DETERMINANTS OF ADDICTION

DETERMINANTS OF ADDICTION

NEUROBIOLOGICAL, BEHAVIORAL, COGNITIVE, AND SOCIOCULTURAL FACTORS

JUSTIN R. YATES
*Experimental Psychology, University of Kentucky Currently a Professor, Psychological Science,
Northern Kentucky University*

Academic Press is an imprint of Elsevier
125 London Wall, London EC2Y 5AS, United Kingdom
525 B Street, Suite 1650, San Diego, CA 92101, United States
50 Hampshire Street, 5th Floor, Cambridge, MA 02139, United States
The Boulevard, Langford Lane, Kidlington, Oxford OX5 1GB, United Kingdom

Copyright © 2023 Elsevier Inc. All rights reserved.

No part of this publication may be reproduced or transmitted in any form or by any means, electronic or mechanical, including photocopying, recording, or any information storage and retrieval system, without permission in writing from the publisher. Details on how to seek permission, further information about the Publisher's permissions policies and our arrangements with organizations such as the Copyright Clearance Center and the Copyright Licensing Agency, can be found at our website: www.elsevier.com/permissions.

This book and the individual contributions contained in it are protected under copyright by the Publisher (other than as may be noted herein).

Notices

Knowledge and best practice in this field are constantly changing. As new research and experience broaden our understanding, changes in research methods, professional practices, or medical treatment may become necessary.

Practitioners and researchers must always rely on their own experience and knowledge in evaluating and using any information, methods, compounds, or experiments described herein. In using such information or methods they should be mindful of their own safety and the safety of others, including parties for whom they have a professional responsibility.

To the fullest extent of the law, neither the Publisher nor the authors, contributors, or editors, assume any liability for any injury and/or damage to persons or property as a matter of products liability, negligence or otherwise, or from any use or operation of any methods, products, instructions, or ideas contained in the material herein.

ISBN: 978-0-323-90578-7

For information on all Academic Press publications visit our website at https://www.elsevier.com/books-and-journals

Publisher: Nikki P. Levy
Acquisitions Editor: Joslyn T. Chaiprasert-Paguio
Editorial Project Manager: Susan Ikeda
Production Project Manager: Swapna Srinivasan
Cover Designer: Vicky Esser Pearson

Typeset by TNQ Technologies

Dedication

To my wife Katy, for your love, support, and patience

    To my sons Landon and Lucas, for being my greatest accomplishments

To my parents Ray and Rita, for raising me to be the best version of myself

    To Joan, for always encouraging me to write a book

In loving memory of Dwayne Sexton

Contents

About the author xi
Preface xiii
Acknowledgements xv
Organization of book xvii

I
Prologue

1. Introduction to addiction: Substance use disorders

Introduction 3
Learning objectives 3
Introduction to addiction and substance use disorders (SUDs) 4
Key aspects of SUDs 4
Commonly used drugs 6
Quiz 1.1 20
Drug combinations 21
Prevalence of SUDs 25
Health and economic impacts of SUDs 27
Translational approach to studying addiction 32
Quiz 1.2 33
Chapter summary 34
Glossary 34
References 35

II
Neurobiological mechanisms of addiction

2. Pharmacological actions of commonly used drugs

Introduction 45
Learning objectives 45
Introduction to the nervous system and neurotransmitters 46
Neurons 47
How neurons communicate with each other 52
Glial cells 55
Quiz 2.1 56
Major neurotransmitters 57
Quiz 2.2 65
Drug mechanisms of action 66
Pharmacological treatments for substance use disorders 71
Quiz 2.3 80
Chapter summary 81
Glossary 81
References 82

3. Neuroanatomical and neurochemical substrates of addiction

Introduction 91
Learning objectives 91
Primary brain structures of the mesocorticolimbic pathway 92
Other brain regions involved in reward 96
Paradigms to study addiction-like behaviors 99
Quiz 3.1 102
Neuroanatomical underpinnings of the addiction process: focus on dopaminergic signaling 102
Quiz 3.2 111
Contribution of other neurotransmitter systems to the addiction process 112
Quiz 3.3 118
Chapter summary 119
Glossary 120
References 120

4. Molecular and cellular mechanisms of addiction

Introduction 133
Learning objectives 133
Heritability of substance use disorders 134

Genes associated with drug addiction in humans 135
Quiz 4.1 139
Genetic approaches to studying addiction in animals 140
Epigenetics and addiction 152
Pharmacogenetic approach to treating addiction 155
Quiz 4.2 155
Cellular changes following drug administration 156
Astrocytes and addiction 164
Quiz 4.3 165
Chapter summary 165
Glossary 166
References 167

III

Behavioral and cognitive mechanisms of addiction

5. Learning mechanisms of addiction: operant conditioning

Introduction 185
Learning objectives 186
Overview of operant conditioning 186
Quiz 5.1 192
Operant processes involved in addiction 194
Using operant conditioning to study addiction 198
Neurobiological differences across self-administration paradigms 208
Quiz 5.2 216
Chapter summary 217
Glossary 218
References 219

6. Learning mechanisms of addiction: Pavlovian conditioning

Introduction 227
Learning objectives 227
Overview of Pavlov's study 228
Principles of Pavlovian conditioning 228
Pavlovian processes involved in addiction 232
Pavlovian-based theories of addiction 239

Quiz 6.1 246
Conditioned place preference 247
Using Pavlovian concepts to treat addiction 251
Quiz 6.2 257
Chapter summary 258
Glossary 258
References 259

7. Attentional and memory processes underlying addiction

Introduction 269
Learning objectives 269
Cognitive model of addiction 270
Drug effects on attention and automaticity 271
Attention to drug warnings 275
Quiz 7.1 276
Drug effects on memory 277
Cognitive impairments as a predictor of drug use problems 281
Modeling relapse-like behavior 282
Conditioned place preference revisited 286
Quiz 7.2 288
Neurobiology of attention and memory: relevance to addiction 288
Cellular mechanisms of learning and addiction 290
Glutamate homeostasis hypothesis of addiction 293
Cognitive-based treatments for addiction 296
Quiz 7.3 300
Chapter summary 300
Glossary 301
References 302

8. Maladaptive decision making and addiction

Introduction 315
Learning objectives 315
Overview of decision making 316
Impulsive decision making 319
Quiz 8.1 325
Risky decision making 326
Maladaptive decision making as a risk factor for substance use disorders 330
Quiz 8.2 332
Shared neuromechanisms of maladaptive decision making and addiction 332
Metacognition, mindfulness, and addiction 336

Treatments for improving drug-induced impairments in decision making 338
Quiz 8.3 340
Chapter summary 341
Glossary 341
References 341

IV
Individual and sociocultural factors linked to addiction

9. Individual differences in addiction: focus on personality traits

Introduction 357
Learning objectives 357
Measuring personality traits 358
Personality factors associated with drug use and addiction 364
Quiz 9.1 368
Impulsivity 369
Attention deficit/hyperactivity disorder 374
Quiz 9.2 377
Chapter summary 378
Glossary 378
References 379

10. Social and sociocultural factors associated with addiction

Introduction 393
Learning objectives 393
Peer and familial influences on drug use 394
Theories of social influence on drug use 396
Social influences on drug use in animal models 399
Quiz 10.1 406
Socioeconomic status (SES) 407
Racial and ethnic differences in addiction 410
Religion/spirituality and addiction 411
Socially based treatments for SUDs 412
Quiz 10.2 418
Chapter summary 419
Glossary 419
References 420

11. Stress and addiction

Introduction 437
Learning objectives 437
Neurobiology of stress 438
Stress and drug use patterns in humans 440
Animal models of stress 444
Quiz 11.1 453
Neuromechanisms mediating relationship between stress and addiction 454
Hedonic allostasis model of addiction revisited 462
Methods for reducing stress 463
Quiz 11.2 466
Chapter summary 466
Glossary 467
References 467

12. Gender and sex differences in addiction

Introduction 487
Learning objectives 487
Differentiating sex from gender 488
Gender differences in drug use patterns 488
Animal studies assessing sex differences in addiction-like behaviors 490
Quiz 12.1 498
Gender/sex as a moderating variable 499
Potential causes of gender/sex differences in drug sensitivity 501
Drug use in sexual/gender minority individuals 509
Quiz 12.2 513
Chapter summary 514
Glossary 514
References 514

V
Epilogue

13. Beyond substance use disorders: Behavioral addictions

Introduction 531
Learning objectives 531
What's considered a behavioral addiction? 532

Comorbid disorders observed in behavioral
    addictions   533
Learning and cognitive mechanisms underlying
    behavioral addictions   534
Quiz 13.1   541
Risk factors associated with behavioral
    addictions   542
Neurobiological underpinnings of behavioral
    addictions   546

Treatment interventions for behavioral
    addictions   552
Quiz 13.2   554
Chapter summary   555
Glossary   556
References   556

Index   571

About the author

Dr. Justin R. Yates is a Professor in the Department of Psychological Science at Northern Kentucky University (NKU). He received his PhD in experimental psychology from the University of Kentucky in 2014, specializing in behavioral neuroscience and psychopharmacology, working under the mentorship of Dr. Michael T. Bardo. His research focuses on elucidating the role of the GluN2B subunit of the *N*-methyl-D-aspartate receptor to maladaptive decision making and addiction-like behaviors in rodents. He has published over 30 research articles and has received funding from the National Institutes of Health to support his research. His research has been recognized at the university, the regional, and the national level. He is a recipient of NKU's Excellence in Research, Scholarship, and Creativity Award and is a Fellow of the Midwestern Psychological Association. He has received three early career awards from the American Psychological Association: the Division six Early Career Award, the Division 25 B.F. Skinner Foundation New Basic Researcher Award, and the Division 28 Young Psychopharmacologist Award.

As a professor at a regional comprehensive university, Dr. Yates' research assistants are undergraduate students. He is passionate about undergraduate research, having served as a Councilor of the Council on Undergraduate Research (CUR) and as Chair of the Psychology Division of CUR. At NKU, he has served as the co-chair of the Celebration of Student Research and Creativity, a yearly event in which students present their experiential learning projects. He has been recognized for his efforts to promote undergraduate research, receiving NKU's Excellence in Undergraduate Research Mentoring Award and the Department of Psychological Science's George Goedel Faculty Mentor Award.

Preface

Although I entered college as a psychology major, I never anticipated that I would become a professor studying addiction, especially a professor using rodent models. As a freshman in 2005, I had no idea what I wanted to do after completing my undergraduate studies. In 2007, I enrolled in a biopsychology course taught by Dr. Walt Isaac, and I became fascinated by the neurobiological underpinnings of learning and behavior. After taking this course, I knew that I wanted to earn my doctorate in biopsychology. To prepare myself for graduate school, I conducted behavioral pharmacology research with Dr. Isaac. Not only was this my first time performing research, but it was my first time handling a rat. I was nervous at first (and admittedly, kind of grossed out), but I loved conducting research. This experience reinforced my desire to earn a doctorate. At this point, Dr. Isaac recommended I apply to the University of Kentucky (UK) as he had earned his PhD from there. When I researched UK's graduate program in Experimental Psychology, I noticed that I could specialize in behavioral neuroscience and psychopharmacology. I was particularly interested in the work of Dr. Mark Fillmore and Dr. Michael Bardo. After interviewing at UK, I was accepted to the program and was invited to work in Dr. Bardo's lab. This is when I began to learn more about psychostimulant addictive-like behaviors and maladaptive decision-making, two areas of research I still pursue today.

As a graduate student, I faced a similar dilemma I experienced as a graduate student: I had no idea what I wanted to do once I earned my degree. The thought of teaching was nauseating as public speaking was terrifying. At UK, all graduate students must serve as a teaching assistant for one semester. To better prepare myself for the teaching assistantship, I enrolled in UK's Preparing Future Faculty (PFF) Program, in which students complete coursework related to college teaching and learning. I took my first class during the Fall 2010 semester and then served as a teaching assistant for a research methods class during the Spring 2011 semester. I did not hate my experience as a teaching assistant, but I did not feel like I had much autonomy in the classroom. After completing the courses for the PFF Program, I taught an introductory psychology course at Northern Kentucky University (NKU). Although still somewhat terrified of speaking in front of large groups, I loved teaching my own course. This experience made me realize I wanted to be a college professor.

Now, as a professor at NKU, I primarily teach courses in research methods, animal learning, biopsychology, and cognitive psychology. However, I have always wanted to develop a course related to my research interests: addiction. Because I study the neurobehavioral mechanisms of addiction, I wanted my course to emphasize the multifaceted nature of addiction. As I looked through prospective textbooks to use in my new course, I found multiple books that discuss the pharmacological actions of drugs, the neuroscience of addiction, and the therapies often used to treat those with an addiction. However, I could not find a text that integrated the neurobiological, the cognitive-behavioral, and the sociocultural determinants of addiction into a single textbook. Providing a comprehensive overview of the various factors linked to addiction is important, especially given the difficulty in treating addiction. At this point, I knew what I had to do: write my own textbook.

I began writing this textbook during the height of the COVID-19 pandemic that affected millions around the world. In addition to learning to teach online, overseeing a research program understaffed due to the pandemic, helping my oldest child navigate kindergarten virtually, and raising a baby (born in April 2020), I had the challenge of writing a book that experts in addiction science could use while making it accessible to nonexperts, particularly undergraduate college students. For experts in the field, I discuss both seminal and recent research studies related to the neurobehavioral mechanisms of addiction. This aspect of the book was easy as I conduct research and understand the addiction field. The more challenging aspect of writing this book was translating the research findings to language that nonexperts could digest. To accomplish this goal, I sent a draft of each chapter to two nonexperts whose opinions I could trust: my parents. My parents do not have a college degree, nor do they have any experience in the addiction field. After they read each chapter, I asked them if the content made sense. Not surprisingly, they had the most difficulty understanding the neurobiology of addiction. As my dad put it "I don't even know how to say half of these words". I hope I have succeeded in applying the technical details of addiction science to situations and experiences most readers will understand.

Acknowledgements

I want to thank my colleague Dr. Cecile Marczinski for serving as my textbook *sensei* as I worked on securing a publishing contract. Her feedback on early drafts of my textbook proposal was extremely helpful. I also want to thank my colleague Dr. Doug Krull for providing additional advice as I worked on my textbook proposal. Many thanks to my editorial project manager Susan Ikeda for making sure I stayed on schedule throughout the writing process. I want to express my gratitude to Swapna Srinivasan for working with me as we revised and formatted the textbook. I also want to acknowledge the National Institutes of Health as some of the work cited in this book was funded by grants P20GM103436 and R15DA047610. Similarly, I want to thank my former research technician Matthew Horchar and the many undergraduate research assistants that have worked in my lab throughout the years.

I want to thank my wife Katy for her unconditional love and support as I spent many late nights and long weekends working to complete this book. I am thankful for my sons Landon and Lucas, whose love and humor helped me stay calm when I started to feel overwhelmed trying to manage this project with my many other responsibilities. I want to thank my parents Ray and Rita for reading drafts of each chapter. Their excitement to read the next chapter helped motivate me to keep writing even when I experienced writer's block or felt like what I was writing was not good.

Organization of book

This book consists of five sections. Section I includes Chapter 1, which serves as a prologue and introduces addiction and substance use disorders (SUDs), as well as terminology related to drug use. Some commonly used drugs are discussed, including their routes of administration, their intoxicating effects, and their corresponding withdrawal symptoms. Additionally, common drug combinations and their effects are discussed. This chapter also details the prevalence rates of SUDs in the United States and across the world, as well as in other psychiatric conditions. The health effects associated with various drug classes are discussed, as well as the economic impact of SUDs to an individual and to society. Specifically, issues related to lost productivity, health costs, and legal fees are covered. This chapter concludes by discussing the translational research approach and the contribution of animal research to understanding the development and maintenance of SUDs.

Section II focuses on the neurobiological mechanisms of drugs and the addiction process. To better understand the neurobiology of addiction, Chapter 2 provides an overview of the nervous system, with a focus on neurons and how they function. The major neurotransmitters are covered in detail as they play a vital role in neuronal neurotransmission and are affected by various drugs. This chapter also discusses the pharmacological treatments for SUDs. Chapter 3 covers the neuromechanisms of addiction at the macro level, focusing on the brain regions and the neurotransmitter systems involved in the addiction process. Chapter 4 discusses the neuromechanisms of addiction at the micro level. Specifically, the genetic and the cellular bases of addiction are covered in detail in this chapter. Chapters 3 and 4 will also detail some of the methods researchers can use to elucidate the neuromechanisms of addiction.

Section III details the behavioral and the cognitive mechanisms of addiction. Chapters 5 and 6 focus on two major forms of associative learning: operant conditioning and Pavlovian conditioning, respectively. Chapters 7 and 8 discuss cognitive factors linked to addiction. Chapter 7 primarily focuses on the relationship between attentional/memory dysfunction and addiction while Chapter 8 discusses maladaptive decision making as a risk factor for addiction. Treatments utilizing behavioral and/or cognitive approaches are introduced in this section, such as cognitive-behavioral therapy, contingency management, and mindfulness training.

Section IV discusses some of the individual and the sociocultural factors of addiction. Chapter 9, introduces the concept of individual differences in addiction risk vulnerability. Chapter 9 focuses on personality traits, which encompass behavioral/cognitive tendencies like impulsivity. Chapter 10 introduces some of the social and sociocultural factors associated with addiction. Specific sociocultural factors discussed in this chapter include socioeconomic status, race and ethnicity, and religion. Addiction treatments utilizing social groups are included in this chapter like family therapy, residential treatment centers, and 12-step programs. Chapter 11 focuses on the contribution of stress to the addiction process. Because many stressors involve social situations (e.g., relationship problems, loss of a loved one, work difficulties, antagonistic interactions, etc.), I chose to include stress in this section. This chapter includes a discussion of the shared neurobiological underpinnings of stress and addiction. Finally, Chapter 12 details how gender/sex moderate drug sensitivity and addiction-like behaviors. Neurobiological and sociocultural factors accounting for gender/sex differences in addiction are covered. This chapter also discusses substance use in gender minority and sexual minority individuals.

Section V includes Chapter 13, which serves as an epilogue as it applies the content of Chapters 2−12 to behavioral addictions. This chapter primarily focuses on gambling disorder, as it is the only formally recognized behavior addiction in the *Diagnostic and Statistical Manual of Mental Disorders*. However, binge eating disorders are also discussed as they share many features of drug addiction.

To help guide one's reading, each chapter includes learning objectives, detailing the most important concepts covered in the chapter. Individuals can test their understanding by completing quizzes embedded in each chapter. Each chapter also includes an Experiment Spotlight, in which I detail the methodology and results of one study. Most chapters include a recent article (published within the past 2 years at the time I wrote the book). I include questions at the end of each Experiment Spotlight to get individuals to think more critically about addiction-related research. For instructors, I have included example exam questions for each chapter. Each exam contains example multiple-choice, true/false, fill-in-the-blank, and short answer/essay questions. Some chapters also contain matching questions. This provides the instructor flexibility as they develop their exams. I personally use a combination of multiple-choice and short answer questions in my exams, but I know some instructors like fill-in-the-blank and true/false questions as well.

SECTION I

Prologue

CHAPTER 1

Introduction to addiction: Substance use disorders

Introduction

> I was a drug addict ... I was a slave to heroin and morphine, experimenting on the side with cocaine, marijuana and opium. In that whole period, I never had a moment of genuine peace or happiness. Those nightmarish intervals of elation or exhilaration which drugs brought me were illusions, I knew even then. - **Leroy Street (p. 3)**[1]

During your lifetime, you may have encountered someone that seemed completely normal. This person may have been friendly, intelligent, talented, calm, or all the above. On the outside, this person seemingly had everything in their life in control. However, this is not always the case as many individuals suffer from some form of **addiction**. These individuals may try to conceal their addiction from everyone they know, including their family and closest friends. Others may acknowledge they have an addiction but have no interest in receiving treatment, whereas others may receive treatment several times before returning to their old habits. There are many individuals that can overcome their addiction; however, avoiding the addictive behavior becomes a life-long battle. This textbook will focus on one specific type of addiction: drug addiction, but I will discuss other behavioral addictions in Chapter 13. For now, the goal of this chapter is to introduce you to the concept of addiction. You will learn about various drug types, their physiological effects when used in isolation or in combination with other drugs, and what happens when an individual quits using a drug. You will also learn about the prevalence of addiction and the health and economic impacts associated with drug use. As you will learn, if you are already not aware, addiction is a serious issue that affects millions of people, their families, and our society.

Learning objectives

By the end of this chapter, you should be able to ...

(1) Define the term addiction and substance use disorder (SUD).
(2) Describe the key aspects of SUDs and differentiate physical dependence from psychological dependence.

(3) Identify the various drugs that are commonly used and describe how they affect an individual.
(4) Explain the health and economic impacts of SUDs.
(5) Understand the role animal research plays in furthering our understanding of SUDs.

Introduction to addiction and substance use disorders (SUDs)

Drug addiction is a severe biomedical disorder characterized by a *compulsive* (uncontrollable) urge to use a substance or substances.[2] When discussing drug addiction, the term **substance use disorder (SUD)** is often used. SUDs involve actions that individuals continuously perform despite potential negative consequences. Importantly, SUDs are not limited to *illicit* (i.e., illegal) drugs; an individual can develop a SUD for numerous drugs. In fact, two of the most common SUDs are for legal substances. The term *drug abuse* is widely used; however, this term is falling out of favor because it often stigmatizes those with a SUD.[3] Remember, individuals with a SUD are people first. Having a SUD does not make someone a "bad" person. Instead of using the term drug abuse, the terms **drug use, drug misuse**, and addiction are commonly used. What is the difference between drug use and drug misuse? The term drug use is often applied to illicit substance use. The term drug misuse refers to the problematic use of legal drugs like prescription medications. The individual may take more of the substance than was prescribed to them (e.g., taking two pills at a time as opposed to one), or they may take the substance in a way that was not intended (e.g., crushing a pill and snorting it instead of swallowing it).

One rumor that needs to be dispelled is that someone can develop an addiction following a single exposure to a substance. Growing up, I heard this "fact" on multiple occasions. The argument I heard was that some drugs have such high misuse potential that someone can get "hooked" immediately. However, addiction does not work like that as the neurobehavioral and cognitive manifestations of addiction occur after prolonged drug use. Thus, saying that someone is addicted to a drug after a single exposure is premature.

Key aspects of SUDs

SUDs are characterized as psychiatric disorders in the *Diagnostic and Statistical Manual of Mental Disorders* (*DSM*) and the *International Classification of Diseases* (*ICD*). The *DSM*, currently in its fifth edition (*DSM-5*), is published by the American Psychiatric Association, a national organization.[2] The *ICD*, currently in its 11th edition (*ICD-11*), is produced by the World Health Organization, an international organization, and is more commonly used outside of the United States.[4] Whereas the *DSM* includes mental disorders only, the *ICD* includes mental and physical disorders. This section will first describe important terms related to SUDs before describing how the *DSM* and the *ICD* define SUDs.

Terminology related to SUDs

Many drugs produce **intoxication**, characterized by alterations in consciousness, cognition, perception, and behavior that are transient. You are probably aware of intoxication if you have ever consumed ethanol (alcohol) or have watched one of your friends drink ethanol.

One behavioral alteration that occurs during ethanol intoxication is increased sociability. Individuals that have consumed ethanol tend to talk more and have increased confidence. These individuals may lack inhibitory control and may act impulsively. At the cognitive level, ethanol intoxication leads to decreased reaction time; that is, an individual will be slower to respond to a question or to notice something that has appeared. We will discuss the intoxicating effects of drugs, including ethanol, in more detail later in this chapter.

With increased substance use, individuals often develop **tolerance** to the effects of the substance. They will need to increase the amount of substance to get the desired effects. Going back to ethanol: as a person develops tolerance, they may need to consume four or five beverages to become intoxicated instead of just two or three drinks. With prolonged substance use, individuals will experience **withdrawal** symptoms when they try to stop using the substance. Withdrawal symptoms vary depending on the substance being used. When specific drugs are covered later in this chapter, I will describe their withdrawal symptoms. For now, just know that withdrawal symptoms are unpleasant, to put it mildly. For some drugs, withdrawal symptoms can be life-threatening! When someone displays tolerance to the drug's effects and experiences withdrawal symptoms when they try to stop using the drug, they are **physically dependent** on the drug. Physical dependence on its own does not mean that someone has a SUD. For example, with pain management, there are individuals that use specific drugs to manage their symptoms. Even if they never misuse the drug, they can still become physically dependent, experiencing withdrawal symptoms if they stop taking the medication.

In contrast to physical dependence, **psychological dependence** is a defining feature of a SUD. Psychological dependence entails different behaviors that can greatly interfere with one's life. These individuals become obsessed with acquiring additional drug and neglect to fulfill other activities, such as spending time at work or with friends and family. When you see the diagnostic criteria of SUDs below, you will notice that many of the criteria center around psychological dependence.

At its heart, SUDs are **relapsing** disorders. Individuals with a SUD are often able to stop using the drug for periods of time. Relapse is the phenomenon in which an individual resumes engaging in a behavior, like drug use, following a period of **abstinence** (period in which the behavior does not occur). Relapse is particularly dangerous because the individual may use a dose of the drug that they frequently used before initiating abstinence. If enough time has passed since the last drug use, the individual may no longer be tolerant to the drug's effects. This can result in a drug **overdose** and potentially death.

Defining SUDs: *Diagnostic and Statistical Manual (DSM-5)*

According to the *DSM-5*, SUDs encompass a "cluster of cognitive, behavioral, and physiological symptoms" (p. 483)[2] and span 11 different criteria:

(1) Taking the substance in larger amounts or for longer than intended.
(2) Wanting to reduce substance use but not managing to do so.
(3) Spending increasing amounts of time procuring, using, or recovering from substance use.
(4) Experiencing cravings/urges to use the substance.
(5) Having difficulty managing work/school and personal responsibilities.

(6) Continuing substance use despite it causing relationship problems.
(7) Abandoning social, occupational, or recreational activities to use the substance.
(8) Continuing substance use despite repeatedly being placed in dangerous situations.
(9) Continuing substance use even when knowing that an existing physical or psychological problem can be exacerbated by substance use.
(10) Needing more of the substance to achieve the desired effect.
(11) Developing withdrawal symptoms.

Notice that points 2-9 revolve around psychological dependence. According to the *DSM-5*, displaying two to three of the criteria listed above meets the definition of a mild SUD while displaying four to five of the criteria is considered a moderate SUD. If individuals display six or more of these criteria, they are considered to have a severe SUD. The *DSM-5* lists 10 different classes of drugs to which the term SUD can be applied, with one of these categories being the catch-all "Other". We will discuss many of the drugs identified in the *DSM-5* later in this chapter.

Defining SUDs: *International Classification of Diseases (ICD-11)*

The *ICD-11* uses two primary diagnoses for SUDs: dependence and harmful use requiring physical or mental harm.[5] To be diagnosed with a SUD, an individual needs to meet three of the six dependence criteria:

(1) Having a strong desire or compulsion to take the substance.
(2) Having difficulty in controlling when to take the substance, how much of the substance to take, and when to stop using the substance.
(3) Developing withdrawal symptoms.
(4) Needing more of the substance to achieve the desired effect.
(5) Neglecting other interests because too much time is spent either taking the substance, obtaining the substance, or recovering from the effects of the substance.
(6) Continuing substance use even when there are harmful consequences.

To meet the category of harmful use, one of the following must be met:

(1) Use of the substance where impairment could be dangerous.
(2) Continued use despite the presence of a physical, psychological, or cognitive problem related to use.
(3) Detrimental behaviors and social problems related to use.
(4) Interpersonal conflict attributed to use. (p. 698).[5]

Although many of the criteria listed in the *ICD-11* are similar to those found in the *DSM-5*, notice that the criteria used for determining the presence of a SUD are different. The *DSM-5* defines SUDs on severity, whereas the *ICD-11* defines SUDs on meeting a minimum number of criteria, thus treating them as an "all-or-nothing" disorder.

Commonly used drugs

According to the Substance Abuse and Mental Health Services Administration (SAMHSA),[6] approximately 164.8 million people 12 years of age or older reported using a drug during the past month, representing approximately 60% of the U.S. population. Drugs can be classified by the effects they have on an individual, but they can also be placed in different categories based on legal definitions. In the United States, drugs are classified by their potential medical uses and their potential for misuse and dependence. The Drug Enforcement Administration (DEA) places drugs in one of five *schedules*, labeled Schedules I through V. Schedule V drugs are those that have legitimate medical purposes and have low misuse potential. Antidiarrheal drugs fall under Schedule V. I do not know of any individuals that go out of their way to use antidiarrheal drugs for fun. As the medical utility decreases and risk of misuse increases, a drug will be placed in the next schedule. For example, Schedule III drugs have medical uses, but these drugs can be easily misused if not carefully taken. Two well-known examples of Schedule III drugs are anabolic steroids and testosterone, which can be misused by bodybuilders and athletes to gain an unfair advantage during competitions. Most of the drugs that will be covered below fall under Schedule I or Schedule II, although some are designated as Schedule IV drugs.

Routes of administration

In addition to understanding how drugs can be classified, you need to understand drug **routes of administration**, or the ways individuals can take a drug. This is important because the way someone takes a drug will influence how quickly it reaches the brain. The speed at which a drug enters the brain is a major determinant of how the individual feels after taking the drug.

The first way an individual can take a drug is *orally*. The oral route involves swallowing the drug so that it enters the stomach. This route of administration is the slowest, with drugs reaching the brain approximately 30–90 min after administration. Additionally, because the drug goes to the stomach, much of it is metabolized and excreted; therefore, not as much of the drug enters the brain. The oral route is abbreviated as p.o., which comes from the Latin words *per os* ("by mouth"). Instead of swallowing a drug, an individual can place the drug inside their mouth, either under the tongue (*sublingual* route) or pressed against the cheek (*buccal* route). These two routes are used with very specific types of drugs and are not as common as the oral route or some of the routes you will read about below.

Instead of ingesting a drug, individuals can place a patch or other device on the skin (known as the *transdermal* route). Like the oral route, this route of administration is slow as the drug must pass through the skin before entering the bloodstream. When I was a graduate student, I volunteered to be a research participant in a study to measure the safety of a drug administered by the transdermal route. I had to wear the transdermal patch for over a week. By the time the researcher removed my patch, my arm looked like an octopus had suctioned itself to me. It was worth the experience as I earned $300, a king's ransom for a broke graduate student.

The oral and transdermal routes deliver the drug at a slower rate compared to the *injection* route. A person can either inject the drug directly into their skin (called subcutaneous injection; s.c.), into their muscles (called intramuscular injection; i.m.), or into a vein (called intravenous injection; i.v.). Because intravenous injections allow the drug to reach the brain in a shorter amount of time compared to other injection methods (often within 30 s), this route of administration is one of the most common methods of administering illicit drugs.

Injecting a drug can be challenging for some, especially intravenous injections. As such, some individuals may prefer to snort the drug into the nasal passage; this is known as the *intranasal* route of administration. Colloquially, this is called "snorting". For some drugs, individuals may prefer to smoke the drug (known as the *inhalation* route). The inhalation route allows the drug to pass through the lungs before reaching the brain. Drugs that are inhaled reach the brain quickly (within 10–20 s). Not surprisingly, drugs that are smoked have some of the highest dependence rates.

Psychostimulants

Psychostimulants, also referred to as stimulants or "uppers", increase arousal and make individuals feel more alert. Stimulants include, but are not limited to, the following drugs: nicotine, cocaine, methamphetamine, and the ADHD medications amphetamine (Adderall, Dexedrine, Vyvanse) and methylphenidate (Ritalin, Concerta).

Nicotine

Nicotine is the main psychoactive compound of the tobacco leaf (Fig. 1.1A) and is found in products such as cigarettes, cigars, snuff, chew (or dip), and snus. Cigarettes are paper-

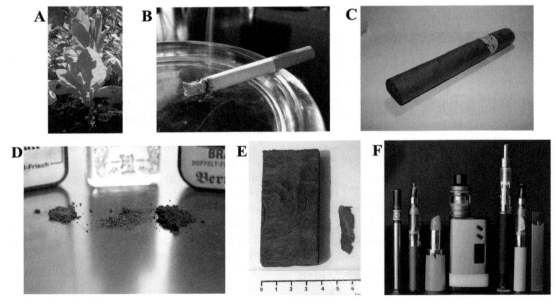

FIGURE 1.1 The tobacco leaf (A) and various tobacco products that contain nicotine: cigarette (B), cigar (C), snuff (D), chew (in plug form) (E), and various e-cigarettes (F). *Panel (A) comes from Wikimedia Commons and was posted by user Hendrick128 (https://commons.wikimedia.org/wiki/File:Burley_Jupiter_2.JPG). Panel (B) comes from Wikimedia Commons and was posted by Tomasz Sienicki (By © 2005 by Tomasz Sienicki [user: tsca, mail: tomasz.sienicki at gmail.com] - Photograph by Tomasz Sienicki/Own work, CC BY-SA 3.0, https://commons.wikimedia.org/w/index.php?curid=172810). Panel (C) comes from Wikimedia Commons and was posted by user 0r14th0 (By 0r14th0 - Own work, CC BY-SA 3.0, https://commons. wikimedia.org/w/index.php?curid=186085). Panel (D) comes from Wikimedia Commons and was posted by user Oimel (By Oimel - Own work, CC BY-SA 3.0, https://commons.wikimedia.org/w/index.php?curid=4382168). Panel (E) comes from Wikimedia Commons and was posted by the Auckland Museum (https://commons.wikimedia.org/wiki/File:Tobacco_plug_(AM_ 1967.79-1).jpg). Panel (F) comes from Wikimedia Commons and was posted by the California Department of Public Health (https://commons.wikimedia.org/wiki/File:Various_types_of_e-cigarettes.jpg).*

wrapped tobacco with a filter at one end (Fig. 1.1B), whereas cigars are cured tobacco traditionally wrapped in leaf tobacco (Fig. 1.1C). Both cigarettes and cigars are smoked. Snuff is a finely cut or powdered tobacco that is also inhaled as individuals pinch some of the tobacco and sniff it (Fig. 1.1D). Snus is a moist form of snuff originating from Sweden that is inserted behind an individual's lip. Chew (dip) is like snuff, but it is made up of loose tobacco leaves that can be packaged as plugs (small, brick-like shapes) (Fig. 1.1E) and twists (braids of leaves). Because chew causes excess saliva production, individuals need to spit the tobacco onto the ground or into a container. Failure to spit the tobacco out can make the user ill. Recently, nicotine has been sold in the form of electronic cigarettes (or e-cigarettes) (Fig. 1.1F). E-cigarettes house a nicotine-containing liquid that is aerosolized when heated. Unlike cigarettes and cigars, e-cigarettes do not contain tobacco.

Each year, the SAMSHA administers the *National Survey on Drug Use and Health* (*NSDUH*) to estimate the prevalence of substance use. According to the *NSDUH*,[6] cigarette smoking is the most common method of tobacco consumption in current users (~80% of tobacco users) as individuals can rapidly get nicotine into the brain. Since 2002, cigarette use has declined in individuals over the age of 12. In 2002, approximately 26% of individuals reported cigarette use, but this percentage decreased to approximately 17.2% in 2018. The largest decreases in cigarette use have been observed in individuals aged 12–17 and in individuals aged 18–25, as the percentage of cigarette users has decreased from 13.0% to 2.7% and from 40.8% to 19.1%, respectively. Individuals aged 26 or older have shown a slower decrease in cigarette use (25.2% in 2002 vs. 18.5% in 2018). As noted by the SAMHSA, caution needs to be taken when discussing these results, as the *NSDUH* does not currently ask individuals about e-cigarette use. Thus, the decrease in cigarette use may be an artifact of individuals that have switched from traditional cigarettes to e-cigarettes. Indeed, in 2020, approximately 19.6% of high school students (3.02 million) and 4.7% of middle school students (550,000) reported using e-cigarettes.[7] One encouraging finding is that these numbers are lower than what was reported in 2019 (4.11 million high school students and 1.24 million middle school students).

Nicotine, specifically in the form of cigarettes, is one of the most used drugs in the world and is legal to use in almost every country, with one exception being the small Himalayan country Bhutan. In the United States, cigarettes are the third most used drug, with an estimated 45.9 million people smoking them during the past month.[6] Despite its high dependence potential, the DEA does not place nicotine in one of its five schedules. However, as of 2019, the United States now requires individuals to be at least 21 years of age to purchase tobacco products. Before the Federal Food, Drug, and Cosmetic Act was amended, most states required individuals be at least 18 years of age to purchase tobacco products. I imagine we will see research reports come out during the next decade detailing how tobacco use rates have been affected, if at all, by this administrative change.

Although nicotine is a stimulant, it does not produce a considerable "high" like the other drugs in this class. Even though nicotine increases blood pressure and heart rate, individuals often report feeling calm after using nicotine products.[8] However, tolerance to nicotine develops rapidly.[9] To compensate for tolerance, individuals often increase the number of cigarettes smoked in a day. Some individuals can smoke two or more packs of cigarettes in a single day.[10] That is over 40 cigarettes! Even though withdrawal symptoms do not pose a serious risk to one's physical health, they are still aversive to an individual as they experience headaches and extreme irritability, and they can experience increased anxiety.[2] Because these withdrawal symptoms can last several days to several weeks,[11] relapse rates are high.[12] The quote "Giving up smoking is the easiest thing in the world. I know because I've done it thousands of times" is attributed to famed writer Mark Twain and illustrates the difficulty of abstaining from nicotine use.

Can one overdose on nicotine? The answer is "yes", but fatal overdoses are rare because only a small portion of nicotine enters the body during smoking.[13] If overdose does occur, individuals may experience initial symptoms such as nausea/vomiting, abdominal pain, salivation, sweating, headache, rapid heart rate, and tremors followed by diarrhea, decreased heart rate/blood pressure, lethargy, and weakness.[14,15] We will cover the long-term health consequences associated with tobacco use later in this chapter.

Schedule II psychostimulants

While nicotine is not scheduled by the DEA, many psychostimulants fall under Schedule II. Derived from the coca leaf (Fig. 1.2A), cocaine is a white powder that is primarily snorted (Fig. 1.2B). However, it can be injected intravenously or ingested as a pill. *Crack cocaine* is a crystalline form of cocaine produced by combing powdered cocaine with baking soda (Fig. 1.2C). Crack is primarily smoked instead of snorted. Cocaine does have a therapeutic purpose as it is an effective topical anesthetic, often used in eye surgery. Unlike cocaine, methamphetamine is not derived from a plant. Instead, it is manufactured in clandestine labs by combining several chemicals. The ability to synthesize methamphetamine has made it a popular drug during the past 30 years as individuals can create it in a home lab using pseudoephedrine, a common ingredient in cold medications (note, I am not going to teach you how to make methamphetamine). Methamphetamine can be injected, snorted, or ingested as a pill, but the most common route of administration is inhalation by smoking. Just as crack is a smokable form of cocaine, *crystal methamphetamine* is a smokable variant of methamphetamine. This form of methamphetamine is known as crystal or glass due to its crystal/glass-like appearance (Fig. 1.2D). Methamphetamine is a Schedule II drug because it can be used to treat ADHD, marketed as Desoxyn. Other Schedule II stimulants used to treat ADHD are methylphenidate (Fig. 1.2E) and amphetamine (Fig. 1.2F). Like Desoxyn, these drugs are prescribed in oral form, although individuals will misuse these drugs by crushing the pills and snorting them or by dissolving the pills in liquid before injecting the solution.

The Schedule II psychostimulants described above increase blood pressure and heart rate just as nicotine, but they can produce intense feelings of *euphoria*, especially when injected or smoked.[16] As stimulants, these drugs also increase energy, alertness, and talkativeness while decreasing appetite.[16] Overdosing on these drugs can be life-threatening. For example, cocaine/methamphetamine overdose is associated with hyperthermia (increased body temperature), increased chest pain, and increased heart rate, which can lead to a heart attack or a stroke.[16–18] Tolerance to the subjective and the physiological effects of these stimulants develops over time. Stimulants like amphetamine were once used as diet pills. Because tolerance develops to the appetite-suppressing effects of stimulants, these drugs are not effective

FIGURE 1.2 The coca plant (A) is used to create cocaine (B). Baking soda can be added to cocaine to create crack cocaine (C). Other psychostimulants include methamphetamine (pictured here as crystal methamphetamine) (D), methylphenidate (E), and amphetamine (F). *Panel (A) comes from Wikimedia Commons and was posted by user Dbotany (https://commons.wikimedia.org/wiki/File:Erythroxylum_novogranatense_var._Novogranatense_(retouched).jpg). Panel (B) comes from Wikimedia Commons and was posted by an employee of the DEA (https://commons.wikimedia.org/wiki/File:CocaineHydrochloridePowder.jpg). Panel (C) comes from Wikimedia Commons and was posted by an employee of the DEA (https://commons.wikimedia.org/wiki/File:Rocks_of_crack_cocaine.jpg). Panel (D) comes from Wikimedia Commons and was posted by user Radspunk (https://commons.wikimedia.org/wiki/File:Crystal_Meth.jpg). Panel (E) comes from Wikimedia Commons and was posted by user Octavio (https://commons.wikimedia.org/wiki/File:Ritalin20mg.jpg). Panel (F) comes from Wikimedia Commons and was posted by user Synesthezia (https://commons.wikimedia.org/wiki/File:Adderall_20mg_capsules.JPG).*

for long-term weight loss. Like nicotine, withdrawal symptoms are not life-threatening. Two major withdrawal symptoms are loss of pleasure (**anhedonia**) and depression.[19] Increased depression can be serious, particularly in those that have major depressive disorder. Other withdrawal symptoms include fatigue, unpleasant dreams, insomnia or hypersomnia, decreased heart rate, slowed movements, increased appetite, and irritability.[2]

Cathinone and synthetic cathinones

Cathinone is a naturally occurring alkaloid found in the shrub *Catha edulis*, better known as khat (Fig. 1.3A). Khat is used more frequently in the Arabian Peninsula and East Africa.[20] Individuals in this part of the world chew khat leaves to stay alert just as many of you reading this book drink caffeinated beverages.[21] Synthetic cathinones are manufactured in labs to mimic the effects of cathinone and are better known as "bath salts". Do not confuse "bath salts" with products like Epsom salt that individuals use during bathing. Those bath salts are not psychoactive drugs like synthetic cathinones. Synthetic cathinones include

FIGURE 1.3 Khat (A) is a naturally derived cathinone whereas synthetic cathinones ("bath salts") are made in laboratories. Unmarked synthetic cathinone in crystalized form (B) and methylone in powder form (C). *Panel (A) comes from Wikimedia Commons and was posted by Trevor Bake (https://commons.wikimedia.org/wiki/File:Khat.jpg). Panel (B) comes from the DEA (https://www.drugabuse.gov/publications/drugfacts/synthetic-cathinones-bath-salts). Panel (C) comes from Wikimedia Commons and was posted by user DMTrott (By DMTrott - Own work. Originally published in The Honest Drug Book [ISBN: 978—0995593602]., CC BY-SA 4.0, https://commons.wikimedia.org/w/index.php?curid=72057204).*

3,4-methylenedioxypyrovalerone (MDPV), 3,4-methylenedioxy-N-methylcathinone (MDMC or methylone), and 4-methyl methcathinone (4-MMC or mephedrone) (Fig. 1.3B and C). Synthetic cathinones come in powder form or as small crystals, and they are often snorted, but they can be ingested orally, smoked, or injected. Due to their stimulant properties, cathinones have many of the same physiological effects as drugs like cocaine, such as increased heart rate and blood pressure.[22,23] At high enough concentrations, synthetic cathinones can lead to paranoia, hallucinations, and agitation.[24] Overdosing on synthetic cathinones can even lead to death.[25] Prolonged use of synthetic cathinones leads to withdrawal symptoms upon cessation of use, including depression, anxiety, tremors, sleep disturbances, and paranoia.[26,27]

Depressants

In contrast to psychostimulants, **depressants** increase sedation and make the user feel calm and relaxed. Colloquially, these drugs are referred to as "downers". Major depressants include ethanol, opioids, and sedative-hypnotics.

Ethanol

Many young adults are familiar with ethanol, as it is the psychoactive drug found in beverages such as beer, wine, and liquor (Fig. 1.4A). The term alcohol is often used synonymously with ethanol, but ethanol is just one type of alcohol. There are other types of alcohol, such as isopropyl alcohol (better known as rubbing alcohol) and methanol, but I

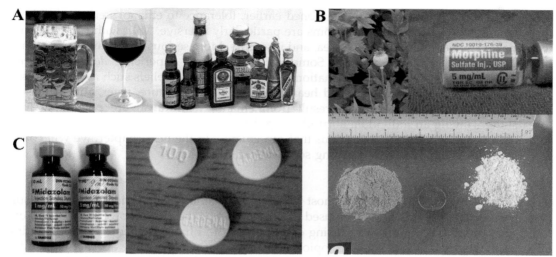

FIGURE 1.4 Various types of depressant drugs. (A) Different types of ethanol, including beer, wine, and liquor. (B) The opium poppy plant, which is used to create opiates like morphine (pictured in an injection vial) and heroin (pictured in powder form). (C) The benzodiazepine midazolam (pictured in an injection vial) and the barbiturate phenobarbital (pictured in pill form). *Panel (A) comes from Wikimedia Commons. The photo of beer was posted by user Usien (By Usien, derivative work Lämpel - Own work, CC BY-SA 3.0, https://commons.wikimedia.org/w/index.php?curid=48714315). The photo of wine was posted by Andre Karwath (By André Karwath aka Aka - Own work, CC BY-SA 3.0, https://commons.wikimedia.org/w/index.php?curid=35268). The photo of liquor was posted by Ralf Roletschek (By Ralf Roletschek - Own work, GFDL 1.2, https://commons.wikimedia.org/w/index.php?curid=58540679). Panel (B) comes from Wikimedia Commons (opium poppy plant and morphine) and the DEA (heroin). The photo of the opium poppy plant was posted by user Dinkum (https://commons.wikimedia.org/wiki/File:Coquelicots_-_Parc_floral_7.JPG). The photo of morphine was posted by user Vaprotan (By Vaprotan - Own work, CC BY-SA 3.0, https://commons.wikimedia.org/w/index.php?curid=9878213). The photo of heroin was posted by the DEA (https://www.dea.gov/factsheets/heroin). Panel (C) comes from the DEA. The photo of midazolam can be found at https://www.dea.gov/factsheets/benzodiazepines. The photo of phenobarbital can be found at https://www.dea.gov/factsheets/barbiturates.*

do not recommend that you drink either of those forms of alcohol as they are more toxic than ethanol, especially methanol.[28] Ethanol comes from the fermentation of yeast, sugars, and starches. For example, wine comes from fermented fruit (primarily grapes), tequila is produced by fermenting the agave plant, and whiskey comes from fermented grains like wheat or barley. Like nicotine, the DEA does not place ethanol into a schedule, but ethanol sales are limited to individuals aged 21 or older. Whereas tobacco use is legal in almost all countries, ethanol use is prohibited in several countries, mostly those that have a large Muslim population (e.g., Afghanistan, Saudi Arabia, and Somalia). In the United States, ethanol is by far the most used drug, as almost 140 million people are estimated to have used ethanol during the past month.[6]

I briefly described some of the intoxicating effects of ethanol earlier in this chapter. Mild intoxication is marked by increased sociability and confidence. Individuals can also experience feelings of *sedation* (reduced irritation or agitation). As ethanol consumption continues, individuals will experience slurred speech and motor difficulties such as the inability to walk in a straight line. If too much ethanol is consumed, respiratory depression can occur, which can lead to coma and death.[29] As mentioned earlier, tolerance to ethanol's effects develops over time. Ethanol withdrawal symptoms are particularly aversive. Mild withdrawal symptoms include anxiety, headaches, nausea, and insomnia. More serious withdrawal symptoms include hallucinations and seizures. Some individuals may experience *delirium tremens* (*DTs*).[30] DTs consist of intense hallucinations and *delusions* (false beliefs such as being persecuted), as well as confusion, increased heart rate and blood pressure, fever, and increased perspiration. The seizures that can result from ethanol withdrawal are potentially life-threatening. Having an individual abruptly stop drinking ethanol is not recommended due to the dangerous withdrawal symptoms that can emerge. In the next chapter, we will cover a pharmacological approach to weaning someone off ethanol.

Opioids

If you have ever had surgery, you most likely were prescribed medication to manage your pain. Historically, **opioids** have been used in this manner as they have powerful *analgesic* effects. They are also notorious for causing constipation.[31,32] I have listened to multiple radio advertisements for drugs that reduce opioid-induced constipation. There are many opioids, but the most well-known ones are heroin, morphine, oxycodone (OxyContin), hydrocodone (Vicodin), hydromorphone (Dilaudid), codeine, and fentanyl. Morphine and codeine are made naturally from the opium poppy plant (Fig. 1.4B). Opioids that are derived from the poppy plant are known as **opiates**. Think of it this way: all opiates are opioids, but not all opioids are opiates. This is analogous to saying that all bourbons are whisky, but not all whiskies are bourbon. Heroin, oxycodone, hydrocodone, and hydromorphone are considered to be semisynthetic opioids. These drugs are created in labs from natural opiates. Fentanyl is a fully synthetic opioid as it is created in a lab without the use of natural opiates. Opioids are primarily Schedule II drugs due to their ability to blunt pain resulting from surgery or from chronic conditions. Heroin is a major exception, as it is classified as a Schedule I drug.

Heroin, morphine, and fentanyl are primarily injected intravenously, whereas prescription opioids such as oxycodone, hydrocodone, hydromorphone, and codeine are taken in pill form. Like ADHD medications, prescription opioids can be misused by being crushed, diluted, and then injected intravenously. Not all individuals start with intravenous injections as this technique can be difficult to execute. Instead, some begin by "skin popping", injecting the opioid under their skin via the subcutaneous route. Because heroin often comes in powder form, individuals may snort or smoke it as opposed to injecting it. When someone uses an opioid, they experience intense euphoria and become less sensitive to pain.[33,34] Over time, individuals develop tolerance to the euphoric and the analgesic effects of opioids. More of the drug is needed to produce the same effects, often leading to dependence. Withdrawal symptoms, although not life-threatening, are extremely unpleasant. Common withdrawal symptoms include muscle aches, runny nose, restlessness, insomnia, diarrhea, nausea, vomiting, increased heart rate, and increased blood pressure.[19] Opioid withdrawal symptoms often resemble the flu. Another major withdrawal symptom is *dysphoria* (general feeling of lousiness).[35] Essentially, individuals going through opioid withdrawal feel terrible. The risk for overdose is extremely high with opioids, which is characterized by respiratory depression, coma, and death. The risk of overdose is particularly high for fentanyl, as it is 50–100 times stronger than morphine,[36] meaning that very small amounts can lead to death.

Sedative-hypnotics

Another major class of depressants is composed of *benzodiazepines* and *barbiturates*. Collectively, these drugs are known as **sedative-hypnotics** and are manufactured by pharmaceutical companies (Fig. 1.4C). These drugs have anxiolytic (i.e., anti-anxiety), hypnotic (i.e., induce sleepiness), and anesthetic effects. The most well-known benzodiazepines are diazepam (Valium), alprazolam (Xanax), and clonazepam (Klonopin), although there are numerous benzodiazepines. Zolpidem (Ambien) is a benzodiazepine-like drug that shares many features of common benzodiazepine drugs. Common barbiturates include amobarbital (Amytal), pentobarbital (Nembutal), and secobarbital (Seconal). Sedative-hypnotics are taken orally in pill form, but they can be misused in similar fashion as prescription stimulants and opioids. Benzodiazepines are Schedule IV drugs, and barbiturates fall anywhere between Schedule IV and Schedule II.

If benzodiazepines and barbiturates are both considered sedative-hypnotics, why are some barbiturates Schedule II drugs and not Schedule IV like benzodiazepines? It comes down to safety. Although both drug classes reduce anxiety and increase sedation, barbiturate overdose is more severe compared to benzodiazepine overdose. At a high enough dose, barbiturates can lead to respiratory depression, coma, and death.[37] Overdosing on benzodiazepines is much more difficult. Explaining why this is the case is beyond the scope of this chapter. As tolerance develops to the sedative/hypnotic effects of benzodiazepines/barbiturates, individuals will increase how much drug they take. Withdrawal symptoms are similar to those discussed for ethanol. Individuals can experience more mild symptoms such as anxiety, insomnia, agitation, and elevated blood pressure, but they can also experience more severe symptoms like DTs.[19] Like ethanol, withdrawal from sedative-hypnotic drugs can be life threatening.

Gamma-hydroxybutyrate (GHB)

The drug gamma-hydroxybutyrate (GHB) (also known as sodium oxybate) is often used in clubs and at raves. When GHB is used in this manner, the DEA categorizes it as a Schedule I drug. However, GHB is sold as the prescription drug Xyrem, which is used to treat the sleep disorder narcolepsy, which is marked by excessive sleepiness and recurring daytime sleep attacks. Xyrem is classified as a Schedule III drug. GHB is manufactured in labs and can be sold as either a powder or in liquid form. The powder form of GHB is commonly dissolved in a beverage and consumed orally. GHB increases euphoria, libido, and tranquility, but it can also cause excessive sweating, nausea, vomiting, headaches, exhaustion, confusion, and clumsiness.[38] At high enough doses, GHB can lead to hallucinations, loss of consciousness, and amnesia. Because GHB can interfere with consciousness and memory, it is considered a "date-rape" drug as it can be placed in alcoholic drinks to subdue an individual. Overdose is possible and potentially fatal as symptoms include sedation, seizures, coma, and respiratory depression.[38] Tolerance will develop to GHB's effects, and withdrawal symptoms include sweating, insomnia, anxiety, and tremors.[38]

Hallucinogens

Hallucinogens are known for their ability to distort a user's perception of sensory events. For example, individuals using a hallucinogen may report seeing vivid colors or may hear things that are not actually there. Hallucinogens can be divided into two major classes: **psychedelic (classic) hallucinogens** and **dissociative hallucinogens**.

Psychedelic hallucinogens

The most well-known psychedelic hallucinogens are D-lysergic acid diethylamide (LSD) (Fig. 1.5A), psilocybin (4-phosphoryloxy-N,N-dimethyltryptamine) (Fig. 1.5B), mescaline (peyote) (Fig. 1.5C), and dimethyltryptamine (DMT). Psychedelic hallucinogens are derived from natural sources. LSD was first synthesized by Dr. Albert Hofmann (1906–2008) in 1938 from ergot, a fungus that grows on grains like rye. Hofmann discovered LSD's hallucinogenic effects in 1943 when he accidently ingested some of the drug.[39] Psilocybin comes from a type of mushroom, and mescaline comes from the peyote cactus. DMT can be synthesized from different types of plants. All psychedelic hallucinogens are classified as Schedule I drugs, which has received criticism as recent evidence has emerged showing that they have therapeutic uses.[40] LSD is currently being investigated as a treatment for alcohol and nicotine use disorders,[41,42] and psychedelic hallucinogens can treat conditions such as posttraumatic stress disorder (PTSD).[43] In addition to treating alcohol and nicotine use disorders, self-report data indicate reduced consumption of drugs like stimulants and opioids following use of psychedelic hallucinogens.[44]

Most psychedelic hallucinogens are ingested orally, although DMT can be inhaled. In addition to hallucinations, psychedelic hallucinogens can increase heart rate and nausea, as well as distorting how one senses time (e.g., time seems to pass very slowly). Depending

FIGURE 1.5 The various hallucinogens. Psychedelic hallucinogens include LSD (A), psilocybin (B), and peyote (C). Dissociative hallucinogens include ketamine (D) and PCP (E). *Panel (A) comes from Wikimedia Commons and was posted by William Rafti of the William Rafti Institute (https://commons.wikimedia.org/wiki/File:Ruby_Slippers_LSD_Sheet.jpg). Panel (B) comes from the DEA (https://www.dea.gov/factsheets/psilocybin). Panel (C) comes from the DEA (https://www.dea.gov/factsheets/mescaline-and-peyote). Panel (D) comes from Wikimedia Commons and was posted by user Doc James (By Doc James - Own work, CC BY-SA 4.0, https://commons.wikimedia.org/w/index.php?curid=80359732). Panel (E) comes from Wikipedia (https://en.wikipedia.org/wiki/Phencyclidine).*

on which specific drug is taken, other effects associated with psychedelic hallucinogens include increased blood pressure, breathing, and body temperature; loss of appetite; dry mouth; sleep problems; uncoordinated movements; excessive sweating; panic; paranoia; and feelings of relaxation.[45–47] Fatal overdose resulting from psychedelic halluincogen use is rare, but these drugs can cause unpleasant hallucinations known as a "bad trip".[48] With prolonged drug use, tolerance can develop to some psychedelic hallucinogen drugs like LSD but not others such as DMT. In contrast to many drugs, psychedelic hallucinogens do not readily produce withdrawal symptoms.[49]

Dissociative hallucinogens

In addition to producing hallucinations, dissociative hallucinogens cause feelings of detachment or *dissociation*. Users feel as if they are viewing themselves from above or afar. The two most widely known dissociative hallucinogens are ketamine (Fig. 1.5D) and phencyclidine (PCP) (Fig. 1.5E), but there are other dissociative hallucinogens like dextromethorphan (a cough suppressant found in over-the-counter cold and cough medicines) and salvia (a type of plant found in Southern Mexico and in Central and South America). Although salvia is naturally occurring, most dissociative hallucinogens are synthesized in a lab. Georges Mion[50] provides an interesting history of PCP and ketamine. PCP was first synthesized by Parke-Davis and Company's laboratory in Detroit, Michigan in 1956 to be used as an anesthetic. In 1962, ketamine was synthesized to replace PCP as an anesthetic. To this day, ketamine is often used as an anesthetic for rodents. Dissociative hallucinogens are either unscheduled by the DEA (dextromethorphan and salvia) or are placed in Schedule II (PCP) or Schedule III (ketamine).

Dissociative hallucinogens are associated with various routes of administration. Ketamine is taken orally (in pill or liquid form) or intranasally. PCP is primarily smoked or injected, but it can be taken in pill form or snorted. Because dextromethorphan is found in over-the-counter medicines, it is taken orally. Salvia can be smoked or consumed orally. Specifically, individuals can consume raw or dried salvia, or they can brew it in a tea. Unlike psychedelic hallucinogens, dissociative hallucinogens readily produce withdrawal symptoms, which include rapid heart rate, stiff muscles, depressed breathing, convulsions, altered body temperature, headaches, and sweating.[51] Dissociative hallucinogens are more likely to lead to fatal overdose compared to psychedelic hallucinogens.[51]

Entactogens (empathogens)

Entactogens (also known as empathogens) primarily increase feelings of empathy and sympathy (hence the name *empath*ogen), although they also exhibit psychostimulant and hallucinogenic properties. The most well-known entactogen is 3,4-methylenedioxymethamphetamine (MDMA; "ecstasy") (Fig. 1.6). Other entactogens include 3,4-methylenedioxyamphetamine (MDA), 3,4-methylenedioxy-N-ethylamphetamine (MDEA), and 3,4-methylenedioxy-N-hydroxyamphetamine (MDOH). It is much easier to refer to these drugs by their abbreviations. These drugs are taken orally in pill form. All entactogens are Schedule I drugs with no approved medical uses, although this designation could change one day as MDMA has recently been tested as a potential treatment for PTSD.[52] Due to their stimulant-like effects and their ability to increase empathy, entactogens are often associated with raves and dance parties. The depressant drug GHB, which you learned about

FIGURE 1.6 The entactogen MDMA. *This image comes from the DEA (https://www.dea.gov/factsheets/ecstasy-or-mdma-also-known-molly).*

earlier, is sometimes categorized with entactogens because it can increase prosocial behavior.[53] Entactogens increase heart rate and blood pressure, while decreasing appetite.[46,54] Overdosing on entactogens can be life-threatening as drugs like MDMA cause hyperthermia and dehydration.[55,56] Tolerance develops to entactogens over time, and withdrawal symptoms for drugs like MDMA include depression, anxiety, insomnia or fatigue, changes in appetite, difficulty concentrating, and headaches.[57]

Cannabis and synthetic cannabinoids

I assume many of you reading this textbook are familiar with cannabis, better known as marijuana, considering approximately 48.2 million individuals have reported using it during the past year.[6] Cannabis comes from the cannabis plant, of which three species are currently recognized: *Cannabis sativa*, *Cannabis indica*, and *Cannabis ruderalis* (Fig. 1.7A and B). The main psychoactive compound found in cannabis is Δ-9-tetrahydrocannabinol (THC). THC is just one of many *cannabinoids* found in cannabis. For example, the cannabinoid cannabidiol (CBD) is often sold in shops as CBD oil. Cannabis is often smoked, but it can be ingested orally as it can be placed in food (known as "edibles"). When someone uses cannabis, they feel euphoric and may experience altered states of mind and time, but they may also have

FIGURE 1.7 Cannabis leaves (A) and buds (B), as well as the synthetic cannabinoid K2. *Panel (A) comes from Wikimedia Commons and was posted by user Chmee2 (By Chmee2 - Own work, CC BY 3.0, https://commons.wikimedia.org/w/index.php?curid=11437037). Panel (B) comes from Wikimedia Commons and was posted by Evan Amos (https://commons.wikimedia.org/wiki/File:Marijuana-Cannabis-Weed-Bud-Gram.jpg). Panel (C) comes from the DEA (https://www.dea.gov/factsheets/spice-k2-synthetic-marijuana).*

difficulty concentrating and can experience impairments in short-term memory. Cannabis is also well known for its ability to stimulate appetite,[58] which is often referred to as the "munchies". At higher doses, cannabis can increase anxiety, paranoia, delusions, and hallucinations.[59] Once upon a time, I was an undergraduate student. One night my roommate and his girlfriend at the time were using cannabis. My roommate's girlfriend started seeing images of melting faces and started to panic. At the time, I knew very little about cannabis and its effects on cognition, so I had no idea what my roommate's girlfriend was experiencing.

One advantage of cannabis is that fatal overdose is extremely rare. However, overdose can still occur, which leads to some of the more severe effects of cannabis described above (e.g., paranoia, delusions, hallucinations). Cannabis tolerance develops with repeated exposure, and withdrawal symptoms manifest when someone tries to stop using cannabis. Individuals can experience feelings of depression, decreased appetite, irritability, insomnia, cold sweats, and mood changes.[19] Like the other drugs we have discussed thus far, people can develop a SUD to cannabis. One common misconception I have heard over the years is that people cannot become "addicted" to cannabis, but that is not true according to the scientific literature. Cannabis use disorder is recognized in the *DSM-5*.

Beyond recreational use, cannabis can be used medically. The fact that cannabis has appetite-stimulating effects is useful for helping improve appetite in individuals diagnosed with acquired immunodeficiency syndrome (AIDS).[60] Cannabis can also reduce nausea and vomiting during chemotherapy for cancer patients.[61] Finally, cannabis can be used to treat chronic pain.[62] Given that cannabis has medical uses, one would assume that the DEA would classify this drug as a Schedule II drug. However, it is considered a Schedule I drug. Why is this the case? From a historical perspective, cannabis was once labeled a "gateway" drug, meaning that its use would introduce individuals to more drugs. When the DEA first started scheduling drugs, cannabis' potential therapeutic effects had not been explored. Efforts have been made to decriminalize cannabis possession and use. Currently, cannabis possession and use are legal in 15 states and the District of Columbia, and most states have provisions that allow for medical use of cannabis. Idaho and South Dakota are the only remaining states in which cannabis is illegal (cannabis use is illegal in Nebraska, but it has been decriminalized).

Although cannabis is derived naturally from the cannabis plant, synthetic cannabinoid use increased during the 2010s (Fig. 1.7C depicts the synthetic cannabinoid K2). These cannabinoids are synthetized in a lab and are often sprayed on dried leaves that can then be smoked. Synthetic cannabinoids can also be sold as a liquid that can be vaporized or used in an e-cigarette. Well-known brands of synthetic cannabinoids included K2, Spice, Joker, Black Mamba, Kush, and Kronic. Like cannabis, synthetic cannabinoids are Schedule I drugs. Although the effects of synthetic cannabinoids are similar to cannabis, overdosing on synthetic cannabinoids is more dangerous compared to overdosing on cannabis, as kidney damage and seizures have been reported.[63-65]

Inhalants

Inhalants consist of substances that are taken by the inhalation method only (known as "huffing"). These substances include solvents, aerosol sprays, and gases. Common inhalants include spray paints, markers, glues, and paint thinners/removers, gasoline, lighter fluid, hair sprays, ether, and nitrous oxide. Because inhalants are found in many household items, they are currently not scheduled by the DEA. Individuals that use inhalants feel euphoric, but they may experience dizziness, slurred speech, and light headedness.[66] Some individuals may experience hallucinations and delusions. Because these products contain harmful chemicals that are not intended for human consumption, inhalant use can be extremely dangerous. Overdose can result in seizures, coma, and death.[66,67] Tolerance occurs with repeated use, and withdrawal symptoms include nausea, loss of appetite, sweating, sleep disturbances, and mood changes.[68]

Quiz 1.1

1. An individual has been prescribed methylphenidate to treat their ADHD symptoms. After taking the drug for several years, they abruptly abstain from further use. The individual reports feeling lousy and more tired than normal. Can we conclude that this individual is psychologically dependent on methylphenidate?
2. Which of the following drugs makes the user feel as if they are viewing themselves from above?
   a. Amphetamine
   b. LSD
   c. Cannabis
   d. PCP

3. What is the major difference between the *DSM* and the *ICD* when classifying SUDs?
4. If someone wants to become intoxicated as quickly as possible, which route of administration should they use?
   a. Inhalation
   b. Intramuscular injection
   c. Oral
   d. Transdermal
5. Imagine a company invents a new drug for weight loss. Even though the drug can help people lose weight, there are reported cases of individuals crushing the pill and snorting it to become intoxicated. Based on this information, into which schedule should the DEA place this new drug?
   a. I
   b. II
   c. IV
   d. V

Answers to quiz 1.1

1. No. This person is exhibiting physical dependence. Physical dependence, on its own, is insufficient evidence for psychological dependence or a diagnosis of a SUD.
2. d — PCP

3. The criteria used for determining the presence of a SUD differs across each publication. The *DSM* defines SUDs on severity, whereas the *ICD* defines SUDs as meeting a minimum number of criteria.
4. a — inhalation
5. b — II

Drug combinations

So far, I have discussed drugs in isolation from one another. However, individuals often engage in **polydrug use**, the use of two or more substances. There are two ways to define polydrug use: concurrently or simultaneously. The term concurrent drug use describes an individual that uses two or more substances independently of one another, and simultaneous drug use refers to an individual that combines two or more substances during a single use. Drug combinations allow the user to either (a) extend the effects of one drug that has already been consumed, (b) enhance the effects of one drug, or (c) reduce the effects of one drug.[69] There are numerous ways that one can combine drugs. For simplicity, I will cover some of the more common drug combinations.

One common drug combination is ethanol and nicotine. As noted by Shiffman and Balabanis,[70] "drinkers smoke and smokers drink" (p. 107). According to the National Institute on Drug Abuse (NIDA),[71] approximately 23.8% of individuals with an alcohol use disorder have a nicotine use disorder, although only 12.9% of individuals with a nicotine use disorder have a co-occurring alcohol use disorder. There are two ways to describe concurrent ethanol and tobacco use. Individuals that consume ethanol are more likely to use tobacco products and vice versa. This is referred to as a between-person interaction. Second, individuals that use both substances may use them at the same time in certain situations. This is known as a situational interaction. Anecdotally, I had a friend in high school that only smoked cigarettes when he drank ethanol at parties. This demonstrates a situational interaction. Not only is cooccurrence of cigarette and ethanol use high, but approximately 20%–34% of individuals that use ethanol also smoke cannabis.[72]

An extremely popular drug combination is ethanol and caffeine. We have not discussed caffeine in this chapter, but caffeine is a drug, a stimulant to be exact. In fact, it is the most widely used drug in the world.[73] I can safely assume that many of you reading this book have had caffeine at some point in your life, whether in the form of coffee, tea, carbonated sodas, energy drinks, or chocolate. At bars, Jägermeister is combined with the energy drink Red Bull to create the Jagerbomb. Up until the end of 2010, the United States allowed the sale of alcoholic drinks premixed with energy drinks (e.g., Four Loco, Sparks, and Tilt). My colleague Dr. Cecile Marczinski studies how combining ethanol and energy drinks affects cognition and behavior. Some of her work has shown that people given alcohol mixed with energy drinks (AmED) report a greater desire to continue drinking compared to those given just ethanol or a placebo drink (no ethanol or energy drink).[74,75] You can read more about these results in the Experiment Spotlight below.

Experiment Spotlight: Desire to drink alcohol is enhanced with high caffeine energy drink mixers (Marczinski et al.[75])

Alcoholic beverages premixed with energy drinks are no longer commercially available in the United States, but young individuals still combine ethanol with energy drinks such as

Red Bull. As noted by Marczinski et al.[75] one physiological effect of caffeine is to increase the consumption of food and beverages. One concern is that combining energy drinks with ethanol can increase one's willingness to continue consuming alcoholic beverages that contain caffeine. The goal of this experiment was to determine if college students have a greater desire to continue drinking an alcoholic beverage combined with an energy drink compared to an alcoholic beverage containing no energy drink. Twenty-six social drinkers aged 21–30 were recruited to participate in this study. Before completing the study, participants were given a general health questionnaire and various surveys to measure impulsivity, personal drinking habits, and caffeine use. Using a *double-blind, placebo-control* study, Marczinski et al.[75] had participants attend six sessions in which they consumed one of the following beverages:

(1) 1.21 mL/kg vodka mixed with 3.63 mL/kg decaffeinated soft drink
(2) 1.21 mL/kg vodka mixed with 3.63 mL/kg energy drink
(3) 1.21 mL/kg vodka mixed with 6.05 mL/kg energy drink
(4) 3.36 mL/kg decaffeinated soft drink
(5) 3.36 mL/kg energy drink
(6) 6.05 mL/kg energy drink

With a double-blind, placebo-control study, both the researcher and the participant do not know which beverage is being consumed during a session. Importantly, the order that each participant consumed each beverage was *counterbalanced*; in other words, some participants consumed the 3.36 mL/kg energy drink first, while others consumed the vodka/energy drink first, and so on. Additionally, each participant was asked to consume each beverage within 5 min. At different time points following consumption of each beverage, participants provided a breath sample and completed the Desire-for-Drug questionnaire.

Compared to each dose of the energy drink and to the decaffeinated soft drink, ethanol administered by itself increased the desire to consume more of the beverage. When combined with either dose of the energy drink, participants reported a greater desire to continue drinking compared to ethanol alone.

Questions to consider:

(1) Why do you think using a double-blind, placebo-control study is important?
(2) Why did the researchers give the participants questionnaires about their health, ethanol use, caffeine use, and impulsivity?
(3) Why is counterbalancing important for a research study?

Answers to questions:

(1) First, if the participant knows which type of drink they are consuming, this can be problematic. If the participant suspects they are receiving an alcoholic beverage mixed with an energy drink, they may state they have a higher desire to continue drinking because this is what he/she anticipates the researchers are trying to discover in their study. This is known as a *demand characteristic* and can ruin the results of a study. In some studies, the researcher needs to avoid knowing what they are giving to the participant. If the researcher knows they are about to give a participant the ethanol/energy drink combination, they may provide subtle cues that could influence the

participants' responses. This is known as *experimenter bias* and can also ruin an experiment. In any study, we want to compare the results of our experimental compound/drug to a control (placebo). In the current experiment, Marczinski et al. compared the desirability of the ethanol/energy drink combination to a decaffeinated soft drink and to energy drinks. The fact that they observed increased desirability to drink following administration of the combined ethanol/energy drink beverage and not following the decaffeinated soft drink or the energy drinks allows the researchers to conclude that the increased desirability is specific to the drink combination and not to some other factor (such as satiating thirst).

(2) The researchers want to ensure that there are no other factors that could influence the results of the study. As you will learn later in this book, impulsivity is a major behavioral/personality factor associated with SUDs. Theoretically, Marczinski et al. could have found that only high impulsive participants responded more favorably to the ethanol/energy drink combination. This would certainly be an interesting finding, but this is something that researchers try to take into account when conducting experiments. Providing a general health survey is important because researchers want to ensure that the participants do not have health issues that could cause them to have an adverse reaction to ethanol or to the energy drink.

(3) In research, counterbalancing is a critical component of research. Because each participant received all six drink types in this experiment, the researchers do not want each person to drink the same drinks in the same order. This could potentially bias the results of the study. If every person started with the decaffeinated soft drink and then had the energy drinks followed by the alcoholic beverages, participants may be able to better detect which drinks they are receiving, which could then lead to the demand characteristics I mentioned above.

One interesting finding reported from Dr. Marczinski's laboratory is that individuals given AmED beverages do not show as many motor control deficits associated with ethanol use compared to individuals given ethanol alone,[76] and individuals given an AmED report feeling less fatigue.[77] Other research has shown that combining ethanol with energy drinks can make someone feel less intoxicated.[78] These findings are important because some may report feeling less impaired by ethanol if consuming AmEDs and may want to attempt to drive home.

Combining ethanol with other depressant drugs such as opioids, benzodiazepines, and barbiturates dramatically increases the risk of overdose death. Because each drug class acts to suppress activity in our brain, combining two or more depressants increases the probability of respiratory failure. Even benzodiazepine overdose can lead to death when combined with ethanol. Former professional wrestler Joan Laurer, better known by the moniker "Chyna", died in 2016 after combining ethanol with opioids and benzodiazepines. Cory Monteith, an actor best known for his role in the show *Glee*, died of a heroin and ethanol overdose in 2013.

Ethanol can also be combined with stimulants such as cocaine and ADHD medications. Individuals combine ethanol with stimulants to mitigate some of the negative side effects of the stimulant, such as anxiety and twitching. In turn, the stimulant counteracts some of the depressant effects of ethanol, like drowsiness. Combining ethanol with cocaine is dangerous. When ethanol and cocaine are metabolized by the liver, cocaethylene is produced. In short, cocaethylene is not good for the body, as it negatively impacts multiple organs, particularly the heart and the liver.[79] Blood pressure can spike, and individuals can suffer a heart attack.[80] Because blood vessels and brain tissue can become damaged, individuals are at heightened risk of developing a stroke or an aneurysm.[81]

One recent trend that has increased in popularity in Europe and other parts of the world is combining cannabis and tobacco and smoking it in a bong (known as a *spliff* in the United States and as a *moke* in other regions). This trend has become so popular that some have estimated that up to 90% of European cannabis users mix cannabis with tobacco.[82] There are some issues related to combining cannabis and tobacco. One study found that smoking tobacco with cannabis can increase symptoms of cannabis dependence.[83] Combining cannabis and tobacco can lead to lung problems in individuals as early as in their mid-30s.[84]

In addition to tobacco, cannabis can be laced with PCP, heroin, LSD, methamphetamine, ketamine, and cocaine, as well as other chemicals like formaldehyde.[85] Cannabis cigarettes laced with other substances are known as "wet" or "fry". Cannabis is popular because of its ability to reduce the effects of other drugs. For example, one reported reason that cannabis is combined with MDMA is to reduce the stimulant effects of MDMA. The stimulant-like effects of MDMA can be aversive to some users. As described by one person:

> What a lot of people hate about E is the speediness from it. Even if it's supposedly pure or whatever … it's what makes your teeth grind and makes you all jittery, and uncomfortable basically. And the weed definitely takes the edge off, like so much, and it helps with the nausea and it helps with all the negative effects, to make it …. the whole thing like pretty pleasurable or as pleasurable as possible. - Hunt et al. (p. 506).[69]

In clubs, entactogens like MDMA are often combined with other drugs. In fact, one study reported that approximately 92% of 18—29-year-old individuals in New York City that use club drugs engage in polydrug use.[86] In addition to cannabis, common MDMA drug combinations include ketamine, cocaine, LSD, GHB, methamphetamine, and ethanol.[86,87] Grov et al.[86] found that approximately half of the sample reported using MDMA and cannabis. While cannabis can blunt the effects of MDMA, combining MDMA with a psychostimulant like cocaine or methamphetamine can enhance and extend the effects of MDMA. However, combining MDMA with psychostimulants is dangerous due to MDMA's stimulant properties. By combining two stimulant-like drugs, the user can easily overdose.

The final combination I want to highlight is known as the "speedball", which consists of cocaine and heroin/morphine, although other stimulants can be used in place of cocaine, and benzodiazepines/barbiturates can substitute opioids (known as a "set up"). Regardless, the speedball is a particularly risky drug combination as the effects of the stimulant subside at a faster rate compared to the opioid.[88] This can lead to increased risk of respiratory depression as an individual may not realize the dose of the opioid is too high. The famous comedian Chris Farley died of an overdose after combining cocaine and morphine. More recently, the actor Phillip Seymour Hoffman, after years of sobriety, died after combining stimulants with heroin and benzodiazepines.

Although polydrug use is popular, extreme caution needs to be taken when combining two or more drugs. As detailed in this section, combining drugs with similar effects can be life-threatening. Combining depressants can lead to respiratory failure and coma, whereas combining stimulants can lead to cardiovascular issues such as heart attack. Even combining drugs with different profiles can be dangerous, as noted above with the speedball.

Prevalence of SUDs

How prevalent are SUDs? Earlier, I provided some statistics from the *NSDUH* regarding ethanol and drug use reported from Americans. Importantly, the number of individuals with a SUD is predicted from the responses of 67,625 participants that completed the *NSDUH*. With a sample size this large, we can accurately predict the number of Americans living with a SUD. This section will first provide some statistics regarding the prevalence of SUDs in the United States.

In the United States

As of 2019, there are approximately 20.4 million individuals aged 12 or older that have been diagnosed with a SUD.[6] An estimated 14.5 million individuals had an alcohol use disorder during the past year, and 8.3 million individuals had a nonalcohol use disorder. Approximately 2.4 million people have a concurrent alcohol use disorder and illicit drug use disorder. The *NSDUH* reports SUD rates across different age groups: 12–17 years, 18–25 years, and 26 years or older. SUDs are most common in 18–25-year-olds (14.1 million vs. 4.5 and 6.7 million 12–17-year-olds and over 26–year-olds, respectively).

The *NSDUH* does not report information on tobacco use disorder. However, the *DSM-V* notes that nicotine dependence is reported in 13% of adults 18 years of age or older. The number of individuals with nicotine dependence decreases across age groups, from 17% of those aged 18–29 to 4% of those aged 65 years or older. Alcohol use disorder is the most prevalent SUD in the United States, with 414,000 adolescents, 3.1 million young adults, and 11 million adults aged 26 or older meeting diagnostic criteria. However, the number of individuals with a current alcohol use disorder has steadily decreased since 2002 for each age group. Unlike alcohol use disorder, the number of those with a cannabis use disorder has remained relatively constant since 2002 (4.3 million vs. 4.8 million). When examining different age groups, cannabis use disorder diagnoses have decreased significantly for those aged 12–17 years (1.1 million in 2002 vs. 699,000 in 2019) but have increased for those aged 26 or older (1.4 million in 2002 vs. 2.2 million in 2019). The number of 18–25-year-old individuals diagnosed with a cannabis use disorder has remained consistent during the past two decades, with an estimated 2 million people.

Prevalence of cocaine use disorder is lower now compared to 20 years ago for all age groups (1.5 million people in 2002 vs. 1 million in 2019), although the number of individuals diagnosed with a cocaine use disorder remained relatively stable during the 2010s. Prescription stimulant use disorder and methamphetamine use disorder have been tracked since 2015. In 2019, approximately 1 million people were diagnosed with a methamphetamine use disorder, whereas 558,000 individuals had a prescription stimulant use disorder.

In contrast to other drugs, heroin use disorder has increased substantially during the past two decades. Since 2002, the number of individuals diagnosed with a heroin use disorder has more than doubled (214,000 in 2002 vs. 438,000 in 2019). The increased prevalence of heroin use disorder seems due to a shift away from prescription opioid use. In 2015, approximately 2 million individuals met the criteria for prescription opioid use disorder, but only 1.4 million people had a prescription opioid use disorder in 2019. This shift away from prescription opioids to heroin may result from higher availability of heroin and decreased costs compared to prescription opioids. The number of individuals with a benzodiazepine/barbiturate use disorder (labeled as prescription tranquilizer or sedative use disorder in the *NSDUH*) has remained consistent since 2015, with approximately 681,000 having a benzodiazepine/barbiturate use disorder.

The prevalence of inhalant use disorder is lower compared to other drug classes, with a rate of 0.02% among all Americans aged 18 or older. Teenagers are most likely to be diagnosed with an inhalant use disorder, with a prevalence rate of 0.4% for all individuals aged 12–17. This makes sense as the chemicals that are used for huffing are easy to obtain and are inexpensive to procure. Like inhalant use disorder, the number of individuals with a hallucinogen use disorder is low. According to the *DSM-5*, the number of individuals with a PCP or a ketamine use disorder is unknown, although the number of individuals reporting using either substance during the past year is estimated to be 1.6–2.5% of the population. When all other hallucinogens are considered, only 0.5% of teenagers aged 12–17 and 0.1% of adults 18 years of age or older meet the criteria for a hallucinogen use disorder.

Across the world

The *Global Burden of Diseases, Injuries, and Risk Factors Study (GBD)* provides some statistics on world-wide SUD rates.[89] This study focuses on SUDs related to the following substances: ethanol, opioids, cocaine, amphetamines, cannabis, and other (any drug not else listed here). Even though some countries restrict the use of ethanol, alcohol use disorder is still the most prevalent SUD in the world, with an estimated 100.4 million cases as of 2016. Opioid use disorder and cannabis use disorder are the most common drug use disorders, with an estimated 26.8 and 22.1 million cases, respectively. All other drugs combined account for less than 14 million diagnoses of SUDs. The prevalence of SUDs varies across geographical areas. Alcohol use disorder is most pronounced in Eastern Europe, with 4,245.6 cases per 100,000 individuals. Australasia (encompassing Australia, New Zealand, New Guinea, and Melanesia) has the highest rates of amphetamine use disorder (574.2 cases per 100,000 people) and "other drug" use (154.6 cases per 100,000 people). North America, specifically high-income areas, has the highest incident rates of cannabis, cocaine, and opioid use disorders (884.3, 524.1, and 1,168.3 cases per 100,000 people, respectively). Because the Islamic faith strictly prohibits ethanol consumption, North Africa and the Middle East, regions with a large Muslim population, have the lowest rates of alcohol use disorder, with only 593 cases per 100,000 people reported.

In other psychiatric conditions

The term **comorbidity** is used to describe two or more disorders/illnesses occurring in the same individual. SUDs are common in individuals with another psychiatric condition. Individuals with schizophrenia, a complex condition characterized by hallucinations, delusions, and disordered thinking, are more likely to use nicotine, ethanol, cannabis, and cocaine.[90] Nicotine use in schizophrenia is particularly high, with approximately 70% of all individuals with schizophrenia using tobacco products.[91] One potential reason for the increased tobacco use in schizophrenia is related to nicotine's cognitive enhancing effects. Because schizophrenia is marked by deficits in cognition, individuals with schizophrenia may use tobacco products to "self-medicate"; that is, they are trying to decrease their symptoms by smoking.[91] Cultural differences have been shown to affect the comorbidity rate between schizophrenia and tobacco use. In Sri Lanka, 30.41% of males and just 1.90% of females with schizophrenia use tobacco products.[92] The lower rates of tobacco use observed in Sri Lanka may be due to the inability of many individuals in this country to afford cigarettes.

Smoking rates are not just higher in individuals with schizophrenia. Over 70% of individuals with bipolar disorder smoke cigarettes,[93] and approximately 40%–60% of individuals with major depressive disorder smoke.[94] Overall, the percentage of individuals with bipolar disorder diagnosed with a SUD is 47.3% while individuals with major depression disorder have high rates of comorbid alcohol use disorder (40.3%) and drug use disorders (17.2%).[95] Borderline personality disorder (BPD), characterized by mood swings, distorted self-image, tumultuous relationships, impulsive behavior, and self-injurious behavior, is diagnosed in approximately 1%–2% of the general population; however, the percentage of individuals with BPD that have a SUD ranges from 9% to 65%.[96] One challenge to treating comorbid mood/personality disorders (e.g., depression/BPD) and SUDs is that individuals with such a diagnosis are more likely to harm themselves and have suicidal ideations.[97]

Individuals with ADHD have higher rates of SUDs, particularly for ethanol, nicotine, cannabis, psychostimulants, sedative-hypnotics, and opioids.[98] Specifically, children with ADHD are three times more likely to develop nicotine dependence, over twice as likely to develop cannabis use disorder and cocaine use disorder, and 1.7 times more likely to develop an alcohol use disorder during adolescence/adulthood.[99] Individuals with ADHD that have a comorbid psychiatric condition such as depression or a disruptive behavior disorder are at greater risk of developing a SUD later in life.[100] Given that the major treatments for ADHD are amphetamine and methylphenidate, two psychostimulants that have misuse potential, there have been concerns about prescribing these medications to those with ADHD. We will return to this discussion in Chapter 9.

Health and economic impacts of SUDs

Some health impacts of substance use have already been mentioned in this chapter. Overdose is a major concern for multiple drugs like opioids, barbiturates, cocaine, and methamphetamine. Drug withdrawal produces physiological symptoms that can make an individual feel sick. Long-term substance use is associated with other health complications that can negatively impact one's quality of life and can ultimately lead to death. In addition to health effects, SUDs have a dramatic economic impact, at both the personal and the societal level. This section will highlight some of the major health and economic impacts of SUDs.

Health impacts

Chronic nicotine use is especially problematic as tobacco products contain chemicals that have been linked to various cancers. Smoking can lead to lung cancer,[101] while smokeless tobacco (e.g., chew and snuff) is associated with oral and throat cancers.[102] Cancer is just one disease that can result from tobacco use. My father smoked for years and now lives with chronic obstructive pulmonary disease (COPD; note, emphysema and chronic bronchitis are two conditions that make up COPD). In COPD, the lungs become inflamed, which restricts airflow from the lungs. Individuals with COPD have difficulty breathing and tend to have a persistent cough. They may also experience wheezing and increased mucus production. Another lung-related disease is asthma. Individuals that smoke tobacco are at higher risk of developing Buerger's disease,[103] a condition that causes blood vessels to swell, leading to pain, tissue damage, and possibly gangrene. Smoking is also linked to heart disease, stroke, gum disease and tooth loss, and diabetes.[104–106] Approximately 16 million Americans live with a disease caused by smoking, and over 480,000 Americans die each year from complications related to smoking.[107] Across the world, approximately 7 million people die each year from smoking-related illnesses.[108]

When e-cigarettes were first introduced, they were marketed as being safer than traditional cigarettes because they did not contain the carcinogens found in cigarettes. However, recent research has found that e-cigarettes are not as safe as they appear. JUUL pods contain as much nicotine as 20 regular cigarettes.[109] Newer devices have larger batteries that can heat the liquid to a higher temperature compared to older models; this can lead to elevated levels of carcinogens like formaldehyde.[110] On the topic of e-cigarette batteries, there have been some reports of injuries resulting from thermal burns and exploding batteries.[111] Smoking e-cigarettes can lead to lung damage, a condition known as e-cigarette, or vaping, product use associated lung injury (or EVALI for short). Patients with EVALI have a cough and difficulty breathing.[112] Another risk associated with e-cigarettes is that the nicotine-containing liquid can be poisonous, especially in young children; fortunately, exposure to liquid nicotine has decreased in young children since 2015 as legislation now requires child-resistant packaging on e-cigarette products.[113] Overall, there are some serious concerns about the use of e-cigarettes, and more research is needed to determine the long-term health consequences of using these products.

Smoking-related illnesses are not isolated to tobacco products. Smoking cannabis carries some of the same health risks as smoking tobacco, as cannabis smoke contains many of the same carcinogens as tobacco smoke.[114] Individuals can experience increased cough and sputum (mucus coughed up from lungs); with long-term use, cannabis smoke can lead to symptoms of bronchitis.[115] If cannabis smoke contains some of the same carcinogens as tobacco smoke, does this mean that smoking cannabis can cause lung cancer? So far, research has not been able to provide clear evidence of the association between smoking cannabis and lung cancer. As noted by Jett et al.[116] one difficulty in ascertaining the contribution of cannabis smoke to lung cancer is finding enough people that are classified as heavy cannabis smokers that do not smoke tobacco. Recall that tobacco use and cannabis use often cooccur. More research is merited in this area, and over the years we should be able to collect enough evidence to determine if smoking cannabis leads to increased risk of lung cancer, especially now that cannabis legalization is spreading across the United States.

Long-term ethanol use causes irreparable damage to certain parts of the body. Because the liver is largely responsible for filtering out toxins (like ethanol), long-term ethanol use can lead to cirrhosis.[117] In cirrhosis, scar tissue known as *fibrosis* forms on the liver as it tries to repair damaged tissue. As this disease progresses, individuals become tired and weak, develop jaundice (yellowing of the skin), bruise easily, and may experience fluid buildup in the abdomen. Cirrhosis can also lead to hepatic encephalopathy, a condition associated with forgetfulness, confusion, and irritability.[118] Cirrhosis is life-threatening, with approximately 1 million people dying from it each year.[119]

In addition to liver damage, chronic ethanol use can damage part of the brain. We will cover neuroanatomy in Chapter 3, but I want to highlight a part of the brain called the *mamillary bodies*. Over time, ethanol consumption leads to damage to this structure. Ethanol in itself does not damage the mamillary bodies. Instead, individuals with alcohol use disorder often fail to get enough vitamin B1 (thiamin/thiamine), which then leads to brain damage.[120] Damage to the mamillary bodies causes *Korsakoff's syndrome*, a disease marked by memory deficits. Individuals with Korsakoff's syndrome may have difficulty learning new information (known as *anterograde amnesia*) and may have difficulty remembering past events (known as *retrograde amnesia*). These individuals may also *confabulate* events that they cannot remember. When someone engages in confabulation, they are not deliberately trying to lie or deceive someone. They legitimately believe the event occurred as described.

You have probably heard the expression "beer belly". There is a reason this term is used, as another health-related consequence of ethanol use is weight gain. Ethanol is calorically dense, with a single beer ranging anywhere from 80 to 220 calories and shots of liquor averaging around 100 calories per shot. Mixed drinks such as pina coladas can have up to 500 calories per drink! That is equivalent to a full meal. Given that a popular drug combination is ethanol and energy drinks, this can be problematic as energy drinks contain high levels of sugar and have between 250 and 300 calories per drink.

Injection of drugs, regardless of drug class, carries several health-related risks. Individuals that share needles are at heightened risk of contracting a sexually transmitted infection (STI) such as human immunodeficiency virus (HIV; leads to AIDS) or hepatitis. Individuals may also discard empty syringes in a public space (e.g., playground) where another person may accidently prick themself with the needle. Because drugs and/or unsterile needles can be contaminated with pathogens,[121] injection sites can become infected, causing abscesses. One health complication that can arise from repeated intravenous injections is endocarditis,[122] inflammation of the interior lining of the heart. If left untreated, endocarditis can damage heart valves and can lead to death.

Psychostimulants often produce repetitious behaviors known as *stereotypies*. Individuals may pick at their skin, resulting in scabs and lesions. Another issue related to psychostimulants is psychosis at high enough doses. Psychosis can lead some to engage in self-injurious behaviors. For example, in 2018, a 21-year-old woman named Kaylee Muthart snorted and injected methamphetamine that had been tainted with another substance. After being awake for nearly 40 straight hours, Kaylee began hallucinating and believed that she had to remove her eyes to save the world. Horrifyingly, Kaylee physically gouged her own eyes out while in the methamphetamine-induced psychosis. Although she survived the incident, Kaylee is now permanently blind.[123]

Speaking of methamphetamine, this stimulant can lead to a condition known as "meth mouth". Over time, individuals that use methamphetamine can lose their teeth and damage their remaining teeth. This occurs for several reasons. One, the intoxicating effects of methamphetamine last longer compared to other psychostimulants; think of Kaylee who was awake for nearly two straight days. During this state of intoxication, I doubt individuals worry about oral hygiene. This can lead to damaged gums, which are important for anchoring our teeth in place. Additionally, stimulant use dries out the mouth. Saliva is important for our gums. The lack of saliva further breaks down gums. Third, individuals on methamphetamine grind their teeth, one of the stereotypies associated with stimulant use, which can crack/damage teeth.

Finally, methamphetamine that is manufactured in home labs contains corrosive materials that damage teeth. Fig. 1.8 shows the damage methamphetamine can have on one's teeth. Although meth mouth is not life-threatening, prolonged stimulant use can have catastrophic effects on the heart. For example, cocaine has been shown to cause inflammation of the heart and can lead to aortic ruptures.[124] Individuals that use stimulants are at heightened risk of stroke.[125] Acclaimed singer Whitney Houston died in 2012 after accidently drowning in her bathtub. Her death was linked to heart disease caused by long-term cocaine use.

Drugs do not just affect the user. Drug use during pregnancy carries significant risks for a developing fetus. Because drugs can cross the placenta, babies can go into withdrawal when they are born.[126] Some drugs can cause irreversible damage to a fetus. For example, ethanol use during pregnancy can lead to *fetal alcohol syndrome (FAS)*. FAS causes physical and mental abnormalities. Physical effects are most prominently observed in the face (Fig. 1.9), as individuals with FAS can have a small head circumference, a thin upper lip, an upturned nose, a smooth philtrum (the skin between the nose and upper lip), and/or epicanthal folds (skin fold that covers inner corner of eye).[127] FAS is also associated with learning disorders, delayed development, poor coordination/balance, attentional deficits, hyperactivity, and impaired decision making.[128] Like ethanol, exposure to cocaine in utero can lead to premature birth, reduced birth weight, decreased head circumference, increased tremors, excessive irritability, hyperalertness, and some slight delays in language development.[129,130]

Since 2020, the world has been affected by the COVID-19 pandemic. The pandemic has had a negative impact on the health of those with a SUD.[131] In the United States, individuals with a SUD are more likely to develop COVID-19 and are more likely to be hospitalized or to die from COVID-19 symptoms,[132] and an increase in the number of individuals starting or increasing substance use has been observed during the pandemic.[133] This has led to an increase in drug overdoses, particularly for opioids, with 40 states reporting increases in overdose deaths during 2020.[131] The increased drug use observed during the pandemic may be attributed to increased stress and anxiety. Many people have faced economic hardships and uncertainty as employers have had to lay off employees in response to lost revenue. Compounding this anxiety is a sense of loneliness as social isolation was implemented to

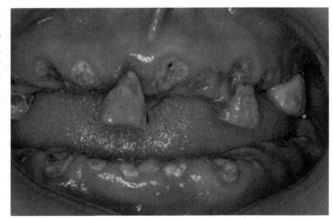

FIGURE 1.8 Long-term methamphetamine use can lead to "meth mouth", characterized by gum disease and damage/loss of teeth. *Image comes from Wang P, Chen X, Zheng L, Guo L, Li X, Shen S. Comprehensive dental treatment for "meth mouth": a case report and literature review. J Formos Med Assoc. 2014;13(11):867–871. https://doi.org/10.1016/j.jfma.2012.01.016. Copyright 2014, with permission from Elsevier.*

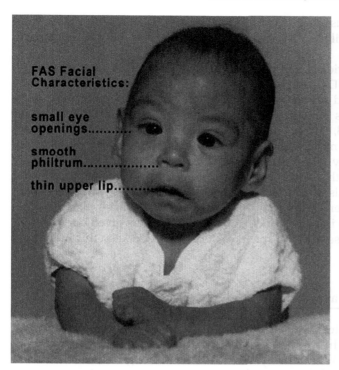

FIGURE 1.9 Image of a child born with fetal alcohol syndrome (FAS). Notice the thin upper lip and smooth philtrum of these children. These are defining physical features of FAS. *Image comes from Wikimedia Commons and was posted by Teresa Kellerman (By Teresa Kellerman - http://www.come-over.to/FAS/fasbabyface.jpg, CC BY-SA 3.0, https://commons.wikimedia.org/w/index.php?curid=4847497).*

reduce the risk of contracting COVID-19. With fewer avenues to alleviate stress/boredom during the pandemic, some turned to drug use or increased their existing drug use. In Chapter 11, you will learn more about how stress and anxiety exacerbate substance use.

Economic impacts

Simply put, drugs are expensive, especially because individuals with a SUD must take the drug frequently, with some drugs being used multiple times a day to avoid withdrawal symptoms. As detailed earlier, individuals that use cigarettes develop tolerance to the effects of nicotine and need to increase the number of cigarettes smoked during the day. In some countries like the United States and Australia, the cost of cigarettes has increased drastically since the early 2000s. As of 2022, here in the state of Kentucky (where I live), the average price of a single pack, which contains 20 cigarettes, is approximately $5.36. This is not that expensive compared to other states. In New York, the average cost of a pack of cigarettes is $10.45, the highest in the country.[134]

Purchasing the drug is not the only cost associated with SUDs. If an individual wants to seek treatment for their SUD, going to a rehabilitation center will cost money. The amount an individual will pay will depend on a few factors, such as the type of treatment intervention and whether the individual has insurance, but the cost can be anywhere between $1,000 and $30,000+.[135] Even if a person does not want to seek treatment, they may have to pay medical bills for conditions related to drug use (e.g., treating an abscess or a heart condition). If an individual faces legal problems for illicit drug use, they can lose their job and incur legal costs.

Not only do SUDs affect an individual, but they also negatively impact a society's economy. In the United States alone, SUDs cost the country more than $700 billion due to lost work productivity, healthcare-related expenses, and crime.[136] A large portion of the healthcare costs stem from tobacco use, as patients can experience multiple forms of cancer and lung diseases like COPD. As of 2018, 30 U.S. states have "three strikes" laws, which means that these states will impose harsh minimum mandatory sentences for those convicted of their third felony. Often, the penalty is a mandatory life sentence. I mention this because some individuals can receive a lengthy prison sentence if their third felony relates to drug possession; we will come back to this in Chapter 5. During the mid-2010s, taxpayers spent approximately $33,000 per prisoner, although this value greatly varies across states (~$15,000 in Alabama vs. ~$70,000 in New York).[137] If an individual is given a mandatory life sentence for a drug-related offense, we will spend a lot of money to keep this individual incarcerated.

Translational approach to studying addiction

To better understand and to better treat addiction, addiction science relies heavily on the **translational approach** to research. The expression "from bench to bedside" encapsulates the translational research approach. What exactly does this mean though? To better understand translational research, I need to define a couple of terms. *Basic research* aims to expand our knowledge on some phenomenon, such as the neurobiological, cognitive-behavioral, or sociocultural processes involved in addiction. I am a basic researcher, as I focus on elucidating the neural mechanisms of impulsive and risky decision-making, concepts you will learn about in Chapter 8. *Applied research* uses the theories generated from basic research to create new treatments, technologies, or procedures. For example, a chemist may synthesize a new compound that a behavioral pharmacologist tests in rats. Imagine that the behavioral pharmacologist shows that this novel compound is efficacious in reducing relapse-like behavior in a rat. This knowledge on its own is interesting, but if we want to help those with a SUD, we need to conduct a clinical study in which individuals volunteer to receive treatments of this new drug. However, before we can conduct this clinical study, additional basic research needs to be conducted to ensure that the drug is safe. Scientists may also want to study how this drug affects certain parts of the brain and/or alters other processes like decision-making.

Whether studying addiction, diabetes, or cancer, animal research plays a critical role in translational research. Animal research has greatly enhanced our knowledge of the neurobehavioral mechanisms of addiction. One major advantage of using animals in research is that we can control their environmental, genetic, and developmental histories, things we cannot ethically or feasibly do with humans. This control allows us to systematically determine which factors are important mediators of SUDs, thus enabling researchers to try to develop novel treatments to battle addiction. Although animal research provides great control, there are individuals that oppose animal research. I can understand why some are uncomfortable with animal research, but some groups engage in destructive acts that can jeopardize the safety and well-being of humans and animals alike. In California, Dr. David Jentsch, an addiction scientist, has had his car bombed on at least two occasions and has had razor blades tainted with HIV sent to his home.[138,139] Other groups have broken into research labs and have released animals.[140] In theory, this seems like a great way to free animals, but this is detrimental to animals that have been raised in a laboratory. These animals have not learned how to hunt for food or how to avoid predators. Realistically, these animals will die shortly after being released into the wild.

To ensure research with animals is conducted as humanely as possible, each institution that engages in animal testing must establish an Institutional Animal Care and Use Committee (IACUC). The goal of IACUC is to oversee all research conducted with vertebrate animals like rodents and birds. Any time a researcher wants to conduct a new study, they must submit a protocol for review. Members of the IACUC committee will review the proposal and will provide feedback and/or concerns about the proposal. One member of the IACUC must be a nonscientist. This requires the researcher to explain the value of their experiment such that a nonexpert can readily understand it. In addition to reviewing proposals, the IACUC will conduct inspections of the animal housing areas and the testing space to ensure animals are being treated ethically. I currently serve on our institution's IACUC, and I can attest to the amount of care and rigor that goes into certifying all animals are treated humanely.

Quiz 1.2

1. Which of the following statements regarding polydrug use is true?
    a. Combining certain drugs can lead to a fatal overdose, even if the drugs are of different classes.
    b. Concurrent drug use refers to individuals that combine two or more drugs.
    c. Ethanol is rarely used in conjunction with other drugs.
    d. Mixing a depressant with a stimulant is safe because the effects of one drug cancel out the effects of the other drug.
2. A researcher is conducting a study to determine the effects of cocaine on locomotor activity in different strains of mice. What type of research is being conducted?
    a. Applied
    b. Basic
    c. Translational
    d. None of the above
3. Which of the following is not a health outcome associated with tobacco use?
    a. Cirrhosis of the liver
    b. Gum disease
    c. Lung cancer
    d. Stroke
4. Which region of the world has the lowest incidence rate of alcohol use disorder?
    a. Eastern Europe

    b. Middle East
    c. North America
    d. South Asia
5. An individual comes into a clinic with skin lesions and missing teeth. While in the clinic, this individual grinds his/her teeth and picks at their skin. Which drug is most likely to cause these symptoms?
    a. Ethanol
    b. Heroin
    c. Methamphetamine
    d. Tobacco

Answers to quiz 1.2

1. a — Combining certain drugs can lead to a fatal overdose, even if the drugs are of different classes.
2. b — Basic
3. a — Cirrhosis of the liver
4. b — Middle East
5. c — Methamphetamine

Chapter summary

Because addiction is a relapsing disorder characterized by compulsive drug seeking, abstaining from drug use is a life-long battle for those living with a SUD. One major challenge for those with nicotine use disorder and/or alcohol use disorder is that both substances are legal in many parts of the world. Due to their availability, treating nicotine use disorder and alcohol use disorder can be difficult. Another challenge to treating SUDs is that individuals often combine two or more drugs and can develop a SUD for more than one substance. Additionally, SUDs are often observed in other clinical populations, such as those with depression or schizophrenia, which further increases the difficulty of successfully treating an individual with a SUD. Given the health and economic impacts associated with SUDs, being able to develop novel behavioral and/or pharmacological treatments is important. Translational research is a valuable tool in the fight against SUDs. By testing both animal subjects and studying individuals with an SUD, we can improve our understanding of addiction.

Glossary

Abstinence — a period in which an individual does not use a substance.
Addiction — a chronic, relapsing disorder characterized by compulsive drug seeking and use despite adverse consequences.
Anhedonia — inability to feel pleasure.
Comorbidity — two or more disorders/illnesses occurring in the same individual.
Depressant — a drug class that reduces activity in the brain, causing feelings of sedation and calmness.
Dissociative hallucinogen — a type of hallucinogen that is known to produce hallucinations, delusions, and feelings of detachment.
Drug misuse — using a substance in a way that is not intended (e.g., crushing a pill and snorting it).
Drug use — term often used to describe the use of illicit drugs.

Hallucinogen — a drug class that produces alterations in one's consciousness, such that individuals may see, hear, or feel things that are not actually present.

Inhalant — a drug class that includes substances that are inhaled only (e.g., aerosol sprays, glues, paint thinners, etc.).

Intoxication — a state in which an individual's physical and/or mental abilities are impaired by a substance.

Opiate — an opioid, such as morphine and codeine, that is derived from the opium poppy plant.

Opioid — a type of depressant that is best known for its analgesic (i.e., pain-reducing) effects.

Overdose — an event in which an individual takes too much of a substance.

Physical dependence — a phenomenon in which a person needs to continue using a substance to avoid experiencing aversive withdrawal symptoms.

Polydrug use — using two or more substances, whether concurrently or simultaneously.

Psychedelic hallucinogen — a type of hallucinogen that is known for making individuals see, hear, or feel things that are not present, as well as making them believe things that are not true. Also referred to as **classic hallucinogen**.

Psychological dependence — a phenomenon in which an individual will engage in harmful activities to procure and use more substance; it is considered to be a defining characteristic of addiction.

Psychostimulant — a drug class that increases activity in the brain, causing increased alertness, arousal, and motor activity. Also called **stimulant**.

Relapse — a return to drug use following a period of abstinence.

Route of administration — the method by which a person gets a substance into their body.

Sedative-hypnotic — a type of depressant that is best known for reducing anxiety and/or increasing drowsiness.

Substance use disorder (SUD) — term often used in place of addiction, which is marked by continued drug use despite adverse consequences.

Tolerance — the phenomenon in which an individual habituates to the effects of a substance over time.

Translational approach — applying basic research (understanding biological or cognitive-behavioral processes) to create new treatments, technologies, or procedures that can help individuals.

Withdrawal — physiological and behavioral symptoms that emerge when an individual stops taking a substance.

Supplementary data related to this chapter can be found online at doi:10.1016/B978-0-323-90578-7.00014-1

References

1. Street L. *I Was a Drug Addict*. Arlington House; 1953.
2. American Psychiatric Association. *Diagnostic and Statistical Manual of Mental Disorders*. 5th ed. 2013. Washington, DC.
3. Botticelli MP, Koh HK. Changing the language of addiction. *JAMA*. 2016;316(13):1361–1362. https://doi.org/10.1001/jama.2016.11874.
4. World Health Organization. *International Statistical Classification of Diseases and Related Health Problems*. 11th ed.; 2019. https://icd.who.int/.
5. Hoffmann NG, Kopak AM. How well do the DSM-5 alcohol use disorder designations map to the ICD-10 disorders? *Alcohol Clin Exp Res*. 2015;39(4):697–701. https://doi.org/10.1111/acer.12685.
6. Substance Abuse and Mental Health Services Administration. *Key Substance Use and Mental Health Indicators in the United States: Results from the 2019 National Survey on Drug Use and Health (HHS Publication No. PEP20-07-01-001, NSDUH Series H-55)*. Rockville, MD: Center for Behavioral Health Statistics and Quality, Substance Abuse and Mental Health Services Administration; 2020. Retrieved from: https://www.samhsa.gov/data/.
7. Wang TW, Neff LJ, Park-Lee E, Ren C, Cullen KA, King BA. E-cigarette use among middle and high school students — United States, 2020. *Morb & Mortal Wkly Rep*; September 9, 2020. Retrieved from: https://www.cdc.gov/mmwr/volumes/69/wr/pdfs/mm6937e1-H.pdf.
8. Donny EC, Houtsmuller E, Stitzer ML. Smoking in the absence of nicotine: behavioral, subjective and physiological effects over 11 days. *Addiction*. 2007;102(2):324–334. https://doi.org/10.1111/j.1360-0443.2006.01670.x.
9. Benowitz NL. Nicotine addiction. *N Engl J Med*. 2010;362(24):2295–2303. https://doi.org/10.1056/NEJMra0809890.
10. Freedman ND, Leitzmann MF, Hollenbeck AR, Schatzkin A, Abnet CC. Cigarette smoking and subsequent risk of lung cancer in men and women: analysis of a prospective cohort study. *Lancet Oncol*. 2008;9(7):649–656. https://doi.org/10.1016/S1470-2045(08)70154-2.

11. McLaughlin I, Dani JA, De Biasi M. Nicotine withdrawal. *Curr. Top. Behav. Neurosci.* 2015;24:99–123. https://doi.org/10.1007/978-3-319-13482-6_4.
12. García-Rodríguez O, Secades-Villa R, Flórez-Salamanca L, Okuda M, Liu SM, Blanco C. Probability and predictors of relapse to smoking: results of the national epidemiologic survey on alcohol and related conditions (NESARC). *Drug Alcohol Depend.* 2013;132(3):479–485. https://doi.org/10.1016/j.drugalcdep.2013.03.008.
13. Mayer B. How much nicotine kills a human? Tracing back the generally accepted lethal dose to dubious self-experiments in the nineteenth century. *Arch Toxicol.* 2014;88(1):5–7. https://doi.org/10.1007/s00204-013-1127-0.
14. Faulkner JM. Nicotine poisoning by absorption through the skin. *JAMA.* 1933;100(21):1664–1665. https://doi.org/10.1001/jama.1933.02740210012005.
15. Tonstad S, Gustavsson G, Kruse E, Walmsley JM, Westin Å. Symptoms of nicotine toxicity in subjects achieving high cotinine levels during nicotine replacement therapy. *Nicotine Tob Res.* 2014;16(9):1266–1271. https://doi.org/10.1093/ntr/ntu076.
16. Asser A, Taba P. Psychostimulants and movement disorders. *Front Neurol.* 2015;6:75. https://doi.org/10.3389/fneur.2015.00075.
17. Callaway CW, Clark RF. Hyperthermia in psychostimulant overdose. *Ann Emerg Med.* 1994;24(1):68–76. https://doi.org/10.1016/s0196-0644(94)70165-2.
18. Westover AN, McBride S, Haley RW. Stroke in young adults who abuse amphetamines or cocaine: a population-based study of hospitalized patients. *Arch Gen Psychiatr.* 2007;64(4):495–502. https://doi.org/10.1001/archpsyc.64.4.495.
19. Lerner A, Klein M. Dependence, withdrawal and rebound of CNS drugs: an update and regulatory considerations for new drugs development. *Brain Commun.* 2019;1(1):fcz025. https://doi.org/10.1093/braincomms/fcz025.
20. Nakajima M, Hoffman R, Alsameai A, Khalil NS, al'Absi M. Development of the khat knowledge, attitudes and perception scale. *Drug Alcohol Rev.* 2018;37(6):802–809. https://doi.org/10.1111/dar.12828.
21. Teni FS, Surur AS, Hailemariam A, et al. Prevalence, reasons, and perceived effects of khat chewing among students of a college in Gondar Town, Northwestern Ethiopia: a cross-sectional study. *Ann Med Health Sci Res.* 2015;5(6):454–460. https://doi.org/10.4103/2141-9248.177992.
22. Getahun W, Gedif T, Tesfaye F. Regular Khat (*Catha edulis*) chewing is associated with elevated diastolic blood pressure among adults in Butajira, Ethiopia: a comparative study. *BMC Publ Health.* 2010;10:390. https://doi.org/10.1186/1471-2458-10-390.
23. Widler P, Mathys K, Brenneisen R, Kalix P, Fisch HU. Pharmacodynamics and pharmacokinetics of khat: a controlled study. *Clin Pharmacol Ther.* 1994;55(5):556–562. https://doi.org/10.1038/clpt.1994.69.
24. Capriola M. Synthetic cathinone abuse. *J Clin Pharmacol.* 2013;5:109–115. https://doi.org/10.2147/CPAA.S42832.
25. La Maida N, Di Trana A, Giorgetti R, Tagliabracci A, Busardò FP, Huestis MA. A review of synthetic cathinone-related fatalities from 2017 to 2020. *Ther Drug Monit.* 2021;43(1):52–68. https://doi.org/10.1097/FTD.0000000000000808.
26. Abdeta T, Tolessa D, Adorjan K, Abera M. Prevalence, withdrawal symptoms and associated factors of khat chewing among students at Jimma University in Ethiopia. *BMC Psychiatr.* 2017;17(1):142. https://doi.org/10.1186/s12888-017-1284-4.
27. Kersten BP, McLaughlin ME. Toxicology and management of novel psychoactive drugs. *J Pharm Pract.* 2015;28(1):50–65. https://doi.org/10.1177/0897190014544814.
28. Gallagher N, Edwards FJ. The diagnosis and management of toxic alcohol poisoning in the emergency department: a review article. *Adv J Emerg Med.* 2019;3(3):e28. https://doi.org/10.22114/ajem.v0i0.153.
29. Lin LA, Bonar EE, Zhang L, Girard R, Coughlin LN. Alcohol-involved overdose deaths in US veterans. *Drug Alcohol Depend.* 2022;230:109196. https://doi.org/10.1016/j.drugalcdep.2021.109196.
30. Schuckit MA. Recognition and management of withdrawal delirium (delirium tremens). *N Engl J Med.* 2014;371(22):2109–2113. https://doi.org/10.1056/NEJMra1407298.
31. Kaufman PN, Krevsky B, Malmud LS, et al. Role of opiate receptors in the regulation of colonic transit. *Gastroenterology.* 1988;94(6):1351–1356. https://doi.org/10.1016/0016-5085(88)90673-7.
32. Prichard D, Norton C, Bharucha AE. Management of opioid-induced constipation. *Br J Nurs.* 2016;25(10):S4–S11. https://doi.org/10.12968/bjon.2016.25.10.S4.

References

33. Kaiko RF, Wallenstein SL, Rogers AG, Grabinski PY, Houde RW. Analgesic and mood effects of heroin and morphine in cancer patients with postoperative pain. *N Engl J Med*. 1981;304(25):1501–1505. https://doi.org/10.1056/NEJM198106183042501.
34. Mirin SM, Meyer RE, McNamee HB. Psychopathology and mood during heroin use: acute vs chronic effects. *Arch Gen Psychiatr*. 1976;33(12):1503–1508. https://doi.org/10.1001/archpsyc.1976.01770120107011.
35. Kanof PD, Handelsman L, Aronson MJ, Ness R, Cochrane KJ, Rubinstein KJ. Clinical characteristics of naloxone-precipitated withdrawal in human opioid-dependent subjects. *J Pharmacol Exp Therapeut*. 1992;260(1):355–363.
36. Vardanyan RS, Hruby VJ. Fentanyl-related compounds and derivatives: current status and future prospects for pharmaceutical applications. *Future Med Chem*. 2014;6(4):385–412. https://doi.org/10.4155/fmc.13.215.
37. Coupey SM. Barbiturates. *Pediatr Rev*. 1997;18(8):260–265. https://doi.org/10.1542/pir.18-8-260.
38. Busardò FP, Jones AW. GHB pharmacology and toxicology: acute intoxication, concentrations in blood and urine in forensic cases and treatment of the withdrawal syndrome. *Curr Neuropharmacol*. 2015;13(1):47–70. https://doi.org/10.2174/1570159X13666141210215423.
39. Doblin RE, Christiansen M, Jerome L, Burge B. The past and future of psychedelic science: an introduction to this issue. *J Psychoact Drugs*. 2019;51(2):93–97. https://doi.org/10.1080/02791072.2019.1606472.
40. Belouin SJ, Henningfield JE. Psychedelics: where we are now, why we got here, what we must do. *Neuropharmacology*. 2018;142:7–19. https://doi.org/10.1016/j.neuropharm.2018.02.018.
41. Fuentes JJ, Fonseca F, Elices M, Farré M, Torrens M. Therapeutic use of LSD in psychiatry: a systematic review of randomized-controlled clinical trials. *Front Psychiatr*. 2020;10:943. https://doi.org/10.3389/fpsyt.2019.00943.
42. Johnson MW, Garcia-Romeu A, Cosimano MP, Griffiths RR. Pilot study of the 5-HT2AR agonist psilocybin in the treatment of tobacco addiction. *J Psychopharmacol*. 2014;28(11):983–992. https://doi.org/10.1177/0269881114548296.
43. Krediet E, Bostoen T, Breeksema J, van Schagen A, Passie T, Vermetten E. Reviewing the potential of psychedelics for the treatment of PTSD. *Int J Neuropsychopharmacol*. 2020;23(6):385–400. https://doi.org/10.1093/ijnp/pyaa018.
44. Garcia-Romeu A, Davis AK, Erowid E, Erowid F, Griffiths RR, Johnson MW. Persisting reductions in cannabis, opioid, and stimulant misuse after naturalistic psychedelic use: an online survey. *Front Psychiatr*. 2020;10:955. https://doi.org/10.3389/fpsyt.2019.00955.
45. Ditman KS, Tietz W, Prince BS, Forgy E, Moss T. Harmful aspects of the LSD experience. *J Nerv Ment Dis*. 1967;145(6):464–474. https://doi.org/10.1097/00005053-196712000-00004.
46. Holze F, Vizeli P, Müller F, et al. Distinct acute effects of LSD, MDMA, and D-amphetamine in healthy subjects. *Neuropsychopharmacology*. 2020;45(3):462–471. https://doi.org/10.1038/s41386-019-0569-3.
47. Schmid Y, Enzler F, Gasser P, et al. Acute effects of lysergic acid diethylamide in healthy subjects. *Biol Psychiatr*. 2015;78(8):544–553. https://doi.org/10.1016/j.biopsych.2014.11.015.
48. Ungerleider JT, Fisher DD, Fuller M, Caldwell A. The "bad trip"–The etiology of the adverse LSD reaction. *Am J Psychiatr*. 1968;124(11):1483–1490. https://doi.org/10.1176/ajp.124.11.1483.
49. Halpern JH, Suzuki J, Huertas PE, Passie T. Hallucinogen abuse and dependence. In: Stolerman IP, Price LH, eds. *Encyclopedia of Psychopharmacology*. Springer; 2014. https://doi.org/10.1007/978-3-642-27772-6_43-2.
50. Mion G. History of anaesthesia: the ketamine story – past, present and future. *Eur J Anaesthesiol*. 2017;34(9):571–575. https://doi.org/10.1097/EJA.0000000000000638.
51. Bey T, Patel A. Phencyclidine intoxication and adverse effects: a clinical and pharmacological review of an illicit drug. *Calif J Emerg Med*. 2007;8(1):9–14.
52. Mithoefer MC, Feduccia AA, Jerome L, et al. MDMA-assisted psychotherapy for treatment of PTSD: study design and rationale for phase 3 trials based on pooled analysis of six phase 2 randomized controlled trials. *Psychopharmacology*. 2019;236:2735–2745. https://doi.org/10.1007/s00213-019-05249-5.
53. Bosch OG, Eisenegger C, Gertsch J, et al. Gamma-hydroxybutyrate enhances mood and prosocial behavior without affecting plasma oxytocin and testosterone. *Psychoneuroendocrinology*. 2015;62:1–10. https://doi.org/10.1016/j.psyneuen.2015.07.167.
54. Vollenweider FX, Gamma A, Liechti M, Huber T. Psychological and cardiovascular effects and short-term sequelae of MDMA ("ecstasy") in MDMA-naïve healthy volunteers. *Neuropsychopharmacology*. 1998;19(4):241–251. https://doi.org/10.1016/S0893-133X(98)00013-X.

55. Costa G, Gołembiowska K. Neurotoxicity of MDMA: main effects and mechanisms. *Exp Neurol.* 2022;347:113894. https://doi.org/10.1016/j.expneurol.2021.113894.
56. Green AR, Cross AJ, Goodwin GM. Review of the pharmacology and clinical pharmacology of 3,4-methylenedioxymethamphetamine (MDMA or "Ecstasy"). *Psychopharmacology.* 1995;119(3):247−260. https://doi.org/10.1007/BF02246288.
57. Cottler LB, Leung KS, Abdallah AB. Test-re-test reliability of DSM-IV adopted criteria for 3,4-methylenedioxymethamphetamine (MDMA) abuse and dependence: a cross-national study. *Addiction.* 2009;104(10):1679−1690. https://doi.org/10.1111/j.1360-0443.2009.02649.x.
58. Kirkham TC. Cannabinoids and appetite: food craving and food pleasure. *Int Rev Psychiatr.* 2009;21(2):163−171. https://doi.org/10.1080/09540260902782810.
59. Hindley G, Beck K, Borgan F, et al. Psychiatric symptoms caused by cannabis constituents: a systematic review and meta-analysis. *Lancet Psychiatr.* 2020;7(4):344−353. https://doi.org/10.1016/S2215-0366(20)30074-2.
60. Beal JE, Olson R, Lefkowitz L, et al. Long-term efficacy and safety of dronabinol for acquired immunodeficiency syndrome-associated anorexia. *J Pain Symptom Manag.* 1997;14(1):7−14. https://doi.org/10.1016/S0885-3924(97)00038-9.
61. Abrams DI, Guzman M. Cannabis in cancer care. *Clin Pharmacol Ther.* 2015;97(6):575−586. https://doi.org/10.1002/cpt.108.
62. Aviram J, Samuelly-Leichtag G. Efficacy of cannabis-based medicines for pain management: a systematic review and meta-analysis of randomized controlled trials. *Pain Phys.* 2017;20(6):E755−E796.
63. Castaneto MS, Gorelick DA, Desrosiers NA, Hartman RL, Pirard S, Huestis MA. Synthetic cannabinoids: epidemiology, pharmacodynamics, and clinical implications. *Drug Alcohol Depend.* 2014;144:12−41. https://doi.org/10.1016/j.drugalcdep.2014.08.005.
64. Schneir AB, Baumbacher T. Convulsions associated with the use of a synthetic cannabinoid product. *J Med Toxicol.* 2012;8(1):62−64. https://doi.org/10.1007/s13181-011-0182-2.
65. Tyndall JA, Gerona R, De Portu G, et al. An outbreak of acute delirium from exposure to the synthetic cannabinoid AB-CHMINACA. *Clin Toxicol.* 2015;53(10):950−956. https://doi.org/10.3109/15563650.2015.1100306.
66. Howard MO, Bowen SE, Garland EL, Perron BE, Vaughn MG. Inhalant use and inhalant use disorders in the United States. *Addiction Sci Clin Pract.* 2011;6(1):18−31.
67. Bowen SE. Two serious and challenging medical complications associated with volatile substance misuse: sudden sniffing death and fetal solvent syndrome. *Subst Use Misuse.* 2011;46(sup1):68−72. https://doi.org/10.3109/10826084.2011.580220.
68. Perron BE, Howard MO, Vaughn MG, Jarman CN. Inhalant withdrawal as a clinically significant feature of inhalant dependence disorder. *Med Hypotheses.* 2009;73(6):935−937. https://doi.org/10.1016/j.mehy.2009.06.036.
69. Hunt GP, Bailey N, Evans K, Moloney M. Combining different substances in the dance scene: enhancing pleasure, managing risk and timing effects. *J Drug Use.* 2009;39(3):495−522. https://doi.org/10.1177/002204260903900303.
70. Shiffman S, Balabanis M. Do drinking and smoking go together? *Alcohol Health Res World.* 1996;20(2):107−110.
71. NIDA. *Common Comorbidities with Substance Use Disorders Research Report: What Are Some Approaches to Diagnosis?*; May 27, 2020. Retrieved from: https://www.drugabuse.gov/publications/research-reports/common-comorbidities-substance-use-disorders/what-are-some-approaches-to-diagnosis.
72. Romaguera A, Torrens M, Papaseit E, Arellano AL, Farré M. Concurrent use of cannabis and alcohol: neuropsychiatric effect consequences. *CNS Neurol Disord - Drug Targets.* 2017;16(5):592−597. https://doi.org/10.2174/1871527316666170419161839.
73. Meredith SE, Juliano LM, Hughes JR, Griffiths RR. Caffeine use disorder: a comprehensive review and research agenda. *J Caffeine Res.* 2013;3(3):114−130. https://doi.org/10.1089/jcr.2013.0016.
74. Marczinski CA, Fillmore MT, Henges AL, Ramsey MA, Young CR. Mixing an energy drink with an alcoholic beverage increases motivation for more alcohol in college students. *Alcohol Clin Exp Res.* 2013;37(2):276−283. https://doi.org/10.1111/j.1530-0277.2012.01868.x/.
75. Marczinski CA, Fillmore MT, Stamates AL, Maloney S. Desire to drink alcohol is enhanced with high caffeine energy drink mixers. *Alcohol Clin Exp Res.* 2016;40(9):1982−1990. https://doi.org/10.1111/acer.13152.
76. Marczinski CA, Fillmore MT, Stamates AL, Maloney S. Alcohol-induced impairment of balance is antagonized by energy drinks. *Alcohol Clin Exp Res.* 2018;42(1):144−152. https://doi.org/10.1111/acer.13521.
77. Marczinski CA, Fillmore MT, Henges AL, Ramsey MA, Young CR. Effects of energy drinks mixed with alcohol on information processing, motor coordination and subjective reports of intoxication. *Exp Clin Psychopharmacol.* 2012;20(2):129−138. https://doi.org/10.1037/a0026136.

78. Forward J, Akhurst J, Bruno R, et al. Nature versus intensity of intoxication: co-ingestion of alcohol and energy drinks and the effect on objective and subjective intoxication. *Drug Alcohol Depend.* 2017;180:292−303. https://doi.org/10.1016/j.drugalcdep.2017.08.013.
79. Andrews P. Cocaethylene toxicity. *J Addict Dis.* 1997;16(3):75−84. https://doi.org/10.1300/J069v16n03_08.
80. Bunn WH, Giannini AJ. Cardiovascular complications of cocaine abuse. *Am Fam Phys.* 1992;46(3):769−773.
81. Farooq MU, Bhatt A, Patel M. Neurotoxic and cardiotoxic effects of cocaine and ethanol. *J Med Toxicol.* 2009;5(3):134−138. https://doi.org/10.1007/BF03161224.
82. Hindocha C, Freeman TP, Ferris JA, Lynskey MT, Winstock AR. No Smoke without Tobacco: a global overview of cannabis and tobacco routes of administration and their association with intention to quit. *Front Psychiatr.* 2016;7:104. https://doi.org/10.3389/fpsyt.2016.00104.
83. Ream GL, Benoit E, Johnson BD, Dunlap E. Smoking tobacco along with marijuana increases symptoms of cannabis dependence. *Drug Alcohol Depend.* 2008;95(3):199−208. https://doi.org/10.1016/j.drugalcdep.2008.01.011.
84. Simon G. *Health Officials: COPD on the Rise*; January 6, 2016. Retrieved from: http://www.govt.lc/news/health-officials-copd-on-the-rise.
85. Gilbert CR, Baram M, Cavarocchi NC. "Smoking wet": respiratory failure related to smoking tainted marijuana cigarettes. *Tex Heart Inst J.* 2013;40(1):64−67.
86. Grov C, Kelly BC, Parsons JT. Polydrug use among club-going young adults recruited through time-space sampling. *Subst Use Misuse.* 2009;44(6):848−864. https://doi.org/10.1080/10826080802484702.
87. Uys JDK, Niesink RJM. Pharmacological aspects of the combined used of 3,4-methylenedioxymethamphetamine (MDMA, ecstasy) and gamma-hydroxybutyric acid (GHB): a review of the literature. *Drug Alcohol Rev.* 2005;24(4):359−368. https://doi.org/10.1080/09595230500295725.
88. NIDA. *Real Teens Ask about Speedballs*; June 26, 2013. Retrieved from: https://archives.drugabuse.gov/blog/post/real-teens-ask-about-speedballs.
89. GBD 2016 Alcohol and Drug Use Collaborators. The global burden of disease attributable to alcohol and drug use in 195 countries and territories, 1990−2016: a systematic analysis for the Global Burden of Disease Study 2016. *Lancet Psychiatr.* 2018;5(12):987−1012. https://doi.org/10.1016/S2215-0366(18)30337-7.
90. Winklbaur B, Ebner N, Sachs G, Thau K, Fischer G. Substance abuse in patients with schizophrenia. *Dialogues Clin Neurosci.* 2006;8(1):37−43. https://doi.org/10.31887/DCNS.2006.8.1/bwinklbaur.
91. Winterer G. Why do patients with schizophrenia smoke? *Curr Opin Psychiatr.* 2010;23(2):112−119. https://doi.org/10.1097/YCO.0b013e3283366643.
92. Wijesundera H, Hanwella R, de Silva VA. Antipsychotic medication and tobacco use among outpatients with schizophrenia: a cross-sectional study. *Ann Gen Psychiatr.* 2014;13(1):7. https://doi.org/10.1186/1744-859X-13-7.
93. Thomson D, Berk M, Dodd S, et al. Tobacco use in bipolar disorder. *Clin Psychopharmacol & Neurosci.* 2015;13(1):1−11. https://doi.org/10.9758/cpn.2015.13.1.1.
94. Kalman D, Morissette SB, George TP. Co-morbidity of smoking in patients with psychiatric and substance use disorders. *Am J Addict.* 2005;14(2):106−123. https://doi.org/10.1080/10550490590924728.
95. Pettinati HM, O'Brien CP, Dundon WD. Current status of co-occurring mood and substance use disorders: a new therapeutic target. *Am J Psychiatr.* 2013;170(1):23−30. https://doi.org/10.1176/appi.ajp.2012.12010112.
96. Trull TJ, Sher KJ, Minks-Brown C, Durbin J, Burr R. Borderline personality disorder and substance use disorders: a review and integration. *Clin Psychol Rev.* 2000;20(2):235−253. https://doi.org/10.1016/s0272-7358(99)00028-8.
97. Pennay A, Cameron J, Reichert T, et al. A systematic review of interventions for co-occurring substance use disorder and borderline personality disorder. *J Subst Abuse Treat.* 2011;41(4):363−373. https://doi.org/10.1016/j.jsat.2011.05.004.
98. De Alwis D, Lynskey MT, Reiersen AM, Agrawal A. Attention-deficit/hyperactivity disorder subtypes and substance use and use disorders in NESARC. *Addict Behav.* 2014;39(8):278−1285. https://doi.org/10.1016/j.addbeh.2014.04.003.
99. Lee SS, Humphreys KL, Flory K, Liu R, Glass K. Prospective association of childhood attention-deficit/hyperactivity disorder (ADHD) and substance use and abuse/dependence: a meta-analytic review. *Clin Psychol Rev.* 2011;31(3):328−341. https://doi.org/10.1016/j.cpr.2011.01.006.

100. Harstad E, Levy S, Committee on Substance Abuse. Attention-deficit/hyperactivity disorder and substance abuse. *Pediatrics*. 2014;134(1):e293−e301. https://doi.org/10.1542/peds.2014-0992.
101. Walser T, Cui X, Yanagawa J, et al. Smoking and lung cancer: the role of inflammation. *Proc Am Thorac Soc*. 2008;5(8):811−815. https://doi.org/10.1513/pats.200809-100TH.
102. Asthana S, Vohra P, Labani S. Association of smokeless tobacco with oral cancer: a review of systematic reviews. *Tobac Preven & Cessat*. 2019;5:34. https://doi.org/10.18332/tpc/112596.
103. Klein-Weigel P, Volz TS, Zange L, Richter J. Buerger's disease: providing integrated care. *J Multidiscip Healthc*. 2016;9:511−518. https://doi.org/10.2147/JMDH.S109985.
104. Borojevic T. Smoking and periodontal disease. *Mater Soc Med*. 2012;24(4):274−276. https://doi.org/10.5455/msm.2012.24.274-276.
105. Campagna D, Alamo A, Di Pino A, et al. Smoking and diabetes: dangerous liaisons and confusing relationships. *Diabetol Metab Syndr*. 2019;11:85. https://doi.org/10.1186/s13098-019-0482-2.
106. West R. Tobacco smoking: health impact, prevalence, correlates and interventions. *Psychol Health*. 2017;32(8):1018−1036. https://doi.org/10.1080/08870446.2017.1325890.
107. U.S. Department of Health and Human Services. *The Health Consequences of Smoking—50 Years of Progress: A Report of the Surgeon General*. U.S. Department of Health and Human Services, Centers for Disease Control and Prevention, National Center for Chronic Disease Prevention and Health Promotion, Office on Smoking and Health; 2014. Retrieved from: https://www.cdc.gov/tobacco/data_statistics/sgr/50th-anniversary/index.htm.
108. World Health Organization. *WHO Report on the Global Tobacco Epidemic*. Author; 2017, 2017.
109. Willett JG, Bennett M, Hair EC, et al. Recognition, use and perceptions of JUUL among youth and young adults. *Tobac Control*. 2018;28(1):115−116. https://doi.org/10.1136/tobaccocontrol-2018-054273.
110. Kosmider L, Sobczak A, Fik M, et al. Carbonyl compounds in electronic cigarette vapors: effects of nicotine solvent and battery output voltage. *Nicotine Tob Res*. 2014;16(10):1319−1326. https://doi.org/10.1093/ntr/ntu078.
111. Tzortzi A, Kapetanstrataki M, Evangelopoulou V, Behrakis P. A systematic literature review of e-cigarette-related illness and injury: not just for the respirologist. *Int J Environ & Publ Health*. 2020;17(7):2248. https://doi.org/10.3390/ijerph17072248.
112. Kalininskiy A, Bach CT, Nacca NE, et al. E-cigarette, or vaping, product use associated lung injury (EVALI): case series and diagnostic approach. *Lancet Respir Med*. 2019;7(12):1017−1026. https://doi.org/10.1016/S2213-2600(19)30415-1.
113. Govindarajan P, Spiller HA, Casvant MJ, Chounthirath T, Smith GA. E-cigarette and liquid nitrogen exposures among young children. *Pediatrics*. 2018;141(5):e20173361. https://doi.org/10.1542/peds.2017-3361.
114. Tashkin DP. Effects of marijuana smoking on the lung. *Ann Am Thor Soc*. 2013;10(3):239−247. https://doi.org/10.1513/AnnalsATS.201212-127FR.
115. Yayan J, Rasche K. Damaging effects of cannabis use on the lungs. *Adv Exp Med Biol*. 2016;952:31−34. https://doi.org/10.1007/5584_2016_71.
116. Jett J, Stone E, Warren G, Cummings KM. Cannabis use, lung cancer, and related issues. *J Thorac Oncol*. 2018;13(4):480−487. https://doi.org/10.1016/j.jtho.2017.12.013.
117. Rehm J, Shield KD. Global alcohol-attributable deaths from cancer, liver cirrhosis, and injury in 2010. *Alcohol Res*. 2013;35(2):174−183.
118. Ferenci P. Hepatic encephalopathy. *Gastroenterol Rep*. 2017;5(2):138−147. https://doi.org/10.1093/gastro/gox013.
119. Asrani SK, Devarbhavi H, Eaton J, Kamath PS. Burden of liver diseases in the world. *J Hepatol*. 2019;70(1):151−171. https://doi.org/10.1016/j.jhep.2018.09.014.
120. Langlais PJ. Alcohol-related thiamine deficiency: impact on cognitive and memory functioning. *Alcohol Health Res World*. 1995;19(2):113−121.
121. Wurcel AG, Merchant EA, Clark RP, Stone DR. Emerging and underrecognized complications of illicit drug use. *Clin Infect Dis*. 2015;61(12):1840−1849. https://doi.org/10.1093/cid/civ689.
122. Rodger L, Shah M, Shojaei E, Hosseini S, Koivu S, Silverman M. Recurrent endocarditis in persons who inject drugs. *Open Forum Infect Dis*. 2019;6(10):ofz396. https://doi.org/10.1093/ofid/ofz396.
123. Narins E. *I'm the Girl Who Clawed Her Own Eyes Out. This is My Story*. Cosmopolitan; March 9, 2018. https://www.cosmopolitan.com/health-fitness/a19179723/kaylee-muthart-eye-gouge-crystal-meth/.
124. Maraj S, Figueredo VM, Lynn Morris D. Cocaine and the heart. *Clin Cardiol*. 2010;33(5):264−269. https://doi.org/10.1002/clc.20746.

125. Fonseca AC, Ferro JM. Drug Nd stroke. *Curr Neurol Neurosci Rep.* 2013;13(2):325. https://doi.org/10.1007/s11910-012-0325-0.
126. Sanlorenzo LA, Stark AR, Patrick SW. Neonatal abstinence syndrome: an update. *Curr Opin Pediatr.* 2018;30(2):182–186. https://doi.org/10.1097/MOP.0000000000000589.
127. Wattendorf DJ, Muenke M. Fetal alcohol spectrum disorders. *Am Fam Phys.* 2005;72(2):279–285.
128. Kodituwakku PW. Neurocognitive profile in children with fetal alcohol spectrum disorders. *Develop Disabil Res. Rev.* 2009;15(3):218–224. https://doi.org/10.1002/ddrr.73.
129. Behnke M, Smith VC, Committee on Substance Abuse, & Committee on Fetus and Newborn. Prenatal substance abuse: short- and long-term effects on the exposed fetus. *Pediatrics.* 2013;131(3):e1009–e1024. https://doi.org/10.1542/peds.2012-3931.
130. Cain MA, Bornick P, Whiteman V. The maternal, fetal, and neonatal effects of cocaine exposure in pregnancy. *Clin Obstet Gynecol.* 2013;56(1):124–132. https://doi.org/10.1097/GRF.0b013e31827ae167.
131. Abramson A. Substance abuse during the pandemic. *Mon Psychol.* March 1, 2021;52(2). Retrieved from: https://www.apa.org/monitor/2021/03/substance-use-pandemic.
132. Wang QQ, Kaelber DC, Xu R, Volkow ND. COVID-19 risk and outcomes in patients with substance use disorders: analyses from electronic records in the United States. *Mol Psychiatr*; 2021. https://doi.org/10.1038/s41380-020-00880-7.
133. Czeisler MÉ, Lane RI, Petrosky E, et al. Mental health, substance use, and suicidal ideation during the COVID-19 pandemic – United States, June 24-30, 2021. *MMWR (Morb Mortal Wkly Rep).* 2020;69:1049–1057. https://doi.org/10.15585/mmwr.mm6932a1external icon.
134. Sales Tax Handbook. *Map of Cigarette Excise Taxes & Price Per Pack*; 2022. https://www.salestaxhandbook.com/cigarette-tax-map.
135. French MT, Popovici I, Tapsell L. The economic costs of substance abuse treatment: updated estimates and cost bands for program assessment and reimbursement. *J Subst Abuse Treat.* 2008;35(4):462–469. https://doi.org/10.1016/j.jsat.2007.12.008.
136. NIDA. *Drugs, Brains, and Behavior: The Science of Addiction*; July 2020. Retrieved from: https://nida.nih.gov/publications/drugs-brains-behavior-science-addiction/introduction.
137. Mai C, Subramanian R. *The Price of Prisons*; 2017. Retrieved from: https://www.vera.org/publications/price-of-prisons-2015-state-spending-trends.
138. Martinez M. *Activist Group Claims to Send AIDS-Tainted Razors to Animal Researcher.* CNN; November 24, 2010. Retrieved from: http://www.cnn.com/2010/US/11/23/california.ucla.threat/index.html.
139. Sample I. *Animal Rights Activists Torch Scientist's Car.* The Guardian; March 18, 2009. Retrieved from: https://www.theguardian.com/science/blog/2009/mar/18/animal-rights-attack-ucla-neuroscientist.
140. Abbott A. Animal-rights activists wreak havoc in Milan laboratory. *Nature.* 2013. https://doi.org/10.1038/nature.2013.12847.

SECTION II

Neurobiological mechanisms of addiction

CHAPTER

2

Pharmacological actions of commonly used drugs

Introduction

Imagine that your neighbor has a known heroin use disorder and that one day you come home to find your neighbor unresponsive in front of their house. You call 911 and tell the dispatcher that your neighbor is not breathing and that you believe your neighbor has overdosed on heroin. The paramedics arrive and spray something into your neighbor's nose. Soon after, your neighbor starts breathing and becomes responsive. You later learn that the paramedics gave your neighbor a drug called *naloxone*. What exactly is naloxone, and how did it revive someone that had overdosed on heroin? To understand what naloxone is and how it works, we need to cover the pharmacological actions of drugs like heroin. But first, a discussion of how our nervous system functions is needed. If you already have a strong background in neuroscience and/or psychopharmacology, most of this chapter will serve as a review of concepts you have learned previously. Regardless, understanding the pharmacological actions of drugs is important as we can develop novel pharmacological treatments for substance use disorders (SUDs). We will conclude this chapter by discussing some of these treatments, including naloxone.

Learning objectives

By the end of this chapter, you should be able to …

(1) Describe what a neuron is and identify the parts of a neuron.
(2) Explain how neurons communicate with one another.
(3) Identify the various glial cells and their functions.
(4) Identify several neurotransmitters and explain how they are affected by commonly used drugs.
(5) Explain the pharmacological treatments that have been used to combat SUDs.
(6) Define harm reduction and discuss the main policies of harm reduction.

Introduction to the nervous system and neurotransmitters

Our **nervous system** is highly complex and serves to (1) transmit sensory information (e.g., information from the outside world) to our brain for further processing and (2) transmit motor information to our muscles to allow us to respond to sensory information correctly. For example, when we walk into a brightly lit room, specialized cells in our eyes send messages about the change in brightness to our brain. The brain then sends a message back to the iris to constrict the pupil; this prevents too much light from further entering our eyes. On a broad scale, our nervous system can be divided into the *central nervous system*, which contains the brain and the spinal cord, and the *peripheral nervous system*, which contains everything else (Fig. 2.1).

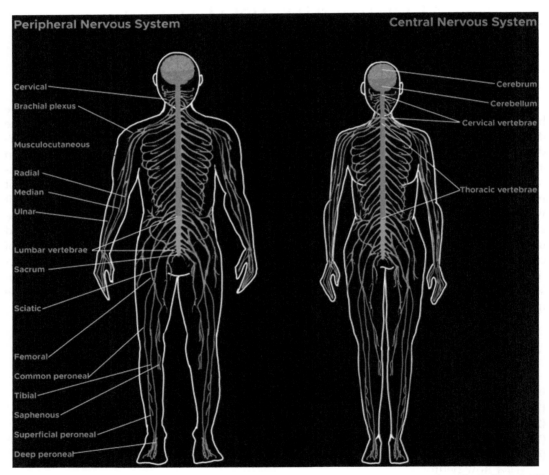

FIGURE 2.1 The central (in red) and peripheral (in orange) nervous systems. *Image comes from Fig. 2 of Ludwig PE, Reddy V, Varacallo M. Neuroanatomy, Central Nervous System (CNS): StatPearls Publishing; 2021. This book is distributed under the terms of the Creative Commons Attribution 4.0 International License (https://creativecommons.org/licenses/by/4.0/).*

The peripheral nervous system can be further fractioned into the somatic nervous system and the autonomic nervous system. The somatic nervous system regulates voluntary control of body movements. The autonomic nervous system, as the name implies, controls automatic processes that we do not consciously control. The autonomic nervous system can be further subdivided into the sympathetic and the parasympathetic nervous systems. The sympathetic nervous system mediates our *fight or flight* response. The sympathetic nervous system stimulates parts of the body necessary for encountering a stressful or dangerous situation, such as increasing heart rate, while inhibiting other activities that are not essential for survival at that point in time like salivation and digestion. The parasympathetic nervous system, on the other hand, is active when we are in a more restive state. This system is colloquially referred to as the *rest and digest* system. Functions such as salivation and digestion increase when the parasympathetic nervous system is more active. We will revisit the sympathetic nervous system in Chapter 11 when we discuss the relationship between stress and addiction.

The nervous system is composed of several types of specialized cells. Before discussing these cells in more detail, I want to first introduce the concept of the **neurotransmitter**. Neurotransmitters are chemicals that act as messengers within the nervous system. There are numerous neurotransmitters in the mammalian nervous system, which control various physiological, behavioral, and cognitive functions. The following list contains some, but certainly not all, of the major neurotransmitters in the nervous system: *acetylcholine, dopamine, endocannabinoids, endogenous opioids, γ-aminobutyric acid (GABA), glutamate, norepinephrine, and serotonin*. We will discuss these neurotransmitter systems in more detail later in this chapter.

One important concept to keep in mind is that neurotransmitters can increase *excitation* or *inhibition* within the nervous system.[1] Excitation means increased activity within part of the nervous system, whereas inhibition means decreased activity. One analogy that is often used to describe excitation and inhibition is a gas pedal and a brake pedal of a car. However, this analogy often oversimplifies the complex effects neurotransmitters can have in the nervous system. To understand how this can be the case, let's discuss the first major cell type of the nervous system: the **neuron**.

Neurons

Neurons are specialized cells that send messages to other cells (Fig. 2.2). There are several types of neurons. The first one is a *sensory neuron*. As its name suggests, a sensory neuron's job is to send sensory information from the external environment to the brain.[2] Going back to the example from earlier, information about brightness gets sent from the eye to the brain through sensory neurons. In addition to sensory neurons, there are *motor neurons*. Motor neurons transmit information from the brain to muscles and glands.[3] In the case of our example, motor neurons send a message to the iris to constrict the pupil. A third type of neuron is called the *interneuron*. Interneurons act as a relay between sensory neurons and/or motor neurons.[4] Hence, these neurons are also called relay neurons. This section will identify the major structures of the neuron and their functions. This knowledge is important for understanding how neurons communicate with each other.

Cell body

Like most cells of an organism, neurons have a *cell body*, sometimes referred to as a soma that contains the *nucleus* and vital *organelles*. The nucleus contains the genetic information of

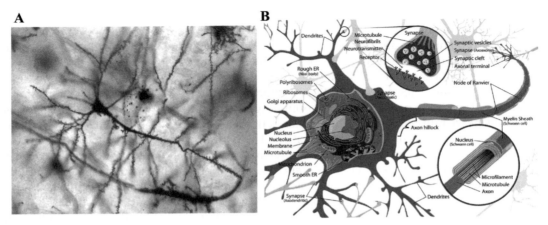

FIGURE 2.2 An image (A) and a schematic (B) of a neuron. *Panel (A) came from Wikimedia Commons and was posted by user MethoxyRoxy (https://commons.wikimedia.org/wiki/File:Pyramidal_hippocampal_neuron_40x.jpg). Panel (B) came from Wikimedia Commons and was posted by user LadyofHats (https://commons.wikimedia.org/wiki/File:Complete_neuron_cell_diagram_en.svg).*

the cell; importantly, genes are responsible for creating *proteins*, highly complex molecules that serve many roles in the body. Organelles help keep the neuron alive. Some of the major organelles are listed below:

Mitochondrion—power source of the cell as it gathers, stores, and releases energy.
Rough endoplasmic reticulum—important for protein synthesis.
Smooth endoplasmic reticulum—important for *lipid* (fat) synthesis.
Golgi apparatus—transports proteins to their destination in the cell.
Lysosome—helps break down waste inside the cell.

Dendrites and receptors

Derived from the Greek word *déndron* ("tree" or "tree branch"), **dendrites** are branch-like projections that originate from the cell body.[5] Dendrites receive messages from surrounding neurons. Dendrites often contain **dendritic spines**, which are small projections that help increase the ease at which neurons receive incoming messages.[6] Dendritic spines come in different forms. For example, some dendritic spines appear as small "stubs" on the dendrite, whereas others have a long, thin projection with a rounded head at the tip[7] (Fig. 2.3). Discussing why there are different dendritic spines is beyond the scope of this chapter. This is a concept that will be covered in more detail in Chapter 7.

Receptors are specialized proteins located on dendrites and the cell body that control the activity of neurons.[8] Some receptors are excitatory, whereas other receptors are inhibitory. The neurotransmitters that bind to their corresponding receptors are called **agonists**. An agonist is a chemical that binds to a receptor and activates it.[9] This activation can lead to either excitation or inhibition of the neuron, so do not think that only excitatory neurotransmitters are agonists. Agonists are not isolated to just neurotransmitters. There are numerous chemicals and drugs that can serve as agonists at different receptors. Later in this chapter, you will learn that some commonly used drugs act as agonists at certain receptors (e.g., nicotine). However, not all drugs act as agonists. Some are **antagonists** at certain receptors, meaning they block the receptor and prevent it from functioning properly[9] (e.g., dissociative hallucinogens).

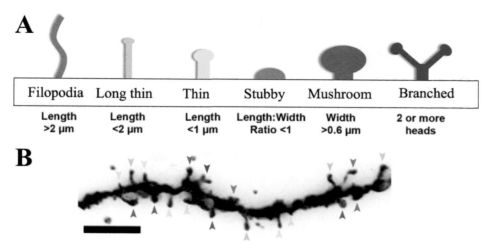

FIGURE 2.3 A schematic showing the various types of dendritic spines (A) and an image of a dendrite highlighting the different types of dendritic spines (B). *This image modified from Fig. 1 of Risher WC, Ustunkaya T, Singh Alvarado J, Eroglu C. Rapid golgi analysis method for efficient and unbiased classification of dendritic spines. PLoS One 2014;9(9):e107591. https://doi.org/10.1371/journal.pone.0107591. This is an open-access article distributed under the terms of the Creative Commons Attribution 4.0 International License (https://creativecommons.org/licenses/by/4.0/).*

To complicate matters even more, receptors can be categorized as either *ionotropic* or *metabotropic* (Fig. 2.4). With an ionotropic receptor, a neurotransmitter binds to a specific site on the receptor.[10] This then causes a channel embedded in the receptor to open, allowing *ions* (charged atoms/molecules) to enter the neuron. If the receptor is excitatory, it will allow the positively charged ion sodium (Na^+) into the neuron. If the receptor is inhibitory, it will allow the negatively charged ion chloride (Cl^-) into the cell. This concept will become more important later in this chapter. Metabotropic receptors, like ionotropic receptors, are either excitatory or inhibitory; however, they are slower compared to ionotropic receptors.[11] When a neurotransmitter binds to a metabotropic receptor, this does not result in the opening of a channel within the receptor. Instead, it will activate an internal part of the receptor called a *guanyl-nucleotide-binding protein* (*G-protein* for short).[12] G-proteins consist of three subunits known as alpha, beta, and gamma. Once activated, the alpha subunit of the G-protein breaks away from the receptor and causes some event to occur within the neuron.

There are three major classes of metabotropic receptors: G_s, G_q, and G_i. G_s and G_q receptors are excitatory.[11] If a G_s metabotropic receptor is activated, it will cause the G-protein to increase *adenyl cyclase* levels in the cell.[13] Adenyl cyclase is an enzyme that causes *cyclic adenosine monophosphate* (*cyclic AMP or cAMP*) levels to increase. cAMP is known as a **second messenger**. Essentially, increased cAMP levels will cause an ion channel to open somewhere on the neuron, allowing positively charged ions into the cell. If cAMP is a second messenger, what is the first messenger? In this case, the first messenger is the neurotransmitter that binds to the metabotropic receptor. Think of adenyl cyclase/cAMP as "middle management". Because cAMP levels must increase before an ion channel opens, this leads to the slower activation associated with metabotropic receptors relative to ionotropic receptors.

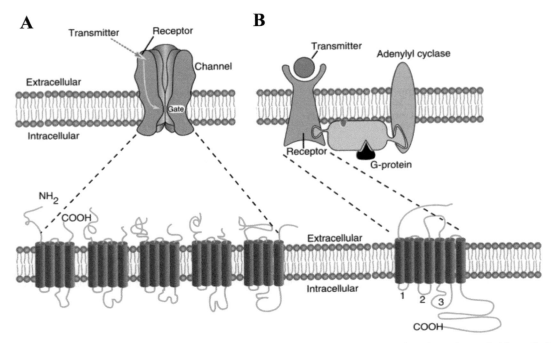

FIGURE 2.4 A comparison of ionotropic (also called ligand-gated ion channel) (A) and metabotropic (also called G-protein-coupled) (B) receptors. In (A), the neurotransmitter binds to a specific site on the ionotropic receptor, causing it to open a channel. The type of receptor (either excitatory or inhibitory) will determine which ions flow across the membrane separating the synapse and the membrane of the neuron. Excitatory ionotropic receptors allow Na^+ or Ca^{++} into the neuron, whereas inhibitory ionotropic receptors allow Cl^- into the neuron. In (B), the neurotransmitter binds to a site on the metabotropic receptor, which then causes a G-protein to break away from the receptor. The G-protein can either increase or decrease adenyl cyclase depending on whether it is excitatory or inhibitory. The result is to open a channel located away from the metabotropic receptor (not pictured on this graph). These channels will allow either positively charged ions (like Ca^{++}) to flow into the neuron (if excitatory) or positively charged ions (like K^+) to flow out of the neuron (if inhibitory). Gq-coupled receptors do not alter adenyl cyclase levels. Instead, they cause a cascade of events to occur in the neuron (not pictured here). *This image comes from Fig. 8.1 reprinted from Squire LR, Berg D, Bloom FE, du Lac S, Ghosh A, Spitzer NC. Fundamental Neuroscience. 4th ed. 164. Copyright 2013, with permission from Elsevier.*

G_q receptors are even more complex than G_s receptors. When activated, G_q receptors do not increase cAMP levels. Instead, the alpha subunit dissociates from the receptor and activates the enzyme *phospholipase C (PLC)*. PLC then cuts a small lipid in the cell membrane called *phosphatidylinositol 4,5-bisphosphate* (PIP_2). PIP_2 is then cleaved into *inositol triphosphate* (IP_3) and *diacylglycerol (DAG)*.[13] IP_3 and DAG are second messengers that cause certain events to occur in the neuron. For simplicity, I will focus on IP_3 for now. IP_3 binds to the endoplasmic reticulum, which causes the positively charged ion calcium (Ca^{++}) to be released into the neuron. The increase in intracellular Ca^{++} is excitatory and can cause alterations to occur to the neuron, such as the insertion of additional receptors.

G_i receptors are inhibitory and act in a manner that is somewhat opposite of G_s receptors.[11] When G_i receptors are activated, they decrease adenyl cyclase and cAMP levels, which ultimately causes positively charged potassium (K^+) ions to be removed from the neuron.[13] If you have been paying attention, you will notice that excitatory receptors allow positively charged ions into the neuron while inhibitory receptors either allow negatively charged ions into the neuron or remove positively charged ions from the neuron. This is a critically important concept for understanding neuronal communication. We will come back to this topic.

Why even have ionotropic and metabotropic receptors? Ionotropic receptors are quickly activated, and their effects last for a short period of time, approximately a fraction of a second. Metabotropic receptors are slower to activate, but their effects last for a longer period, several seconds to a few minutes. In some cases, quick transmission is important, so ionotropic receptors are needed. Our skeletal muscles are controlled by an ionotropic receptor, allowing us to quickly move our limbs. Likewise, our visual and auditory systems are controlled by ionotropic receptors. However, in many cases, slower transmission is preferable. For example, a specific metabotropic receptor controls our heart rate. Slower transmission allows us to maintain a stable heart rate throughout the day. This is not to say that metabotropic receptors cannot be used for processes that are required immediately. An inhibitory metabotropic receptor controls parasympathetic input to the heart, but an excitatory metabotropic receptor causes heart rate to increase when we encounter a stressful situation.

Another important concept is that some neurotransmitters bind to both ionotropic and metabotropic receptors, and some neurotransmitter systems are composed of excitatory and inhibitory receptors. This means that a single neurotransmitter can be either excitatory or inhibitory, depending on the receptor to which it binds. Why are some neurotransmitters both excitatory and inhibitory? Given the sheer number of physiological and cognitive functions neurotransmitters serve, allowing them to stimulate or inhibit certain processes increases flexibility. We do not need thousands of neurotransmitters to ensure that our muscles work properly or to allow us to think or to feel emotions.

Axon

Connected to the cell body is the **axon**, a long tube-like projection that is responsible for sending an electrical signal to the end of the neuron. This electrical signal will be covered in more detail in the next section. In mammalian species, most axons are covered by the *myelin sheath*. The myelin sheath is a fatty tissue that helps increase the speed at which the axon can transmit signals to the end of the neuron, which increases the speed of neuronal communication.[14] The importance of the myelin sheath is easily understood when one considers the disease multiple sclerosis (MS). MS is characterized by a wide range of symptoms, including, but not limited to, vision problems, cognitive issues like depression, muscle weakness, and tingling sensations or numbness in the extremities. These symptoms emerge because of the breakdown of myelin sheath.[15]

Inside the axon are specialized structures called *microtubules*.[16] Microtubules help provide structural support to the neuron, and they act as a transport system. After the Golgi apparatus has transported newly formed proteins from the cell body to the edge of the axon, the microtubules take the proteins and transport them to the end of the neuron, which we need to discuss next.

Terminal buttons and transporters

The **terminal buttons** (also called terminal boutons, axon terminals, or axon boutons) are projections that originate from the axon. As described above, receptors on dendrites can receive either excitatory or inhibitory messages. The terminal button is the region of the neuron that sends these excitatory/inhibitory messages to the next neuron.[17] Located at the terminal button are proteins called **transporters** (Fig. 2.5). When a terminal button releases a chemical message to the next neuron, there is often an excess of neurotransmitter that needs to be "recycled" for future use. Think of it this way: neurons want to be "eco-friendly" and do not want to waste resources when there is no need to do so. Transporters are responsible for sending this excess neurotransmitter back into the presynaptic terminal in a process known as *re-uptake*.[18]

How neurons communicate with each other

So far, you have learned about the various structures of the neuron, but we need to tie everything together by explaining how neurons communicate with each other. To do so, I need to describe, in detail, what the axon does to send a signal to the terminal button, and I need to describe how the terminal buttons transmit their messages to surrounding neurons.

Action potentials

If a neuron receives enough excitation, it will send an **action potential** (electrical signal) down the length of the axon.[19] Before discussing the action potential, I first need to describe a neuron's **resting membrane potential**[20] (Fig. 2.6). What exactly does this mean? If we stick one electrode inside a neuron and one electrode outside of a neuron, we would notice that the inside of the neuron is more negatively charged than the outside of the neuron. Why is this

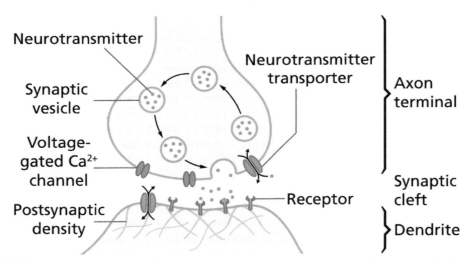

FIGURE 2.5 An image of a terminal button of a presynaptic neuron, a synapse, and a postsynaptic dendrite. On the presynaptic terminal is the transporter, which is responsible for taking excess neurotransmitter back into the terminal button. *This image came from Wikimedia Commons and was posted by Thomas Splettstoesser (https://commons.wikimedia.org/wiki/File:SynapseSchematic_en.svg).*

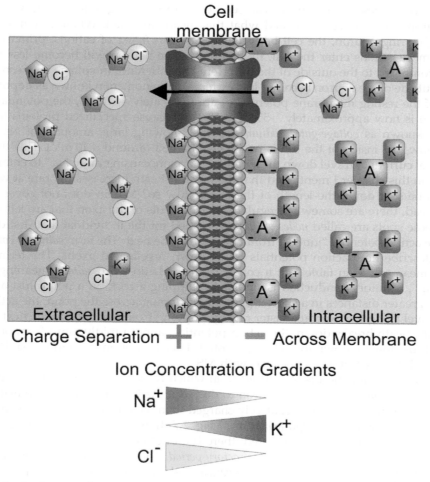

FIGURE 2.6 A schematic of a neuron at resting membrane potential. The concentration of sodium (Na$^+$) and chloride (Cl$^-$) ions is higher in the extracellular space (outside) compared to the intracellular space (inside) of the neuron. Conversely, the concentration of potassium (K$^+$) ions and anions is higher in the intracellular space compared to the extracellular space. Due to the large number of anions located inside the neuron, the resting membrane potential is approximately −70 mV. *This image came from Wikimedia Commons and was posted by user Synaptitude (https://commons.wikimedia.org/wiki/File:Basis_of_Membrane_Potential2.png).*

the case? Inside and outside of the cell are positively charged ions (known as *cations*) and negatively charged proteins (known as *anions*). The inside of the neuron contains numerous anions and K$^+$ ions. Even though K$^+$ is positively charged, there are more anions on the inside compared to K$^+$ ions. The outside of the neuron contains negatively charged chloride (Cl$^-$) ions, but there are much higher concentrations of Na$^+$. Because there are more Na$^+$ ions on the outside of the cell than there are K$^+$ ions on the inside, the net result is a negatively charged inside. The resting membrane potential of a neuron ranges anywhere from −30 to −90 mV (millivolts).

As previously described, neurons can send excitatory messages or inhibitory messages to other neurons, but I never explained what this means in detail. When a neuron is excited by a neighboring neuron, the cell body will receive an *influx* of cations, primarily Na^+ or Ca^{++}. If more cations enter the cell, the interior of the neuron will become less negatively charged compared to the outside of the cell. Here, the difference in voltage between the inside and the outside of the neuron becomes less extreme, a phenomenon known as **depolarization**. Instead of the resting membrane potential of approximately -70 mV, the potential inside of the neuron is now approximately -55 mV. This will cause specialized proteins to open on the axon, known as *voltage-gated sodium channels*, allowing large amounts of Na^+ into the axon.[19] Now, the inside of the cell is positively charged (around $+40$ mV). This then causes an electrical current to travel down the length of the axon, causing more Na^+ to enter the axon.

Earlier in this chapter I mentioned that the myelin sheath increases the rate at which messages can be sent down the length of the axon. How so? Myelin does not cover the entire axon. Instead, there are somewhat regularly spaced spots of the axon that are left unmyelinated.[21] These spots are called *nodes of Ranvier*, named for the individual that first discovered them. The action potential "jumps" from one node to the next. The term *saltatory conduction* is used to describe how action potentials travel down myelinated axons. The term saltatory does not mean salt as in table salt; it comes from the Latin word *saltare*, meaning "to hop" or "to leap". Saltatory conduction is similar to skipping a rock on a pond. Just as the rock can cover greater distances in a shorter period by skipping across the pond, the action potential can travel along the axon at a faster rate by skipping from one node to the next. In MS, as the myelin sheath breaks down, the action potential cannot rapidly jump from one node to the next, thus slowing down the action potential. This disruption in neurotransmission leads to the long list of symptoms associated with MS.

Following depolarization is *repolarization*, in which K^+ ions are removed from the axon.[19] This causes the voltage inside the neuron to decrease. As K^+ ions continue to exit the axon, the neuron becomes even more negatively charged than what it was at resting membrane potential. This is known as **hyperpolarization**.[19] Fig. 2.7 shows the events of an action potential.

After an action potential has occurred, there is a brief period in which another action potential cannot occur.[19] This is called a *refractory period*. There are two refractory periods. The first one is known as the *absolute refractory period*, and the second one is known as the *relative refractory period*. During the absolute refractory period, the neuron will not fire another action potential regardless of how much that neuron is excited. This occurs because the Na^+ ion channels of the axon become desensitized, meaning they will not open even if there is additional excitation. During the relative refractory period, the neuron can generate another action potential if it is sufficiently stimulated.

Synapses and chemical transmission

When the action potential reaches the terminal button, this causes Ca^{++} to enter the cell.[22] Inside of the terminal buttons are special sacs called **synaptic vesicles** (Fig. 2.4 shows synaptic vesicles). These vesicles contain different neurotransmitters. When Ca^{++} enters the terminal buttons, it causes these vesicles to start moving toward the edge of the terminal button. The vesicles open, releasing neurotransmitter into the space between neurons, which is the **synapse**.[23] Importantly, neurons do not contact each other. Instead, the neurotransmitter

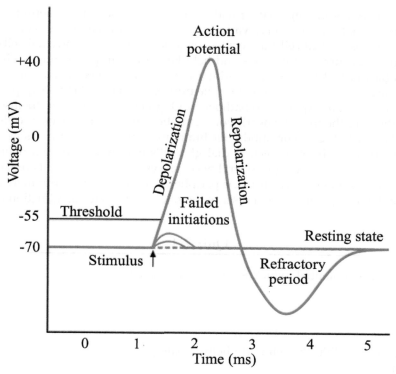

FIGURE 2.7 A schematic showing the course of an action potential. The resting membrane potential shows the voltage at −70 mV. If enough excitation occurs, the axon becomes depolarized as Na^+ ions enter the neuron. This causes the voltage to become positively charged. If there is not enough excitation (failed initiations), depolarization does not occur. Following depolarization is repolarization, in which K^+ ions are removed from the axon. As K^+ ions continue to be removed from the axon, hyperpolarization occurs. During hyperpolarization, the voltage becomes more negatively charged compared to the resting membrane potential. The refractory period also occurs during this time, in which another action potential either cannot occur at all (absolute refractory period) or can only occur with increased stimulation (relative refractory period). Eventually, the neuron returns to its resting membrane potential. This image comes from Wikimedia Commons and was originally posted by user Chris 73 and later updated by user Diberri (https://commons.wikimedia.org/wiki/File:Action_potential.svg). The author of this textbook edited the text to make it less blurry.

crosses the synapse and binds to receptors located on the dendrite of a neighboring neuron. Depending on which neurotransmitter is released and to which receptor it binds, it will either excite or inhibit the next neuron. If it is excitatory, it will cause an action potential to occur in the next neuron (assuming there is sufficient stimulation), thus sending the message further along in the nervous system.

Glial cells

Now that you have a general understanding of how neurons work, it is now time to turn our attention to a different group of cells that play important contributions to our nervous system. **Glial cells** are nonneuronal cells located within the central nervous system and the peripheral nervous system that provide supporting functions to neurons.[24] There are different types of glial cells that serve various roles. Within the central nervous system, there are four major types of glial cells: astrocytes, oligodendrocytes, ependymal cells, and radial glia.[24-26] Astrocytes surround neurons and help keep them in place and also help form the **blood–brain barrier (BBB)**. The BBB is primarily made up of specialized cells call *endothelial cells*, but they are supported by astrocytes. The purpose of the BBB is to help protect the central nervous system from exposure to pathogens. Oligodendrocytes wrap themselves around axons; thus, they are the myelin sheath you read about earlier! Ependymal cells help create *cerebrospinal fluid* (CSF). CSF surrounds our brain and spinal cord and acts as a cushion to protect these structures from impact. Radial glia cells help newly formed neurons migrate to where they need to go. In the peripheral nervous system, Schwann cells replace oligodendrocytes.[27] There are other glial cells in the peripheral nervous system, but there is no need to cover them in detail right now. I will discuss the role of astrocytes in addiction in Chapter 4.

Quiz 2.1

1. During an action potential, which ion enters the axon?
   a. Ca^{++}
   b. Cl^-
   c. K^+
   d. Na^+
2. Which structure within the cell body helps create proteins?
   a. Lysosome
   b. Mitochondrion
   c. Rough endoplasmic reticulum
   d. Smooth endoplasmic reticulum
3. Why are dendritic spines important?
4. What is the difference between an ionotropic receptor and a metabotropic receptor?
5. Which of the following is a function of astrocytes?
   a. Act as the myelin sheath of axons
   b. Help create CSF
   c. Make up part of the BBB
   d. Transport immature neurons to their location

Answers to quiz 2.1

1. d — Na^+
2. c — Rough endoplasmic reticulum
3. Dendritic spines allow dendrites to receive incoming messages more easily from surrounding neurons.
4. When an agonist binds to an ionotropic receptor, a channel pore located within the receptor will open, allowing ions to enter the neuron (e.g., Na^+ if excitatory or Cl^- if inhibitory). When an agonist binds to a metabotropic receptor, a G-protein will break away from the receptor. This G-protein can do one of several things, depending on the

type of receptor. If the receptor is excitatory, the G-protein will increase adenyl cyclase levels or will activate a second messenger, which will cause a channel to open located somewhere else on the postsynaptic neuron or will lead to the insertion of additional receptors. If the receptor is inhibitory, the G-protein will decrease adenyl cyclase levels, leading to the opening of a channel to remove K^+ from the neuron.
5. c — Make up part of the BBB

Major neurotransmitters

Understanding the major functions of each neurotransmitter, how each one is synthesized and degraded, and the various receptors/transporters of each neurotransmitter is important as these details will help increase understanding of how drugs exert their effects and how certain medications can be used to reduce drug use. We will begin with the first neurotransmitter to be discovered: acetylcholine.

Acetylcholine

Although acetylcholine was first isolated in 1913 by Sir Henry Dale (1875—1968), its functions as a neurotransmitter were not uncovered until 1921 by Dr. Otto Loewi (1873—1961).[28] Loewi conducted an experiment with two frog hearts. One heart was still attached to the *vagus nerve* and was placed in a chamber filled with saline (salt water). The second heart was placed in a separate chamber, such that saline from the first chamber could flow into it. Loewi applied an electrical current to the vagus nerve of the first heart and noticed that the heart's beating slowed. After a short delay, he noticed that the beating of the heart in the second chamber also slowed. Loewi correctly hypothesized that stimulation of the vagus nerve causes a chemical to flow into the fluid, thus decreasing heart rate. At the time, Loewi called this chemical "Vagusstoff", but we now know it as acetylcholine. Dale and Loewi received the Nobel Prize in 1936 for their contributions in neurotransmission.

Due to Loewi's work, we know that acetylcholine is important for controlling multiple processes in the peripheral nervous system, such as heart rate. One of its most important roles is controlling skeletal muscles.[29] The drug Botox (botulinum toxin) gives people smooth-looking skin by blocking cholinergic receptors (note: the term cholinergic is used to describe the acetylcholine system). By blocking these receptors, the muscles in the face become paralyzed. Caution needs to be taken when using Botox because overdose can lead to paralysis of limbs. Acetylcholine is the major neurotransmitter released in the parasympathetic nervous system. Within the central nervous system, acetylcholine is important for keeping us alert and sustaining our attention, and it has been shown to be an important mediator of learning and memory.[29] Alzheimer's disease, a neurocognitive disorder characterized by memory loss, is associated with damage to cholinergic neurons.[30]

Where does acetylcholine come from? Acetylcholine is made from two things: the *precursor* choline and acetyl-coenzyme A.[31] Choline is a nutrient we obtain from our diet. Our livers can make choline, but only at very low levels. Foods high in choline include proteins like beef, chicken, fish, and liver; egg yolks; broccoli; soy; dairy; and cauliflower. The enzyme choline acetyltransferase converts choline and acetyl-coenzyme A into acetylcholine. This conversion occurs in the presynaptic terminal, where the newly synthesized acetylcholine is packaged into vesicles. Following an action potential, acetylcholine is released into the synapse where it can bind to one of two major types of receptors: the ionotropic *nicotinic receptor* or various metabotropic *muscarinic receptors*.[32]

Nicotinic receptors are composed of five subunits that form to create a channel.[33,34] In total, there are seven different alpha subunits (α_2, α_3, α_4, α_5, α_6, α_7, α_9, and α_{10}; there is an α_8 subunit that has been found in avian species but not in mammals) and three beta subunits (β_2, β_3, and β_4). The most common nicotinic subunit compositions consist of five α_7 or two α_4 and two β_2, with an additional subunit, typically a third β_2 or an α_5.[35] In the case of the *homomeric* α_7 receptors, acetylcholine binds at the site where two subunits come into contact. Thus, acetylcholine can bind to five different sites on these receptors. In the case of the *heteromeric* α_4/β_2 receptors, acetylcholine binds to where an α_4 subunit meets with a β_2 subunit. Fig. 2.8 shows the binding site of acetylcholine to various nicotinic receptor subtypes, as well as the distributions of each receptor subtype in the rat brain.

There are five types of muscarinic receptors,[36] simply labeled M_1-M_5. M_1, M_3, and M_5 receptors are excitatory and are coupled to G_q proteins, whereas M_2 and M_4 receptors are inhibitory (G_i-coupled). In Loewi's famous experiment, vagus nerve stimulation caused acetylcholine to bind to M_2 receptors in the heart. Because these receptors are inhibitory, they cause heart rate to decrease. The G_i-coupled M_2 and M_4 receptors are located on both the postsynaptic neuron and on the terminal buttons of the presynaptic neuron. What happens when acetylcholine binds to presynaptic M_2/M_4 receptors? Activation of these receptors acts as a *negative feedback loop*; that is, the presynaptic neuron is "commanded" to stop releasing acetylcholine into the synapse.[37] This can be accomplished by stopping the conversion of acetylcholine from choline and acetyl-coenzyme A or from preventing acetylcholine release into the synapse. Receptors located on the presynaptic neuron (such as M_2 receptors) are often referred to as **autoreceptors**. Autoreceptors are sensitive to the neurotransmitter that is released from that neuron. In the case of the M_2 autoreceptor, these receptors will respond to the acetylcholine that is released from that neuron. In contrast, postsynaptic muscarinic receptors will respond to acetylcholine released from multiple neurons. As such, these receptors are known as **heteroreceptors**. Fig. 2.9 provides an example of a synapse that contains an autoreceptor and a heteroreceptor.

After acetylcholine has been released from the presynaptic neuron, it needs to be removed from the synapse to prevent acetylcholine from overstimulating receptors. Overstimulation of cholinergic neurons is problematic as it can cause cramps, muscular weakness, and even paralysis. Not only does the postsynaptic membrane contain nicotinic/muscarinic receptors, but it contains an enzyme called acetylcholinesterase. Acetylcholinesterase *hydrolyzes* (fancy way of saying chemical breakdown due to reaction with water) acetylcholine into choline and acetic acid.[38] The choline is taken back up into the presynaptic neuron where it can be converted into acetylcholine once again while we excrete acetic acid.

Monoamines

Monoamines make up a neurotransmitter class that is characterized by the shape of its chemical structure. Specifically, monoamines contain a single amine group connected to an aromatic ring by a two-carbon chain. Monoamines can be subdivided into two major classes:

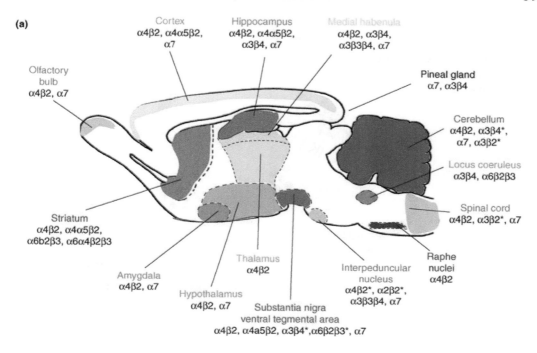
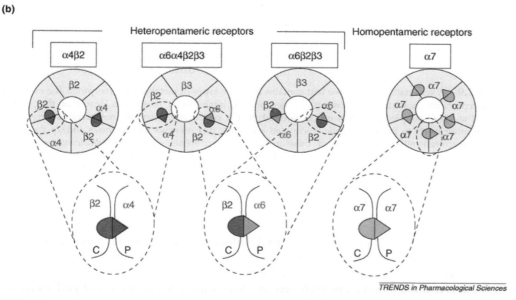

FIGURE 2.8 (A) Distribution of various nicotinic subtypes in the rat brain. (B) Schematic of how nicotinic subunits can be arranged. Notice that with the α7 homopentameric receptor subtype, acetylcholine can bind to five different sites. However, with heteropentameric receptor subtypes, acetylcholine binds to two sites only. *Image comes from Fig. 1 of reprinted from Gotti C, Zoli M, Clementi F, Brain nicotinic acetylcholine receptors: native subtypes and their relevance. Trends in Pharmacological Sciences. 2006;27(9):482–491. Copyright 2006, with permission from Elsevier.*

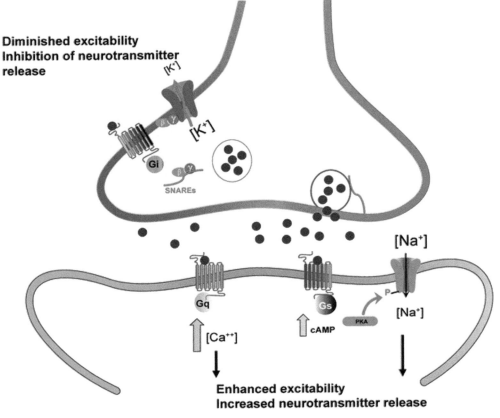

FIGURE 2.9 A schematic of a synapse that contains the three types of metabotropic receptors. The excitatory G_q and G_s receptors are located on the postsynaptic neuron. The inhibitory G_i receptor is located on the presynaptic neuron, although G_i-coupled receptors can be located on the postsynaptic neuron (not pictured here). When a neurotransmitter binds to the presynaptic G_i-coupled receptors, this prevents the synthesis and/or the release of additional neurotransmitter. Activation of these autoreceptors prevents *soluble N-ethylmaleimide-sensitive factor attachment protein receptors (SNAREs)* from priming the vesicle to release more neurotransmitter into the synaptic space. *Image comes from Fig. 1 of reprinted by permission from Roth BL. Molecular pharmacology of metabotropic receptors targeted by neuropsychiatric drugs. Spring. Nat.* Nat Struc Mol Biol. 2019. Copyright 2019.

catecholamines and indolamines.[39,40] Catecholamines contain the neurotransmitters dopamine, norepinephrine, and epinephrine, whereas indolamines contain 5-hydroxytryptophan (5-HT; also known as serotonin) and melatonin.

Collectively, monoamines are involved in many different physiological and mental processes. Dopamine is largely involved in reward learning, movement, executive functioning (i.e., self-control), arousal, motivation, lactation, and nausea.[41] Parkinson's disease, characterized by excessive shaking, results from destruction of a group of dopaminergic neurons.[42] Schizophrenia, a neurocognitive disorder marked by hallucinations and delusions, is hypothesized to stem from excess dopamine levels in the brain.[43] Norepinephrine and epinephrine play a vital role in the fight or flight response.[44] As acetylcholine is the major neurotransmitter of the parasympathetic nervous system, norepinephrine is the major neurotransmitter of the sympathetic nervous system. Therefore, norepinephrine prepares us to respond to a stressful situation. Norepinephrine and epinephrine are also involved in vigilance, wakefulness, attention, and memory.[45] Serotonin helps control mood, sleep, appetite, bowel movements, and nausea.[46] In fact, most of the serotonin in the body is found in the gastrointestinal tract, explaining its contribution to processes like appetite. Even though most of our serotonin is in our gastrointestinal tract, it plays an important role in cognition, as major depressive disorder is believed to result from decreased serotonin, as well as norepinephrine, levels in the brain.[47]

Catecholamines are synthesized from the amino acid tyrosine.[48] Recall from an introductory biology course that amino acids are the basic building blocks of proteins. Our body naturally produces tyrosine from phenylalanine, but we also get tyrosine from our diet, as high-protein foods (such as chicken, turkey, fish, milk, yogurt, cheese, peanuts, and almonds) contain tyrosine. Fun fact: tyrosine comes from the Greek word *týros*, meaning cheese, the food in which tyrosine was first discovered. In the neuron, the enzyme tyrosine hydroxylase first converts tyrosine into L-3,4-dihydroxyphenylalanine (L-DOPA).[48] L-DOPA is then converted into dopamine by the enzyme DOPA decarboxylase (sometimes referred to as aromatic L-amino acid decarboxylase or AADC).[48] Interestingly, dopamine is the precursor to norepinephrine, as the enzyme dopamine β-hydroxylase (DβH) is responsible for converting the former into the latter.[49] Finally, the enzyme phenylethanolamine N-methyltransferase converts norepinephrine into epinephrine.[50]

Indolamines follow a different synthesis pathway, with the precursor being the amino acid tryptophan.[51] Unlike tyrosine, our bodies do not naturally produce tryptophan; thus, it is an essential amino acid. Foods high in tryptophan include high-protein foods (similar to those high in tyrosine), egg whites, chocolate, and oats. Growing up, you may have heard that eating a lot of turkey during Thanksgiving makes you sleepy due to the high tryptophan levels found in it, although this is not necessarily true. We most likely get tired from eating large amounts of food during the day. Tryptophan is first converted into 5-hydroxy-L-tryptophan (5-HTP) by the enzyme tryptophan hydroxylase.[51] The enzyme 5-hydroxytryptophan decarboxylase (also referred to as AADC) converts 5-HTP into serotonin.[51] Serotonin then serves as a precursor to melatonin. There are a couple of steps needed to get from serotonin to melatonin, but I will spare you those details as we will not discuss melatonin in this book.

Dopamine receptors are subdivided into two major classes: dopamine D_1-like and dopamine D_2-like.[41,52] All dopamine receptors are metabotropic. The D_1-like receptor class consists of D_1 and D_5 receptors, which are G_s-coupled. The D_2-like receptor class consists of D_2, D_3, and D_4 receptors, and they are G_i-coupled. Just like the cholinergic system, the excitatory D_1-like receptors are located on the postsynaptic neuron only, but D_2-like receptors are located on both dendrites of the postsynaptic neuron and on the presynaptic neuron. Similar to dopamine receptors, all noradrenergic receptors are metabotropic. They can be subdivided into three major types: α_1, α_2, and β. The α_1 and β subtypes are stimulatory and are heteroreceptors, with α_1 receptors being G_q-coupled and β being G_s-coupled. The α_2 receptor is G_i-coupled and can be either an autoreceptor or a heteroreceptor.

Serotonergic receptors are more complicated than dopaminergic and noradrenergic receptors. Thus far, 14 different 5-HT receptors have been identified, which are categorized into seven different families (5-HT$_1$-5HT$_7$).[53] 5-HT receptors are metabotropic, except for the 5-HT$_3$ receptor. Most 5-HT receptors are excitatory, with the exceptions of the 5-HT$_1$ and 5-HT$_5$ families. The 5-HT$_1$ and 5-HT$_5$ families primarily serve as autoreceptors, although some 5-HT$_1$ receptors are heteroreceptors in certain brain regions.

Excess monoamines are sent back into the presynaptic neuron through transporters. Each of the major monoamines has their own transporter: the dopamine transporter (DAT), the norepinephrine transporter (NET), and the serotonin transporter (SERT).[54] Next, some of the excess neurotransmitter can be *metabolized* (broken down) by certain enzymes. One such enzyme is **monoamine oxidase (MAO)**.[55] MAO is bound to the mitochondria in the neuron and can be differentiated into MAO-A and MAO-B. MAO metabolizes dopamine, norepinephrine, and serotonin. In addition to MAO, catechol-*O*-methyltransferase (COMT) breaks down catecholamines (e.g., dopamine and norepinephrine).[56] Like MAO, there are two variants of COMT: soluble form (S-COMT) and membrane-bound form (MB-COMT).[57] S-COMT is found in the *cytosol* (intracellular fluid of the cell); essentially, S-COMT floats around in the neuron. MB-COMT, as the name suggests, is bound to a membrane; in this case, it is bound to the rough endoplasmic reticulum. Why are there two different enzymes devoted to breaking down dopamine and norepinephrine? An important consideration is that MAO and COMT are found in different parts of the brain and in different parts of the body. For example, COMT is heavily found in our liver, not in the brain.

For the monoamines that have not been metabolized by MAO or COMT, they can reenter the vesicle through the **vesicular monoamine transporter 2 (VMAT-2)**.[58] VMAT-2 functions similarly to transporters like DAT and NET. Instead of removing excess neurotransmitter from the synapse, VMAT-2 removes neurotransmitter from the cytosol. This allows the neuron to reuse some of the monoamines that have already been synthesized from tyrosine/tryptophan.

Glutamate

Glutamate is the major excitatory neurotransmitter in the brain and is involved in learning and memory.[59] Glutamate is also involved in sensation, as many neurons transmitting information about vision, hearing, and pain release glutamate.[60-63] Glutamate is synthesized from the amino acid glutamine via the enzyme glutaminase.[64] This conversion largely occurs in the presynaptic neuron. Our bodies can produce glutamate, thus making it a nonessential amino acid. However, we can get glutamate from our diet in the form of monosodium glutamate, better known as MSG. If you have ever eaten Ramen noodles, you are quite familiar with MSG as the seasoning packets contain MSG. You may have previously learned that our taste buds can be divided into sweet, sour, salty, and bitter. However, a fifth taste is now often discussed: *umami*. This refers to a meaty, savory taste, and this taste is driven by MSG.[65]

Glutamate can bind to either metabotropic or ionotropic receptors, with both types being subdivided even further.[66] There are three major classes of metabotropic receptors: Group I, Group II, and Group III. The Group I family is composed of mGluR$_1$ and mGluR$_5$, which are excitatory (G$_q$-coupled) and located on the postsynaptic neuron. The Group II family consists of mGluR$_2$ and mGluR$_3$, and the Group III family contains mGluR$_4$, mGluR$_6$, mGluR$_7$, and mGluR8. Even though glutamate is the major excitatory neurotransmitter in the brain, the Group II and Group III families are inhibitory, not excitatory. The Group II and Group III metabotropic receptors can function as heteroreceptors or as autoreceptors.

Ionotropic receptors can be subdivided into α-amino-3-hydroxy-5-methyl-4-isoxazolepropionic acid (AMPA), N-methyl-D-aspartate (NMDA), and kainate. We will not discuss kainate receptors as they can be difficult to differentiate from AMPA receptors.[66] All three receptors are composed of four subunits that surround a channel that opens once it has been stimulated. In the case of AMPA receptors, the subunits are abbreviated as GluA1, GluA2, GluA3, and GluA4.[67] Most AMPA receptors contain two GluA2 subunits, with the two other subunits consisting of combinations of GluA1, GluA3, and GluA4. Glutamate can bind to any of the four AMPA receptor subunits. Once activated, the AMPA receptor allows Na^+ to enter the neuron, thus depolarizing it.

NMDA receptors are much more complex than AMPA receptors. All NMDA receptors have two GluN1 subunits. These subunits are required to make the NMDA receptor functional. The other two subunits are labeled as GluN2 and are composed of combinations of GluN2A, GluN2B, GluN2C, and GluN2D. Glutamate binds to the GluN2 subunit.[66] One characteristic that makes NMDA receptors unique is that it requires a *co-agonist* to function. The neurotransmitter glycine must bind to the GluN1 subunits to allow glutamate to activate the receptor.[68] Another unique characteristic of NMDA receptors is that they are dependent on AMPA receptors to function. At resting membrane potential, NMDA receptors are blocked by a positively charged magnesium (Mg^{++}) ion.[66] Remember, the inside of a neuron is more negatively charged compared to the outside of the cell. This attracts the Mg^{++} ion, which tries to enter through the channel of an NMDA receptor. However, the Mg^{++} ion is too large to fit through the NMDA receptor, so it stays trapped inside the channel. Even if NMDA receptors are stimulated by both glycine and glutamate, nothing can happen as other positively charged ions like Na^+ cannot enter the neuron because the Mg^{++} ion blocks the receptor. This is where AMPA receptors play an important role in NMDA receptor activity. As glutamate stimulates surrounding AMPA receptors, this allows more Na^+ to enter the neuron. As the concentration of Na^+ increases, the neuron becomes less positively charged. This depolarization causes the Mg^{++} ion to dissociate from the receptor channel. Now, Na^+ and Ca^{++} can enter the NMDA receptor when it is stimulated.

In the nerve terminal, glutamate is packaged inside vesicles by the vesicular glutamate transporter (VGLUT)[69] and is released into the synapse once Ca^{++} enters the terminal following an action potential. Like monoamines, some excess glutamate can be removed from the synapse via a family of transporters known as the excitatory amino acid transporters (EAATs).[70] There are five different EAATs, but for simplicity, we will not discuss each one in detail. All you need to know is that glutamate is typically removed from the synapse via an EAAT into a neighboring astrocyte. From there, the enzyme glutamine synthetase metabolizes glutamate back into glutamine.

GABA

GABA is the major inhibitory neurotransmitter in the nervous system.[71] It is largely responsible for regulating the release of other neurotransmitters, but it plays specific roles in neuron growth and sleep.[71] Even though GABA is the major inhibitory neurotransmitter in the mammalian brain, it is synthesized from glutamate, the major excitatory neurotransmitter.[72] The enzyme glutamic acid decarboxylase is responsible for this conversion, which occurs in the presynaptic neuron. Following this conversion, GABA is stored in synaptic vesicles via the vesicular GABA transporter (VGAT). Once GABA is released into the synapse, it can bind to either $GABA_A$ or $GABA_B$ receptors.[73,74] $GABA_A$ receptors are ionotropic, containing five subunits. There are approximately 16 different possible subunits: six alpha, four beta, four gamma, one delta, and one epsilon. Although there are numerous potential combinations that can occur, most $GABA_A$ receptors are composed of two alpha subunits, two beta subunits, and one gamma subunit. GABA binds to where the alpha and beta subunits meet. Once GABA has activated the receptor, the ion channel opens, allowing Cl^- to enter the cell. This, in turn, inhibits the neuron. $GABA_B$ receptors are metabotropic and are inhibitory like the $GABA_A$ receptor. These receptors are located on both the pre- and the postsynaptic neurons. From the synapse, excess GABA is removed via VGAT and is metabolized into glutamate by GABA transaminase.[72]

Endogenous opioids

You may be familiar with endorphins, which are associated with pain reduction. Endorphins, along with enkephalins and dynorphins, are *endogenous opioids*. The term endogenous means found within the body. These endogenous opioids are derived from larger *polypeptides*, which are long chains of amino acids. Endorphins and enkephalins are formed when the polypeptide proopiomelanocortin (POMC) is cleaved. POMC also gives rise to melanocyte-stimulating hormone (α-MSH) and adrenocorticotropic hormone (ACTH). α-MSH is important for controlling satiety and sexual behavior and regulates the movement of melanin, the chemical that determines our skin color. ACTH is responsible for secreting "stress" hormones such as cortisol, which you will learn more about in Chapter 11. It is amazing that POMC can be cleaved to create peptides with vastly different responsibilities in the body.

Endogenous opioids bind to one of four opioid receptors: mu, kappa, delta, and nociceptin opioid peptide (also known as nociceptin/orphanin FQ). They are inhibitory metabotropic receptors that are differentially sensitive to the endogenous opioids. Endorphins primarily bind to mu receptors, dynorphins bind to kappa receptors, enkephalins bind to delta receptors, and nociceptin binds to nociceptin opioid peptide receptors. Like the other G_i-coupled receptors described in this chapter, opioid receptors are located on the pre- and the postsynaptic neuron. Because opioids are peptides, not neurotransmitters, there is one major difference in the opioid system. Unlike neurotransmitters such as dopamine, glutamate, and GABA that can be repackaged into vesicles, there are no transporters that can remove excess opioids from the synapse. Once released into the synapse, the endogenous opioids bind to their respective receptors before they are quickly metabolized.

Endocannabinoids

The term endocannabinoid means endogenous cannabinoid. Endocannabinoids have been implicated in appetite, pain suppression, motor control, and cognition.[75] The two most well-known endocannabinoids are anandamide (coming from the Sanskrit word meaning "bliss") and 2-arachidonoylglycerol (2-AG). Anandamide and 2-AG were not discovered until 1992 and 1995, respectively.[76–78] The endocannabinoid system is vastly different from other neurotransmitter systems for several reasons. First, endocannabinoids are not stored in vesicles. They are synthesized and released on demand. When Ca^{++} enters the postsynaptic neuron, the enzyme transacylase converts phosphatidylethanolamine into N-acyl-phosphatidylethanolamine (NAPE), which is then converted into anandamide by the enzyme phospholipase D after it cleaves NAPE. The synthesis pathway for 2-AG is different, but I will not cover it here.[75] Second, endocannabinoids are not synthesized in the presynaptic neuron, but in the postsynaptic neuron. Third, because endocannabinoids are released from the postsynaptic neuron, they bind to endocannabinoid receptors located on the presynaptic neuron. There are no endocannabinoid receptors located on postsynaptic neurons. This type of neurotransmission is referred to as *retrograde signaling*. There are two types of cannabinoid receptors, both of which are inhibitory: CB_1 and CB_2 receptors.[75] CB_1 receptors are found primarily in the central nervous system, whereas CB_2 are located in the peripheral nervous system. When endocannabinoids bind to CB receptors, they inhibit neurotransmitter release from the presynaptic neuron. Excess anandamide is degraded primarily by the enzyme fatty acid amide hydrolase (FAAH) into ethanolamine,[79] but it can also be oxidized by cyclooxygenase-2 (COX-2) into prostamide E_2. Fun fact: COX-2 inhibitors are often used as pain medications (e.g., ibuprofen).

Quiz 2.2

1. Which neurotransmitter system is largely involved in reward learning?
    a. Acetylcholine
    b. Dopamine
    c. GABA
    d. Serotonin
2. Which of the following receptors is inhibitory?
    a. Dopamine D_1
    b. Glutamate NMDA
    c. Muscarinic M_2
    d. Serotonin 5-HT_3
3. What features make the endocannabinoid system unique compared to other neurotransmitter systems?
4. Which of the following receptors is ionotropic?
    a. Dopamine D_2
    b. $GABA_A$
    c. $mGluR_5$
    d. Serotonin 5-HT_{2A}
5. Describe what happens to dopamine after it has been released into the synapse.

Answers to quiz 2.2

1. b — Dopamine
2. c — Muscarinic M_2

3. Endocannabinoids are synthesized on demand and are not stored in synaptic vesicles. Also, endocannabinoids are synthesized in the postsynaptic neuron and bind to receptors located on the presynaptic neuron only.
4. b — $GABA_A$
5. Dopamine can bind to one of two types of receptors: dopamine D_1-like and dopamine D_2-like. A portion of the excess dopamine will be metabolized by catechol-O-methyltransferase (COMT), whereas some of the excess dopamine can be taken back into the presynaptic neuron via the DAT. Following re-uptake, some of the dopamine will be repackaged into synaptic vesicles, whereas some will be metabolized by MAO.

Drug mechanisms of action

Before discussing the mechanisms of action of commonly used drugs, I need to first differentiate the terms **pharmacokinetics** and **pharmacodynamics**.[80] Pharmacokinetics refers to how a drug is absorbed into the body (i.e., route of administration), its distribution through the body, its metabolism, and its excretion. Pharmacodynamics concerns the effects of the drug on the body and the mechanisms that control these effects. In Chapter 1, you read about the effects various drugs have on the body, but you did not learn *how* drugs are able to exert these effects. Let's now discuss these pharmacodynamic mechanisms of action in more detail. Each drug you have learned about increases the neurotransmitter dopamine. Recall that dopamine is largely involved in reward learning, so this finding is not surprising. Drugs can either stimulate dopamine release directly or stimulate dopamine release indirectly (i.e., they target other neurotransmitter systems that then allow dopamine to be released). I will begin this section by discussing drugs that directly stimulate the release of dopamine: psychostimulants.

Psychostimulants

Psychostimulants such as cocaine, amphetamine, and methamphetamine directly interact with the monoaminergic system to increase dopamine, as well as other monoamine, levels[81]; however, each psychostimulant drug has a unique way of altering neurotransmitter levels. Cocaine is a DAT, a NET, and a SERT inhibitor.[82] That is, cocaine blocks each of these transporters, preventing excess dopamine, norepinephrine, and serotonin from being taken back up into the presynaptic neuron. This causes monoamine levels to build up in the synapse, allowing these neurotransmitters to stimulate their respective receptors repeatedly. Amphetamine does not block DAT; instead, it reverses the flow of DAT, as well as NET. In a sense, amphetamine causes DAT/NET to work in reverse. When amphetamine is introduced to the synapse, it causes DAT/NET to release the dopamine that is in the presynaptic neuron into the synapse.[83] Additionally, amphetamine is effective at preventing MAO from metabolizing monoamines. Normally, MAO metabolizes dopamine that has not been packaged into a vesicle, but amphetamine stops this process from occurring. This allows even more dopamine to spill out of DAT into the synaptic space. Methamphetamine's mechanism of action is similar to amphetamine. However, methamphetamine is also a potent VMAT-2 reverser.[84]

It causes dopamine/norepinephrine to spill out of the vesicle into the nerve terminal. Because MAO is also inhibited by methamphetamine, dopamine and norepinephrine levels increase in the terminal, and these neurotransmitters are eventually sent into the synapse through DAT/NET. Notice that the end result is the same: there are increased levels of monoamines in the synaptic space. Fig. 2.10 shows how psychostimulants increase dopamine levels.

While classified as an entactogen, MDMA's mechanisms of action are similar to those of psychostimulants. Like cocaine, MDMA inhibits DAT, NET, and SERT, although it shows greater inhibition of NET and SERT compared to DAT. The greater activity of MDMA at SERT may explain some of its hallucinogenic effects, as increased serotonin levels are linked to hallucinations. MDMA can also act as a VMAT-2 inhibitor and an MAO inhibitor.[85]

Although nicotine is a stimulant, its effects on dopaminergic activity are different compared to drugs like cocaine and amphetamines. Nicotine is an agonist at acetylcholine nicotinic receptors. Because nicotinic receptors are excitatory, nicotine administration depolarizes the neuron, thus sending messages to neighboring dopaminergic neurons.[86] In addition to its effects on dopaminergic neurons, nicotine stimulates nicotinic receptors located on glutamatergic neurons.[87] Stimulating these glutamatergic neurons causes them to release glutamate, which then binds to AMPA/NMDA receptors found on dopamine neurons. This also depolarizes the neuron, leading to the generation of an action potential.

Depressants

In contrast to psychostimulants, depressants do not directly target monoaminergic systems. Ethanol's mechanism of action has been harder to identify compared to other drugs. One of my former professors once described ethanol as being "promiscuous" as it interacts with most major neurotransmitter systems at high enough doses. In the case of glutamate,

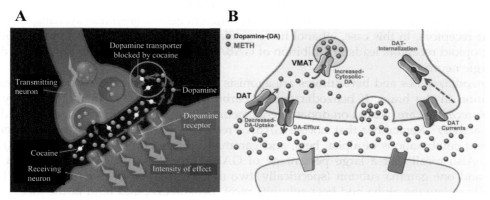

FIGURE 2.10 Schematic showing how cocaine (A) and methamphetamine (B) increase dopamine levels in the synapse. Cocaine is a potent dopamine transporter (DAT) inhibitor. Methamphetamine reverses the flow of DAT and reverses the flow of the vesicular monoamine transporter 2 (VMAT). *Panel (A) modified from NIDA. How Does Cocaine Produce Its Effects?; June 11, 2020; Retrieved from: https://www.drugabuse.gov/publications/research-reports/cocaine/how-does-cocaine-produce-its-effects. Panel (B) comes from Fig. 1 of reprinted from Sambo DO, Lebowitz JJ, Khoshbouei H. The sigma-1 receptor as a regulator of dopamine neurotransmission: a potential therapeutic target for methamphetamine addiction. Pharmacol Ther. 2018;186:152−167, Copyright 2018.*

the hypothesis is that ethanol blocks ionotropic glutamate receptors.[88] This makes intuitive sense as ionotropic glutamate receptors are responsible for excitatory neurotransmission, and ethanol increases feelings of sedation as consumption increases. Ethanol also *potentiates* (increases) $GABA_A$ receptor functioning.[89] Basically, ethanol increases the ability of $GABA_A$ receptors to inhibit the neuron. Ethanol's ability to reduce excitatory transmission and increase inhibitory transmission makes excessive ethanol consumption particularly dangerous. If someone overdoses on ethanol, too much inhibition occurs in the brain, eventually causing regions involved in breathing and heart rate to stop working. This can lead to death.

If ethanol potentiates GABAergic neurons, how can it increase dopamine levels? To answer this question, let's first turn our attention to another depressant drug class: opioids. You already know that endogenous opioids bind to various opioid receptors. Opioid drugs like heroin are agonists at the mu opioid receptor. The ability of opioids to bind to mu receptors accounts for their analgesic effects. Interestingly, when heroin enters the brain, it is converted into 6-monoacetylmorphine.[90] The major difference between heroin and morphine is that heroin is more fat soluble (*lipophilic*), meaning that it dissolves in fat more easily compared to morphine. So, why does this matter? Drugs that are more lipophilic pass the BBB more easily compared to other drugs. Therefore, heroin enters the brain much faster than morphine. Recall, drugs that enter the brain faster tend to have higher misuse potential.

How exactly do opioids increase dopamine levels? In certain parts of the brain, dopamine-containing neurons are inhibited by GABAergic neurons. Some of these GABAergic neurons contain mu opioid receptors on their dendrites. When opioids bind to these receptors, they inhibit the inhibitory GABAergic neuron. Think of it this way: in mathematics, you learned that multiplying a negative number to a negative number results in a positive number. The same principle applies here. If an inhibitory neuron is inhibited itself, the result is excitation. This is known as *disinhibition*. Because the dopamine neurons are no longer inhibited by GABA, they can be stimulated, causing release of dopamine into the synapse. This brings us back to ethanol. Ethanol interacts with many different neurotransmitter systems, including opioid receptors. In this case, ethanol has a similar mechanism of action as opioids. Stimulating opioid receptors leads to inhibition of GABAergic neurons that normally inhibit dopaminergic neurons.[91]

Benzodiazepines and barbiturates are agonists at the $GABA_A$ receptor,[92] thus increasing inhibition in the brain. If benzodiazepines/barbiturates increase GABAergic activity and GABA receptors are located on dopaminergic neurons, shouldn't these drugs decrease dopamine activity? Remember that not all $GABA_A$ receptors are the same as they can contain different combinations of alpha, beta, and gamma (and sometimes delta and epsilon) subunits. Also recall that a large percentage of $GABA_A$ receptors consist of two alpha, two beta, and one gamma subunit (specifically, two α_1, two β_2, one γ_2). Whereas GABA binds at a site where the alpha and beta subunits meet, benzodiazepines bind at the junction between the alpha and gamma subunits. In a particular part of the brain, GABAergic neurons contain the $2\alpha_1:2\beta_2:1\gamma_2$ GABA receptors. The dopaminergic neurons in the same brain region contain GABA receptors with a different combination—$2\alpha_3:2\beta_2:1\gamma_2$. Benzodiazepines/barbiturates will bind to both types of GABA receptors; however, the inhibitory effects are more pronounced on GABAergic neurons compared to dopaminergic neurons. As a result, the GABAergic neurons that inhibit dopaminergic neurons become inhibited.[93] The result is similar to what is observed when ethanol or opioids are administered: increased dopamine release by disinhibition (Fig. 2.11).

Hallucinogens

Psychedelic hallucinogens exert their effects primarily on serotonin receptors as agonists at $5-HT_{2A}$ receptors.[94] The agonistic effects at these receptors are believed to lead to the hallucinogenic effects of these drugs. Although evidence has shown that LSD can stimulate dopamine D_1-like receptors,[95] most other psychedelic hallucinogens have little activity at dopamine receptors. Because $5-HT_{2A}$ receptors stimulate dopamine release[96] as they are excitatory, the rewarding effects of psychedelic hallucinogens appear to be mediated by activation of these serotonergic receptors.

Dissociative hallucinogens such as ketamine and PCP primarily target the glutamatergic system by blocking ionotropic glutamate receptors, specifically NMDA receptors.[97] After AMPA receptors have depolarized the neuron, thus allowing the Mg^{++} ion to be displaced from the NMDA receptor channel, ketamine/PCP are able to enter the channel and occupy a site near where the Mg^{++} ion once occupied. Because ketamine and PCP cannot bind to the NMDA receptor until after it has been activated, they are considered **uncompetitive antagonists**. One theory as to how dissociative hallucinogens increase dopamine levels is by inhibiting NMDA receptors located on GABAergic neurons. If the GABAergic

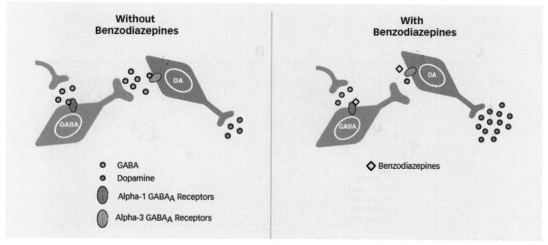

FIGURE 2.11 A schematic showing how benzodiazepines increase dopamine levels. When there are no benzodiazepines in someone's nervous system, GABA can bind to Alpha-1 $GABA_A$ receptors (composed of $2\alpha_1:2\beta_2:1\gamma_2$) located on GABAergic neurons and to Alpha-3 $GABA_A$ receptors (composed of $2\alpha_3:2\beta_2:1\gamma_2$) located on dopaminergic neurons. When GABA binds to Alpha-3 $GABA_A$ receptors, dopamine release is decreased. Benzodiazepines selectively bind to Alpha-1 $GABA_A$ receptors without binding to Alpha-3 $GABA_A$ receptors. Because the GABAergic neuron is inhibited, this allows the dopaminergic neuron to release more dopamine. *This image comes from NIDA. Well-Known Mechanism Underlies Benzodiazepines' Addictive Properties. April 19, 2012; Retrieved from: https://archives.drugabuse.gov/news-events/nida-notes/2012/04/well-known-mechanism-underlies-benzodiazepines-addictive-properties.*

neuron is inhibited, it cannot function properly. Therefore, these neurons can no longer inhibit glutamatergic neurons that are responsible for stimulating dopaminergic neurons[98] (Fig. 2.12).

Cannabis

Cannabis, more specifically tetrahydrocannabinol (THC), is a cannabinoid that binds to CB receptors and activates them. So far, you have learned about agonists such as nicotine and opioids. THC is also an agonist, but it is a **partial agonist**. A partial agonist, as the name suggests, shows partial *efficacy* at a receptor (i.e., they have reduced ability to activate the receptor). The ability of THC to increase dopamine levels is similar to how opioids increase dopamine levels: they inhibit GABAergic neurons that inhibit dopaminergic neurons.[99]

Inhalants

To elucidate the mechanisms of action of inhalants, toluene is used in experiments as this chemical is often used as a solvent in paint thinners, permanent markers, and some glues. Like ethanol, toluene has multiple mechanisms of action, including stimulating 5-HT$_{2A}$ receptors[100] and GABA receptors,[101,102] while inhibiting NMDA receptors.[102,103] The rewarding effects of inhalants are most likely due to 5-HT$_{2A}$ receptor-mediated dopamine release, like psychedelic hallucinogens.

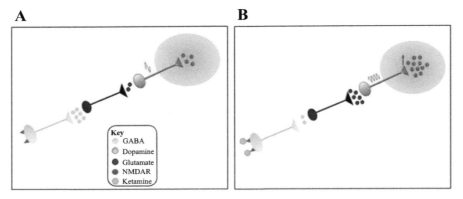

FIGURE 2.12 Potential mechanism by which ketamine increases dopamine release. In (A), activation of NMDA receptors located on GABAergic neurons causes the release of GABA that then binds to GABA receptors located on glutamatergic neurons. This then decreases the amount of glutamate sent to dopaminergic neurons. The result is to decrease dopamine release. In (B), ketamine blocks the NMDA receptors on the GABAergic neuron, which then disinhibits the glutamate neuron, causing increased glutamate release. This causes increased dopamine release. *This image modified from Fig. 5 of Kokkinou M, Ashok AH, Howes OD. The effects of ketamine on dopaminergic function: meta-analysis and review of the implications for neuropsychiatric disorders. Molr Psyc. 2017;23:59—69. https://doi.org/10.1038/mp.2017.190. This is an open-access article distributed under the terms of the Creative Commons Attribution 4.0 International License (https://creativecommons.org/licenses/by/4.0/).*

Pharmacological treatments for substance use disorders

Knowing how drugs exert their effects in the brain enables us to develop medications that can either act as a substitute for the drug or block the "high" individuals feel when taking the drug. Pharmacological treatments for addiction generally fall into two major categories: receptor antagonist therapy and **agonist treatment**. Receptor antagonist therapy entails providing an individual with a pharmacotherapy that blocks the effects of the drug. The goal of agonist treatment is to replace a drug that poses high health risks with a similar drug that has fewer risks. Agonist treatment is sometimes referred to as substitution treatment or as agonist replacement therapy.[104] There are concerns about using the term substitution to describe agonist treatment as it often gives the impression that one drug is simply being substituted for another. This terminology can increase stigma and slow down progress in developing novel medications for treating SUDs.[105] To avoid potential stigma, some have called for the use of the term agonist treatment.[105]

Agonist treatment is one strategy underlying **harm reduction**. Harm reduction refers to a set of policies aimed at reducing the negative social and/or physical consequences of drug use. In addition to agonist treatment, some policies related to harm reduction include:

1. Needle exchange programs: individuals that inject drugs can exchange their used needles for a sterile needle. This program helps reduce the risk of contracting a sexually transmitted infection (STI), and it helps decrease the likelihood of an individual discarding their used syringe in a public space.
2. Supervised consumption services: individuals can use drugs (licit or illicit) in a safe environment in the presence of trained staff. This allows staff to ensure the individual is using the drug in a safe manner. The goal of drug consumption rooms is to prevent overdose deaths. One vital tenet of harm reduction is to keep the individual alive.
3. Good Samaritan laws: individuals using illicit drugs will not be arrested if they call emergency services if they witness an overdose.
4. Providing individuals with naloxone: You encountered naloxone in the Introduction to this chapter, and you will read more about naloxone below. As you have already learned, this drug can be used to reverse opioid overdose. Naloxone is easy to administer and can save someone's life. In fact, one of my former students saved her neighbor's life by using naloxone, which provided the inspiration for the Introduction to this chapter.

Since 2008, the Harm Reduction International has released a biennial *Global State of Harm Reduction* report focusing on harm reduction policies implemented globally. The most recent report was published in 2020[106] (at least at the time this book was written). This report identifies if a country:

1. explicitly includes harm reduction in national policy,
2. provides at least one needle exchange program,
3. provides at least one opioid agonist treatment program,
4. provides at least one supervised consumption service,
5. distributes naloxone to peers,
6. provides opioid agonist treatment in at least one prison, and
7. provides a needle exchange program in at least one prison.

Currently, 87 countries have a national harm reduction policy, with most of these countries offering at least one needle exchange program (86 countries) and at least one opioid agonist treatment (84 countries, a decrease of two since 2018). However, only 59 countries offer opioid agonist treatment in at least one prison, and only 10 offer a needle exchange program in at least one prison. The number of countries that provide supervised consumption services and peer distribution of naloxone is low (12 and 16 countries, respectively). Canada is the only country in the world that has a national harm reduction policy and offers all of the services described above. Conversely, there are over 60 countries that do not implement any harm reduction policies; these countries are located primarily in Latin America/the Caribbean, Oceania, the Middle East, and Africa. For those that are curious, the United States offers most harm reduction services, with the major exception of needle exchange programs within prisons. In 2021, New York became the first state to implement supervised consumption.

Some medications used to treat SUDs do not act as receptor antagonists or agonists. Instead, they may prevent certain enzymes from exerting their effects, or they may inhibit other proteins such as transporters. Now, let's cover the major pharmacotherapies that have been used to combat various SUDs. This section will cover medications that have been tested in humans. Some are approved by the Food and Drug Administration (FDA), whereas others have been used experimentally only. There are numerous compounds that have been tested experimentally in animals. To cover each one of these compounds is far beyond the scope of this chapter. The next chapter will provide further discussion of experiments that have been conducted with animal subjects to further elucidate the neuromechanisms of addiction.

Treating nicotine use disorder

The most common FDA-approved therapy for tobacco dependence comes in the form of nicotine gum (Nicorette; approved in 1984) or patches (Nicoderm or Nicotrol; approved in 1996 for sale without the need of a prescription). Nicotine gum and patches are a form of agonist treatment. Individuals receive the nicotine they get from tobacco products without being exposed to the carcinogens associated with cigarettes, cigars, snuff, etc. In addition to gum and transdermal patches, nicotine agonist treatment can come in the form of a sublingual tablet, an oral inhaler, a nasal spray, or a lozenge. Nicotine agonist treatment can help individuals abstain from tobacco use for a period of at least 6 months[107]; however, individuals often stop using nicotine agonist treatment prematurely for a variety of reasons, including side effects, cost of treatment, and the feeling that they should be able to quit on their own.[108] Although nicotine agonist treatment can help individuals abstain from tobacco use, abstinence rates are low. Only 15.8% of individuals given nicotine gum reduce their daily cigarette use by 50%, compared to 6.7% of those given placebo.[109] Abstinence rates for those treated with nicotine patches are low as well, with one study reporting that 22% of those treated with nicotine patches maintained abstinence for at least 6 months, compared to 12% treated with placebo patches.[110] These low abstinence rates highlight the need for additional treatment options for combating nicotine use disorder.

In 2006, the FDA approved the drug varenicline (Chantix and Champix), a partial agonist at nicotinic receptors. Because varenicline is a partial agonist at nicotinic receptors, an individual gets some of the effects of nicotine without experiencing withdrawal symptoms. Varenicline is more effective than placebo and nicotine patches in helping participants remain abstinent without increasing neuropsychiatric adverse events.[111,112] When combined with counseling, varenicline can increase the rate at which adolescents abstain from tobacco use, although abstinence rates are similar between those treated with varenicline and those treated with placebo by the end of treatment.[113] Increased abstinence can be achieved if varenicline is combined with nicotine agonist treatment.[114,115] While varenicline is more effective in reducing tobacco use relative to other pharmacological treatments, long-term abstinence rates are relatively low at between 18% and 30%.[116] The Experiment Spotlight below details an attempt to increase the number of individuals willing to try varenicline to reduce tobacco use.

Experiment Spotlight: A pilot randomized clinical trial of remote varenicline sampling to promote treatment engagement and smoking cessation (Carpenter et al.[117])

You just read that long-term abstinence rates following varenicline treatment are relatively low. Low long-term abstinence rates are observed across most, if not all, pharmacological treatments for SUDs. However, gauging long-term abstinence in a real-world setting is somewhat difficult because fewer than 25% of individuals use pharmacotherapies to treat a SUD. Carpenter et al.[117] wanted to pilot a technique called medication sampling to determine if this intervention increases the number of individuals willing to try varenicline. Between November 2018 and April 2020, 99 smokers from South Carolina either received a 4-week mailed sample of varenicline or did not receive varenicline (control group). In order to participate in the study, participants had to meet 11 criteria. Some of these criteria included being a daily smoker (25+ days in a month smoking 5+ cigarettes during the day) for at least 1 year, showing some interest in quitting smoking, having a primary care doctor, and owning a smartphone or having regular access to email. Participants that received varenicline were given minimal suggestive instructions and were given a letter they could provide to their healthcare provider if they wished to continue receiving treatment.

Participants received a survey link at four different time points: ~14 days, ~1 month, ~2 months, ~4 months. Participants received money for completing each survey and could earn up to $160 for completing all four surveys. Participants were asked about their varenicline use, their motivation and confidence to stop smoking, their attitudes toward varenicline, and their cigarette use. Participants were also asked about any adverse events that occurred while taking varenicline.

Individuals given varenicline primarily used the medication during the first 2 weeks of the study. Whereas 54% and 66% of participants reported using varenicline at least once during weeks 1 and 2, respectively, only 36% and 24% of participants used varenicline during weeks 3 and 4, respectively. Averaged across the experiment, 76% of the participants provided with varenicline tried it at least once, and over half of the sample (58%) used it regularly. Motivation to quit smoking was consistent in those given varenicline but decreased in those in the control group. Confidence to quit smoking increased in the varenicline group but decreased somewhat in the control group. Overall, 18% of those given varenicline reported abstinence at the 4-month follow-up compared to just 6% of those in the control group. Additionally, at the 4-month follow-up, 42% of those that sampled varenicline reduced their daily cigarette use by 50%, compared to 12% of individuals in the control group.

Questions to consider:

(1) Why do you think the researchers required participants to have a primary care physician and access to a smartphone/email?
(2) After reading the results of this study, do you think this approach should be implemented on a wider scale? Why or why not?
(3) In the current study, the researchers asked participants about their cigarette use but did not collect urine samples to confirm abstinence. Why is this a limitation to the study?

Answers to questions:

(1) The researchers wanted participants to have a primary care physician so they could talk to them about continuing varenicline treatment if they wished to do so. Because this experiment involved emailing participants follow-up surveys, it was imperative they could access those surveys, whether on their smartphone or through their email.
(2) On one hand, the results are somewhat promising in the sense that individuals given varenicline reported a greater decrease in cigarette use and were more confident about quitting smoking. On the other hand, because many participants did not try the varenicline (~25% of the experimental group never used the varenicline sent to them, and only slightly more than half used it on a consistent basis), there are some concerns about the cost-effectiveness of this program. Mailing medications to numerous people could be cost prohibitive.
(3) Because the researchers have to trust the participants' word, there is no way to verify that participants abstained from cigarette use. By collecting urine samples, the researchers can determine how many participants abstained from cigarette use during a period of time.

Bupropion (Zyban), which has been used as an antidepressant (under the name Wellbutrin) since the mid-1980s, was approved by the FDA in 1997 as a smoking cessation aid. Bupropion blocks the reuptake of dopamine and norepinephrine[118] and acts as a **noncompetitive antagonist** at nicotinic receptors.[119] A noncompetitive antagonist blocks the receptor and cannot be removed by increasing the concentration of the agonist (in this case, nicotine). Because bupropion is able to block nicotinic receptors, it prevents nicotine from exerting its effects; this, in turn, decreases nicotine cravings. Similar to varenicline, approximately 20%–30% of individuals given bupropion treatment stop smoking and remain abstinent for 1 year following treatment.[120,121] In an effort to increase abstinent rates, combination therapies have been used. Combining nicotine agonist treatment with bupropion does not increase abstinence rates compared to either treatment administered alone[120]; however, combining varenicline with bupropion increases smoking cessation compared to varenicline monotherapy.[122] One challenge to using bupropion as a smoking cessation aid is its side effect profile. Individuals treated with bupropion may experience insomnia, nausea/vomiting, and dizziness. Seizures have been reported, but this is extremely rare.[123]

Rimonabant (Acomplia in Europe and Zimulti in the United States) is a CB_1 **inverse agonist** that was used primarily for weight loss but was also tested as a smoking cessation aid. An inverse agonist is a chemical that binds to the same receptor as an agonist but produces a physiological response that is opposite of the agonist. Why target the cannabinoid system to treat tobacco use disorder? The answer to this question is that the endocannabinoid system is known to alter the rewarding effects of nicotine.[124] Early work showed some promise as rimonabant increased abstinence rates compared to placebo treatment.[125] Unfortunately, rimonabant was discontinued in Europe in the late 2000s and was never approved by the FDA in the United States due to increased anxiety, depression, and suicidal ideations.[126]

One experimental treatment that has been used in smokers is N-acetylcysteine. I will not provide much detail about N-acetylcysteine's mechanism of action right now, but know that it plays an important role in regulating extracellular glutamate concentrations.[127] N-acetylcysteine is used clinically to treat acetaminophen (Tylenol) overdose[128] and can be used to loosen thick mucus associated with chronic obstructive pulmonary disease (COPD).[129] As a smoking cessation aid, some mixed results have been observed. One study found that N-acetylcysteine decreases the number of cigarettes smoked,[130] but another study failed to observe reductions in cigarette use following treatment.[131] One double-blind, placebo-controlled study found that those treated with N-acetylcysteine reported less craving for nicotine and were more likely to abstain from nicotine use,[132] although others have reported no alterations in craving.[130,131,133]

Treating alcohol use disorder

One treatment that has been used to reduce ethanol use is disulfiram (Antabuse). Disulfiram is neither an agonist nor an antagonist at receptors affected by ethanol. Instead, it blocks an enzyme involved in the metabolism of ethanol.[134] Normally, when we consume ethanol, the enzyme alcohol dehydrogenase breaks down ethanol into acetaldehyde. The enzyme aldehyde dehydrogenase then converts acetaldehyde into acetic acid, which we then excrete. Disulfiram's mechanism of action is to inhibit aldehyde dehydrogenase, causing toxic levels of acetaldehyde to build up in the bloodstream. Excess acetaldehyde can lead to headache, nausea, vomiting, flushing of the skin, and sweating. The goal of disulfiram is to prevent individuals from consuming ethanol because they know they will become ill if they do so.

Dr. Helge Kragh provides an account of the long and interesting history of disulfiram.[135] Disulfiram was first synthesized in 1881 by a German chemist. The synthesis of disulfiram did not gain much attention outside of organic chemistry until the early 1900s when rubber factories started using it to accelerate the rubber hardening process. In the late 1930s, a plant physician noticed that some workers became ill after drinking ethanol, and he hypothesized that their illness was related to disulfiram; however, he never tested this hypothesis. Finally, in the 1940s, the Danish physician Dr. Erik Jacobsen (1903—1985) tested the effects of disulfiram on himself and noted the unpleasant experiences he had when consuming ethanol. By the late 1940s, Jacobsen and some collaborators had published the first papers suggesting disulfiram be used as a treatment for ethanol dependence. In 1951, disulfiram became the first FDA-approved drug to treat alcohol use disorder.

Considering disulfiram has been used since the 1950s, one would assume that this is the best treatment option for those with an alcohol use disorder. The answer to this question depends on how you define how disulfiram should decrease ethanol use. If individuals know that they will receive disulfiram, this treatment reduces ethanol use to a greater extent compared to placebo; however, if individuals are not told that they have received disulfiram, this treatment performs no better than placebo.[136] One argument for disulfiram's efficacy as a pharmacotherapy relates to the individual's expectancies regarding what will happen if they drink ethanol after taking disulfiram as opposed to its physiological actions. If individuals in both an experimental group and a control group are not told what treatment they have been given, each participant may assume they have been given disulfiram and will avoid drinking ethanol. This may account for why "blinded" studies (i.e., participants are unaware of the treatment they have received) fail to show that disulfiram is more efficacious compared to placebo.

One major problem associated with disulfiram is compliance. Early research found that only 7% of individuals treated with disulfiram continued taking it after 1 year.[137] Granted, compliance is a challenge for many medications, but compliance tends to be worse for disulfiram compared to other pharmacotherapies for alcohol use disorder. In one study with veterans, individuals took disulfiram on only 41.3% of days that the medication was available, compared to 44.7%–54.6% for other treatments.[138] Given the compliance issues with disulfiram, other potential pharmacotherapies for ethanol dependence have been tested.

In 1994, the FDA approved the use of naltrexone (as ReVia) to treat alcohol use disorder, which can be administered either orally or as an injectable. You will see naltrexone again when we discuss treatments for opioid use disorders because it is a mu opioid receptor **competitive antagonist**. A competitive antagonist, unlike a noncompetitive antagonist, can be removed from the binding site if enough agonist is added to the synapse. Because ethanol has actions on opioid receptors, naltrexone's ability to block the rewarding effects of ethanol should not be surprising. Laboratory studies have shown that naltrexone reduces ethanol consumption and cravings for ethanol.[139] Evidence has also shown that naltrexone reduces relapse to drinking.[140] However, there is little evidence that naltrexone can lead to long-term abstinence,[141] although requiring individuals to abstain from ethanol use before the beginning of naltrexone treatment can increase abstinence rates.[142]

Another antagonist that is used to help individuals with alcohol use disorder is acamprosate (Campral), approved by the FDA in 2004. Acamprosate is primarily an NMDA receptor antagonist, although it can also potentiate GABA receptor functioning. Acamprosate is not typically used to prevent drinking; it is more commonly used to help combat withdrawal symptoms that can emerge following abrupt ethanol cessation. This is important as ethanol withdrawal can be potentially life-threatening due to the onset of seizures. Getting someone to wean off ethanol is more important than having them quit abruptly. Not surprisingly, individuals treated with acamprosate are not as likely to remain abstinent compared to those treated with disulfiram; yet, acamprosate decreases ethanol cravings more than disulfiram.[143] Acamprosate is a good option for those that are beginning the process of quitting ethanol consumption, but it is not ideal as a long-term treatment.

Although not approved by the FDA for treating alcohol use disorder, some researchers have tested the efficacy of the drug modafinil. Modafinil is a psychostimulant used to treat the sleep disorder narcolepsy (as Provigil), but has a different pharmacological profile compared to stimulants such as cocaine, methamphetamine, and ADHD medications. Modafinil's mechanism of action is not completely understood, but it has been shown to be a weak DAT inhibitor.[144] As modafinil can increase dopamine levels, some have hypothesized that modafinil can decrease ethanol consumption. Modafinil does not reduce the number of heavy drinking days in individuals, but it can prolong relapse in individuals that have increased impulsivity. One problem with using modafinil as a treatment for alcohol use disorder is that this drug increases drinking and impairs abstinence in individuals with low levels of impulsivity.[145]

Topiramate (Trokendi or Topamax) is approved by the FDA to treat seizures and to prevent migraine headaches but has not been approved to treat SUDs. Its mechanism of action is to inhibit voltage-gated Na^+ channels.[146] Recall that during an action potential, Na^+ channels open, allowing more Na^+ to enter the neuron. Topiramate blocks this process, which decreases excitation in the nervous system. This drug has shown promise in treating alcohol use disorder, as individuals given topiramate reduce the number of drinks they consume and experience more days of abstinence.[147] Topiramate also decreases ethanol craving.[148] One drawback to topiramate is that it can lead to mild cognitive impairment. For example, when military veterans with a traumatic brain injury and alcohol use disorder were treated with topiramate, they experienced impairments in verbal fluency and working memory.[149] Fortunately, these impairments are observed for a short period of time. To minimize these negative side effects, the dose of topiramate is slowly increased over time.

Treating opioid use disorders

The primary treatment for opioid use disorder is agonist treatment. The drug methadone (Dolophine, Methadose) is an agonist at mu opioid receptors and was first used in the 1940s to withdraw individuals from heroin. In the 1960s, clinical studies showed that methadone could be used as a form of maintenance therapy for opioid use disorders.[150] In 1972, the FDA approved the use of methadone for treating opioid use disorders. Methadone is often taken orally in tablet form (Dolophine) or in liquid form (Methadose). As discussed in Chapter 1, injecting drugs such as heroin is associated with several health risks, such as contracting HIV/hepatitis or developing an abscess at the injection site. Additionally, by getting individuals to reduce drug injections, they are less likely to leave syringes in public areas where someone can prick themselves. Methadone is an agonist like heroin, but it has lower misuse potential, most likely because methadone does not rapidly increase dopamine levels.[151]

In addition to methadone, the drug buprenorphine (Subutex, Probuphine), first approved in 2002, can be used to replace heroin. Buprenorphine is a partial agonist at mu opioid receptors. Similar to varenicline's use for tobacco dependence, buprenorphine is useful in weaning individuals off of heroin without causing them to experience withdrawal symptoms. Overall, both methadone and buprenorphine decrease opioid use,[152,153] although buprenorphine has become the medication of choice for most patients.[152] Long-term abstinence following methadone maintenance is low, with approximately 9%—21% of individuals reporting no heroin use compared to 10%—19% of individuals not treated with methadone.[154] A more recent study reported an abstinence rate of 31%.[155]

Although the FDA approved the use of naltrexone for alcohol use disorder in 1994, naltrexone (as Vivitrol) was approved in 1984 to help treat opioid addiction. Ideally, naltrexone should be given after an individual has detoxed from opioids, as giving the individual naltrexone will precipitate withdrawal symptoms. Because incarcerated individuals are more likely to have an opioid use disorder,[156] one study gave half of the criminal justice offenders in the sample naltrexone and half their usual treatment. Individuals given naltrexone took longer to relapse compared to those given their usual treatment (10.5 vs. 5.0 weeks), but after approximately 1 year, opioid use was similar across each group.[157]

Like naltrexone, naloxone (Narcan) is a mu opioid receptor antagonist. However, its main purpose is to help treat those that have overdosed on an opioid. Naloxone is also used in conjunction with buprenorphine in the form of Suboxone. Suboxone is an ingenious drug that is taken sublingually (i.e., the individual places a film underneath their tongue). When taken as prescribed, the individual receives the partial agonist buprenorphine to prevent withdrawal symptoms. Yet, if an individual tries to misuse Suboxone by injecting it, the naloxone blocks mu opioid receptors, preventing the person from feeling the effects of the buprenorphine. If someone takes the drug as prescribed, wouldn't the naloxone still block opioid receptors? When taken orally, naloxone gets broken down in the stomach and does not reach the brain. Like I said, an ingenious drug. Even though the addition of naloxone to buprenorphine may, at first glance, lead to compliance issues, treatment retention appears to be similar for buprenorphine/naloxone compared to buprenorphine or methadone.[158]

Treating psychostimulant use disorders

Currently, there are no FDA-approved medications for psychostimulant use disorders. Similar to using methadone in place of heroin, studies have examined if replacing cocaine with either *d*-amphetamine or methylphenidate can reduce cocaine use. Results of laboratory experiments and clinical trials have provided support that amphetamine reduces cocaine use.[159] Mixed results have been obtained with methylphenidate. Some studies have found that methylphenidate reduces the rewarding effects of cocaine[160,161] while others have found that methylphenidate is an ineffective pharmacotherapy for psychostimulant use disorders.[162,163] Even though amphetamine can reduce cocaine use, ADHD medications do not appear to be efficacious in treating methamphetamine use disorder.[164,165] Amphetamine reduces methamphetamine withdrawal symptoms and cravings,[164] suggesting that amphetamine can be useful in weaning someone off of methamphetamine. Given that amphetamine and methylphenidate have misuse potential, there are some concerns about using these drugs as treatment for cocaine/methamphetamine use disorder. The same logic used in methadone replacement therapy for heroin use disorder can be applied here. Cocaine (as crack) and methamphetamine are often smoked, which can be problematic for one's lungs. The manufacturing of methamphetamine can be dangerous, and individuals can experience the same complications from injecting stimulants as with opioids.

Not only can disulfiram be used to treat alcohol use disorder, but it has been used experimentally to decrease cocaine use in individuals.[166,167] How can a drug that primarily targets ethanol metabolism decrease cocaine use in people? Like ethanol, disulfiram has multiple mechanisms of action. In addition to inhibiting aldehyde dehydrogenase, disulfiram affects the metabolism of dopamine.[166] When dopamine is metabolized by MAO, the first metabolite created is dihydroxyphenylacetaldehyde (DOPAL), which is then converted by aldehyde dehydrogenase into 3,4-dihydroxypenylacetic acid (DOPAC). Because DOPAL is toxic to dopaminergic terminals, one hypothesis is that the buildup of this metabolite decreases dopamine release. This can then decrease the rewarding effects of cocaine. There is one limitation to this hypothesis: individuals treated with disulfiram often report experiencing a greater "high" after taking cocaine[168] or amphetamine.[169] One potential reason for this paradoxical finding is that disulfiram prevents DβH from converting dopamine into norepinephrine,[166] thus increasing dopamine levels. By increasing dopamine levels via DβH inhibition, disulfiram acts as a form of agonist treatment for cocaine dependence.

Due to its stimulant-like profile, some have tested modafinil's ability to treat psychostimulant use disorders. One double-blind placebo-controlled study found that modafinil failed to decrease the percentage of days individuals abstained from cocaine use during the 12-week treatment period, but modafinil reduced cravings for cocaine. Additionally, modafinil decreased the percentage of days individuals abstained from cocaine in individuals without a comorbid alcohol use disorder.[170] Although modafinil has some beneficial effects on cocaine use disorder, other studies have not observed similar reductions in methamphetamine craving or use.[171,172]

A few clinical trials have tested the efficacy of topiramate in treating cocaine use disorder. When combined with another therapy, topiramate reduces short-term cocaine use (3 weeks of continuous abstinence).[173] Individuals given topiramate are more likely to provide cocaine-free urine samples compared to individuals treated with placebo,[174,175] but the overall percentage of cocaine-free urine samples for the topiramate-treated individuals is low (16.6% urinary cocaine-free weeks compared to 5.8% for placebo-treated individuals). Topiramate does not appear to be efficacious in treating crack cocaine use in individuals with comorbid alcohol use disorder, although individuals treated with topiramate are more likely to be abstinent during the final 3 weeks of treatment.[176] Due to the limited number of clinical trials that have tested topiramate, more work is needed to determine if this drug is a viable treatment for psychostimulant use disorders.

Treating cannabis use disorder

Just as there are no FDA-approved medications for psychostimulant use disorders, there are no approved pharmacotherapies for cannabis use disorder. Before the CB_1 inverse agonist rimonabant was withdrawn from further clinical testing, one study determined if rimonabant could decrease the subjective effects of cannabis (e.g., how high one feels after taking a drug). Participants were less likely to report feeling intoxicated following cannabis use when taking rimonabant.[177] Agonist treatment has also been tested as an option for cannabis use disorder. In one case study, two individuals were given dronabinol, a synthetic form of THC; at the time the study was published, one of the individuals had been abstinent for 4 years, whereas the other individual had been abstinent for over 6 months.[178] Dronabinol decreases the amount of cannabis individuals are willing to use[179]; however, in a double-blind, placebo-control study, dronabinol treatment did not increase 2-week abstinence rates, although it increased treatment retention rates.[180]

Instead of targeting the CB_1 receptor, other researchers have tested the effects of naltrexone on cannabis use. Interestingly, one study found that naltrexone blunted the intoxicating effects of cannabis in current smokers but enhanced these intoxicating effects in nonusers.[181] Unfortunately, individuals with heavy cannabis use were more likely to report liking cannabis' effects when given naltrexone, bringing into question the utility of naltrexone treatment for cannabis use disorder. Because participants received acute treatment[181] (i.e., received a single administration of one of several doses of naltrexone), a follow-up study treated participants with a single dose of naltrexone for 16 days.[182] As this treatment regimen more closely mirrors what someone would experience in real life, this study could better determine the efficacy of naltrexone treatment for cannabis use. Indeed, when individuals were given naltrexone maintenance, they reported decreased positive effects and were less likely to use cannabis. In a more recent study, individuals given naltrexone reported using cannabis for fewer days during the week.[183] With a sample size of only 12 participants, more work is needed to determine if naltrexone can be used to treat cannabis use disorder.

In adolescents, N-acetylcysteine has shown promise for reducing cannabis use. Adolescents treated with N-acetylcysteine are more likely to provide cannabis-free urine samples (40.9% of samples) compared to those treated with placebo (27.2%).[184] Not only can N-acetylcysteine decrease cannabis use in adolescents, but it can also reduce the consumption of ethanol in this population.[185] N-acetylcysteine may be able to reduce cannabis use in adolescents, but there is no evidence that it is effective in treating cannabis use disorder in adults.[186] The reason for the discrepancy between adolescents and adults has not been elucidated, but as noted by Gray et al., one potential explanation is that adolescent brains are still developing and may be more sensitive to the effects of N-acetylcysteine.[186]

Quiz 2.3

1. Which drug acts as an NMDA receptor antagonist?
   a. Cocaine
   b. Heroin
   c. Peyote
   d. Phencyclidine
2. Which drug is typically used to reverse opioid overdose?
   a. Buprenorphine
   b. Naloxone
   c. Naltrexone
   d. Suboxone
3. Which of the following pharmacological treatments is an example of agonist treatment?
   a. Bupropion
   b. Disulfiram
   c. Methadone
   d. Rimonabant
4. Which country provides all seven harm reduction policies as outlined by the Harm Reduction International?
   a. Canada
   b. Finland
   c. the United Kingdom
   d. the United States
5. Which of the following pharmacological treatments has NOT been tested for psychostimulant use disorder?
   a. Acamprosate
   b. Amphetamine
   c. Disulfiram
   d. Topiramate

Answers quiz 2.3

1. d — Phencyclidine
2. b — Naloxone
3. c — Methadone
4. a — Canada
5. a — Acamprosate

Chapter summary

The purpose of this chapter was to introduce some of the pharmacological actions of commonly used drugs. Although drugs have different targets, they share the ability to increase dopamine levels in the central nervous system. Given the complexity of neurons and neurotransmitter systems, drugs do not have to directly target the dopaminergic system to increase dopamine levels. Some medications have shown some promise in reducing drug use; however, there are no pharmacotherapies that can magically cure someone of their SUD (if there were, I most likely would not be writing this book right now). In fact, long-term abstinence rates are generally low, even for the more promising pharmacotherapies. This highlights the difficulty of treating addiction and shows that medications alone are not necessarily sufficient in helping someone overcome a SUD. As such, you will learn about other treatment interventions in subsequent chapters. You may now know that drugs increase dopamine levels, but this does not explain why some individuals develop a SUD. Having a fundamental understanding of how drugs affect neurotransmitter systems is important for the next chapter as we dive deeper into the neuromechanisms of addiction. You will learn how dopamine, as well as other neurotransmitters, within specific brain regions mediates distinct aspects of the addiction process.

Glossary

Action potential – an electrical impulse that travels the length of a neuron's axon, which then allows the release of neurotransmitters into the synapse.
Agonist – a chemical that binds to a receptor and activates it.
Agonist treatment – a treatment approach in which individuals are given a medication that mimics the drug to which they are dependent; this allows the individual to slowly wean off the substance without experiencing aversive withdrawal symptoms.
Antagonist – a chemical that binds to a receptor and blocks it.
Autoreceptor – receptor often located on presynaptic neuron that is sensitive to a neurotransmitter released from the same neuron.
Axon – a long, slender projection that extends from the cell body of a neuron to its terminal buttons.
Blood-brain barrier (BBB) – highly selective semipermeable border that separates blood from the extracellular fluid of the central nervous system.
Competitive antagonist – an antagonist that can be removed from the binding site if enough agonist is placed into the synapse.
Dendrite – a tree-like projection extending from the cell body of a neuron that is responsible for receiving messages from nearby neurons.
Dendritic spine – a small projection located on a dendrite.
Depolarization – the process by which positively charged ions (primarily sodium) begin to enter the axon, thus making the inside of the neuron less negatively charged compared to its resting membrane potential.
Glial cell – nonneuronal cell located in the central nervous system and the peripherhal nervous system that provides many support functions for neurons.
Harm reduction – a treatment principle that focuses on minimizing the negative social and physical outcomes associated with drug use.

Heteroreceptor — receptor often located on postsynaptic neuron that is sensitive to a neurotransmitter released from multiple neurons.
Hyperpolarization — the process by which an overabundance of positively charged ions (primarily potassium) exit the neuron, thus making the inside of the neuron more negatively charged compared to its resting membrane potential.
Inverse agonist — a chemical that binds to the same site as an agonist but exerts the opposite effect as the agonist.
Monoamine oxidase (MAO) — an enzyme that metabolizes the monoamines dopamine, norepinephrine, epinephrine, serotonin, and melatonin, as well as other chemicals with an amine structure.
Nervous system — a network of neurons that controls our physiological, behavioral, and cognitive functions.
Neuron — a specialized cell that uses a combination of electrical impulses and chemical transmission to communicate with other cells.
Neurotransmitter — a chemical that binds to a receptor to either excite or inhibit a neuron.
Noncompetitive antagonist — an antagonist that cannot be removed from the binding site even as the agonist concentration increases.
Partial agonist — a chemical that binds to a receptor and activates it to a lesser extent compared to an agonist.
Pharmacodynamics — refers to the effects of a drug on physiological, behavioral, and cognitive processes.
Pharmacokinetics — refers to how a drug moves through the body and includes the absorption, metabolism, and excretion of the drug.
Receptor — a protein that recognizes specific molecules that bind to it.
Resting membrane potential — the electrical potential difference across the membrane of a neuron when it is in a nonactivated state.
Second messenger — an intracellular signaling molecule activated by a first messenger (e.g., a neurotransmitter) that is responsible for activating certain physiological changes, such as opening an ion channel.
Synapse — the space between two neurons.
Synaptic vesicle — a sac located within the terminal buttons of a neuron that stores neurotransmitters.
Terminal button — the region of the neuron where the axon ends; neurotransmitters are often synthesized and stored in this part of the neuron.
Transporter — a protein that is responsible for removing excess neurotransmitter from the synapse and sending it back into the terminal buttons.
Uncompetitive antagonist — an antagonist that can only bind to the receptor when it is in an activated state.
Vesicular monoamine transporter 2 (VMAT-2) — a protein that is responsible for placing neurotransmitters into the synaptic vesicles.

Supplementary data related to this chapter can be found online at doi:10.1016/B978-0-323-90578-7.00008-6

References

1. He H-Y, Cline HT. What is excitation/inhibition and how is it regulated? A case of the elephant and the wisemen. *J Exp Neurosci*. 2019;13. https://doi.org/10.1177/1179069519859371, 1179069519859371.
2. Crawford LK, Caterina MJ. Functional anatomy of the sensory nervous system: updates from the neuroscience bench. *Toxicol Pathol*. 2020;48(1):174–189. https://doi.org/10.1177/0192623319869011.
3. Stifani N. Motor neurons and the generation of spinal motor neuron diversity. *Front Cell Neurosci*. 2014;8:293. https://doi.org/10.3389/fncel.2014.00293.
4. Kepecs A, Fishell G. Interneuron cell types are fit to function. *Nature*. 2014;505(7483):318–326. https://doi.org/10.1038/nature12983.
5. Rollenhagen A, Lübke JHR. Dendrites: a key structural element of neurons. In: Pfaff DW, ed. *Neuroscience in the 21st Century*. Springer; 2013. https://doi.org/10.1007/978-1-4614-1997-6_11.
6. Gipson CD, Olive MF. Structural and functional plasticity of dendritic spines - root or result of behavior? *Gene Brain Behav*. 2017;16(1):101–117. https://doi.org/10.1111/gbb.12324.
7. Peters A, Kaiserman-Abramof IR. The small pyramidal neuron of the rat cerebral cortex. The perikaryon, dendrites and spines. *Am J Anat*. 1970;127(4):321–355. https://doi.org/10.1002/aja.1001270402.
8. Stone DK. Receptors: structure and function. *Am J Med*. 1998;105(3):244–250. https://doi.org/10.1016/s0002-9343(98)00221-6.
9. Kenakin T. Agonists, partial agonists, antagonists, inverse agonists and agonist/antagonists? *Trends Pharmacol Sci*. 1987;8(11):423–426. https://doi.org/10.1016/0165-6147(87)90229-X.

10. Sakimura K. Ionotropic receptor. In: Binder MD, Hirokawa N, Windhorst U, eds. *Encyclopedia of Neuroscience*. Springer; 2009. https://doi.org/10.1007/978-3-540-29678-2_2596.
11. Roth BL. Molecular pharmacology of metabotropic receptors targeted by neuropsychiatric drugs. *Nat Struct Mol Biol*. 2019;26(7):535–544. https://doi.org/10.1038/s41594-019-0252-8.
12. Wacker D, Stevens RC, Roth BL. How ligands illuminate GPCR molecular pharmacology. *Cell*. 2017;170(3):414–427. https://doi.org/10.1016/j.cell.2017.07.009.
13. Syrovatkina V, Alegre KO, Dey R, Huang XY. Regulation, signaling, and physiological functions of G-proteins. *J Mol Biol*. 2016;428(19):3850–3868. https://doi.org/10.1016/j.jmb.2016.08.002.
14. Williamson JM, Lyons DA. Myelin dynamics throughout life: an ever-changing landscape? *Front Cell Neurosci*. 2018;12:424. https://doi.org/10.3389/fncel.2018.00424.
15. Ghasemi N, Razavi S, Nikzad E. Multiple sclerosis: pathogenesis, symptoms, diagnoses and cell-based therapy. *Cell J*. 2017;19(1):1–10. https://doi.org/10.22074/cellj.2016.4867.
16. Dent EW, Baas PW. Microtubules in neurons as information carriers. *J Neurochem*. 2014;129(2):235–239. https://doi.org/10.1111/jnc.12621.
17. Reichardt LF, Kelly RB. A molecular description of nerve terminal function. *Annu Rev Biochem*. 1983;52:871–926. https://doi.org/10.1146/annurev.bi.52.070183.004255.
18. Focke PJ, Wang X, Larsson HP. Neurotransmitter transporters: structure meets function. *Structure*. 2013;21(5):694–705. https://doi.org/10.1016/j.str.2013.03.002.
19. Kress GJ, Mennerick S. Action potential initiation and propagation: upstream influences on neurotransmission. *Neuroscience*. 2009;158(1):211–222. https://doi.org/10.1016/j.neuroscience.2008.03.021.
20. Moore JW, Cole KS. Resting and action potentials of the squid giant axon in vivo. *J Gen Physiol*. 1960;43(5):961–970. https://doi.org/10.1085/jgp.43.5.961.
21. Salzer JL. Clustering sodium channels at the node of Ranvier: close encounters of the axon-glia kind. *Neuron*. 1997;18(6):843–846. https://doi.org/10.1016/s0896-6273(00)80323-2.
22. Südhof TC. Calcium control of neurotransmitter release. *Cold Spring Harbor Perspect Biol*. 2012;4(1):a011353. https://doi.org/10.1101/cshperspect.a011353.
23. Südhof TC, Malenka RC. Understanding synapses: past, present, and future. *Neuron*. 2008;60(3):469–476. https://doi.org/10.1016/j.neuron.2008.10.011.
24. Jäkel S, Dimou L. Glial cells and their function in the adult brain: a journey through the history of their ablation. *Front Cell Neurosci*. 2017;11:24. https://doi.org/10.3389/fncel.2017.00024.
25. Howard BM, Mo Z, Filipovic R, Moore AR, Antic SD, Zecevic N. Radial glia cells in the developing human brain. *Neuroscientist*. 2008;14(5):459–473. https://doi.org/10.1177/1073858407313512.
26. Jiménez AJ, Domínguez-Pinos MD, Guerra MM, Fernández-Llebrez P, Pérez-Fígares JM. Structure and function of the ependymal barrier and diseases associated with ependyma disruption. *Tissue Barriers*. 2014;2:e28426. https://doi.org/10.4161/tisb.28426.
27. Jessen KR, Mirsky R, Lloyd AC. Schwann cells: development and role in nerve repair. *Cold Spring Harbor Perspect Biol*. 2015;7(7):a020487. https://doi.org/10.1101/cshperspect.a020487.
28. Tansey EM. Henry Dale and the discovery of acetylcholine. *Comptes Rendus Biol*. 2006;329(5–6):419–425. https://doi.org/10.1016/j.crvi.2006.03.012.
29. Brown DA. Acetylcholine and cholinergic receptors. *Brain & Neurosci Adv*. 2019;3. https://doi.org/10.1177/2398212818820506, 2398212818820506.
30. Ferreira-Vieira TH, Guimaraes IM, Silva FR, Ribeiro FM. Alzheimer's disease: targeting the cholinergic system. *Curr Neuropharmacol*. 2016;14(1):101–115. https://doi.org/10.2174/1570159x13666150716165726.
31. Blusztajn JK, Wurtman RJ. Choline and cholinergic neurons. *Science*. 1983;221(4611):614–620. https://doi.org/10.1126/science.6867732.
32. Picciotto MR, Higley MJ, Mineur YS. Acetylcholine as a neuromodulator: cholinergic signaling shapes nervous system function and behavior. *Neuron*. 2012;76(1):116–129. https://doi.org/10.1016/j.neuron.2012.08.036.
33. Gotti C, Zoli M, Clementi F. Brain nicotinic acetylcholine receptors: native subtypes and their relevance. *Trends Pharmacol Sci*. 2006;27(9):482–491. https://doi.org/10.1016/j.tips.2006.07.004.
34. Picciotto MR, Caldarone BJ, King SL, Zachariou V. Nicotinic receptors in the brain. Links between molecular biology and behavior. *Neuropsychopharmacology*. 2000;22(5):451–465. https://doi.org/10.1016/S0893-133X(99)00146-3.
35. Dani JA. Neuronal nicotinic acetylcholine receptor structure and function and response to nicotine. *Int Rev Neurobiol*. 2015;124:3–19. https://doi.org/10.1016/bs.irn.2015.07.001.
36. Wess J. Novel insights into muscarinic acetylcholine receptor function using gene targeting technology. *Trends Pharmacol Sci*. 2003;24(8):414–420. https://doi.org/10.1016/S0165-6147(03)00195-0.

37. Slutsky I, Silman I, Parnas I, Parnas H. Presynaptic M(2) muscarinic receptors are involved in controlling the kinetics of ACh release at the frog neuromuscular junction. *J Physiol*. 2001;536(Pt 3):717−725. https://doi.org/10.1111/j.1469-7793.2001.00717.x.
38. Dvir H, Silman I, Harel M, Rosenberry TL, Sussman JL. Acetylcholinesterase: from 3D structure to function. *Chem Biol Interact*. 2010;187(1−3):10−22. https://doi.org/10.1016/j.cbi.2010.01.042.
39. Goldstein DS. Catecholamines 101. *Clin Auton Res*. 2010;20(6):331−352. https://doi.org/10.1007/s10286-010-0065-7.
40. Rus A, Molina F, Del Moral ML, Ramírez-Expósito MJ, Martínez-Martos JM. Catecholamine and indolamine pathway: a case-control study in fibromyalgia. *Biol Res Nurs*. 2018;20(5):577−586. https://doi.org/10.1177/1099800418787672.
41. Missale C, Nash SR, Robinson SW, Jaber M, Caron MG. Dopamine receptors: from structure to function. *Physiol Rev*. 1998;78(1):189−225. https://doi.org/10.1152/physrev.1998.78.1.189.
42. Meder D, Herz DM, Rowe JB, Lehéricy S, Siebner HR. The role of dopamine in the brain - lessons learned from Parkinson's disease. *Neuroimage*. 2019;190:79−93. https://doi.org/10.1016/j.neuroimage.2018.11.021.
43. McCutcheon RA, Krystal JH, Howes OD. Dopamine and glutamate in schizophrenia: biology, symptoms and treatment. *World Psychiatr*. 2020;19(1):15−33. https://doi.org/10.1002/wps.20693.
44. Wood SK, Valentino RJ. The brain norepinephrine system, stress and cardiovascular vulnerability. *Neurosci Biobehav Rev*. 2017;74(Pt B):393−400. https://doi.org/10.1016/j.neubiorev.2016.04.018.
45. Berridge CW. Noradrenergic modulation of arousal. *Brain Res Rev*. 2008;58(1):1−17. https://doi.org/10.1016/j.brainresrev.2007.10.013.
46. Berger M, Gray JA, Roth BL. The expanded biology of serotonin. *Annu Rev Med*. 2009;60:355−366. https://doi.org/10.1146/annurev.med.60.042307.110802.
47. Hasler G. Pathophysiology of depression: do we have any solid evidence of interest to clinicians? *World Psychiatr*. 2010;9(3):155−161. https://doi.org/10.1002/j.2051-5545.2010.tb00298.x.
48. Fernstrom JD, Fernstrom MH. Tyrosine, phenylalanine, and catecholamine synthesis and function in the brain. *J Nutr*. 2007;137(6 Suppl 1):1539S−1548S. https://doi.org/10.1093/jn/137.6.1539S.
49. Gonzalez-Lopez E, Vrana KE. Dopamine beta-hydroxylase and its genetic variants in human health and disease. *J Neurochem*. 2020;152(2):157−181. https://doi.org/10.1111/jnc.14893.
50. Mahmoodi N, Harijan RK, Schramm VL. Transition-state analogues of phenylethanolamine *N*-methyltransferase. *J Am Chem Soc*. 2020;142(33):14222−14233. https://doi.org/10.1021/jacs.0c05446.
51. Richard DM, Dawes MA, Mathias CW, Acheson A, Hill-Kapturczak N, Dougherty DM. L-tryptophan: basic metabolic functions, behavioral research and therapeutic indications. *Int J Tryptophan Res*. 2009;2:45−60. https://doi.org/10.4137/ijtr.s2129.
52. Mishra A, Singh S, Shukla S. Physiological and functional basis of dopamine receptors and their role in neurogenesis: possible implication for Parkinson's disease. *J Exp Neurosci*. 2018;12. https://doi.org/10.1177/1179069518779829, 1179069518779829.
53. Nichols DE, Nichols CD. Serotonin receptors. *Chem Rev*. 2008;108(5):1614−1641. https://doi.org/10.1021/cr078224o.
54. Lin Z, Canales JJ, Björgvinsson T, et al. Monoamine transporters: vulnerable and vital doorkeepers. *Prog Mol Biol & Transl Sci*. 2011;98:1−46. https://doi.org/10.1016/B978-0-12-385506-0.00001-6.
55. Gaweska H, Fitzpatrick PF. Structures and mechanism of the monoamine oxidase family. *Biomol Concepts*. 2011;2(5):365−377. https://doi.org/10.1515/BMC.2011.030.
56. Axelrod J, Senoh S, Witkop B. O-Methylation of catechol amines in vivo. *J Biol Chem*. 1958;233(3):697−701.
57. Männistö PT. Catechol O-methyltransferase: characterization of the protein, its gene, and the preclinical pharmacology of COMT inhibitors. *Adv Pharmacol*. 1998;42:324−328. https://doi.org/10.1016/s1054-3589(08)60755-3.
58. Guillot TS, Miller GW. Protective actions of the vesicular monoamine transporter 2 (VMAT2) in monoaminergic neurons. *Mol Neurobiol*. 2009;39(2):149−170. https://doi.org/10.1007/s12035-009-8059-y.
59. Riedel G, Platt B, Micheau J. Glutamate receptor function in learning and memory. *Behav Brain Res*. 2003;140(1−2):1−47. https://doi.org/10.1016/s0166-4328(02)00272-3.
60. Brandstätter JH, Koulen P, Wässle H. Diversity of glutamate receptors in the mammalian retina. *Vis Res*. 1998;38(10):1385−1397. https://doi.org/10.1016/s0042-6989(97)00176-4.
61. Dingledine R, Conn PJ. Peripheral glutamate receptors: molecular biology and role in taste sensation. *J Nutr*. 2000;130(4S Suppl):1039S−1042S. https://doi.org/10.1093/jn/130.4.1039S.

62. Nordang L, Cestreicher E, Arnold W, Anniko M. Glutamate is the afferent neurotransmitter in the human cochlea. *Acta Otolaryngol*. 2000;120(3):359–362. https://doi.org/10.1080/000164800750000568.
63. Pereira V, Goudet C. Emerging trends in pain modulation by metabotropic glutamate receptors. *Front Mol Neurosci*. 2019;11:464. https://doi.org/10.3389/fnmol.2018.00464.
64. Yelamanchi SD, Jayaram S, Thomas JK, et al. A pathway map of glutamate metabolism. *J Cell Commun & Signal*. 2016;10(1):69–75. https://doi.org/10.1007/s12079-015-0315-5.
65. Kondoh T, Torii K. Brain activation by umami substances via gustatory and visceral signaling pathways, and physiological significance. *Biol Pharm Bull*. 2008;31(10):1827–1832. https://doi.org/10.1248/bpb.31.1827.
66. Ozawa S, Kamiya H, Tsuzuki K. Glutamate receptors in the mammalian central nervous system. *Prog Neurobiol*. 1998;54(5):581–618. https://doi.org/10.1016/s0301-0082(97)00085-3.
67. Bettler B, Mulle C. Review: neurotransmitter receptors. II. AMPA and kainate receptors. *Neuropharmacology*. 1995;34(2):123–139. https://doi.org/10.1016/0028-3908(94)00141-e.
68. Johnson JW, Ascher P. Glycine potentiates the NMDA response in cultured mouse brain neurons. *Nature*. 1987;325(6104):529–531. https://doi.org/10.1038/325529a0.
69. Du X, Li J, Li M, et al. Research progress on the role of type I vesicular glutamate transporter (VGLUT1) in nervous system diseases. *Cell Biosci*. 2020;10:26. https://doi.org/10.1186/s13578-020-00393-4.
70. Magi S, Piccirillo S, Amoroso S, Lariccia V. Excitatory amino acid transporters (EAATs): glutamate transport and beyond. *Int J Mol Sci*. 2019;20(22):5674. https://doi.org/10.3390/ijms20225674.
71. Ngo DH, Vo TS. An Updated review on pharmaceutical properties of gamma-aminobutyric acid. *Molecules*. 2019;24(15):2678. https://doi.org/10.3390/molecules24152678.
72. Martin DL, Tobin AJ. Mechanisms controlling GABA synthesis and degradation in the brain. In: Martin DL, Olsen RW, eds. *GABA in the Nervous System*. Lippincott Williams & Wilkins; 2000:25–41.
73. Evenseth L, Gabrielsen M, Sylte I. The GABA$_B$ receptor-structure, ligand binding and drug development. *Molecules*. 2020;25(13):3093. https://doi.org/10.3390/molecules25133093.
74. Sigel E, Steinmann ME. Structure, function, and modulation of GABA(A) receptors. *J Biol Chem*. 2012;287(48):40224–40231. https://doi.org/10.1074/jbc.R112.386664.
75. Zou S, Kumar U. Cannabinoid receptors and the endocannabinoid system: signaling and function in the central nervous system. *Int J Mol Sci*. 2018;19(3):833. https://doi.org/10.3390/ijms19030833.
76. Devane WA, Hanus L, Breuer A, et al. Isolation and structure of a brain constituent that binds to the cannabinoid receptor. *Science*. 1992;258(5090):1946–1949. https://doi.org/10.1126/science.1470919.
77. Mechoulam R, Ben-Shabat S, Hanus L, et al. Identification of an endogenous 2-monoglyceride, present in canine gut, that binds to cannabinoid receptors. *Biochem Pharmacol*. 1995;50(1):83–90. https://doi.org/10.1016/0006-2952(95)00109-D.
78. Sugiura T, Kondo S, Sukagawa A, et al. 2-Arachidonoylglycerol: a possible endogenous cannabinoid receptor ligand in brain. *Biochem Biophys Res Commun*. 1995;215(1):89–97. https://doi.org/10.1006/bbrc.1995.2437.
79. Dainese E, Oddi S, Simonetti M, et al. The endocannabinoid hydrolase FAAH is an allosteric enzyme. *Sci Rep*. 2020;10(1):5903. https://doi.org/10.1038/s41598-020-62514-w.
80. Lertora J, Vanevski K. Introduction to pharmacokinetics and pharmacodynamics. In: Thoene J, ed. *Small Molecule Therapy for Genetic Disease*. Cambridge University Press; 2010:35–54. https://doi.org/10.1017/CBO9780511777905.004.
81. Faraone SV. The pharmacology of amphetamine and methylphenidate: relevance to the neurobiology of attention-deficit/hyperactivity disorder and other psychiatric comorbidities. *Neurosci Biobehav Rev*. 2018;87:255–270. https://doi.org/10.1016/j.neubiorev.2018.02.001.
82. Kuhar MJ, Ritz MC, Boja JW. The dopamine hypothesis of the reinforcing properties of cocaine. *Trends Neurosci*. 1991;14(7):299–302. https://doi.org/10.1016/0166-2236(91)90141-g.
83. Rothman RB, Baumann MH, Dersch CM, et al. Amphetamine-type central nervous system stimulants release norepinephrine more potently than they release dopamine and serotonin. *Synapse*. 2001;39(1):32–41. https://doi.org/10.1002/1098-2396(20010101)39:1<32::AID-SYN5>3.0.CO;2-3.
84. Kish SJ. Pharmacologic mechanisms of crystal meth. *Can Med Assoc J*. 2008;178(13):1679–1682. https://doi.org/10.1503/cmaj.071675.
85. Kalant H. The pharmacology and toxicology of "ecstasy" (MDMA) and related drugs. *Can Med Assoc J*. 2001;165(7):917–928.
86. Pidoplichko VI, DeBiasi M, Williams JT, Dani JA. Nicotine activates and desensitizes midbrain dopamine neurons. *Nature*. 1997;390(6658):401–404. https://doi.org/10.1038/37120.

87. Schilström B, Nomikos GG, Nisell M, Hertel P, Svensson TH. N-methyl-D-aspartate receptor antagonism in the ventral tegmental area diminishes the systemic nicotine-induced dopamine release in the nucleus accumbens. *Neuroscience*. 1998;82(3):781–789. https://doi.org/10.1016/s0306-4522(97)00243-1.
88. Nagy J, Kolok S, Boros A, Dezso P. Role of altered structure and function of NMDA receptors in development of alcohol dependence. *Curr Neuropharmacol*. 2005;3(4):281–297. https://doi.org/10.2174/157015905774322499.
89. Wallner M, Olsen RW. Physiology and pharmacology of alcohol: the imidazobenzodiazepine alcohol antagonist site on subtypes of GABAA receptors as an opportunity for drug development? *Br J Pharmacol*. 2008;154(2):288–298. https://doi.org/10.1038/bjp.2008.32.
90. Inturrisi CE, Schultz M, Shin S, Umans JG, Angel L, Simon EJ. Evidence from opiate binding studies that heroin acts through its metabolites. *Life Sci*. 1983;33(Suppl 1):773–776. https://doi.org/10.1016/0024-3205(83)90616-1.
91. Cowan MS, Lawrence AJ. The role of opioid-dopamine interactions in the induction and maintenance of ethanol consumption. *Prog Neuro Psychopharmacol Biol Psychiatr*. 1999;23(7):1171–1212. https://doi.org/10.1016/S0278-5846(99)00060-3.
92. Study RE, Barker JL. Cellular mechanisms of benzodiazepine action. *JAMA*. 1982;247(15):2147–2151.
93. Tan KR, Rudolph U, Lüscher C. Hooked on benzodiazepines: GABA$_A$ receptor subtypes and addiction. *Trends Neurosci*. 2011;34(4):188–197. https://doi.org/10.1016/j.tins.2011.01.004.
94. González-Maeso J, Weisstaub NV, Zhou M, et al. Hallucinogens recruit specific cortical 5-HT(2A) receptor-mediated signaling pathways to affect behavior. *Neuron*. 2007;53(3):439–452. https://doi.org/10.1016/j.neuron.2007.01.008.
95. Watts VJ, Mailman RB, Lawler CP, Neve KA, Nichols DE. LSD and structural analogs: pharmacological evaluation at D_1 dopamine receptors. *Psychopharmacology*. 1995;118:401–409. https://doi.org/10.1007/BF02245940.
96. Alex KD, Pehek EA. Pharmacological mechanisms of serotonergic regulation of dopamine neurotransmission. *Pharmacol Ther*. 2007;113(2):296–320. https://doi.org/10.1016/j.pharmthera.2006.08.004.
97. Morris BJ, Cochran SM, Pratt JA. PCP: from pharmacology to modelling schizophrenia. *Curr Opin Pharmacol*. 2005;5(1):101–106. https://doi.org/10.1016/j.coph.2004.08.008.
98. Kokkinou M, Ashok AH, Howes OD. The effects of ketamine on dopaminergic function: meta-analysis and review of the implications for neuropsychiatric disorders. *Mol Psychiatr*. 2017;23:59–69. https://doi.org/10.1038/mp.2017.190.
99. Lupica CR, Riegel AC. Endocannabinoid release from midbrain dopamine neurons: a potential substrate for cannabinoid receptor antagonist treatment of addiction. *Neuropharmacology*. 2005;48(8):1105–1116. https://doi.org/10.1016/j.neuropharm.2005.03.016.
100. Rivera-García MT, López-Rubalcava C, Cruz SL. Preclinical characterization of toluene as a non-classical hallucinogen drug in rats: participation of 5-HT, dopamine and glutamate systems. *Psychopharmacology*. 2015;232(20):3797–3808. https://doi.org/10.1007/s00213-015-4041-8.
101. Beckstead MJ, Weiner JL, Mihic SJ. Glycine and gamma-aminobutyric acid (A) receptor function is enhanced by inhaled drugs of abuse. *Mol Pharmacol*. 2000;57(6):1199–1205.
102. Nimitvilai S, You C, Arora DS, et al. Differential effects of toluene and ethanol on dopaminergic neurons of the ventral tegmental area. *Front Neurosci*. 2016;10:434. https://doi.org/10.3389/fnins.2016.00434.
103. Cruz SL, Mirshahi T, Thomas B, Balster RL, Woodward JJ. Effects of the abused solvent toluene on recombinant N-methyl-D-aspartate and non-N-methyl-D-aspartate receptors expressed in Xenopus oocytes. *J Pharmacol Exp Therapeut*. 1998;286(1):334–340.
104. Sordo L, Barrio G, Bravo MJ, et al. Mortality risk during and after opioid substitution treatment: systematic review and meta-analysis of cohort studies. *BMJ*. 2017;357:j1550. https://doi.org/10.1136/bmj.j1550.
105. Samet JH, Fiellin DA. Opioid substitution therapy — time to replace the term. *Lancet*. 2015;385(9977):P1508–P1509. https://doi.org/10.1016/S0140-6736(15)60750-4.
106. Reduction International H. *Global State of Harm Reduction 2020*. 2020. London.
107. Moore D, Aveyard P, Connock M, Wang D, Fry-Smith A, Barton P. Effectiveness and safety of nicotine replacement therapy assisted reduction to stop smoking: systematic review and meta-analysis. *BMJ*. 2009;338:b1024. https://doi.org/10.1136/bmj.b1024.
108. Balmford J, Borland R, Hammond D, Cummings KM. Adherence to and reasons for premature discontinuation from stop-smoking medications: data from the ITC Four-Country Survey. *Nicotine Tob Res*. 2011;13(2):94–102. https://doi.org/10.1093/ntr/ntq215.

109. Batra A, Klingler K, Landeldt B, Friederich HM, Westin A, Danielsson T. Smoking reduction treatment with 4-mg nicotine gum: a double-blind, randomized, placebo-controlled study. *Clin Pharmacol & Ther.* 2005;78(6):689−696. https://doi.org/10.1016/j.clpt.2005.08.019.
110. Schuurmans MM, Diacon AH, van Biljon X, Bolliger CT. Effect of pre-treatment with nicotine patch withdrawal symptoms and abstinence rates in smokers subsequently quitting with the nicotine patch: a randomized controlled trial. *Addiction.* 2004;99(5):634−640. https://doi.org/10.1111/j.1360-0443.2004.00711.x.
111. Anthenelli RM, Benowitz NL, West R, et al. Neuropsychiatric safety and efficacy of varenicline, bupropion, and nicotine patch in smokers with and without psychiatric disorders (EAGLES): a double-blind, randomised, placebo-controlled clinical trial. *Lancet.* 2016;387(10037):2507−2520. https://doi.org/10.1016/S0140-6736(16)30272-0.
112. Evins AE, Benowitz NL, West R, et al. Neuropsychiatric safety and efficacy of varenicline, bupropion, and nicotine patch in smokers with psychotic, anxiety, and mood disorders in the EAGLES trial. *J Clin Psychopharmacol.* 2019;39(2):108−116. https://doi.org/10.1097/JCP.0000000000001015.
113. Gray KM, Baker NL, McClure EA, et al. Efficacy and safety of varenicline for adolescent smoking cessation: a randomized clinical trial. *JAMA Pediatr.* 2019;173(12):1146−1153. https://doi.org/10.1001/jamapediatrics.2019.3553.
114. Chang P-H, Chiang C-H, Ho W-C, Wu P-Z, Tsai J-S, Guo F-R. Combination therapy of varenicline with nicotine replacement therapy is better than varenicline alone: a systematic review and meta-analysis of randomized controlled trials. *BMC Publ Health.* 2015;15:689. https://doi.org/10.1186/s12889-015-2055-0.
115. Koegelenberg CFN, Noor F, Bateman ED, et al. Efficacy of varenicline combined with nicotine replacement therapy vs. varenicline alone for smoking cessation: a randomized clinical trial. *JAMA.* 2014;312(2):155−161. https://doi.org/10.1001/jama.2014.7195.
116. Jordan CJ, Xi Z-X. Discovery and development of varenicline for smoking cessation. *Expet Opin Drug Discov.* 2018;13(7):671−683. https://doi.org/10.1080/17460441.2018.1458090.
117. Carpenter MJ, Gray KM, Wahlquist AE, et al. *A Pilot Randomized Clinical Trial of Remote Varenicline Sampling to Promote Treatment Engagement and Smoking Cessation.* Nicotine & Tobacco Research; 2020. https://doi.org/10.1093/ntr/ntaa241. ntaa241.
118. Warner C, Shoaib M. How does bupropion work as a smoking cessation aid? *Addiction Biol.* 2005;10(3):219−231. https://doi.org/10.1080/13556210500222670.
119. Costa R, Oliveira NG, Dinis- Oliveira RJ. Pharmacokinetic and pharmacodynamic of bupropion: integrative overview of relevant clinical and forensic aspects. *Drug Metabol Rev.* 2019;51(3):292−313. https://doi.org/10.1080/03602532.2019.1620763.
120. Stapleton J, West R, Hajek P, et al. Randomized trial of nicotine replacement therapy (NRT), bupropion and NRT plus bupropion for smoking cessation: effectiveness in clinical practice. *Addiction.* 2013;108(12):2193−2201. https://doi.org/10.1111/add.12304.
121. Wilkes S. The use of bupropion SR in cigarette smoking cessation. *Int J Chronic Obstr Pulm Dis.* 2008;3(1):45−53. https://doi.org/10.2147/copd.s1121.
122. Vogeler T, McClain C, Evoy EE. Combination bupropion SR and varenicline for smoking cessation: asystematic review. *Am J Drug Alcohol Abuse.* 2016;42(2):129−139. https://doi.org/10.3109/00952990.2015.1117480.
123. Settle Jr EC. Bupropion sustained release: side effect profile. *J Clin Psychiatr.* 1998;59(Suppl 4):32−36.
124. Merritt LL, Martin BR, Walters C, Lichtman AH, Damaj MI. The endogenous cannabinoid system modulates nicotine reward and dependence. *J Pharmacol Exp Therapeut.* 2008;326(2):483−492. https://doi.org/10.1124/jpet.108.138321.
125. Cahill K, Ussher M. Cannabinoid type 1 receptor antagonists (rimonabant) for smoking cessation. *Cochrane Database Syst Rev.* 2007;(3):CD005353. https://doi.org/10.1002/14651858.CD005353.pub2.
126. King A. Neuropsychiatric adverse effects signal the end of the line for rimonabant. *Nat Rev Cardiol.* 2010;7:602. https://doi.org/10.1038/nrcardio.2010.148.
127. Pedre B, Barayeu U, Ezerina D, Dick TP. The mechanism of action of N-acetylcysteine (NAC): the emerging role of H_2S and sulfane sulfur species. *Pharmacol Ther.* 2021;228:107916. https://doi.org/10.1016/j.pharmthera.2021.107916.
128. Heard KJ. Acetylcysteine for acetaminophen poisoning. *N Engl J Med.* 2008;359(3):285−292. https://doi.org/10.1056/NEJMct0708278.

129. Dekhuijzen PN, van Beurden WJ. The role for N-acetylcysteine in the management of COPD. *Int J Chronic Obstr Pulm Dis*. 2006;1(2):99−106. https://doi.org/10.2147/copd.2006.1.2.99.
130. Knackstedt LA, LaRowe S, Mardikian P, et al. The role of cystine-glutamate exchange in nicotine dependence in rats and humans. *Biol Psychiatr*. 2009;65(10):841−845. https://doi.org/10.1016/j.biopsych.2008.10.040.
131. Schulte M, Goudriaan AE, Kaag AM, et al. The effect of N-acetylcysteine on brain glutamate and gamma-aminobutyric acid concentrations and on smoking cessation: a randomized, double-blind, placebo-controlled trial. *J Psychopharmacol*. 2017;31(10):1377−1379. https://doi.org/10.1177/0269881117730660.
132. Froeliger B, McConnell PA, Stankeviciute N, McClure EA, Kalivas PW, Gray KM. The effects of N-Acetylcysteine on frontostriatal resting-state functional connectivity, withdrawal symptoms and smoking abstinence: a double-blind, placebo-controlled fMRI pilot study. *Drug Alcohol Depend*. 2015;156:234−242. https://doi.org/10.1016/j.drugalcdep.2015.09.021.
133. Schmaal L, Berk L, Hulstijn KP, Cousijn J, Wiers RW, van den Brink W. Efficacy of N-acetylcysteine in the treatment of nicotine dependence: a double-blind placebo-controlled pilot study. *Eur Addiction Res*. 2011;17(4):211−216. https://doi.org/10.1159/000327682.
134. Cederbaum AI. Alcohol metabolism. *Clin Liver Dis*. 2012;16(4):667−685. https://doi.org/10.1016/j.cld.2012.08.002.
135. Kragh H. From disulfiram to Antabuse: the invention of a drug. *Bulletin for the History of Chemistry*. 2008;33(2):82−88.
136. Skinner MD, Lahmek P, Pham H, Aubin H-J. Disulfiram efficacy in the treatment of alcohol dependence: a meta-analysis. *PLoS One*. 2014;9(2):e87366. https://doi.org/10.1371/journal.pone.0087366.
137. Ludwig AM, Levine J, Stark LH. *LSD and Alcoholism*. Springfield: Charles C. Thomas; 1970.
138. Walker JR, Korte JE, McRae-Clark AL, Hartwell KJ. Adherence across FDA-approved medications for alcohol use disorder in a veterans administration population. *J Stud Alcohol Drugs*. 2019;80(5):572−577. https://doi.org/10.15288/jsad.2019.80.572.
139. Hendershot CS, Wardell JD, Samokhvalov AV, Rehm J. Effects of naltrexone on alcohol self-administration and craving: meta-analysis of human laboratory studies. *Addiction Biol*. 2017;22(6):1515−1527. https://doi.org/10.1111/adb.12425.
140. Donoghue K, Elzerbi C, Saunders R, Whittington C, Pilling S, Drummond C. The efficacy of acamprosate and naltrexone in the treatment of alcohol dependence, Europe versus the rest of the world: a meta-analysis. *Addiction*. 2015;110(6):920−930. https://doi.org/10.1111/add.12875.
141. Cheng H-Y, McGuinness LA, Elbers RG, et al. Treatment interventions to maintain abstinence from alcohol in primary care: systematic review and network meta-analysis. *BMJ*. 2020;371:m3934. https://doi.org/10.1136/bmj.m3934.
142. O'Malley SS, Garbutt JC, Gastfriend DR, Dong Q, Kranzler HR. Efficacy of extended-release naltrexone in alcohol-dependent patients who are abstinent before treatment. *J Clin Psychopharmacol*. 2007;27(5):507−512. https://doi.org/10.1097/jcp.0b013e31814ce50d.
143. de Sousa A, de Sousa A. An open randomized study comparing disulfiram and acamprosate in the treatment of alcohol dependence. *Alcohol Alcohol*. 2005;40(6):545−548. https://doi.org/10.1093/alcalc/agh187.
144. Hashemian SM, Farhadi T. A review on modafinil: the characteristics, function, and use in critical care. *J Drug Assess*. 2020;9(1):82−86. https://doi.org/10.1080/21556660.2020.1745209.
145. Joos L, Goudriaan AE, Schmaal L, et al. Effect of modafinil on impulsivity and relapse in alcohol dependent patients: a randomized, placebo-controlled trial. *Eur Neuropsychopharmacol*. 2013;23(8):948−955. https://doi.org/10.1016/j.euroneuro.2012.10.004.
146. Mula M, Cavanna AE, Monaco F. Psychopharmacology of topiramate: from epilepsy to bipolar disorder. *Neuropsychiatric Dis Treat*. 2006;2(4):475−488. https://doi.org/10.2147/nedt.2006.2.4.475.
147. Johnson BA, Ait-Daoud N, Bowden CL, et al. Oral topiramate for treatment of alcohol dependence: a randomised controlled trial. *Lancet*. 2003;361(9370):1677−1685. https://doi.org/10.1016/S0140-6736(03)13370-3.
148. Wetherill RR, Spilka N, Jagannathan K, et al. Effects of topiramate on neural responses to alcohol cues in treatment-seeking individuals with alcohol use disorder: preliminary findings form a randomized, placebo-controlled trial. *Neuropsychopharmacology*. 2021;46:1414−1420. https://doi.org/10.1038/s41386-021-00968-w.
149. Pennington DL, Bielenberg J, Lasher B, et al. A randomized pilot trial of topiramate for alcohol use disorder in veterans with traumatic brain injury: effects on alcohol use, cognition, and post-concussive symptoms. *Drug Alcohol Depend*. 2020;214:108149. https://doi.org/10.1016/j.drugalcdep.2020.108149.

150. Dole VP, Nyswander M. A medical treatment for diacetylmorphine (heroin) addiction. A clinical trial with methadone hydrochloride. *JAMA*. 1965;193(8):646–650. https://doi.org/10.1001/jama.1965.03090080008002.
151. Jordan CJ, Cao J, Newman AH, Xi Z-X. Progress in agonist therapy for substance use disorders: lessons learned from methadone and buprenorphine. *Neuropharmacology*. 2019;158:107609. https://doi.org/10.1016/j.neuropharm.2019.04.015.
152. Koehl JL, Zimmerman DE, Bridgeman PJ. Medications for management of opioid use disorder. *Am J Health Syst Pharm*. 2019;76(15):1097–1103. https://doi.org/10.1093/ajhp/zxz105.
153. Ling W, Wesson DR. Clinical efficacy of buprenorphine: comparisons to methadone and placebo. *Drug Alcohol Depend*. 2003;70(2):S49–S57. https://doi.org/10.1016/S0376-8716(03)00059-0.
154. Maddux JF, Desmond DP. Methadone maintenance and recovery from opioid dependence. *Am J Drug Alcohol Abuse*. 1992;18(1):63–74. https://doi.org/10.3109/00952999209001612.
155. Jones S, Jack B, Kirby J, Wilson TL, Murphy PN. Methadone-assisted opiate withdrawal and subsequent heroin abstinence: the importance of psychological preparedness. *Am J Addict*. 2021;30(1):11–20. https://doi.org/10.1111/ajad.13062.
156. Substance Abuse and Mental Health Services Administration. *Key Substance Use and Mental Health Indicators in the United States: Results from the 2019 National Survey on Drug Use and Health (HHS Publication No. PEP20-07-01-001, NSDUH Series H-55)*. Rockville, MD: Center for Behavioral Health Statistics and Quality, Substance Abuse and Mental Health Services Administration; 2020. Retrieved from: https://www.samhsa.gov/data/.
157. Lee JD, Friedman PD, Kinlock TW, et al. Extended-release naltrexone to prevent opioid relapse in criminal justice offenders. *N Engl J Med*. 2016;374(13):1232–1242. https://doi.org/10.1056/NEJMoa1505409.
158. Dalton K, Butt N. Does the addition of naloxone in buprenorphine/naloxone affect retention in treatment in opioid replacement therapy?: a systematic review and meta-analysis. *J Addict Nurs*. 2019;30(4):254–260. https://doi.org/10.1097/JAN.0000000000000308.
159. Rush CR, Stoops WW. Agonist replacement therapy for cocaine dependence: a translational review. *Future Med Chem*. 2012;4(2):245–265. https://doi.org/10.4155/fmc.11.184.
160. Collins SL, Levin FR, Foltin RW, Kleber HD, Evans SM. Response to cocaine alone and in combination with methylphenidate in cocaine abusers with ADHD. *Drug Alcohol Depend*. 2006;82(2):158–167. https://doi.org/10.1016/j.drugalcdep.2005.09.003.
161. Levin FR, Evans SM, Brooks DJ, Garawi F. Treatment of cocaine dependent treatment seekers with adult ADHD: double-blind comparison of methylphenidate and placebo. *Drug Alcohol Depend*. 2007;87(1):20–29. https://doi.org/10.1016/j.drugalcdep.2006.07.004.
162. Grabowski J, Roache JD, Schmitz JM, Rhoades H, Creson D, Korszun A. Replacement medication for cocaine dependence: methylphenidate. *J Clin Psychopharmacol*. 1997;17(6):485–488. https://doi.org/10.1097/00004714-199712000-00008.
163. Schubiner H, Saules KK, Arfken CL, et al. Double-blind placebo-controlled trial of methylphenidate in the treatment of adult ADHD patients with comorbid cocaine dependence. *Exp Clin Psychopharmacol*. 2002;10(3):286–294. https://doi.org/10.1037//1064-1297.10.3.286.
164. Galloway GP, Buscemi R, Coyle JR, et al. A randomized, placebo-controlled trial of sustained-release dextroamphetamine for treatment of methamphetamine addiction. *Clin Pharmacol & Ther*. 2011;89(2):276–282. https://doi.org/10.1038/clpt.2010.307.
165. Miles SW, Sheridan J, Russell B, et al. Extended-release methylphenidate for treatment of amphetamine/methamphetamine dependence: a randomized. Double-blind, placebo-controlled trial. *Addiction*. 2013;108(7):1279–1286. https://doi.org/10.1111/add.12109.
166. Gaval-Cruz M, Weinshenker D. Mechanisms of disulfiram-induced cocaine abstinence: antabuse and cocaine relapse. *Mol Interv*. 2009;9(4):175–187. https://doi.org/10.1124/mi.9.4.6.
167. McCance-Katz EF, Kosten TR, Jatlow P. Chronic disulfiram treatment effects on intranasal cocaine administration: initial results. *Biol Psychiatr*. 1998;43(7):540–543. https://doi.org/10.1016/S0006-3223(97)00506-4.
168. Sofuoglu M, Poling J, Waters A, Sewell A, Hill K, Kosten T. Disulfiram enhances subjective effects of dextroamphetamine in humans. *Pharmacol, Biochem Behav*. 2008;90(3):394–398. https://doi.org/10.1016/j.pbb.2008.03.021.
169. Anderson AL, Reid MS, Li S-H, et al. Modafinil for the treatment of cocaine dependence. *Drug Alcohol Depend*. 2009;104(1–2):133–139. https://doi.org/10.1016/j.drugalcdep.2009.04.015.

170. Anderson AL, Li S-H, Biswas K, et al. Modafinil for the treatment of methamphetamine dependence. *Drug Alcohol Depend*. 2012;120(1−3):135−141. https://doi.org/10.1016/j.drugalcdep.2011.07.007.
171. Johnson BA, Ait-Daoud N, Wang X-Q, et al. Topiramate for the treatment of cocaine addiction: a randomized clinical trial. *JAMA Psychiatr*. 2013;70(12):1338−1346. https://doi.org/10.1001/jamapsychiatry.2013.2295.
172. Heinzerling KG, Swanson A-N, Kim S, et al. Randomized, double-blind, placebo-controlled trial of modafinil for the treatment of methamphetamine dependence. *Drug Alcohol Depend*. 2010;109(1−3):20−29. https://doi.org/10.1016/j.drugalcdep.2009.11.023.
173. Kampman KM, Pettinati H, Lynch KG, Spratt K, Wierzbicki MR, O'Brien CP. A double-blind, placebo-controlled trial of topiramate for the treatment of comorbid cocaine and alcohol dependence. *Drug Alcohol Depend*. 2013;133(1):94−99. https://doi.org/10.1016/j.drugalcdep.2013.05.026.
174. Kampman KM, Pettinati H, Lynch KG, et al. A pilot trial of topiramate for the treatment of cocaine dependence. *Drug Alcohol Depend*. 2004;75(3):233−240. https://doi.org/10.1016/j.drugalcdep.2004.03.008.
175. Huestis MA, Boyd SJ, Heishman SJ, Preston KL. Single and multiple doses of rimonabant antagonize acute effects of smoked cannabis in male cannabis users. *Psychopharmacology*. 2007;194(4):505−515. https://doi.org/10.1007/s00213-007-0861-5.
176. Levin FR, Kleber HD. Use of dronabinol for cannabis dependence: two case reports and review. *Am J Addict*. 2008;17(2):161−164. https://doi.org/10.1080/10550490701861177.
177. Schilenz NJ, Lee DC, Stitzer ML, Vandrey R. The effect of high-dose dronabinol (oral THC) maintenance on cannabis self-administration. *Drug Alcohol Depend*. 2018;187:254−260. https://doi.org/10.1016/j.drugalcdep.2018.02.022.
178. Levin FR, Mariani JJ, Brooks DJ, Pavlicova M, Cheng W, Nunes EV. Dronabinol for the treatment of cannabis dependence: a randomized, double-blind, placebo-controlled trial. *Drug Alcohol Depend*. 2011;116(1−3):142−150. https://doi.org/10.1016/j.drugalcdep.2010.12.010.
179. Haney M. Opioid antagonism of cannabinoid effects: differences between marijuana smokers and nonmarijuana smokers. *Neuropsychopharmacology*. 2007;32(6):1391−1403. https://doi.org/10.1038/sj.npp.1301243.
180. Cooper ZD, Haney M. Opioid antagonism enhances marijuana's effects in heavy marijuana smokers. *Psychopharmacology*. 2010;211(2):141−148. https://doi.org/10.1007/s00213-010-1875-y.
181. Haney M, Ramesh D, Glass A, Pavlicova M, Bedi G, Cooper ZD. Naltrexone maintenance decreases cannabis self-administration and subjective effects in daily cannabis smokers. *Neuropsychopharmacology*. 2015;40(11):2489−2498. https://doi.org/10.1038/npp.2015.108.
182. Notzon DP, Kelly MA, Choi CJ, et al. Open-label pilot study of injectable naltrexone for cannabis dependence. *Am J Drug Alcohol Abuse*. 2018;44(6):619−627. https://doi.org/10.1080/00952990.2017.1423321.
183. Gray KM, Carpenter MJ, Baker NL, et al. A double-blind randomized controlled trial of N-acetylcysteine in cannabis-dependent adolescents. *Am J Psychiatr*. 2012;169(8):805−812. https://doi.org/10.1176/appi.ajp.2012.12010055.
184. Squeglia LM, Baker NL, McClure EA, Tomko RL, Adisetiyo V, Gray KM. Alcohol use during a trial of N-acetylcysteine for adolescent marijuana cessation. *Addict Behav*. 2016;63:172−177. https://doi.org/10.1016/j.addbeh.2016.08.001.
185. Gray KM, Sonne SC, McClure EA, et al. A randomized placebo-controlled trial of N-acetylcysteine for cannabis use disorder in adults. *Drug Alcohol Depend*. 2017;177:249−257. https://doi.org/10.1016/j.drugalcdep.2017.04.020.
186. Tomko RL, Baker NL, Hood CO, et al. Depressive symptoms and cannabis use in a placebo-controlled trial of N-Acetylcysteine for adult cannabis use disorder. *Psychopharmacology*. 2020;237(2):479−490. https://doi.org/10.1007/s00213-019-05384-z.

CHAPTER 3

Neuroanatomical and neurochemical substrates of addiction

Introduction

I once had a conversation with a neighbor about my research. He claimed that individuals with a substance use disorder (SUD) "lack willpower" and that they "choose" to have an addiction. Unfortunately, this is an opinion I have heard from multiple individuals that are unfamiliar with the research that has examined the neurobiological bases of SUDs. The previous chapter covered the pharmacological actions of commonly used drugs. Drugs increase dopamine levels, whether directly or indirectly. However, drug-induced dopamine release does not fully explain why individuals develop a SUD. This chapter will discuss some of the neurobiological determinants of addiction. As you are about to learn, long-term drug use causes numerous changes to occur in the brain, which greatly impact how individuals respond to drug-paired cues and how individuals make decisions. These neural adaptations can lead to intense drug cravings, thus precipitating relapse. This chapter can be considered a "Part I" of the neuromechanisms of addiction as I will focus on the neurobiological determinants of addiction at a macro level. That is, the major brain structures and neurotransmitters that mediate the addiction process will be discussed in this chapter. The next chapter will focus on the molecular and the cellular mechanisms involved in addiction. After reading these next two chapters, I hope that you can see why addiction is now considered a biomedical disease and not simply the result of a personal failing.

Learning objectives

By the end of this chapter, you should be able to …

(1) Identify some of the behavioral, neuroanatomical, and neurochemical techniques used to study addiction.
(2) Discuss the specific functions of brain structures related to reward.
(3) Explain the neural circuitry that mediates distinct stages of the addiction cycle.

Primary brain structures of the mesocorticolimbic pathway

Many of the neurotransmitters that you learned about in the previous chapter follow certain *pathways* in the brain. Cell bodies in one region of the brain project axons to another part of the brain. Because commonly used drugs increase dopamine levels, I will focus on dopamine's pathways for now. There are four major dopaminergic pathways (Fig. 3.1). Two of these pathways will be covered as they are directly linked to addiction. The *mesolimbic* pathway connects the ventral tegmental area (VTA) to the *ventral striatum*, which contains the nucleus accumbens (NAc). Rewards like food, sex, and drugs increase dopamine levels in this pathway.[1] The *mesocortical* pathway connects the VTA to the cerebral cortex. Because the mesolimbic and the mesocortical pathways both involve the VTA and are closely related, they are often referred together as the mesocorticolimbic pathway, or informally as the "reward pathway" of the brain.

Not surprisingly, the brain structures that make up the mesocorticolimbic pathway play a pivotal role in processing and responding to rewards. The major brain structures of the mesocorticolimbic pathway will be discussed in more detail below, as well as structures that have direct connections with these structures. For now, I will discuss how these brain regions respond to natural rewards. This will allow you to learn how these structures function normally and how they are altered by drug use.

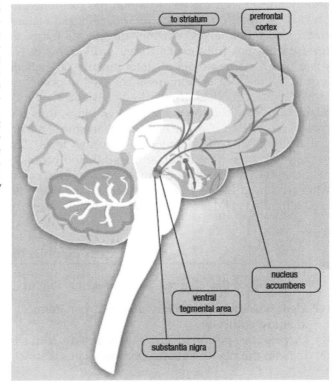

FIGURE 3.1 A schematic highlighting the major dopaminergic pathways. The nigrostriatal pathway originates in the substantia nigra and ends in the striatum (more specifically, the dorsal striatum). The mesolimbic pathway originates in the ventral tegmental area (VTA) and ends in the nucleus accumbens (NAc). The mesocortical pathway also originates in the VTA and ends in the frontal cortex. The tuberinfundibular pathway is shown but is not labeled. This pathway is confined to the hypothalamus, which is located anterior to the VTA (just to the right of the VTA in this image). *Images comes from the National Institute on Drug Abuse (https://www.drugabuse.gov/publications/research-reports/methamphetamine/are-people-who-misuse-methamphetamine-risk-contracting-hivaids-hepatitis-b-c).*

Ventral tegmental area (VTA)

The VTA is one brain region where dopaminergic cell bodies originate. Recall that when excitatory receptors on a neuron are stimulated, this causes an action potential to propagate down the length of the axon, which is then followed by release of neurotransmitter into the synapse. Dopaminergic neurons in the VTA show two different types of "firing" patterns.[2] The first type is known as **tonic firing**. Here, neurons engage in slow, asynchronous firing; in other words, neurons do not fire in unison. Tonic firing is somewhat like a pacemaker, leading to low concentrations of dopamine in the mesocorticolimbic pathway. **Phasic firing** is characterized by several populations of neurons firing in quick succession. This causes dopamine levels to increase for a short period of time. Having two different firing patterns is proposed to underly reward learning. Before encountering something that is rewarding, dopaminergic neurons engage in tonic firing. When we encounter something that is rewarding, like food or sex, phasic firing of intra-VTA dopamine neurons occurs.[3] The increased dopamine enables us to remember that what we just experienced is pleasurable and that we should repeat it in the future. Reward learning serves an important biological function: it keeps us alive! We need to eat and stay hydrated; we also need to procreate to ensure the survival of our species. One interesting finding is that dopamine neurons in the VTA are more sensitive to rewards we have not yet experienced or are not anticipating on receiving. Read the Experiment Spotlight to learn more about how VTA dopamine neurons respond to rewards.

Experiment Spotlight: Dopaminergic neurons report an error in the temporal prediction of reward during learning (Hollerman & Schultz[4])

This study is older compared to the others you will read throughout this book, but it is a seminal piece of work that showed how dopamine neurons in the VTA respond to rewards. Using two monkeys, Hollerman and Schultz[4] implanted electrodes into the VTA and into the substantia nigra. The monkeys saw two images appear on a computer screen, one on the left side of the screen and one on the right side of the screen. The pictures varied randomly between the left and the right side of the screen. Initially, monkeys were exposed to just one picture. Each time the monkey touched a lever below the image, it received some juice. Eventually, a second image was added. Touching the lever under this image did not result in delivery of juice. After the monkeys learned to consistently respond to the correct image, they received pairs of images they had never seen before. Using trial and error, the monkeys had to learn which novel image led to juice delivery.

What Hollerman and Schultz[4] found in this study was remarkable. Dopaminergic neurons in the VTA, as well as in the substantia nigra, were activated early in training when monkeys were more likely to select the wrong image and received juice reward at unpredictable times. Once monkeys learned the task and were able to consistently select the correct image, dopaminergic activity greatly diminished. When the juice was delivered at an unexpected time, dopaminergic activity increased dramatically. Conversely, if the juice reward was omitted when it was expected, dopaminergic activity decreased substantially.

The results of this study suggest that dopamine neurons are responsible for *error signaling*. That is, when we receive an unexpected reward for doing something (e.g., your mom unexpectedly gives you $20 after showing up to her house), dopamine neurons in the VTA respond. This allows us to remember this information in the future, so we can respond accordingly. If you always receive $20 for showing up to your mom's house, there is no need for the dopamine neurons to respond because there is nothing new to learn. You have already learned the association between going to your mom's house and receiving money.

Questions to consider:
(1) Although this study used juice as the reward, how do you think these results apply to commonly used drugs?
(2) Hollerman and Schultz used two monkeys in their study. Is the use of just two monkeys in the current experiment a design flaw? Why or why not?
(3) How do we know that the dopaminergic activity is specific to the VTA? What control could be incorporated to show specificity of dopaminergic activity concerning reward presentation/cues that predict rewards?

Answers to questions:
(1) Drugs, like juice, are rewarding to animals. The results of this study can be used to make predictions about how drug-paired stimuli activate intra-VTA dopaminergic neurons. As you will learn, cues paired with drugs activate dopaminergic neurons.
(2) Not necessarily. One thing to keep in mind is that although only two monkeys were used, the "subjects" in this study were individual neurons located in the VTA and the substantia nigra. Hollerman and Schultz recorded the activity of numerous neurons in these brain regions across both animal subjects. Because many tests were performed, Hollerman and Shultz obtained a cornucopia of data for this study.
(3) An important control in a study like this one is recording dopaminergic activity outside of the VTA and/or recording activity from nondopaminergic neurons. In fact, Hollerman and Schultz recorded activity from nondopaminergic neurons located in the substantia nigra. These neurons did not respond like the dopaminergic ones located in the VTA or in the substantia nigra. These results provide further evidence that dopaminergic neurons are important for error signaling.

Not only does the VTA respond to unexpected rewards, but it responds to cues (**stimuli**) that are consistently paired with the reward. In another seminal study conducted by Dr. Wolfram Schultz,[5] monkeys were given access to juice following presentation of a light and a sound. Initially, dopaminergic neurons fired in response to the juice delivery. However, over repeated pairings of the light/sound and juice, dopaminergic neurons began responding as soon as the light/sound were presented but did not fire following presentation of juice. In Chapter 6, you will learn more about this type of learning. For now, just know that this research is important because it shows that things in the environment can become paired with rewards, and these pairings elicit a strong physiological response in the VTA. This will become critically important as we continue discussing addiction, especially relapse.

How does responding to reward-paired cues relate to us? There are many stimuli in our environment that become paired with rewards. Aromas become paired with specific foods. If you smell cookies baking in the oven, the olfactory cues signal that food delivery is imminent. As such, dopaminergic neurons begin firing when you smell the cookies, but not when you eat them. If you are expecting a text from someone that you like, you may get excited when you hear/feel the notification go off on your phone. Here, the dopaminergic neurons in the VTA respond to the cue that signals delivery of the message that you have been anticipating.

Nucleus accumbens (NAc)

The NAc is often colloquially referred to as the pleasure center of the brain. Activation of the VTA leads to dopamine release in the NAc. Because neurons in the VTA respond to cues that become paired with rewards, dopamine release in the NAc is important for assigning **incentive salience** to these cues.[6] Incentive salience is a form of attention that motivates us to approach some object or perceived outcome. Imagine that you really enjoy coffee and that you wake up one morning to the smell of someone brewing coffee. Because you have learned to associate the smell of coffee to the coffee itself, this olfactory cue has high incentive salience. The smell may motivate you to get some coffee even if you had not planned on brewing some for yourself that day.

Although dopamine is released in the NAc following VTA stimulation, dopaminergic neurons are not located in this region. Instead, the NAc contains a high concentration of special GABAergic neurons called **medium spiny neurons** (**MSN**; also known as medium projection neurons), which account for 90%–95% of the neurons in the NAc. Most MSNs either have D_1-like or D_2-like receptors, although some have both receptor subtypes. At baseline, MSNs containing D_2-like receptors are more active compared to MSNs containing D_1-like receptors.[7] Evidence suggests that D_1-containing MSNs play a critical role in reward learning while D_2-containing MSNs are more important for aversive learning.[8] For example, learning to associate pleasurable odors with foods that you like involves activation of D_1-containing MSNs, whereas learning to associate malodorous scents with something (e.g., the stench of a skunk with the skunk itself) involves the activation of D_2-containing MSNs.

The NAc can be subdivided into two major regions: the NAc shell (outer portion) and the NAc core (inner portion). Each subregion of the NAc has different functions and distinct projections to other areas of the brain. The NAc shell is important for mediating the pleasure we experience when we receive a reward. For those that enjoy coffee, drinking a cup of java will activate the NAc shell. The shell is also important for attributing incentive salience to cues that become paired with the reward (as discussed above). The NAc shell projects GABAergic neurons back to the VTA, which ultimately decreases the amount of dopamine that is released in the NAc core. This acts as a negative feedback loop, similar to what you learned about in the previous chapter when I discussed autoreceptors on terminal buttons of presynaptic neurons.

The NAc core is important for establishing motor functions related to obtaining a reward. Going back to our example of coffee: after you have learned that coffee is rewarding, neurons in the NAc core encode the behavioral processes necessary for acquiring coffee in the future. These processes could include operating a coffee pot or French press or driving to your favorite coffee shop. GABAergic neurons originating from the NAc core project to a region of the basal ganglia known as the ventral pallidum. Given the basal ganglia's role in movement, communication between the NAc core and the ventral pallidum is important for directing actions toward reward-paired cues. Being able to coordinate actions to acquire rewards like food is evolutionarily advantageous as it allows a species to survive.

Prefrontal cortex (PFC)

In addition to the NAc, dopaminergic neurons originating from the VTA project to the cerebral cortex. Dopamine plays an important role in the *prefrontal cortex* (*PFC*), the most anterior region of the frontal lobe. Like the NAc, the PFC can be fractioned into multiple areas. At a broad level, the PFC can be divided into the orbitofrontal cortex (OFC) and the medial prefrontal cortex (mPFC). The OFC can be further subdivided into the lateral OFC (lOFC) and the medial OFC (mOFC), while the mPFC can be fractioned into the infralimbic region (IL), and the prelimbic region (PL). Collectively, the PFC is important for the *motivational* aspects of rewards. Just because something is rewarding does not always mean we should engage in the behavior that provides us access to that reward. Personally, I love cheesecake; however, eating it constantly is not in my best interest. I could gain a lot of weight, leading to several health complications. Our PFC engages in a risk/benefit analysis when dealing with rewards. The PFC helps us maintain self-control.

Other brain regions involved in reward

While the VTA, the NAc, and the PFC are important mediators of reward, and thus addiction, there are numerous brain regions involved in the sequalae of addiction. Fig. 3.2 shows a network of brain structures that are involved in the addiction process. The contributions of the regions depicted in Fig. 3.2 will be discussed below. As mentioned above, I will discuss how these regions mediate the rewarding aspects of natural rewards.

Dorsal striatum

The dorsal striatum is composed of the caudate nucleus and the putamen. The dorsal striatum and the globus pallidus make up the basal ganglia. The primary function of the basal ganglia is to control voluntary movements. However, they also mediate **habit formation**. We are creatures of habit. We often follow a similar routine during the day. After we wake up, some of us will make coffee and/or eat breakfast before going to work/school. We often take the same route to work/school each day. Driving becomes a habit, to the point where we often zone out while driving. Habits are important because we do not have to exert much cognitive effort when planning our actions. We just do them. How is this related to rewards? By having a routine that we follow, this makes obtaining rewards easier. Do you find yourself eating the same few meals over the span of several weeks? Sticking to the few meals that we know we like reduces the cognitive load on the frontal cortex, which allows it to focus on other tasks, such as studying for an upcoming exam. Having too many options can be cognitively overwhelming. Imagine trying to pick a meal at a restaurant with a large selection, or imagine trying to select a movie/show to watch on your favorite streaming service. The formation of habits around rewards is believed to occur due to the connections between the ventral striatum and the dorsal striatum. The NAc core controls dopamine release in the dorsal striatum via its connection with the substantia nigra.

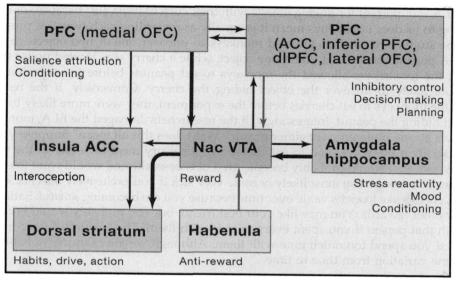

FIGURE 3.2 The neural circuity of addiction. The ventral tegmental area (VTA) and the nucleus accumbens (Nac) mediate the rewarding effects of drugs. The insula and the anterior cingulate cortex (ACC) are involved in *interoception*, how we feel soon after taking a drug. The dorsal striatum receives information from both the Nac and the insula/ACC. With repeated drug use, the dorsal striatum drives compulsive drug seeking. At this point, drug use becomes a habitual action. The VTA and the Nac receive inputs from the habenula, which normally inhibits the VTA when expected reward is not delivered or when something aversive happens. With repeated drug use, the habenula becomes less effective at inhibiting the VTA. The prefrontal cortex (PFC), particularly the medial orbitofrontal cortex (OFC), is important for attributing incentive salience to cues paired with drugs. Over time, stimuli paired with drug use (e.g., syringe) elicit intense cravings by activating the VTA/NAc. Parts of the PFC, such as the ACC, the inferior PFC, the dorsolateral PFC (dlPFC), and the lateral OFC, have a reciprocal connection with the VTA. This part of the brain is important for decision making and impulse control. Like the habenula, repeated drug use weakens the control of the PFC over VTA activity. Finally, reward structures have reciprocal connections with the amygdala and the hippocampus, which are important for memory, conditioning (like the medial OFC), and stress reactivity. During the addiction process, the limbic system becomes heavily involved. Drug-paired stimuli trigger "memories" of drug use, which can elicit cravings. Experiencing stress can activate the reward pathway, further increasing cravings. *Image modified from Fig. 5 of Volkow ND, Morales M. The brain on drugs: from reward to addiction. Cell. 2015;162(4):712–725. https://doi.org/10.1016/j.cell.2015.07.046. Copyright 2015, with permission from Elsevier.*

Amygdala

The amygdala (coming from a Latin word meaning "almond"), as part of the *limbic system*, is heavily involved in fear and emotional memories, but this brain structure plays an important role in reward. The amygdala is composed of many subregions, each having distinct functions. Specific to reward, we will focus on two such subregions: the central nucleus of the amygdala (CeA) and the basolateral amygdala (BLA). The CeA, like the VTA, responds to objects that have been paired with rewards. If a light is turned on before food is delivered to a rat, the rat will approach the light because the light becomes associated with food, thus acquiring rewarding properties. However, rats with a damaged CeA fail to approach the light after it has been paired with food although they still approach the food.[9]

The BLA is important for linking objects with their current *value*. Just because something is rewarding to us does not always mean it is equally as rewarding each time it is presented to us. In one study,[10] researchers allowed monkeys to uncover one of two objects to reveal a reward. A peanut was located under one object, while a cherry was located under the other object. If the researchers allowed the monkeys to eat peanuts before the experiment, they were more likely to uncover the object hiding the cherry. Conversely, if the researchers allowed the monkeys to eat cherries before the experiment, they were more likely to uncover the object hiding the peanut. Interestingly, if the researchers damaged the BLA, monkeys did not show a strong preference for either reward. What does this all mean? Suppose you enjoy cheesecake like I do. If you had to eat cheesecake for every meal each day of the week for several weeks, you would probably become tired of cheesecake and would want to eat something else. You would also most likely become very sick if you exclusively ate cheesecake! In this case, cheesecake loses its value over time because you are becoming *satiated*. Satiation can occur for other rewards. You may like your best friend, but you probably would become irritated with that person if you spent every minute with them. The "value" of your friend can decrease if you spend too much time with them. Although we are creatures of habit, we do enjoy some variation from time to time.

Anterior cingulate cortex and insula

The anterior cingulate cortex (ACC) is just posterior to the PFC and encodes information regarding action-outcome associations. If an animal performs a certain action and receives food for engaging in this action, the ACC helps store this information. Similarly, if an animal engages in a behavior but does not receive reinforcement for that action, the ACC will encode that information. In one study, monkeys were given two objects and had to move just one object to receive food. After monkeys learned this task, the contingencies were reversed. The animal then had to displace the other object to receive food. Monkeys learned to switch their preference to receive food. However, monkeys with ACC lesions were impaired in reversal learning. Similar results are observed when monkeys must reverse a behavioral action, such as pulling a joystick or turning a joystick.[11]

The insula is in the *lateral fissure* of the cerebral cortex. The insula is proposed to mediate numerous processes, including processing sensory information, responding to *novelty* (i.e., how "new" something appears to be), and interpreting emotional experiences. Directly related to rewards, the insula contributes to *interoceptive awareness*, our awareness of events occurring in the body, a process also controlled by the ACC. Imagine how you feel when you are thirsty or when you need to use the restroom. Your throat may feel dry, or you may feel pressure in your bladder. These are interoceptive cues we use to guide certain behaviors, such as getting a glass of water or using the restroom.

Hippocampus

Like the amygdala, the hippocampus is part of the limbic system and is important for learning and memory; specifically, it helps consolidate short-term memories into long-term memories. The hippocampus is also important for spatial learning. The term hippocampus comes from a Greek word meaning "seahorse" because this structure resembles a seahorse (genus *Hippocampus*). Due to its role in memory, the hippocampus is hypothesized to respond to the novelty of an object. What's interesting is that we frequently engage in habits, but we are sensitive to novel objects; even infants as young as 10 weeks old will spend more time gazing at a novel object compared to a familiar one.[12] Sensitivity to novelty may be an evolutionary adaptation as we are able to explore unknown objects to determine if they are beneficial or harmful to us.

Lateral habenula

The habenula makes up part of the *epithalamus*. Although divided into a medial habenula and lateral habenula, the latter has received recent attention for its involvement in reward processing. Specifically, the lateral habenula is activated when we receive something aversive or do not receive a reward that is expected; this activation ultimately leads to inhibition of dopaminergic neurons in the VTA. The lateral habenula sends glutamatergic projections to a brain region known as the rostromedial tegmental nucleus, which is the posterior portion of the VTA. Activation of this region causes GABA release in the VTA, which inhibits the dopaminergic cell bodies located in the VTA.

Paradigms to study addiction-like behaviors

Now that you have some basic knowledge of the brain regions involved in reward, we need to discuss the ways researchers study addiction-like behaviors. There are multiple paradigms that are used to model distinct aspects of addiction. Just a few major paradigms will be briefly discussed below. These paradigms will be discussed in greater detail in Chapters 5–7.

In the **drug self-administration** paradigm (Fig. 3.3A), animals often undergo catheter implantation surgery before being placed in a special testing chamber. The animal is allowed to press a lever or stick their nose into an aperture (collectively, things such as levers and apertures are known as *manipulanda*). A drug of interest is then infused into the animal's jugular or femoral vein. Animals often continue responding on manipulanda to receive additional drug infusions. Drug self-administration studies can be conducted in humans, as individuals are allowed to respond on a key to earn access to cigarette puffs, pills, or intranasal sprays of a drug.[13] Drug self-administration measures the direct *reinforcing* effects of a drug. In other words, this paradigm determines if animals/humans will continue engaging in a behavior to earn access to the drug. In real life, individuals engage in certain behaviors to gain access to drugs, such as paying another individual. If drugs were not reinforcing, individuals would not be willing to spend their money to use them. You will learn more about reinforcement and drug self-administration in Chapter 5.

Drug self-administration can be used to measure drug-seeking behavior. In one type of drug-seeking procedure, animals respond on a manipulandum a certain number of times to receive access to a stimulus such as a light (e.g., they may have to respond 10 times to receive access to the light). No drug is administered at this point. To receive the drug, the animal must respond on the manipulandum after waiting a certain amount of time (e.g., 15 min). When the animal receives the drug infusion, they also receive presentation of the

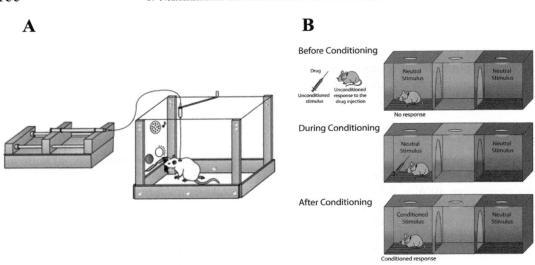

FIGURE 3.3 (A) A schematic of the drug self-administration paradigm. A syringe is attached to a pump that pushes drug through tubing that connects to a catheter port attached to the rat. The drug then flows into a catheter implanted into a vein (typically right jugular) after the rat responds on some manipulandum (lever in this image). Other stimuli can be presented when the rat receives drug reinforcement, such as lights and/or sounds. The number of responses should increase during the first few sessions as the animal learns to associate its responses with drug reinforcement. Eventually, the number of responses will plateau. (B) A schematic of the CPP paradigm. Each compartment is associated with different stimuli, such as visual and tactile cues. In this image, the dark blue compartment has rod floors, the light blue compartment has a solid floor, and the green compartment has a mesh floor. During conditioning sessions, a rat is isolated to either end compartment following injection of either the drug of interest or vehicle. Exposure to drug/vehicle occurs on alternating sessions. During the final test session, the rat is allowed to explore all three compartments in a drug-free state. *Panel (A) is modified from Fig. 2 of Watterson LR, Oliver MF. Synthetic cathinones and their rewarding and reinforcing effects in rodents. Adv Neurosci. 2014;2014:209875. https://doi.org/10.1155/2014/209875. This is an open-access article distributed under the terms of the Creative Commons Attribution 4.0 International License (https://creativecommons.org/licenses/by/4.0/). Panel (B) comes from Fig. 1 of McKendrick G, Graziane NM. Drug-induced conditioned place preference and its practical use in substance use disorder research. Front Behav Neurosci. 2020;14:582147. https://doi.org/10.3389/fnbeh.2020.582147. This is an open-access article distributed under the terms of the Creative Commons Attribution 4.0 International License (https://creativecommons.org/licenses/by/4.0/).*

light. This is known as a *second-order schedule of reinforcement*. What happens is that animals will increase their rate of responding to gain access to the drug-paired cue even though these responses do not increase how much drug they earn. The increase in responding is interpreted as an increase in drug-seeking behavior. Drug seeking can also be measured by delivering an aversive stimulus (e.g., foot shock) each time an animal receives drug reinforcement. If animals continue responding for the drug despite receiving the aversive stimulus, they are engaging in drug-seeking behavior, specifically *compulsive* drug seeking. Measuring compulsive drug seeking is valuable as addiction is characterized by compulsive drug use. Individuals with a SUD will continue using the drug despite negative consequences like overdose.

Another way to measure drug reward is with **conditioned place preference (CPP)** (Fig. 3.3B), in which animals associate the effects of a specific drug with an environmental context. Most CPP chambers consist of two or three compartments that are separated by guillotine doors. Each compartment provides distinct visual, tactile, and/or olfactory cues. Unlike drug self-administration, animals do not work to receive the drug. The researcher injects the drug into the animal and isolates it to one of the CPP compartments. On alternating sessions, animals receive an injection of vehicle (think of placebo) and are placed in a different compartment. Eventually, the animal is allowed to explore each compartment in a drug-free state. If the drug is rewarding, the animal will spend more time in the drug-paired compartment. Thus, CPP measures the *conditioned rewarding* effects of a drug instead of the direct reinforcing effects of a drug. Given that multiple stimuli become paired with drugs, CPP models this aspect of drug use by pairing the drug with specific environmental contexts. Granted, environmental cues can be paired with drug administration in self-administration experiments, but one large advantage of CPP over self-administration is that surgeries do not need to be performed on animals. CPP will be discussed in more detail in Chapter 6.

To measure relapse, researchers often use the **reinstatement** model. Reinstatement can be used in drug self-administration or in CPP studies. In drug self-administration, after animals have learned to respond on a manipulandum to earn drug infusions, they will receive test sessions in which they no longer receive the drug infusion for each response. This is known as **extinction** and mimics an individual that is abstaining from drug use. Once the animal decreases their responding to a sufficient level, the researcher can either inject the animal with a small amount of the drug they previously self-administered, they can turn on lights and/or present sounds that were previously paired with each drug infusion, or they can subject the animal to a stressful event. The researcher then measures how much the animal responds on the previously drug-paired manipulandum. Importantly, animals never receive drug infusions for responding on the manipulandum during the reinstatement test. Increased responding suggests that the animal is showing relapse-like behavior. In CPP, extinction can be accomplished different ways; however, the result is to give animals multiple sessions in which the drug is no longer paired with the original environment. Eventually, the animal will spend equal amounts of time in the previously drug-paired and vehicle-paired compartments. Researchers can then inject the animal with a small amount of the drug or subject them to stress before measuring how much time the animal spends in each compartment. Relapse-like behavior occurs when the animal spends significantly more time in the drug-paired compartment during the reinstatement test.

Closely related to reinstatement is the **incubation of craving** effect, although this is primarily measured via drug self-administration. To measure incubation of craving, animals are first trained to self-administer a drug. Animals are not actively trained to associate responding with the absence of drug delivery (i.e., extinction training) like with reinstatement. Instead, after completing the final drug self-administration session, animals spend time in their home cage for a certain number of days. Animals are then placed back in the testing chamber and are presented with cues that were previously paired with drug administration. The researcher can measure how much the animal starts responding on the drug-paired manipulandum. Just as in reinstatement, animals never receive drug infusions for responding on the manipulandum. Animals receive several test sessions as they go through withdrawal. Increased responses throughout the withdrawal period are interpreted as incubation of craving. Reinstatement and incubation of craving will be a major focus of Chapter 7.

Quiz 3.1

1. A researcher trains a rat to respond on a lever to earn an infusion of cocaine. Each time the rat earns an infusion of cocaine, two lights turn on. After 2 weeks, the researcher stops reinforcing the rat when it responds on the lever. The lights never turn on at this point. After an additional 2 weeks, the researcher places the rat in the testing chamber and turns the lights on. The researcher notices that the rat starts pressing the lever even though it does not earn any cocaine for doing so. Which behavioral paradigm has the researcher just used to study addiction?
    a. CPP
    b. Incubation of craving
    c. Reinstatement
    d. Second-order schedule
2. Out of the following regions, where do most dopaminergic cell bodies originate?
    a. Dorsal striatum
    b. Frontal cortex
    c. NAc
    d. VTA
3. Which brain region plays an important role in evaluating the current value of a stimulus?
    a. Basolateral nucleus of the amygdala
    b. Central nucleus of the amygdala
    c. NAc shell
    d. VTA
4. Which neurotransmitter is associated with MSNs?
    a. Acetylcholine
    b. Dopamine
    c. GABA
    d. Glutamate

Quiz 3.1 answers

1. c — Reinstatement
2. d — VTA
3. a — Basolateral nucleus of the amygdala
4. c — GABA

Neuroanatomical underpinnings of the addiction process: focus on dopaminergic signaling

In addition to identifying which brain regions mediate different aspects of drug addiction, this section will spotlight how dopaminergic activity in these regions drives addiction-like behaviors. Later, you will learn about the contribution of other neurotransmitter systems to addiction. First, I want to briefly highlight some of the techniques researchers can use to elucidate the neuroanatomical and neurochemical underpinnings of addiction.

In humans, imaging methods such as functional magnetic resonance imaging (fMRI) and positron emission tomography (PET) can be used to determine which brain regions are active when an individual uses a drug or encounters cues that have been paired with a drug. In animals, researchers often perform *lesion* studies to determine the contribution of a brain region on behavior, such as drug self-administration. Typically, the NMDA receptor agonist quinolinic acid is infused into a brain region. Overstimulation of NMDA receptors causes too much Ca^{++} to enter the neuron. Ca^{++} is important for numerous processes in the neuron, including the release of neurotransmitter from presynaptic vesicles. However, too much Ca^{++} is detrimental to the neuron, leading to *excitotoxicity*. An alternative to lesions is to temporarily inactivate a brain region using the GABA receptor agonists baclofen and muscimol.

The techniques above can inform us about which brain regions control certain behaviors, but they do not provide any information about the underlying neural mechanisms of these behaviors. We know that drugs increase dopamine levels, but *how* do we know this? One technique that is used to measure neurotransmitter release is **microdialysis** (Fig. 3.4A). A probe inserted into the brain collects fluid which is then analyzed at a later point. An alternative to microdialysis is **voltammetry** (Fig. 3.4B). Instead of placing a microdialysis probe into the brain, an electrode is inserted. Whereas microdialysis involves the physical removal of dialysate from the brain through a probe, voltammetry detects changes in current due to *oxidation* (loss of electrons) or *reduction* (gain of electrons). Depending on what type of reaction occurs, researchers can determine if certain neurotransmitters like dopamine are being released at a given point in time. Microdialysis captures tonic firing as it collects analyte over the span of several minutes, but it is less effective for measuring phasic firing. Voltammetry has greater temporal resolution, meaning that it can measure neurotransmitter release that occurs at the subsecond level (i.e., it can measure phasic firing). If one is interested in phasic release of a neurotransmitter, voltammetry is optimal.

Although microdialysis and voltammetry are useful for quantifying neurotransmitter activity in a brain region, they do not identify which receptors are important for controlling behaviors. There are a few approaches one can use to address this issue. First, **behavioral pharmacology** experiments entail administering a drug that targets a specific receptor, transporter, or enzyme. These drugs can be injected *systemically* (through the circulatory system such that it affects the entire body), but these injections do not provide information about neuroanatomic specificity. Alternatively, drugs can be infused directly into the brain, known as *microinfusions* or *microinjections*. Microinfusions allow one to determine if receptors in a specific brain region control a behavior of interest. In Chapter 4, you will learn about additional techniques researchers can use to elucidate the neuromechanisms of addiction.

Initiation of drug use

Given the importance of the VTA and the NAc to reward learning, research has determined if these regions directly mediate the **acquisition** of drug self-administration or drug CPP. As expected, lesions to the VTA disrupt the development of ethanol or cocaine CPP,[14,15] while damage to the NAc greatly reduces drug self-administration.[16] The amygdala also plays a role in the development of CPP, as amygdala-lesioned animals show impaired CPP.[14,17] Recall

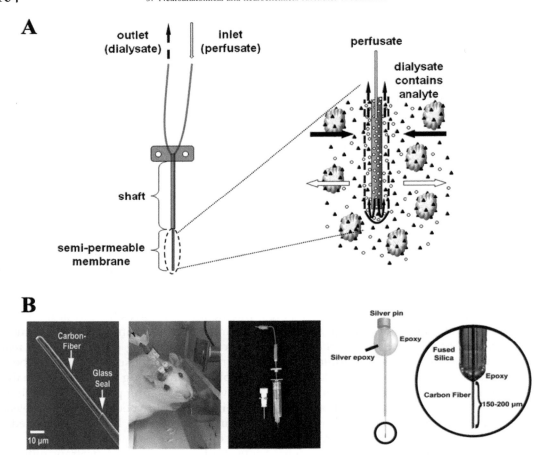

FIGURE 3.4 (A) A schematic of a microdialysis probe. The shaft and the semi-permeable membrane are inserted into a brain region of interest through a guide cannula that has been surgically implanted into the animal's brain and has been secured to the skull with something like dental acrylic. The researcher can infuse a fluid through the inlet (usually something like artificial cerebrospinal fluid). This is known as the perfusate. Flushing the fluid through the inlet will cause neurotransmitters to flow into the semi-permeable membrane. The fluid exits the microdialysis probe through the outlet. The fluid that comes out is now the dialysate, which contains the analyte (essentially, what the researcher is interested in measuring). The dialysate is run through a special machine that detects levels of various neurotransmitters or their metabolites. (B) The left panel shows the carbon-fiber tipped electrode that is implanted into an animal's brain for a voltammetry experiment. The middle panel shows a rat implanted with two guide cannulae and a stimulating electrode located posterior to the guide cannulae. The guide cannulae can be opened to allow for the insertion of a reference electrode and recording electrode. The guide cannula is presented in the right panel, as well as a *micromanipulator*, which helps insert the microelectrode into the guide cannula. The image on the far right shows a schematic of a microelectrode. *Panel (A) Image comes from Wikimedia Commons and was posted by user StS83 (https://commons.wikimedia.org/wiki/File:Schematic_illustration_of_a_microdialysis_probe.png). Panel (B) comes from Fig. 1 of Rodeberg NT, Sandberg SG, Johnson JA, Phillips PEM, Wightman RM. Hitchhiker's guide to voltammetry: acute and chronic electrodes for in vivo fast-scan cyclic voltammetry. ACS Chem Neurosci. 2017;8(2):221−234. https://doi.org/10.1021/acschemneuro.6b00393. Further permissions related to the material excerpted should be directed to the ACS.*

that the BLA is important for linking objects with their current value and that the CeA is associated with pairing stimuli with a reinforcer. While lesioning the BLA impairs drug CPP,[17] inactivating the CeA, but not the BLA, prevents the acquisition of cocaine self-administration.[18] Because the CeA is important for approach toward drug-paired stimuli, damaging this area most likely decreases drug self-administration by diminishing an animal's propensity to approach cues paired with drug delivery (e.g., lever). In contrast to structures like the VTA, the NAc, and the amygdala, OFC-lesioned animals acquire cocaine self-administration at a faster rate compared to control animals.[19] This finding makes intuitive sense considering the OFC is important for decision making and self-control. Individuals with impaired decision making may not consider the long-term ramifications of their drug use.

At the neurochemical level, most drugs increase both tonic firing and phasic firing of intra-VTA dopaminergic neurons. One characteristic that distinguishes drugs from natural rewards is that they cause much higher levels of dopamine to be released in the NAc. In humans, greater dopamine release is associated with increased self-reported feelings of euphoria. This potentially explains why users often feel euphoria when initially taking a drug. The increased dopamine release observed following drug administration does not necessarily mean that dopamine is directly involved in the pleasure of drug taking. Instead, dopamine seems to be important for getting us to repeat pleasurable experiences as we discussed with natural rewards earlier in the chapter. This then promotes the individual to continue using the drug.

Dopamine D_1-like receptors are necessary for the acquisition of drug self-administration. Research has shown that D_1-like receptors in the NAc shell, but not the NAc core, mediate the early stages of cocaine self-administration, but not heroin self-administration.[20] Although intra-NAc shell D_1-like receptors do not appear to be involved in the acquisition of heroin self-administration,[21] systemic administration of a D_1-like receptor antagonist, as well as a D_2-like receptor antagonist, prevents the acquisition of morphine reward as measured with CPP.[22] However, there are some inconsistencies in the literature regarding the role of D_2-like receptors in the acquisition of drug-taking behavior. Whereas a D_2-like receptor antagonist prevents acquisition of morphine CPP, a D_2-like receptor agonist blocks the acquisition of ethanol self-administration.[23] Despite this discrepancy, a plethora of research has shown that dopaminergic activity in the NAc is critical for the initiation of drug-taking behavior.

Maintenance of drug-taking behavior

After some individuals have acquired a drug-taking response, they will often continue using the drug. This is known as **maintenance**. As such, the drug self-administration paradigm is primarily used to model maintenance of drug-taking behavior. Somewhat consistent with research examining the acquisition of drug self-administration, lesions to the NAc core, but not the shell, decrease drug self-administration,[24] whereas OFC and mPFC lesions increase self-administration.[19,25] Earlier in this chapter, you learned that the NAc core is important for establishing motor functions related to obtaining rewards. As a person continues to use a drug, the NAc core is instrumental in encoding information about the physical processes associated with drug use like pouring an alcoholic drink, lighting a cigarette, or preparing a syringe. Because the frontal cortex is important for decision making/self-control, animals with impaired functioning of the OFC/mPFC may not be able to inhibit themselves from taking the drug even when drug use leads to negative consequences. I will return to this concept when discussing the neuromechanisms of compulsive drug seeking.

As with acquisition, maintenance of drug self-administration is largely mediated by dopaminergic activity in the reward pathway. D_1-like and D_2-like receptors in the NAc contribute to self-administration of multiple drugs.[26–28] In addition to the NAc, dopaminergic activity in the amygdala contributes to maintenance of drug self-administration, as blocking D_1-like receptors in this region produces an increase in cocaine self-administration.[29,30] This last result is somewhat inconsistent with what lesion studies have shown regarding the role of the amygdala in maintaining drug self-administration. One possibility for this discrepancy is that lesions to a brain region affect multiple neurotransmitter systems. Although blocking dopamine D_1-like receptors in the amygdala increases drug self-administration, there could be other neurotransmitter systems within this region that may have opposing roles in controlling maintenance of drug use. Indeed, as you will learn, numerous neurotransmitters are released in the amygdala.

Animals that have increased D_2-like receptor sensitivity self-administer more cocaine compared to animals with decreased D_2-like receptor sensitivity.[31] D_2-like receptor sensitivity is determined by injecting rats with the drug quinpirole, which acts as an agonist at D_2-like receptors, before measuring locomotor activity (i.e., the animal's movement). Rats that show greater locomotor activity following quinpirole injections are classified as having increased D_2-like receptor sensitivity. At first glance, this result seems odd because D_2-like receptors are inhibitory and often act as autoreceptors that decrease dopamine transmission. Why would animals that have increased D_2-like receptor sensitivity show addictive-like behaviors? With prolonged drug use, compensatory changes occur in the brain. You will read about many of these changes in the next chapter. One change I want to highlight here is that drugs like cocaine can increase the sensitivity of D_2-like receptors, meaning that less dopamine is needed to activate these receptors.[32] This effect is primarily observed in the NAc shell and the dorsal striatum, suggesting that the transition to drug maintenance requires striatal D_2-like receptors.

Sensitization to the behavioral effects of drugs

With continued drug use, some drugs can cause **sensitization** to occur for certain physiological actions of the drug. Animals will show a greater response to the drug than it originally did. Take psychostimulants as an example. If a rat is injected repeatedly with a drug like cocaine or methamphetamine, they will become more hyperactive compared to when they were first injected. This is known as *locomotor sensitization* (Fig. 3.5). Stimulants are not the only drug class that can cause locomotor sensitization. Repeated injections of ethanol,[33] opioids,[34,35] and dissociative hallucinogens[36,37] produce locomotor sensitization.

Sensitization is a complex behavior that is controlled by several brain regions. Lesions to the NAc shell or the dorsal PFC can prevent or attenuate locomotor sensitization.[38,39] However, lesions to the NAc core or to the amygdala potentiate locomotor sensitization.[40] This result is particularly interesting because lesions to the NAc core and the amygdala tend to decrease the conditioned rewarding and direct reinforcing effects of drugs. These results highlight that locomotor sensitization and drug reward are mediated by different cortical circuits. Unlike the other brain regions described here, there are inconsistencies concerning the

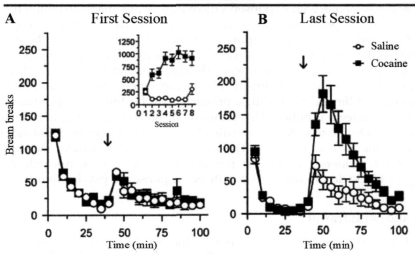

FIGURE 3.5 Locomotor activity (measured in photobeam breaks) following saline (*open circles*) or cocaine (*black squares*) injections during the first session (A) and during the last session (session 8) (B). The arrow at 40 min indicates when each animal was injected. Notice that on session 8, animals show much higher locomotor activity following cocaine injection compared to session 1. The insert in panel (A) shows total locomotor activity across each session. Cocaine-induced activity increases across the first few sessions before reaching a stable level. The increased cocaine-induced hyperactivity across sessions is indicative of locomotor sensitization. *Image modified from Fig. 1 of Ferrario CR, Li X, Wang X, Reimers JM, Uejima JL, Wolf ME. The role of glutamate receptor redistribution in locomotor sensitization to cocaine. Neuropsychopharmacology. 2010;35(3):818–833. https://doi.org/10.1038/npp.2009.190 to make the font less blurry. Copyright 2010.*

role of the mPFC to locomotor sensitization. Some research has shown that lesions to this region disrupt locomotor sensitization[41] while other research has shown the opposite effect of mPFC lesions.[42] More work is needed to uncover the precise role of the mPFC in behavioral sensitization. These inconsistencies could be related to the composition of the mPFC. The mPFC is subdivided into the IL and the PL regions, which form distinct connections with other parts of the brain. Theoretically, if one lesions the mPFC, the lesion could affect the PL more than the IL, whereas another individual could lesion the IL to a greater extent than the PL. These differences could potentially explain the discrepancies reported concerning the role of the mPFC in sensitization.

Dopaminergic activity in the dorsal striatum is important for the development of sensitization. Sensitization can increase the *efficacy* of dopamine for D_1-like receptors in this region. In other words, sensitization increases how well dopamine activates these receptors. Yet, sensitization decreases *potency* of D_1-like receptor agonists; that is, a higher concentration of the same drug is needed to exert its effects at the receptor.[43] Dopaminergic activity in the ventral striatum also mediates behavioral sensitization. Repeated morphine administration increases dopamine release in the NAc core but decreases dopamine release in the NAc shell.[44] Given the connection between the NAc core and the dorsal striatum, these results suggest that repeated stimulation of D_1-like receptors in the NAc core leads to alterations to dopamine receptor functioning in the dorsal striatum as described above.

One important function of the basal ganglia is to inhibit actions that we do not need to do. If you think about someone with Parkinson's disease, they are unable to inhibit the shaking in their hands that makes performing many actions difficult. The inability to stop shaking results from decreased dopaminergic inputs from the substantia nigra to the basal ganglia. As part of the basal ganglia, the dorsal striatum controls two pathways related to movement: the direct pathway and the indirect pathway. The direct pathway is responsible for initiating movements, whereas the indirect pathway inhibits movements that are not of importance to us at that moment in time. Certain drugs, primarily psychostimulants, produce stereotypies like skin picking and teeth grinding. This primarily results because increased dopamine preferentially binds to D_1-like receptors in the direct pathway but does not bind as easily to D_2-like receptors in the indirect pathway. The result is that motor activity increases, which is difficult to inhibit. Evidence to support this assertion comes from research showing that D_1-like receptor blockade can suppress sensitized responses to psychostimulants.[45]

Tolerance and withdrawal

Even though we show sensitization to some of the physiological effects of drugs, we often develop tolerance to their rewarding effects. Repeated drug administration lowers baseline dopamine levels in the NAc[46] and elicits less dopamine release over time.[47] In the next chapter, we will explore the neural adaptations that occur with repeated drug use that can account for drug tolerance. For now, we will focus more on withdrawal, as withdrawal symptoms are observed upon drug cessation in individuals that have developed tolerance to the reinforcing effects of the drug. During drug withdrawal, further reductions in dopamine levels are observed in the NAc.[48] Dopaminergic activity is attenuated in other regions as well, such as the substantia nigra and the VTA.[49] Withdrawal can also alter dopaminergic control of PFC excitation. For example, in saline-treated rats, blocking D_2-like receptors increases neuronal excitation in the PFC; however, in rats going through cocaine withdrawal, inhibition of D_2-like receptors fails to alter neuronal activity in the PFC.[50] VTA stimulation decreases PFC activity in rats experiencing cocaine withdrawal but increases PFC activity in saline-treated animals.[50]

The hypoactive dopaminergic state observed during withdrawal can help explain the dysphoria individuals experience upon drug cessation. The negative mood state individuals in withdrawal experience is a major motivating factor for continued drug use. If withdrawal symptoms are mediated by reduced dopaminergic activity, one intuitive treatment to counteract drug withdrawal is the administration of dopamine D_1-like receptor agonists or D_2-like receptor antagonists. Indeed, D_1-like receptor agonists and D_2-like receptor antagonists are capable of ameliorating morphine withdrawal symptoms.[51,52] Paradoxically, research has shown that dopamine D_1-receptor antagonists or D_2-like agonists can *diminish* withdrawal symptoms to drugs like morphine.[52,53] These results seem to suggest that other neurotransmitter systems may interact with the dopaminergic system to influence drug withdrawal symptoms.

Craving and drug seeking

When individuals are exposed to images of drug cues, such as an ashtray, several brain regions are activated. Brain structures associated with vision (like the occipital cortex), the cingulate gyrus, the mPFC, and the dorsal striatum are commonly activated. In individuals that have been deprived of cigarettes, areas like the occipital cortex and the mPFC are more activated compared to individuals that have recently had cigarettes.[13] Neurochemically, dopamine release is elevated in the ventral striatum, the dorsal striatum, the PFC, the amygdala, and the hippocampus when individuals are presented with drug-paired stimuli.[54,55] Recall that the PFC encodes information about the motivational aspects of a reward. As someone continues to use the drug, the motivational value of the drug increases at the expense of other rewards. This can lead to increased craving for the drug when the individual is not currently using it.

As mentioned earlier, because the PFC no longer functions properly, individuals are more likely to engage in compulsive drug seeking; that is, individuals will continue using the drug despite its harmful consequences. In the case of someone with methamphetamine use disorder, they will continue to use methamphetamine even after losing their teeth or damaging their skin due to the increased stereotypies associated with skin picking. This issue is exacerbated by the finding that drug use becomes a habitual action regulated by the dorsal striatum. For some individuals, drug use becomes part of one's routine. Individuals may smoke a cigarette at the same times each day (e.g., right after waking up, after a meal, etc.). Individuals with alcohol use disorder become accustomed to drinking a beverage with each meal or each time they are with certain peers.

In animals, inactivating the NAc core or the VTA decreases drug seeking, but only when animals are in a drug-free state.[56,57] Conversely, inactivating or lesioning the NAc core can *increase* cocaine drug-seeking after animals have been recently exposed to cocaine.[56,58] As suggested by Di Ciano and Everitt,[56] this discrepancy may occur because the increased dopamine levels in the NAc following drug self-administration may offset the effects of temporary inactivation. Selective inactivation of the dorsal striatum or lesions to the OFC suppress drug seeking,[59,60] whereas lesions to the BLA increase cocaine seeking.[61] The BLA and the CeA differentially initiate and maintain striatal-mediated drug seeking, as the BLA is responsible for the recruitment of striatal control of cocaine seeking whereas the CeA is more important for maintaining long-term cocaine seeking.[62]

Dopaminergic activity in several brain regions influences drug seeking. Cues that become paired with drug self-administration elicit dopamine release in the NAc core, but not in the NAc shell.[47] Animals that engage in drug-seeking behavior have increased dopamine release in the dorsal striatum.[63] Conversely, blocking dopamine receptors in the BLA reduces cocaine-seeking behavior.[64] These results highlight that increased dopaminergic activity in regions such as the NAc core, the dorsal striatum, and the BLA influence drug-seeking behavior.

Relapse

Dopaminergic activity plays an important role in relapse. Using a CPP task, Calipari et al. showed that NAc D_1-containing MSNs become more activated during a reinstatement test, whereas activity of D_2-containing MSNs does not change.[7] This finding suggests that activation of D_1-like receptors in the reward pathway is necessary for relapse to occur. Evidence to support this claim comes from the finding that systemic administration of dopamine D_1-like receptor agonists induces reinstatement,[65] whereas dopamine D_1-like receptor antagonists prevent relapse-like behavior.[66–69] Although D_1-like receptors are required for relapse to occur, D_2-like receptors mediate relapse as well.[67,68]

Lesion/inactivation studies have identified the involvement of many brain regions in relapse-like behavior, including the VTA, the NAc (both core and shell regions), the OFC, the PL region of mPFC, the CeA, the hippocampus, the lateral habenula, the posterior pareventricular nucleus of the thalamus, the substantia nigra, and the basal ganglia.[61,70–74] Microinjection studies have shown that dopamine receptors located in some of these brain regions mediate relapse-like behavior. Stimulating intra-NAc D_1-like receptors increases drug-seeking behavior,[75] whereas blocking D_1-like receptors in the VTA, the NAc, the caudate putamen, the mPFC, the OFC, the hippocampus, or the amygdala decreases reinstatement to different drugs.[76–80] Additionally, stimulating D_2-like receptors in the NAc or the dorsal striatum can increase relapse-like behavior,[75] while blocking these receptors can prevent reinstatement.[76]

One interesting observation is that stimulating both D_1-like and D_2-like receptors increases relapse-like behavior while blocking these receptors inhibits relapse-like behavior. At first glance, these results seem paradoxical. How so? Stimulating D_1-like receptors increases dopamine levels. The fact that stimulating these receptors potentiates relapse-like behavior makes intuitive sense. However, stimulating D_2-like receptors should decrease dopamine levels as these receptors are inhibitory and largely serve as autoreceptors. Stimulating D_2-like receptors should have a similar effect as blocking D_1-like receptors. Another paradoxical finding is that stimulating D_1-like receptors can *increase* relapse-like behavior,[65] whereas stimulating D_2-like receptors can *decrease* relapse-like behavior.[81] An important consideration is that there are multiple ways to measure reinstatement in animals. The way in which reinstatement is tested can influence the effects of D_1-like and D_2-like receptors on relapse-like behavior. We will come back to this issue in Chapter 7.

Dopamine depletion hypothesis of addiction

During initial drug use, dopamine levels increase substantially; however, drugs become less effective at increasing dopamine levels with prolonged drug use. Over time, dopamine levels in the mesocorticolimbic pathway decrease. Dr. Charles Dackis and Dr. Mark Gold originally developed the **dopamine depletion hypothesis of addiction** to explain cocaine addiction.[82] This hypothesis argues that depleted dopamine levels in the reward pathway lead to dysphoria, which then motivate the individual to use drugs to replenish the depleted dopamine. Because drug use is the cause of dopamine reduction, individuals must continue using the drug to try to maintain a somewhat normal level of dopamine. This leads to drug seeking and ultimately addiction. This hypothesis can also be used to explain why individuals with SUDs do not derive much pleasure from other rewards.

The dopamine depletion hypothesis has high face validity. However, there are some challenges to this hypothesis. For one, as noted by Nutt et al., the dopamine depletion hypothesis can be used to explain psychostimulant use disorders because these drugs act directly on the dopaminergic system, but it does not fully account for other SUDs.[83] Reduced dopamine release is not commonly observed in cannabis use disorder. Nutt et al. argue that dopamine is just one component of addiction, particularly psychostimulant addiction.[83] As you will soon learn, numerous neurotransmitter systems are involved in the direct reinforcing and the conditioned rewarding effects of drugs, and they play important roles in drug-seeking and relapse-like behaviors.

Quiz 3.2

1. Out of the following, which brain structure is most important for the initiation of drug use?
   a. Amygdala
   b. Dorsal striatum
   c. NAc
   d. Orbital frontal cortex
2. With continued drug use, individuals often develop _____ to the reinforcing effects of stimulants and develop _____ to the locomotor-stimulant effects of a drug.
   a. sensitization/sensitization
   b. sensitization/tolerance
   c. tolerance/sensitization
   d. tolerance/tolerance
3. Which of the following statements about the contribution of dopamine to addiction is true?
   a. Only dopamine D_1-like receptors are involved in relapse-like behavior.
   b. Prolonged drug use leads to a hyperdopaminergic state in the brain.
   c. Stimulating dopamine D_1-like receptors decreases drug self-administration in animals.
   d. The transition from occasional drug use to habitual drug use is largely mediated by the dorsal striatum.
4. What result is most likely to occur following damage to the basolateral nucleus of the amygdala?
   a. Decreased drug relapse-like behavior
   b. Decreased drug self-administration
   c. Increased drug CPP
   d. Increased drug-seeking behavior
5. Why is the finding that both dopamine D_1-like and D_2-like antagonists decrease relapse-like behavior paradoxical?

Answers to quiz 3.2

1. c — NAc
2. c — tolerance/sensitization
3. d — The transition from occasional drug use to habitual drug use is largely mediated by the dorsal striatum.
4. d — Increased drug-seeking behavior
5. Because D_1-like receptors are excitatory and D_2-like receptors are inhibitory, the finding that blocking both receptor subtypes can decrease relapse-like behaviors suggests that increasing *and* decreasing dopaminergic activity can be effective in relapse prevention.

Contribution of other neurotransmitter systems to the addiction process

Dopamine may be the neurotransmitter most linked to addiction, but it certainly is not the only one that contributes to the development of SUDs. Because neurotransmitter systems often work in tandem with one another and can influence the release of other neurotransmitters, it should not be surprising to learn that multiple neurotransmitter systems play critical roles in the addiction process.

GABA

In the last chapter, you learned that some drugs increase dopamine levels indirectly by inhibiting GABAergic neurons that inhibit dopaminergic neurons, and some drugs exert their effects by binding to GABAergic neurons (benzodiazepines and barbiturates) or by potentiating the inhibitory effects of GABA (ethanol). The $GABA_A$ and $GABA_B$ receptor agonists muscimol and baclofen are often used to inactivate brain regions to test their contribution to addiction-like behaviors. With that said, if dopamine release in the mesocorticolimbic pathway is necessary for addiction-like behaviors to develop, one logical step for treating SUDs would be to administer GABA receptor agonists or GABA transaminase inhibitors. GABA transaminase breaks GABA back down to glutamate. These pharmacological agents should be able to decrease dopamine release following drug administration, thus blunting the rewarding and/or reinforcing effects of the drug. This is exactly what happens. Stimulating $GABA_B$ receptors attenuates the rewarding effects of drugs[84–86] and reduces drug-seeking behavior.[87] Furthermore, increasing GABA concentration in the VTA prevents the acquisition of heroin self-administration and decreases the maintenance of heroin self-administration.[88] Increasing hippocampal or amygdalar GABA levels decreases morphine CPP.[89,90] Conversely, blocking GABAergic receptors increases the rewarding properties of drugs. Specifically, $GABA_A$ receptor antagonists potentiate the conditioned rewarding effects of ethanol and morphine.[90,91]

GABAergic activity is also important for relapse-like behavior, as stimulating GABA receptors diminishes reinstatement.[85,86,92] One somewhat paradoxical finding has been reported concerning the role of GABA receptors in relapse. Blocking $GABA_B$, but not $GABA_A$, receptors in the VTA prevents reinstatement of cocaine seeking.[93] As proposed by Blacktop et al., this seemingly contradictory effect may result from inhibition of GABAergic neurons in the VTA.[93] Normally, stimulating $GABA_B$ receptors inhibits VTA neurons that send inhibitory messages to the NAc. By blocking these receptors, VTA neurons can send inhibitory signals to the NAc, thus limiting the release of dopamine.

Because $GABA_A$ receptors are ionotropic with five subunits, research has determined which subunits are required for addiction-like phenotypes to emerge. The α_1 and α_5 subunits mediate ethanol self-administration,[94,95] and the α_5 subunit is involved in the reinstatement of ethanol-seeking.[94] Hippocampal α_5 subunits influence an animal's self-administration of ethanol.[96] These studies have shown that impairing α_1- or α_5-containing $GABA_A$ receptors *decreases* ethanol self-administration, perhaps by a similar mechanism as $GABA_B$ receptor antagonists described above. More research is needed to determine if the α_1 and α_5 subunits mediate ethanol-related behaviors only or generalize to other drug classes. While research has shown a role for the α_1 and the α_5 $GABA_A$ subunit to ethanol use, the α_2 subunit does not appear to mediate drug self-administration or relapse-like behavior.[97]

Glutamate

Not only do drugs increase dopamine levels, but they increase glutamate levels by various mechanisms. Some drugs interact directly with NMDA receptors, such as ethanol, ketamine, and PCP. Because the glutamatergic system is composed of several ionotropic receptors and several families of metabotropic receptors, numerous studies have attempted to elucidate the specific contribution of glutamate to the addiction process.

Blocking ionotropic glutamate receptors attenuates dopamine release following drug administration.[98] Directly related to this finding, systemic administration of AMPA/NMDA receptor antagonists decreases the conditioned rewarding effects of various drugs[99] and attenuates the direct reinforcing effects of drugs.[100] Considering that the NMDA receptor antagonists ketamine and PCP have high misuse potential, targeting specific subunits of the NMDA receptor may provide avenues for treating SUDs. To this end, GluN2B-selective antagonists can block the conditioned rewarding effects of drugs.[101,102]

Like ionotropic glutamate receptors, metabotropic glutamate receptors (mGluRs) are known to influence the reinforcing/conditioned rewarding effects of drugs. Blocking excitatory Group I metabotropic receptors attenuates the reinforcing and the conditioned rewarding effects of cocaine,[103,104] while stimulating inhibitory Group II metabotropic glutamate receptors reduces cocaine self-administration.[105] Group I mGluRs are not just important for mediating the reinforcing/rewarding effects of drugs; they are also involved in the locomotor-stimulant effects of drugs[106] and behavioral sensitization.[107]

Glutamatergic activity in the VTA has been directly implicated in the development of drug CPP.[108] Additionally, NMDA receptors in the NAc and AMPA receptors in the NAc core, but not in the NAc shell, are important for drug self-administration.[109,110] Glutamatergic activity in the NAc core contributes to locomotor sensitization. In one study, rats received daily injections of cocaine for 1 week. Animals that developed locomotor sensitization had greater glutamate release in the NAc core compared to rats that did not develop locomotor sensitization and to rats that never received cocaine.[111] AMPA receptors, but not NMDA receptors, in the NAc core control drug-seeking behavior.[112]

AMPA/NMDA receptor antagonism inhibits drug-induced reinstatement.[99,113,114] AMPA/NMDA receptors in the NAc play a critical role in relapse.[115,116] NMDA receptors in the amygdala and AMPA receptors in the dorsal striatum also contribute to drug relapse.[117,118] To complicate matters even more, Group I and Group II mGluRs contribute to relapse-like behavior as well. Systemic administration of mGluR$_5$ antagonists attenuates reinstatement to cocaine.[119] Additionally, mGluR$_{2/3}$ agonists prevent reinstatement of cocaine-seeking[120] and attenuate the incubation of cocaine craving,[121] an effect that is also observed following microinjection of an mGluR$_{2/3}$ agonist into the amygdala. Hippocampal mGluR$_1$ and dorsal striatal mGluR$_5$ are known to mediate relapse-like behavior, as blocking these receptors reduces reinstatement.[122,123]

I know this is a lot of information to keep straight. Collectively, these results show that increased glutamatergic transmission is required for several behaviors related to addiction as antagonizing excitatory glutamatergic receptors, whether NMDA/AMPA or Group I mGluRs, decreases drug self-administration and relapse-like behavior. You are going to learn much more about glutamate's influence on addiction throughout this book, especially when we discuss the cellular mechanisms of learning that promote the development of addictive-like behaviors.

Serotonin

As dopaminergic neurons follow certain pathways, the other monoamine neurotransmitters have their own pathways. While dopaminergic cell bodies largely originate from the VTA or the substantia nigra, serotonergic cell bodies originate from a region of the brain known as the raphe nuclei. These nuclei are located in the brain stem; axons extending from these cell bodies travel to numerous areas of the brain, including the frontal cortex, the VTA, the striatum, the amygdala, and the hippocampus, just to name a few. Because serotonin is released in multiple regions associated with addiction, one can logically argue that this neurotransmitter underlies addiction-like behaviors. Indeed, increasing serotonin levels attenuates the acquisition of heroin self-administration[124] and maintenance of heroin self-administration.[125] Serotonin transporter (SERT) inhibitors, which increase serotonin levels, decrease reinstatement.[126] Whereas elevated forebrain serotonin levels decrease drug seeking, reduced serotonin levels in the frontal cortex increase drug seeking.[127] Depleting serotonin levels also increases relapse-like behavior.[128]

The serotonergic system is highly complex, with multiple receptor subtypes. The 5-HT_1 and the 5-HT_2 receptors have received the most attention. Although 5-HT_1 receptor antagonists decrease the reinforcing effects of drugs and prevent relapse-like behavior,[129] they fail to decrease MDMA drug self-administration.[130] These drugs may be ineffective for MDMA self-administration because this drug increases serotonin levels in the brain; thus, increasing serotonin levels via 5-HT_1 receptor blockade may not be effective in altering serotonin levels when someone is using MDMA. One interesting finding is that stimulating 5-HT_{1B} receptors enhances the reinforcing effects of cocaine but decreases psychostimulant reinstatement.[131] How can this be the case? 5-HT_{1B} receptors are inhibitory and can be found on GABAergic neurons that project from the VTA to the NAc. By stimulating 5-HT_{1B} receptors, this can inhibit GABAergic neurons, which then causes dopamine levels to increase in the NAc.[132] This could explain why 5-HT_{1B} receptor agonists increase the reinforcing effects of a drug. More work is needed to understand why 5-HT_{1B} receptor agonists decrease reinstatement. Perhaps animals given a 5-HT_{1B} receptor agonist are more sensitive to the drug itself than the drug-paired cues. Because the animal does not receive the drug during a reinstatement test, it may "give up" at a faster rate as it learns that it will not receive drug at that point.

Serotonin 5-HT_{2C} receptor agonists consistently decrease addiction-like behaviors,[126,133–135] while 5-HT_{2C} antagonists potentiate these behaviors.[136] 5-HT_{2C} receptors in the mPFC and the CeA play a prominent role in relapse, as directly stimulating these receptors decreases cocaine reinstatement.[137,138] The efficacy of 5-HT_{2C} receptor agonists may stem from their ability to decrease drug-induced changes in dopamine release in the NAc.[133] While 5-HT_{2C} receptor agonists decrease addictive-like behaviors, 5-HT_{2A} receptor *antagonists* are effective at decreasing relapse-like behavior,[130,134,136] but not to MDMA-paired cues.[130] 5-HT_{2A} receptor blockade in the mPFC appears to be a major locus for the prevention of relapse.[139] The ability of 5-HT_{2A} receptor antagonists to attenuate reinstatement may result from their ability to decrease dopamine release in the dorsal striatum as opposed to the NAc.[140]

Like the glutamatergic system, there is a lot to unpack here given the complexity of the serotonergic system. Overall, increasing serotonin levels decreases the misuse potential of drugs while decreasing serotonin levels can potentiate their misuse potential. The one notable exception is that decreasing serotonin levels via the inhibitory 5-HT_{1B} receptor decreases relapse-like behavior.

Norepinephrine

In the section above, I mentioned that increasing serotonin levels attenuates the acquisition of heroin self-administration. In the study cited above, the researchers used a mixed SERT/NET inhibitor[124]; thus, the decreased heroin self-administration could be mediated more by norepinephrine than serotonin. Like the serotonergic system, noradrenergic neurons project to several brain regions implicated in addiction, such as the amygdala, the cingulate cortex, the hippocampus, the frontal cortex, and the striatum. Although SERT/NET reuptake inhibitors do not affect heroin CPP,[141] the NET inhibitor atomoxetine decreases relapse-like behavior to heroin.[142]

Recall from the previous chapter that noradrenergic α_1 receptors are excitatory as they are coupled to G_q metabotropic receptors. These receptors are known to regulate dopamine transmission in the NAc shell.[143] Because α_1 receptors help increase dopamine release, blocking these receptors decreases the rewarding properties of drugs.[144–146] Conversely, α_1 receptor agonists enhance addiction-like behaviors.[147] In contrast to α_1 receptors, adrenergic α_2 receptors are inhibitory G_i-coupled metabotropic receptors. Research examining the contribution of the adrenergic system to addiction has largely focused on α_2 receptors. Adrenergic α_2 agonists have shown promise in decreasing the reinforcing and the conditioned reward effects of drugs[148,149] and attenuating relapse-like behavior.[150] In addition to reducing the reinforcing effects of drugs, α_2 receptor agonists can ameliorate withdrawal symptoms in animals and in humans. In rats, α_2 receptor agonists decrease withdrawal symptoms associated with morphine cessation.[151] In humans, the α_2 receptor agonist guanfacine decreases cannabis withdrawal symptoms[152] and helps reduce relapse,[153] suggesting that targeting the α_2 receptor can help individuals transition away from drug use and to help keep them abstinent for longer periods of time. Activation of α_2 receptors in the CeA seems largely responsible for the beneficial effects of α_2 receptor agonists on relapse-like behaviors.[154] Although α_2 receptor agonists decrease the reinforcing effects of drugs, caution is needed if using them as a therapeutic for individuals with SUDs as chronic administration of an α_2 receptor agonist can increase the reinforcing effects of cocaine.[155]

While α_2 receptor agonists decrease addictive-like behaviors, α_2 receptor antagonists increase drug self-administration.[156] Experimentally, the α_2 receptor antagonist yohimbine is often used to precipitate relapse-like behavior in animals. Because the noradrenergic system is involved in stress, I will discuss the use of yohimbine in animal models of relapse in Chapter 11.

Endogenous opioids

To briefly recap the endogenous opioid system, there are four major receptor subtypes (mu, kappa, delta, and nociceptin opioid peptide) that are G_i-coupled. Each receptor subtype has distinct functions beyond analgesia. For example, mu opioid receptors are associated with the euphoria individuals experience after taking an opioid,[157] while kappa receptors are involved in dysphoria and the aversive effects of opioids.[158] Stimulating kappa receptors can cause hallucinations/dissociation.[159] Delta receptors appear to regulate mood, as delta receptor agonists have antidepressant-like effects.[160]

Given that the mu opioid receptor antagonist naltrexone is used to treat SUDs beyond opioids, the endogenous opioid system has received considerable attention in addiction research. Not only do drugs increase dopamine levels, but stimulants and ethanol increase endogenous opioid levels.[161,162] You learned in Chapter 1 that cigarettes and ethanol are often used concurrently. Preclinical evidence has shown that nicotine administration increases ethanol self-administration, a phenomenon that is mediated by mu opioid receptors but not kappa opioid receptors.[163] Consistent with clinical studies examining the efficacy of naltrexone for the treatment of SUDs, preclinical studies have shown that mu opioid receptor antagonists attenuate drug-induced changes in dopamine release[164] and decrease drug self-administration of various drug classes.[164–166] Mu opioid receptors in the VTA and the NAc play an important role in drug reinforcement, as directly stimulating or blocking these receptors increases and decreases drug self-administration, respectively.[167,168]

Mu opioid receptor antagonists are not just beneficial for attenuating the reinforcing effects of drugs, but they can be used to decrease stereotypic behaviors. In mice, methamphetamine often produces a stereotypy that is marked by increased biting (analogous to an individual that picks at his or her skin excessively). This stereotypy decreases following administration of a mu opioid receptor antagonist.[169] Mu opioid receptors also appear to be involved in the locomotor-stimulant effects of ethanol, as blocking these receptors decreases ethanol-induced increases in locomotor activity.[170]

Concerning drug-seeking behavior and relapse, a selective mu receptor antagonist reduces ethanol-seeking behavior and ethanol intake,[171] and it attenuates cocaine and heroin drug seeking.[172] Not surprisingly, mu opioid receptor antagonists attenuate opioid reinstatement.[173] Mu opioid receptors are also important mediators of reinstatement of ethanol- and cocaine-seeking behavior,[174,175] particularly mu opioid receptors located in the NAc,[176] primarily the shell subregion.[177]

Opioids increase dopamine levels by binding to inhibitory mu opioid receptors on GABAergic neurons, thus disinhibiting dopaminergic neurons in the reward pathway. However, one finding that seems counterintuitive at first is that mu opioid receptor agonists can decrease drug self-administration in animals[178] and in humans.[179] If an agonist for these receptors is directly infused into the VTA, rats self-administer less cocaine.[167] How can mu opioid receptor agonists, which increase dopamine levels, reduce drug self-administration? Because mu opioid receptor agonists increase dopamine levels in the reward pathway, there may not be much need for the animal to work to receive infusions of a drug such as cocaine. There is only so much dopamine that can be released during a self-administration session. In the last chapter, you learned about agonist treatments (e.g., methadone). In a sense, treating animals with a mu opioid receptor agonist acts as a form of agonist treatment that can be extended to other drug classes. While a mu opioid receptor agonist decreases cocaine self-administration, a low dose of the same mu opioid receptor agonist *potentiates* self-administration of a low dose of cocaine.[167]

Mu opioid receptors are not the only ones that control drug use. In addition to being a mu opioid receptor antagonist, naltrexone is a kappa opioid receptor partial agonist. As such, kappa opioid receptors have also been implicated in the reinforcing effects of drugs as stimulating these receptors decreases heroin-induced increases in intra-NAc dopamine release[180] and decreases oxycodone[181] and cocaine self-administration.[182] Whereas stimulating kappa receptors blocks the reinforcing effects of opioids, antagonizing delta opioid receptors attenuates opioid reward[183,184] and blunts opioid withdrawal symptoms.[185] Delta opioid receptor antagonists are also efficacious in decreasing CPP to ethanol and psychostimulants.[186,187] The effects of delta opioid receptor antagonists are dependent on which subpopulation of delta opioid receptors is blocked. If a delta opioid receptor antagonist is microinfused into the NAc, rats self-administer less cocaine; however, if this antagonist is infused directly into the VTA, rats self-administer more cocaine.[188] The increased cocaine self-administration observed following intra-VTA blockade of delta opioid receptors may result from disinhibition of the mPFC.[188]

Endocannabinoids

There are two major endocannabinoids that you learned about previously: anandamide and 2-arachidonoylglycerol (2-AG). Endocannabinoids and exogenous cannabinoids like THC bind to one of two cannabinoid receptors: CB_1 and CB_2. Both receptors are metabotropic and inhibitory. Not only do cannabinoids increase dopamine levels in the reward pathway,[189] but drugs can differentially alter the levels of anandamide and 2-AG in the brain. For example, Caillé et al. used microdialysis to detect anandamide and 2-AG levels in the NAc shell following self-administration of ethanol, heroin, and cocaine. Whereas ethanol increased 2-AG levels without altering anandamide levels, heroin increased anandamide levels while decreasing 2-AG levels.[190] Cocaine failed to alter levels of either anandamide or 2-AG. Like heroin, other opioids such as morphine and fentanyl increase anandamide levels while decreasing 2-AG levels in the NAc shell.[191,192]

The endocannabinoid system influences the reinforcing effects of drugs. Recall that the CB_1 antagonist rimonabant was used experimentally to treat cannabis and nicotine use disorders, although this drug was discontinued. At the preclinical level, research has consistently shown that decreasing endocannabinoid levels or blocking CB_1 receptors attenuates drug reinforcement and reinstatement.[193–197] Conversely, increasing endocannabinoid levels potentiates the reinforcing effects of drugs.[198] One interesting finding is that increasing endocannabinoid levels can reduce multiple symptoms of opioid withdrawal. This finding aligns with other evidence showing that cannabinoids are effective at attenuating pain. Thus, cannabinoids may be useful for getting someone through opioid withdrawal.

In the reward pathway, direct injections of a CB_1 receptor antagonist into either the NAc or the VTA decreases ethanol self-administration.[199,200] Direct blockade of intra-VTA CB_1 receptors also reduces nicotine self-administration,[201] and CB_1 receptor antagonism in the NAc decreases heroin self-administration.[202] Taken together, the behavioral pharmacology studies highlighted here implicate the endocannabinoid system as an important mediator of addiction. Overall, decreasing endocannabinoid levels decreases the misuse potential of drugs.

Acetylcholine

Cell bodies of cholinergic neurons originate from two areas of the brain: the nuclei of the basal forebrain and the brainstem. The nuclei of the basal forebrain consist of several structures, including the NAc. Collectively, axons of cholinergic neurons originating from the nuclei of the basal forebrain travel to the cerebral cortex and to the amygdala. Axons originating from the brainstem, specifically the pedunculopontine tegmental nucleus and the laterodorsal pontine tegmentum, extend primarily to the thalamus and the nuclei of the basal forebrain, as well as the VTA. You have already learned that nicotine, which binds to nicotinic receptors, increases dopamine levels. Mechanistically, this occurs because nicotine stimulates neurons that then stimulate dopaminergic neurons in the VTA.

If increased acetylcholine release is associated with reinforcement, one potential therapeutic for SUDs should attempt to decrease acetylcholine release. To this end, the nicotinic receptor antagonist bupropion is approved by the FDA as a smoking cessation medication. Experimentally, administration of nicotinic receptor antagonists impairs acquisition of nicotine self-administration and decreases nicotine self-administration during maintenance.[203] Nicotinic receptors have been implicated in the rewarding and the reinforcing effects of other drugs, as nicotinic receptor antagonists block drug self-administration.[100,204] Evidence has suggested that nicotinic receptor activation in the VTA, the NAc, the hippocampus, the amygdala, and/or the lateral habenula is required for drug reinforcement.[205-207]

Muscarinic receptors also contribute to addiction-like behaviors. Recall that $M_1/M_3/M_5$ receptors are excitatory, and M_2 and M_4 receptors are inhibitory. Additional evidence for the role of muscarinic receptors in addiction vulnerability comes from behavioral pharmacology experiments. M_5 receptor antagonists decrease drug self-administration[208] while stimulating $M_{2/4}$ receptors decreases drug self-administration.[209] One interesting observation is that microinfusion of the drug atropine, which non-selectively blocks each muscarinic receptor subtype, into the NAc core decreases acquisition of drug self-administration,[205] whereas an M_1 receptor agonist infused into the NAc shell decreases drug self-administration.[210] These results imply that muscarinic receptors, particularly the M_1 subtype, in the core and shell regions of the NAc may differentially mediate the reinforcing effects of drugs.

Both nicotinic and muscarinic receptors contribute to relapse-like behavior, as antagonizing nicotinic receptors or stimulating M_1 or M_4 receptors can impair reinstatement.[211,212] Both nicotinic and muscarinic receptors in the VTA contribute to reinstatement,[206] particularly early during withdrawal.[213] Muscarinic receptors in the NAc and the amygdala also mediate relapse-like behavior.[214,215]

Quiz 3.3

1. Which of the following results from administration of a Group I mGluR antagonist?
   a. Decreased locomotor sensitization
   b. Increased CPP
   c. Increased drug self-administration
   d. Increased reinstatement
2. Which of the following drugs would be best for blocking relapse-like behavior?
   a. Atropine (muscarinic M_1 and M_2 receptor antagonist)
   b. Bicuculline ($GABA_A$ receptor antagonist)
   c. Trazadone (serotonin $5-HT_{2A}$ receptor antagonist)
   d. Yohimbine (noradrenergic α_2 antagonist)

3. Increasing _____ levels typically decreases addiction-like behaviors.
   a. acetylcholine
   b. endocannabinoid
   c. norepinephrine
   d. serotonin
4. Where do noradrenergic cell bodies originate in the brain?
   a. Interpeduncular nucleus
   b. Locus coeruleus
   c. Raphe nuclei
   d. VTA
5. How are noradrenergic α_1 receptor antagonists capable of reducing drug use?

Answers to quiz 3.3

1. a — Decreased locomotor sensitization
2. b — Bicuculline (GABA$_A$ receptor antagonist)
3. d — serotonin
4. b — Locus coeruleus
5. Stimulation of α_1 receptors increases dopamine levels. By antagonizing these receptors, drugs cannot effectively increase dopamine levels, blunting their reinforcing effects.

Chapter summary

Even though dopaminergic activity in the mesocorticolimbic pathway controls distinct stages of addiction, addiction is mediated by many neurotransmitter systems and by other brain regions outside of the reward pathway. Overall, evidence suggests that decreasing dopamine, glutamate, norepinephrine, opioid, endocannabinoid, and acetylcholine levels decreases the reinforcing effects of drugs, while increasing GABA and serotonin levels decreases the reinforcing effects of drugs. Considering multiple neurotransmitters govern addiction, I hope you now understand why treating SUDs is incredibly difficult. At face value, giving an individual a dopamine D$_1$-like receptor antagonist seems like a good idea for preventing relapse. However, these drugs fail to prevent glutamatergic activity in areas like the amygdala or the hippocampus from precipitating drug-seeking behavior. In fact, one study found that the dopamine receptor antagonist flupenthixol increases relapse rates to ethanol in individuals.[216] Understanding some of the neuromechanisms of addiction has hopefully shed some light as to why there is no pharmacotherapy that is 100% effective in treating SUDs. However, there is still more to learn concerning the neuromechanisms of addiction. This chapter primarily focused on the contribution of neurotransmitter systems in several brain regions on addiction-like behaviors. One caveat to behavioral pharmacology studies is that many drugs have actions on other receptors. This can make isolating the specific contributions of a neurotransmitter system to addiction difficult. In the next chapter, you will learn more about the molecular and the cellular mechanisms controlling addiction. As such, you will learn about additional techniques that researchers can use to elucidate the neuromechanisms of addiction.

Glossary

Acquisition — the initial learning of emitting a response to receive a reward (as in self-administration) or the initial learning of the association between two stimuli (as in CPP).

Behavioral pharmacology — a branch of science that aims to study drug effects on behavior.

Conditioned place preference (CPP) — a research paradigm in which an animal learns to associate one environmental context with a drug and another environmental context with the absence of drug. Animals will eventually approach the drug-paired context even when in a drug-free state.

Dopamine depletion hypothesis of addiction — a hypothesis that posits that addiction-like behaviors such as drug seeking emerge because the dopaminergic system is in a hypoactive state (i.e., not enough dopamine in the brain).

Drug self-administration — a research paradigm in which animals and humans are allowed to respond on some manipulandum (e.g., key or lever) to receive access to a drug.

Extinction — the process by which an animal learns that responding on a manipulandum (e.g., a lever) or approaching an environmental context previously paired with drug no longer gives them access to the drug. This causes responding or approach behavior to decrease.

Habit formation — the establishment of routines that we follow on a daily or near daily basis.

Incentive salience — form of attention that motivates us to approach some object or perceived outcome.

Incubation of craving — an animal model of addiction in which an animal is exposed to drug-paired cues after a certain period of forced abstinence; responses on a manipulandum previously associated with drug increase in the presence of drug-paired cues even though no drug is delivered. These responses increase as the withdrawal period increases.

Maintenance — the continuation of drug self-administration after animals have acquired the behavior.

Medium spiny neuron (MSN) — a type of GABAergic neuron located in the NAc.

Microdialysis — a neuroscience technique in which a semipermeable probe is inserted into an animal's brain to collect analyte. This technique allows for the quantification of neurotransmitter release that occurs over several minutes.

Phasic firing — a pattern of neuronal action potentials characterized by groups of neurons firing in rapid succession.

Reinstatement — an animal model of relapse in which an animal, following a period of extinction, is either exposed to a stimulus previously paired with a drug, given a priming injection of the previously self-administered drug, or is subjected to a stressor.

Sensitization — the process by which an organism shows an increased physiological response to a drug.

Stimulus — a thing or event that evokes a reaction from any part of an organism. Things such as lights, sounds, odors, and colors are considered stimuli.

Tonic firing — a pattern of action potentials characterized by slow, asynchronous firing.

Voltammetry — a neuroscience technique in which a probe inserted into the brain of an animal detects oxidation (removal of electron) or reduction (addition of electron) after an electrical stimulus has been applied. This technique allows for the quantification of neurotransmitter release at the subsecond level.

Supplementary data related to this chapter can be found online at doi:10.1016/B978-0-323-90578-7.00001-3

References

1. Alcaro A, Huber R, Panksepp J. Behavioral functions of the mesolimbic dopaminergic system: an affective neuroethological perspective. *Brain Res Rev*. 2007;56(2):283–321. https://doi.org/10.1016/j.brainresrev.2007.07.014.
2. Dreyer JK, Herrik KF, Berg RW, Hounsgaard JD. Influence of phasic and tonic dopamine release on receptor activation. *J Neurosci*. 2010;30(42):14273–14283. https://doi.org/10.1523/JNEUROSCI.1894-10.2010.
3. Heien ML, Wightman RM. Phasic dopamine signaling during behavior, reward, and disease states. *CNS Neurol Disord - Drug Targets*. 2006;5(1):99–108. https://doi.org/10.2174/187152706784111605.
4. Hollerman JR, Schultz W. Dopaminergic neurons report an error in the temporal prediction of reward during learning. *Nat Neurosci*. 1998;1:304–309. https://doi.org/10.1038/1124.
5. Schultz W, Dayan P, Montague R. A neural substrate of prediction and reward. *Science*. 1997;275(5306):1593–1599. https://doi.org/10.1126/science.275.5306.1593.
6. Berridge KC. From prediction error to incentive salience: mesolimbic computation of reward motivation. *Eur J Neurosci*. 2012;35(7):1124–1143. https://doi.org/10.1111/j.1460-9568.2012.07990.x.

7. Calipari ES, Bagot RC, Purushothaman I, et al. In vivo imaging identifies temporal signature of D1 and D2 medium spiny neurons in cocaine reward. *Proceedings of the National Academy of the United States of America*. 2016;113(10):2726–2731. https://doi.org/10.1073/pnas.1521238113.
8. Kravitz AV, Tye LD, Kreitzer AC. Distinct roles for direct and indirect pathwat striatal neurons in reinforcement. *Nat Neurosci*. 2012;15(6):816–818. https://doi.org/10.1038/nn.3100.
9. Gallagher M, Graham PW, Holland PC. The amygdala central nucleus and appetitive Pavlovian conditioning: lesions impair one class of conditioned behavior. *J Neurosci*. 1990;10(6):1906–1911. https://doi.org/10.1523/JNEUROSCI.10-06-01906.1990.
10. Málková L, Gaffan D, Murray EA. Excitotoxic lesions of the amygdala fail to produce impairment in visual learning for auditory secondary reinforcement but interfere with reinforcer devaluation effects in rhesus monkeys. *J Neurosci*. 1997;17(15):6011–6020. https://doi.org/10.1523/JNEUROSCI.17-15-06011.1997.
11. Chudasama Y, Daniels TE, Gorrin DP, Rhodes SEV, Rudebeck PH, Murray EA. The role of the anterior cingulate cortex in choices based on reward value and reward contingency. *Cerebr Cortex*. 2013;23(12):2884–2898. https://doi.org/10.1093/cercor/bhs266.
12. Wetherford MJ, Cohen LB. Developmental changes in infant visual preferences for novelty and familiarity. *Child Dev*. 1973;44(3):416–424.
13. Engelmann JM, Versace F, Robinson JD, et al. Neural substrates of smoking cue reactivity: a meta-analysis of fMRI studies. *Neuroimage*. 2012;60(1):252–262. https://doi.org/10.1016/j.neuroimage.2011.12.024.
14. Gremel CM, Cunningham CL. Roles of the nucleus accumbens and amygdala in the acquisition and expression of ethanol-conditioned behavior in mice. *J Neurosci*. 2008;28(5):1076–1084. https://doi.org/10.1523/JNEUROSCI.4520-07.2008.
15. Ouachikh O, Dieb W, Durif F, Hafidi A. Differential behavioral reinforcement effects of dopamine receptor agonists in the rat with bilateral lesion of the posterior ventral tegmental area. *Behav Brain Res*. 2013;252(1):24–31. https://doi.org/10.1016/j.bbr.2013.05.042.
16. Singer G, Wallace M. Effects of 6-OHDA lesions in the nucleus accumbens on the acquisition of self injection of heroin under schedule and non schedule conditions in rats. *Pharmacol, Biochem Behav*. 1984;20(5):807–809. https://doi.org/10.1016/0091-3057(84)90204-1.
17. Fuchs RA, Weber SM, Rice HJ, Neisewander JL. Effects of excitotoxic lesions of the basolateral amygdala on cocaine-seeking behavior and cocaine conditioned place preference in rats. *Brain Res*. 2002;929(1):15–25. https://doi.org/10.1016/s0006-8993(01)03366-2.
18. Warlow SM, Robinson MJF, Berridge KC. Optogenetic central amygdala stimulation intensifies and narrows motivation for cocaine. *J Neurosci*. 2017;37(35):8330–8348. https://doi.org/10.1523/JNEUROSCI.3141-16.2017.
19. Grakalic I, Panlilio LV, Quiroz C, Schindler CW. Effects of orbitofrontal cortex lesions on cocaine self-administration. *Neuroscience*. 2010;165(2):313–324. https://doi.org/10.1016/j.neuroscience.2009.10.051.
20. Pisanu A, Lecca D, Valentini V, et al. Impairment of acquisition of intravenous cocaine self-administration by RNA-interference of dopamine D1-receptors in the nucleus accumbens shell. *Neuropharmacology*. 2015;89:398–411. https://doi.org/10.1016/j.neuropharm.2014.10.018.
21. Gerrits MA, Ramsey NF, Wolterink G, van Ree JM. Lack of evidence for an involvement of nucleus accumbens dopamine D1 receptors in the initiation of heroin self-administration in the rat. *Psychopharmacology*. 1994;114(3):486–494. https://doi.org/10.1007/BF02249340.
22. Le Merrer J, Gavello-Baudy S, Galey D, Cazala P. Morphine self-administration into the lateral septum depends on dopaminergic mechanisms: evidence from pharmacology and Fos neuroimaging. *Behav Brain Res*. 2007;180(2):203–217. https://doi.org/10.1016/j.bbr.2007.03.014.
23. Rodd ZA, Melendez RI, Bell RL, et al. Intracranial self-administration of ethanol within the ventral tegmental area of male Wistar rats: evidence for involvement of dopamine neurons. *J Neurosci*. 2004;24(5):1050–1057. https://doi.org/10.1523/JNEUROSCI.1319-03.2004.
24. Alderson HL, Parkinson JA, Robbins TW, Everitt BJ. The effects of excitotoxic lesions of the nucleus accumbens core or shell regions on intravenous heroin self-administration in rats. *Psychopharmacology*. 2001;153(4):455–463. https://doi.org/10.1007/s002130000634.
25. Schenk S, Horger BA, Peltier R, Shelton K. Supersensitivity to the reinforcing effects of cocaine following 6-hydroxydopamine lesions to the medial prefrontal cortex in rats. *Brain Res*. 1991;543(2):227–235. https://doi.org/10.1016/0006-8993(91)90032-q.

26. Hodge CW, Samson HH, Chappelle AM. Alcohol self-administration: further examination of the role of dopamine receptors in the nucleus accumbens. *Alcohol Clin Exp Res.* 1997;21(6):1083−1091. https://doi.org/10.1111/j.1530-0277.1997.tb04257.x.
27. Maldonado R, Robledo P, Chover AJ, Caine SB, Koob GF. D1 dopamine receptors in the nucleus accumbens modulate cocaine self-administration in the rat. *Pharmacol, Biochem Behav.* 1993;45(1):239−242. https://doi.org/10.1016/0091-3057(93)90112-7.
28. Phillips GD, Robbins TW, Everitt BJ. Bilateral intra-accumbens self-administration of d-amphetamine: antagonism with intra-accumbens SCH-23390 and sulpiride. *Psychopharmacology.* 1994;114(3):477−485. https://doi.org/10.1007/BF02249339.
29. Caine SB, Heinrichs SC, Coffin VL, Koob GF. Effects of the dopamine D-1 antagonist SCH 23390 microinjected into the accumbens, amygdala or striatum on cocaine self-administration in the rat. *Brain Res.* 1995;692(1−2):47−56. https://doi.org/10.1016/0006-8993(95)00598-k.
30. McGregor A, Roberts DC. Dopaminergic antagonism within the nucleus accumbens or the amygdala produces differential effects on intravenous cocaine self-administration under fixed and progressive ratio schedules of reinforcement. *Brain Res.* 1993;624(1−2):245−252. https://doi.org/10.1016/0006-8993(93)90084-z.
31. Merritt KE, Bachtell RK. Initial d2 dopamine receptor sensitivity predicts cocaine sensitivity and reward in rats. *PLoS One.* 2013;8(11):e78258. https://doi.org/10.1371/journal.pone.0078258.
32. Bailey A, Metaxas A, Yoo JH, McGee T, Kitchen I. Decrease of D_2 receptor binding but increase in D_2-stimulated G-protein activation, dopamine transporter binding and behavioural sensitization in brains of mice treated with a chronic escalating dose 'binge' cocaine administration paradigm. *Eur J Neurosci.* 2008;28(4):759−770. https://doi.org/10.1111/j.1460-9568.2008.06369.x.
33. Lister RG. The effects of repeated doses of ethanol on exploration and its habituation. *Psychopharmacology.* 1987;92(1):78−83. https://doi.org/10.1007/BF00215483.
34. Kvello AMS, Andersen JM, Boix F, Mørland J, Bogen IL. The role of 6-acetylmorphine in heroin-induced reward and locomotor sensitization in mice. *Addiction Biol.* 2020;25(2):e12727. https://doi.org/10.1111/adb.12727.
35. Zarrindast M-R, Heidari-Darvishani A, Rezayof A, Fathi-Azarbaijani F, Jafari-Sabet M, Hajizadeh-Moghaddam A. Morphine-induced sensitization in mice: changes in locomotor activity by prior scheduled exposure to GABAA receptor agents. *Behav Pharmacol.* 2007;18(4):303−310. https://doi.org/10.1097/FBP.0b013e3282186baa.
36. Galvano JP, Manhães AC, Carvalho-Nogueira ACC, Silva JM, Filgueiras CC, Abreu-Villaça Y. Profiling of behavioral effects evoked by ketamine and the role of 5HT 2 and D 2 receptors in ketamine-induced locomotor sensitization in mice. *Prog Neuro Psychopharmacol Biol Psychiatr.* 2020;97:109775. https://doi.org/10.1016/j.pnpbp.2019.109775.
37. Johnson KM, Phillips M, Wang C, Kevetter GA. Chronic phencyclidine induces behavioral sensitization and apoptotic cell death in the olfactory and piriform cortex. *J Neurosci Res.* 1998;52(6):709−722. https://doi.org/10.1002/(SICI)1097-4547(19980615)52:6<709::AID-JNR10>3.0.CO;2-U.
38. Brenhouse HC, Montalto S, Stellar JR. Electrolytic lesions of a discrete area within the nucleus accumbens shell attenuate the long-term expression, but not early phase, of sensitization to cocaine. *Behav Brain Res.* 2006;170(2):219−223. https://doi.org/10.1016/j.bbr.2006.02.029.
39. Pierce RC, Reeder DC, Hicks J, Morgan ZR, Kalivas PW. Ibotenic acid lesions of the dorsal prefrontal cortex disrupt the expression of behavioral sensitization to cocaine. *Neuroscience.* 1998;82(4):1103−1114. https://doi.org/10.1016/s0306-4522(97)00366-7.
40. Kelsey JE, Gerety LP, Guerriero RM. Electrolytic lesions of the nucleus accumbens core (but not the medial shell) and the basolateral amygdala enhance context-specific locomotor sensitization to nicotine in rats. *Behav Neurosci.* 2009;123(3):577−588. https://doi.org/10.1037/a0015573.
41. Li Y, Hu XT, Berney TG, et al. Both glutamate receptor antagonists and prefrontal cortex lesions prevent induction of cocaine sensitization and associated neuroadaptations. *Synapse.* 1999;34(3):169−180. https://doi.org/10.1002/(SICI)1098-2396(19991201)34:3<169::AID-SYN1>3.0.CO;2-C.
42. Banks KE, Gratton A. Possible involvement of medial prefrontal cortex in amphetamine-induced sensitization of mesolimbic dopamine function. *Eur J Pharmacol.* 1995;282(1−3):157−167. https://doi.org/10.1016/0014-2999(95)00306-6.
43. Goutier W, O'Connor JJ, Lowry JP, McCreary AC. The effect of nicotine induced behavioral sensitization on dopamine D1 receptor pharmacology: an in vivo and ex vivo study in the rat. *Eur Neuropsychopharmacol.* 2015;25(6):933−943. https://doi.org/10.1016/j.euroneuro.2015.02.008.

44. Cadoni C, Di Chiara G. Reciprocal changes in dopamine responsiveness in the nucleus accumbens shell and core and in the dorsal caudate-putamen in rats sensitized to morphine. *Neuroscience*. 1999;90(2):447–455. https://doi.org/10.1016/s0306-4522(98)00466-7.
45. McDougall SA, Rudberg KN, Veliz A, et al. Importance of D1 and D2 receptor stimulation for the induction and expression of cocaine-induced behavioral sensitization in preweanling rats. *Behav Brain Res*. 2017;326:226–236. https://doi.org/10.1016/j.bbr.2017.03.001.
46. Maisonneuve IM, Ho A, Kreek MJ. Chronic administration of a cocaine "binge" alters basal extracellular levels in male rats: an in vivo microdialysis study. *J Pharmacol Exp Therapeut*. 1995;272(2):652–657.
47. Ito R, Dalley JW, Howes SR, Robbins TW, Everitt BJ. Dissociation in conditioned dopamine release in the nucleus accumbens core and shell in response to cocaine cues and during cocaine-seeking behavior in rats. *J Neurosci*. 2000;20(19):7489–7495. https://doi.org/10.1523/JNEUROSCI.20-19-07489.2000.
48. Rossetti ZL, Hmaidan Y, Gessa GL. Marked inhibition of mesolimbic dopamine release: a common feature of ethanol, morphine, cocaine and amphetamine abstinence in rats. *Eur J Pharmacol*. 1992;221(2–3):227–234. https://doi.org/10.1016/0014-2999(92)90706-a.
49. Lee TH, Gao WY, Davidson C, Ellinwood EH. Altered activity of midbrain dopamine neurons following 7-day withdrawal from chronic cocaine abuse is normalized by D2 receptor stimulation during the early withdrawal phase. *Neuropsychopharmacology*. 1999;21(1):127–136. https://doi.org/10.1016/S0893-133X(99)00011-1.
50. Nogueira L, Kalivas PW, Lavin A. Long-term neuroadaptations produced by withdrawal from repeated cocaine treatment: role of dopaminergic receptors in modulating cortical excitability. *J Neurosci*. 2006;26(47):12308–12313. https://doi.org/10.1523/JNEUROSCI.3206-06.2006.
51. Chartoff EH, Mague SD, Barhight MF, Smith AM, Carlezon Jr WA. Behavioral and molecular effects of dopamine D_1 receptor stimulation during naloxone-precipitated morphine withdrawal. *J Neurosci*. 2006;26(24):6450–6457. https://doi.org/10.1523/JNEUROSCI.0491-06.2006.
52. Rodríguez-Arias M, Pinazo J, Miñarro J, Stinus L. Effects of SCH 23390, raclopride, and haloperidol on morphine withdrawal-induced aggression in male mice. *Pharmacol, Biochem Behav*. 1999;64(1):123–130. https://doi.org/10.1016/s0091-3057(99)00067-2.
53. Rodgers HM, Lim S-A, Yow J, et al. Dopamine D1 or D3 receptor modulators prevent morphine tolerance and reduce opioid withdrawal symptoms. *Pharmacol, Biochem Behav*. 2020;194:172935. https://doi.org/10.1016/j.pbb.2020.172935.
54. Fotros A, Casey KF, Larcher K, et al. Cocaine cue-induced dopamine release in amygdala and hippocampus: a high-resolution PET [^{18}F]fallypride study in cocaine dependent participants. *Neuropsychopharmacology*. 2013;38(9):1780–1788. https://doi.org/10.1038/npp.2013.77.
55. Milella MS, Fotros A, Gravel P, et al. Cocaine cue-induced dopamine release in the human prefrontal cortex. *J Psychiatry Neurosci*. 2016;41(5):322–330. https://doi.org/10.1503/jpn.150207.
56. Di Ciano P, Everitt BJ. Contribution of the ventral tegmental area to cocaine-seeking maintained by a drug-paired conditioned stimulus in rats. *Eur J Neurosci*. 2004;19(6):1661–1667. https://doi.org/10.1111/j.1460-9568.2004.03232.x.
57. Ito R, Robbins TW, Everitt BJ. Differential control over cocaine-seeking behavior by nucleus accumbens core and shell. *Nat Neurosci*. 2004;7(4):389–397. https://doi.org/10.1038/nn1217.
58. Hutcheson DM, Parkinson JA, Robbins TW, Everitt BJ. The effects of nucleus accumbens core and shell lesions on intravenous heroin self-administration and the acquisition of drug-seeking behaviour under a second-order schedule of heroin reinforcement. *Psychopharmacology*. 2001;153(4):464–472. https://doi.org/10.1007/s002130000635.
59. Jonkman S, Pelloux Y, Everitt BJ. Differential roles of the dorsolateral and midlateral striatum in punished cocaine seeking. *J Neurosci*. 2012;32(13):4645–4650. https://doi.org/10.1523/JNEUROSCI.0348-12.2012.
60. Hutcheson DM, Everitt BJ. The effects of selective orbitofrontal cortex lesions on the acquisition and performance of cue-controlled cocaine seeking in rats. *Ann N Y Acad Sci*. 2003;1003:410–411. https://doi.org/10.1196/annals.1300.038.
61. Pelloux Y, Murray JE, Everitt BJ. Differential roles of the prefrontal cortical subregions and basolateral amygdala in compulsive cocaine seeking and relapse after voluntary abstinence in rats. *Eur J Neurosci*. 2013;38(7):3018–3026. https://doi.org/10.1111/ejn.12289.
62. Murray JE, Belin-Rauscent A, Simon M, et al. Basolateral and central amygdala differentially recruit and maintain dorsolateral striatum-dependent cocaine-seeking habits. *Nat Commun*. 2015;6:10088. https://doi.org/10.1038/ncomms10088.

63. Ito R, Dalley JW, Robbins TW, Everitt BJ. Dopamine release in the dorsal striatum during cocaine-seeking behavior under the control of a drug-associated cue. *J Neurosci.* 2002;22(14):6247–6253. https://doi.org/10.1523/JNEUROSCI.22-14-06247.2002.
64. Di Ciano P, Everitt BJ. Direct interactions between the basolateral amygdala and nucleus accumbens core underlie cocaine-seeking behavior by rats. *J Neurosci.* 2004;24(32):7167–7173. https://doi.org/10.1523/JNEUROSCI.1581-04.2004.
65. Graham DL, Hoppenot R, Hendryx A, Self DW. Differential ability of D1 and D2 dopamine receptor agonists to induce and modulate expression and reinstatement of cocaine place preference in rats. *Psychopharmacology.* 2007;191(3):719–730. https://doi.org/10.1007/s00213-006-0473-5.
66. Alleweireldt AT, Weber SM, Kirschner KF, Bullock BL, Neisewander JL. Blockade or stimulation of D1 dopamine receptors attenuates cue reinstatement of extinguished cocaine-seeking behavior in rats. *Psychopharmacology.* 2002;159(3):284–293. https://doi.org/10.1007/s002130100904.
67. Khroyan TV, Barrett-Larimore RL, Rowlett JK, Spealman RD. Dopamine D1- and D2-like receptor mechanisms in relapse to cocaine-seeking behavior: effects of selective antagonists and agonists. *J Pharmacol Exp Therapeut.* 2000;294(2):680–687.
68. Liu X, Jernigen C, Gharib M, Booth S, Caggiula AR, Sved AF. Effects of dopamine antagonists on drug cue-induced reinstatement of nicotine-seeking behavior in rats. *Behav Pharmacol.* 2010;21(2):153–160. https://doi.org/10.1097/FBP.0b013e328337be95.
69. Liu Y, Jean-Richard-Dit-Bressel P, Yau JO-Y, et al. The mesolimbic dopamine activity signatures of relapse to alcohol-seeking. *J Neurosci.* 2020;40(33):6409–6427. https://doi.org/10.1523/JNEUROSCI.0724-20.2020.
70. Bianchi PC, de Oliveira PEC, Palombo P, et al. Functional inactivation of the orbitofrontal cortex disrupts context-induced reinstatement of alcohol seeking in rats. *Drug Alcohol Depend.* 2018;186:102–112. https://doi.org/10.1016/j.drugalcdep.2017.12.045.
71. Gabriele A, See RE. Lesions and reversible inactivation of the dorsolateral caudate-putamen impair cocaine-primed reinstatement to cocaine-seeking in rats. *Brain Res.* 2011;1417:27–35. https://doi.org/10.1016/j.brainres.2011.08.030.
72. Gill MJ, Ghee SM, Harper SM, See RE. Inactivation of the lateral habenula reduces anxiogenic behavior and cocaine seeking under conditions of heightened stress. *Pharmacol, Biochem Behav.* 2013;111:24–29. https://doi.org/10.1016/j.pbb.2013.08.002.
73. Matzeu A, Weiss F, Martin-Fardon R. Transient inactivation of the posterior paraventricular nucleus of the thalamus blocks cocaine-seeking behavior. *Neurosci Lett.* 2015;608:34–39. https://doi.org/10.1016/j.neulet.2015.10.016.
74. Rogers JL, See RE. Selective inactivation of the ventral hippocampus attenuates cue-induced and cocaine-primed reinstatement of drug-seeking in rats. *Neurobiol Learn Mem.* 2007;87(4):688–692. https://doi.org/10.1016/j.nlm.2007.01.003.
75. Costa Campos R, Dias C, Darlot F, Cador M. Double dissociation between actions of dopamine D1 and D2 receptors of the ventral and dorsolateral striatum to produce reinstatement of cocaine seeking behavior. *Neuropharmacology.* 2020;172:108113. https://doi.org/10.1016/j.neuropharm.2020.108113.
76. Assar N, Mahmoudi D, Farhoudian A, Hasan Farhadi M, Fatahi Z, Haghparast A. D1- and D2-like dopamine receptors in the CA1 region of the hippocampus are involved in the acquisition and reinstatement of morphine-induced conditioned place preference. *Behav Brain Res.* 2016;312:394–404. https://doi.org/10.1016/j.bbr.2016.06.061.
77. Berglind WJ, Case JM, Parker MP, Fuchs RA, See RE. Dopamine D1 or D2 receptor antagonism within the basolateral amygdala differentially alters the acquisition of cocaine-cue associations necessary for cue-induced reinstatement of cocaine-seeking. *Neuroscience.* 2006;137(2):699–706. https://doi.org/10.1016/j.neuroscience.2005.08.064.
78. Bossert JM, Poles GC, Wihbey KA, Koya E, Shaham Y. Differential effects of blockade of dopamine D1-family receptors in nucleus accumbens core or shell on reinstatement of heroin seeking induced by contextual and discrete cues. *J Neurosci.* 2007;27(46):12655–12663. https://doi.org/10.1523/JNEUROSCI.3926-07.2007.
79. Cosme CV, Gutman AL, Worth WR, LaLumiere RT. D1, but not D2, receptor blockade within the infralimbic and medial orbitofrontal cortex impairs cocaine seeking in a region-specific manner. *Addiction Biol.* 2018;23(1):16–27. https://doi.org/10.1111/adb.12442.

80. Farahimanesh S, Moradi M, Nazari-Serenjeh F, Zarrabian S, Haghparast A. Role of D1-like and D2-like dopamine receptors within the ventral tegmental area in stress-induced and drug priming-induced reinstatement of morphine seeking in rats. *Behav Pharmacol.* 2018;29(5):426—436. https://doi.org/10.1097/FBP.0000000000000381.
81. Thiel KJ, Wenzel JM, Pentkowski NS, Hobbs RJ, Alleweireldt AT, Neisewander JL. Stimulation of dopamine D2/D3 but not D1 receptors in the central amygdala decreases cocaine-seeking behavior. *Behav Brain Res.* 2010;214(2):386—394. https://doi.org/10.1016/j.bbr.2010.06.021.
82. Dackis CA, Gold MS. New concepts in cocaine addiction: the dopamine depletion hypothesis. *Neurosci Biobehav Rev.* 1985;9(3):469—477. https://doi.org/10.1016/0149-7634(85)90022-3.
83. Nutt DJ, Lingford-Hughes A, Erritzoe D, Stokes PRA. The dopamine theory of addiction: 40 years of highs and lows. *Nat Rev Neurosci.* 2015;16(5):305—312. https://doi.org/10.1038/nrn3939.
84. Walker BM, Koob GF. The gamma-aminobutyric acid-B receptor agonist baclofen attenuates responding for ethanol in ethanol-dependent rats. *Alcohol Clin Exp Res.* 2007;31(1):11—18. https://doi.org/10.1111/j.1530-0277.2006.00259.x.
85. Campbell UC, Lac ST, Carroll ME. Effects of baclofen on maintenance and reinstatement of intravenous cocaine self-administration in rats. *Psychopharmacology.* 1999;143(2):209—214. https://doi.org/10.1007/s002130050937.
86. Paterson NE, Froestl W, Markou A. Repeated administration of the GABAB receptor agonist CGP44532 decreased nicotine self-administration, and acute administration decreased cue-induced reinstatement of nicotine-seeking in rats. *Neuropsychopharmacology.* 2005;30(1):119—128. https://doi.org/10.1038/sj.npp.1300524.
87. Di Ciano P, Everitt BJ. The GABA(B) receptor agonist baclofen attenuates cocaine- and heroin-seeking behavior by rats. *Neuropsychopharmacology.* 2003;28(3):510—518. https://doi.org/10.1038/sj.npp.1300088.
88. Xi ZX, Stein EA. Increased mesolimbic GABA concentration blocks heroin self-administration in the rat. *J Pharmacol Exp Therapeut.* 2000;294(2):613—619.
89. Zarrindast M-R, Ahmadi S, Haeri-Rohani A, Rezayof A, Jafari M-R, Jafari-Sabet M. GABA(A) receptors in the basolateral amygdala are involved in mediating morphine reward. *Brain Res.* 2004;1006(1):49—58. https://doi.org/10.1016/j.brainres.2003.12.048.
90. Zarrindast M-R, Massoudi R, Sepehri H, Rezayof A. Involvement of GABA(B) receptors of the dorsal hippocampus on the acquisition and expression of morphine-induced place preference in rats. *Physiol Behav.* 2006;87(1):31—38. https://doi.org/10.1016/j.physbeh.2005.08.041.
91. Chester JA, Cunningham CL. GABA(A) receptors modulate ethanol-induced conditioned place preference and taste aversion in mice. *Psychopharmacology.* 1999;144(4):363—372. https://doi.org/10.1007/s002130051019.
92. Spano MS, Fattore L, Fratta W, Fadda P. The GABAB receptor agonist baclofen prevents heroin-induced reinstatement of heroin-seeking behavior in rats. *Neuropharmacology.* 2007;52(7):1555—1562. https://doi.org/10.1016/j.neuropharm.2007.02.012.
93. Blacktop JM, Vranjkovic O, Mayer M, Van Hoof M, Baker DA, Mantsch JR. Antagonism of GABA-B but not GABA-A receptors in the VTA prevents stress- and intra-VTA CRF-induced reinstatement of extinguished cocaine seeking in rats. *Neuropharmacology.* 2016;102:197—206. https://doi.org/10.1016/j.neuropharm.2015.11.013.
94. Chandler CM, Reeves-Darby J, Jones SA, et al. α5GABA A subunit-containing receptors and sweetened alcohol cue-induced reinstatement and active sweetened alcohol self-administration in male rats. *Psychopharmacology.* 2019;236(6):1797—1806. https://doi.org/10.1007/s00213-018-5163-6.
95. Holtyn AF, Tiruveedhula VVN, P B, Stephen MR, Cook JM, Weerts EM. Effects of the benzodiazepine GABA$_A$ α1-preferring antagonist 3-isopropoxy-β-carboline hydrochloride (3-ISOPBC) on alcohol seeking and self-administration in baboons. *Drug Alcohol Depend.* 2017;170:25—31. https://doi.org/10.1016/j.drugalcdep.2016.10.036.
96. June HL, Harvey SC, Foster KL, et al. GABA(A) receptors containing (alpha)5 subunits in the CA1 and CA3 hippocampal fields regulate ethanol-motivated behaviors: an extended ethanol reward circuitry. *J Neurosci.* 2001;21(6):2166—2177. https://doi.org/10.1523/JNEUROSCI.21-06-02166.2001.
97. Dixon CI, Halbout B, King SL, Stephens DN. Deletion of the GABAA α2-subunit does not alter self administration of cocaine or reinstatement of cocaine seeking. *Psychopharmacology.* 2014;231(13):2695—2703. https://doi.org/10.1007/s00213-014-3443-3.
98. Pap A, Bradberry CW. Excitatory amino acid antagonists attenuate the effects of cocaine on extracellular dopamine in the nucleus accumbens. *J Pharmacol Exp Therapeut.* 1995;274(1):127—133.

99. Maldonado C, Rodríguez-Arias M, Castillo A, Aguilar MA, Miñarro J. Effect of memantine and CNQX in the acquisition, expression and reinstatement of cocaine-induced conditioned place preference. *Prog Neuro Psychopharmacol Biol Psychiatr*. 2007;31(4):932−939. https://doi.org/10.1016/j.pnpbp.2007.02.012.
100. Blokhina EA, Kashkin VA, Zvartau EE, Danysz W, Bespalov AY. Effects of nicotinic and NMDA receptor channel blockers on intravenous cocaine and nicotine self-administration in mice. *Eur Neuropsychopharmacol*. 2005;15(2):219−225. https://doi.org/10.1016/j.euroneuro.2004.07.005.
101. Suzuki T, Kato H, Tsuda M, Suzuki H, Misawa M. Effects of the non-competitive NMDA receptor antagonist ifenprodil on the morphine-induced place preference in mice. *Life Sci*. 1999;64(12):PL151−156. https://doi.org/10.1016/s0024-3205(99)00036-3.
102. Yates JR, Campbell HL, Hawley LL, Horchar MJ, Kappesser JL, Wright MR. Effects of the GluN2B-selective antagonist Ro 63-1908 on acquisition and expression, of methamphetamine conditioned place preference in male and female rats. *Drug Alcohol Depend*. 2021;225:108785.
103. Lee B, Platt DM, Rowlett JK, Adewale AS, Spealman RD. Attenuation of behavioral effects of cocaine by the Metabotropic Glutamate Receptor 5 Antagonist 2-Methyl-6-(phenylethynyl)-pyridine in squirrel monkeys: comparison with dizocilpine. *J Pharmacol Exp Therapeut*. 2005;312(3):1232−1240. https://doi.org/10.1124/jpet.104.078733.
104. Yu F, Zhong P, Liu X, Sun D, Gao H-Q, Liu Q-S. Metabotropic glutamate receptor I (mGluR1) antagonism impairs cocaine-induced conditioned place preference via inhibition of protein synthesis. *Neuropsychopharmacology*. 2013;38(7):1308−1321. https://doi.org/10.1038/npp.2013.29.
105. Adewale AS, Platt DM, Spealman RD. Pharmacological stimulation of group ii metabotropic glutamate receptors reduces cocaine self-administration and cocaine-induced reinstatement of drug seeking in squirrel monkeys. *J Pharmacol Exp Therapeut*. 2006;318(2):922−931. https://doi.org/10.1124/jpet.106.105387.
106. Chiamulera C, Epping-Jordan MP, Zocchi A, et al. Reinforcing and locomotor stimulant effects of cocaine are absent in mGluR5 null mutant mice. *Nat Neurosci*. 2001;4(9):873−874. https://doi.org/10.1038/nn0901-873.
107. Dravolina OA, Danysz W, Bespalov AY. Effects of group I metabotropic glutamate receptor antagonists on the behavioral sensitization to motor effects of cocaine in rats. *Psychopharmacology*. 2006;187(4):397−404. https://doi.org/10.1007/s00213-006-0440-1.
108. Harris GC, Aston-Jones G. Critical role for ventral tegmental glutamate in preference for a cocaine-conditioned environment. *Neuropsychopharmacology*. 2003;28(1):73−76. https://doi.org/10.1038/sj.npp.1300011.
109. Rassnick S, Pulvirenti L, Koob GF. Oral ethanol self-administration in rats is reduced by the administration of dopamine and glutamate receptor antagonists into the nucleus accumbens. *Psychopharmacology*. 1992;109(1−2):92−98. https://doi.org/10.1007/BF02245485.
110. Suto N, Ecke LE, Wise RA. Control of within-binge cocaine-seeking by dopamine and glutamate in the core of nucleus accumbens. *Psychopharmacology*. 2009;205(3):431−439. https://doi.org/10.1007/s00213-009-1553-0.
111. Pierce RC, Bell K, Duffy P, Kalivas PW. Repeated cocaine augments excitatory amino acid transmission in the nucleus accumbens only in rats having developed behavioral sensitization. *J Neurosci*. 1996;16(4):1550−1560. https://doi.org/10.1523/JNEUROSCI.16-04-01550.1996.
112. Di Ciano P, Everitt BJ. Dissociable effects of antagonism of NMDA and AMPA/KA receptors in the nucleus accumbens core and shell on cocaine-seeking behavior. *Neuropsychopharmacology*. 2001;25(3):341−360. https://doi.org/10.1016/S0893-133X(01)00235-4.
113. Bäackström P, Hyytiä P. Involvement of AMPA/kainate, NMDA, and mGlu5 receptors in the nucleus accumbens core in cue-induced reinstatement of cocaine seeking in rats. *Psychopharmacology*. 2007;192(4):571−580. https://doi.org/10.1007/s00213-007-0753-8.
114. Gipson CD, Reissner KJ, Kupchik YM, et al. Reinstatement of nicotine seeking is mediated by glutamatergic plasticity. *Proc Natl Acad Sci USA*. 2013;110(22):9124−9129. https://doi.org/10.1073/pnas.1220591110.
115. Cornish JL, Duffy P, Kalivas PW. A role for nucleus accumbens glutamate transmission in the relapse to cocaine-seeking behavior. *Neuroscience*. 1999;93(4):1359−1367. https://doi.org/10.1016/s0306-4522(99)00214-6.
116. LaLumiere RT, Kalivas PW. Glutamate release in the nucleus accumbens core is necessary for heroin seeking. *J Neurosci*. 2008;28(12):3170−3177. https://doi.org/10.1523/JNEUROSCI.5129-07.2008.
117. Feltenstein MW, See RE. NMDA receptor blockade in the basolateral amygdala disrupts consolidation of stimulus-reward memory and extinction learning during reinstatement of cocaine-seeking in an animal model of relapse. *Neurobiol Learn Mem*. 2007;88(4):435−444. https://doi.org/10.1016/j.nlm.2007.05.006.
118. Vanderschuren LJMJ, Di Ciano P, Everitt BJ. Involvement of the dorsal striatum in cue-controlled cocaine seeking. *J Neurosci*. 2005;25(38):8665−8670. https://doi.org/10.1523/JNEUROSCI.0925-05.2005.

119. Kumaresan V, Yuan M, Yee J, et al. Metabotropic glutamate receptor 5 (mGluR5) antagonists attenuate cocaine priming- and cue-induced reinstatement of cocaine seeking. *Behav Brain Res*. 2009;202(2):238–244. https://doi.org/10.1016/j.bbr.2009.03.039.
120. Baptista MAS, Martin-Fardon R, Weiss F. Preferential effects of the metabotropic glutamate 2/3 receptor agonist LY379268 on conditioned reinstatement versus primary reinforcement: comparison between cocaine and a potent conventional reinforcer. *J Neurosci*. 2004;24(20):4723–4727. https://doi.org/10.1523/JNEUROSCI.0176-04.2004.
121. Lu L, Uejima JL, Gray SM, Bossert JM, Shaham Y. Systemic and central amygdala injections of the mGluR(2/3) agonist LY379268 attenuate the expression of incubation of cocaine craving. *Biol Psychiatr*. 2007;61(5):591–598. https://doi.org/10.1016/j.biopsych.2006.04.011.
122. Simmons DL, Mandt BH, Ng CMC, et al. Low- and high-cocaine locomotor responding rats differ in reinstatement of cocaine seeking and striatal mGluR5 protein expression. *Neuropharmacology*. 2013;75:347–355. https://doi.org/10.1016/j.neuropharm.2013.08.001.
123. Xie X, Ramirez DR, Lasseter HC, Fuchs RA. Effects of mGluR1 antagonism in the dorsal hippocampus on drug context-induced reinstatement of cocaine-seeking behavior in rats. *Psychopharmacology*. 2010;208(1):1–11. https://doi.org/10.1007/s00213-009-1700-7.
124. Magalas Z, De Vry J, Tzschentke TM. The serotonin/noradrenaline reuptake inhibitor venlafaxine attenuates acquisition, but not maintenance, of intravenous self-administration of heroin in rats. *Eur J Pharmacol*. 2005;528(1–3):103–109. https://doi.org/10.1016/j.ejphar.2005.10.038.
125. Higgins GA, Wang Y, Corrigall WA, Sellers EM. Influence of 5-HT3 receptor antagonists and the indirect 5-HT agonist, dexfenfluramine, on heroin self-administration in rats. *Psychopharmacology*. 1994;114(4):611–619. https://doi.org/10.1007/BF02244992.
126. Rüedi-Bettschen D, Spealman RD, Platt DM. Attenuation of cocaine-induced reinstatement of drug seeking in squirrel monkeys by direct and indirect activation of 5-HT2C receptors. *Psychopharmacology*. 2015;232(16):2959–2968. https://doi.org/10.1007/s00213-015-3932-z.
127. Pelloux Y, Dilleen R, Economidou D, Theobald D, Everitt BJ. Reduced forebrain serotonin transmission is causally involved in the development of compulsive cocaine seeking in rats. *Neuropsychopharmacology*. 2012;37(11):2505–2514. https://doi.org/10.1038/npp.2012.111.
128. Tran-Nguyen LT, Bellew JG, Grote KA, Neisewander JL. Serotonin depletion attenuates cocaine seeking but enhances sucrose seeking and the effects of cocaine priming on reinstatement of cocaine seeking in rats. *Psychopharmacology*. 2001;157(4):340–348. https://doi.org/10.1007/s002130100822.
129. Schenk S. Effects of the serotonin 5-HT(2) antagonist, ritanserin, and the serotonin 5-HT(1A) antagonist, WAY 100635, on cocaine-seeking in rats. *Pharmacol, Biochem Behav*. 2000;67(2):363–369. https://doi.org/10.1016/s0091-3057(00)00377-4.
130. Schenk S, Foote J, Aronsen D, et al. Serotonin antagonists fail to alter MDMA self-administration in rats. *Pharmacol, Biochem Behav*. 2016;148:38–45. https://doi.org/10.1016/j.pbb.2016.06.002.
131. Pentkowski NS, Acosta JI, Browning JR, Hamilton EC, Neisewander JL. Stimulation of 5-HT(1B) receptors enhances cocaine reinforcement yet reduces cocaine-seeking behavior. *Addiction Biol*. 2009;14(4):419–430. https://doi.org/10.1111/j.1369-1600.2009.00162.x.
132. Yan Q-S, Zheng S-Z, Yan S-E. Involvement of 5-HT1B receptors within the ventral tegmental area in regulation of mesolimbic dopaminergic neuronal activity via GABA mechanisms: a study with dual-probe microdialysis. *Brain Res*. 2004;1021(1):82–91. https://doi.org/10.1016/j.brainres.2004.06.053.
133. Berro LF, Perez Diaz M, Maltbie E, Howell LL. Effects of the serotonin 2C receptor agonist WAY163909 on the abuse-related effects and mesolimbic dopamine neurochemistry induced by abused stimulants in rhesus monkeys. *Psychopharmacology*. 2017;234(17):2607–2617. https://doi.org/10.1007/s00213-017-4653-2.
134. Fletcher PJ, Rizos Z, Sinyard J, Tampakeras M, Higgins GA. The 5-HT2C receptor agonist Ro60-0175 reduces cocaine self-administration and reinstatement induced by the stressor yohimbine, and contextual cues. *Neuropsychopharmacology*. 2008;33(6):1402–1412. https://doi.org/10.1038/sj.npp.1301509.
135. Neelakantan H, Holliday ED, Fox RG, et al. Lorcaserin suppresses oxycodone self-administration and relapse vulnerability in rats. *ACS Chem Neurosci*. 2017;8(5):1065–1073. https://doi.org/10.1021/acschemneuro.6b00413.
136. Fletcher PJ, Grottick AJ, Higgins GA. Differential effects of the 5-HT(2A) receptor antagonist M100907 and the 5-HT(2C) receptor antagonist SB242084 on cocaine-induced locomotor activity, cocaine self-administration and cocaine-induced reinstatement of responding. *Neuropsychopharmacology*. 2002;27(4):576–586. https://doi.org/10.1016/S0893-133X(02)00342-1.

137. Pentkowski NS, Duke FD, Weber SM, et al. Stimulation of medial prefrontal cortex serotonin 2C (5-HT(2C)) receptors attenuates cocaine-seeking behavior. *Neuropsychopharmacology*. 2010;35(10):2037−2048. https://doi.org/10.1038/npp.2010.72.
138. Pockros-Burgess LA, Pentowski NS, Der-Ghazarian T, Neisewander JL. Effects of the 5-HT2C receptor agonist CP809101 in the amygdala on reinstatement of cocaine-seeking behavior and anxiety-like behavior. *Int J Neuropsychopharmacol*. 2014;17(11):1751−1762. https://doi.org/10.1017/S1461145714000856.
139. Pockros LA, Pentowski NS, Swinford SE, Neisewander JL. Blockade of 5-HT2A receptors in the medial prefrontal cortex attenuates reinstatement of cue-elicited cocaine-seeking behavior in rats. *Psychopharmacology*. 2011;213(2−3):307−320. https://doi.org/10.1007/s00213-010-2071-9.
140. Murnane KS, Winschel J, Schmidt KT, et al. Serotonin 2A receptors differentially contribute to abuse-related effects of cocaine and cocaine-induced nigrostriatal and mesolimbic dopamine overflow in nonhuman primates. *J Neurosci*. 2013;33(33):13367−13374. https://doi.org/10.1523/JNEUROSCI.1437-13.2013.
141. Tzschentke TM, Magalas Z, De Vry J. Effects of venlafaxine and desipramine on heroin-induced conditioned place preference in the rat. *Addiction Biol*. 2006;11(1):64−71. https://doi.org/10.1111/j.1369-1600.2006.00009.x.
142. Economidou D, Dalley JW, Everitt BJ. Selective norepinephrine reuptake inhibition by atomoxetine prevents cue-induced heroin and cocaine seeking. *Biol Psychiatr*. 2011;69(3):266−274. https://doi.org/10.1016/j.biopsych.2010.09.040.
143. Mitrano DA, Schroeder JP, Smith Y, et al. Alpha-1 adrenergic receptors are localized on presynaptic elements in the nucleus accumbens and regulate mesolimbic dopamine transmission. *Neuropsychopharmacology*. 2012;37(9):2161−2172. https://doi.org/10.1038/npp.2012.68.
144. Greenwall TN, Walker BM, Cottone P, Zorrilla EP, Koob GF. The alpha1 adrenergic receptor antagonist prazosin reduces heroin self-administration in rats with extended access to heroin administration. *Pharmacol, Biochem Behav*. 2009;91(3):295−302. https://doi.org/10.1016/j.pbb.2008.07.012.
145. Forget B, Wertheim C, Mascia P, Pushparaj A, Goldberg SR, Le Foll B. Noradrenergic alpha1 receptors as a novel target for the treatment of nicotine addiction. *Neuropsychopharmacology*. 2010;35(8):1751−1760. https://doi.org/10.1038/npp.2010.42.
146. Verplaetse TL, Rasmussen DD, Froehlich JC, Czachowski CL. Effects of prazosin, an α1-adrenergic receptor antagonist, on the seeking and intake of alcohol and sucrose in alcohol-preferring (P) rats. *Alcohol Clin Exp Res*. 2012;36(5):881−886. https://doi.org/10.1111/j.1530-0277.2011.01653.x.
147. Barnaba Solecki W, Szklarczyk K, Pradel K, Kwiatkowska K, Dobrzański G, Przewłocki R. Noradrenergic signaling in the VTA modulates cocaine craving. *Addiction Biol*. 2018;23(2):596−609. https://doi.org/10.1111/adb.12514.
148. Czoty PW, Nader MA. Effects of the α-2 adrenergic receptor agonists lofexidine and guanfacine on food-cocaine choice in socially housed cynomolgus monkeys. *J Pharmacol Exp Therapeut*. 2020;375(1):193−201. https://doi.org/10.1124/jpet.120.266007.
149. Samini M, Kardan A, Mehr SE. Alpha-2 agonists decrease expression of morphine-induced conditioned place preference. *Pharmacol, Biochem Behav*. 2008;88(4):403−406. https://doi.org/10.1016/j.pbb.2007.09.013.
150. Lê AD, Funk D, Juzytsch W, et al. Effect of prazosin and guanfacine on stress-induced reinstatement of alcohol and food seeking in rats. *Psychopharmacology*. 2011;218(1):89−99. https://doi.org/10.1007/s00213-011-2178-7.
151. Tierney C, Nadaud D, Koenig-Berard E, Stinus L. Effects of two alpha 2 agonists, rilmenidine and clonidine, on the morphine withdrawal syndrome and their potential addictive properties in rats. *Am J Cardiol*. 1988;61(7):D35−D38. https://doi.org/10.1016/0002-9149(88)90462-6.
152. Haney M, Cooper ZD, Bedi G, et al. Guanfacine decreases symptoms of cannabis withdrawal in daily cannabis smokers. *Addiction Biol*. 2019;24(4):707−716. https://doi.org/10.1111/adb.12621.
153. Haney M, Hart CL, Vosburg SK, Comer SD, Collins Reed S, Foltin RW. Effects of THC and lofexidine in a human laboratory model of marijuana withdrawal and relapse. *Pscyhopharmacology*. 2008;197(1):157−168. https://doi.org/10.1007/s00213-007-1020-8.
154. Yamada H, Bruijnzeel AW. Stimulation of α2-adrenergic receptors in the central nucleus of the amygdala attenuates stress-induced reinstatement of nicotine seeking in rats. *Neuropharmacology*. 2011;60(2−3):303−311. https://doi.org/10.1016/j.neuropharm.2010.09.013.
155. Kohut SJ, Fivel PA, Mello NK. Differential effects of acute and chronic treatment with the α2-adrenergic agonist, lofexidine, on cocaine self-administration in rhesus monkeys. *Drug Alcohol Depend*. 2013;133(2):593−599. https://doi.org/10.1016/j.drugalcdep.2013.07.032.

156. Bertholomey ML, Verplaetse TL, Czachowski CL. Alterations in ethanol seeking and self-administration following yohimbine in selectively bred alcohol-preferring (p) and high alcohol drinking (had-2) rats. *Behav Brain Res.* 2013;238:252−258. https://doi.org/10.1016/j.bbr.2012.10.030.
157. Al-Hasani R, Bruchas MR. Molecular mechanisms of opioid receptor-dependent signaling and behavior. *Anesthesiology.* 2011;115(6):1363−1381. https://doi.org/10.1097/ALN.0b013e318238bba6.
158. Lalanne L, Ayranci G, Kieffer BL, Lutz PE. The kappa opioid receptor: from addiction to depression, and back. *Front Psychiatr.* 2014;5:170. https://doi.org/10.3389/fpsyt.2014.00170.
159. Clark SD, Abi-Dargham A. The role of dynorphin and the kappa opioid receptor in the symptomatology of schizophrenia: a review of the evidence. *Biol Psychiatr.* 2019;86(7):502−511. https://doi.org/10.1016/j.biopsych.2019.05.012.
160. Jutkiewicz EM. The antidepressant -like effects of delta-opioid receptor agonists. *Mol Interv.* 2006;6(3):162−169. https://doi.org/10.1124/mi.6.3.7.
161. Colasanti A, Searle GE, Long CJ, et al. Endogenous opioid release in the human brain reward system induced by acute amphetamine administration. *Biol Psychiatr.* 2012;72(5):371−377. https://doi.org/10.1016/j.biopsych.2012.01.027.
162. Mitchell JM, O'Neill JP, Janabi M, Marks SM, Jagust WJ, Fields HL. Alcohol consumption induces endogenous opioid release in the human orbitofrontal cortex and nucleus accumbens. *Sci Transl Med.* 2012;4(116):116ra6. https://doi.org/10.1126/scitranslmed.3002902.
163. Domi E, Xu L, Pätz M, et al. Nicotine increases alcohol self-administration in male rats via a μ-opioid mechanism within the mesolimbic pathway. *Br J Pharmacol.* 2020;177(19):4516−4531. https://doi.org/10.1111/bph.15210.
164. Cunningham JI, Todtenkopf MS, Dean RL, et al. Samidorphan, an opioid receptor antagonist, attenuates drug-induced increases in extracellular dopamine concentrations and drug self-administration in male Wistar rats. *Pharmacol, Biochem Behav.* 2021;204:173157. https://doi.org/10.1016/j.pbb.2021.173157.
165. Liu X, Jernigen C. Activation of the opioid μ1, but not δ or κ, receptors is required for nicotine reinforcement in a rat model of drug self-administration. *Prog Neuro Psychopharmacol Biol Psychiatr.* 2011;35(1):146−153. https://doi.org/10.1016/j.pnpbp.2010.10.007.
166. Maguire DR, Gerak LR, Sanchez JJ, et al. Effects of acute and repeated treatment with methocinnamox, a mu opioid receptor antagonist, on fentanyl self-administration in rhesus monkeys. *Neuropsychopharmacology.* 2020;45(12):1986−1993. https://doi.org/10.1038/s41386-020-0698-8.
167. Corrigall WA, Coen KM, Adamson KL, Chow BL. The mu opioid agonist DAMGO alters the intravenous self-administration of cocaine in rats: mechanisms in the ventral tegmental area. *Psychopharmacology.* 1999;141(4):428−435. https://doi.org/10.1007/s002130050853.
168. Cornish JL, Lontos JM, Clemens KJ, McGregor IS. Cocaine and heroin ('speedball') self-administration: the involvement of nucleus accumbens dopamine and mu-opiate, but not delta-opiate receptors. *Psychopharmacology.* 2005;180(1):21−32. https://doi.org/10.1007/s00213-004-2135-9.
169. Kitanaka J, Kitanaka N, Hall FS, et al. The selective μ opioid receptor antagonist β-funaltrexamine attenuates methamphetamine-induced stereotypical biting in mice. *Brain Res.* 2013;1522:88−98. https://doi.org/10.1016/j.brainres.2013.05.027.
170. Arias C, Molina JC, Spear NE. Differential role of mu, delta and kappa opioid receptors in ethanol-mediated locomotor activation and ethanol intake in preweanling rats. *Physiol Behav.* 2010;99(3):348−354. https://doi.org/10.1016/j.physbeh.2009.11.012.
171. Giuliano C, Peña-Oliver Y, Goodlett CR, et al. Evidence for a long-lasting compulsive alcohol seeking phenotype in rats. *Neuropsychopharmacology.* 2018;43(4):728−738. https://doi.org/10.1038/npp.2017.105.
172. Giuliano C, Robbins TW, Wille DR, Bullmore ET, Everitt BJ. Attenuation of cocaine and heroin seeking by μ-opioid receptor antagonism. *Psychopharmacology.* 2013;227(1):137−147. https://doi.org/10.1007/s00213-012-2949-9.
173. Bossert JM, Hoots JK, Fredriksson I, et al. Role of mu, but not delta or kappa, opioid receptors in context-induced reinstatement of oxycodone seeking. *Eur J Neurosci.* 2019;50(3):2075−2085. https://doi.org/10.1111/ejn.13955.
174. Ciccocioppo R, Martin-Fardon R, Weiss F. Effect of selective blockade of mu(1) or delta opioid receptors on reinstatement of alcohol-seeking behavior by drug-associated stimuli in rats. *Neuropsychopharmacology.* 2002;27(3):391−399. https://doi.org/10.1016/S0893-133X(02)00302-0.

175. Gerrits MAFM, Kuzmin AV, van Ree JM. Reinstatement of cocaine-seeking behavior in rats is attenuated following repeated treatment with the opioid receptor antagonist naltrexone. *Eur Neuropsychopharmacol.* 2005;15(3):297−303. https://doi.org/10.1016/j.euroneuro.2004.11.004.
176. Simmons D, Self DW. Role of mu- and delta-opioid receptors in the nucleus accumbens in cocaine-seeking behavior. *Neuropsychopharmacology.* 2009;34(8):1946−1957. https://doi.org/10.1038/npp.2009.28.
177. Richard JM, Fields HL. Mu-opioid receptor activation in the medial shell of nucleus accumbens promotes alcohol consumption, self-administration and cue-induced reinstatement. *Neuropharmacology.* 2016;108:14−23. https://doi.org/10.1016/j.neuropharm.2016.04.010.
178. Negus SS, Mello NK. Effects of mu-opioid agonists on cocaine- and food-maintained responding and cocaine discrimination in rhesus monkeys: role of mu-agonist efficacy. *J Pharmacol Exp Therapeut.* 2002;300(3):1111−1121. https://doi.org/10.1124/jpet.300.3.1111.
179. Foltin RW, Fischman MW. Effects of methadone or buprenorphine maintenance on the subjective and reinforcing effects of intravenous cocaine in humans. *J Pharmacol Exp Therapeut.* 1996;278(3):1153−1164.
180. Xi ZX, Fuller SA, Stein EA. Dopamine release in the nucleus accumbens during heroin self-administration is modulated by kappa opioid receptors: an in vivo fast-cyclic voltammetry study. *J Pharmacol Exp Therapeut.* 1998;284(1):151−161.
181. Zamarripa CA, Naylor JE, Huskinson SL, Townsend EA, Prisinzano TE, Freeman KB. Kappa opioid agonists reduce oxycodone self-administration in male rhesus monkeys. *Psychopharmacology.* 2020;237(5):1471−1480. https://doi.org/10.1007/s00213-020-05473-4.
182. Glick SD, Visker KE, Maisonneuve IM. Effects of cyclazocine on cocaine self-administration in rats. *Eur J Pharmacol.* 1998;357(1):9−14. https://doi.org/10.1016/s0014-2999(98)00548-2.
183. Martin TJ, Kim SA, Cannon DG, et al. Antagonism of delta(2)-opioid receptors by naltrindole-5′-isothiocyanate attenuates heroin self-administration but not antinociception in rats. *J Pharmacol Exp Therapeut.* 2000;294(3):975−982.
184. Billa SK, Xia Y, Morón JA. Disruption of morphine-conditioned place preference by a delta2-opioid receptor antagonist: study of mu-opioid and delta-opioid receptor expression at the synapse. *Eur J Neurosci.* 2010;32(4):625−631. https://doi.org/10.1111/j.1460-9568.2010.07314.x.
185. Yang P-P, Yeh T-K, Loh HH, Law P-Y, Wang Y, Tao P-L. Delta-opioid receptor antagonist naltrindole reduces oxycodone addiction and constipation in mice. *Eur J Pharmacol.* 2019;852:265−273. https://doi.org/10.1016/j.ejphar.2019.04.009.
186. Matsuzawa S, Suzuki T, Misawa M, Nagase H. Different roles of mu-, delta- and kappa-opioid receptors in ethanol-associated place preference in rats exposed to conditioned fear stress. *Eur J Pharmacol.* 1999;368(1):9−16. https://doi.org/10.1016/s0014-2999(99)00008-4.
187. Suzuki T, Mori T, Tsuji M, Misawa M, Hagase H. The role of delta-opioid receptor subtypes in cocaine- and methamphetamine-induced place preferences. *Life Sci.* 1994;55(17):PL339−PL344. https://doi.org/10.1016/0024-3205(94)00774-8.
188. Ward SJ, Roberts DCS. Microinjection of the delta-opioid receptor selective antagonist naltrindole 5′-isothiocyanate site specifically affects cocaine self-administration in rats responding under a progressive ratio schedule of reinforcement. *Behav Brain Res.* 2007;182(1):140−144. https://doi.org/10.1016/j.bbr.2007.05.003.
189. Fadda P, Scherma M, Spano MS, et al. Cannabinoid self-administration increases dopamine release in the nucleus accumbens. *Neuroreport.* 2006;17(15):1629−1632. https://doi.org/10.1097/01.wnr.0000236853.40221.8e.
190. Caillé S, Alvarez-Jaimes L, Polis I, Stouffer DG, Parsons LH. Specific alterations of extracellular endocannabinoid levels in the nucleus accumbens by ethanol, heroin, and cocaine self-administration. *J Neurosci.* 2007;27(14):3695−3702. https://doi.org/10.1523/JNEUROSCI.4403-06.2007.
191. Sustkova-Fisero M, Charalambous C, Havlickova T, et al. Alterations in rat accumbens endocannabinoid and GABA content during fentanyl treatment: the role of ghrelin. *Int J Mol Sci.* 2017;18(11):2486. https://doi.org/10.3390/ijms18112486.
192. Sustkova-Fisero M, Jerabek P, Havlickova T, Syslova K, Kacer P. Ghrelin and endocannabinoids participation in morphine-induced effects in the rat nucleus accumbens. *Psychopharmacology.* 2016;233(3):469−484. https://doi.org/10.1007/s00213-015-4119-3.
193. Adamczyk P, Miszkiel J, McCreary AC, Filip M, Papp M, Przegaliński E. The effects of cannabinoid CB1, CB2 and vanilloid TRPV1 receptor antagonists on cocaine addictive behavior in rats. *Brain Res.* 2012;1444:45−54. https://doi.org/10.1016/j.brainres.2012.01.030.

References

194. Economidou D, Mattioli L, Cifani C, et al. Effect of the cannabinoid CB1 receptor antagonist SR-141716A on ethanol self-administration and ethanol-seeking behaviour in rats. *Psychopharmacology*. 2006;183(4):394–403. https://doi.org/10.1007/s00213-005-0199-9.
195. He X-H, Jordan CJ, Vemuri K, et al. Cannabinoid CB 1 receptor neutral antagonist AM4113 inhibits heroin self-administration without depressive side effects in rats. *Acta Pharmacol Sin*. 2019;40(3):365–373. https://doi.org/10.1038/s41401-018-0059-x.
196. Nawata Y, Kitaichi K, Yamamoto T. Prevention of drug priming- and cue-induced reinstatement of MDMA-seeking behaviors by the CB1 cannabinoid receptor antagonist AM251. *Drug Alcohol Depend*. 2016;160:76–81. https://doi.org/10.1016/j.drugalcdep.2015.12.016.
197. Schindler CW, Redhi GH, Vemuri K, et al. Blockade of nicotine and cannabinoid reinforcement and relapse by a cannabinoid cb1-receptor neutral antagonist am4113 and inverse agonist rimonabant in squirrel monkeys. *Neuropsychopharmacology*. 2016;41(9):2283–2293. https://doi.org/10.1038/npp.2016.27.
198. Solinas M, Panlilio LV, Tanda G, Makriyannis A, Matthews SA, Goldberg SR. Cannabinoid agonists but not inhibitors of endogenous cannabinoid transport or metabolism enhance the reinforcing efficacy of heroin in rats. *Neuropsychopharmacology*. 2005;30(11):2046–2057. https://doi.org/10.1038/sj.npp.1300754.
199. Malinen H, Hyytiä P. Ethanol self-administration is regulated by CB1 receptors in the nucleus accumbens and ventral tegmental area in alcohol-preferring AA rats. *Alcohol Clin Exp Res*. 2008;32(11):1976–1983. https://doi.org/10.1111/j.1530-0277.2008.00786.x.
200. Ramesh D, Ross GR, Schlosburg JE, et al. Blockade of endocannabinoid hydrolytic enzymes attenuates precipitated opioid withdrawal symptoms in mice. *J. Pharmacol. Exp. Therapeut*. 2011;339(1):173–185. https://doi.org/10.1124/jpet.111.181370.
201. Simonnet A, Cador M, Caille S. Nicotine reinforcement is reduced by cannabinoid CB1 receptor blockade in the ventral tegmental area. *Addiction Biol*. 2013;18(6):930–936. https://doi.org/10.1111/j.1369-1600.2012.00476.x.
202. Caillé S, Parsons LH. Cannabinoid modulation of opiate reinforcement through the ventral striatopallidal pathway. *Neuropsychopharmacology*. 2006;31(4):804–813. https://doi.org/10.1038/sj.npp.1300848.
203. Madsen HB, Koghar HS, Pooters T, Massalas JS, Drago J, Lawrence AJ. Role of α4- and α6-containing nicotinic receptors in the acquisition and maintenance of nicotine self-administration. *Addiction Biol*. 2015;20(3):500–512. https://doi.org/10.1111/adb.12148.
204. Solinas M, Scherma M, Fattore L, et al. Nicotinic alpha 7 receptors as a new target for treatment of cannabis abuse. *J Neurosci*. 2007;27(21):5615–5620. https://doi.org/10.1523/JNEUROSCI.0027-07.2007.
205. Crespo JA, Sturm K, Saria A, Zernig G. Activation of muscarinic and nicotinic acetylcholine receptors in the nucleus accumbens core is necessary for the acquisition of drug reinforcement. *J Neurosci*. 2006;26(22):6004–6010. https://doi.org/10.1523/JNEUROSCI.4494-05.2006.
206. Kuzmin A, Jerlhag E, Liljequist S, Engel J. Effects of subunit selective nACh receptors on operant ethanol self-administration and relapse-like ethanol-drinking behavior. *Psychopharmacology*. 2009;203(1):99–108. https://doi.org/10.1007/s00213-008-1375-5.
207. Zarrindast M-R, Meshkani J, Rezayof A, Beigzadeh R, Rostami P. Nicotinic acetylcholine receptors of the dorsal hippocampus and the basolateral amygdala are involved in ethanol-induced conditioned place preference. *Neuroscience*. 2010;168(2):505–513. https://doi.org/10.1016/j.neuroscience.2010.03.019.
208. Gunter BW, Gould RW, Bubser M, McGowan KM, Lindsley CW, Jones CK. Selective inhibition of M 5 muscarinic acetylcholine receptors attenuates cocaine self-administration in rats. *Addiction Biol*. 2018;23(5):1106–1116. https://doi.org/10.1111/adb.12567.
209. Rasmussen T, Sauerberg P, Nielsen EB, et al. Muscarinic receptor agonists decrease cocaine self-administration rates in drug-naive mice. *Eur J Pharmacol*. 2000;402(3):241–246. https://doi.org/10.1016/s0014-2999(00)00442-8.
210. Mark GP, Kinney AE, Grubb MC, et al. Injection of oxotremorine in nucleus accumbens shell reduces cocaine but not food self-administration in rats. *Brain Res*. 2006;1123(1):51–59. https://doi.org/10.1016/j.brainres.2006.09.029.
211. Polston JE, Pritchett CE, Sell EM, Glick SD. 18-Methoxycoronaridine blocks context-induced reinstatement following cocaine self-administration in rats. *Pharmacol, Biochem Behav*. 2012;103(1):83–94. https://doi.org/10.1016/j.pbb.2012.07.013.
212. Stoll K, Hart R, Lindsley CW, Thomsen M. Effects of muscarinic M_1 and M_4 acetylcholine receptor stimulation on extinction and reinstatement of cocaine seeking in male mice, independent of extinction learning. *Psychopharmacology*. 2018;235(3):815–827. https://doi.org/10.1007/s00213-017-4797-0.

213. Solecki W, Wickham RJ, Behrens S, et al. Differential role of ventral tegmental area acetylcholine and N-methyl-D-aspartate receptors in cocaine-seeking. *Neuropharmacology*. 2013;75:9—18. https://doi.org/10.1016/j.neuropharm.2013.07.001.
214. See RE, McLaughlin J, Fuchs RA. Muscarinic receptor antagonism in the basolateral amygdala blocks acquisition of cocaine-stimulus association in a model of relapse to cocaine-seeking behavior in rats. *Neuroscience*. 2003;117(2):477—483. https://doi.org/10.1016/s0306-4522(02)00665-6.
215. Yee J, Famous KR, Hopkins TJ, McMullen MC, Pierce RC, Schmidt HD. Muscarinic acetylcholine receptors in the nucleus accumbens core and shell contribute to cocaine priming-induced reinstatement of drug seeking. *Eur J Pharmacol*. 2011;650(2—3):596—604. https://doi.org/10.1016/j.ejphar.2010.10.045.
216. Walter H, Ramskogler K, Semler B, Lesch OM, Platz W. Dopamine and alcohol relapse: D_1 and D_2 antagonists increase relapse rates in animal studies and in clinical trials. *J Biomed Sci*. 2001;8(1):83—88. https://doi.org/10.1159/000054017.

CHAPTER 4

Molecular and cellular mechanisms of addiction

Introduction

When discussing complex behaviors, one question that has dominated Psychology for years is whether these behaviors result from nature or from nurture. This is aptly called the "Nature-Nurture Question". Are behaviors hardwired from birth, or are they learned from others and from personal experiences? Addiction is no exception. Is someone born with a predisposition to develop a substance use disorder (SUD), or do they learn to acquire the behaviors associated with addiction over time? Here is a thought experiment: imagine there are two identical twins, and one develops a SUD. What is the likelihood that the other twin will also develop a SUD? Is this twin doomed to addiction? If genetics are the sole determining factor of addiction, does this mean we could identify genetic markers that put someone at risk of developing a SUD? Could we alter someone's genetics to reduce the risk of developing a SUD? This chapter will primarily focus on the molecular (genetic) basis of addiction and will address some of these questions. Because genes encode things such as receptors, this chapter will also discuss the cellular bases of addiction.

Learning objectives

By the end of this chapter, you should be able to …

(1) Discuss the heritability of addiction and identify specific genes implicated in various SUDs.
(2) Describe the molecular mechanisms of addiction.
(3) Explain how epigenetics influences sensitivity to drug reinforcement.
(4) Describe the pharmacogenetic approach to treating SUDs.
(5) Identify some of the major cellular changes that occur because of substance use.
(6) Describe the cellular mechanisms underlying gene expression and addiction vulnerability.
(7) Discuss the role of astrocytes in addiction.

Heritability of substance use disorders

To determine if individuals are naturally more likely to develop a SUD, we can measure the **heritability** of SUDs. In layman's terms, heritability measures how much of a trait comes from our parents. Heritability ranges from 0 to 1 (or 0%—100%), with values closer to 0 indicating that a trait results primarily from the environment and values closer to 1 (100%) indicating that the trait is influenced primarily by genetics. When reading the literature, you may see that the heritability of a trait is 0.5 (or 50%). This does not mean that 50% of the trait is caused by genetic factors, and 50% is caused by environmental factors. Instead, this means that 50% of the variability (differences) in a trait within the population is due to genetic differences among individuals. For example, the heritability of height is approximately 0.8. Not everyone is the same height. I am 5 feet, 6 inches tall, and my dad and mom are 5 feet, 7 inches and 5 feet, 3 inches tall, respectively. Notice that I am not the same height as my parents. This is probably true for many of you reading this book. Also, some of you reading this book may be shorter than me, while some of you are most certainly taller than me. Notice that the heritability of height is not 1.0. Remember, a heritability of 0.8 means that the differences in individuals' heights are largely due to genetic differences across these individuals. Two individuals can have parents that are nearly the same height, but they will not necessarily be the same height. My wife's parents are nearly the same height as my parents, but I am taller than my wife.

There are a few techniques researchers can use to isolate the heritability of a SUD from other causes like family upbringing, social influences, or learning. Researchers can conduct family, twin, or adoption studies. In family studies, one can determine if an individual with a SUD has family members, such as parents or siblings, that also have a SUD. In twin studies, researchers can use twins that are either *monozygotic* (i.e., identical) or *dizygotic* (i.e., fraternal). In adoption studies, researchers can examine children of substance-dependent parents that were adopted by nonsubstance-dependent individuals. If addiction is heritable, one would expect to see the following across these studies: (a) individuals with a SUD should be more likely to have family members that have a SUD compared to those without a SUD; (b) higher concordance of SUDs in monozygotic twins compared to dizygotic twins because monozygotic twins share 100% of their genes while dizygotic twins share 50% of their genes; and (c) individuals with substance-dependent parents should have a greater risk of developing a SUD, even if raised by nondependent individuals.

Overall, evidence shows that SUDs are highly heritable. For example, the heritability of alcohol use disorder is approximately 43%—53%.[1] Twin studies indicate that the heritability of SUDs ranges from 39% (for hallucinogenic drugs) to 72% (for cocaine).[2] Additionally, individuals adopted during infancy whose biological parents smoke have higher drug use.[3] Although SUDs have high heritability, family/adoption studies do not isolate which specific genes are associated with addiction. Furthermore, despite the high heritability rates of various SUDs, research shows a vital role of the environment in addiction. In the adoption study I just mentioned, Keyes et al. also found that individuals whose adoptive parents smoke are at increased risk of developing a SUD.[3] We will discuss the role of environmental factors on addiction later in this book. For now, let's start highlighting some of the major genes associated with addiction.

Genes associated with drug addiction in humans

With advances in technology, researchers are now able to examine an individual's entire genome at once. In a **genome-wide association study**, researchers recruit numerous participants and scan their genomes to find potential genetic variations (**polymorphisms**) associated with a disease, including SUDs. Doing an experiment like this was extremely difficult before the early 2000s. With the completion of the Human Genome Project and the International HapMap Project (note, HapMap is short for **haplotype** map), genome-wide association studies can be conducted with greater ease. Given that our DNA (deoxyribonucleic acid) is approximately 99.5% identical to all other individuals, research has focused on trying to determine which genetic changes are responsible for the development of physical and psychiatric conditions. One polymorphism of particular interest is a **single-nucleotide polymorphism (SNP)**. A SNP is simply a substitution of a single *nucleotide* at a specific position in the genome[4] (Fig. 4.1). Imagine most individuals have a nucleotide of adenine-thymine at a certain position of DNA located on chromosome 7. If some individuals have a guanine-cytosine nucleotide at this position, this is considered a SNP. There are multiple SNPs that can occur within a gene. Detailing the specific SNPs associated with each SUD is beyond the scope of this book. Instead, I will highlight some of the major genes that have been implicated in SUDs.

Genes encoding for dopamine receptors

Because dopamine has received extensive attention in SUDs, as we covered in the previous chapter, examining genes that encode dopamine receptors is a logical first step in unraveling the genetic basis of addiction. Each dopamine receptor is encoded by a separate gene: conveniently named *DRD1*, *DRD2*, *DRD3*, *DRD4*, and *DRD5*. Based on the gene names, it should be obvious which dopamine receptors are associated with each gene. *DRD1* and *DRD5* encode the dopamine D_1-like receptors D_1 and D_5, and *DRD2-DRD4* encode the dopamine D_2-like receptors D_2, D_3, and D_4.

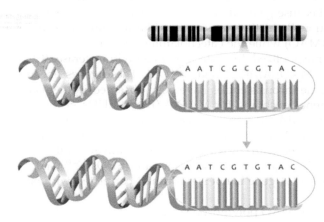

FIGURE 4.1 An image showing an example of a single nucleotide polymorphism. Notice that cytosine has been replaced by thymine. *The image comes from Wikimedia Commons and was posted by the NHS National Genetics and Genomics Education Center (https://commons.wikimedia.org/wiki/File: Single_nucleotide_polymorphism_substitution_ mutation_diagram_-_cytosine_to_thymine.png).*

Although many behavioral pharmacology experiments have examined the contribution of D_1-like receptors to multiple aspects of addiction, genome-wide association studies have not examined the link between the *DRD1* or the *DRD5* genes to addiction as much as one would expect. Somewhat surprisingly, polymorphisms of genes encoding for D_1-like receptors have not been consistently linked to psychostimulant dependence.[5] A haplotype of the *DRD1* gene has been associated with ethanol dependence,[6] and certain SNPs of the *DRD1* gene have been linked to heroin dependence.[7] More specifically, individuals with one of two *DRD1* gene SNPs are more likely to develop rapid heroin dependence following the first use.[8]

The *DRD2* gene is associated with crack cocaine use in women,[9] but there have been some inconsistencies when examining the contribution of the *DRD2* and the *DRD3* genes to psychostimulant dependence. Some studies have failed to detect significant correlations between these genes and psychostimulant dependence[10,11] while others have strongly implicated these genes in cocaine dependence.[12] *DRD4* gene polymorphisms have been tied to methamphetamine dependence.[13]

Beyond cocaine and methamphetamine, polymorphisms in the *DRD2*, the *DRD3*, and the *DRD4* genes are associated with smoking behavior,[14] smoking frequency,[15] and craving for nicotine.[16] Genes encoding for D_2-like receptors are important mediators of opioid dependence[17] (but see[18]) and cue-elicited craving for heroin.[19,20] In particular, polymorphisms in the *DRD3* gene are associated with the early onset of heroin dependence.[21] Finally, genetic association studies have shown that D_2-like receptors are involved in cannabis dependence.[22] Overall, genetic studies have shown that D_2-like receptors are more consistently linked to SUDs compared to D_1-like receptors. This finding is interesting given that behavioral pharmacology experiments have consistently implicated D_1-like receptors as an important locus of addiction-like behaviors. Later in this chapter, we will discuss how long-term drug use alters D_2-like receptor expression in the brain. As you will learn, drugs cause widespread changes in the number of D_2-like receptors in the brain.

Genes encoding for monoamine clearance

Perhaps genetic association studies fail to detect significant correlations between D_1-like receptor encoding genes and drug use because genes that encode for monoamine clearance are more important determinants of addiction. Recall that monoamines like dopamine can be metabolized by either monoamine oxidase (MAO) while the catecholamines dopamine, norepinephrine, and epinephrine can also be metabolized by catechol-O-methyltransferase (COMT). There are two genes associated with MAO: *MAO-A* and *MAO-B*. These genes encode for MAO-A and MAO-B, respectively. Some inconsistencies have been observed in the literature concerning the association between genes encoding for MAO-A/B and SUDs. Some studies have shown that polymorphisms in the *MAO-A* gene are associated with increased tobacco use[23] and adolescent ethanol consumption.[24] Additionally, some evidence has shown that polymorphisms in the *MAO-B* gene are associated with smoking behavior.[23] However, other studies have found no association between *MAO* gene polymorphisms and SUDs.[25,26] One meta-analysis found that, collectively, the *MAO-A* gene is not significantly correlated with ethanol dependence.[27]

Similar to what has been reported with MAO polymorphisms, the influence of *COMT* gene polymorphisms on SUDs is not consistent across or even within drug classes. Some research has implicated alterations in the *COMT* gene to nicotine, ethanol, cannabis, and methamphetamine dependence,[28–31] smoking cessation,[32] and heroin relapse.[33] Additionally, a specific polymorphism of the *COMT* gene has been linked to increased treatment dropout rates for opioid dependence.[34] Conversely, *COMT* gene polymorphisms are not associated with synthetic cannabinoid use disorder,[35] opioid addiction,[36] or with Type 2 alcohol use disorder (marked by early onset antisocial behaviors).[37] Although one study found a link between the *COMT* gene and nicotine dependence, another study failed to detect such a relationship.[38]

Monoamines can be cleared from the synapse by transporters and repackaged into vesicles. The *SLC6A3* gene encodes for the dopamine transporter (DAT). Thus, this gene is also called the *DAT1* gene. Up until now, the genes that have been covered have names that map onto the name of the protein. Why is the gene that encodes for DAT called *SLC6A3*? SLC6A3 is an abbreviation for **S**olute **C**arrier family six member 3, which is another name for DAT. To go on a brief tangent: the term solute means a solution, so solute carriers are just proteins that move solutions across a membrane. There are 66 solute carrier families, with family six being composed of transporters for various neurotransmitters and amino acids, among a few other things. With that said, polymorphisms of the *SLC6A3* gene have been associated with cocaine use[11] and crack cocaine use.[39] *SLC6A3* polymorphisms can also affect how individuals respond to pharmacotherapies for SUDs. Individuals with genetically higher DAT levels have better treatment outcomes when using disulfiram (Antabuse) to treat cocaine dependence.[40] While this gene is associated with cocaine dependence, the *SLC6A3* gene does not appear to be an important mediator of nonstimulant use disorders, as polymorphisms in this gene are not associated with ethanol dependence[41] or heroin dependence,[42] as well as cue-elicited craving for heroin.[19]

Given that the *SLC6A3* gene is not consistently linked to nonstimulant drug dependence, other research has focused on the gene that encodes for the serotonin transporter (SERT), the *SLC6A4* gene. In contrast to the *SLC6A3* gene, polymorphisms of the *SLC6A4* gene are not consistently linked to cocaine dependence,[43,44] but there is some evidence that a specific polymorphism of this gene is associated with faster escalated cocaine use.[45] Like the *SLC6A3* gene, the *SLC6A4* gene is not a significant predictor of alcohol use disorder,[46] although a polymorphism of the *SLC6A4* leads to better treatment outcomes when using disulfiram to treat cocaine dependence.[47] Likewise, *SLC6A4* gene polymorphisms are associated with treatment outcomes for ethanol dependence, but not cocaine dependence, in African Americans.[48] Polymorphisms of the *SLC6A4* gene have been implicated in heroin dependence, but only in White individuals.[49] *SLC6A4* gene polymorphisms have also been associated with adolescent smoking[50] and risk of tobacco dependence,[51] but these polymorphisms are not associated with treatment response to nicotine replacement therapy.[52] Overall, genes that regulate monoamine clearance are a mixed bag. Some of the work detailed here is promising, but there are several inconsistencies that raise serious questions about the utility of targeting these genes. One thing I would like to point out is that racial differences appear to have a modulatory role when examining genetic associations with addiction. Chapter 10 will discuss some of the sociocultural factors associated with addiction, including race.

Genes encoding for acetylcholine nicotinic receptor

Because nicotine binds to acetylcholine nicotinic receptors, which activate dopaminergic neurons in the ventral tegmental area (VTA), there has been much interest in determining the contribution of genes that encode the nicotinic receptor to addiction, particularly nicotine use disorder. Nicotinic receptors are composed of five subunits, and there are approximately 17 distinct subunits. Each subunit is encoded by a different gene: *CHRNA1-CHRNA10* (alpha subunits), *CHRNB1-CHRNB4* (beta subunits), *CHNRD* (delta subunit), *CHRNE* (epsilon subunit), and *CHRNG* (gamma subunit). Not surprisingly, polymorphisms in various nicotinic receptor genes are implicated in nicotine dependence. The *CHRNA2* gene appears to be an important mediator of the subjective effects individuals experience when smoking a cigarette.[53] The *CHRNA2*, *CHRNA3*, *CHRNA4*, *CHRNA5*, *CHRNA6*, *CHRNA7*, *CHNRB1*, and *CHRNB3* genes are all associated with smoking behavior and nicotine dependence,[54–61] and the *CHRNA2* gene is important for initial subjective responses to nicotine.[62] This is important because individuals that have a pleasurable experience after smoking a cigarette are going to be more likely to continue smoking while individuals that have a less pleasurable initial experience may not want to continue smoking. For example, the first time I smoked a cigarette, I thought it was absolutely disgusting. As such, I rarely had a desire to try cigarettes in the future. One last thing I want to mention concerning nicotinic receptor genes and nicotine dependence is that women lacking a certain *allele* (copy) of the *CHRNA5* gene are more sensitive to smoking-related images, as determined by greater activation of the hippocampus and the dorsal striatum.[63] This finding is relevant as increased sensitivity to drug-paired cues could precipitate relapse. Thus, women missing this allele may be at risk of developing nicotine use disorder.

Beyond nicotine, SNPs in the *CHRNA3*, *CHRNA5*, and *CHRNA6* genes are associated with ethanol dependence.[64,65] Like nicotine, the *CHRNA2* gene mediates the initial subjective response to ethanol.[62] The shared genetics of subjective responses to nicotine and ethanol could partially explain why ethanol and nicotine use cooccur. Polymorphisms in the *CHRNA5* gene are associated with cocaine/crack cocaine use disorder,[65,66] as well as slower transition from first cocaine use to cocaine use disorder.[67] Variants of the *CHRNB3* gene are associated with cocaine dependence and ethanol dependence[64,68] while a SNP for the *CHRNB4* gene is associated with shorter time to relapse in cocaine users.[67] Given the high consistency across studies, genes encoding for nicotinic receptors are important for various SUDs, not just nicotine dependence.

Genes encoding for ethanol metabolism

There are two ethanol metabolizing genes: *ADH1B* and *ALDH2*. *ADH1B* encodes for alcohol dehydrogenase, and *ALDH2* encodes for aldehyde dehydrogenase. As covered in Chapter 2, ethanol is first metabolized by alcohol dehydrogenase into acetaldehyde before being metabolized into acetate by aldehyde dehydrogenase. As expected, variations in both genes affect ethanol consumption and ethanol dependence.[69] Beyond alcohol use disorder, research has shown that individuals with a particular allele for the *ALDH2* gene are less likely to develop heroin dependence,[18] although interactions between the *ALDH2* gene and the *ADH1B* gene (as well as *DRD2* gene) modify the protective effects of this allele.[70] Specific alleles of the *ALDH2* gene also predict smoking cessation.[71] These genetic studies corroborate behavioral pharmacology studies examining the use of disulfiram to treat various SUDs.

Summary of human gene association studies

Given the complexity of drug addiction, there is no single gene that determines if one is at increased risk of developing a SUD. To complicate matters, individual genes may not be associated with drug use, but gene interactions may act as a predisposing factor for dependence. For example, Costa-Mallen et al. found that polymorphisms in the *MAO-B* gene or the *DRD2* gene failed to predict smoking behavior; however, men with a certain *MAO-B* polymorphism and a certain *DRD2* polymorphism were more likely to smoke.[72] To complicate matters more, an allele that is a risk factor for one SUD can be protective against a different SUD. A specific SNP for the *CHRNA5* gene is associated with increased nicotine dependence but is associated with decreased cocaine dependence.[58] One important consideration is that some genes implicated in SUDs have not been widely *replicated*. Replication refers to repeating an experiment and is an important foundation of research. If a single study examining the link between one gene and addiction is conducted, there is always the chance that the researchers detect a false positive. Think of it this way: if I had to shoot a basketball from midcourt, I would rarely get the ball into the basket (I am not good at basketball). There is the chance I get lucky and make this shot, albeit the frequency of such an event would be extremely low. A similar phenomenon occurs in research in which a study happens to yield significant correlations. Theoretically, if we were to replicate many of these genome-wide association studies, the significant correlations observed in the original study may not hold up. The reason I mention this is that one comprehensive review found that the only genes consistently linked to SUDs are the *DRD2*, the *CHRNA5/CHRNA3/CHRNB4*, the *ALH1B/ALDH2*, and the *ankyrin repeat and kinase domain containing 1 (ANKK1)* (associated with D_2-like receptor expression).[54] This finding highlights the need for experimental replications.

Quiz 4.1

1. Which gene encodes for alcohol dehydrogenase?
    a. *ADH1B*
    b. *ALDH2*
    c. *SLC6A3*
    d. *SLC6A4*
2. The heritability of cocaine use disorder is approximately 72%. What does this mean?
    a. 72 out of 100 individuals will develop cocaine use disorder at some point in their life.
    b. Genetic factors explain 72% of the variability in cocaine dependence across individuals.
    c. Individuals have a 72% chance of inheriting cocaine use disorder if one of their parents has cocaine use disorder.
    d. None of the above.
3. Which of the following genes has NOT been consistently linked to SUDs?
    a. *ALDH2*
    b. *CHRNA3*
    c. *DRD1*
    d. *DRD2*

Answers to quiz 4.1

1. a — *ADH1B*
2. b — Genetic factors explain 72% of the variability in cocaine dependence across individuals.
3. c — *DRD1*

Genetic approaches to studying addiction in animals

Because gene association studies are correlative by nature, researchers cannot control the environmental histories of participants. For example, individuals with a polymorphism in the *MAO-A* gene that have experienced sexual abuse are more likely to consume ethanol during adolescence.[24] If researchers do not know the life history of participants, they may erroneously conclude that a gene is not associated with addiction. Yet, this may not be the case as the association between a gene and a behavior may be mediated by an environmental factor. Another issue that complicates gene association studies is that race or ethnicity can modulate the associations between a gene and dependence-like behaviors. When discussing the *SLC6A3* gene, I cited one study that failed to detect an association between a polymorphism of this gene and heroin dependence.[42] This study was conducted using Chinese participants. However, one study conducted in India reported a significant association between the *SLC6A3* gene and heroin dependence.[73] This is interesting because these collective results seem to suggest that Chinese individuals may be protected against heroin dependence even if they have alleles of the *SLC6A3* that put others at risk. To better elucidate the genetic basis of addiction, preclinical studies are useful for isolating cause and effect. In other words, studies with animals can better enable us to determine which genes are responsible for addiction-like behaviors.

Measuring drug-induced changes in transcription factor expression

Proteins called **transcription factors** bind to DNA and control gene expression by either initiating or inhibiting *transcription*, the process of converting a segment of DNA into RNA (ribonucleic acid). There are numerous transcription factors, with several being implicated in dependence-related behaviors.[74] This section will cover just one major class of transcription factors: the Fos family. The Fos family is composed of several individual transcription factors: c-Fos, FosB and its truncated splice variants delta FosB (ΔFosB) and Δ2ΔFosB, Fos-related antigen 1 (FRA1), and FRA2.[75] Most drug addiction studies have focused on c-Fos and FosB/ΔFosB. There are multiple stimuli that can activate c-Fos. Researchers are interested in c-Fos because its activation occurs very quickly and lasts for a short period of time; thus, it is considered an **immediate early gene**. Expression of c-Fos provides an indirect measure of neuronal activity in a brain region as it is often activated during action potentials. In short, numerous drugs increase c-Fos levels in structures associated with addiction, such as the nucleus accumbens (NAc), the frontal cortex, and the dorsal striatum,[76,77] as well as in other areas of the brain connected to structures in the reward pathway, like the hypothalamus and the hippocampus.[78] c-Fos expression is also observed during drug withdrawal[79] and during reinstatement of drug-seeking behavior.[80]

Like c-Fos, FosB/ΔFosB levels increase in several brain regions implicated in addiction following exposure to multiple drugs, such as cocaine,[81] methamphetamine,[82] ethanol,[83] morphine,[84] and nicotine.[85] Furthermore, drug withdrawal and relapse-like behavior increase FosB/ΔFosB levels in the dorsal striatum and the ventral striatum.[80,86] In the NAc, ΔFosB has received considerable attention in addiction research. While drugs often produce transient increases in Fos transcription levels, chronic drug administration leads to accumulating levels of ΔFosB without altering other Fos family transcription expression.[87] Furthermore, overexpression of ΔFosB increases both cocaine conditioned place preference (CPP) and morphine CPP[88,89] and increases cocaine self-administration.[90] These results show that during distinct stages of the addiction cycle, transcription of DNA into RNA increases. This will become more relevant when we discuss the cellular mechanisms of addiction near the end of this chapter.

While c-Fos and FosB/ΔFosB provide information about brain activation, one limitation is that this information does not always inform one of the specific neurotransmitter systems at play during drug self-administration or during withdrawal and relapse. To learn more about which neurotransmitter systems are involved in addiction-like behaviors, other genetic approaches need to be considered.

Measuring drug-induced changes in mRNA expression

Messenger RNA (mRNA) is involved in the synthesis of proteins like receptors and transporters. Measuring drug-induced changes in mRNA expression can allow us to better isolate the genetic underpinnings of SUDs.

Drugs can either **upregulate** (increase) or **downregulate** (decrease) mRNA expression. Covering the effects of every drug class on each neurotransmitter system is not the purpose of this section. Many drugs increase D_1-like and D_2-like receptor mRNA expression in the NAc and in the dorsal striatum[91,92] (but see[93]) and psychostimulant administration leads to increased dopamine D_1-like receptor mRNA expression in the frontal cortex.[91] However, extended self-administration of cocaine leads to decreased D_2-like mRNA expression in the medial prefrontal cortex (mPFC), the orbitofrontal cortex (OFC), and the cingulate cortex, a finding that is not observed in animals given limited access to cocaine during the day.[92] The decreased D_2-like mRNA expression in the frontal cortex is not isolated to psychostimulants, as ethanol decreases D_2-like mRNA expression in the frontal cortex.[93] In short, drugs greatly affect mRNA expression of dopaminergic receptors in the mesocorticolimbic pathway. The implications of these findings will be discussed later in this chapter. For now, I will briefly highlight how drugs can alter mRNA expression of nondopaminergic neurons.

Concerning the serotonergic system, cocaine increases 5-HT_{1B} mRNA expression in the NAc shell and in the dorsal striatum.[94,95] Chronic ethanol administration also increases 5-HT_{1B} mRNA in the dorsal striatum.[96] Acute injections of nicotine increase 5-HT_{1A} mRNA expression in the cerebral cortex and in the hippocampus, although chronic nicotine administration decreases mRNA expression.[97] Ethanol and cannabinoid administration increases 5-HT_{1A} mRNA expression in the hippocampus[98,99] (but see[100]) The findings reported here are mostly consistent with behavioral pharmacology experiments showing that stimulation of 5-HT$_1$ receptors increases the rewarding effects of drugs.

Opioid receptor gene expression is also altered following drug exposure, as cocaine increases mRNA expression of mu, kappa, and delta opioid receptor subtypes in the dorsal striatum and in the PFC.[101–103] Nicotine also increases mu opioid and kappa opioid receptor mRNA expression.[104,105] While acute drug administration increases opioid receptor mRNA expression, chronic drug treatment, whether with cocaine or with ethanol, decreases kappa receptor mRNA expression.[106] Recall that opioid receptors play an important role in the rewarding properties of drugs. Stimulation of opioid receptors on GABAergic neurons allows dopaminergic neurons to become disinhibited, thus increasing dopamine levels in the NAc.

In addition to the alterations described above, ethanol administration affects NMDA receptor and CB$_1$ mRNA levels. GluN1 subunit mRNA expression is increased in the dorsal striatum of rats[107] and in the frontal cortex of primates.[108] Ethanol causes widespread changes in CB$_1$ receptor expression, decreasing expression in the striatum, the central nucleus of the amygdala (CeA), and part of the hypothalamus.[109] The effects of ethanol on NMDA receptor mRNA levels are somewhat consistent with the effects of phencyclidine (PCP) on AMPA receptor mRNA levels in the PFC.[110] Like ethanol, opioids decrease CB$_1$ receptor mRNA in the dorsal striatum and in the NAc.[111] The increased expression of NMDA/AMPA mRNA is consistent with the findings that activation of ionotropic glutamate receptors is important for the stimulation of dopaminergic neurons in the mesocorticolimbic pathway. However, drug effects on CB$_1$ mRNA expression are inconsistent with what one would expect to observe. So far, drugs upregulate mRNA expression of receptors that increase drug reward when stimulated (e.g., D$_1$-like, 5-HT$_{1B}$, mu opioid). Although stimulation of CB$_1$ receptors increases the rewarding effects of drugs, drugs often downregulate mRNA CB$_1$ receptor expression. We will return to this later in the chapter.

Drug withdrawal also affects mRNA expression in the brain. Specifically, cocaine withdrawal increases kappa opioid receptor expression,[112] but decreases D$_1$-like and 5-HT$_{1B}$ expression in the dorsal striatum.[95,113] Cocaine withdrawal causes a transient decrease in D$_2$-like mRNA expression in the NAc and in the dorsal striatum, which returns to baseline levels within 24 h.[114] Withdrawal from morphine increases CB$_1$ mRNA in the dorsal striatum,[111] but decreases 5-HT$_{1A}$ mRNA expression in the raphe nuclei.[115] Similarly, ethanol withdrawal decreases 5-HT$_{1A}$ mRNA expression in the hippocampus.[100] Notice that withdrawal often alters mRNA expression in the opposite direction compared to following drug administration.

The main takeaway is that various drug classes cause widespread changes in mRNA expression in the reward pathway. While this information is important, measuring mRNA expression on its own does not provide the complete story of the genetic basis of addiction. To better determine the molecular basis of addiction, researchers can manipulate gene expression in animals to determine if specific mRNA mediates addiction-like behaviors.

Effects of altered gene expression on addiction-like behaviors

As highlighted at the end of the last chapter, one weakness of behavioral pharmacology studies is that drugs often have more than one mechanism of action. For example, some glutamatergic drugs have actions on serotonergic receptors. This can make interpreting the results of behavioral pharmacology experiments difficult. One way to circumvent this issue is to use animals that have been genetically modified such that they are missing a receptor of interest. Typically, mice are used in this type of research as their genome is remarkably similar to humans and are fairly inexpensive. Numerous studies have examined how genetically altering specific receptors affects drug reward/reinforcement. Overall, reducing/knocking out receptors that stimulate dopamine release, whether directly or indirectly, attenuates drug reinforcement. These receptors include:

— Dopamine D_1/D_5 receptors
— Serotonin 5-HT_1 receptors
— Norepinephrine α_1 and $\beta_{1/2}$ receptors
— Acetylcholine nicotinic and muscarinic $M_1/M_3/M_5$ receptors
— Glutamate AMPA/NMDA receptors and Group I mGluRs
— Opioid mu/kappa/delta receptors
— Cannabinoid CB_1/CB_2 receptors

Concerning cocaine, mice lacking either the D_1-like receptor, the noradrenergic α_1 receptor, the muscarinic M_5 receptor, the $mGluR_5$, the mu and delta opioid receptors, or the CB_1 receptor fail to learn to respond for cocaine infusions or self-administer less cocaine.[116–122] Selective reduction of D_1-like receptors in the NAc shell decreases the reinforcing effects of cocaine.[123] Deletion of the nicotinic β_2 subunit or the muscarinic M_1 receptor decreases cocaine CPP[124,125] while CB_1 deficient mice self-administer less MDMA.[126]

Unsurprisingly, removing nicotinic receptor α_4, α_6, or β_2 subunits prevents nicotine self-administration.[127] Similarly, NMDA receptor deletion from dopaminergic neurons prevents nicotine CPP.[128] Research has shown differential roles for the α_7 and the β_2 subunits on nicotine reward. Mice lacking the β_2 subunit show an initial decrease in preference for nicotine that increases across weeks, whereas mice lacking the α_7 subunit show an initial preference for nicotine that decreases across time.[129] Studies have shown that the nicotinic α_4 receptor is required for nicotine CPP.[130] Mu or delta deficient mice show decreased preference for nicotine,[131,132] and CB_1 knockout mice fail to develop nicotine or THC CPP.[133,134]

Mice lacking the GluN2A subunit of the NMDA receptor do not show a preference for ethanol[135] or engage in dependence-like ethanol drinking.[136] Furthermore, selective deletion of the GluN2B subunit from cortical interneurons decreases ethanol consumption.[137] Deleting the mu or the delta opioid receptor decreases preference for ethanol.[138] CB_1 receptor knockout mice show reduced ethanol drug self-administration and CPP.[139]

Mice lacking mu or delta opioid receptors show decreased preference for opioids.[140] Reducing the number of GluN2B-containing neurons in the NAc also prevents morphine CPP[141] while increasing GluN2B expression in the frontal cortex potentiates morphine CPP.[142] CB_1 knockout mice fail to self-administer morphine or develop morphine CPP.[143,144] Noradrenergic α_1- and M_1/M_5-lacking mice show blunted morphine CPP.[118,124,145]

In addition to blunting drug reinforcement/reward, knockout models have shown that selective deletion of certain receptors can attenuate the locomotor-stimulant effects of drugs. D_1 receptor and noradrenergic α_1 receptor knockout mice fail to develop locomotor sensitization to both stimulants and opioids,[118,146–148] and D_5 knockout mice show a blunted response to the locomotor-stimulant effects of cocaine.[149] The GABA α_2 subunit influences an animal's response to the acute locomotor effects of ethanol. Animals lacking the α_2 subunits are more likely to fall and take longer to right themselves following ethanol administration.[150]

In contrast to receptors that increase dopamine levels, studies have shown that deleting receptors that decrease dopamine release increases addiction-like behaviors. Mice lacking D_2-like receptors acquire cocaine self-administration at a faster rate than mice with these receptors and self-administer more cocaine.[151] In particular, D_2-like autoreceptors play a critical role in the acquisition of drug-taking behavior, as mice lacking these autoreceptors acquire cocaine self-administration at a faster rate and display enhanced cocaine CPP and cocaine-seeking behaviors.[152,153] Furthermore, mice lacking D_2-like receptors show increased behavioral responses to D_1-like receptor agonists.[154] Dobbs et al. argue that this hypersensitivity promotes behaviors such as drug self-administration and locomotor sensitization.[154] Why? Remember D_2-like receptors are inhibitory and work to reduce dopamine synthesis and/or dopamine release. Without this regulatory mechanism in place, D_1-like receptors can be stimulated more frequently. Outside of the dopaminergic system, mice lacking muscarinic M_4 receptors self-administer more ethanol and cocaine compared to mice with these receptors,[155,156] and mGluR$_2$ knockout mice are more sensitive to the conditioned rewarding effects of cocaine.[157]

Interestingly, D_2-like knockout mice fail to develop locomotor sensitization to psychostimulants[158] as D_1-like receptor knockout mice. However, if D_2 autoreceptors are deleted, mice become more hyperactive following cocaine administration,[152] which mirrors the finding that these animals are more sensitive to cocaine reward. Selective deletion of mGluR$_2$ increases the locomotor-stimulant effects of cocaine as well.[157]

Although there is high concordance between behavioral pharmacology experiments and genetic alteration studies, there are some major discrepancies. One major discrepancy concerns 5-HT$_{1B}$ receptors. In the previous chapter, you read that stimulating these receptors potentiates the reinforcing effects of drugs. Thus, removing these receptors should attenuate drug reinforcement. Although 5-HT$_{1B}$ knockout mice do not develop cocaine CPP,[159] they are more sensitive to the acute locomotor-stimulant effects of cocaine,[160] acquire cocaine self-administration at a faster rate[161] and are more motivated to self-administer cocaine.[162] Another paradox is that pharmacological blockade of 5-HT$_3$ receptors reduces addiction-like phenotypes, but *overexpression* of the 5-HT$_3$ receptor blunts the conditioned rewarding effects of cocaine.[163] Serotonin is not the only monoamine where discordant findings have been observed. D_2-like knockout mice self-administer less ethanol and morphine[164,165] while **transgenic** mice overexpressing either the α_3, α_5, or the β_4 nicotinic subunits drink less ethanol compared to wildtype controls.[166] Furthermore, even though GluN2B-selective antagonists are efficacious in reducing the reinforcing effects of drugs, deleting the GluN2B subunit from D_1-like receptors enhances cocaine CPP 4 weeks following the initial preference test.[167] Another discrepancy associated with the glutamatergic system is that deletion of mGluR$_2$ decreases drug-seeking behavior and relapse-like behavior.[168] One notable exception has been observed with the opioid system: mice with decreased delta opioid receptors consume more ethanol than mice with these receptors.[169] These findings provide further justification for the use of genetic models to elucidate the neuromechanisms of addiction.

Optogenetics

Today, there are methods that allow researchers to isolate which specific neurons in a brain region are involved in certain behaviors. In **optogenetics**, neurons are genetically modified such that they express light-sensitive ion channels. To understand how optogenetics works, a brief discussion of how the visual system functions is warranted. When light enters our eyes, it activates photoreceptors by breaking the bond between *opsin* and *retinal* proteins. Put simply, there are special receptors in our eye that respond to light instead of neurotransmitters like dopamine or glutamate. There are different types of opsin proteins. For example, *channelrhodopsin* is found in green algae and serves to move the algae in response to light. Specifically, channelrhodopsin responds to blue light, causing cell excitation. Another type of opsin is *halorhodopsin*, which is sensitive to yellow/green light. Halorhodopsin causes Cl^- to enter cells; thus, this rhodopsin is inhibitory.

How exactly does optogenetics work? The animal needs to have one of the various opsin genes inserted into select neurons. This requires the researcher to engage in genetic engineering. If the researcher tries to inject the opsin gene of interest into a brain region, the neurons will not express the light-sensitive ion channels. The researcher needs to couple the opsin gene with something called a *promoter*. A promoter is a sequence of DNA that activates a gene; the promoter allows transcription to occur. Next, the researcher has two options on how to insert the opsin gene and the promoter into the animal. First, the researcher can insert the opsin gene and the promoter into a viral vector and inject this vector into the brain region of interest. Second, the researcher can inject the gene/promoter into a zygote (fertilized egg). During the experiment, animals are connected to a fiber optic cable. This cable transmits a specific-colored light to activate the neurons that now express the opsin gene of interest.[170] See Fig. 4.2 for a schematic of how optogenetics works and to see an image of a rat connected to a fiber optic cable.

As a relatively newer neuroscience technique, not nearly as many optogenetic studies have been conducted as behavioral pharmacology or receptor knockout/knock-in studies. So far, researchers have found that optogenetic stimulation of D_2-containing medium spiny neurons (MSNs) in the NAc suppresses cocaine self-administration,[171] and stimulation of GABAergic neurons in the NAc, particularly those that project to the rostromedial tegmental nucleus, disrupts cocaine CPP.[172] Inhibiting NAc neurons that project to the ventral pallidum can block the reinstatement of cocaine self-administration.[173] Optogenetic inhibition of tyrosine hydroxylase-containing neurons (remember, tyrosine hydroxylase is required for dopamine synthesis) in the VTA also decreases relapse-like behaviors.[174] Collectively, these studies confirm what many behavioral pharmacology experiments have shown previously: decreasing dopamine neurotransmission in the mesolimbic pathway reduces the reinforcing effects of drugs, particularly psychostimulants.

Outside of the NAc/VTA, optogenetic inhibition of the CeA decreases cocaine self-administration, while stimulating these neurons, but not intra-basolateral amygdala (BLA) neurons, increases preference for cocaine over sucrose.[175] These studies highlight that increased activity of the CeA increases the reinforcing effects of drugs, possibly by activating NAc neurons.

In the frontal cortex, selectively inactivating the prelimbic (PL) region of the mPFC increases cocaine self-administration but impairs reinstatement of cocaine-seeking behavior;

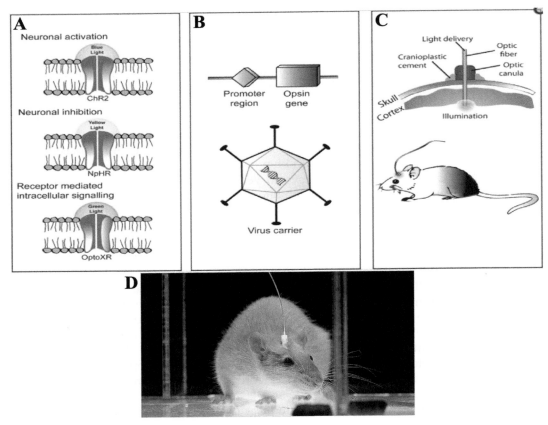

FIGURE 4.2 (A) Schematic showing the main concept of optogenetics. If channelrhodopsin 2 receptors (ChR2) are inserted into the brain, they will be activated by blue light, which ultimately excites neurons containing these receptors. If halorhodopsin (NpHR) receptors are inserted into the brain, they will be activated by yellow light, which will cause neuronal inhibition. If OptoXR receptors are inserted into the brain, they will be activated by green light. Activation of these receptors mimics G protein-coupled receptors. For example, there are opsin G protein-coupled receptors that resemble β_2 noradrenergic receptors (G_s) and α_2 noradrenergic receptors (G_q) (B) Schematic showing what needs to be inserted into a viral vector. In addition to adding the opsin gene of interest (whether ChR2, NpHR, or OptoXR), a promoter needs to be added to the viral vector. This allows the gene of interest to be expressed in specific neurons. (C) After the viral vector has been injected into the animal's brain, the animal is connected to a fiber optic cable that extends into the animal's brain. Light emitted from the cable will activate neurons that express the opsin gene. (D) A photo of a rat connected to a fiber optic cable. Notice a blue light is turned on; thus, this researcher is causing neuronal excitation by stimulating ChR2-containing neurons. *Panels (A–C) come from Fig. 1 of Pama EAC, Colzato LS, Hommel B. Optogenetics as a neuromodulation tool in cognitive neuroscience. Front Psychol. 2013;4:610. https://doi.org/10.3389/fpsyg.2013.00610. This is an open-access article distributed under the terms of the Creative Commons Attribution 4.0 International License (https://creativecommons.org/licenses/by/4.0/). Panel (D) comes from Wikimedia Commons and was posted by Dr. Karl Deisseroth (via his website https://web.stanford.edu/group/dlab/media/layout/frontrat.png).*

however, this effect is isolated to animals that show elevated baseline levels of cocaine self-administration.[176] Other research has shown that inactivating the PL region reduces methamphetamine relapse-like behavior.[177] Collectively, this research shows a dissociable role of the PL mPFC in addiction. The finding that inactivation of the PL region increases drug self-administration makes sense as we have discussed the importance of the frontal cortex on self-control. If the PFC is important for self-control, why does inactivating this region *decrease* relapse-like behavior? Wouldn't these animals be more likely to reinstate drug seeking? As you will learn in Chapter 7, the PFC is also an important locus for memory, especially drug-paired memories. By inactivating areas associated with memory, cues previously paired with drugs become less effective at triggering drug-associated memories. Indeed, optogenetic inhibition of hippocampal neurons, which are involved in memory, prevents cocaine CPP,[178] and inhibition of NAc core-projecting BLA neurons, as well as NAc core-projecting PL neurons/NAc shell-projecting infralimbic (IL) neurons, decreases cocaine-seeking behavior and reinstatement.[179,180]

Concerning depressant drugs, optogenetic inhibition of specific neurons in the dorsal striatum and in the substantia nigra decreases ethanol self-administration and reinstatement of ethanol self-administration.[181] The establishment of morphine CPP appears to be largely driven by connections between the VTA and the dorsal raphe nuclei (DRN). Optogenetic activation of neurons that connect the rostral region of the VTA to the DRN produces an aversion to morphine, whereas activation of the caudal VTA-DRN pathway increases preference for morphine.[182] Glutamatergic neurons projecting from the NAc core to the insular cortex are important for the reinstatement of morphine CPP, as inhibiting these neurons suppresses reinstatement.[183]

Although optogenetics is an exciting method for elucidating the neural circuitry underlying addiction, this technique has one major disadvantage. To successfully activate or inhibit a specific region, the light emitted to the brain via the optic fiber needs to be stimulated with a certain frequency. Determining which frequency is effective can be difficult as multiple experiments are needed to optimize this parameter. An alternative to the traditional optogenetic approach is to use *stable-step function opsins* (SSFOs). With an SSFO, a light can be briefly illuminated, which causes long-lasting depolarizations to occur in targeted neurons. This then increases activity in the brain region of interest. Because these neurons remain highly active for an extended period (approximately 30 min), animals do not need to be connected to a fiber optic cable during behavioral testing. Using the SFFO technique, Müller Ewald et al. showed that stimulating activity in the IL region of the mPFC attenuates relapse-like behavior.[184] Due to the high costs associated with optogenetics research, SSFOs will allow more researchers to better study the neural circuitry of addiction. The one downside to SSFOs is that this technique is only amenable for procedures that last 30 min or less, which limits what kind of procedures can be used to measure drug reinforcement.

Chemogenetics

One major challenge to conducting optogenetics research is cost.[185] A less expensive alternative to optogenetics is **chemogenetics**. Like optogenetics, chemogenetics targets neurons in a specific neural circuit. Instead of using light to stimulate photosensitive receptors that have been inserted into the animal, researchers can chemically engineer receptors that are then inserted into subpopulations of neurons. For simplicity, I will discuss the most commonly used chemogenetic technique: **designer receptor exclusively activated by a designer drug (DREADD)**. The first DREADDs were developed in Dr. Bryan Roth's lab at the University of North Carolina.[186] Armbuster et al.[186] were able to modify acetylcholine muscarinic receptors such that they only responded to an otherwise pharmacologically inert drug: clozapine-N-oxide (CNO). These genetically engineered receptors are known as hM3Dq and hM4Di. hM3Dq is a modified form of the human M_3 muscarinic receptor. Recall that M_3 muscarinic receptors are G_q-coupled; thus, they are excitatory. hM4Di is a modified form of the human M_4 muscarinic receptor. These receptors are inhibitory as they are G_i-coupled. CNO can activate either hM3Dq or hM4Di. If a researcher is interested in activating a subpopulation of receptors, they will pair hM3Dq with something known as a fluorescent tag (helps detect which neurons express the DREADD) before inserting the DREADD into a viral vector. This viral vector is then microinjected into the brain. On certain test days, the researcher can inject the animal with CNO to activate those receptors. Fig. 4.3 shows a schematic of how DREADDs work. Read the Experiment Spotlight to learn more about how a chemogenetic experiment is conducted.

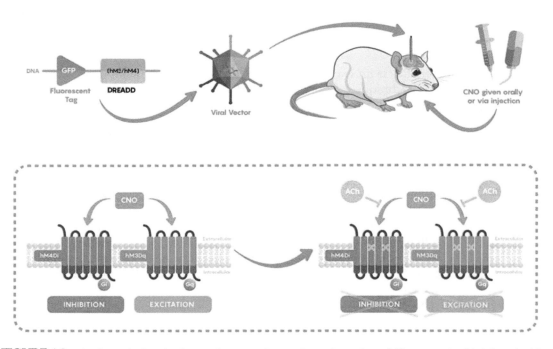

FIGURE 4.3 A schematic showing how a chemogentic experiment is conducted. First, an animal is injected with a viral vector that contains the DREADD of interest (usually either hM3Dq or hM4Di) and a fluorescent tag, which allows the researcher to see where the DREADDs were inserted in the brain. The animal is then systemically treated with a drug that selectively targets the DREADDs (such as CNO). If neurons contain hM3Dq, administration of CNO will stimulate those receptors, causing the neuron to become excited. If neurons contain hM4Di, administration of CNO will stimulate those receptors, causing the neuron to become inhibited. *Image comes from Fig. 1 of Section 3.4 of Ju W, Santos A, Freeman A, Daniele E. Neuroscience: Canadian. 1st ed. eCampusOntario; 2018. This book is distributed under the terms of the Creative Commons Attribution 4.0 International License (https://creativecommons.org/licenses/by/4.0/).*

Experiment Spotlight: Chemogenetic manipulations of ventral tegmental area dopamine neurons reveal multifaceted roles in cocaine abuse (Mahler et al.[187])

Although studies have examined the contribution of the VTA to addiction-like behaviors, Mahler et al.[187] wanted to determine the precise role of intra-VTA dopaminergic G-coupled metabotropic receptors on the reinstatement of cocaine seeking. This study performed additional experiments, but I will focus on the effects of DREADD activation on the reinstatement of cocaine seeking. Mahler et al. used male transgenic rats that expressed tyrosine hydroxylase (TH) and wildtype control rats. Both TH-expressing and wildtype controls were implanted with indwelling jugular catheters and received an infusion of a viral vector containing mCherry-tagged hM4Di (G_i-coupled), mCherry-tagged hM3Dq (G_q-coupled), or rM3Ds (G_s-coupled) into the ventromedial midbrain. Rats were then trained to self-administer cocaine. Next, rats learned that responding no longer led to cocaine reinforcement. Eventually, Mahler et al. tested rats for reinstatement of cocaine seeking-behavior with one of four methods: (1) they presented the rats with cues that were previously paired with cocaine infusions; (2) they gave the rats a systemic injection of cocaine; (3) they gave the rats a systemic injection of the noradrenergic α_2 receptor antagonist yohimbine and exposed them to cues associated with cocaine; (4) they injected the rats with CNO (1 or 10 mg/kg) or vehicle (control for CNO). Before the other reinstatement tests, the researchers administered CNO (1 or 10.0 mg/kg) or vehicle.

Mahler et al. found that chemogenetic stimulation of G_q-coupled receptors increased the reinstatement of cocaine-seeking behavior across all four reinstatement conditions. Interestingly, stimulating G_s-coupled receptors modestly increased cue-induced reinstatement but did not reliably alter reinstatement in the other conditions. Finally, stimulating G_i-coupled receptors decreased reinstatement in rats given a cocaine injection or a yohimbine injection before the reinstatement test. Mahler et al. also found that these effects were observed in the transgenic rats only. See the figure on the next page for the results of the experiment.

Questions to consider:

(1) Why did the researchers use transgenic rats expressing TH?
(2) What do you think mCherry is? Why is it included with the DREADDs?
(3) What do the results of this experiment indicate about intra-VTA dopaminergic G-protein-coupled receptors?
(4) What could potentially explain why stimulating G_i-coupled receptors failed to decrease cue-induced reinstatement of cocaine seeking?
(5) How can the researchers ensure that DREADD-induced alterations in reinstatement are specific to the VTA and not to any surrounding regions (like the substantia nigra)?

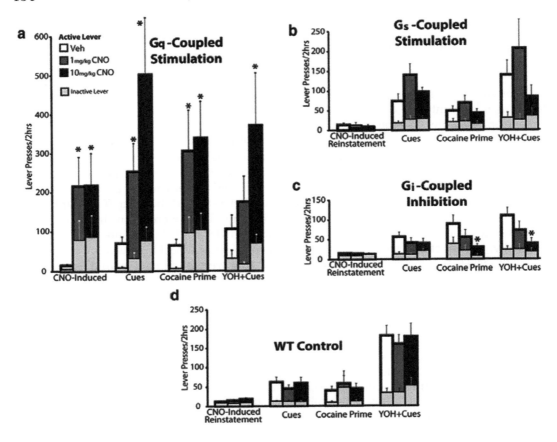

The image is adapted from Fig. 3 of Mahler SV, Brodnik ZD, Cox BM, et al. Chemogenetic manipulations of ventral tegmental area dopamine neurons reveal multifaceted roles in cocaine abuse. J Neurosci. 2019;39(3):503–518. https://doi.org/10.1523/JNEUROSCI.0537-18.2018. This article is distributed under the terms of the Creative Commons Attribution 4.0 International License (https://creativecommons.org/licenses/by/4.0/).

Answers to questions:

(1) Because the researchers are interested in dopaminergic neurons, they need to ensure that the DREADDs are expressed in these neurons. Because TH is involved in the synthesis of dopamine, using transgenic mice that overexpress TH allows them to better target these neurons.

(2) mCherry is a fluorescent tag that allows researchers to visualize where the DREADDs are localized. Looking at the figure on the next page, you will see that mCherry-tagged neurons appear red, the color of *cherries*.

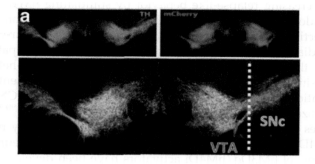

The image is adapted from Fig. 2 of Mahler SV, Brodnik ZD, Cox BM, et al. Chemogenetic manipulations of ventral tegmental area dopamine neurons reveal multifaceted roles in cocaine abuse. J Neurosci. 2019;39(3):503—518. https://doi.org/10.1523/JNEUROSCI.0537-18.2018. This article is distributed under the terms of the Creative Commons Attribution 4.0 International License (https://creativecommons.org/licenses/by/4.0/).

(3) G_q-coupled receptors located on dopaminergic VTA neurons are more important for the reinstatement of drug-seeking behavior compared to G_s-coupled receptors. Remember from Chapter 2 that G_s-coupled and G_q-coupled receptors have different mechanisms even though they are both excitatory. You will learn more about these differences later in this chapter. These results suggest that G_q-coupled receptors seem to be the primary drivers of addiction-like behaviors.

(4) If you look at panel (C) of the figure on the previous page, notice that baseline responding (following vehicle [Veh] treatment) is lower for the "cues" group compared to the "cocaine prime" and "YOH + cues" groups. Stimulating G_i-coupled receptors slightly lowered responding in the "cues" group, but this was not a statistically significant difference. Because the response is already so low at baseline, the researchers will have a lot of difficulties seeing further decreases in behavior, regardless of which manipulation they use. In research, this is called a *floor effect*. It just means that using cues to reinstate cocaine-seeking was not as effective in rats that received the hM4Di receptor compared to rats that received either the hM3Dq or the rM3Ds receptor.

(5) In an experiment like this, it is important to image brain regions that are proximal to the brain region of interest. Mahler et al. examined the substantia nigra in addition to the VTA. Look at the top left corner of the figure at the top of the page. You will see a brain section that shows where TH-expressing neurons are located (green color). The VTA appears as a circular structure while the substantia nigra looks like a column jutting out at an angle. As already discussed, mCherry tells the researcher where the DREADDs have been incorporated into neurons. Notice that you see very little red in the substantia nigra. The bottom image shows colabeling of both TH and mCherry. If both TH and mCherry are present in a neuron, it will appear yellow. Notice that the colabeling occurs almost exclusively in the VTA. Indeed, in their manuscript, Mahler et al. note that over 70% of VTA neurons expressed DREADDs, whereas fewer than 10% of substantia nigra neurons did.

In addition to decreasing relapse-like behavior by stimulating hM4Di receptors, ethanol self-administration decreases following hM4Di DREADD injection into the insular cortex and CNO microinjection into the NAc core, suggesting that the insular-ventral striatum circuit is a critical mediator of ethanol sensitivity.[188] The glutamatergic pathway between the insular cortex and the CeA plays an important role in methamphetamine relapse-like behavior, as decreasing activity in this pathway attenuates reinstatement.[189]

One major limitation of using CNO to activate DREADDs is that CNO can be reversed metabolized into clozapine, which binds to non-DREADD receptors.[190] This is problematic because any changes in behavior following CNO administration may not be mediated by the hM3Dq or hM4Di receptors that have been inserted into the neurons of interest. To address this issue, novel hM3Dq/hM4Di activators have been developed that do not have the issues associated with CNO, but research examining the effects of these novel compounds on addiction-like behaviors is not available at the time of this writing.

Epigenetics and addiction

Epigenetics is the study of heritable changes in gene expression that do not involve changes to the DNA sequence. Another way to phrase this is that epigenetics refers to changes in **phenotype** that do not result from changes in **genotype**. The term "epi" means "above", "outside of", or "around". Thus, the term epigenetics means "above genetics". Even though the DNA sequence is unaltered, changes to the physical structure of DNA can occur. These physical changes affect how cells "read" genes. This, in turn, influences which genes are expressed. Epigenetics is the reason why our skin cells differ from our neurons. The DNA for these cells is identical. However, the instructions for making these cells differ.

Although this chapter has focused on the genetic basis of addiction, one important consideration is that the environment can greatly impact our genes. This is a major area of focus in epigenetics. Environmental events can effectively cause gene expression to occur, and this gene expression can be passed down to our offspring. Going back to the gene association studies that were covered earlier in this chapter, one study found that the *SLC6A3* gene influences substance use during adulthood, but this effect is only observed in individuals that tried ethanol or cigarette smoking during adolescence.[191]

There are two major structural changes that can occur to DNA. I will briefly cover what these changes are, and I will discuss how these changes impact addiction-like behaviors.

DNA methylation

The first way that DNA can be physically altered is by adding a methyl group (one carbon atom bonded to three hydrogen atoms—CH_3) to either a cytosine or an adenine base. In animals, only cytosine can be methylated. This process is known as **DNA methylation** and serves primarily to suppress gene transcription, although it can activate gene transcription. Because gene transcription is inhibited, protein synthesis cannot occur. Fig. 4.4 shows a schematic of DNA methylation. Think of methylation as a light switch. As a light switch is either in the "on" or in the "off" position, DNA methylation causes the gene to either be "activated" or "silenced".

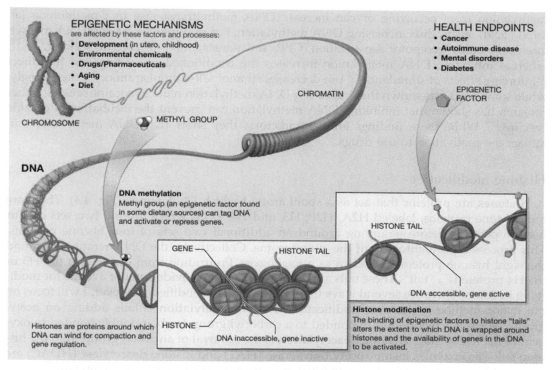

FIGURE 4.4 Image showing two major epigenetic changes that can occur to DNA. First, a methyl group can bind to DNA, which can activate or repress a gene (turn it on or off). Second, an acetyl group (referred to as an "epigenetic factor" in the image) can be added or removed from a histone tail. This influences how tightly DNA is wrapped around histones. When DNA is wrapped around the histone, it is inaccessible. Conversely, when DNA is unwrapped, it can be used. *Image comes from the National Institutes of Health (http://commonfund.nih.gov/epigenomics/figure).*

Drugs are known to alter DNA methylation. Self-administration of cocaine is associated with the methylation of several genes in the mPFC.[192] Research has shown that prolonged access to cocaine differentially alters DNA methylation; rats given extended access to cocaine show decreased DNA methylation in the mPFC, but rats given limited access to cocaine have increased DNA methylation in the mPFC.[193] In the NAc, extended access to cocaine increases DNA methylation of the *GLT-1* gene, corresponding to decreased *GLT-1* gene expression.[194] This gene encodes for the glutamate transporter 1, which removes excess glutamate from the synapse. If there are fewer glutamate transporters, more glutamate can stimulate dopaminergic neurons in areas like the VTA. Rats that show ethanol dependence-like behavior have increased DNA methylation in the mPFC and in the NAc compared to rats that have not developed features of dependence.[195]

Because drugs can alter DNA methylation, one interesting research question that has been addressed is if altering DNA methylation can reduce the risk of developing a SUD. Experimentally, researchers can infuse a DNA methyltransferase inhibitor to prevent DNA methylation from occurring or can increase DNA methyltransferase or L-methionine (an amino acid) levels, thus increasing DNA methylation. For example, increasing DNA methylation attenuates locomotor sensitization, CPP, and reinstatement of cocaine seeking,[196–198] whereas inhibiting DNA methylation increases the conditioned rewarding and the direct reinforcing effects of stimulants,[196] but decreases ethanol self-administration.[195] Interestingly, while some work has shown that increasing DNA methylation reduces cocaine seeking, other research has shown that inhibiting DNA methylation can prevent the incubation of cocaine craving.[199] While these findings are contradictory, they show that DNA methylation can impact the motivation to use drugs.

Histone modification

Histones are proteins that act as a spool around which DNA winds (Fig. 4.4). There are four histone proteins, labeled H2A, H2B, H3, and H4. DNA wraps around two sets of four histone proteins before wrapping around an additional two sets of four histone proteins. This process repeats throughout the chromosome. Collectively, the DNA wrapping around the eight histone proteins is known as a *nucleosome*. Protruding from one end of the H3 or the H4 protein is a "tail". These tails are important as they provide a major avenue for modifying histones. There are several ways that histones can be modified; however, I will focus on the major method of histone modification. **Histone acetylation** entails adding an acetyl group, which is a methyl group bonded to a carbonyl group (carbon double bonded to oxygen) to the histone tail. **Histone deacetylation** is the removal of an acetyl group from the histone tail. The enzymes histone acetyltransferase (HAT) and histone deacetylase (HDAC) are responsible for histone acetylation and histone deacetylation, respectively. Acetylation causes the DNA that is wrapped around the histones to become more relaxed while deacetylation causes DNA to wrap more tightly around the histones. If DNA is tightly wrapped around the histone, gene expression is repressed. Whereas DNA methylation is often equated to a light switch, histones are often compared to volume knobs. Gene expression can be finely adjusted as DNA unwraps itself from the histone.

As with DNA methylation, researchers can test the contribution of histone acetylation/deacetylation by administering compounds that inhibit HAT/HDAC. Most addiction-related studies have focused on HDAC inhibitors. I will highlight just some of the effects HDAC inhibitors have on drug use-related behaviors. HDAC inhibitors decrease drug self-administration,[200] impair drug CPP,[201] and attenuate reinstatement of drug-seeking behavior.[200,202] Not all studies have shown a beneficial effect of HDAC inhibitors on addiction-like behaviors. Direct infusions of HDAC inhibitors into the NAc shell increase cocaine self-administration.[203] Additionally, HDAC inhibitors fail to alter heroin self-administration and even potentiate reinstatement of heroin seeking.[204] HDAC inhibitors also enhance cocaine CPP[205] and morphine CPP.[202] Because many studies administer HDAC inhibitors systemically, this can lead to histone modification throughout multiple brain regions. To better determine the impact of histone modification on addiction, local injection of these inhibitors into the brain is merited. Despite these discrepancies, targeting epigenetic factors is a newer and more exciting approach for unraveling the genetic underpinnings of addiction.

Pharmacogenetic approach to treating addiction

In Chapter 2, you learned about some of the major pharmacotherapies for treating SUDs. Unfortunately, pharmacotherapies rarely produce long-term abstinence (more than 1 year) and reduce drug use in a small percentage of individuals (between 20% and 30%). One reason this could be is that genetic variations may control how well one responds to specific pharmacotherapies. **Pharmacogenetics** is an approach in which physicians consider one's genetic makeup when prescribing a medication. In the context of addiction, studies have revealed that individuals with certain polymorphisms respond better to pharmacological treatments for SUDs. You read about some of these findings earlier in this chapter. Recall that individuals with a genetic variant of the *SLC6A3* gene respond better to disulfiram for the treatment of cocaine dependence. To provide another example, individuals with a polymorphism of the mu-opioid receptor gene show decreased ethanol use and have lower relapse rates when treated with naltrexone compared to those without this polymorphism, and individuals with a polymorphism of the cytochrome P450 2A6 enzyme respond better to nicotine replacement therapies.[206] Cytochrome P450 2A6 helps metabolize nicotine. Individuals who metabolize nicotine faster do not respond as well as to nicotine agonist treatment. This phenomenon could occur because individuals metabolize nicotine too quickly to maintain a stable plasma level of nicotine, which can lead to unpleasant withdrawal symptoms. As we continue understanding more about the genetic basis of addiction, we may see more of these individualized treatment regimens to help individuals with a SUD.

Quiz 4.2

1. What is the major limitation to using CNO to activate DREADDs?
2. What happens when neurons in the PFC are inhibited via optogenetics?
   a. Decreased drug self-administration
   b. Decreased reinstatement of drug-seeking
   c. Increased behavioral sensitization
   d. Increased incubation of craving
3. Which drug increases ΔFosB?
   a. Ethanol
   b. Methamphetamine
   c. Nicotine
   d. All of the above
4. What is the difference between optogenetics and stable-step function opsins?
5. Which of the following statements is true?
   a. The same drug can either increase or decrease DNA methylation.
   b. HDAC inhibitors are effective at treating opioid use disorders.
   c. Inhibiting DNA methylation increases ethanol consumption.
   d. None of the above

Answers to quiz 4.2

1. CNO can be reversed metabolized back to clozapine, which has actions on other receptors, such as dopaminergic and serotonergic receptors.
2. b — Decreased reinstatement of drug seeking
3. d — All of the above
4. In optogenetics, an animal is connected to a fiber optic cable during the entire session, whereas in stable-step function opsin experiments, animals are given a single exposure to light.
5. a — The same drug can either increase or decrease DNA methylation.

Cellular changes following drug administration

In addition to learning about the genetic basis of addiction, you have previously learned how drugs exert their effects in the brain, as well as the neuroanatomical structures and neurotransmitter systems implicated in addiction. Not only do drugs alter neurotransmitter levels like dopamine in multiple brain regions, but they can drastically alter the structure of neurons. This section will discuss some of the major cellular changes that occur following acute exposure and chronic exposure to drugs.

Changes to receptor levels

One question you might have is "haven't we already discussed how drugs affect receptors?". Earlier, you learned about drug-induced changes to receptor mRNA expression. Because mRNA is used to create proteins in a process known as *translation*, mRNA levels do not perfectly correlate with protein expression. Theoretically, the changes that have been observed in mRNA expression following drug exposure may not *translate* (terrible joke, I know) to changes in receptor expression. As such, studies often examine protein expression instead of mRNA expression.

How drugs affect dopamine receptor levels

Like mRNA expression, drugs drastically alter receptor density and receptor expression. Cocaine can upregulate dopamine D_1-like receptors in the cerebral cortex and in the striatum as early as 3 days following initial use.[207] If animals are treated several times each day with cocaine, robust behavioral sensitization occurs with parallel increases in D_1-like receptor expression in the NAc, the ventral pallidum, and the substantia nigra following 14 days of treatment.[208] While Alburges et al. observed no differences in D_2-like receptor expression following cocaine exposure,[207] Sousa et al. found that early withdrawal from cocaine increases D_2-like receptors in the striatum.[209]

Although psychostimulants like cocaine upregulate dopamine receptors following initial drug exposure, prolonged drug use causes the brain to compensate for the repeated influx of dopamine in the NAc. Not only do drugs produce less dopamine release over time, but long-term changes in receptor expression are observed. Prolonged cocaine self-administration increases D_1-like receptor expression in the NAc and in the substantia nigra in rats.[210] Like animals, humans that use methamphetamine have increased D_1-like receptor expression in the NAc[211]; however, individuals with a SUD often show decreased D_2-like receptor expression in the NAc and in the PFC.[211,212] Overall, these results are consistent with mRNA expression studies. Drugs tend to upregulate D_1-like mRNA and protein expression in striatal regions, but they can decrease D_2-like expression in the PFC. This brings up a larger question: what is the purpose of up/downregulation of receptors? Think of it this way: imagine you turn your TV on, and the sound is way too loud. You will compensate by turning the volume down. Similarly, if your NAc and PFC receive too much dopaminergic input, these regions compensate by "turning down" how much dopamine is released into these areas. Because less dopamine is released following chronic drug administration (think of the dopamine depletion hypothesis), D_1-like receptors increase to try to depolarize dopaminergic neurons. Conversely, D_2-like receptors decrease because there is less dopamine to control. These compensatory mechanisms can help explain why individuals feel a blunted drug effect. This neuroadaptation can also explain why those with SUD often experience anhedonia. Because there is less dopamine in the NAc, natural rewards are not as enticing to someone with a SUD.

Decreased dopamine D_2-like receptor levels in the striatum have a profound effect on PFC functioning. Dr. Nora Volkow, currently the director of the National Institute on Drug Abuse (NIDA), previously found that decreased D_2-like receptor levels in the striatum are associated with decreased functioning of the PFC.[213] As discussed previously, decreased PFC functioning can lead to impairments in decision-making that can promote further drug use.

Withdrawal from drugs can cause further alterations in dopamine receptor density/expression. Cocaine withdrawal leads to a downregulation of D_1-like receptors and a downregulation of D_2-like receptors, an effect that lasts up to 30 days following the last treatment,[214] although no changes in D_1-like or D_2-like expression are observed following methamphetamine self-administration.[215] Ethanol has the opposite effect as cocaine, as increased D_1-like and D_2-like receptor density is observed in the NAc, the amygdala, and the dorsal striatum during withdrawal.[216]

How drugs affect receptor levels of other neurotransmitter systems

With specific drug classes, compensatory changes outside of the dopaminergic system are observed. Typically, drugs that act as an agonist at a receptor will promote the downregulation of those receptors after prolonged use. For example, chronic administration of opioids causes downregulation of mu-opioid receptors.[217] This leads to tolerance to the analgesic effects of morphine. Conversely, antagonists tend to cause upregulation of receptors. Because ethanol interferes with NMDA receptor functioning, these receptors are upregulated to compensate for being inhibited. Ethanol administration also causes downregulation of GABA receptors. These compensatory changes account for the tolerance that develops to ethanol's sedative effects. Earlier, you read that depressant-like drugs downregulate CB_1 receptor mRNA expression. Drugs like opioids and ethanol bind to inhibitory receptors (opioid and GABA receptors), and CB_1 receptors are inhibitory. The increased inhibition in the brain may cause CB_1 mRNA expression to decrease as a compensatory response. One notable exception to the "rule" about down/upregulation is nicotine. Even though nicotine is an agonist at nicotinic receptors, prolonged nicotine exposure leads to an upregulation of nicotinic receptors.[218] Other exceptions have been observed as well, such as the downregulation of 5-HT$_{2A}$ receptors following antagonist treatments.[219] Treat this as a general rule of thumb, not as a law.

Drugs that do not directly target a specific neurotransmitter system can cause alterations in receptor expression. Multiple drugs cause widespread changes to glutamatergic receptors. For example, cocaine administration increases GluN1 subunit, GluN2A subunit, mGluR$_5$, and Group II mGluR expression in the NAc, the dorsal striatum, or in the hippocampus.[220–222] Drugs also drastically alter the endogenous opioid system.[223] Drugs increase mu-opioid receptor expression in the NAc and in the dorsal striatum.[224] In addition to its effects on glutamatergic and GABAergic receptors, ethanol downregulates CB$_1$ receptors[225] just as it downregulates CB$_1$ receptor mRNA expression. Similarly, cocaine use downregulates CB$_1$ in the PFC of both humans and mice.[226] However, extended access to cocaine increases CB$_1$ expression in the NAc core and in the amygdala.[227] Reinstatement of cocaine seeking upregulates CB$_1$ receptors in the PFC but downregulates these receptors in the VTA.[228]

As with chronic drug use, drug withdrawal can change the number of receptors across neurotransmitter systems in various regions of the brain. Cocaine withdrawal increases NMDA GluN1 expression in the dorsal striatum[229] and GluN2A subunit expression in the PFC and mGluR$_5$ expression in the hippocampus but decreases mGluR$_5$ expression in the PFC. Morphine withdrawal upregulates CB$_1$ receptors in the NAc.[230]

The information provided here is by no means an exhaustive list of all the drug-induced changes in receptor density/expression that results from drug administration and/or drug withdrawal. The point of this section is to further emphasize that drugs have widespread effects on the brain and can alter neurotransmitter systems that they do not directly target. This increases the difficulty of treating individuals with a SUD. Blocking dopaminergic receptors may be able to decrease short-term drug use, but neural adaptations can occur in other neurotransmitter systems in response to drug cessation, which can promote relapse-like behavior.

Effects of altered receptor levels during abrupt drug cessation

You now know that chronic drug use can lead to alterations in the number of receptors in the brain. Now, imagine someone that abruptly stops using a drug. Cessation of drug use can lead to aversive withdrawal symptoms. The cause of these withdrawal symptoms is related to the compensatory changes that have occurred previously while someone was using the drug repeatedly. In the case of prolonged opioid use, the downregulation of opioid receptors has a huge effect on one's digestive system. Because opioid receptors are largely located in our digestive tract, decreased opioid receptor expression leads to diarrhea, which is often characteristic of opioid withdrawal. In Chapter 1 you learned that withdrawal from ethanol and benzodiazepines/barbiturates can be life-threatening. The reason has to do with the compensatory changes that have occurred to GABA receptors and to NMDA receptors as in the case of ethanol. The use of these depressants leads to a downregulation of GABA receptors, which act to inhibit neuronal activity in the brain. Ethanol also upregulates NMDA receptors. As long as the individual continues using these substances, the alterations that have occurred to these receptors will not harm the individual. However, when the individual abruptly stops taking these depressants, there is no longer enough inhibition in the brain to counteract excitatory neurotransmission due to there being fewer GABA receptors. Over excitation in the brain can lead to seizures and death. This is why individuals with alcohol use disorder are often given acamprosate. This drug helps block NMDA receptors, which allows the individual to safely stop using ethanol. To wean someone from benzodiazepines/barbiturates, they can be given anticonvulsant drugs that target the GABAergic system.[231]

Alterations to dendritic spines

In addition to mRNA and protein expression, drugs alter the number of dendritic spines. Interestingly, the direction of these changes is not the same across drug classes as you have seen with drug-induced changes to mRNA/protein expression. Whereas depressant-type drugs such as ethanol and morphine tend to decrease the number of dendritic spines,[232] psychostimulant drugs increase the number of dendritic spines.[233] Drug-induced alterations in dendritic spine density can persist even after discontinuation of drug use. In rodents, increased dendritic spine density is observed in MSNs in the NAc and in the dorsal striatum 3.5 months following the last amphetamine injection[233] while ethanol withdrawal increases dendritic spine density in the frontal cortex.[234] It is important to note that drugs may decrease dendritic spine density in one region but may increase dendritic spine density in a separate region. For example, Robinson et al. found that morphine decreases the number of dendritic spines in the NAc shell and in part of the hippocampus but increases the number of dendritic spines in the OFC.[232] Morphine can also increase dendritic spine density in the NAc core.[235]

Drugs do not just alter the number of dendritic spines; they can alter the **morphology** of dendritic spines. Amphetamine increases dendritic spine length in the mPFC, but not in the NAc.[236] When an animal goes into withdrawal following chronic administration of cocaine, the spine head diameter increases. If the animal is given an injection of cocaine following withdrawal, the spine head diameter increases compared to when the animal was in withdrawal. Interestingly, as the effects of cocaine dissipate, the dendritic spine head diameter decreases dramatically and becomes smaller than baseline levels (i.e., before the animal was exposed to the drug).[237] Fig. 4.5 shows how dendritic spine head morphology changes during withdrawal from cocaine and following an acute injection of cocaine. Nonstimulant drugs also alter dendritic spine morphology. Morphine increases the number of thin spines and decreases the number of stubby spines in the NAc shell.[238] The relevance of altered dendritic spine numbers and morphology will become more apparent in Chapter 7 when we discuss the cellular mechanisms underlying learning and memory.

Changes to transporter expression and function

So far, this chapter has primarily focused on the contribution of receptors to addiction. However, DAT is critically important for SUDs. Remember that most psychostimulant drugs directly target monoaminergic transporters, particularly DAT. Not surprisingly, drugs can cause changes to DAT expression and function. I first need to differentiate transporter expression from function. Expression indicates if the transporter is located on the cell membrane or if it is located inside of the presynaptic terminal (these are *internalized* transporters).[239] Researchers often analyze both types of expression because increases in internalized transporter expression without alterations in membrane-bound transporter expression will not largely

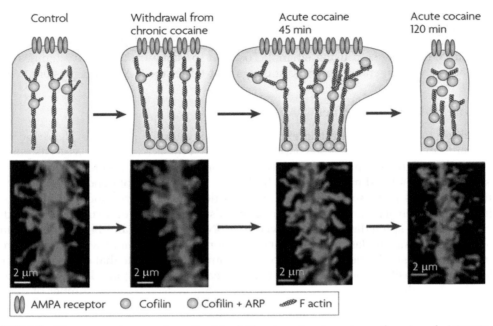

FIGURE 4.5 The top row shows a schematic of dendritic spines at various stages of cocaine administration. The bottom row shows representative images of dendritic spines. The leftmost image shows dendritic spines that have not been exposed to cocaine. The second image shows dendritic spines following withdrawal from repeated cocaine administration. Notice that the spine head is larger, and there is a corresponding upregulation of AMPA receptors. The third image shows what happens to dendritic spines 45 min following an acute injection of cocaine after animals have gone through withdrawal. This causes dendritic spine heads to enlarge in size, and there is a further increase in AMPA receptor upregulation. The rightmost image shows that dendritic spine head size decreases in size 2 h following an acute injection of cocaine. There is a corresponding downregulation of AMPA receptors. Note, F actin is a type of microfilament that helps provide structure to the dendritic spine. Cofilin/cofilin + ARP are proteins that help control actin formation. *Image comes from Fig. 3 of Kalivas PW. The glutamate homeostasis hypothesis of addiction.* Nat Rev Neurosci. *2009;10(8):561−572. https://doi.org/10.1038/nrn2515. Copyright 2009.*

affect how much neurotransmitter is taken back into the presynaptic neuron. Function refers to how quickly the transporter can remove excess neurotransmitters from the synapse. Researchers can also measure the *affinity* of the neurotransmitter to that transporter.[240] In other words, how well is the neurotransmitter "attracted" to that transporter? For example, dopamine should have high affinity for DAT while having a lower affinity for other transporters. With those definitions out of the way, you can now learn about some of the research examining the effects of drugs on transporter expression/function and how altering expression and/or function can influence the reinforcing effects of drugs.

Studies examining the effects of cocaine on DAT expression have provided contrasting results. An early study showed that cocaine self-administration leads to an upregulation of DAT in the dorsal striatum, the NAc, the VTA, and the substantia nigra.[210] However, more recent studies have found reduced DAT expression in the NAc.[241] Cocaine exposure produces tolerance to cocaine-induced inhibition of DAT,[241] meaning DAT can take up more dopamine even after cocaine is administered. This alteration could partially account for why repeated drug use is less effective at producing euphoria even when DAT is downregulated. Given that ADHD medications like methylphenidate are structurally similar to cocaine, one would think the administration of these drugs would affect DAT to a similar extent as cocaine. Yet, this is not always the case. While Calipari et al.[241] observed decreased DAT expression and function following cocaine self-administration, methylphenidate increased DAT expression and function in the NAc. Although cocaine decreases DAT function in the NAc, animals given extended access to cocaine show *increased* DAT function in the NAc,[242] as well as in the PFC and in the OFC.[243] These results suggest that cocaine can exert biphasic effects on DAT function in the NAc depending on how much drug exposure animals have. This biphasic effect could reflect neuroadaptations that occur with prolonged drug exposure. Another thing to consider is that alterations in DAT function may be dependent on the brain region. Cocaine self-administration leads to increased DAT function and SERT availability in the dorsal striatum.[244,245]

Psychostimulant withdrawal profoundly affects DAT expression. Cocaine withdrawal upregulates DAT mRNA expression in the VTA,[246] and methamphetamine withdrawal upregulates DAT expression in the striatum 72 h after the last self-administration session.[215] Alterations in DAT during withdrawal are not isolated to psychostimulants. Morphine withdrawal also increases DAT expression.[247]

Individual differences in DAT functioning can be used to make predictions about an animal's drug sensitivity. Rats that have faster uptake of dopamine through DAT in the dorsal striatum self-administer more cocaine and show increased reinstatement of cocaine-seeking behavior.[248] Given the importance of DAT to psychostimulant reinforcement, research has utilized genetic and pharmacological manipulations of this transporter to clarify its role in the reinforcing effects of drugs like cocaine. DAT knockout mice show decreased drug self-administration.[249] However, it is important to note that DAT knockout mice still self-administer cocaine[250] and develop cocaine CPP.[251] Furthermore, DAT knockout mice develop greater morphine CPP.[252] Instead of deleting DAT, researchers can pharmacologically inhibit this transporter. Administration of DAT inhibitors can decrease cocaine self-administration.[253] The problem with using DAT inhibitors as pharmacotherapies for cocaine use disorder is that they can have high misuse potential as animals self-administer DAT inhibitors other than cocaine.[253] Recently, a novel DAT inhibitor, JJC8-016, was shown to inhibit cocaine reinforcement and reinstatement of cocaine seeking without displaying any misuse potential on its own.[254]

In addition to mediating the reinforcing effects of drugs, DAT influences the locomotor-stimulant effects of drugs, as DAT knockout mice fail to show increased locomotor activity following acute administration and fail to develop locomotor sensitization to repeated cocaine injections.[255] DAT appears to be especially important for the development of locomotor sensitization. Norepinephrine transporter (NET) knockout mice show a blunted locomotor response to acute cocaine, but unlike DAT knockout mice, they develop locomotor sensitization to repeated cocaine injections.[255]

Although this section focused exclusively on DAT, I do want to quickly highlight that research has examined the contribution of SERT to addiction-like behaviors. SERT knockout mice fail to self-administer MDMA,[256] which is not surprising as MDMA exerts most of its effects through the serotonergic system. However, the beneficial effect of SERT deletion on drug self-administration appears to be specific for MDMA, as SERT knockout rats self-administer *more* drugs compared to control rats.[257] In a randomized control trial, participants given the SERT inhibitor escitalopram failed to reduce their choice for cocaine,[258] but citalopram decreases consumption of ethanol in those with alcohol use disorder.[259]

Cellular pathways associated with addiction

In this chapter, you have learned that drugs can alter mRNA expression, protein expression, and dendritic spine density and morphology. However, you have not learned *how* drugs produce these alterations. To understand the "how", you need to understand cellular signaling pathways. You previously learned that many receptors are metabotropic, including all monoaminergic receptors (except for 5-HT$_3$ receptors), opioid receptors, GABA$_B$ receptors, cannabinoid receptors, and certain glutamate receptors (Groups I-III mGluRs). Recall that metabotropic receptors are coupled to a G protein that dissociates from the receptor when it is stimulated. The cellular pathways for each G protein-coupled receptor type differ. G$_s$ receptors stimulate adenylyl cyclase, which then stimulates cyclic adenosine monophosphate (cAMP). While increased cAMP levels cause ion channels located on the cell membrane to open, cAMP has another important function. It activates a **protein kinase**, specifically protein kinase A (PKA). Protein kinases are enzymes that modify other proteins by adding a phosphate group to them. This is known as *phosphorylation* and is the mechanism by which proteins are activated.[260] Once activated, PKA will start a chain reaction in which multiple proteins are phosphorylated (this is known as a *phosphorylation cascade*). Long story short: this will ultimately lead to phosphorylation of cAMP response element-binding protein (CREB), a specific transcription factor in the nucleus of the neuron. You learned about the Fos transcription family earlier. Phosphorylated CREB (pCREB) causes the transcription of DNA into mRNA. Translation then can occur, leading to increased protein expression. This protein expression can lead to upregulation of receptors or structural changes to occur in the neuron.

G$_q$-coupled receptors do not directly target adenylyl cyclase. Instead, G$_q$-coupled receptors activate phospholipase C (PLC), which eventually gets cleaved into inositol triphosphate (IP$_3$) and diacylglycerol (DAG). DAG activates a different protein kinase: protein kinase C (PKC). Like PKA, PKC activates transcription factors, leading to altered receptor expression and morphological changes in the neuron. IP$_3$ causes Ca^{++} to be released from the endoplasmic reticulum; the increased Ca^{++} causes cAMP formation, leading to increased pCREB levels. However, the influx of Ca^{++} also activates something called *calmodulin* (somewhat of a portmanteau of the words "CAlcium", "MODULation", and "proteIN"). Located in the cytoplasm, calmodulin detects increases in intracellular Ca^{++} levels. Because Ca^{++} activates calmodulin, Ca^{++} is a second messenger. Calmodulin then activates Ca^{++}/calmodulin-dependent protein kinase II (CaMKII). CaMKII leads to phosphorylation of CREB and other transcription factors.[261]

What about ionotropic receptors? Let's use NMDA receptors as an example. Stimulation of NMDA receptors allows Ca^{++} to enter the cell. In addition to depolarizing the neuron, the increased Ca^{++} works in the same way as described above for G$_q$-coupled metabotropic receptors. Ca^{++} influx activates calmodulin, which then activates CaMKII.[261] Fig. 4.6 shows a simplified schematic of the signaling pathways described here. Keep in mind the description

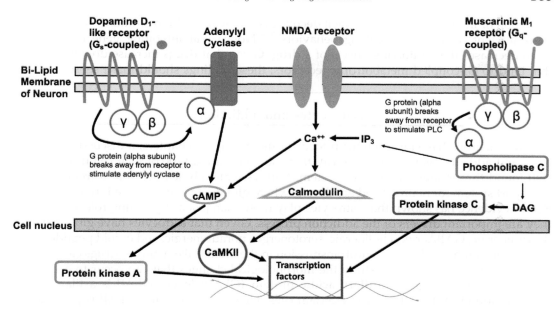

FIGURE 4.6 A simple schematic showing how receptor activation can lead to activation of transcription factors that lead to gene expression in the cell nucleus of the neuron. Stimulation of G_s-coupled metabotropic receptors (D_1-like receptor located on the left side of the figure) causes the alpha subunit to break away from the receptor. This activates adenylyl cyclase, which activates cyclic AMP (cAMP). cAMP activates protein kinase A, resulting in the activation of transcription factors like CREB. Stimulation of the NMDA receptor (located in the center of the figure) causes Ca^{++} ions to enter the neuron. Ca^{++} stimulates cAMP, which leads to the activation of protein kinase A. Ca^{++} also activates calmodulin. Calmodulin stimulates CaMKII, which activates transcription factors. Stimulation of G_q-coupled metabotropic receptors (muscarinic M_1 receptor located on the right side of the figure) causes the alpha subunit to break away from the receptor. Instead of activating adenyl cyclase, the alpha subunit stimulates phospholipase C. Phospholipase C is then cleaved into DAG and IP_3. IP_3 activity causes an influx of Ca^{++}, resulting in either cAMP or calmodulin activation. DAG activates protein kinase C, which then activates transcription factors. *Figure was created by the author using Microsoft PowerPoint.*

of signaling pathways provided here is incredibly simplified. There are numerous kinases, transcription factors, and signaling pathways that I have not covered. At least an entire chapter (if not an entire book) could be devoted to the numerous transcription factors and signaling pathways. Having a basic understanding of signaling pathways is important when you learn about learning mechanisms underlying addiction. We will come back to these cellular pathways as they are critically important for learning and memory.

To investigate the specific role of signaling pathways on addiction, researchers can inhibit the formation of protein kinases or CaMKII. For example, PKA inhibition decreases cocaine self-administration, while stimulating PKA activity increases self-administration.[262] Selective PKA inhibition in the amygdala reduces cocaine-seeking behavior/reinstatement.[263,264] Inhibiting PKC activity does not impair initial cocaine CPP, but it does impair CPP when animals are tested at a later point in time.[265] Furthermore, PKC inhibition attenuates the reinstatement of cocaine CPP.[266] Interestingly mice lacking a specific form of PKC are more sensitive to the reinforcing and the conditioning rewarding effects of morphine.[267] CaMKII reduction or inhibition in the NAc shell decreases cocaine self-administration[203] and prevents reinstatement of morphine-seeking behaviors.[268] Notice that inhibiting gene transcription often decreases the reinforcing effects of a drug. Given that drugs increase dopamine levels that bind to G_s-coupled metabotropic receptors, these findings should make sense.

Astrocytes and addiction

While neurons have been the main focus of the past three chapters, a discussion of astrocytes and addiction is warranted. Recall that astrocytes are a type of glial cell (nonneuronal cell) found throughout the nervous system. You have already learned that astrocytes act as a support system for neurons and are the site where glutamate is synthesized from glutamine. Recent evidence has shown that astrocytes play more than just a supporting role for neurons; they are important drivers of the addiction process. Given that astrocytes have glutamatergic, dopaminergic, GABAergic, cholinergic, serotonergic, noradrenergic, and opioid peptide receptors,[269] they highly influence neurotransmission in the brain. Like neurons, drugs can alter the morphology of astrocytes. To study the morphological changes that can occur to astrocytes, researchers measure **glial fibrillary acidic protein (GFAP)** levels. GFAP is a filament-like structure found in the cytoplasm of certain cells, including astrocytes. GFAP is proposed to aid in astrocyte-neuron communication. As such, drug-induced changes in GFAP expression following drug exposure can be a potential mechanism underlying SUDs. Most drugs increase GFAP expression. For example, methamphetamine increases GFAP in the striatum and in the hippocampus,[270,271] nicotine increases GFAP in the spinal cord,[272] morphine increases GFAP levels in the VTA, the striatum, and the PFC,[273,274] and ethanol increases GFAP in the amygdala.[275] However, there are some exceptions. Cocaine self-administration decreases GFAP expression in the NAc core,[276] and morphine decreases GFAP in the hippocampus.[274]

In addition to drug-induced alterations in GFAP expression in the mesocorticolimbic pathway, drug withdrawal changes GFAP levels. While long-term cocaine withdrawal (3 weeks) upregulates GFAP levels in the PFC and in the NAc,[277] short-term nicotine withdrawal (3–10 days) decreases GFAP in the hippocampus.[278] Reinstatement of nicotine seeking is also associated with decreased GFAP expression in the NAc core.[279]

Early-life drug use can lead to long-lasting changes to astrocyte morphology that persist into adulthood. Adolescent exposure to cannabis leads to elevated GFAP levels in the hippocampus, but decreased levels in the cerebellum, during adulthood.[280,281] Adolescent ethanol consumption in rats causes increased GFAP expression in the hippocampus, the striatum, and the frontal cortex.[282] Amazingly, pregnant rats exposed to ethanol have offspring that show increased GFAP in the hippocampus and in the striatum.[283]

Because drugs greatly affect astrocytes, recent research has focused on the contribution of astrocytes to addiction-like behaviors. Manipulating astrocyte functioning can greatly impact the reinforcing effects of drugs in animals and can influence relapse-like behavior. Using DREADDs to express hM3Dq on glial cells, Nwachukwu et al. recently found that stimulating astrocytes in the amygdala decreases ethanol consumption.[275] Somewhat similarly, activation of G_q-coupled DREADDs located on astrocytes in the NAc core inhibits reinstatement of psychostimulant-seeking behavior.[284] This research highlights the need to consider glial cells in addition to neurons when examining the neurobiological basis of addiction.

Quiz 4.3

1. Which of the following statements about drug-induced changes to dopamine receptors is true?
   a. Acute psychostimulant treatment causes downregulation of D_1-like receptors.
   b. Chronic psychostimulant treatment downregulates D_1-like receptor expression.
   c. Intra-PFC dopamine receptors are unaffected by prolonged psychostimulant administration.
   d. Withdrawal from psychostimulants upregulates D_1-like and D_2-like receptors.
2. What effect does chemogenetic stimulation of glial cells have on addiction-like behaviors?
   a. Decreased reinstatement of psychostimulant drug-seeking
   b. Increased ethanol consumption
   c. Increased relapse-like behavior
   d. None of the above
3. What does DAG stimulate?
   a. Calcium
   b. Phospholipase C
   c. Protein kinase A
   d. Protein kinase C
4. How do drugs alter dendritic spine density?
   a. Both stimulants and depressants decrease dendritic spine density.
   b. Both stimulants and depressants increase dendritic spine density.
   c. Depressants tend to decrease the number of dendritic spines, whereas stimulants increase them.
   d. Stimulants tend to decrease the number of dendritic spines, whereas depressants increase them.

Answers to quiz 4.3

1. d — Withdrawal from psychostimulants upregulates D_1-like and D_2-like receptors.
2. a — Decreased reinstatement of psychostimulant drug seeking.
3. d — Protein kinase C
4. c — Depressants tend to decrease the number of dendritic spines, whereas stimulants increase them.

Chapter summary

In this chapter, you learned about the molecular and the cellular mechanisms controlling addiction-like behaviors. Several genes have been implicated in SUDs, mainly those that encode dopamine D_2 receptors, several nicotinic receptor subunits, and enzymes that metabolize ethanol. Unsurprisingly, drugs can greatly affect gene expression, which can ultimately affect protein expression. Experimentally altering mRNA expression can potentiate or attenuate the reinforcing effects of drugs. Although SUDs are highly heritable, there is not a perfect association between our genes and propensity for drug use. As such, we need to start considering other factors in addiction. For those that are not as enamored with neurobiology as I am, I have some good news. Starting in the next chapter, you will begin reading about the behavioral and the learning mechanisms underlying SUDs. However, I have some bad news … we are not done discussing neurobiology in this textbook. Because neurobiology is closely linked to behavior and cognition, we cannot discuss one without discussing the other.

Glossary

Chemogenetics — a technique in which large molecules (such as proteins) are engineered to interact with previously unrecognized small molecules; also see designer receptor exclusively activated by a designer drug.

Designer receptor exclusively activated by a designer drug (DREADD) — a chemogenetic approach in which a modified receptor (typically a modified acetylcholine muscarinic M_3 or M_4 receptor) is inserted into the brain and then later activated by a drug that normally is inert in the brain.

DNA methylation — the addition of a methyl group to a DNA molecule, specifically a cytosine or an adenine base.

Downregulation — a decrease in a cellular component, such as RNA or protein.

Epigenetics — refers to phenotypic changes that do not result from changes in genotype (i.e., alterations to the DNA sequence).

Genome-wide association study — a research technique in which genetic variations observed in individuals are correlated with certain physical and/or behavioral traits.

Genotype — the genetic constitution of an organism.

Glial fibrillary acidic protein (GFAP) — filament-like structure found in the cytoplasm of cells, particularly glial cells, that are involved in multiple processes in the central nervous system, such as astrocyte-neuron communication.

Haplotype — a group of alleles that are inherited together from a single parent; can also refer to a set of linked single nucleotide polymorphisms that are inherited across generations.

Heritability — a statistic that indicates how much variability in a phenotype can be explained by genetic variability in a population.

Histone — type of protein that acts as a spool around which DNA winds.

Histone acetylation — process of adding an acetyl group to a histone tail.

Histone deacetylation — process of removing an acetyl group from a histone tail.

Immediate early gene — a gene that is activated quickly and for a short period of time in response to certain stimuli. c-Fos is an example of an immediate early gene.

Messenger RNA (mRNA) — a single-stranded molecule of RNA that is read by a ribosome to create a protein.

Morphology — the physical appearance of structures within an organism; for example, the physical features of a neuron.

Optogenetics — a technique in which neurons can be controlled by activating light-sensitive ion channels that have been introduced by a viral vector.

Pharmacogenetics — a medical approach in which physicians consider one's genetic makeup when prescribing a medication.

Phenotype — observable characteristics or traits of an organism that result from our genotype and from the environment.

Polymorphism — occurrence of two or more different forms of a phenotype.

Protein kinase — an enzyme that modifies other proteins by adding a phosphate group to them.

Single-nucleotide polymorphism (SNP) — substitution of a single nucleotide at a specific position in the genome.

Transcription factor — a protein that controls gene expression by binding to DNA to either initiate or inhibit the process of copying RNA from DNA.

Transgenic — often used to refer to genetically modified animals; these animals have had a segment of DNA from one organism inserted into their DNA.

Upregulation — an increase in a cellular component, such as RNA or protein.

Supplementary data related to this chapter can be found online at doi:10.1016/B978-0-323-90578-7.00004-9

References

1. Verhulst B, Neale MC, Kendler KS. The heritability of alcohol use disorders: a meta-analysis of twin and adoption studies. *Psychol Med*. 2015;45(5):1061–1072. https://doi.org/10.1017/S0033291714002165.
2. Ducci F, Goldman D. The genetic basis of addictive disorders. *Psychiatr Clin North Am*. 2012;35(2):495–519. https://doi.org/10.1016/j.psc.2012.03.010.
3. Keyes M, Legrand LN, Iacono WG, McGue M. Parental smoking and adolescent problem behavior: an adoption study of general and specific effects. *Am J Psychiatr*. 2008;165(10):1338–1344. https://doi.org/10.1176/appi.ajp.2008.08010125.
4. Gray IC, Campbell DA, Spurr NK. Single nucleotide polymorphisms as tools in human genetics. *Human Mol Genet*. 2000;9(16):2403–2408. https://doi.org/10.1093/hmg/9.16.2403.
5. Liu H-C, Chen C-K, Leu S-J, Wu H-T, Lin S-K. Association between dopamine receptor D1 A-48G polymorphism and methamphetamine abuse. *Psychiatr & Clin Neurosci*. 2006;60(2):226–231. https://doi.org/10.1111/j.1440-1819.2006.01490.x.
6. Batel P, Houchi H, Daoust M, Ramoz N, Naassila M, Gorwood P. A haplotype of the DRD1 gene is associated with alcohol dependence. *Alcohol: Clin & Exper Res*. 2008;32(4):567–572. https://doi.org/10.1111/j.1530-0277.2008.00618.x.
7. Liu JH, Zhong HJ, Dang J, Peng L, Zhu YS. Single-nucleotide polymorphisms in dopamine receptor D1 are associated with heroin dependence but not impulsive behavior. *Genet & Mol Res*. 2015;14(2):4041–4050. https://doi.org/10.4238/2015.April.27.19.
8. Peng S, Du J, Jiang H, et al. The dopamine receptor D1 gene is associated with the length of interval between first heroin use and onset of dependence in Chinese Han heroin addicts. *J Neural Trans*. 2013;120(11):1591–1598. https://doi.org/10.1007/s00702-013-1029-6.
9. Stolf AR, Cupertino RB, Müller D, et al. Effects of DRD2 splicing-regulatory polymorphism and DRD4 48 bp VNTR on crack cocaine addiction. *J Neural Trans*. 2019;126(2):193–199. https://doi.org/10.1007/s00702-018-1946-5.
10. Fernàndez-Castillo N, Ribasés M, Roncero C, Casas M, Gonzalvo B, Cormand B. Association study between the DAT1, DBH and DRD2 genes and cocaine dependence in a Spanish sample. *Psychiatr Genet*. 2010;20(6):317–320. https://doi.org/10.1097/YPG.0b013e32833b6320.
11. Tsai S-J, Cheng C-Y, Shu L-RR, et al. No association for D2 and D4 dopamine receptor polymorphisms and methamphetamine abuse in Chinese males. *Psychiatr Genet*. 2002;12(1):29–33. https://doi.org/10.1097/00041444-200203000-00004.
12. Moyer RA, Wang D, Papp AC, et al. Intronic polymorphisms affecting alternative splicing of human dopamine D2 receptor are associated with cocaine abuse. *Neuropsychopharmacology*. 2011;36(4):753–762. https://doi.org/10.1038/npp.2010.208.
13. Li T, Chen C-k, Hu X, et al. Association analysis of the DRD4 and COMT genes in methamphetamine abuse. *Am J Med Genet Part B Neuropsychiatr Genet*. 2004;129B(1):120–124. https://doi.org/10.1002/ajmg.b.30024.
14. Pérez-Rubio G, Ramírez-Venegas A, Díaz VN, et al. Polymorphisms in HTR2A and DRD4 predispose to smoking and smoking quantity. *PLoS One*. 2017;12(1):e0170019. https://doi.org/10.1371/journal.pone.0170019.
15. Vandenbergh DJ, O'Connor RJ, Grant MD, et al. Dopamine receptor genes (DRD2, DRD3 and DRD4) and gene-gene interactions associated with smoking-related behaviors. *Addic Biol*. 2007;12(1):106–116. https://doi.org/10.1111/j.1369-1600.2007.00054.x.
16. Harrell PT, Lin H-Y, Park JY, Blank MD, Drobes DJ, Evans DE. Dopaminergic genetic variation moderates the effect of nicotine on cigarette reward. *Psychopharmacology*. 2016;233(2):351–360. https://doi.org/10.1007/s00213-015-4116-6.
17. Kotler M, Cohen H, Segman R, et al. Excess dopamine D4 receptor (D4DR) exon III seven repeat allele in opioid-dependent subjects. *Mol Psychiatr*. 1997;2(3):251–254. https://doi.org/10.1038/sj.mp.4000248.
18. Wang T-Y, Lee S-Y, Chen S-L, et al. The aldehyde dehydrogenase 2 gene is associated with heroin dependence. *Drug & Alcohol Depend*. 2012;120(1–3):220–224. https://doi.org/10.1016/j.drugalcdep.2011.06.008.
19. Li Y-F, Shao C-H, Zhang D, et al. The effect of dopamine D2, D5 receptor and transporter (SLC6A3) polymorphisms on the cue-elicited heroin craving in Chinese. *Am J Med Genet Part B: Neuropsychiatr Genet*. 2006;141B(3):269–273. https://doi.org/10.1002/ajmg.b.30264.
20. Shao C, Li Y, Jiang K, et al. Dopamine D4 receptor polymorphism modulates cue-elicited heroin craving in Chinese. *Psychopharmacology*. 2006;186(2):185–190. https://doi.org/10.1007/s00213-006-0375-6.

21. Kuo S-C, Yeh Y-W, Chen C-Y, et al. DRD3 variation associates with early-onset heroin dependence, but not specific personality traits. *Progr Neuro-Psychopharmacol & Biol Psychiatr.* 2014;51:1–8. https://doi.org/10.1016/j.pnpbp.2013.12.018.
22. Nacak M, Isir AB, Balci SO, Pehlivan S, Benlier N, Aynacioglu S. Analysis of dopamine D2 receptor (DRD2) gene polymorphisms in cannabinoid addicts. *J For Sci.* 2012;57(6):1621–1624. https://doi.org/10.1111/j.1556-4029.2012.02169.x.
23. Tilli EM, Mitiushkina NV, Sukhovskaya OA, Imyanitov EN, Hirvonen AP. The genotypes and methylation of MAO genes as factors behind smoking behavior. *Pharmacogenet & Genom.* 2017;27(11):394–401. https://doi.org/10.1097/FPC.0000000000000304.
24. Nilsson KW, Comasco E, Åslund C, Nordquist N, Leppert J, Oreland L. MAOA genotype, family relations and sexual abuse in relation to adolescent alcohol consumption. *Add Biol.* 2011;16(2):347–355. https://doi.org/10.1111/j.1369-1600.2010.00238.x.
25. Chien C-C, Lin C-H, Chang Y-Y, Lung F-W. Association of VNTR polymorphisms in the MAOA promoter and DRD4 exon 3 with heroin dependence in male Chinese addicts. *World J Biol Psychiatr.* 2010;11(2 Pt 2):409–416. https://doi.org/10.3109/15622970903304459.
26. Gokturk C, Schultze S, Nilsson KW, von Knorring L, Oreland L, Hallman J. Serotonin transporter (5-HTTLPR) and monoamine oxidase (MAOA) promoter polymorphisms in women with severe alcoholism. *Archiv Women's Mental Health.* 2008;11(5–6):347–355. https://doi.org/10.1007/s00737-008-0033-6.
27. Forero DA, López-León S, Shin HD, Park BL, Kim D-J. Meta-analysis of six genes (BDNF, DRD1, DRD3, DRD4, GRIN2B and MAOA) involved in neuroplasticity and the risk for alcohol dependence. *Drug & Alcohol Depend.* 2015;149:259–263. https://doi.org/10.1016/j.drugalcdep.2015.01.017.
28. Beuten J, Payne TJ, Ma JZ, Li MD. Significant association of catechol-O-methyltransferase (COMT) haplotypes with nicotine dependence in male and female smokers of two ethnic populations. *Neuropsychopharmacology.* 2006;31(3):675–684. https://doi.org/10.1038/sj.npp.1300997.
29. Erjavec G,N, Sviglin KN, Perkovic MN, Muck-Seler D, Jovanovic T, Pivac N. Association of gene polymorphisms encoding dopaminergic system components and platelet MAO-B activity with alcohol dependence and alcohol dependence-related phenotypes. *Progr Neuro-Psychopharmacol & Biol Psychiatr.* 2014;54:321–327. https://doi.org/10.1016/j.pnpbp.2014.07.002.
30. Isir ABB, Oguzkan S, Nacak M, Gorucu S, Dulger HE, Arslan A. The catechol-O-methyl transferase Val158Met polymorphism and susceptibility to cannabis dependence. *Am J For Med & Pathol.* 2008;29(4):320–322. https://doi.org/10.1097/PAF.0b013e3181847e56.
31. Jugurnauth SK, Chen C-K, Barnes MR, et al. *Pharmacogenet & Genom.* 2011;21(11):731–740. https://doi.org/10.1097/FPC.0b013e32834a53f9.
32. Li S, Wang Q, Pan L, et al. The association of dopamine pathway gene score, nicotine dependence and smoking cessation in a rural male population of Shandong, China. *Am J Addic.* 2016;25(6):493–498. https://doi.org/10.1111/ajad.12421.
33. Su H, Li Z, Du J, et al. Predictors of heroin relapse: personality traits, impulsivity, COMT gene Val158met polymorphism in a 5-year prospective study in Shanghai, China. *Am J Med Genet Part B: Neuropsychiatr Genet.* 2015;168(8):712–719. https://doi.org/10.1002/ajmg.b.32376.
34. Crist RC, Li J, Doyle GA, Gilbert A, Dechairo BM, Berrettini WH. Pharmacogenetic analysis of opioid dependence treatment dose and dropout rate. *Am J Drug & Alcohol Abuse.* 2018;44(4):431–440. https://doi.org/10.1080/00952990.2017.1420795.
35. Pehlivan S, Aytac HM, Kurnaz S, Pehlivan M, Aydin PC. Evaluation of COMT (rs4680), CNR2 (rs2501432), CNR2 (rs2229579), UCP2 (rs659366), and IL-17 (rs763780) gene variants in synthetic cannabinoid use disorder patients. *J Addic Dis.* 2020;38(4):495–505. https://doi.org/10.1080/10550887.2020.1787770.
36. Oosterhuis BE, LaForge KS, Prodnikov D, et al. Catechol-O-methyltransferase (COMT) gene variants: possible association of the Val158Met variant with opiate addiction in Hispanic women. *Am J Med Genet Part B: Neuropsychiatr Genet.* 2008;147B(6):793–798. https://doi.org/10.1002/ajmg.b.30716.
37. Hallikainen T, Lachman H, Saito T, et al. Lack of association between the functional variant of the catechol-o-methyltransferase (COMT) gene and early-onset alcoholism associated with severe antisocial behavior. *Am J Med Genet Part A.* 2000;96(3):348–352. https://doi.org/10.1002/1096-8628(20000612)96:3<348::aid-ajmg22>3.0.co;2-z.

38. Mutschler J, Abbruzzese E, von der Goltz C, et al. Lack of association of a functional catechol-O-methyltransferase gene polymorphism with risk of tobacco smoking: results from a multicenter case-control study. *Nicotine & Tobacco Res.* 2013;15(7):1322–1327. https://doi.org/10.1093/ntr/nts334.
39. Stolf AR, Szobot CM, Halpern R, et al. Crack cocaine users show differences in genotype frequencies of the 3' UTR variable number of tandem repeats of the dopamine transporter gene (DAT1/SLC6A3). *Neuropsychobiology.* 2014;70(1):44–51. https://doi.org/10.1159/000365992.
40. Kampangkaew JP, Spellicy CJ, Nielsen EM, et al. Pharmacogenetic role of dopamine transporter (SLC6A3) variation on response to disulfiram treatment for cocaine addiction. *Am J Addic.* 2019;28(4):311–317. https://doi.org/10.1111/ajad.12891.
41. van der Zwaluw CS, Engels RCME, Buitelaar J, Verkes RJ, Franke B, Scholte RHJ. Polymorphisms in the dopamine transporter gene (SLC6A3/DAT1) and alcohol dependence in humans: a systematic review. *Pharmacogenetics.* 2009;10(5):853–866. https://doi.org/10.2217/pgs.09.24.
42. Hou Q-F, Li S-B. Potential association of DRD2 and DAT1 genetic variation with heroin dependence. *Neurosci Lett.* 2009;464(2):127–130. https://doi.org/10.1016/j.neulet.2009.08.004.
43. Patkar AA, Berrettini WH, Hoehe M, et al. Serotonin transporter (5-HTT) gene polymorphisms and susceptibility to cocaine dependence among African-American individuals. *Addic Biol.* 2001;6(4):337–345. https://doi.org/10.1080/13556210020077064.
44. Tristán-Noguero A, Fernàndez-Castillo N, Roncero C, et al. Lack of association between the LPR and VNTR polymorphisms of the serotonin transporter gene and cocaine dependence in a Spanish sample. *Psychiatr Res.* 2013;210(3):1287–1289. https://doi.org/10.1016/j.psychres.2013.09.004.
45. Yuferov V, Butelman ER, Randesi M, et al. Association of serotonin transporter (sert) polymorphisms with opioid dependence and dimensional aspects of cocaine use in a Caucasian cohort of opioid users. *Neuropsychiatr Dis & Treat.* 2021;17:659–670. https://doi.org/10.2147/NDT.S286536.
46. Villalba K, Attonito J, Mendy A, Devieux JG, Gasana J, Dorak TM. A meta-analysis of the associations between the SLC6A4 promoter polymorphism (5HTTLPR) and the risk for alcohol dependence. *Psychiatr Genet.* 2015;25(2):47–58. https://doi.org/10.1097/YPG.0000000000000078.
47. Nielsen DA, Harding MJ, Hamon SC, Huang W, Kosten TR. Modifying the role of serotonergic 5-HTTLPR and TPH2 variants on disulfiram treatment of cocaine addiction: a preliminary study. *Genes, Brain & Behav.* 2012;11(8):1001–1008. https://doi.org/10.1111/j.1601-183X.2012.00839.x.
48. Mannelli P, Patkar AA, Murray HW, et al. Polymorphism in the serotonin transporter gene and response to treatment in African American cocaine and alcohol-abusing individuals. *Addic Biol.* 2005;10(3):261–268. https://doi.org/10.1080/13556210500235540.
49. Lin P-Y, Wu Y-S. Association between serotonin transporter gene polymorphisms and heroin dependence: a meta-analytic study. *Neuropsychiatr Dis & Treat.* 2016;12:3061–3067. https://doi.org/10.2147/NDT.S120786.
50. Gerra G, Garofano L, Zaimovic A, et al. Association of the serotonin transporter promoter polymorphism with smoking behavior among adolescents. *Am J Med Genet Part B: Neuropsychiatr Genet.* 2005;135B(1):73–78. https://doi.org/10.1002/ajmg.b.30173.
51. Pizzo de Castro MR, Maes M, Guembarovski RL, et al. A SLC6A4STin2 VNTR genetic polymorphism is associated with tobacco use disorder, but not with successful smoking cessation or smoking characteristics: a case control study. *BMC Genet.* 2014;15:78. https://doi.org/10.1186/1471-2156-15-78.
52. David SP, Munafò MR, Murphy MFG, Walton RT, Johnstone EC. The serotonin transporter 5-HTTLPR polymorphism and treatment response to nicotine patch: follow-up of a randomized controlled trial. *Nicotine & Tobacco Res.* 2007;9(2):225–231. https://doi.org/10.1080/14622200601078566.
53. Hoft NR, Stitzel JA, Hutchison KE, Ehringer MA. CHRNB2 promoter region: association with subjective effects to nicotine and gene expression differences. *Genes, Brain & Behav.* 2011;10(2):176–185. https://doi.org/10.1111/j.1601-183X.2010.00650.x.
54. Bühler KM, Giné E, Echeverry-Alzate V, Calleja-Conde J, de Fonseca FR, López-Moreno JA. Common single nucleotide variants underlying drug addiction: more than a decade of research. *Addic Biol.* 2015;20(5):845–871. https://doi.org/10.1111/adb.12204.
55. Cameli C, Bacchelli E, De Paola M, et al. Genetic variation in CHRNA7 and CHRFAM7A is associated with nicotine dependence and response to varenicline treatment. *Eur J Human Genet.* 2018;26(12):1824–1831. https://doi.org/10.1038/s41431-018-0223-2.

56. Culverhouse RC, Johnson EO, Breslau N, et al. Multiple distinct CHRNB3-CHRNA6 variants are genetic risk factors for nicotine dependence in African Americans and European Americans. *Addiction*. 2014;109(5):814–822. https://doi.org/10.1111/add.12478.
57. Feng Y, Niu T, Xing H, et al. A common haplotype of the nicotine acetylcholine receptor alpha 4 subunit gene is associated with vulnerability to nicotine addiction in men. *Am J Human Genet*. 2004;75(1):112–121. https://doi.org/10.1086/422194.
58. Grucza RA, Wang JC, Stitzel JA, et al. A risk allele for nicotine dependence in CHRNA5 is a protective allele for cocaine dependence. *Biol Psychiatr*. 2008;64(11):922–929. https://doi.org/10.1016/j.biopsych.2008.04.018.
59. Han S, Yang B-Z, Kranzler HR, Oslin D, Anton R, Gelernter J. Association of CHRNA4 polymorphisms with smoking behavior in two populations. *Am J Med Genet Part B: Neuropsychiatr Genet*. 2011;156B(4):421–429. https://doi.org/10.1002/ajmg.b.31177.
60. Lou XY, Ma JZ, Payne TJ, Beuten J, Crew KM, Li MD. Gene-based analysis suggests association of the nicotinic acetylcholine receptor beta1 subunit (CHRNB1) and M1 muscarinic acetylcholine receptor (CHRM1) with vulnerability for nicotine dependence. *Human Genet*. 2006;120(3):381–389. https://doi.org/10.1007/s00439-006-0229-7.
61. Wang S, D van der Vaart A, Xu Q, et al. Significant associations of CHRNA2 and CHRNA6 with nicotine dependence in European American and African American populations. *Human Genet*. 2014;133(5):575–586. https://doi.org/10.1007/s00439-013-1398-9.
62. Ehringer MA, Clegg HV, Collins AC, et al. Association of the neuronal nicotinic receptor beta2 subunit gene (CHRNB2) with subjective responses to alcohol and nicotine. *Am J Med Genet Part B: Neuropsychiatr Genet*. 2007;144B(5):596–604. https://doi.org/10.1002/ajmg.b.30464.
63. Janes AC, Smoller JW, David SP, et al. Association between CHRNA5 genetic variation at rs16969968 and brain reactivity to smoking images in nicotine dependent women. *Drug & Alcohol Depend*. 2012;120(1–3):7–13. https://doi.org/10.1016/j.drugalcdep.2011.06.009.
64. Hoft NR, Corley RP, McQueen MB, Huizinga D, Menard S, Ehringer MA. SNPs in CHRNA6 and CHRNB3 are associated with alcohol consumption in a nationally representative sample. *Genes, Brain & Behav*. 2009;8(6):631–637. https://doi.org/10.1111/j.1601-183X.2009.00495.x.
65. Sherva R, Kranzler HR, Yu Y, et al. Variation in nicotinic acetylcholine receptor genes is associated with multiple substance dependence phenotypes. *Neuropsychopharmacology*. 2010;35(9):1921–1931. https://doi.org/10.1038/npp.2010.64.
66. Aroche AP, Rovaris DL, Grevet EH, et al. Association of CHRNA5 gene variants with crack cocaine addiction. *NeuroMol Med*. 2020;22(3):384–390. https://doi.org/10.1007/s12017-020-08596-1.
67. Forget B, Icick R, Robert J, et al. Alterations in nicotinic receptor alpha5 subunit gene differentially impact early and later stages of cocaine addiction: a translational study in transgenic rats and patients. *Progr Neurobiol*. 2021;197:101898. https://doi.org/10.1016/j.pneurobio.2020.101898.
68. Haller G, Kapoor M, Budde J, et al. Rare missense variants in CHRNB3 and CHRNA3 are associated with risk of alcohol and cocaine dependence. *Human Mol Genet*. 2014;23(3):810–819. https://doi.org/10.1093/hmg/ddt463.
69. Edenberg HJ, Gelernter J, Agrawal A. Genetics of alcoholism. *Curr Psychiatr Rep*. 2019;21(4):26. https://doi.org/10.1007/s11920-019-1008-1.
70. Wang TY, Lee SY, Chen SL, et al. The ADH1B and DRD2 gene polymorphism may modify the protective effect of the ALDH2 gene against heroin dependence. *Progr Neuro-Psychopharmacol & Biol Psychiatr*. 2013;43:134–139. https://doi.org/10.1016/j.pnpbp.2012.12.011.
71. Masaoka H, Gallus S, Ito H, et al. Aldehyde dehydrogenase 2 polymorphism is a predictor of smoking cessation. *Nicotine & Tobacco Res*. 2017;19(9):1087–1094. https://doi.org/10.1093/ntr/ntw316.
72. Costa-Mallen P, Costa LG, Checkoway H. Genotype combinations for monoamine oxidase-B intron 13 polymorphism and dopamine D2 receptor TaqIB polymorphism are associated with ever-smoking status among men. *Neurosci Lett*. 2005;385(2):158–162. https://doi.org/10.1016/j.neulet.2005.05.035.
73. Koijam AS, Hijam AC, Singh AS, et al. Association of dopamine transporter gene with heroin dependence in an Indian subpopulation from Manipur. *J Mol Neurosci*. 2021;71(1):122–136. https://doi.org/10.1007/s12031-020-01633-5.
74. Bali P, Kenny PJ. Transcriptional mechanisms of drug addiction. *Dialogs Clin Neurosci*. 2019;21(4):379–387. https://doi.org/10.31887/DCNS.2019.21.4/pkenny.

75. Tulchinsky E. Fos family members: regulation, structure and role in oncogenic transformation. *Histol & Histopathol*. 2000;15(3):921–928. https://doi.org/10.14670/HH-15.921.
76. Graybiel AM, Moratalla R, Robertson HA. Amphetamine and cocaine induce drug-specific activation of the c-fos gene in striosome-matrix compartments and limbic subdivisions of the striatum. *Proc Nat Acad Sci USA*. 1990;87(17):6912–6916. https://doi.org/10.1073/pnas.87.17.6912.
77. Hope B, Kosofsky B, Hyman SE, Nestler EJ. Regulation of immediate early gene expression and AP-1 binding in the rat nucleus accumbens by chronic cocaine. *Proc Nat Acad Sci USA*. 1992;89(13):5764–5768. https://doi.org/10.1073/pnas.89.13.5764.
78. Zoeller RT, Fletcher DL. A single administration of ethanol simultaneously increases c-fos mRNA and reduces c-jun mRNA in the hypothalamus and hippocampus. *Molecul Brain Res*. 1994;24(1–4):185–191. https://doi.org/10.1016/0169-328X(94)90131-7.
79. Matsumoto I, Leah J, Shanley B, Wilce P. Immediate early gene expression in the rat brain during ethanol withdrawal. *Mol & Cell Neurosci*. 1993;4(6):485–491. https://doi.org/10.1006/mcne.1993.1060.
80. Rubio FJ, Liu Q-R, Li X, et al. Context-induced reinstatement of methamphetamine seeking is associated with unique molecular alterations in Fos-expressing dorsolateral striatum neurons. *J Neurosci*. 2015;35(14):5625–5639. https://doi.org/10.1523/JNEUROSCI.4997-14.2015.
81. Gao P, Limpens JHW, Spijker S, Vanderschuren LJMJ, Voorn P. Stable immediate early gene expression patterns in medial prefrontal cortex and striatum after long-term cocaine self-administration. *Addic Biol*. 2017;22(2):354–368. https://doi.org/10.1111/adb.12330.
82. Liu Q-R, Rubio FJ, Bossert JM, et al. Detection of molecular alterations in methamphetamine-activated Fos-expressing neurons from a single rat dorsal striatum using fluorescence-activated cell sorting (FACS). *J Neurochem*. 2014;128(1):173–185. https://doi.org/10.1111/jnc.12381.
83. Rivera P, Silva-Peña D, Blanco E, et al. Oleoylethanolamide restores alcohol-induced inhibition of neuronal proliferation and microglial activity in striatum. *Neuropharmacology*. 2019;146:184–197. https://doi.org/10.1016/j.neuropharm.2018.11.037.
84. Muller DL, Unterwald EM. D1 dopamine receptors modulate deltaFosB induction in rat striatum after intermittent morphine administration. *J Pharmacol & Exper Ther*. 2005;314(1):148–154. https://doi.org/10.1124/jpet.105.083410.
85. Venebra-Muñoz A, Corona-Morales A, Santiago-García J, Melgarejo-Gutíerrez M, Caba M, García-García F. Enriched environment attenuates nicotine self-administration and induces changes in ΔFosB expression in the rat prefrontal cortex and nucleus accumbens. *Neuroreport*. 2014;25(9):688–692. https://doi.org/10.1097/WNR.0000000000000157.
86. Li X, Davis IR, Lofaro OM, Zhang J, Cimbro R, Rubio FJ. Distinct gene alterations between Fos-expressing striatal and thalamic neurons after withdrawal from methamphetamine self-administration. *Brain & Behav*. 2019;9(9):e01378. https://doi.org/10.1002/brb3.1378.
87. Nestler EJ. Transcriptional mechanisms of drug addiction. *Clin Psychopharmacol & Neurosci*. 2012;10(3):136–143. https://doi.org/10.9758/cpn.2012.10.3.136.
88. Ohnishi YN, Ohnishi YH, Vialou V, et al. Functional role of the N-terminal domain of ΔFosB in response to stress and drugs of abuse. *Neuroscience*. 2015;284:165–170. https://doi.org/10.1016/j.neuroscience.2014.10.002.
89. Zachariou V, Bolanos CA, Selley DE, et al. An essential role for DeltaFosB in the nucleus accumbens in morphine action. *Nature Neurosci*. 2006;9(2):205–211. https://doi.org/10.1038/nn1636.
90. Colby CR, Whisler K, Steffen C, Nestler EJ, Self DW. Striatal cell type-specific overexpression of DeltaFosB enhances incentive for cocaine. *J Neurosci*. 2003;23(6):2488–2493. https://doi.org/10.1523/JNEUROSCI.23-06-02488.2003.
91. Laurier LG, Corrigall WA, George SR. Dopamine receptor density, sensitivity and mRNA levels are altered following self-administration of cocaine in the rat. *Brain Res*. 1994;634(1):31–40. https://doi.org/10.1016/0006-8993(94)90255-0.
92. Briand LA, Flagel SB, Garcia-Fuster MJ, et al. Persistent alterations in cognitive function and prefrontal dopamine D2 receptors following extended, but not limited, access to self-administered cocaine. *Neuropsychopharmacology*. 2008;33(12):2969–2980. https://doi.org/10.1038/npp.2008.18.
93. Rotter A, Biermann T, Amato D, et al. Glucocorticoid receptor antagonism blocks ethanol-induced place preference learning in mice and attenuates dopamine D2 receptor adaptation in the frontal cortex. *Brain Res Bull*. 2012;88(5):519–524. https://doi.org/10.1016/j.brainresbull.2012.05.007.
94. Hoplight BJ, Vincow ES, Neumaier JF. Cocaine increases 5-HT1B mRNA in rat nucleus accumbens shell neurons. *Neuropharmacology*. 2007;52(2):444–449. https://doi.org/10.1016/j.neuropharm.2006.08.013.

95. Neumaier JF, McDevitt RA, Polis IY, Parsons LH. Acquisition of and withdrawal from cocaine self-administration regulates 5-HT mRNA expression in rat striatum. *J Neurochem.* 2009;111(1):217−227. https://doi.org/10.1111/j.1471-4159.2009.06313.x.
96. Nevo I, Langlois X, Laporte AM, et al. Chronic alcoholization alters the expression of 5-HT1A and 5-HT1B receptor subtypes in rat brain. *Eur J Pharmacol.* 1995;281(3):229−239. https://doi.org/10.1016/0014-2999(95)00238-g.
97. Kenny PJ, File SE, Rattray M. Nicotine regulates 5-HT(1A) receptor gene expression in the cerebral cortex and dorsal hippocampus. *Eur J Neurosci.* 2001;13(6):1267−1271. https://doi.org/10.1046/j.0953-816x.2001.01501.x.
98. Burnett EJ, Grant KA, Davenport AT, Hemby SE, Friedman DP. The effects of chronic ethanol self-administration on hippocampal 5-HT1A receptors in monkeys. *Drug & Alcohol Depend.* 2014;136:135−142. https://doi.org/10.1016/j.drugalcdep.2014.01.002.
99. Zavitsanou K, Wang H, Dalton VS, Nguyen V. Cannabinoid administration increases 5HT1A receptor binding and mRNA expression in the hippocampus of adult but not adolescent rats. *Neuroscience.* 2010;169(1):315−324. https://doi.org/10.1016/j.neuroscience.2010.04.005.
100. Kinoshita H, Jessop DS, Roberts DJ, Hishida S, Harbuz MS. Chronic ethanol administration and withdrawal decreases 5-HT1A mRNA, but not 5-HT4 expression in the rat hippocampus. *Pharmacol & Toxicol.* 2003;93(2):100−102. https://doi.org/10.1034/j.1600-0773.2003.930208.x.
101. Mantsch JR, Yuferov V, Mathieu-Kia A-M, Ho A, Kreek MJ. Effects of extended access to high versus low cocaine doses on self-administration, cocaine-induced reinstatement and brain mRNA levels in rats. *Psychopharmacology.* 2004;175(1):26−36. https://doi.org/10.1007/s00213-004-1778-x.
102. Sun H, Luessen DJ, Kind KO, Zhang K, Chen R. Cocaine self-administration regulates transcription of opioid peptide precursors and opioid receptors in rat caudate putamen and prefrontal cortex. *Neuroscience.* 2020;443:131−139. https://doi.org/10.1016/j.neuroscience.2020.07.035.
103. Zhang Y, Schlussman SD, Rabkin J, Butelman ER, Ho A, Kreek MJ. Chronic escalating cocaine exposure, abstinence/withdrawal, and chronic re-exposure: effects on striatal dopamine and opioid systems in C57BL/6J mice. *Neuropharmacology.* 2013;67:259−266. https://doi.org/10.1016/j.neuropharm.2012.10.015.
104. Isola R, Zhang H, Tejwani GA, Neff NH, Hadjiconstantinou M. Acute nicotine changes dynorphin and prodynorphin mRNA in the striatum. *Psychopharmacology.* 2009;201(4):507−516. https://doi.org/10.1007/s00213-008-1315-4.
105. Walters CL, Cleck JN, Kuo YC, Blendy JA. Mu-opioid receptor and CREB activation are required for nicotine reward. *Neuron.* 2005;46(6):933−943. https://doi.org/10.1016/j.neuron.2005.05.005.
106. Rosin A, Lindholm S, Franck J, Georgieva J. Downregulation of kappa opioid receptor mRNA levels by chronic ethanol and repetitive cocaine in rat ventral tegmentum and nucleus accumbens. *Neurosci Lett.* 1999;275(1):1−4. https://doi.org/10.1016/s0304-3940(99)00675-8.
107. Raeder H, Holter SM, Hartmann AM, Spanagel R, Moller HJ, Rujescu D. Expression of N-methyl-d-aspartate (NMDA) receptor subunits and splice variants in an animal model of long-term voluntary alcohol self-administration. *Drug & Alcohol Depend.* 2008;96(1−2):16−21. https://doi.org/10.1016/j.drugalcdep.2007.12.013.
108. Acosta G, Hasenkamp W, Daunais JB, Friedman DP, Grant KA, Hemby SE. Ethanol self-administration modulation of NMDA receptor subunit and related synaptic protein mRNA expression in prefrontal cortical fields in cynomolgus monkeys. *Brain Res.* 2010;1318:144−154. https://doi.org/10.1016/j.brainres.2009.12.050.
109. Oliva JM, Ortiz S, Pérez-Rial S, Manzanares J. Time dependent alterations on tyrosine hydroxylase, opioid and cannabinoid CB1 receptor gene expressions after acute ethanol administration in the rat brain. *Eur Neuropsychopharmacol.* 2008;18(5):373−382. https://doi.org/10.1016/j.euroneuro.2007.09.001.
110. Tomita H, Hikiji M, Fujiwara Y, Akiyama K, Otsuki S. Changes in dopamine D2 and GluR-1 glutamate receptor mRNAs in the rat brain after treatment with phencyclidine. *Acta Medica Okayama.* 1995;49(2):61−68. https://doi.org/10.18926/AMO/30393.
111. Navarro M, Carrera MR, Fratta W, et al. Functional interaction between opioid and cannabinoid receptors in drug self-administration. *J Neurosci.* 2001;21(14):5344−5350. https://doi.org/10.1523/JNEUROSCI.21-14-05344.2001.
112. Caputi FF, Caffino L, Candeletti S, Fumagalli F, Romualdi P. Short-term withdrawal from repeated exposure to cocaine during adolescence modulates dynorphin mRNA levels and BDNF signaling in the rat nucleus accumbens. *Drug & Alcohol Depend.* 2019;197:127−133. https://doi.org/10.1016/j.drugalcdep.2019.01.006.
113. Svensson P, Hurd YL. Specific reductions of striatal prodynorphin and D1 dopamine receptor messenger RNAs during cocaine abstinence. *Brain Res Mol Brain Res.* 1998;56(1−2):162−168. https://doi.org/10.1016/s0169-328x(98)00041-2.

114. Przewłocka B, Lasoń W. Adaptive changes in the proenkephalin and D2 dopamine receptor mRNA expression after chronic cocaine in the nucleus accumbens and striatum of the rat. *Eur Neuropsychopharmacol.* 1995;5(4):465–469. https://doi.org/10.1016/0924-977x(95)80005-m.
115. Lunden JW, Kirby LG. Opiate exposure and withdrawal dynamically regulate mRNA expression in the serotonergic dorsal raphe nucleus. *Neuroscience.* 2013;254:160–172. https://doi.org/10.1016/j.neuroscience.2013.08.071.
116. Caine SB, Thomsen M, Gabriel KI, et al. Lack of self-administration of cocaine in dopamine D1 receptor knockout mice. *J Neurosci.* 2007;27(48):13140–13150. https://doi.org/10.1523/JNEUROSCI.2284-07.2007.
117. Chiamulera C, Epping-Jordan MP, Zocchi A, et al. Reinforcing and locomotor stimulant effects of cocaine are absent in mGluR5 null mutant mice. *Nat Neurosci.* 2001;4(9):873–874. https://doi.org/10.1038/nn0901-873.
118. Drouin C, Darracq L, Trovero F, et al. Alpha1b-adrenergic receptors control locomotor and rewarding effects of psychostimulants and opiates. *J Neurosci.* 2002;22(7):2873–2884. https://doi.org/10.1523/JNEUROSCI.22-07-02873.2002.
119. Gutiérrez-Cuesta J, Burokas A, Mancino S, Kummer S, Martín-García E, Maldonado R. Effects of genetic deletion of endogenous opioid system components on the reinstatement of cocaine-seeking behavior in mice. *Neuropsychopharmacology.* 2014;39(13):2974–2988. https://doi.org/10.1038/npp.2014.149.
120. Mathon DS, Lesscher HMB, Gerrits MAFM, et al. Increased gabaergic input to ventral tegmental area dopaminergic neurons associated with decreased cocaine reinforcement in mu-opioid receptor knockout mice. *Neuroscience.* 2005;130(2):359–367. https://doi.org/10.1016/j.neuroscience.2004.10.002.
121. Soria G, Mendizábal V, Touriño C, et al. Lack of CB1 cannabinoid receptor impairs cocaine self-administration. *Neuropsychopharmacology.* 2005;30(9):1670–1680. https://doi.org/10.1038/sj.npp.1300707.
122. Thomsen M, Woldbye DP, Wörtwein G, Fink-Jensen A, Wess J, Caine SB. Reduced cocaine self-administration in muscarinic M5 acetylcholine receptor-deficient mice. *J Neurosci.* 2005;25(36):8141–8149. https://doi.org/10.1523/JNEUROSCI.2077-05.2005.
123. Pisanu A, Lecca D, Valentini V, et al. Impairment of acquisition of intravenous cocaine self-administration by RNA-interference of dopamine D1-receptors in the nucleus accumbens shell. *Neuropharmacology.* 2015;89:398–411. https://doi.org/10.1016/j.neuropharm.2014.10.018.
124. Carrigan KA, Dykstra LA. Behavioral effects of morphine and cocaine in M1 muscarinic acetylcholine receptor-deficient mice. *Psychopharmacology.* 2007;191(4):985–993. https://doi.org/10.1007/s00213-006-0671-1.
125. Zachariou V, Caldarone BJ, Weathers-Lowin A, et al. Nicotine receptor inactivation decreases sensitivity to cocaine. *Neuropsychopharmacology.* 2001;24(5):576–589. https://doi.org/10.1016/S0893-133X(00)00224-4.
126. Touriño C, Ledent C, Maldonado R, Valverde O. CB1 cannabinoid receptor modulates 3,4-methylenedioxymethamphetamine acute responses and reinforcement. *Biol Psychiatr.* 2008;63(11):1030–1038. https://doi.org/10.1016/j.biopsych.2007.09.003.
127. Pons S, Fattore L, Cossu G, et al. Crucial role of alpha4 and alpha6 nicotinic acetylcholine receptor subunits from ventral tegmental area in systemic nicotine self-administration. *J Neurosci.* 2008;28(47):12318–12327. https://doi.org/10.1523/JNEUROSCI.3918-08.2008.
128. Wang LP, Li F, Shen X, Tsien JZ. Conditional knockout of NMDA receptors in dopamine neurons prevents nicotine-conditioned place preference. *PloS One.* 2010b;5(1):e8616. https://doi.org/10.1371/journal.pone.0008616.
129. Levin ED, Petro A, Rezvani AH, et al. Nicotinic alpha7- or beta2-containing receptor knockout: effects on radial-arm maze learning and long-term nicotine consumption in mice. *Behav Brain Res.* 2009;196(2):207–213. https://doi.org/10.1016/j.bbr.2008.08.048.
130. Sanjakdar SS, Maldoon PP, Marks MJ, et al. Differential roles of α6β2* and α4β2* neuronal nicotinic receptors in nicotine- and cocaine-conditioned reward in mice. *Neuropsychopharmacology.* 2015;40(2):350–360. https://doi.org/10.1038/npp.2014.177.
131. Berrendero F, Kieffer BL, Maldonado R. Attenuation of nicotine-induced antinociception, rewarding effects, and dependence in mu-opioid receptor knock-out mice. *J Neurosci.* 2002;22(24):10935–10940. https://doi.org/10.1523/JNEUROSCI.22-24-10935.2002.
132. Berrendero F, Plaza-Zabala A, Galeote L, et al. Influence of δ-opioid receptors in the behavioral effects of nicotine. *Neuropsychopharmacology.* 2012;37(10):2332–2344. https://doi.org/10.1038/npp.2012.88.
133. Castañé A, Valjent E, Ledent C, Parmentier M, Maldonado R, Valverde O. Lack of CB1 cannabinoid receptors modifies nicotine behavioural responses, but not nicotine abstinence. *Neuropharmacology.* 2002;43(5):857–867. https://doi.org/10.1016/s0028-3908(02)00118-1.

134. Ghozland S, Matthes HWD, Simonin F, Filliol D, Kieffer BL, Maldonado R. Motivational effects of cannabinoids are mediated by mu-opioid and kappa-opioid receptors. *J Neurosci.* 2002;22(3):1146−1154. https://doi.org/10.1523/JNEUROSCI.22-03-01146.2002.
135. Boyce-Rustay JM, Holmes A. Ethanol-related behaviors in mice lacking the NMDA receptor NR2A subunit. *Psychopharmacology.* 2006;187(4):455−466. https://doi.org/10.1007/s00213-006-0448-6.
136. Jury NJ, Radke AK, Pati D, et al. NMDA receptor GluN2A subunit deletion protects against dependence-like ethanol drinking. *Behav Brain Res.* 2018;353:124−128. https://doi.org/10.1016/j.bbr.2018.06.029.
137. Radke AK, Jury NJ, Delpire E, Nakazawa K, Holmes A. Reduced ethanol drinking following selective cortical interneuron deletion of the GluN2B NMDA receptors subunit. *Alcohol.* 2017;58:47−51. https://doi.org/10.1016/j.alcohol.2016.07.005.
138. Roberts AJ, McDonald JS, Heyser CJ, et al. mu-Opioid receptor knockout mice do not self-administer alcohol. *J Pharmacol & Exper Ther.* 2000;293(3):1002−1008.
139. Thanos PK, Dimitrakakis ES, Rice O, Gifford A, Volkow ND. Ethanol self-administration and ethanol conditioned place preference are reduced in mice lacking cannabinoid CB1 receptors. *Behav Brain Res.* 2005;164(2):206−213. https://doi.org/10.1016/j.bbr.2005.06.021.
140. Matthes HW, Maldonado R, Simonin F, et al. Loss of morphine-induced analgesia, reward effect and withdrawal symptoms in mice lacking the mu-opioid-receptor gene. *Nature.* 1996;383(6603):819−823. https://doi.org/10.1038/383819a0.
141. Kao JH, Huang EY, Tao PL. NR2B subunit of NMDA receptor at nucleus accumbens is involved in morphine rewarding effect by siRNA study. *Drug & Alcohol Depend.* 2011;118(2−3):366−374. https://doi.org/10.1016/j.drugalcdep.2011.04.019.
142. Li Y, Ping X, Yu P, et al. Over-expression of the GluN2B subunit in the forebrain facilitates the acquisition of morphine-related positive and aversive memory in rats. *Behav Brain Res.* 2016;311:416−424. https://doi.org/10.1016/j.bbr.2016.05.039.
143. Cossu G, Ledent C, Fattore L, et al. Cannabinoid CB1 receptor knockout mice fail to self-administer morphine but not other drugs of abuse. *Behav Brain Res.* 2001;118(1):61−65. https://doi.org/10.1016/s0166-4328(00)00311-9.
144. Martin M, Ledent C, Parmentier M, Maldonado R, Valverde O. Cocaine, but not morphine, induces conditioned place preference and sensitization to locomotor responses in CB1 knockout mice. *Eur J Neurosci.* 2000;12(11):4038−4046. https://doi.org/10.1046/j.1460-9568.2000.00287.x.
145. Basile AS, Fedorova I, Zapata A, et al. Deletion of the M5 muscarinic acetylcholine receptor attenuates morphine reinforcement and withdrawal but not morphine analgesia. *Proc Nat Acad Sci USA.* 2002;99(17):11452−11457. https://doi.org/10.1073/pnas.162371899.
146. Becker A, Grecksch G, Kraus J, et al. Loss of locomotor sensitisation in response to morphine in D1 receptor deficient mice. *Naunyn-Schmiedeberg's Archiv Pharmacol.* 2001;363(5):562−568. https://doi.org/10.1007/s002100100404.
147. Karasinska JM, George SR, Cheng R, O'Dowd BF. Deletion of dopamine D1 and D3 receptors differentially affects spontaneous behaviour and cocaine-induced locomotor activity, reward and CREB phosphorylation. *Eur J Neurosci.* 2005;22(7):1741−1750. https://doi.org/10.1111/j.1460-9568.2005.04353.x.
148. Karlsson RM, Hefner KR, Sibley DR, Holmes A. Comparison of dopamine D1 and D5 receptor knockout mice for cocaine locomotor sensitization. *Psychopharmacology.* 2008;200(1):117−127. https://doi.org/10.1007/s00213-008-1165-0.
149. Elliot EE, Sibley DR, Katz JL. Locomotor and discriminative-stimulus effects of cocaine in dopamine D5 receptor knockout mice. *Psychopharmacology.* 2003;169(2):161−168. https://doi.org/10.1007/s00213-003-1494-y.
150. Dixon CI, Walker SE, King SL, Stephens DN. Deletion of the gabra2 gene results in hypersensitivity to the acute effects of ethanol but does not alter ethanol self administration. *PloS One.* 2012;7(10):e47135. https://doi.org/10.1371/journal.pone.0047135.
151. Caine SB, Negus SS, Mello NK, et al. Role of dopamine D2-like receptors in cocaine self-administration: studies with D2 receptor mutant mice and novel D2 receptor antagonists. *J Neurosci.* 2002;22(7):2977−2988. https://doi.org/10.1523/JNEUROSCI.22-07-02977.2002.
152. Bello EP, Mateo Y, Gelman DM, et al. Cocaine supersensitivity and enhanced motivation for reward in mice lacking dopamine D2 autoreceptors. *Nat Neurosci.* 2011;14(8):1033−1038. https://doi.org/10.1038/nn.2862.

153. Holroyd KB, Arover MF, Fuino RL, et al. Loss of feedback inhibition via D2 autoreceptors enhances acquisition of cocaine taking and reactivity to drug-paired cues. *Neuropsychopharmacology*. 2015;40(6):1495–1509. https://doi.org/10.1038/npp.2014.336.
154. Dobbs LK, Kaplan AR, Bock R, et al. D1 receptor hypersensitivity in mice with low striatal D2 receptors facilitates select cocaine behaviors. *Neuropsychopharmacology*. 2019;44(4):805–816. https://doi.org/10.1038/s41386-018-0286-3.
155. de la Cour C, Sørensen G, Wortwein G, et al. Enhanced self-administration of alcohol in muscarinic acetylcholine M4 receptor knockout mice. *Eur J Pharmacol*. 2015;746:1–5. https://doi.org/10.1016/j.ejphar.2014.10.050.
156. Schmidt LS, Thomsen M, Weikop P, et al. Increased cocaine self-administration in M4 muscarinic acetylcholine receptor knockout mice. *Psychopharmacology*. 2011;216(3):367–378. https://doi.org/10.1007/s00213-011-2225-4.
157. Morishima Y, Miyakawa T, Furuyashiki T, Tanaka Y, Mizuma H, Nakanishi S. Enhanced cocaine responsiveness and impaired motor coordination in metabotropic glutamate receptor subtype 2 knockout mice. *Proc Nat Acad Sci USA*. 2005;102(11):4170–4175. https://doi.org/10.1073/pnas.0500914102.
158. Solís O, García-Sanz P, Martín AB, et al. Behavioral sensitization and cellular responses to psychostimulants are reduced in D2R knockout mice. *Addic Biol*. 2021;26(1):e12840. https://doi.org/10.1111/adb.12840.
159. Belzung C, Scearce-Levie K, Barreau S, Hen R. Absence of cocaine-induced place conditioning in serotonin 1B receptor knock-out mice. *Pharmacol Biochem & Behav*. 2000;66(1):221–225. https://doi.org/10.1016/s0091-3057(00)00238-0.
160. Castanon N, Scearce-Levie K, Lucas JJ, Rocha B, Hen R. Modulation of the effects of cocaine by 5-HT1B receptors: a comparison of knockouts and antagonists. *Pharmacol Biochem & Behav*. 2000;67(3):559–566. https://doi.org/10.1016/s0091-3057(00)00389-0.
161. Rocha BA, Ator R, Emmett-Oglesby MW, Hen R. Intravenous cocaine self-administration in mice lacking 5-HT1B receptors. *Pharmacol Biochem & Behav*. 1997;57(3):407–412. https://doi.org/10.1016/s0091-3057(96)00444-3.
162. Rocha BA, Scearce-Levie K, Lucas JJ, et al. Increased vulnerability to cocaine in mice lacking the serotonin-1B receptor. *Nature*. 1998;393(6681):175–178. https://doi.org/10.1038/30259.
163. Allan AM, Galindo R, Chynoweth J, Engel SR, Savage DD. Conditioned place preference for cocaine is attenuated in mice over-expressing the 5-HT(3) receptor. *Psychopharmacology*. 2001;158(1):18–27. https://doi.org/10.1007/s002130100833.
164. Elmer GI, Pieper JO, Rubinstein M, Low MJ, Grandy DK, Wise RA. Failure of intravenous morphine to serve as an effective instrumental reinforcer in dopamine D2 receptor knock-out mice. *J Neurosci*. 2002;22(10):RC224. https://doi.org/10.1523/JNEUROSCI.22-10-j0004.2002.
165. Risinger FO, Freeman PA, Rubinstein M, Low MJ, Grandy DK. Lack of operant ethanol self-administration in dopamine D2 receptor knockout mice. *Psychopharmacology*. 2000;152(3):343–350. https://doi.org/10.1007/s002130000548.
166. Gallego X, Ruiz-Medina J, Valverde O, et al. Transgenic over expression of nicotinic receptor alpha 5, alpha 3, and beta 4 subunit genes reduces ethanol intake in mice. *Alcohol*. 2012;46(3):205–215. https://doi.org/10.1016/j.alcohol.2011.11.005.
167. Joffe ME, Turner BD, Delpire E, Grueter BA. Genetic loss of GluN2B in D1-expressing cell types enhances long-term cocaine reward and potentiation of thalamo-accumbens synapses. *Neuropsychopharmacology*. 2018;43(12):2383–2389. https://doi.org/10.1038/s41386-018-0131-8.
168. Yang H-J, Zhang H-Y, Bi G-H, He Y, Gao J-T, Xi Z-X. Deletion of type 2 metabotropic glutamate receptor decreases sensitivity to cocaine reward in rats. *Cell Rep*. 2017;20(2):319–332. https://doi.org/10.1016/j.celrep.2017.06.046.
169. Roberts AJ, Gold LH, Polis I, et al. Increased ethanol self-administration in delta-opioid receptor knockout mice. *Alcohol: Clin & Exper Res*. 2001;25(9):1249–1256. https://doi.org/10.1097/00000374-200109000-00002.
170. Deisseroth K. Optogenetics. *Nat Methods*. 2011;8(1):26–29. https://doi.org/10.1038/nmeth.f.324.
171. Bock R, Shin JH, Kaplan AR, et al. *Nat Neurosci*. 2013;16(5):632–638. https://doi.org/10.1038/nn.3369.
172. Weitz M, Khayat A, Yaka R. GABAergic projections to the ventral tegmental area govern cocaine-conditioned reward. *Addic Biol*. 2021;26(4):e13026. https://doi.org/10.1111/adb.13026.
173. Stefanik MT, Kupchik YM, Brown RM, Kalivas PW. Optogenetic evidence that pallidal projections, not nigral projections, from the nucleus accumbens core are necessary for reinstating cocaine seeking. *J Neurosci*. 2013;33(24):13654–13662. https://doi.org/10.1523/JNEUROSCI.1570-13.2013.

174. Liu Y, Jean-Richard-Dit-Bressel P, Yau JO-Y, et al. The mesolimbic dopamine activity signatures of relapse to alcohol-seeking. *J Neurosci*. 2020;40(33):6409−6427. https://doi.org/10.1523/JNEUROSCI.0724-20.2020.
175. Warlow SM, Robinson MJF, Berridge KC. Optogenetic central amygdala stimulation intensifies and narrows motivation for cocaine. *J Neurosci*. 2017;37(35):8330−8348. https://doi.org/10.1523/JNEUROSCI.3141-16.2017.
176. Martín-García E, Courtin J, Renault P, et al. Frequency of cocaine self-administration influences drug seeking in the rat: optogenetic evidence for a role of the prelimbic cortex. *Neuropsychopharmacology*. 2014;39(10):2317−2330. https://doi.org/10.1038/npp.2014.66.
177. Cordie R, McFadden LM. Optogenetic inhibition of the medial prefrontal cortex reduces methamphetamine-primed reinstatement in male and female rats. *Behav Pharmacol*. 2019;30(6):506−513. https://doi.org/10.1097/FBP.0000000000000485.
178. Zhou Y, Yan E, Cheng D, et al. The projection from ventral ca1, not prefrontal cortex, to nucleus accumbens core mediates recent memory retrieval of cocaine-conditioned place preference. *Front Behav Neurosci*. 2020;14:558074. https://doi.org/10.3389/fnbeh.2020.558074.
179. Puaud M, Higuera-Matas A, Brunault P, Everitt BJ, Belin D. The basolateral amygdala to nucleus accumbens core circuit mediates the conditioned reinforcing effects of cocaine-paired cues on cocaine seeking. *Biol Psychiatr*. 2021;89(4):356−365. https://doi.org/10.1016/j.biopsych.2020.07.022.
180. Stefanik MT, Kupchik YM, Kalivas PW. Optogenetic inhibition of cortical afferents in the nucleus accumbens simultaneously prevents cue-induced transient synaptic potentiation and cocaine-seeking behavior. *Brain Struc & Func*. 2016;221(3):1681−1689. https://doi.org/10.1007/s00429-015-0997-8.
181. Hellard ER, Binette A, Zhuang X, et al. Optogenetic control of alcohol-seeking behavior via the dorsomedial striatal circuit. *Neuropharmacology*. 2019;155:89−97. https://doi.org/10.1016/j.neuropharm.2019.05.022.
182. Li Y, Li C-Y, Xi W, et al. Rostral and caudal ventral tegmental area gabaergic inputs to different dorsal raphe neurons participate in opioid dependence. *Neuron*. 2019b;101(4):748−761. https://doi.org/10.1016/j.neuron.2018.12.012.
183. Zhang R, Jia W, Wang Y, et al. A glutamatergic insular-striatal projection regulates the reinstatement of cue-associated morphine-seeking behavior in mice. *Brain Res Bull*. 2019;152:257−264. https://doi.org/10.1016/j.brainresbull.2019.07.023.
184. Müller Ewald VA, De Corte BJ, Gupta SC, et al. Attenuation of cocaine seeking in rats via enhancement of infralimbic cortical activity using stable step-function opsins. *Psychopharmacology*. 2019;236(1):479−490. https://doi.org/10.1007/s00213-018-4964-y.
185. Shen Y, Campbell RE, Côté DC, Paquet ME. Challenges for therapeutic applications of opsin-based optogenetic tools in humans. *Front Neural Circ*. 2020;14:41. https://doi.org/10.3389/fncir.2020.00041.
186. Armbuster BN, Li X, Pausch MH, Herlitze S, Roth BL. Evolving the lock to fit the key to create a family of G protein-coupled receptors potently activated by an inert ligand. *Proc Nat Acad Sci USA*. 2007;104(12):5163−5168. https://doi.org/10.1073/pnas.0700293104.
187. Mahler SV, Brodnik ZD, Cox BM, et al. Chemogenetic manipulations of ventral tegmental area dopamine neurons reveal multifaceted roles in cocaine abuse. *J Neurosci*. 2019;39(3):503−518. https://doi.org/10.1523/JNEUROSCI.0537-18.2018.
188. Jaramillo AA, Van Voorhies K, Randall PA, Besheer J. Silencing the insular-striatal circuit decreases alcohol self-administration and increases sensitivity to alcohol. *Behav Brain Res*. 2018;348:74−81. https://doi.org/10.1016/j.bbr.2018.04.007.
189. Venniro M, Caprioli D, Zhang M, et al. The anterior insular cortex → central amygdala glutamatergic pathway is critical to relapse after contingency management. *Neuron*. 2017;96(2):414−427. https://doi.org/10.1016/j.neuron.2017.09.024.
190. Manvich DF, Webster KA, Foster SL, et al. The DREADD agonist clozapine N-oxide (CNO) is reverse-metabolized to clozapine and produces clozapine-like interoceptive stimulus effects in rats and mice. *Sci Rep*. 2018;8(1):3840. https://doi.org/10.1038/s41598-018-22116-z.
191. Schmid B, Blomeyer D, Becker K, et al. The interaction between the dopamine transporter gene and age at onset in relation to tobacco and alcohol use among 19-year-olds. *Addic Biol*. 2009;14(4):489−499. https://doi.org/10.1111/j.1369-1600.2009.00171.x.
192. Baker-Andresen D, Zhao Q, Li X, et al. Persistent variations in neuronal DNA methylation following cocaine self-administration and protracted abstinence in mice. *Neuroepigenetics*. 2015;4:1−11. https://doi.org/10.1016/j.nepig.2015.10.001.

193. Ploense KL, Li X, Baker-Andresen D, et al. Prolonged-access to cocaine induces distinct Homer2 DNA methylation, hydroxymethylation, and transcriptional profiles in the dorsomedial prefrontal cortex of male Sprague-Dawley rats. *Neuropharmacology*. 2018;143:299–305. https://doi.org/10.1016/j.neuropharm.2018.09.029.
194. Kim R, Sepulveda-Orengo MT, Healey KL, Williams EA, Reissner KJ. Regulation of glutamate transporter 1 (GLT-1) gene expression by cocaine self-administration and withdrawal. *Neuropharmacology*. 2018;128:1–10. https://doi.org/10.1016/j.neuropharm.2017.09.019.
195. Barbier E, Tapocik JD, Juergens N, et al. DNA methylation in the medial prefrontal cortex regulates alcohol-induced behavior and plasticity. *J Neurosci*. 2015;35(15):6153–6164. https://doi.org/10.1523/JNEUROSCI.4571-14.2015.
196. LaPlant Q, Vialou V, Covington III HE, et al. Dnmt3a regulates emotional behavior and spine plasticity in the nucleus accumbens. *Nat Neurosci*. 2010;13(9):1137–1143. https://doi.org/10.1038/nn.2619.
197. Tian W, Zhao M, Li M, et al. Reversal of cocaine-conditioned place preference through methyl supplementation in mice: altering global dna methylation in the prefrontal cortex. *PLoS One*. 2012;7(3):e33435. https://doi.org/10.1371/journal.pone.0033435.
198. Wright KN, Hollis F, Duclot F, et al. Methyl supplementation attenuates cocaine-seeking behaviors and cocaine-induced c-Fos activation in a DNA methylation-dependent manner. *J Neurosci*. 2015;35(23):8948–8958. https://doi.org/10.1523/JNEUROSCI.5227-14.2015.
199. Massart R, Barnea R, Dikshtein Y, et al. Role of DNA methylation in the nucleus accumbens in incubation of cocaine craving. *J Neurosci*. 2015;35(21):8042–8058. https://doi.org/10.1523/JNEUROSCI.3053-14.2015.
200. Jeanblanc J, Lemoine S, Jeanblanc V, Alaux-Cantin S, Naassila M. The class I-specific HDAC inhibitor ms-275 decreases motivation to consume alcohol and relapse in heavy drinking rats. *Int J Neuropsychopharmacol*. 2015;18(9):pyv029. https://doi.org/10.1093/ijnp/pyv029.
201. Pastor V, Host L, Zwiller J, Bernabeu R. Histone deacetylase inhibition decreases preference without affecting aversion for nicotine. *J Neurochem*. 2011;116(4):636–645. https://doi.org/10.1111/j.1471-4159.2010.07149.x.
202. Wang Y, Lai J, Cui H, et al. Inhibition of histone deacetylase in the basolateral amygdala facilitates morphine context-associated memory formation in rats. *J Mol Neurosci*. 2015;55(1):269–278. https://doi.org/10.1007/s12031-014-0317-4.
203. Wang L, Lv Z, Hu Z, et al. Chronic cocaine-induced H3 acetylation and transcriptional activation of CaMKIIalpha in the nucleus accumbens is critical for motivation for drug reinforcement. *Neuropsychopharmacology*. 2010;35(4):913–928. https://doi.org/10.1038/npp.2009.193.
204. Chen W-S, Xu W-J, Zhu H-Q, et al. Effects of histone deacetylase inhibitor sodium butyrate on heroin seeking behavior in the nucleus accumbens in rats. *Brain Res*. 2016;1652:151–157. https://doi.org/10.1016/j.brainres.2016.10.007.
205. Raybuck JD, McCleery EJ, Cunningham CL, Wood MA, Lattal KM. The histone deacetylase inhibitor sodium butyrate modulates acquisition and extinction of cocaine-induced conditioned place preference. *Pharmacol Biochem & Behav*. 2013;106:109–116. https://doi.org/10.1016/j.pbb.2013.02.009.
206. Sturgess JE, George TP, Kennedy JL, Heinz A, Müller DJ. Pharmacogenetics of alcohol, nicotine and drug addiction treatments. *Addic Biol*. 2011;16(3):357–376. https://doi.org/10.1111/j.1369-1600.2010.00287.x.
207. Unterwood EM, Ho A, Rubenfeld JM, Kreek MJ. Time course of the development of behavioral sensitization and dopamine receptor up-regulation during binge cocaine administration. *J Pharmacol & Exper Ther*. 1994;270(3):1387–1396.
208. Alburges ME, Narang N, Wamsley JK. Alterations in the dopaminergic receptor system after chronic administration of cocaine. *Synapse*. 1993;14(4):314–323. https://doi.org/10.1002/syn.890140409.
209. Sousa FC, Gomes PB, Macêdeo DS, Marinho MM, Viana GS. Early withdrawal from repeated cocaine administration upregulates muscarinic and dopaminergic D2-like receptors in rat neostriatum. *Pharmacol Biochem & Behav*. 1999;62(1):15–20. https://doi.org/10.1016/s0091-3057(98)00142-7.
210. Tella SR, Ladenheim B, Andrews AM, Goldberg SR, Cadet JL. Differential reinforcing effects of cocaine and GBR-12909: biochemical evidence for divergent neuroadaptive changes in the mesolimbic dopaminergic system. *J Neurosci*. 1996;16(23):7416–7427. https://doi.org/10.1523/JNEUROSCI.16-23-07416.1996.
211. Worsely JN, Moszcynska A, Falardeau P, et al. Dopamine D1 receptor protein is elevated in nucleus accumbens of human, chronic methamphetamine users. *Mol Psychiatr*. 2000;5(6):664–672. https://doi.org/10.1038/sj.mp.4000760.

212. Volkow ND, Wang GJ, Fowler JS, et al. Cocaine uptake is decreased in the brain of detoxified cocaine abusers. *Neuropsychopharmacology*. 1996;14(3):159−168. https://doi.org/10.1016/0893-133X(95)00073-M.
213. Volkow ND, Fowler JS, Wang GJ, et al. Decreased dopamine D2 receptor availability is associated with reduced frontal metabolism in cocaine abusers. *Synapse*. 1993;14(2):169−177. https://doi.org/10.1002/syn.890140210.
214. Macêdo DS, Correia EE, Vasconcelos SMM, Aguiar LMV, Viana GSB, Sousa FCF. Cocaine treatment causes early and long-lasting changes in muscarinic and dopaminergic receptors. *Cell & Mol Neurobiol*. 2004;24(1):129−136. https://doi.org/10.1023/b:cemn.0000012718.08443.60.
215. D'Arcy C, Luevano JE, Miranda-Arango M, et al. Extended access to methamphetamine self-administration up-regulates dopamine transporter levels 72 hours after withdrawal in rats. *Behav Brain Res*. 2016;296:125−128. https://doi.org/10.1016/j.bbr.2015.09.010.
216. Sari Y, Bell RL, Zhou FC. Effects of chronic alcohol and repeated deprivations on dopamine D1 and D2 receptor levels in the extended amygdala of inbred alcohol-preferring rats. *Alcohol: Clin & Exper Res*. 2006;30(1):46−56. https://doi.org/10.1111/j.1530-0277.2006.00010.x.
217. Sim-Selley LJ, Selley DE, Vogt LJ, Childers SR, Martin TJ. Chronic heroin self-administration desensitizes μ opioid receptor-activated G-proteins in specific regions of rat brain. *J Neurosci*. 2000;20(12):4555−4562. https://doi.org/10.1523/JNEUROSCI.20-12-04555.2000.
218. Govind AP, Vezina P, Green WN. Nicotine-induced upregulation of nicotinic receptors: underlying mechanisms and relevance to nicotine addiction. *Biochem Pharmacol*. 2009;78(7):756−765. https://doi.org/10.1016/j.bcp.2009.06.011.
219. Gray JA, Roth BL. Paradoxical trafficking and regulation of 5-HT(2A) receptors by agonists and antagonists. *Brain Res Bull*. 2001;56(5):441−451. https://doi.org/10.1016/s0361-9230(01)00623-2.
220. Beveridge TJR, Smith HR, Nader MA, Porrino LJ. Group II metabotropic glutamate receptors in the striatum of non-human primates: dysregulation following chronic cocaine self-administration. *Neurosci Lett*. 2011;496(1):15−19. https://doi.org/10.1016/j.neulet.2011.03.077.
221. Pomierny-Chamiolo L, Miszkiel J, Frankowska M, et al. Withdrawal from cocaine self-administration and yoked cocaine delivery dysregulates glutamatergic mGlu5 and NMDA receptors in the rat brain. *Neurot Res*. 2015;27(3):246−258. https://doi.org/10.1007/s12640-014-9502-z.
222. Smaga I, Wydra K, Frankowska M, Fumagalli F, Sanak M, Filip M. Cocaine self-administration and abstinence modulate NMDA receptor subunits and active zone proteins in the rat nucleus accumbens. *Molecules*. 2020;25(15):3480. https://doi.org/10.3390/molecules25153480.
223. Trigo JM, Martin-García E, Berrendero F, Robledo P, Maldonado R. The endogenous opioid system: a common substrate in drug addiction. *Drug & Alcohol Depend*. 2010;108(3):183−194. https://doi.org/10.1016/j.drugalcdep.2009.10.011.
224. Rosin A, Kitchen I, Georgieva J. Effects of single and dual administration of cocaine and ethanol on opioid and ORL1 receptor expression in rat CNS: an autoradiographic study. *Brain Res*. 2003;978(1−2):1−13. https://doi.org/10.1016/s0006-8993(03)02674-x.
225. Basavarajappa BS, Hungund BL. Role of the endocannabinoid system in the development of tolerance to alcohol. *Alcohol & Alcohol*. 2005;40(1):15−24. https://doi.org/10.1093/alcalc/agh111.
226. Álvaro-Bartolomé M, García-Sevilla JA. Dysregulation of cannabinoid CB1 receptor and associated signaling networks in brains of cocaine addicts and cocaine-treated rodents. *Neuroscience*. 2013;247:294−308. https://doi.org/10.1016/j.neuroscience.2013.05.035.
227. Orio L, Edwards S, George O, Parsons LH, Koob GF. A role for the endocannabinoid system in the increased motivation for cocaine in extended-access conditions. *J Neurosci*. 2009;29(15):4846−4857. https://doi.org/10.1523/JNEUROSCI.0563-09.2009.
228. Bystrowska B, Frankowska M, Smaga I, Niedzielska-Andres E, Pomierny-Chamioło L, Filip M. Cocaine-induced reinstatement of cocaine seeking provokes changes in the endocannabinoid and N-acylethanolamine levels in rat brain structures. *Molecules*. 2019;24(6):1125. https://doi.org/10.3390/molecules24061125.
229. Loftis JM, Janowsky A. Cocaine treatment- and withdrawal-induced alterations in the expression and serine phosphorylation of the NR1 NMDA receptor subunit. *Psychopharmacology*. 2002;164(4):349−359. https://doi.org/10.1007/s00213-002-1209-9.
230. Yuan WX, Heng LJ, Ma J, et al. Increased expression of cannabinoid receptor 1 in the nucleus accumbens core in a rat model with morphine withdrawal. *Brain Res*. 2013;1531:102−112. https://doi.org/10.1016/j.brainres.2013.07.047.

231. Brett J, Murnion B. Management of benzodiazepine misuse and dependence. *Austr Prescriber.* 2015;38(5):152−155. https://doi.org/10.18773/austprescr.2015.055.
232. Robinson TE, Gorny G, Savage VR, Kolb B. Widespread but regionally specific effects of experimenter- versus self-administered morphine on dendritic spines in the nucleus accumbens, hippocampus, and neocortex of adult rats. *Synapse.* 2002;46(4):271−279. https://doi.org/10.1002/syn.10146.
233. Li Y, Kolb B, Robinson TE. The location of persistent amphetamine-induced changes in the density of dendritic spines on medium spiny neurons in the nucleus accumbens and caudate-putamen. *Neuropsychopharmacology.* 2003;28(6):1082−1085. https://doi.org/10.1038/sj.npp.1300115.
234. McGuier NS, Padula AE, Lopez MF, Woodward JJ, Mulholland PJ. Withdrawal from chronic intermittent alcohol exposure increases dendritic spine density in the lateral orbitofrontal cortex of mice. *Alcohol.* 2015;49(1):21−27. https://doi.org/10.1016/j.alcohol.2014.07.017.
235. Kobrin KL, Moody O, Arena DT, Moore CF, Heinrichs SC, Kaplan GB. Acquisition of morphine conditioned place preference increases the dendritic complexity of nucleus accumbens core neurons. *Addic Biol.* 2016;21(6):1086−1096. https://doi.org/10.1111/adb.12273.
236. Heijtz RD, Kolb B, Forssberg H. Can a therapeutic dose of amphetamine during pre-adolescence modify the pattern of synaptic organization in the brain? *Eur J Neurosci.* 2003;18(12):3394−3399. https://doi.org/10.1046/j.0953-816x.2003.03067.x.
237. Kalivas PW. The glutamate homeostasis hypothesis of addiction. *Nat Rev Neurosci.* 2009;10(8):561−572. https://doi.org/10.1038/nrn2515.
238. Geoffroy H, Canestrelli C, Marie N, Noble F. Morphine-induced dendritic spine remodeling in rat nucleus accumbens is corticosterone dependent. *Int J Neuropsychopharmacol.* 2019;22(6):394−401. https://doi.org/10.1093/ijnp/pyz014.
239. Melikian HE. Neurotransmitter transporter trafficking: endocytosis, recycling, and regulation. *Pharmacol & Ther.* 2004;104(1):17−27. https://doi.org/10.1016/j.pharmthera.2004.07.006.
240. Rao A, Sorkin A, Zahniser NR. Mice expressing markedly reduced striatal dopamine transporters exhibit increased locomotor activity, dopamine uptake turnover rate, and cocaine responsiveness. *Synapse.* 2013;67(10):668−677. https://doi.org/10.1002/syn.21671.
241. Calipari ES, Ferris MJ, Melchior JR, et al. Methylphenidate and cocaine self-administration produce distinct dopamine terminal alterations. *Addic Biol.* 2014;19(2):145−155. https://doi.org/10.1111/j.1369-1600.2012.00456.x.
242. Oleson EB, Talluri S, Childers SR, et al. Dopamine uptake changes associated with cocaine self-administration. *Neuropsychopharmacology.* 2009;34(5):1174−1184. https://doi.org/10.1038/npp.2008.186.
243. McIntosh S, Howell L, Hemby SE. Dopaminergic dysregulation in prefrontal cortex of rhesus monkeys following cocaine self-administration. *Front Psychiatr.* 2013;4:88. https://doi.org/10.3389/fpsyt.2013.00088.
244. Banks ML, Czoty PW, Gage HD, et al. Effects of cocaine and MDMA self-administration on serotonin transporter availability in monkeys. *Neuropsychopharmacology.* 2008;33(2):219−225. https://doi.org/10.1038/sj.npp.1301420.
245. Ramamoorthy S, Samuvel DJ, Balasubramaniam A, See RE, Jayanthi LD. Altered dopamine transporter function and phosphorylation following chronic cocaine self-administration and extinction in rats. *Biochem & Biophys Res Commun.* 2010;391(3):1517−1521. https://doi.org/10.1016/j.bbrc.2009.12.110.
246. Arroyo M, Baker WA, Everitt BJ. Cocaine self-administration in rats differentially alters mRNA levels of the monoamine transporters and striatal neuropeptides. *Brain Res Mol Brain Res.* 2000;83(1−2):107−120. https://doi.org/10.1016/s0169-328x(00)00205-9.
247. García-Pérez D, Núñez C, Laorden ML, Milanés MV. Regulation of dopaminergic markers expression in response to acute and chronic morphine and to morphine withdrawal. *Addic Biol.* 2016;21(2):374−386. https://doi.org/10.1111/adb.12209.
248. Shaw JK, Pamela Alonso I, Lewandowski SI, et al. Individual differences in dopamine uptake in the dorsomedial striatum prior to cocaine exposure predict motivation for cocaine in male rats. *Neuropsychopharmacology.* 2021;46(10):1757−1767. https://doi.org/10.1038/s41386-021-01009-2.
249. Thomsen M, Hall FS, Uhl GR, Caine SB. Dramatically decreased cocaine self-administration in dopamine but not serotonin transporter knock-out mice. *J Neurosci.* 2009;29(4):1087−1092. https://doi.org/10.1523/JNEUROSCI.4037-08.2009.
250. Rocha BA, Fumagalli F, Gainetdinov RR, et al. Cocaine self-administration in dopamine-transporter knockout mice. *Nat Neurosci.* 1998;1(2):132−137. https://doi.org/10.1038/381.

251. Sora I, Wichems C, Takahashi N, et al. Cocaine reward models: conditioned place preference can be established in dopamine- and in serotonin-transporter knockout mice. *Proc Nat Acad Sci USA*. 1998;95(13):7699−7704. https://doi.org/10.1073/pnas.95.13.7699.
252. Spielewoy C, Gonon F, Roubert C, et al. Increased rewarding properties of morphine in dopamine-transporter knockout mice. *Eur J Neurosci*. 2000;12(5):1827−1837. https://doi.org/10.1046/j.1460-9568.2000.00063.x.
253. Lindsey KP, Wilcox KM, Votaw JR, et al. Effects of dopamine transporter inhibitors on cocaine self-administration in rhesus monkeys: relationship to transporter occupancy determined by positron emission tomography neuroimaging. *J Pharmacol & Exper Ther*. 2004;309(3):959−969. https://doi.org/10.1124/jpet.103.060293.
254. Zhang HY, Bi GH, Yang HJ, et al. The novel modafinil analog, JJC8-016, as a potential cocaine abuse pharmacotherapeutic. *Neuropsychopharmacology*. 2017;42(9):1871−1883. https://doi.org/10.1038/npp.2017.41.
255. Mead AN, Rocha BA, Donovan DM, Katz JL. Intravenous cocaine induced-activity and behavioural sensitization in norepinephrine-, but not dopamine-transporter knockout mice. *Eur J Neurosci*. 2002;16(3):514−520. https://doi.org/10.1046/j.1460-9568.2002.02104.x.
256. Trigo JM, Renoir T, Lanfumey L, et al. 3,4-methylenedioxymethamphetamine self-administration is abolished in serotonin transporter knockout mice. *Biol Psychiatr*. 2007;62(6):669−679. https://doi.org/10.1016/j.biopsych.2006.11.005.
257. Caffino L, Mottarlini F, Van Reijmersdal B, et al. The role of the serotonin transporter in prefrontal cortex glutamatergic signaling following short- and long-access cocaine self-administration. *Addic Biol*. 2021;26(2):e12896. https://doi.org/10.1111/adb.12896.
258. Verrico CD, Haile CN, Mahoney 3rd JJ, Thompson-Lake DG, Newton TF, De La Garza 2nd R. Treatment with modafinil and escitalopram, alone and in combination, on cocaine-induced effects: a randomized, double blind, placebo-controlled human laboratory study. *Drug & Alcohol Depend*. 2014;141:72−78. https://doi.org/10.1016/j.drugalcdep.2014.05.008.
259. Naranjo CA, Poulos CX, Bremner KE, Lanctôt KL. Citalopram decreases desirability, liking, and consumption of alcohol in alcohol-dependent drinkers. *Clin Pharmacol & Ther*. 1992;51(6):729−739. https://doi.org/10.1038/clpt.1992.85.
260. Ardito F, Giuliani M, Perrone D, Troiano G, Lo Muzio L. The crucial role of protein phosphorylation in cell signaling and its use as targeted therapy (Review). *Int J Mol Med*. 2017;40(2):271−280. https://doi.org/10.3892/ijmm.2017.3036.
261. Brini M, Calì T, Ottolini D, Carafoli E. Neuronal calcium signaling: function and dysfunction. *Cell & Mol Life Sci*. 2014;71(15):2787−2814. https://doi.org/10.1007/s00018-013-1550-7.
262. Self DW, Genova LM, Hope BT, Barnhart WJ, Spencer JJ, Nestler EJ. Involvement of cAMP-dependent protein kinase in the nucleus accumbens in cocaine self-administration and relapse of cocaine-seeking behavior. *J Neurosci*. 1998;18(5):1848−1859. https://doi.org/10.1523/JNEUROSCI.18-05-01848.1998.
263. Arguello AA, Hodges MA, Wells AM, Lara 3rd H, Xie X, Fuchs RA. Involvement of amygdalar protein kinase A, but not calcium/calmodulin-dependent protein kinase II, in the reconsolidation of cocaine-related contextual memories in rats. *Psychopharmacology*. 2014;231(1):55−65. https://doi.org/10.1007/s00213-013-3203-9.
264. Sanchez H, Quinn JJ, Torregrossa MM, Taylor JR. Reconsolidation of a cocaine-associated stimulus requires amygdalar protein kinase A. *J Neurosci*. 2010;30(12):4401−4407. https://doi.org/10.1523/JNEUROSCI.3149-09.2010.
265. Lai YT, Fan HY, Cherng CG, Chiang CY, Kao GS, Yu L. Activation of amygdaloid PKC pathway is necessary for conditioned cues-provoked cocaine memory performance. *Neurobiol Learn & Memory*. 2008;90(1):164−170. https://doi.org/10.1016/j.nlm.2008.03.006.
266. Ortinski PI, Briand LA, Pierce RC, Schmidt HD. Cocaine-seeking is associated with PKC-dependent reduction of excitatory signaling in accumbens shell D2 dopamine receptor-expressing neurons. *Neuropharmacology*. 2015;92:80−89. https://doi.org/10.1016/j.neuropharm.2015.01.002.
267. Newton PM, Kim JA, McGeehan AJ, et al. Increased response to morphine in mice lacking protein kinase C epsilon. *Genes, Brain & Behav*. 2007;6(4):329−338. https://doi.org/10.1111/j.1601-183X.2006.00261.x.
268. Liu Z, Zhang JJ, Liu XD, Yu LC. Inhibition of CaMKII activity in the nucleus accumbens shell blocks the reinstatement of morphine-seeking behavior in rats. *Neurosci Lett*. 2012;518(2):167−171. https://doi.org/10.1016/j.neulet.2012.05.003.

269. Haydon PG, Carmignoto G. Astrocyte control of synaptic transmission and neurovascular coupling. *Physiol Rev*. 2006;86(3):1009−1031. https://doi.org/10.1152/physrev.00049.2005.
270. Gonçalves J, Baptista S, Martins T, et al. Methamphetamine-induced neuroinflammation and neuronal dysfunction in the mice hippocampus: preventive effect of indomethacin. *Eur J Neurosci*. 2010;31(2):315−326. https://doi.org/10.1111/j.1460-9568.2009.07059.x.
271. Robson MJ, Turner RC, Naser ZJ, et al. SN79, a sigma receptor antagonist, attenuates methamphetamine-induced astrogliosis through a blockade of OSMR/gp130 signaling and STAT3 phosphorylation. *Exper Neurol*. 2014;254:180−189. https://doi.org/10.1016/j.expneurol.2014.01.020.
272. Hawkins JL, Denson JE, Miley DR, Durham PL. Nicotine stimulates expression of proteins implicated in peripheral and central sensitization. *Neuroscience*. 2015;290:115−125. https://doi.org/10.1016/j.neuroscience.2015.01.034.
273. Goins EC, Bajic D. Astrocytic hypertrophy in the rat ventral tegmental area following chronic morphine differs with age. *J Neurol & Neurorehabilit Res*. 2018;3(1):14−21.
274. Łupina M, Tarnowski M, Baranowska-Bosiacka I, et al. SB-334867 (an orexin-1 receptor antagonist) effects on morphine-induced sensitization in mice-A view on receptor mechanisms. *Mol Neurobiol*. 2018;55(11):8473−8485. https://doi.org/10.1007/s12035-018-0993-0.
275. Nwachukwu KN, Evans WA, Sides TR, Trevisani P, Davis A, Marshall SA. Chemogenetic manipulation of astrocytic signaling in the basolateral amygdala reduces binge-like alcohol consumption in male mice. *J Neurosci Res*. 2021;99(8):1957−1972. https://doi.org/10.1002/jnr.24841.
276. Scofield MD, Li H, Siemsen BM, et al. Cocaine self-administration and extinction leads to reduced glial fibrillary acidic protein expression and morphometric features of astrocytes in the nucleus accumbens core. *Biol Psychiatr*. 2016;80(3):207−215. https://doi.org/10.1016/j.biopsych.2015.12.022.
277. Bowers MS, Kalivas PW. Forebrain astroglial plasticity is induced following withdrawal from repeated cocaine administration. *Eur J Neurosci*. 2003;17(6):1273−1278. https://doi.org/10.1046/j.1460-9568.2003.02537.x.
278. Xu Z, Seidler FJ, Tate CA, Garcia SJ, Slikker Jr W, Slotkin TA. Sex-selective hippocampal alterations after adolescent nicotine administration: effects on neurospecific proteins. *Nicotine & Tobacco Res*. 2003;5(6):955−960. https://doi.org/10.1080/14622200310001615321.
279. Namba MD, Kupchik YM, Spencer SM, et al. Accumbens neuroimmune signaling and dysregulation of astrocytic glutamate transport underlie conditioned nicotine-seeking behavior. *Addic Biol*. 2020;25(5):e12797. https://doi.org/10.1111/adb.12797.
280. Suárez I, Bodega G, Fernández-Ruiz JJ, Ramos JA, Rubio M, Fernáandez B. Reduced glial fibrillary acidic protein and glutamine synthetase expression in astrocytes and Bergmann glial cells in the rat cerebellum caused by delta(9)-tetrahydrocannabinol administration during development. *Develop Neurosci*. 2002;24(4):300−312. https://doi.org/10.1159/000066744.
281. Zamberletti E, Gabaglio M, Grilli M, et al. Long-term hippocampal glutamate synapse and astrocyte dysfunctions underlying the altered phenotype induced by adolescent THC treatment in male rats. *Pharmacol Res*. 2016;111:459−470. https://doi.org/10.1016/j.phrs.2016.07.008.
282. Evrard SG, Duhalde-Vega M, Tagliaferro P, Mirochnic S, Caltana LR, Brusco A. A low chronic ethanol exposure induces morphological changes in the adolescent rat brain that are not fully recovered even after a long abstinence: an immunohistochemical study. *Exper Neurol*. 2006;200(2):438−459. https://doi.org/10.1016/j.expneurol.2006.03.001.
283. Ramos AJ, Evrard SG, Tagliaferro P, Tricárico MV, Brusco A. Effects of chronic maternal ethanol exposure on hippocampal and striatal morphology in offspring. *Annals New York Acad Sci*. 2002;965:343−353. https://doi.org/10.1111/j.1749-6632.2002.tb04176.x.
284. Siemsen BM, Reichel CM, Leong KC, et al. Effects of methamphetamine self-administration and extinction on astrocyte structure and function in the nucleus accumbens core. *Neuroscience*. 2019;406:528−541.

SECTION III

Behavioral and cognitive mechanisms of addiction

CHAPTER 5

Learning mechanisms of addiction: operant conditioning

Introduction

The first section of this book covered the neurobiological bases of addiction. We know that drugs increase dopamine levels in the mesocorticolimbic pathway, which accounts for their rewarding properties. If the neurobiological changes that occur following prolonged drug use are solely responsible for addiction, we would have an easier time treating substance use disorders (SUDs). However, medications used to treat the various SUDs often fail to lead to permanent abstinence. This section will cover the behavioral and the cognitive factors associated with addiction. This chapter will specifically focus on operant conditioning, a type of learning in which an organism associates a behavioral response with some type of **reinforcer**. Before defining operant conditioning, I want you to read the following quote from an individual named Ashley:

> At the beginning stages of my drug use, I was just having fun on the weekends with my friends. But in just a short time, I found myself using drugs throughout the week by myself. And on the weekend, I found myself taking more and more substances that would have a huge impact on my mental health. I didn't realize that I had a problem until I found myself becoming dependent on the drugs that I would take. Withdrawals became an issue and my mental health began to plummet. My mind was consumed by my drug habits and they were becoming more of an issue the more I would take.[1]

This quote highlights the role that reinforcement has on the addiction process. What you will learn in this chapter is that reinforcement is not always enjoyable. In the case of Ashley, he initially used drugs because he was "having fun on the weekends". Over time, he found drug use to be less fun as he became physically dependent on the drugs and had to deal with the aversive side effects of drug withdrawal. One important aspect of drug use is that liking a drug is not necessarily the same thing as wanting/needing a drug.

Reinforcement is not the only aspect of operant conditioning involved in drug addiction. We also must consider how drug use and *punishment* are interrelated. The following is a case report about an individual named Tom who used heroin:

One day, Tom called his father from a bus station. When his father arrived there, he was shocked by Tom's appearance. Tom was black and blue and very confused. His father took him to an urgent care center, where he was diagnosed with rhabdomyolysis, which was caused by a loss of consciousness after Tom injected a mixture of heroin and another toxin and then lay on his arm for at least 24 hours ... One night, Tom was with a friend in a motel in Texas "shooting up" heroin and cocaine ... Tom went to the bathroom and fell on the floor. The other boy who was with him put him in the shower to try to revive him. Tom vomited all over himself ... He died later that night (pp. 44, 47).[2]

This tragic story highlights the punishment that can occur from drug use. In Tom's case, he ultimately died from a drug overdose. However, there are other ways in which drug use can punish one's behavior, which will be covered in this chapter in more detail.

Learning objectives

By the end of this chapter, you should be able to ...

(1) Explain what operant conditioning is and describe the role reinforcement and punishment have on the addiction process.
(2) Identify ways in which researchers can use operant conditioning to study addiction in a laboratory setting.
(3) Compare and contrast treatment options that utilize operant conditioning in clinical populations.

Overview of operant conditioning

Operant conditioning (also known as instrumental learning or R–S learning) is a form of learning in which an organism comes to associate certain behaviors with specific outcomes.[3] More simply, we learn that our actions have consequences. Those consequences can dramatically alter our behavior in the future. To fully understand operant conditioning and its contribution to drug addiction, a brief history lesson is needed.

B.F. Skinner and the operant conditioning chamber

Dr. Burrhus Frederick (B.F.) Skinner (1904–1990) was a behavioral psychologist who emphasized that the external environment shapes who we are,[4] and thus, is considered to be the father of operant conditioning. As a *radical behaviorist*, Skinner did not believe in mental processes. Instead, he wanted to focus on what was physically observable: behavior. To better test the role of the environment on behavior, Skinner designed a special apparatus—the operant conditioning chamber (see Fig. 5.1 for a modern-day version of the operant chamber). In an operant conditioning chamber, an animal, such as a rat or a pigeon, can engage in some behavior like pressing a lever or stepping on a treadle to receive a reinforcer (e.g., food, water, etc.). The reason we use the term operant conditioning to describe this type of learning is because the organism *operates* something in its environment. For example, the rat operates the lever to receive food. The most impressive thing about Skinner's work is that he designed

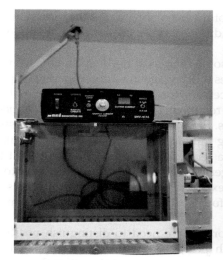 

FIGURE 5.1 The inside of a modern operant conditioning chamber. At the top left of the chamber is a house light that provides illumination and can be used to signal the beginning of an experimental session or an individual trial within a session. Two levers are located on the right side of the chamber. Often, only one of these levers is associated with reinforcement. In between each lever is a food tray, which is where the animal will go to receive a food pellet. Above each lever is a stimulus light that can be used in multiple ways, such as signaling a time-out period in which further responding will not result in reinforcement or signaling reinforcement delivery. The image on the right shows a closer view of the levers, the food tray, and the stimulus lights. One advantage of operant conditioning chambers is that they can be easily modified to include nose poke apertures instead of levers or to include additional levers on the left side of the chamber; also, tone generators can be added to provide an auditory stimulus for the animal. In the image on the left, one can see the tether that is connected to a rat for drug self-administration experiments. The black box on top of the operant conditioning chamber is a shock generator that can be used to measure compulsive drug seeking. *Images were taken by the author in his lab.*

the operant conditioning chamber while he was a graduate student at Harvard University. I was lucky if I could stay awake during my graduate stats course!

By creating the operant conditioning chamber, Skinner could empirically test how certain stimuli strengthen or weaken behaviors. Although Skinner did not study drug addiction, his study of behavior and the development of the operant conditioning chamber have played a critical role in our understanding of the development and maintenance of SUDs. We will now turn our attention to the first major component of operant conditioning: reinforcement.

Reinforcement

A reinforcer is anything that increases the probability or frequency of a behavior occurring or increases the strength of a behavior. For example, if a hungry rat is placed in an operant conditioning chamber and is allowed to press a lever to earn food, receiving food will reinforce the rat's lever-press behavior by increasing the frequency at which (or increasing the probability that) the rat presses the lever. Skinner could also increase how hard the animal had to press the lever to receive reinforcement. Animals learn to exert more force on the lever, thus demonstrating an increase in the strength of the behavior.[4] Using the term reinforcer on its own does not tell the entire story. There are different types of reinforcers that we need to discuss.

There are **primary reinforcers** which fulfill some biological need, such as food, water, or sex. Not surprisingly, animals of different species will engage in various operant responses to earn access to these reinforcers as they are necessary for survival. Because animals do not need to learn that primary reinforcers are, well, reinforcing, they are also called *unconditioned reinforcers*. There are also **secondary reinforcers**. Secondary reinforcers do not satisfy a biological need and are not necessarily reinforcing initially, but animals will eventually work to receive access to them.[5] In the case of a rat, a secondary reinforcer can be a light that becomes paired with food. Even if food is no longer delivered, rats will press a lever to receive access to the light. In humans, the prototypic example of a secondary reinforcer is money. Money can be used to purchase primary reinforcers like food and water (in some places one can even buy sex).

In addition to primary and secondary reinforcers, we need to distinguish **positive reinforcement** from **negative reinforcement**.[6] Both positive reinforcement and negative reinforcement *increase* behavior. In positive reinforcement, an **appetitive stimulus** is *added* to the animal's environment to increase the frequency/strength of a behavior. In the case of an animal, food is added to the environment following a behavioral response. In our case, an employer will give us money to continue coming to work. As much as I love my job, I would spend my time looking for a new one if my university stopped paying me. With negative reinforcement, something aversive is removed from the animal's environment to increase the frequency/strength of a behavior. Animals will learn to press a lever to avoid receiving an aversive foot shock. Imagine you have a roommate that nags you to do the dishes. Eventually, you may do the dishes just to stop hearing your roommate nag. The next time your roommate starts to nag, you will be more likely to perform the behavior of washing the dishes.

Two important concepts related to reinforcement need to be discussed: **contingency** and **contiguity**.[7] Contingency refers to how reliably/consistently a reinforcer is delivered following a specific behavior. If you want to train your dog to sit but give it treats randomly for performing other behaviors such as staring at you or laying on the couch, you are going to have a more difficult time getting your dog to sit on command. Instead, you want to consistently reinforce your dog for sitting, and for sitting only. Once your dog has learned this behavior, you can start teaching it the next trick. Contiguity refers to how much time elapses between the occurrence of the behavior and the delivery of reinforcement. If your dog sits, but you wait 5 min to reinforce the behavior, your dog is going to have a difficult time connecting the behavior with the outcome of receiving a treat. At this point, the dog just associates whatever it is doing at that moment in time with reinforcement. Immediate reinforcement is important for learning! This is why online quizzes can be a valuable tool for students. Students can get near immediate reinforcement as they quickly learn how well they performed as opposed to having to wait 1–2 weeks to get exam scores back.

Another important concept to know is that there are certain instances in which a reinforcer is more or less effective compared to other instances. Overall, this concept is referred to as a **motivating operation**.[8] Motivating operations can be broken down into establishing operations and abolishing operations. An establishing operation is anything that increases the effectiveness of a reinforcer. A slice of pizza is more appealing when one is hungry compared to being completely full. Here, hunger is an establishing operation for food reinforcement. In animal research, researchers often restrict how much food an animal can eat during the day to motivate it to respond during an operant task. An abolishing operation is anything that decreases the effectiveness of a reinforcer. As hunger can be an establishing operation for food reinforcement, satiation can be an abolishing operation for the same reinforcer. Additionally, if someone is extremely wealthy, they may be less willing to respond on a computer key to earn a few dollars compared to someone that is struggling to make ends meet. In this example, wealth can serve as an abolishing operation for behaviors maintained by monetary reinforcement.

We need to cover one last concept related to reinforcement: **discriminative stimuli**. A discriminative stimulus is a special cue that tells an organism when reinforcement is available or unavailable.[9] For example, a rat can be trained to respond on a lever when a light is on. Each time the light turns on, responding on the lever leads to food delivery. When the light turns off, responding on the same lever will produce no reinforcement. In real life, a red light on a vending machine signals that the machine is broken; thus, reinforcement is unavailable even if someone inserts money into the machine. Thus, the discriminative stimulus is the presence/absence of a red light.

Schedules of reinforcement

As mentioned earlier, B.F. Skinner designed the operant conditioning chamber to study how the environment can influence behavior. In particular, these chambers are useful because they allow one to study **schedules of reinforcement**. These schedules describe the rules needed for an animal to receive reinforcement.[10] The four primary schedules of reinforcement are *fixed ratio (FR), variable ratio (VR), fixed interval (FI),* and *variable interval (VI)*. Each of these schedules produces their own schedule effect, a characteristic pattern of responding. Fig. 5.2 compares the schedule effects of the FR, VR, FI, and VI schedules of reinforcement.

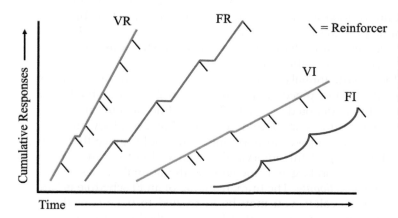

FIGURE 5.2 Schematic showing the typical pattern of responding associated with the four main schedules of reinforcement. For simplicity, each schedule of reinforcement occupies a different space on the x-axis. *This image was created by the author using Microsoft PowerPoint.*

In an FR schedule of reinforcement, an animal responds on some manipulandum (whether a lever or a nose poke aperture) a certain number of times to receive reinforcement. For example, in an FR five (FR 5) schedule of reinforcement, a rat will not receive a reinforcer until after it has pressed on the lever five times. In real life, you may reinforce yourself (by taking a break, playing a game, etc.) after you have read 10 pages of your textbook (example of an FR 10 schedule). FR schedules produce a high rate of responding followed by a *postreinforcement pause* (*PRP*). Let's break this down. Imagine we are training a rat to respond on an FR 10 schedule of reinforcement. Once the animal starts pressing the lever to meet this requirement, it will not stop responding until it earns the reinforcer. In other words, the rat will press the lever 10 times in quick succession before stopping. Once the animal receives the reinforcer, it will take a short break before it resumes responding. This break is the PRP. For you, you may read 10 pages of your textbook before taking a short break. As you read those 10 pages, you are unlikely to stop reading (unless what you are reading is incredibly boring; hopefully this is not the case right now). After reading, you may need a break in order to rest your eyes.

In a VR schedule of reinforcement, an animal must respond nth number of times before receiving reinforcement. Here, the animal does not know the exact number of times it has to respond before receiving reinforcement. To illustrate, imagine during one trial, an animal has to respond 10 times to receive a reinforcer. However, on the next trial, the animal has to respond 15 times. Finally, the animal has to respond five times. Averaged together, the animal is responding on a VR 10 schedule of reinforcement. For humans, the prototypic example of a VR schedule is playing on a slot machine. If you play on a slot machine for an extended period of time, you will notice that you win at random times. You may have to play 30 times before winning any money, but you may then win after playing 11 additional times. Because the animal cannot predict when they will receive reinforcement (because it is variable, or random), it will respond on the manipulandum at a very high rate and will rarely exhibit PRPs. Why are PRPs so rare in the VR schedule? This is related to the unpredictable nature of the VR schedule. In the FR schedule, the animal takes a break because it knows that it can go back to the lever and press on it five times (if FR 5 schedule) to receive another reinforcer. In the VR schedule, the rat does not know if it needs to press the lever just five times or 55 times. To maximize the amount of food it can earn, the rat will take very short breaks after earning a reinforcer to continue pressing the lever. Think of gambling: using a VR schedule ensures that people keep playing even after they have won some money. Casinos are trying to take your money, not give you money, after all.

In an FI schedule of reinforcement, animals must wait a certain amount of time before a response leads to reinforcement. A rat or pigeon may have to wait 20 s before the first lever press leads to food delivery. This is denoted as an FI 20″ schedule of reinforcement (the quotation mark denotes seconds). Animals are decent at timing; however, they are not perfect. If you were to watch the animal respond in real time, you would notice that they do not press on the lever at the beginning of this interval (i.e., you will not see them respond after 3–5 s). Instead, the rate of responding will increase as the animal approaches 20 s. This pattern of responding is called *scalloping*. The human analog of an FI schedule is baking. If following a recipe (say, for cheesecake), you may be instructed to bake something for 55 min. Here, your responding is going to be dictated by a 55′ schedule of reinforcement (the single apostrophe denotes minutes). During the first few minutes, you are not going to check the oven because you know the food will not be close to being ready. Because you do not want to burn your food, you may check the oven starting at 50 min and will continue checking as you approach 55 min. After the 55-minute interval ends, you can remove the cheesecake from the oven. Realistically, you may pull the cheesecake out before 55 min in order to avoid overbaking it, but this example highlights scalloping.

The VI schedule of reinforcement is a combination of the FI and the VR schedule in the sense that it randomizes how long an animal must wait before responding leads to reinforcement. In a VI 20″ schedule of reinforcement, an animal may have to wait 30 s before responding leads to reinforcement on one trial, but may only have to wait 10 s during the next trial. Checking your final grades at the end of the semester operates on a VI schedule of reinforcement. You have a general idea of when your professor will enter final grades, but you are not entirely sure when that moment will occur. Because of the uncertainty associated with the VI schedule, you will check your grades at a steady rate (i.e., you may check your grades first thing in the morning, after lunch, and then before bed) until your professor has posted the grade online. Unlike the FI schedule, scalloping is not observed in the VI schedule. Instead, this schedule produces a steady, but fairly low, rate of responding. In the VR schedule, animals respond at very high rates because they do not know how many responses are needed to receive reinforcement. In the VI schedule, only the first response following some interval leads to reinforcement. There is no need for the animal to continuously press the lever, just as there is no need for you to check your final grades every 30 s.

Although we have covered four major schedules of reinforcement here, there are numerous schedules of reinforcement that are used to study behavior in animals. We will cover some of the more complex schedules of reinforcement when we discuss how we can use operant conditioning to measure the reinforcing effects of drugs in animals. For now, let's turn our attention to the second major outcome of a behavior: punishment.

Punishment

A **punisher** is anything that decreases the probability or frequency of a behavior occurring or decreases the strength of a behavior. Like reinforcement, we need to differentiate **positive punishment** from **negative punishment**.[6] With positive punishment, something aversive is added to the environment to decrease the probability of a behavior occurring. For rats, a positive punisher can be a mild electric foot shock. If a rat presses a lever and gets shocked, they are going to be less likely to press the lever again in the future. In humans, a parent spanking their child is a form of positive punishment. With negative punishment, something appetitive is removed to decrease the frequency of a behavior. Omitting reinforcement when a response is made is a form of negative punishment. A real-life example of negative punishment is someone having their license suspended for driving under the influence of ethanol/drugs. By removing a person's ability to drive, they should be less likely to drink ethanol and/or use drugs and drive in the future.

The principles of contingency and contiguity apply to punishment as they do reinforcement. If you want to punish a behavior, you need to consistently apply the punisher to the problematic behavior (contingency). Pretend someone uses a shock collar to minimize excessive barking (note: I am not condemning nor condoning the use of a shock collar on one's dog). In order for the shock collar to be effective, you must reliably use it when your dog "misbehaves" during the early stages of learning. If you sporadically use the shock collar, your dog will have difficulty learning to associate its excessive barking with delivery of shock. You also need to punish the behavior immediately for punishment to be effective (contiguity). For example, I have seen individuals try to spank their dog after coming home to find it has soiled the carpet or try to rub the dog's nose in its mess. This is problematic because too much time has elapsed since the dog soiled the carpet. At this point, the dog has no idea why it is being punished and will associate the punishment with whatever it is doing at that moment in time. This can lead to general suppression of behavior. What I mean by this: the dog may just stop doing whatever it is doing when it sees you because it is afraid it is going to get spanked or have its nose rubbed in something.

Just as motivating operations influence the effectiveness of a reinforcer, they can alter how well a punisher alters someone's behavior. Ideally, a ticket should decrease someone's willingness to go above the speed limit while driving. However, if this were the case, we would see few individuals speeding on the highway. I doubt there has ever been a time where I have been on the road and have NOT seen anyone going excessively above the speed limit, and I am not talking about going just 5 miles per hour above the limit. There are multiple reasons why someone may not alter their driving behavior even after receiving a ticket. One such possibility is something we covered earlier: wealth. A wealthy individual may not be sensitive to the ticket because the punishment is inconsequential to them. Here, wealth acts as an abolishing operation for the effectiveness of the speeding ticket. Scandinavian countries such as Finland have implemented a system in which fines are based on an individual's income. Individuals that make more money will pay larger fines for infractions. One individual was forced to pay $103,000 for going 15 miles per hour over the speed limit.[11]

One area I see students get tripped up is differentiating negative reinforcement from positive punishment. My students often erroneously equate the two. As described above, reinforcers, whether positive or negative, increase the occurrence of a behavior. Conversely, punishers, whether positive or negative, decrease behavior. The terms positive and negative merely refer to something being added to or removed from the animal's environment (see Fig. 5.3 for a chart comparing positive/negative reinforcement and punishment).

Quiz 5.1

1. Which schedule of reinforcement produces the highest rate of responding?
    a. FI
    b. FR
    c. VI
    d. VR
2. A rat receives a reinforcer after every 30 responses. What schedule of reinforcement is being used?
    a. FI
    b. FR
    c. VI
    d. VR

3. After helping your friend move, they give you a gift card to your favorite restaurant. What type of reinforcer have you just received?
4. A rat has previously learned that responding on a lever leads to food. Now, when the rat presses the lever, nothing happens. Omitting the food is a _____ _____.
   a. negative punisher
   b. negative reinforcer
   c. positive punisher
   d. positive reinforcer
5. Why did B.F. Skinner develop the operant conditioning chamber?

Answers to quiz 5.1

1. d — VR
2. b — FR
3. Secondary reinforcer. The gift card is not intrinsically reinforcing. However, because you have learned over time that gift cards (like money) can be used to receive primary reinforcers, it becomes a nice gift indeed.
4. a — negative punisher
5. Skinner wanted to determine how animals (including humans) operate on their environment and how corresponding changes in the environment can influence our future behaviors.

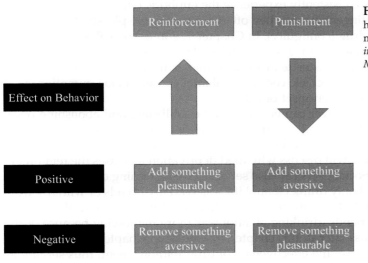

FIGURE 5.3 Schematic showing how positive and negative reinforcement/punishment affect behavior. *This image was created by the author using Microsoft PowerPoint.*

Operant processes involved in addiction

Now that we have a fundamental understanding of operant conditioning, it is time to apply principles of reinforcement and punishment to SUDs.

Positive reinforcement

Drugs are highly reinforcing in both humans and animals. Animals are willing to emit a response on a manipulandum to receive an infusion of a drug such as cocaine or heroin. Although drugs do not satisfy a biological need, they are often considered primary reinforcers because little learning is needed to associate the euphoric effects of the drug to the drug itself. This statement needs to be qualified as individuals may need multiple exposures to cigarettes and/or ethanol before developing a preference for them. Recall from Chapter 3 that the ventral tegmental area (VTA) and the nucleus accumbens (NAc) are the brain structures most often associated with reinforcement. By using operant conditioning, we have been able to further elucidate our understanding of the neurobiological basis of SUDs. Using microdialysis, Pettit and Justice Jr. found that dopamine levels increase in the NAc following cocaine administration.[12] When discussing the neuromechanisms of addiction, many of the studies I cited used operant conditioning procedures.

Not surprisingly, positive reinforcement is important for the initiation of drug use. Because drugs are positively reinforcing, it encourages a person to continue taking the drug. With drugs, we can see the concepts of contingency and contiguity in action. Recall that contingency refers to how reliably the reinforcer follows the behavior, and contiguity refers to how much time passes between the behavior and reinforcement delivery. During the early stages of drug use, individuals consistently experience the euphoric effects of the drug after they have administered it (contingency), and they often experience the pleasurable effects of the drug soon after taking the drug (contiguity). In Chapter 1, you learned about the various routes of administration associated with different drugs. Drugs associated with routes of administration that reach the brain faster tend to have higher misuse potential (e.g., cigarettes, cannabis, methamphetamine, crack cocaine). Ethanol is one major exception as it has high misuse rates despite a slower onset of action.

The effectiveness of a drug reinforcer can be altered by establishing and abolishing operations just like any other reinforcer. In rats, food deprivation increases self-administration of cocaine[13] and heroin.[14] Thus, food deprivation serves as an establishing operation for both food and drug reinforcement. We will discuss why food deprivation increases the reinforcing effects of drugs in Chapter 11. Social peers can also serve as an establishing operation in some contexts. When I was younger, I only drank ethanol if I was with my friends or with my wife. On my own, I did not enjoy the taste of ethanol enough to drink it if I was home alone eating dinner or watching TV. With friends, drinking ethanol was more reinforcing because of the social context. As we will discuss later in this chapter, as well as in Chapter 10, certain peer interactions can be used to decrease the effectiveness of drug reinforcement, thus serving as an abolishing operation.

Knowing that drugs are reinforcing does not explain why some individuals develop a SUD whereas others do not. Perhaps individuals predisposed to developing a SUD shows greater *reward sensitivity*. Some work has shown increased reward sensitivity in those with cocaine use disorder and alcohol use disorder[15–17] and in adolescent girls that misuse ethanol.[18] However, other studies have shown decreased reward sensitivity in those with a SUD.[19,20] What could account for these discrepancies? Increased reward sensitivity may be more important for the initiation of drug use. Individuals that score higher in reward sensitivity feel more stimulated after consuming ethanol, an effect that takes longer to diminish.[21] High reward sensitivity is also associated with greater positive subjective effects following *d*-amphetamine administration.[22,23] Recall from Chapter 3 that as individuals continue using drugs, compensatory changes occur in the mesocorticolimbic pathway, resulting in decreased dopamine release in response to both drug and nondrug reinforcers. A recent study found that decreased activation of the ventral striatum is predictive of substance use in young adults.[24] Neuroadaptations in the mesocorticolimbic pathway do not fully explain the discrepancies observed across studies regarding reward sensitivity and addiction vulnerability. Individuals that have used cocaine for over 10 years show increased reward sensitivity compared to controls.[15] These discrepancies do highlight that positive reinforcement, although important for the initiation of drug use, is not the reason why individuals develop a SUD. We need to turn our focus to the role of negative reinforcement on the addiction process.

Negative reinforcement

As an individual continues using a drug, they do not experience the same efflux of dopamine in the NAc as they originally did, nor do they experience the same "high" as they did when initially taking the drug. One may ask: "if the person no longer likes taking the drug, why do they continue using it?". If an individual attempts to stop using the drug, they experience withdrawal symptoms. These withdrawal symptoms are extremely unpleasant; thus, the person takes more drug to escape the aversive side effects of drug withdrawal. This is analogous to an individual taking acetaminophen or ibuprofen to alleviate a headache. You may not particularly enjoy taking these medications, but you do so to get rid of something unpleasant. Because taking a drug like heroin quickly (contiguity) and consistently (contingency) alleviates withdrawal symptoms, an individual's continued drug use is driven by negative reinforcement.

Opponent-process theory of addiction

The process by which drugs transition from positive reinforcers to negative reinforcers is described by the **opponent-process theory of addiction**. The opponent-process theory was originally developed by the German physiologist Dr. Karl Hering (1834–1918) to explain color vision. Hering argued that our color vision arises from three pairs of opponent colors: red-green, yellow-blue, and white-black.[25] So, what does this have to do with substance use? Hering's theory was extended to drug addiction by Dr. Richard Solomon (1918–1995). According to the opponent-process theory of addiction, drug effects can be represented by two states that reflect different (or opponent) processes[26] (Fig. 5.4). State A represents the *hedonic*, or pleasurable, effects of the drug, whereas State B reflects the aversive effects of the

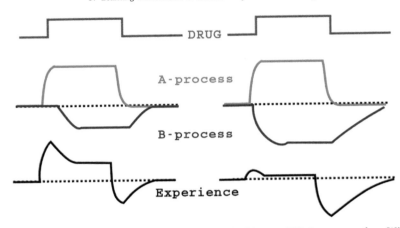

FIGURE 5.4 A schematic of the opponent process theory of addiction. *This image came from Wikimedia commons and was posted by Andy Wills (https://commons.wikimedia.org/wiki/File:Opponent_process_theory_of_drug_addiction.svg).*

drug. Early during drug use, the drug will cause a large increase in State A. As the individual goes into acute withdrawal, State B will begin to increase; however, the peak of State B will not be as high as the peak of State A. This mirrors what we discussed above: during the early stages of addiction, the positive reinforcing effects increase the likelihood that an individual continues using a drug, and they often outweigh the negative effects of the drug. Over time, as the individual develops tolerance to the drug, using the drug produces little increases in State A. Now, coming off the drug produces a much higher peak in State B, reflecting worsening withdrawal symptoms. Continued drug use occurs because individuals try to minimize the peak of State B, which reflects negative reinforcement.

Here is a concrete example of how opponent-process theory can explain a SUD. Imagine a young college student goes to a party and consumes ethanol. This individual is going to experience the disinhibitory effects of ethanol (e.g., become more sociable, engage in more risky behaviors, etc.) and may end up having a "good time" (State A). The person may experience some negative effects of ethanol consumption (e.g., vomiting, waking up with a hangover, etc.) (State B). Even though this person wakes up feeling lousy, they may think to themselves "Last night was really fun. I can't wait until the next time I can do that again". Over time, this individual will develop tolerance to the effects of ethanol, needing to drink more to experience the same pleasurable feelings they received initially. With prolonged drinking, if the individual abruptly stops drinking, they may experience severe withdrawal symptoms. Remember, ethanol withdrawal can be fatal! Now, the aversive effects of ethanol (State B) outweigh the hedonic effects one receives from drinking (State A). The person now drinks in order to avoid withdrawal symptoms.

The opponent-process theory reiterates an important concept when discussing SUDs. Liking and wanting a drug are not the same thing.[27] In the context of operant conditioning, liking is analogous to the positive reinforcing effects of a drug. The person enjoys taking the drug because they like what the drug does to them, whether making them feel euphoric or giving them more energy. Wanting is more akin to needing the drug to escape withdrawal symptoms (e.g., negative reinforcement), although as you will learn in the next chapter, wanting is more than just merely escaping withdrawal symptoms. For now, know that just because someone wants a drug does not mean that they like the drug. This is a common occurrence with individuals struggling with a SUD. Individuals with SUDs often talk about needing the drug just to be able to make it through the day.

Hedonic allostasis model of addiction

Dr. George Koob, currently the director of the National Institute on Alcohol Abuse and Alcoholism and a professor at The Scripps Research Institute, expanded the opponent process theory of addiction to include neurobiological changes that occur as an individual develops a SUD.[28] Before discussing Dr. Koob's model, we need to cover a couple of terms. The first one is *homeostasis*. Many of you are probably familiar with this term and have a general understanding of what homeostasis is. Homeostasis is an equilibrium that organisms try to maintain.[29] For instance, our body temperature is relatively constant. Just a change of a few degrees in either direction can be problematic. We also maintain a constant level of oxygen in our blood and keep blood pressure as consistent as possible. *Allostasis* is like homeostasis in the sense that the goal is to maintain consistency within an organism.[30] The difference between these processes is the way by which consistency is achieved. With homeostasis, a system tries to maintain consistency indefinitely. With allostasis, a system maintains stability by allowing things to change internally. Here is one example to differentiate the two systems. In warm-blooded animals, our body temperature is very consistent, regardless of what is happening outside. Our body temperature reflects a homeostatic process. Cold-blooded animals have body temperatures that change based on the external temperature. Although the animal's body temperature has changed, the end result helps the animal survive. This is an allostatic process.

What does allostasis have to do with addiction? Koob and Le Moal proposed the **hedonic allostasis model of addiction** to describe how the brain compensates for changes in the hedonic value of drugs with prolonged drug use.[28] With repeated drug use, there is less dopamine release in the mesocorticolimbic pathway, and there is a downregulation of dopamine D_2-like receptors in the prefrontal cortex (PFC). During initial drug use, the increased dopamine levels correspond to State A. As the individuals goes into acute withdrawal, decreasing dopamine levels are observed. This coincides with State B. According to Koob and Le Moal, the neuroadaptations that occur because of repeated drug use are an example of allostasis.[28] The organism must reset its *hedonic set point*. Think of hedonic set point as a baseline level of "happiness" that we experience from day to day. With chronic drug use, the dopamine system becomes dysregulated; as such, individuals often feel dysphoric as they go into withdrawal. Furthermore, the hedonic set point increases, such that it takes more drug to reach that set point, corresponding to tolerance. Continued drug use continues to increase the hedonic set point. Even when an individual stops using the drug, the hedonic set point remains higher than baseline levels. This may be a major factor that leads to relapse. We will return to this model when we discuss the role of stress on addiction as Koob posits that stress hormones play a critical role in hedonic allostasis.[31]

Punishment

Although most of the focus up to this point has centered around reinforcement, we need to discuss the relationship between punishment and drug addiction. Importantly, when someone uses a substance, they risk various types of punishment. An individual can contract a sexually transmitted infection (STI), such as hepatitis or human immunodeficiency virus (HIV), from sharing needles or can experience physical distress after taking the drug, such as a seizure. There is the risk of being imprisoned for using an illicit substance or losing one's license for driving while under the influence of ethanol or drugs. The ultimate risk that someone takes when using a drug is overdosing and dying, as you read about in the Introduction to this chapter.

The question becomes, once again, "why does someone continue using the drug?". During the early stages of drug use, the positive reinforcing effects of the drug appear to outweigh the risk of punishment. This is similar to the opponent-process theory of addiction as State A is larger than State B. Indeed, there is some evidence that individuals with a SUD are less sensitive to punishment[32–35] (but see[17]). An individual during the initial stages of heroin use disorder may enjoy the effects of heroin so much that they fail to consider the potential life-changing events that can occur to them by sharing a needle. Another thing to consider is that some of the aversive effects of the drug (mainly withdrawal symptoms) come later in time compared to the reinforcing effects of the drug, although death can come very quickly if someone administers too much! The fact that the reinforcing effects of the drug are more contiguous to the drug-taking experience relative to the punishing effects of the drug may be a contributing factor as to why people continue using the drug during the initial stages of addiction. Over time, as tolerance to the drug develops, the negative reinforcing effects start to outweigh punishment. Here, the person using heroin feels so awful during the onset of withdrawal that they will do anything to procure the drug, even if they must engage in illegal and/or unsafe activities to gain access to it. Here, withdrawal symptoms serve as a powerful establishing operation for continued drug use. Granted, there are many other potential reasons why someone continues using a drug despite its negative consequences. In Chapter 4, we discussed the genetic basis of addiction, and in Chapter 9 we will discuss personality traits that are associated with increased addiction vulnerability (e.g., impulsivity), just to name a couple of examples.

Using operant conditioning to study addiction

There are two major paradigms that utilize operant conditioning that allow us to measure the direct reinforcing effects or discriminative stimulus-like effects of a drug. We will first discuss drug self-administration, a tool widely used to study SUDs in both animals and humans.

Drug self-administration

In Chapter 3, you briefly learned about drug self-administration. Fig. 5.5 shows an image of an animal performing a self-administration task. Research has shown that animals will self-administer a wide variety of drugs, including cocaine, methamphetamine, methylphenidate,

FIGURE 5.5 An image of a rat responding for drug reinforcement. Here, the rat sticks its nose into an aperture to receive reinforcement. *Image was taken by the author of this textbook.*

d-amphetamine, heroin, morphine, THC, pentobarbital, phencyclidine, ketamine, ethanol, nicotine, and synthetic cathinones.[36–46] As you are about to find out, there are multiple self-administration tasks that one can use to measure the reinforcing effects of a drug. We will start off by discussing the most simplistic drug self-administration paradigm.

Short-access model

The first task allows an animal to respond on a manipulandum according to an FR schedule of reinforcement, most often a simple FR 1 schedule (although other schedules like the VI are used), during a short period of time during a session.[47] These **short-access** sessions usually last between 1 and 3 h and have high face validity; that is, they appear to adequately model aspects of human addiction. Some experiments will test various doses of a drug to determine how animals respond to lower/higher drug doses. Research typically shows that drug self-administration across doses follows an inverted U shape (see Fig. 5.6 for an example). All this means is that animals respond most to intermediate doses of a drug compared to lower or higher doses. Low doses are not self-administered as much because they are not as reinforcing; high doses are not self-administered as much because animals reach the optimal blood concentration of the drug at a faster rate; thus, there is no need for the animal to continue responding at that point. Another way to phrase this is that high doses of a drug cause an animal to reach its hedonic set point faster.

To prevent an animal from overdosing, a couple of strategies can be incorporated into the experimental session. First, a *time-out period* (typically 20 s) is enforced following delivery of each reinforcer. This means that the animal can respond as many times as it wants during these 20 s, but they will not receive another infusion until the time-out period ends.[48] This ensures the animals do not earn too many infusions in quick succession. Second, the number of reinforcers can be capped, particularly early in training.[49] Once the animal has reached the maximum number of drug infusions, the experimental session ends.

In the case of ethanol, rats typically do not receive intravenous infusions after emitting a response (but see[44]). Instead, ethanol flows into a small cup which is then consumed by the animal. One challenge to studying ethanol self-administration is that rodents tend to have a taste aversion to ethanol,[50,51] although there are some strains of rats/mice that will willingly consume ethanol.[52,53] If using a strain of rodent that does not naturally prefer the taste of

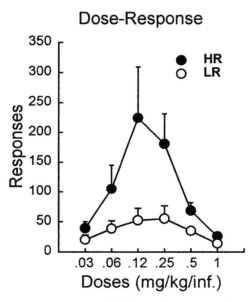

FIGURE 5.6 Example of a dose response curve. In this experiment, animals were classified as high responders (HR) or low responders (LRs) for novelty (determined by measuring how much time animals spent moving around in a novel environment) and then allowed to self-administer various doses of cocaine. Notice that responses for cocaine follow an inverted U shape for each group, although it is more pronounced for HR animals. Note, novelty seeking will be discussed in Chapter 9. *Graph modified from Fig. 4B of Piazza PV, Deroche-Gamonent V, Rouge-Pont F, Le Moal M. Vertical shifts in self-administration dose-response functions predict a drug-vulnerable phenotype predisposed to addiction. J Neurosci. 2000;20(11):4226–4232. https://doi.org/10.1523/JNEUROSCI.20-11-04226.2000 Copyright 2000 Society for Neuroscience.*

ethanol, a couple of methods can be used to increase responding for ethanol. One, a sucrose-fading procedure can be used.[51] Rodents naturally consume sweet foods like sucrose. In the sucrose-fading procedure, ethanol is mixed with sucrose; over time, the amount of sucrose added to the ethanol is reduced (or faded) until only ethanol remains. Although this may appear like "cheating" to get the animal to self-administer ethanol, this technique provides a nice model of human drinking behavior. Adolescents/young adults often begin ethanol consumption by mixing liquor with a sweetener such as soda or Kool-Aid. Eventually, individuals transition to beer or to shots of liquor. Another technique used to increase ethanol self-administration is to allow animals to consume ethanol in their home cage before testing them.[50] This allows the animal to acquire the taste of ethanol, making it less likely that they avoid it during testing. The analogy here is that some people need to acquire the taste of beer before they enjoy it. For example, I absolutely hated the taste of beer the first time I tried it. I had to start by drinking a sweetened alcoholic beverage (see above). After I had tried nonsweetened beer a few times, I started to enjoy the taste more. When I tried a more hoppy beer (like an India Pale Ale [IPA]) for the first time, I detested it. Now, IPAs are my favorite type of beer.

Long-access model

Some have argued that the short-access model allows one to measure recreational use of a drug only and does not fully model the addiction process.[54] To fully capture the transition to addictive-like behaviors, one self-administration paradigm that can be used is the **long-access (extended access) model**,[54,55] which was developed by Dr. Koob. In the long-access model, animals are allowed to self-administer a drug for an extended period of time, usually 6 h during a single session. What researchers have found is that animals dramatically increase the amount of drug they self-administer across sessions, a phenomenon known as *escalation* (Fig. 5.7). Escalation has been observed with several drugs, including cocaine,[56–59] methamphetamine,[60–62] d-amphetamine,[63] methylphenidate,[64] heroin,[65–67] and oxycodone.[68] Interestingly, research has shown that animals do not show escalation of nicotine self-administration,[69] but one study found that escalation of nicotine was observed in animals given brief periods of abstinence of 24–48 h between sessions.[70] Although the long-access model has often been used to model escalation, some studies have shown that short-access models can adequately model addictive-like behaviors.[71,72] For example, Allain and Samaha[71] tested two groups of rats. One group self-administered cocaine during 6-h sessions (long-access group), and one group self-administered cocaine during 2-h sessions (short-access group). Even though the long-access group showed escalation of cocaine consumption only, both groups developed behaviors that are characteristic of the addiction process, including relapse-like behavior. Still, the long-access model is a popular self-administration paradigm.

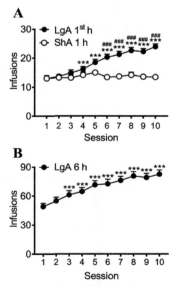

FIGURE 5.7 Example of escalation in animals given extended access to methamphetamine. In this study, rats self-administered methamphetamine for either 1 h (ShA group) or for 6 h (LgA group). Across a 10-day period, the number of infusions significantly increased for the LgA group, but not for the ShA group during the first hour of self-administration (A). When examining the entire 6-hour period across 10 days, animals significantly increase their responding for methamphetamine (B). *Graph modified from Fig. 2 of Tunstall BJ, Ho CP, Cao J, et al. Atypical dopamine transporter inhibitors attenuate compulsive-like methamphetamine self-administration in rats. Neuropharmacology. 2018;131:96–103. https://doi.org/10.1016/j.neuropharm.2017.12.006. Copyright 2018, with permission from Elsevier.*

Progressive ratio schedule

In addition to using simple schedules of reinforcement to study addiction, researchers often use a **progressive ratio (PR)** schedule of reinforcement[73] that was originally developed by Dr. William Hodos.[74] How a PR schedule works is that the response requirement needed to receive reinforcement increases either (1) after the animal receives each reinforcer or (2) at the beginning of each daily session. For example, in the first situation, at the beginning of the experimental session, an animal may need to respond on the lever just once to receive a drug infusion. However, to receive the next drug infusion, the animal may need to respond twice. Then, they will need to respond four times and then eight times and so on to receive subsequent infusions. In the second situation, the response requirement could be set to an FR 1 on the first day, increase to an FR 2 on the second day, increase to an FR 4 on the third day, and so on. The main variable of interest in a PR experiment is the *break point*, the response requirement at which animals no longer respond for the reinforcer. A PR schedule is used to assess how effective a reinforcer is, that is, how hard an animal is willing to work to receive drug infusions; thus, the term **reinforcer efficacy** is often used in association with PR schedules. Fig. 5.8 shows some data from an experiment using a PR schedule. Panlilio et al.[75] tested rats that either had been previously treated with THC or treated with vehicle (no THC) in a PR schedule for cocaine reinforcement. At each dose of cocaine tested (0.1, 0.3, 1.0 mg/kg/infusion), THC significantly decreased the number of responses for cocaine, meaning that THC reduced the reinforcing efficacy of cocaine in these rats.

Threshold procedure

Another drug self-administration task that can be used is called the **threshold procedure**.[76] In the threshold procedure, the amount of drug delivered following completion of a response requirement (typically FR 1) decreases within a session. This effectively lowers the dose of the drug that the animal receives (Fig. 5.9). This task is closely related to *behavioral economics*, the psychology of how we make economic decisions. If you have ever taken a course in economics, you probably learned about elastic and inelastic goods. Elastic goods are items that we are willing to avoid if the price of such good increases too much. Luxury items or

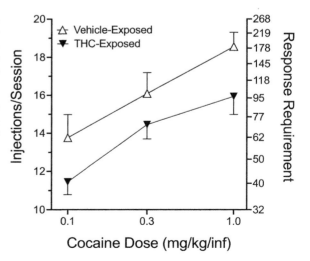

FIGURE 5.8 Data from an experiment using a PR schedule of reinforcement to measure the reinforcing efficacy of cocaine for rats treated either with THC (*black triangles*) or vehicle (*white triangles*). The right *y*-axis shows the last response requirement (measure of break point). *Graph recreated from Fig. 3A of Panlilio LV, Solinas M, Matthews SA, Goldberg SR. Previous exposure to THC alters the reinforcing efficacy and anxiety-related effects of cocaine in rats. Neuropsychopharmacology. 2007;32(3):646–657. https://doi.org/10.1038/sj.npp.1301109. Copyright 2007, with permission from Elsevier.*

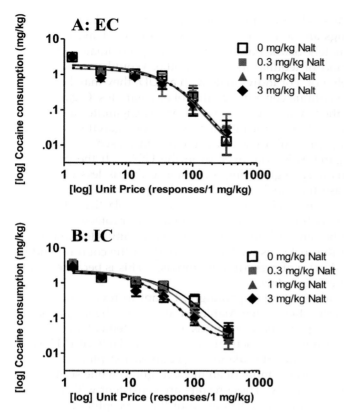

FIGURE 5.9 Animals were raised either in an enriched condition (EC) (A) or in an isolated condition (IC) (B) and then trained to respond for cocaine reinforcement in a threshold procedure in which the dose of cocaine decreased across the session. Rats were treated with several doses of naltrexone before responding for cocaine. Naltrexone decreased responding for cocaine in IC rats, but not in EC rats. In behavioral economic terms, naltrexone makes cocaine more of an elastic good in IC rats. *Graph modified from Fig. 1 of Hofford RS, Beckmann JS, Bardo MT. Rearing environment differentially modulates cocaine self-administration after opioid pretreatment: a behavioral economic analysis. Drug Alcohol Depend. 2016;167:89—94. https://doi.org/10.1016/j.drugalcdep.2016.07.026. Copyright 2016, with permission from Elsevier.*

activities such as going to the movies or sporting events are elastic goods. Inelastic goods are things that we will continue to purchase, even as the price continues to increase. Common inelastic goods include gasoline for vehicles and certain food staples. How does elasticity of a good relate to the threshold procedure? By decreasing the amount of drug delivered across time, we are increasing the "price" of that drug. For example, imagine we train animals to respond on a lever to earn 0.75 mg/kg/infusion of cocaine. Initially, the syringe pump that delivers the cocaine is set to deliver the cocaine across a 5.9-s interval, which ensures delivery of 0.1 mL of solution, thus allowing us to deliver the intended dose of 0.75 mg/kg/infusion. As the session progresses, the duration of the infusion period decreases, which causes the pump to deliver less than 0.1 mL of drug. This then causes the dose of the drug to decrease correspondingly. In order to earn the same amount of drug that they initially earned, the animal has to respond more frequently on the lever. This is the price of the reinforcer. For some animals, cocaine is an elastic good. If the price increases too steeply in a short amount of time, they will stop responding entirely. For other animals, cocaine is an inelastic good. They will continue responding for the drug regardless of its current price. When I was a graduate student, my mentors and I found that animals raised in an enriched environment (e.g., raised in groups with access to toys) treated cocaine as an elastic good, whereas animals raised in isolation treated it more as an inelastic good[77].

The concept of behavioral economics is highly applicable to humans and SUDs. For individuals that do not have a SUD, drugs are an elastic good. Think of it this way. Imagine a bottle of liquor goes from approximately $20 a bottle (this is the approximate price of a 750 mL bottle of a very popular brand of tequila) to $500 a bottle. Would you be willing to pay $500 for that bottle? Most likely not. For individuals with a SUD, they may be willing to spend large amounts of money to continue consuming ethanol or cigarettes. Cigarettes, in particular, are an inelastic good for those with nicotine use disorder. Cigarette prices have increased substantially over the years. Adjusted for inflation, a pack of cigarettes cost $2.29 in the 1950s. In the 1990s, the average price of a pack of cigarettes increased to $3.24. The average price of a single pack of cigarettes is now over $6 in the USA.[78] If the price of a drug increases too much, individuals may begin using a substance that is less expensive. For example, if ethanol prices increase too much, an individual may switch to a cheaper benzodiazepine. This is a form of *cross-price elasticity*. This cross-price elasticity has been observed in certain parts of the country battling the heroin epidemic. As a professor in Northern Kentucky, I have seen the impact heroin has had on the local community. Heroin use has increased dramatically in the state of Kentucky, as well as other parts of the country, and its increase can be linked to the increased price of prescription opioids.[79] This is because heroin is a cheaper opioid compared to prescription drugs.

Before moving on to the next self-administration paradigm, I want to revisit the experiment spotlight from the previous chapter. Recall that Mahler et al.[80] used chemogenetics to study the contribution of intra-VTA dopaminergic neurons to reinstatement of cocaine-seeking behavior. This study also determined how activating hM3Dq, hM4Di, or rM3Ds receptors controls economic demand of cocaine. Stimulating G_q-coupled receptors decreases demand elasticity of cocaine (i.e., makes cocaine more of an inelastic good) while stimulating G_i-coupled receptors increases demand elasticity. These results are consistent with what we discussed in the previous chapter concerning the contribution of these receptors to relapse-like behavior. That is, stimulation of excitatory receptors located on VTA dopaminergic neurons increases addiction-like behaviors.

Choice procedure

Although the self-administration techniques we have covered so far are important for increasing our understanding of the reinforcing effects of drugs, they do not always provide a realistic view of drug use. In real life, people have to make choices between reinforcers. You and your friends have to decide if you would rather go to the movies or go play miniature golf. For the traditional student, one must decide if they would rather go to class or stay home and watch TV or play video games. Some of my students must decide if they would rather go to class or go to work! We constantly have to make choices between **alternative reinforcers**. How can we model this with an animal? In a **concurrent schedule** of reinforcement, one manipulandum is associated with drug reinforcement, and one manipulandum is associated with a different reinforcer, typically food. For example, Thomsen et al.[81] allowed rats to respond on one lever to receive cocaine infusions and on a different lever to receive vanilla flavored Ensure protein drink. Thomsen et al. found that rats showed exclusive preference for the liquid food when no cocaine was delivered but showed increasing preference for cocaine as the drug dose increased.[81] Someone with a SUD must decide between using the drug or engaging in other activities, such as spending time with family or going to school. When we cover operant-based therapies for SUDs, we will discuss how alternative reinforcement can be used to our advantage.

Second-order schedule

As you have already learned, a second-order schedule of reinforcement is often used to measure drug-seeking behavior. As related to drug use, this procedure is most associated with Dr. Steve Goldberg (1941–2014), who spent most of his career with the Intramural Research Program of the National Institute on Drug Abuse (NIDA). Second-order schedules have also been used extensively by Dr. Barry Everitt, a professor at the University of Cambridge. Recall that in the second-order schedule animals are first trained to respond on a manipulandum for drug reinforcement that is paired with cues such as lights. Animals do not receive an infusion of drug until the first response after a certain period (e.g., 15 min). Animals can gain access to drug-paired cues according to a different schedule of reinforcement, typically an FR schedule. Across multiple drugs, animals will continue to respond for the drug-paired cues[82,83] (Fig. 5.10). This paradigm is useful because it shows how powerful drug-paired stimuli are at motivating an animal to seek drug reinforcement. In the next chapter, we will continue discussing how stimuli become paired with drugs and how they contribute to drug seeking.

Compulsive drug seeking

As discussed earlier, humans will often continue taking a drug despite its negative consequences. This is known as *compulsive drug seeking* and can be modeled in animals using a drug self-administration procedure that combines a punisher (often foot shock) with drug delivery. Specifically, rats that have had extended access to cocaine will continue to respond for the

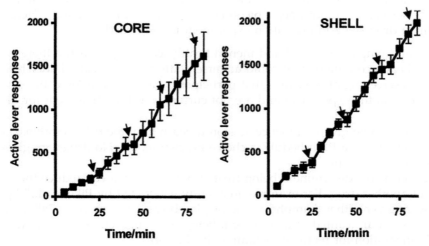

FIGURE 5.10 Data from an experiment using a second-order schedule of reinforcement to measure cocaine-seeking behavior. Rats received a 90-min session in which cocaine was delivered according to an FI 20′ schedule of reinforcement and drug-paired stimuli were presented after 10 responses (FR 10). Arrows indicate when rats received each cocaine infusion. During a 90-min session, rats could earn a maximum of four cocaine infusions. Even though the maximum number of infusions was low, rats in both experimental groups (rats were implanted with guide cannulae into either the NAc core or the NAc shell for microdialysis) responded over 1,500 times during the session! *Graph comes from Fig. 3 of Ito R, Dalley JW, Howes SR, Robbins TW, Everitt BJ. Dissociation in conditioned dopamine release in the nucleus accumbens core and shell in response to cocaine cues and during cocaine-seeking behavior in rats. J Neurosci. 2000;20(19):7489–7495. https://doi.org/10.1523/JNEUROSCI.20-19-07489.2000. Copyright (2000) Society for Neuroscience.*

drug even when they receive a foot shock, an effect that is not observed in rats that have limited experience with cocaine or in rats self-administering sucrose.[84] Instead of shocking an animal when responding for drug, an aversive agent can be added to the drug. Holtz and Carroll[85] first trained adolescent and adult rats to self-administer cocaine. Following training, Holtz and Carroll added histamine to the cocaine solution. If you are familiar with allergies, histamine is what is released when we experience an allergic reaction. In my case, when I am in the lab working with rats, I experience increased histamine release. I do not think anyone has ever said "allergies are awesome! I love having itchy, watery eyes and a runny nose!". Histamine not only negatively affects us, but it affects other animals as well. In the Holtz and Carroll study,[85] adding histamine to the cocaine decreased the number of responses for cocaine. However, histamine did not completely block cocaine responding, further showing that animals will continue to use drugs even when aversive consequences are linked to the behavior.

Human drug self-administration

Thus far, we have discussed drug self-administration in animals. Drug self-administration experiments are not limited to animals such as rodents and nonhuman primates. Many studies have used humans as participants. I want to emphasize that human studies can only use participants that are currently using a drug and are not actively seeking treatment for a SUD (doing so would be unethical). Like animal studies, different drug self-administration tasks can be used with human participants, ranging from FR schedules to PR schedules to concurrent schedules.[86] Studies have had human participants self-administer the following drugs: cocaine,[87–91] *d*-amphetamine,[92,93] methamphetamine,[92] ethanol,[94,95] heroin,[96,97] morphine,[98] nicotine,[99–101] and pentobarbital.[102] Here is how a typical human drug self-administration experiment is conducted:

(1) Participants complete physical and mental health screening to ensure there are no underlying health conditions that could adversely affect the individual after receiving the drug. This screening also ensures that participants have recently used the drug that will be used in the experiment and are not currently seeking treatment for their drug use.
(2) Participants can be given a practice session to learn how the behavioral task(s) will work, whether it is a PR schedule, a choice procedure, etc. No drugs are given to the participants at this time.
(3) Some experiments determine if being treated with one drug can reduce the reinforcing effects of another drug. For example, in a study conducted by Pike et al.,[103] participants were given either *d*-amphetamine or placebo at 7:00 a.m. or at 7:00 p.m. for 7 days. Pike et al. wanted to determine if being on *d*-amphetamine reduced subsequent self-administration for methamphetamine (it did not).
(4) Participants complete the operant task to earn drug reinforcers. The drug is delivered either in pill form, intranasally, or intravenously. In the case of the Pike et al.[103] study, participants received intranasal methamphetamine.
(5) Participants are given a questionnaire to rate how they feel about the drug (e.g., do they like the drug, and would they be willing to take the drug again?).
(6) Participants have physiological measures taken, such as heart rate, blood pressure, temperature, etc. This allows researchers to determine if the drug has a physiological effect on the participants. For example, stimulants like methamphetamine should increase

heart rate. This also ensures the safety of the participant as the researchers and nursing staff can closely monitor how someone reacts to the drug.
(7) Other behavioral/cognitive tasks can be administered to the participant to determine how the drug alters constructs such as impulsivity, risky decision making, etc.
(8) Following experimental procedures, participants are debriefed, meaning they are told what the purpose of the experiment was.

This chapter has covered procedures related to the acquisition and maintenance of drug self-administration. In Chapter 7, we will discuss additional operant conditioning procedures that are used to study memory processes related to addiction, such as reinstatement, a model of relapse. For now, let's shift our focus from drug self-administration to a different type of operant procedure: drug discrimination.

Drug discrimination

In a **drug discrimination** experiment (Fig. 5.11), an animal is initially trained to respond on a manipulandum for food reinforcement in a drug-induced state. For example, a rat will receive an injection of cocaine and will then be placed in the operant chamber. Responses on one manipulandum, such as the lever located on the left side of the chamber, will lead to reinforcement. During a separate session, the rat will receive an injection of vehicle and will be trained to respond on the right lever. During discrimination training, the acute drug effects serve as a discriminative stimulus that signals the availability of reinforcement. The list of drugs that can serve as a discriminative stimulus is long, including psychostimulants, opioids, hallucinogens, THC, ethanol, and nicotine; even other drug classes have been used as discriminative stimuli, such as anxiolytics and antipsychotics.[104] Once the animal has learned this discrimination, the researcher can inject the animal with a different drug like

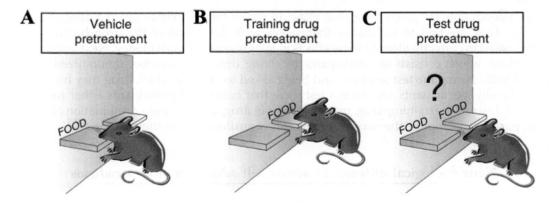

FIGURE 5.11 A schematic of the drug discrimination paradigm. *This image modified from Fig. 1 of Solinas M, Panlilio LV, Justinova Z, Yasar S, Goldberg SR. Using drug-discrimination techniques to study the abuse-related effects of psychoactive drugs in rats. Nat Protoc. 2006;1(3):1194—1206. https://doi.org/10.1038/nprot.2006.167. Copyright 2006.*

amphetamine. The researcher will then measure how much the animal responds on the lever associated with cocaine.

There are three possible outcomes of a drug discrimination experiment. First, if the drugs are very similar to one another, the animal will respond as if it had received the original training drug. This is called *stimulus substitution*. Animals initially trained with cocaine will show a similar pattern of responding if they are injected with *d*-amphetamine.[105] Second, the new drug may cause the animal to respond on the manipulandum associated with the original training drug, but not to the same extent as the training drug. This is called *partial substitution*. For example, animals originally trained to respond under the influence of THC show some responding on the same lever following administration of diazepam, a benzodiazepine.[106] Finally, some drugs do not produce any responses on the manipulandum paired with the original training drug. Just because Drug A substitutes or partially substitutes for Drug B, this does not mean that Drug B will substitute for Drug A. In the partial substitution example, I mentioned that diazepam partially substitutes for THC. In contrast, THC does not substitute for diazepam.[106] Weird, huh?

So, why should we care about drug discrimination? The drug discrimination paradigm is important because it allows researchers to determine if a novel drug shares characteristics with a known drug. Imagine a scenario in which a pharmaceutical company is trying to develop a novel pharmacotherapy for obesity. One assay the researchers can use is drug discrimination to determine if animals respond to the potential weight loss drug as they do with a drug such as cocaine or methamphetamine. If animals respond in a similar manner when a drug is substituted in place of the weight loss drug, there will be concerns about the misuse potential of this novel drug, which will then increase the need to conduct additional experiments such as drug self-administration. Prescribing medications with high misuse potential such as opioids has become a major problem in this country, particularly as individuals switch to cheaper alternatives like heroin, as mentioned above when discussing cross-price elasticity.

As drug self-administration can be conducted in animals and humans, drug discrimination experiments can be conducted across various species. In human drug discrimination experiments, participants are given a *sampling phase*, in which they are either given a drug or given placebo. They are told to familiarize themselves with the effects of the drug because they will be asked to correctly identify these effects in later sessions to earn money. Next is the *acquisition phase*, which consists of participants receiving drug and placebo, randomized across several trials during the test sessions, and being asked to identify which drug they have been given. Finally, participants are given a *test phase* that consists of participants either receiving different doses of the training drug, receiving a novel drug, or receiving a combination of drugs. Here, participants receive the maximum reward, regardless of how they respond.[107]

Neurobiological differences across self-administration paradigms

In the previous chapters, I discussed the neurobiological constructs of addiction in a simplified manner. Given that the self-administration procedures described in the previous section measure different aspects of addiction-like behaviors, determining if the neurobiological substrates underlying performance in each task differ is important for further clarifying the neuromechanisms of addiction. I am not going to detail all of the neurobiological findings that have been observed across each self-administration paradigm. I do want to emphasize that there is generally high concordance across procedures. Specifically, drugs that decrease drug reinforcement as assessed in short-access models tend to decrease drug reinforcement as measured with PR schedules or with extended access models. However, some discrepancies have been observed across experimental procedures. Blocking D_1-like receptors increases cocaine self-administration when an FR 1 schedule of reinforcement is used but decreases cocaine self-administration in a PR schedule.[108] D_1-like receptor antagonists decrease self-administration of higher doses of cocaine,[109] showing that the dose of the drug greatly influences pharmacological manipulations of self-administration behavior. Additionally, a D_3 receptor antagonist has been shown to decrease cocaine-seeking behavior as assessed in a second-order schedule without affecting cocaine self-administration using an FR 1 schedule of reinforcement.[110] This result suggests that D_3 receptors are important for assigning incentive salience to cues paired with the drug as opposed to mediating the direct reinforcing effects of the drug.

In addition to the discrepancies observed in behavioral pharmacology experiments, knockdown/knock-in models have shown discrepancies between FR and PR schedules. Reducing the number of D_2-like receptors in the VTA does not affect the acquisition of cocaine self-administration or the total number of cocaine infusions earned in an FR schedule, but it potentiates cocaine self-administration in a PR schedule.[111] Another study found that overexpression of serotonin $5-HT_6$ receptors decreases cocaine self-administration when FR schedules are used but does not affect the reinforcing efficacy of cocaine as measured using a PR schedule.[112] Finally, drug-induced changes in receptor expression are sometimes task dependent. For example, escalation of cocaine self-administration decreases AMPA receptor expression in the PFC, an effect that is not observed in animals given short access to cocaine.[113] This last result suggests that dysregulated drug use is associated with an altered glutamatergic system. This is a topic we will come back to in Chapter 7. Overall, these results highlight the utility in assessing potential pharmacotherapies for SUDs using a variety of self-administration procedures. A potential pharmacotherapy may reduce self-administration of a drug, but it may not prevent escalation of drug-taking behavior or impair the ability of secondary reinforcers to drive drug-seeking behavior.

Using operant principles to treat addiction

Now that we have covered the pivotal role operant conditioning has on the development of drug addiction, we now need to discuss how we can use operant conditioning to help treat those with a SUD.

Contingency management

Earlier in this chapter I introduced the concept of alternative reinforcers. Someone with a SUD must decide if they would rather spend their time and money on drugs as opposed to doing other activities like spending time with family/friends, going to school/work, or working on a hobby. As a teenager, I worked at a grocery store, and I witnessed some customers that did not have enough money to buy cigarettes and food/essential items. These customers had to make a decision between competing reinforcers. In my experience, most (if not all) customers I encountered would either remove the cigarettes or ask for a cheaper brand so they could get everything they needed. However, this is not always the case, as demonstrated by a study published in Australia. When participants were interviewed, some discussed how their neighbors would choose cigarettes over essential goods:

> You see that all the time, especially in our area where we are … you see kids not eating and parents are smoking (p. 603).[114]

> Yes friends of mine will not get their groceries so they can get their smokes … they get behind in their bills (p. 603).[114]

How can we use alternative reinforcement to help treat those with a SUD? One form of therapy is **contingency management**. In contingency management, an individual with a SUD receives money or a food voucher if they provide a urine-free sample. Contingency management reduces use of the following drugs: nicotine,[115,116] ethanol,[117,118] stimulants,[119–121] opioids,[122,123] and cannabis.[124,125] See the Experiment Spotlight below for a detailed description of how a contingency management study works.

Experiment Spotlight: A contingency management method for 30-day abstinence in nontreatment seeking young adult cannabis users (Schuster et al.[124])

Behind ethanol, cannabis is the most frequently used drug in young adults, excluding caffeine. Although cannabis has multiple health benefits, there are still concerns about its long-term use as it may negatively impact the developing brain. As such, Schuster et al.[124] wanted to use contingency management to reduce cannabis use in a sample of young adults. They recruited 38 adults aged 18–25 who reported using cannabis at least weekly and were not currently seeking treatment for their cannabis use. These participants were screened for psychiatric conditions, mood and anxiety disorders, and attention-deficit/hyperactivity disorder (ADHD). During a span of 4 weeks, participants visited the lab on seven occasions (four during the first week and one during each subsequent week) and provided a urine sample. Each time a participant visited the lab, they earned money on an increasing scale. They received $5 during the initial visit but received $35 during the seventh visit. If the participant provided a drug-free urine sample, they received additional money on an increasing scale (e.g., $35 after the second visit and $105 following the seventh visit). Following the 4-week contingency management phase, participants were no longer paid for abstaining from cannabis use. After 2 additional weeks, participants were invited to come back to the lab for a follow-up visit. Results showed that 34 of the 38 participants (89.5%) abstained from cannabis use during the 4-week contingency management phase. Out of these 34 participants, 33 (97.1%) returned at the 2-week follow-up appointment. When alternative reinforcement was no longer delivered for cannabis abstinence, Schuster et al. found that 31 out of 33 (93.9%) of the participants had resumed using cannabis.[124]

Questions to consider:

(1) Why were participants screened for psychiatric conditions, mood/anxiety disorders, and ADHD?
(2) Why do you think participants had to show up to the lab four times during the first week but only once per week following the first week?
(3) What are some limitations (weaknesses) to the study that you can think of?
(4) Overall, do you think contingency management is an effective treatment for cannabis use disorder? Why or why not?

Answers to questions:

(1) Because there is often high comorbidity between SUDs and other psychiatric conditions (such as depression and ADHD), researchers often record this information to determine if their sample is representative of what is observed in the general population. In the current experiment, the researchers found that 48.6% of the participants had a life-time diagnosis of a psychiatric disorder and that 45.9% of the participants reported having a life-time diagnosis of depression.
(2) By meeting with the participants more frequently during the first week, the researchers were able to better monitor participants to ensure they were meeting abstinence requirements.
(3) Some potential limitations:
   (a) The sample size of 38 is small. It can be difficult to generalize these results to all young adults in this country.
   (b) There was no control group. In research, comparing an intervention (in this case, contingency management) to another group of individuals that does not receive the intervention allows one to draw more definitive conclusions about the effectiveness of the intervention. Ideally, the researchers should have recruited an additional 38 participants. These participants would not receive contingency management during the course of the experiment. The researchers could then show that these individuals do not stop using cannabis, which would provide additional support for the use of contingency management.
   (c) Providing each participant money on an increasing scale could be cost-prohibitive if trying to treat numerous people.
(4) In the short-term, contingency management appears to be effective. However, contingency management failed to produce long-lasting alterations in behavior, as most of the participants resumed cannabis use once alternative reinforcement was no longer provided.

After reading the Experiment Spotlight, one thing that you should have immediately noticed is that most of the participants resumed drug taking once alternative reinforcement was no longer provided. This effect has been observed with drugs other than cannabis.[126] Although contingency management often fails to prevent individuals from relapsing, it is an effective treatment strategy for reducing drug use in the short term. In order to promote long-term abstinence, other treatment strategies are needed. Let's cover a few other treatment options that incorporate operant principles.

Community reinforcement approach

Dr. Nathan Azrin (1930—2013), a former student of B.F. Skinner, wanted to apply principles of behaviorism to the treatment of SUDs, particularly alcohol use disorder. Azrin argued that individuals struggling with alcohol use disorder need to receive reinforcement from the community (in the form of family, friends, and coworkers) for abstaining from ethanol. Additionally, Azrin wanted individuals to discover activities that did not revolve around ethanol they could enjoy. These principles underly the **community reinforcement approach (CRA)**. Below are the basic techniques used in CRA[127]:

(1) *Functional Analysis of Substance Use*: a therapist can interview the client to discover when he/she is most likely to use drugs (essentially, identify possible triggers) and identify the positive and the negative consequences of the drug use. For example, a therapist may discover that someone may smoke cannabis every time they feel stressed. The clinician would then work with the client to understand why he/she feels stressed and then work on strategies to reduce stress that do not rely on cannabis use.

(2) *Sobriety Sampling*: clients agree to try short-term abstinence. The client may agree to stop smoking cannabis for a period of 1 week. Encouraging short-term abstinence is important because trying to enact long-term abstinence may seem overly daunting to an individual, which will cause compliance to decrease.

(3) *CRA Treatment Plan*: clients are asked to complete the Happiness Scale. Clients rate how unhappy/happy, on a scale of 1—10, they are with different aspects of their life, such as drug use, job/education progress, social life, spiritual life, etc. Clients then see in which areas they are unhappy and can develop a plan to address these issues.

(4) *Behavioral Skills Training*: clients are taught three major skills—problem-solving, communication, and drug refusal. Developing problem-solving skills is a necessity as individuals need to learn how to deal with issues without resorting to drug use. Similarly, communication skills are important as individuals need to be able to talk to loved ones about any stressors they are facing, as well as to openly discuss cravings. Learning drug refusal skills is a no brainer. Individuals need to be able to "say no to drugs" to maintain abstinence.

(5) *Job Skills Training*: clients receive training on how to find and keep a job that they will enjoy. By having a job, this reduces how much time an individual can spend using a drug. Importantly, if the individual likes their job, this job can serve as alternative reinforcement to drug-taking behaviors.

(6) *Social and Recreational Counseling*: clients are instructed to try new social/recreational activities (e.g., join an intramural sports team). Similar to the job skills training described above, if someone is able to find activities that they enjoy, they will have less time to use drugs, and they can meet people that can help them abstain from drug use.

(7) *Relapse Prevention*: by completing the Functional Analysis from Step 1, clients learn how to identify early warning signs that may lead to relapse. Clients can recruit someone to keep track of potential early warning signs. For example, if someone identifies that they drink ethanol when they get into an argument with their partner, they can

work on ways to better cope with the argument. This step can be used in conjunction with developing communication skills.
(8) *Relationship Counseling*: clients work on improving interaction with their partner. If the client's relationship with their partner is strong, the partner can serve as a powerful alternative reinforcer to drug use. The partner can help the client identify early warning signs of relapse-like behavior.

As noted by Miller et al., there are two major factors that influence how well CRA works.[128] First, the therapist needs to be optimistic and energetic, and he/she must consistently reinforce the client, even if they make minor progress. The therapist also needs to be personal as opposed to treating each interaction as a negotiation. If the client is ready to start treatment, Miller et al. emphasize that treatment needs to begin as soon as possible, not in 1 month or more.[128] Related to timing, during the early stages of treatment, more sessions should be scheduled within a short period of time in order to maximize therapy. This principle is similar to what was observed in the Schuster et al. contingency management study.[124]

Although CRA was originally used to treat alcohol use disorder,[129] it has been successfully applied to other SUDs[130–134] and has been used in various settings (e.g., inpatient vs. outpatient, rural vs. urban[127]) and with different groups of people (e.g., homeless individuals[135,136]). One major advantage of CRA is that it can be combined with other therapies to improve treatment outcomes. Numerous studies have examined the impact of combined CRA and contingency management on SUDs. Overall, results indicate that combined CRA/contingency management reduces drug use to a greater extent compared to other treatment options,[137,138] although this is not always the case. In one study, CRA/contingency management did not significantly decrease cocaine use in pregnant women/women with young children compared to a 12-step program combined with contingency management[139] (note, you will learn more about 12-step programs in Chapter 10).

Since its inception in the 1970s, CRA has been modified to be used in conjunction with other therapies (as you read above) or to be used with special populations. Below, I will discuss two such special populations: (1) those that refuse to seek treatment and (2) adolescents.

Community reinforcement and family training (CRAFT)

There are some individuals with a SUD that refuse to seek treatment. The goal of community reinforcement and family training (CRAFT) is to engage unmotivated individuals. Additionally, CRAFT aims to help family members and friends develop strategies to help the affected person and to help themselves cope during the course of treatment. The CRAFT intervention is composed of two phases. During the first phase, family and friends of the affected individual are taught skills for modifying the individual's behaviors related to drug use and to enhance treatment engagement. Specifically, family and friends are encouraged to

(1) [raise] awareness of negative consequences caused by [the individual's] drug use and possible personal benefits of treatment,
(2) [learn] specific strategies for preventing dangerous situations,

(3) [use] contingency management training to reinforce the [individual's] nonusing behaviors and to extinguish drug use,
(4) [use] social skill training to improve relationship communication and problem-solving skills,
(5) [plan] … activities that interfere and compete with the [individual's] drug use,
(6) [practice] strategies to interfere with actual and potential [individual's] drug use, and,
(7) [prepare] to initiate treatment when the [individual] appears ready, and [support] the [individual] once treatment has begun (pp. 295–296).[140]

In addition to the training described above, family and friends receive training to improve their quality of life outside of the relationship with the individual. This first phase of training lasts approximately 6 months or 12 sessions.

During the second phase, motivated individuals receive community reinforcement. Before receiving community reinforcement, individuals are given motivational enhancement therapy (MET) for a few sessions to increase their motivation to seek treatment. During the community reinforcement phase individuals:

(1) have their drug use analyzed,
(2) develop treatment goals,
(3) follow sobriety sampling,
(4) receive drug refusal training,
(5) receive social skills training, including communication skills and problem-solving skills,
(6) receive social and recreation counseling, and,
(7) receive relapse prevention training.[140]

Like the first phase, treatment lasts for either 6 months or 12 sessions (with two emergency sessions).

In an early CRAFT study, Myers et al.[140] had 62 concerned significant others (CSOs) participate in the intervention. Overall, 74% of the CSOs were able to engage their loved one to seek treatment. Myers et al.[140] also found that individuals that participated in treatment had higher abstinence rates for both illicit drugs and ethanol compared to those that were unengaged. Although CRAFT is typically administered to individuals, it can be used in a group setting[141]; additionally, individuals can complete a self-directed workbook to increase flexibility.[142] Even though CRAFT can be delivered in various formats, a recent meta-analysis found highly variable treatment entry rates depending on which modality is used.[143] Specifically, Archer et al.[143] found that entry rates ranged from 12.5% to 71% for individual settings and from 13.3% to 40% for self-directed workbooks. In contrast, the most successful outcomes (77%–86% entry rate) were observed with a multimodal approach, meaning that individuals attend individual and group sessions. Thus, a therapist using CRAFT to treat an individual with a SUD should consider a multimodal approach that is tailor-made for each individual.

Adolescent community reinforcement approach

The adolescent community reinforcement approach (A-CRA) is a variant of the CRA that focuses on adolescents and young adults. Because the focus is on younger individuals, some modifications had to be made to the original CRA, such as including categories related to friends and school in the Happiness Scale and simplifying the communications skills training procedure.[127] The most consequential difference between the A-CRA and the CRA is that caregivers are more involved in the A-CRA. Caregivers attend several sessions, some of which that they attend alone in order to learn about parenting styles and some that they attend with the adolescent/young adult seeking treatment. The caregivers are exposed to some of the same training that the adolescent has received, such as communication skills.

Although research has provided evidence for the efficacy of A-CRA, it does not always lead to the best outcomes for individuals with SUDs. For example, Davis et al.[144] divided male and female adolescent/young adults with an opioid use disorder into several groups. One group received A-CRA as described above. Another group received MET combined with cognitive-behavioral therapy (CBT; you will learn more about CBT in Chapter 7). A third group received CBT alone. Finally, one group received the treatment that they currently had before enrolling in the study (e.g., 12-step program). Whereas treatment type did not alter outcomes for adolescent females, adolescent males and young adults (regardless of sex) had poorer outcomes following A-CRA compared to other treatment options. As with CRAFT, these results highlight the need to customize treatment for adolescents as one size does not fit all.

Use of punishment to combat substance use disorders?

Thus far, we have discussed how reinforcement can be used to help treat individuals with SUDs. Can punishment be used as effectively as reinforcement for reducing substance use? According to the seminal work of Kahneman and Tversky,[145] people tend to be more sensitive to losses, a form of punishment, relative to gains. One would expect that an individual would be just as upset about losing $1,000 as they would be happy to receive $1,000. However, individuals tend to be more upset about losing something. Think of it this way: imagine your professor adds 10 points to your grade for showing up to class early. You may be happy about this outcome. But, imagine your professor deducts 10 points from your grade for showing up late. You would most likely feel more distraught about losing these points. When I bring this hypothetical situation up to my students, I can see how outraged they are about the mere thought of having points deducted from their grade, whereas they barely react to the possibility of having points added to their grade.

In the United States, individuals that are caught with illicit drugs are often sentenced to jail or prison, a form of punishment that theoretically should outweigh the reinforcing effects of the drug. It is highly important to note that approximately 80% of incarcerated individuals in the United States have a SUD.[146] The idea is that if someone is incarcerated, they will be forced to abstain from drug-taking behaviors. Although loss sensitivity can be applied to economic decision making (i.e., we tend to be risk averse when making decisions), the assumption that imprisoning someone will decrease future drug use is not supported by the literature. Most studies have found that a significant percentage of individuals resume drug use following incarceration, ranging from approximately 20%–80%,[147–150] suggesting that incarceration alone does not adequately treat those with a SUD. One reason why incarceration fails to prevent relapse is because a large majority (up to 80%) of incarcerated individuals do not receive treatment while in prison/jail.[151] Also, recall that the incubation of craving model of relapse is somewhat analogous to an individual that has been incarcerated. Craving for drugs increases during protracted withdrawal. These results emphasize the need of using positive reinforcement to treat those with SUDs instead of merely punishing the behavior.

Another problem with incarceration is the use of mandatory sentencing requirements. You may know these better as the "three strikes" laws. Some individuals can receive a life sentence if their third felony relates to drug possession. In 2014, a 26-year-old man named Chris Young was sentenced to life in prison after being arrested for possessing cocaine and crack cocaine with the intent of distributing these substances. Young's previous arrests were drug possession charges. Although Young was not a violent individual, the state of Tennessee's habitual offender law was automatically triggered following his third felony. Even the federal judge that sentenced Young to life in prison felt the sentence was too harsh. Eventually, Young's life sentence was commuted by former president Donald Trump in 2020.[152]

Quiz 5.2

1. When someone initially starts using a drug, the drug acts as a _____ reinforcer; however, over time, the drug acts as a _____ reinforcer.
2. What is the major disadvantage of using contingency management to treat SUDs?
3. An individual is told they can earn puffs on a cigarette if they hit the space key on a keyboard. The first time the individual hits the key, they earn access to the cigarette. To earn another puff on the cigarette, the person now has the hit the key two times. Then, they must hit the key four times, followed by eight times, 16 times, etc. Which schedule of reinforcement is being used?
    a. Long-access
    b. PR
    c. Threshold procedure
    d. VR
4. A rat has been trained to respond on the left lever in an operant chamber to earn food after it has been injected with heroin. The rat is also trained to respond on the right lever following saline injections. Imagine the researcher now tests different drugs to measure how much the animal responds on the heroin-paired lever. For each drug listed below, decide if the rat will respond as if it were given heroin:
    a. Cocaine
    b. Methamphetamine
    c. Morphine
    d. Oxycodone

e. Phencyclidine (PCP)
   f. THC
5. When would a therapist want to use Community Reinforcement and Family Training (CRAFT) instead of CRA?

Answers to quiz 5.2

1. positive, negative
2. Individuals often revert to drug use when alternative reinforcement (e.g., money/food voucher) is no longer provided.
3. b — PR
4. Cocaine — the rat would show very little responding on the heroin-paired lever because cocaine, as a psychostimulant, produces effects that are opposite to those caused by heroin.
   Methamphetamine — see response above for cocaine.
   Morphie — the rat would respond as if it were given heroin. Heroin and morphine are both opioids. Remember, heroin is converted into 6-monoacetylmorphine when it enters the brain.
   Oxycodone — as with morphine, responses would be high on the heroin-paired lever because oxycodone is an opioid.
   PCP — it is a dissociative hallucinogen that produces a different behavioral response compared to analgesics like heroin. PCP can produce analgesia, but this occurs at extremely high doses. Responding on this lever would most likely be low.
   THC — there will be at least partial substitution here. THC has analgesic effects and acts to stimulate cannabinoid receptors, which are inhibitory (just as mu opioid receptors are inhibitory).
5. CRAFT is primarily used when dealing with individuals that are resistant to going to therapy for their SUD.

Chapter summary

After reading this chapter, you should now know that drugs are powerful reinforcers. During initial drug use, drugs serve as positive reinforcers; however, over time, drug use becomes negatively reinforcing as individuals try to escape the aversive symptoms of withdrawal. Although B.F. Skinner was not an addiction researcher, his contributions to psychology have profoundly affected how we study SUDs. With the development of the operant conditioning chamber, researchers were and are still able to tightly control the environment to study how drugs serve as reinforcers in animals. Drug self-administration and drug discrimination experiments continue to be used and are often combined with neuroscience techniques to unravel the neurobiology underlying SUDs. Not only can operant conditioning be used to study the addiction process, but it can be used to help individuals decrease their drug use in the form of contingency management or community reinforcement. Even though we are at the end of this chapter, we are not finished discussing the contributions of operant conditioning to addiction research. Later, we will examine how operant conditioning can be used to study relapse-like behavior, and you will learn how peers can reinforce drug-taking behaviors and can serve as discriminative stimuli that signal drug availability.

Glossary

Alternative reinforcement — a stimulus (usually activity) that competes with a different stimulus. For example, spending time with friends is an alternative reinforcer that competes with drug use.
Appetitive stimulus — a positive reinforcer that an animal will approach. Depriving an animal of this reinforcer will make them work to earn access to it (e.g., food is an appetitive stimulus for hungry rats).
Community reinforcement approach (CRA) — a type of therapy that emphasizes the importance of external reinforcement (e.g., whether from family, friends, coworkers or from extracurricular activities) in treating SUDs.
Concurrent schedule of reinforcement — a schedule of reinforcement in which an animal chooses between two or more reinforcers. Each reinforcer is associated with a different manipulandum. For example, a rat can respond on one lever to receive an infusion of drug or can respond on a different lever to receive food.
Contiguity — refers to how much times passes between the occurrence of a behavior and the delivery of a reinforcer.
Contingency — refers to how reliably or consistently a reinforcer is delivered following the occurrence of a behavior.
Contingency management — a type of therapy in which individuals are provided alternative reinforcers (e.g., money) to abstain from drug use.
Discriminative stimulus — a cue that signals the availability/unavailability of reinforcement.
Drug discrimination — an operant procedure in which animals are trained to respond on one manipulandum while under the influence of a specific drug and to respond on a different manipulandum when given vehicle.
Hedonic allostasis model of addiction — posits that chronic drug use leads to dysregulation of the reward circuitry.
Long-access model — a drug self-administration paradigm in which animals are allowed to respond for a drug during an extended period of time during a single session (e.g., 6 h); also known as the **extended access model**.
Motivating operation — anything that changes the effectiveness of a consequence, whether reinforcement or punishment.
Negative punishment — removing something pleasurable from an animal's environment to decrease the probability of a behavior occurring (e.g., revoking someone's license for driving under the influence of ethanol/drugs).
Negative reinforcement — removing something aversive from an animal's environment to increase the probability of a behavior occurring (e.g., an individual continues to use drugs to avoid withdrawal symptoms).
Operant conditioning — a form of learning in which animals learn to associate behaviors with their outcomes.
Opponent-process theory of addiction — theory stating that drug addiction arises from a drug's transition from a positive reinforcer to a negative reinforcer. More specifically, a drug has hedonic effects (State A) and aversive effects (State B). During early drug use, State A is more pronounced than State B; however, over time, State B is greater than State A.
Positive punishment — adding something aversive to an animal's enivornment to decrease the probability of a behavior occurring (e.g., spanking a child).
Positive reinforcement — adding something pleasurable to an animal's enivornment to increase the probability of a behavior occurring (e.g., an employer paying an employee to show up to work).
Primary reinforcer — something that fulfills a biological need, such as food, water, and sex.
Progressive ratio (PR) schedule of reinforcement — a schedule of reinforcement in which the response requirement needed to receive reinforcement increases after each delivery of the reinforcer.
Punisher — a stimulus that decreases the likelihood of a behavior occurring in the future.
Reinforcer — a stimulus that increases the likelihood of a behavior occurring in the future.
Reinforcer efficacy — how effective a reinforcer is in maintaining a specific response.
Schedule of reinforcement — the rules governing how an animal must respond to receive reinforcement.
Secondary reinforcer — a stimulus that is not innately reinforcing but becomes reinforcing over time due to its association with primary reinforcers (e.g., money is a secondary reinforcer for humans).
Short-access model — animals are given a relatively short amount of time to earn reinforcers (typically around 1–3 h).
Threshold procedure — a drug self-administration paradigm in which the amount of drug delivered decreases across the experimental session.

Supplementary data related to this chapter can be found online at https://educate.elsevier.com/9780323905787.

References

1. Betel Australia. *Ash: Interview with A Former Drug Addict*; n.d. https://www.betel.org.au/ash-interview-with-a-former-addict/.
2. Mauro T. The many victims of substance abuse. *Psychiatry*. 2007;4(9):43–51.
3. Staddon JE, Cerutti DT. Operant conditioning. *Annu Rev Psychol*. 2003;54:115–144. https://doi.org/10.1146/annurev.psych.54.101601.145124.
4. Skinner BF. *The Behavior of Organisms: An Experimental Analysis*. Appleton-Century; 1938.
5. Beck SM, Locke HS, Savine AC, Jimura K, Braver TS. Primary and secondary rewards differentially modulate neural activity dynamics during working memory. *PLoS One*. 2010;5(2):e9251. https://doi.org/10.1371/journal.pone.0009251.
6. Scott HK, Jain A, Cogburn M. *Behavior Modification*. StatPearls Publishing; 2021.
7. Elsner B, Hommel B. Contiguity and contingency in action-effect learning. *Psychol Res*. 2004;68(2–3):138–154. https://doi.org/10.1007/s00426-003-0151-8.
8. Laraway S, Snycerski S, Michael J, Poling A. Motivating operations and terms to describe them: some further refinements. *J Appl Behav Anal*. 2003;36(3):407–414. https://doi.org/10.1901/jaba.2003.36-407.
9. Michael J. The discriminative stimulus or S(D). *Behav Analyst*. 1980;3(1):47–49. https://doi.org/10.1007/BF03392378.
10. Ferster CB, Skinner BF. *Schedules of Reinforcement*. Prentice-Hall; 1957.
11. Pinsker J. *Finland, Home of The $103,000 Speeding Ticket*. Atlantic; March 12, 2015. https://www.theatlantic.com/business/archive/2015/03/finland-home-of-the-103000-speeding-ticket/387484/.
12. Pettit HO, Justice Jr JB. Dopamine in the nucleus accumbens during cocaine self-administration as studied by in vivo microdialysis. *Pharmacol Biochem Behav*. 1989;34(4):899–904. https://doi.org/10.1016/0091-3057(89)90291-8.
13. Campbell UC, Carroll ME. Effects of ketoconazole on the acquisition of intravenous cocaine self-administration under different feeding conditions in rats. *Psychopharmacology*. 2001;154(3):311–318. https://doi.org/10.1007/s002130000627.
14. Carroll ME, Campbell UC, Heidman P. Ketoconazole suppresses food restriction-induced increases in heroin self-administration in rats: sex differences. *Exp Clin Psychopharmacol*. 2001;9(3):307–316. https://doi.org/10.1037/1064-1297.9.3.307.
15. Balconi M, Finocchiaro R, Campanella S. Reward sensitivity, decisional bias, and metacognitive deficits in cocaine drug addiction. *J Addiction Med*. 2014;8(6):399–406. https://doi.org/10.1097/ADM.0000000000000065.
16. Mellick W, Tolliver BK, Brenner H, Prisciandaro JJ. Delay discounting and reward sensitivity in a 2 × 2 study of bipolar disorder and alcohol dependence. *Addiction*. 2019;114(8):1369–1378. https://doi.org/10.1111/add.14625.
17. Mellick W, Prisciandaro JJ, Brenner H, Brown D, Tolliver BK. A multimethod examination of sensitivity to reward and sensitivity to punishment in bipolar disorder and alcohol dependence: results from a 2 × 2 factorial design. *Psychopathology*. 2021;54(2):70–77. https://doi.org/10.1159/000512661.
18. Loxton NJ, Dawe S. Alcohol abuse and dysfunctional eating in adolescent girls: the influence of individual differences in sensitivity to reward and punishment. *Int J Eat Disord*. 2001;29(4):455–462. https://doi.org/10.1002/eat.1042.
19. Joyner KJ, Bowyer CB, Yancey JR, et al. Blunted reward sensitivity and trait disinhibition interact to predict substance use problems. *Clin Psychol Sci*. 2019;7(5):1109–1124. https://doi.org/10.1177/2167702619838480.
20. Volkow ND, Wang GJ, Telang F, et al. Profound decreases in dopamine release in striatum in detoxified alcoholics: possible orbitofrontal involvement. *J Neurosci*. 2007;27(46):12700–12706. https://doi.org/10.1523/JNEUROSCI.3371-07.2007.
21. Morris DH, Treloar H, Tsai CL, McCarty KN, McCarthy DM. Acute subjective response to alcohol as a function of reward and punishment sensitivity. *Addict Behav*. 2016;60:90–96. https://doi.org/10.1016/j.addbeh.2016.03.012.
22. Kirkpatrick MG, Johanson CE, de Wit H. Personality and the acute subjective effects of d-amphetamine in humans. *J Psychopharmacol*. 2013;27(3):256–264. https://doi.org/10.1177/0269881112472564.
23. White TL, Lott DC, de Wit H. Personality and the subjective effects of acute amphetamine in healthy volunteers. *Neuropsychopharmacology*. 2006;31(5):1064–1074. https://doi.org/10.1038/sj.npp.1300939.

24. Bart CP, Nusslock R, Ng TH, et al. Decreased reward-related brain function prospectively predicts increased substance use. *J Abnorm Psychol.* 2021;130(8):886–898. https://doi.org/10.1037/abn0000711.
25. Hering E. *Zur lehre vom lichtsinne sechs mittheilungen an die kaiserliche akademie der wissenschaften in wien.* Wien: Carl Gerold's Sohn; 1878.
26. Solomon RL. The opponent-process theory of acquired motivation: the costs of pleasure and the benefits of pain. *Am Psychol.* 1980;35(8):691–712. https://doi.org/10.1037//0003-066x.35.8.691.
27. Berridge KC, Robinson TE. Liking, wanting, and the incentive-sensitization theory of addiction. *Am Psychol.* 2016;71(8):670–679. https://doi.org/10.1037/amp0000059.
28. Koob GF, Le Moal M. Drug addiction, dysregulation of reward, and allostasis. *Neuropsychopharmacology.* 2001;24(2):97–129. https://doi.org/10.1016/S0893-133X(00)00195-0.
29. Kotas ME, Medzhitov R. Homeostasis, inflammation, and disease susceptibility. *Cell.* 2015;160(5):816–827. https://doi.org/10.1016/j.cell.2015.02.010.
30. McEwen BS, Wingfield JC. The concept of allostasis in biology and biomedicine. *Horm Behav.* 2003;43(1):2–15. https://doi.org/10.1016/s0018-506x(02)00024-7.
31. George O, Le Moal M, Koob GF. Allostasis and addiction: role of the dopamine and corticotropin-releasing factor systems. *Physiol Behav.* 2012;106(1):58–64. https://doi.org/10.1016/j.physbeh.2011.11.004.
32. Duehlmeyer L, Hester R. Impaired learning from punishment of errors in smokers: differences in dorsolateral prefrontal cortex and sensorimotor cortex blood-oxygen-level dependent responses. *NeuroImage. Clinical.* 2019;23:101819. https://doi.org/10.1016/j.nicl.2019.101819.
33. Hammond CJ, Krishnan-Sarin S, Mayes LC, Potenza MN, Crowley MJ. Associations of cannabis- and tobacco-related problem severity with reward and punishment sensitivity and impulsivity in adolescent daily cigarette smokers. *Int J Ment Health Addiction.* 2021;19(6):1963–1979. https://doi.org/10.1007/s11469-020-00292-2.
34. Hester R, Bell RP, Foxe JJ, Garavan H. The influence of monetary punishment on cognitive control in abstinent cocaine-users. *Drug Alcohol Depend.* 2013;133(1):86–93. https://doi.org/10.1016/j.drugalcdep.2013.05.027.
35. Potts GF, Bloom EL, Evans DE, Drobes DJ. Neural reward and punishment sensitivity in cigarette smokers. *Drug Alcohol Depend.* 2014;144:245–253. https://doi.org/10.1016/j.drugalcdep.2014.09.773.
36. Balster RL, Schuster CR. A comparison of d-amphetamine, l-amphetamine, and methamphetamine self-administration in rhesus monkeys. *Pharmacol Biochem Behav.* 1973;1(1):67–71. https://doi.org/10.1016/0091-3057(73)90057-9.
37. Balster RL, Woolverton WL. Continuous-access phencyclidine self-administration by rhesus monkeys leading to physical dependence. *Psychopharmacology.* 1980;70(1):5–10. https://doi.org/10.1007/BF00432363.
38. Broadbear JH, Winger G, Woods JH. Self-administration of fentanyl, cocaine and ketamine: effects on the pituitary-adrenal axis in rhesus monkeys. *Psychopharmacology.* 2004;176(3–4):398–406. https://doi.org/10.1007/s00213-004-1891-x.
39. Deneau G, Yanagita T, Seevers MH. Self-administration of psychoactive substances by the monkey. *Psychopharmacologia.* 1969;16(1):30–48. https://doi.org/10.1007/BF00405254.
40. Johanson CE, Schuster CR. A choice procedure for drug reinforcers: cocaine and methylphenidate in the rhesus monkey. *J Pharmacol Exp Therapeut.* 1975;193(2):676–688.
41. Lang WJ, Latiff AA, McQueen A, Singer G. Self administration of nicotine with and without a food delivery schedule. *Pharmacol, Biochem Behav.* 1977;7(1):65–70. https://doi.org/10.1016/0091-3057(77)90012-0.
42. Lemaire GA, Meisch RA. Pentobarbital self-administration in rhesus monkeys: drug concentration and fixed-ratio size interactions. *J Exp Anal Behav.* 1984;42(1):37–49. https://doi.org/10.1901/jeab.1984.42-37.
43. Marusich JA, Gay EA, Stewart DA, Blough BE. Sex differences in inflammatory cytokine levels following synthetic cathinone self-administration in rats. *Neurotoxicology.* 2022;88:65–78. https://doi.org/10.1016/j.neuro.2021.11.002.
44. Smith SG, Davis WM. Intravenous alcohol self-administration in the rat. *Pharmacol Res Commun.* 1974;6(4):397–402. https://doi.org/10.1016/s0031-6989(74)80039-1.
45. van Ree JM, Slangen JL, de Wied D. Intravenous self-administration of drugs in rats. *J Pharmacol Exp Therapeut.* 1978;204(3):547–557.
46. Weeks JR. Experimental morphine addiction: method for automatic intravenous injections in unrestrained rats. *Science.* 1962;138(3537):143–144. https://doi.org/10.1126/science.138.3537.143.
47. Panlilio LV, Goldberg SR. Self-administration of drugs in animals and humans as a model and an investigative tool. *Addiction.* 2007;102(12):1863–1870. https://doi.org/10.1111/j.1360-0443.2007.02011.x.

48. Corrigall WA, Coen KM. Nicotine maintains robust self-administration in rats on a limited-access schedule. *Psychopharmacology*. 1989;99(4):473–478. https://doi.org/10.1007/BF00589894.
49. Strickland JC, Abel JM, Lacy RT, et al. The effects of resistance exercise on cocaine self-administration, muscle hypertrophy, and BDNF expression in the nucleus accumbens. *Drug Alcohol Depend*. 2016;163:186–194. https://doi.org/10.1016/j.drugalcdep.2016.04.019.
50. Blegen MB, da Silva E Silva D, Bock R, Morisot N, Ron D, Alvarez VA. Alcohol operant self-administration: investigating how alcohol-seeking behaviors predict drinking in mice using two operant approaches. *Alcohol*. 2018;67:23–36. https://doi.org/10.1016/j.alcohol.2017.08.008.
51. Tolliver GA, Sadeghi KG, Samson HH. Ethanol preference following the sucrose-fading initiation procedure. *Alcohol*. 1988;5(1):9–13. https://doi.org/10.1016/0741-8329(88)90036-5.
52. Ciccocioppo R. Genetically selected alcohol preferring rats to model human alcoholism. *Current Topics in Behavioral Neurosciences*. 2013;13:251–269. https://doi.org/10.1007/7854_2012_199.
53. Crabbe JC. Sensitivity to ethanol in inbred mice: genotypic correlations among several behavioral responses. *Behav Neurosci*. 1983;97(2):280–289. https://doi.org/10.1037//0735-7044.97.2.280.
54. Ahmed SH, Koob GF. Transition from moderate to excessive drug intake: change in hedonic set point. *Science*. 1998;282(5387):298–300. https://doi.org/10.1126/science.282.5387.298.
55. Spanagel R, Holter SM, Allingham K, Landgraf R, Zieglgansberger W. Acamprosate and alcohol: I. Effects on alcohol intake following alcohol deprivation in the rat. *Eur J Pharmacol*. 1996;305(1–3):39–44. https://doi.org/10.1016/0014-2999(96)00174-4.
56. Allain F, Bouayad-Gervais K, Samaha A-N. High and escalating levels of cocaine intake are dissociable from subsequent incentive motivation for the drug in rats. *Psychopharmacology*. 2018;235(1):317–328. https://doi.org/10.1007/s00213-017-4773-8.
57. Cocker PJ, Rotge J-Y, Daniel M-L, Belin-Rauscent A, Belin D. Impaired decision making following escalation of cocaine self-administration predicts vulnerability to relapse in rats. *Addiction Biol*. 2020;25(3):e12738. https://doi.org/10.1111/adb.12738.
58. Guillem K, Ahmed SH, Peoples LL. Escalation of cocaine intake and incubation of cocaine seeking are correlated with dissociable neuronal processes in different accumbens subregions. *Biol Psychiatr*. 2014;76(1):31–39. https://doi.org/10.1016/j.biopsych.2013.08.032.
59. Mandt BH, Copenhagen LI, Zahniser NR, Allen RM. Escalation of cocaine consumption in short and long access self-administration procedures. *Drug Alcohol Depend*. 2015;149:166–172. https://doi.org/10.1016/j.drugalcdep.2015.01.039.
60. Anker JJ, Baron TR, Zlebnik NE, Carroll ME. Escalation of methamphetamine self-administration in adolescent and adult rats. *Drug Alcohol Depend*. 2012;124(1–2):149–153. https://doi.org/10.1016/j.drugalcdep.2012.01.004.
61. Le Cozannet R, Markou A, Kuczenski R. Extended-access, but not limited-access, methamphetamine self-administration induces behavioral and nucleus accumbens dopamine response changes in rats. *Eur J Neurosci*. 2013;38(10):3487–3495. https://doi.org/10.1111/ejn.12361.
62. Rogers JL, De Santis S, See RE. Extended methamphetamine self-administration enhances reinstatement of drug seeking and impairs novel object recognition in rats. *Psychopharmacology*. 2008;199(4):615–624. https://doi.org/10.1007/s00213-008-1187-7.
63. Gipson CD, Bardo MT. Extended access to amphetamine self-administration increases impulsive choice in a delay discounting task in rats. *Psychopharmacology*. 2009;207(3):391–400. https://doi.org/10.1007/s00213-009-1667-4.
64. Marusich JA, Beckmann JS, Gipson CD, Bardo MT. Methylphenidate as a reinforcer for rats: contingent delivery and intake escalation. *Exp Clin Psychopharmacol*. 2010;18(3):257–266. https://doi.org/10.1037/a0019814.
65. Greenwell TN, Funk CK, Cottone P, et al. Corticotropin-releasing factor-1 receptor antagonists decrease heroin self-administration in long- but not short-access rats. *Addiction Biol*. 2009;14(2):130–143. https://doi.org/10.1111/j.1369-1600.2008.00142.x.
66. Picetti R, Caccavo JA, Ho A, Kreek MJ. Dose escalation and dose preference in extended-access heroin self-administration in Lewis and Fischer rats. *Psychopharmacology*. 2012;220(1):163–172. https://doi.org/10.1007/s00213-011-2464-4.

67. Towers EB, Tunstall BJ, McCracken ML, Vendruscolo LF, Koob GF. Male and female mice develop escalation of heroin intake and dependence following extended access. *Neuropharmacology*. 2019;151:189−194. https://doi.org/10.1016/j.neuropharm.2019.03.019.
68. Zhang Y, Mayer-Blackwell B, Schlussman SD, et al. Extended access oxycodone self-administration and neurotransmitter receptor gene expression in the dorsal striatum of adult C57BL/6J mice. *Psychopharmacology*. 2014;231(7):1277−1287. https://doi.org/10.1007/s00213-013-3306-3.
69. Paterson NE, Markou A. Prolonged nicotine dependence associated with extended access to nicotine self-administration in rats. *Psychopharmacology*. 2004;173(1−2):64−72. https://doi.org/10.1007/s00213-003-1692-7.
70. Cohen A, Koob GF, George O. Robust escalation of nicotine intake with extended access to nicotine self-administration and intermittent periods of abstinence. *Neuropsychopharmacology*. 2012;37(9):2153−2160. https://doi.org/10.1038/npp.2012.67.
71. Allain F, Samaha A-N. Revisiting long-access versus short-access cocaine self-administration in rats: intermittent intake promotes addiction symptoms independent of session length. *Addiction Biol*. 2019;24(4):641−651. https://doi.org/10.1111/adb.12629.
72. Beckmann JS, Gipson CD, Marusich JA, Bardo MT. Escalation of cocaine intake with extended access in rats: dysregulated addiction or regulated acquisition? *Psychopharmacology*. 2012;222(2):257−267. https://doi.org/10.1007/s00213-012-2641-0.
73. Richardson NR, Roberts DC. Progressive ratio schedules in drug self-administration studies in rats: a method to evaluate reinforcing efficacy. *J Neurosci Methods*. 1996;66(1):1−11. https://doi.org/10.1016/0165-0270(95)00153-0.
74. Hodos W. Progressive ratio as a measure of reward strength. *Science*. 1961;134(3483):943−944. https://doi.org/10.1126/science.134.3483.943.
75. Panlilio LV, Solinas M, Matthews SA, Goldberg SR. Previous exposure to THC alters the reinforcing efficacy and anxiety-related effects of cocaine in rats. *Neuropsychopharmacology*. 2007;32(3):646−657. https://doi.org/10.1038/sj.npp.1301109.
76. Oleson EB, Richardson JM, Roberts DCS. A novel IV cocaine self-administration procedure in rats: differential effects of dopamine, serotonin, and GABA drug pre-treatments on cocaine consumption and maximal price paid. *Psychopharmacology*. 2011;214(2):567−577. https://doi.org/10.1007/s00213-010-2058-6.
77. Yates JR, Bardo MT, Beckmann JS. Environmental enrichment and drug value: a behavioral economic analysis in male rats. *Addiction Biol*. 2019;24(1):65−75. https://doi.org/10.1111/adb.12581.
78. Sauter MB. Price of a pack of cigarettes through the decades. https://247wallst.com/special-report/2019/06/19/price-of-a-pack-of-cigarettes-through-the-decades/3/; June 19, 2019.
79. Al-Tayyib AA, Koester S, Riggs P. Prescription opioids prior to injection drug use: comparisons and public health implications. *Addict Behav*. 2017;65:224−228. https://doi.org/10.1016/j.addbeh.2016.08.016.
80. Mahler SV, Brodnik ZD, Cox BM, et al. Chemogenetic manipulations of ventral tegmental area dopamine neurons reveal multifaceted roles in cocaine abuse. *J Neurosci*. 2019;39(3):503−518. https://doi.org/10.1523/JNEUROSCI.0537-18.2018.
81. Thomsen M, Barrett AC, Negus SS, Caine SB. Cocaine versus food choice procedure in rats: environmental manipulations and effects of amphetamine. *J Exp Anal Behav*. 2013;99(2):211−233. https://doi.org/10.1002/jeab.15.
82. Goldberg SR. Comparable behavior maintained under fixed-ratio and second-order schedules of food presentation, cocaine injection or d-amphetamine injection in the squirrel monkey. *J Pharmacol Exp Therapeut*. 1973;186(1):18−30.
83. Whitelaw RB, Markou A, Robbins TW, Everitt BJ. Excitotoxic lesions of the basolateral amygdala impair the acquisition of cocaine-seeking behaviour under a second-order schedule of reinforcement. *Psychopharmacology*. 1996;127(3):213−224.
84. Vanderschuren LJMJ, Everitt BJ. Drug seeking becomes compulsive after prolonged cocaine self-administration. *Science*. 2004;305(5686):1017−1019. https://doi.org/10.1126/science.1098975.
85. Holtz NA, Carroll ME. Cocaine self-administration punished by intravenous histamine in adolescent and adult rats. *Behav Pharmacol*. 2015;26(4):393−397. https://doi.org/10.1097/FBP.0000000000000136.
86. Jones JD, Comer SD. A review of human drug self-administration procedures. *Behav Pharmacol*. 2013;24(0):384−395. https://doi.org/10.1097/FBP.0b013e3283641c3d.
87. Donny EC, Bigelow GE, Walsh SL. Choosing to take cocaine in the human laboratory: effects of cocaine dose, inter-choice interval, and magnitude of alternative reinforcement. *Drug Alcohol Depend*. 2003;69(3):289−301. https://doi.org/10.1016/s0376-8716(02)00327-7.

88. Foltin RW, Fischman MW. Self-administration of cocaine by humans: choice between smoked and intravenous cocaine. *J Pharmacol Exp Therapeut*. 1992;261(3):841–849.
89. Foltin RW, Fischman MW. Effects of buprenorphine on the self-administration of cocaine by humans. *Behav Pharmacol*. 1994;5(1):79–89. https://doi.org/10.1097/00008877-199402000-00009.
90. Lynch WJ, Sughondhabriom A, Pittman B, et al. A paradigm to investigate the regulation of cocaine self-administration in human cocaine users: a randomized trial. *Psychopharmacology*. 2006;185(3):306–314. https://doi.org/10.1007/s00213-006-0323-5.
91. Stoops WW, Lile JA, Glaser PE, Hays LR, Rush CR. Intranasal cocaine functions as reinforcer on a progressive ratio schedule in humans. *Eur J Pharmacol*. 2010;644(1–3):101–105. https://doi.org/10.1016/j.ejphar.2010.06.055.
92. Kirkpatrick MG, Gunderson EW, Johanson C-E, Levin FR, Foltin RW, Hart CL. Comparison of intranasal methamphetamine and d-amphetamine self-administration by humans. *Addiction*. 2012;107(4):783–791. https://doi.org/10.1111/j.1360-0443.2011.03706.x.
93. Vansickel AR, Stoops WW, Rush CR. Human sex differences in D-amphetamine self-administration. *Addiction*. 2010;105(4):727–731. https://doi.org/10.1111/j.1360-0443.2009.02858.x.
94. Barrett SP, Tichauer M, Leyton M, Pihl RO. Nicotine increases alcohol self-administration in non-dependent male smokers. *Drug Alcohol Depend*. 2006;81(2):192–204. https://doi.org/10.1016/j.drugalcdep.2005.06.009.
95. Mello NK, Mendelson JH. Operant analysis of drinking patterns of chronic alcoholics. *Nature*. 1965;206:43–46. https://doi.org/10.1038/206043a0.
96. Comer SD, Collins ED, Fischman MW. Choice between money and intranasal heroin in morphine-maintained humans. *Behav Pharmacol*. 1997;8(8):677–690. https://doi.org/10.1097/00008877-199712000-00002.
97. Comer SD, Collins ED, MacArthur RB, Fischman MW. Comparison of intranasal and intravenous heroin self-administration by morphine-maintained humans. *Psychopharmacology*. 1999;143(4):327–338. https://doi.org/10.1007/s002130050956.
98. Lamb RJ, Preston KL, Schindler CW, et al. The reinforcing and subjective effects of morphine in post-addicts: a dose-response study. *J Pharmacol Exp Therapeut*. 1991;259(3):1165–1173.
99. Griffiths RR, Bigelow GE, Liebson IA. Facilitation of human tobacco self-administration by ethanol: a behavioral analysis. *J Exp Anal Behav*. 1976;25(3):279–292. https://doi.org/10.1901/jeab.1976.25-279.
100. Griffiths RR, Rush CR, Puhala KA. Validation of the multiple-choice procedure for investigating drug reinforcement in humans. *Exp Clin Psychopharmacol*. 1996;4(1):97–106. https://doi.org/10.1037/1064-1297.4.1.97.
101. Sofuoglu M, Yoo S, Hill KP, Mooney M. Self-administration of intravenous nicotine in male and female cigarette smokers. *Neuropsychopharmacology*. 2008;33(4):715–720. https://doi.org/10.1038/sj.npp.1301460.
102. Griffiths RR, Troisi JR, Silverman K, Mumford GK. Multiple-choice procedure: an efficient approach for investigating drug reinforcement in humans. *Behav Pharmacol*. 1993;4(1):3–13.
103. Pike E, Stoops WW, Hays LR, Glaser PEA, Rush CR. Methamphetamine self-administration in humans during D-amphetamine maintenance. *J Clin Psychopharmacol*. 2014;34(6):675–681. https://doi.org/10.1097/JCP.0000000000000207.
104. Young R. Chapter 3. Drug discrimination. In: Buccafusco JJ, ed. *Methods of Behavior Analysis in Neuroscience*. 2nd ed. Boca Raton: CRC Press/Taylor & Francis; 2009. https://www.ncbi.nlm.nih.gov/books/NBK5225/.
105. Colpaert FC, Niemegeers CJE, Janssen PAJ. Discriminative stimulus properties of cocaine and *d*-amphetamine, and antagonism by haloperidol: a comparative study. *Neuropharmacology*. 1978;17(11):937–942. https://doi.org/10.1016/0028-3908(78)90135-1.
106. Wiley JL, Martin BR. Effects of SR141716A on diazepam substitution for Δ^9-tetrahydrocannabinol in rat drug discrimination. *Pharmacol, Biochem Behav*. 1999;64(3):519–522. https://doi.org/10.1016/S0091-3057(99)00130-6.
107. Bolin BL, Alcorn III JL, Reynolds AR, Lile JA, Rush CR. Human drug discrimination: a primer and methodological review. *Exp Clin Psychopharmacol*. 2016;24(4):214–228. https://doi.org/10.1037/pha0000077.
108. Ranaldi R, Wise RA. Blockade of D1 dopamine receptors in the ventral tegmental area decreases cocaine reward: possible role for dendritically released dopamine. *J Neurosci*. 2001;21(15):5841–5846. https://doi.org/10.1523/JNEUROSCI.21-15-05841.2001.
109. Hubner CB, Moreton JE. Effects of selective D1 and D2 dopamine antagonists on cocaine self-administration in the rat. *Psychopharmacology*. 1991;105(2):151–156. https://doi.org/10.1007/BF02244301.

110. Di Ciano P, Underwood RJ, Hagan JJ, Everitt BJ. Attenuation of cue-controlled cocaine-seeking by a selective D3 dopamine receptor antagonist SB-277011-A. *Neuropsychopharmacology*. 2003;28(2):329−338. https://doi.org/10.1038/sj.npp.1300148.
111. de Jong JW, Roelofs TJ, et al. Reducing ventral tegmental dopamine D2 receptor expression selectively boosts incentive motivation. *Neuropsychopharmacology*. 2015;40(9):2085−2095. https://doi.org/10.1038/npp.2015.60.
112. Brodsky M, Gibson AW, Smirnov D, Nair SG, Neumaier JF. Striatal 5-HT6 receptors regulate cocaine reinforcement in a pathway-selective manner. *Neuropsychopharmacology*. 2016;41(9):2377−2387. https://doi.org/10.1038/npp.2016.45.
113. Sun WL, Zelek-Molik A, McGinty JF. Short and long access to cocaine self-administration activates tyrosine phosphatase STEP and attenuates GluN expression but differentially regulates GluA expression in the prefrontal cortex. *Psychopharmacology*. 2013;229(4):603−613. https://doi.org/10.1007/s00213-013-3118-5.
114. Guillaumier A, Bonevski B, Paul C. 'Cigarettes are priority': a qualitative study of how Australian socioeconomically disadvantaged smokers respond to rising cigarette prices. *Health Educ Res*. 2015;30(4):599−608. https://doi.org/10.1093/her/cyv026.
115. Rash CJ, Petry NM, Alessi SM. A randomized trial of contingency management for smoking cessation in the homeless. *Psychol Addict Behav*. 2018;32(2):141−148. https://doi.org/10.1037/adb0000350.
116. Stitzer ML, Bigelow GE. Contingent reinforcement for reduced breath carbon monoxide levels: target-specific effects on cigarette smoking. *Addict Behav*. 1985;10(4):345−349. https://doi.org/10.1016/0306-4603(85)90030-9.
117. Dougherty DM, Lake SL, Hill-Kapturczak N, et al. Using contingency management procedures to reduce at-risk drinking in heavy drinkers. *Alcohol Clin Exp Res*. 2015;39(4):743−751. https://doi.org/10.1111/acer.12687.
118. Petry NM, Martin B, Cooney JL, Kranzler HR. Give them prizes, and they will come: contingency management for treatment of alcohol dependence. *J Consult Clin Psychol*. 2000;68(2):250−257. https://doi.org/10.1037//0022-006x.68.2.250.
119. Higgins ST, Budney AJ, Bickel WK, Foerg FE, Donham R, Badger GJ. Incentives improve outcome in outpatient behavioral treatment of cocaine dependence. *Arch Gen Psychiatr*. 1994;51(7):568−576. https://doi.org/10.1001/archpsyc.1994.03950070060011.
120. Johnson MW, Bruner NR, Johnson PS, Silverman K, Berry MS. Randomized controlled trial of d-cycloserine in cocaine dependence: effects on contingency management and cue-induced cocaine craving in a naturalistic setting. *Exp Clin Psychopharmacol*. 2020;28(2):157−168. https://doi.org/10.1037/pha0000306.
121. Shoptaw S, Huber A, Peck J, Yang X, Liu J, Dang J, Roll J, Shapiro B, Rotheram-Fuller E, Ling W. Randomized, placebo-controlled trial of sertraline and contingency management for the treatment of methamphetamine dependence. *Drug Alcohol Depend*. 2006;85(1):12−18. https://doi.org/10.1016/j.drugalcdep.2006.03.005.
122. Carroll KM, Ball SA, Nich C, et al. Targeting behavioral therapies to enhance naltrexone treatment of opioid dependence. *Arch Gen Psychiatr*. 2001;58(8):755−761. https://doi.org/10.1001/archpsyc.58.8.755.
123. Jarvis BP, Holtyn AF, DeFulio A, et al. The effects of extended-release injectable naltrexone and incentives for opiate abstinence in heroin-dependent adults in a model therapeutic workplace: a randomized trial. *Drug Alcohol Depend*. 2019;197:220−227. https://doi.org/10.1016/j.drugalcdep.2018.12.026.
124. Schuster RM, Hanly A, Gilman J, Budney A, Vandrey R, Evins AE. A contingency management method for 30-days abstinence in non-treatment seeking yound adult cannabis users. *Drug Alcohol Depend*. 2016;167:199−206. https://doi.org/10.1016/j.drugalcdep.2016.08.622.
125. Stanger C, Ryan SR, Scherer EA, Norton GE, Budney AJ. Clinic- and home-based contingency management plus parent training for adolescent cannabis use disorders. *J Am Acad Child Adolesc Psychiatr*. 2015;54(6):445−453.e2. https://doi.org/10.1016/j.jaac.2015.02.009.
126. Silverman K, Higgins ST, Brooner RK, et al. Sustained cocaine abstinence in methadone maintenance patients through voucher-based reinforcement therapy. *Arch Gen Psychiatr*. 1996;53(5):409−415. https://doi.org/10.1001/archpsyc.1996.01830050045007.
127. Meyers RJ, Roozen HG, Smith JE. The community reinforcement approach: an update of the evidence. *Alcohol Res Health*. 2011;33(4):380−388.
128. Miller RJ, Meyers RJ, Hiller-Sturmhöfel S. The community-reinforcement approach. *Alcohol Res Health*. 1999;23(2):116−121.
129. Hunt GM, Azrin NH. A community-reinforcement approach to alcoholism. *Behav Res Ther*. 1973;11(1):91−104. https://doi.org/10.1016/0005-7967(73)90072-7.

130. Abbott PJ, Weller SB, Delaney HD, Moore BA. Community reinforcement approach in the treatment of opiate addicts. *Am J Drug Alcohol Abuse*. 1998;24(1):17−30. https://doi.org/10.3109/00952999809001696.
131. Azrin NH, McMahon PT, Donohue B, et al. Behavior therapy for drug abuse: a controlled treatment outcome study. *Behav Res Ther*. 1994;32(8):857−866. https://doi.org/10.1016/0005-7967(94)90166-x.
132. Azrin NH, Donohue B, Teichner GA, Crum T, Howell J, DeCato LA. A controlled evaluation and description of individual-cognitive problem solving and family-behavior therapies in dually diagnosed conduct-disordered and substance-dependent youth. *J Child Adolesc Subst Abuse*. 2001;11(1):1−43. https://doi.org/10.1300/J029v11n01_01.
133. Roozen HG, Van Beers SEC, Weevers HJA, et al. Effects on smoking cessation: naltrexone combined with a cognitive behavioral treatment based on the community reinforcement approach. *Subst Use Misuse*. 2006;41(1):45−60. https://doi.org/10.1080/10826080500318665.
134. Secades-Villa R, Sánchez-Hervás E, Zacarés-Romaguera F, García-Rodríguez O, Santonja-Gómez FJ, García-Fernández G. Community reinforcement approach (CRA) for cocaine dependence in the Spanish public health system: 1 year outcome. *Drug Alcohol Rev*. 2011;30(6):606−612. https://doi.org/10.1111/j.1465-3362.2010.00250.x.
135. Slesnick N, Guo X, Brakenhoff B, Bantchevska D. A comparison of three interventions for homeless youth evidencing substance use disorders: results of a randomized clinical trial. *J Subst Abuse Treat*. 2015;54:1−13. https://doi.org/10.1016/j.jsat.2015.02.001.
136. Zhang J, Slesnick N. Substance use and social stability of homeless youth: a comparison of three interventions. *Psychol Addict Behav*. 2018;32(8):873−884. https://doi.org/10.1037/adb0000424.
137. Garcia-Rodriguez O, Secades-Villa R, Higgins ST, et al. Effects of voucher-based intervention on abstinence and retention in an outpatient treatment for cocaine addiction: a randomized controlled trial. *Exp Clin Psychopharmacol*. 2009;17(3):131−138. https://doi.org/10.1037/a0015963.
138. Higgins ST, Delaney DD, Budney AJ, et al. A behavioral approach to achieving initial cocaine abstinence. *Am J Psychiatr*. 1991;148(9):1218−1224. https://doi.org/10.1176/ajp.148.9.1218.
139. Schottenfeld RS, Moore B, Pentalon MV. Contingency management with community reinforcement approach or twelve-step facilitation drug counseling for cocaine dependent pregnant women or women with young children. *Drug Alcohol Depend*. 2011;118(1):48−55. https://doi.org/10.1016/j.drugalcdep.2011.02.019.
140. Meyers RJ, Miller WR, Hill DE, Tonigan JS. Community reinforcement and family training (CRAFT): engaging unmotivated drug users in treatment. *J Subst Abuse*. 1998;10(3):291−308. https://doi.org/10.1016/S0899-3289(99)00003-6.
141. Meyers RJ, Miller WR, Smith JE, Tonigan JS. A randomized trial of two methods for engaging treatment-refusing drug users through concerned significant others. *J Consult Clin Psychol*. 2002;70(5):1182−1185. https://doi.org/10.1037/0022-006X.70.5.1182.
142. Manuel JK, Austin JL, Miller WR, et al. Community Reinforcement and Family Training: a pilot comparison of group and self-directed delivery. *J Subst Abuse Treat*. 2012;43(1):129−136. https://doi.org/10.1016/j.jsat.2011.10.020.
143. Archer M, Harwood H, Stevelink S, Rafferty L, Greenberg N. Community reinforcement and family training and rates of treatment entry: a systematic review. *Addiction*. 2020;115(6):1024−1037. https://doi.org/10.1111/add.14901.
144. Davis JP, Prindle JJ, Eddie D, Pedersen ER, Dumas TM, Christie NC. Addressing the opioid epidemic with behavioral interventions for adolescents and young adults: a quasi-experimental design. *J Consult Clin Psychol*. 2019;87(10):941−951. https://doi.org/10.1037/ccp0000406.
145. Kahneman D, Tversky A. Prospect theory: an analysis of decision under risk. *Econometrica*. 1979;47(2):263−291. https://doi.org/10.2307/1914185.
146. Belenko S, Peugh J. Estimating drug treatment needs among state prison inmates. *Drug Alcohol Depend*. 2005;77(3):269−281. https://doi.org/10.1016/j.drugalcdep.2004.08.023.
147. Chamberlain A, Nyamu S, Aminawung J, Wang EA, Shavit S, Fox AD. Illicit substance use after release from prison among formerly incarcerated primary care patients: a cross-sectional study. *Addiction Sci Clin Pract*. 2019;14(1):7. https://doi.org/10.1186/s13722-019-0136-6.
148. Inciardi JA, Martin SS, Butzin CA. Five-year outcomes of therapeutic community treatment of drug-involved offenders after release from prison. *Crime Delinquen*. 2004;50(1):88−107. https://doi.org/10.1177/0011128703258874.

149. Kinlock TW, Gordon MS, Schwartz RP, O'Grady KE. A study of methadone maintenance for male prisoners: 3-month postrelease outcomes. *Crim Justice Behav*. 2008;35(1):34–47. https://doi.org/10.1177/0093854807309111.
150. Malouf E, Stuewig J, Tangney J. Self-control and jail inmates' substance misuse post-release: meditation by friends' substance use and moderation by age. *Addict Behav*. 2012;37(11):1198–1204. https://doi.org/10.1016/j.addbeh.2012.05.013.
151. Chandler R, Fletcher B, Volkow ND. Treating drug abuse and addiction in the criminal justice system: improving public health and safety. *JAMA, J Am Med Assoc*. 2009;301(2):183–190. https://doi.org/10.1001/jama.2008.976.
152. Bruer W, Gallagher D. This former prisoner had an unlikely supporter: the judge who sentence him. *CNN*; February 1, 2021. https://www.cnn.com/2021/02/01/us/chris-young-freed-with-help-from-judge-kevin-sharp/index.html.

CHAPTER 6

Learning mechanisms of addiction: Pavlovian conditioning

Introduction

> I used heroin for around 5 years ... Life, in general, was completely miserable ... the dependency was unbearable. Feeling like I need a substance to live ... was such a gross feeling.[1]

Andrew Warwick was first introduced to heroin at 20 years of age and initially used it to help self-medicate his bipolar disorder. At first, he enjoyed the effects of heroin, describing it as "slightly euphoric".[1] With continued drug use, Andrew noticed that he no longer enjoyed using the drug as he felt miserable. However, in Andrew's words, he felt like he needed heroin to survive. In the last chapter, you learned about the role of operant conditioning in maintaining drug-taking behaviors. In this chapter, you will learn about another learning phenomenon that drives continued drug use despite its negative consequences: **Pavlovian conditioning**. Specifically related to this form of learning, you will learn how certain stimuli in our environment become paired with drug use. These cues can elicit strong drug cravings even when the individual is not particularly interested in using the drug. You will learn more about the dissociation between "liking" and "wanting" (or "needing") a drug. Andrew is not alone when stating that he felt like he needed heroin to live. When many individuals with substance use disorders (SUDs) are interviewed, a common theme is that they no longer enjoy using the drug, but they feel they need the drug to function during the day.

Learning objectives

By the end of this chapter, you should be able to ...

(1) Differentiate unconditioned stimuli/responses from conditioned stimuli/responses.
(2) Discuss the influence of Pavlovian conditioning on operant conditioning.
(3) Explain how Pavlovian conditioning contributes to addiction.
(4) Describe the compensatory response and the incentive sensitization theories of addiction.

(5) Explain how paradigms such as Pavlovian conditioned approach and conditioned place preference are used to study addiction.
(6) Compare and contrast treatment options that utilize Pavlovian conditioning in clinical populations.

Overview of Pavlov's study

Dr. Ivan Pavlov (1849—1936) is one of the most heavily discussed figures in Psychology, but funnily enough, he was not a psychologist. He was a physiologist interested in studying digestion. In fact, he received the Nobel Prize in the Physiology or Medicine category in 1904 for his work on the digestive system. In his work, Pavlov often used dogs as subjects. Pavlov would make an incision in the dog's cheek, which would allow him to externalize the salivary gland. This then allowed Pavlov to collect salivary secretions when dogs were presented with different types of food, as well as other substances like marbles or sand. During his studies, Pavlov noticed something strange—dogs started to salivate when they heard Pavlov's assistant walking toward them to deliver food. Pavlov, being the good scientist that he was, decided to conduct a study to investigate what he originally called "psychic secretions".[2]

In what is now one of the most famous Psychology experiments ever conducted, Pavlov first presented dogs food by itself. Not surprisingly, the dogs salivated at the sight of the food. Next, Pavlov presented a metronome, a device that produces an audible clicking sound at regular intervals, by itself. The metronome failed to elicit salivation in the dogs. Pavlov then presented the metronome before presenting the food. Because dogs had access to the food, they started salivating again. Pavlov repeated the metronome-food pairings across multiple sessions. Eventually, Pavlov presented the metronome by itself. The dogs began to salivate at the sight/sound of the metronome! The dogs had learned to associate the metronome with the food.[2] Learning to associate two stimuli with each other is now known as Pavlovian (or classical) conditioning. Pavlovian conditioning is also known as stimulus-stimulus (S—S) learning.

The key distinction between Pavlovian conditioning and operant conditioning is the type of association that is formed. In Pavlovian conditioning, organisms learn to associate two stimuli together, regardless of the behavior that is occurring. In operant conditioning, an organism learns to associate a specific behavior with an outcome. Even though these two phenomena measure distinct aspects of learning, they often work in tandem to influence behaviors. We will come back to this topic later in this chapter. But first, I need to cover some basic principles of Pavlovian conditioning.

Principles of Pavlovian conditioning

In Pavlov's study, food naturally produced saliva in the dog. Stimuli that elicit a natural, unlearned response from an organism are known as **unconditioned stimuli (US)**. In other words, animals do not have to learn to produce saliva when presented with food. Salivating in response to food is an example of an **unconditioned response (UR)**. Unconditioned responses are those that are unlearned or occur naturally in response to a specific stimulus.

Before conditioning, the metronome was a *neutral* stimulus that did not trigger a response from the animal. Because the animal associated the metronome with food over repeated pairings, the metronome eventually elicited salivation. The metronome became a **conditioned stimulus (CS)**. Salivating in response to the metronome is a **conditioned response (CR)**. Unlike URs, learning must occur for a CR to manifest. The dogs in Pavlov's experiment learned that the presence of the metronome meant that they were about to eat; therefore, they started salivating in anticipation of the food. Fig. 6.1 shows a depiction of Pavlov's seminal study.

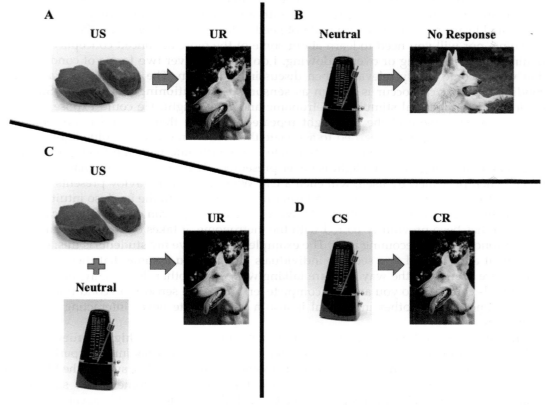

FIGURE 6.1 Process of Pavlovian conditioning. (A) Presentation of an unconditioned stimulus (US) such as food elicits a natural response, or unconditioned response (UR), from an animal such as a dog. (B) The presentation of a neutral stimulus like a metronome does not elicit a response from the organism other than an orienting response. (C) When the US is paired with the neutral stimulus, the dog will show a UR. The UR occurs because of the presence of the US. (D) After repeated pairings of the US and the neutral stimulus, the neutral stimulus becomes a conditioned stimulus (CS) that elicits a conditioned response (CR). *The figure was created by the author using Microsoft PowerPoint. The image of the steaks was obtained through Wikimedia Commons and was posted by Venison Steaks (https://commons.wikimedia.org/wiki/File:Venison_Steaks.jpg). The image of the drooling dog was obtained through Wikimedia Commons and was posted by Ildar Sagdejev (https://commons.wikimedia.org/wiki/File:2008-08-12_White_German_Shepherd_drooling_in_the_sun.jpg). The image of the metronome was obtained through Wikimedia Commons and was posted by Vincent Quach (https://commons.wikimedia.org/wiki/File:Metronome_Nikko.jpg). The image of the dog playing with the ball was obtained through Pexels and was posted by Daniel Bendig (https://www.pexels.com/photo/photography-of-a-dog-biting-green-tennis-ball-931876).*

Based on his findings, Pavlov developed a theory of conditioning that posited that CSs take on the same properties as the US and that CRs are functionally the same as URs.[2] Through conditioning, the metronome took on the same properties as the food, which is why dogs salivated in response to the food and to the metronome following conditioning. This is known as the **stimulus substitution theory of conditioning**.

Pavlovian conditioning gets more complicated than what was presented above. In my Animal Learning course, students often assume the section on Pavlovian conditioning will be easy because they have heard the story of Pavlov several times before enrolling in my class. To their surprise, they soon realize that Pavlovian conditioning is so much more than dogs drooling when presented with a metronome. Luckily for you, this is not an Animal Learning course, so you will not need to learn the mathematical theories of conditioning, nor will you need to learn about some of the more advanced concepts of conditioning such as blocking or overshadowing. I do want to cover two forms of conditioning that will become highly relevant when discussing addiction. The first type of Pavlovian conditioning that can occur is known as **sensory preconditioning**[3] (Fig. 6.2A). Suppose Pavlov had two neutral stimuli: a metronome and a blue light. He could expose dogs to both the metronome and the blue light repeatedly. Notice that no food is presented at this point. After dogs have learned to associate the metronome with the blue light, Pavlov could begin conditioning the dog to associate the metronome with the food. During this stage of conditioning, the blue light is never presented. By now, you know that the metronome, when presented by itself, will elicit a CR. What happens if Pavlov presents the blue light by itself? The blue light has never been directly paired with the food, so intuition says that the blue light should do nothing. However, the blue light can elicit a CR. Because the blue light has been previously paired with the metronome, it takes on the same features as the metronome, thus becoming a CS. The example I like to give my students is this: imagine you are at a party, and you see two individuals talking. You assume that these two individuals are friends by the way they are talking with one another. Eventually, one of these individuals comes up to you and is a complete jerk. Through sensory preconditioning, you may assume that the other individual is also a jerk, despite never interacting with this individual.

Another advanced type of Pavlovian conditioning that can occur is **higher-order (second-order) conditioning**[4] (Fig. 6.2B). Imagine after Pavlov completed his initial experiment, he wanted to begin pairing the metronome with another neutral stimulus, such as the blue light I used in the sensory preconditioning example above. The metronome at this stage is an established CS. As Pavlov begins pairing the CS with the blue light, food is never presented. Eventually, the blue light is presented on its own. Amazingly, the light can elicit a CR even though it was never paired with food. Like sensory preconditioning, the blue light, through its pairings with the metronome, begins to take on the same psychophysical properties as the metronome, thus eliciting a CR. To give you another example: imagine that you go through a bad breakup with someone. Each time you now see this person, you may have feelings of anger or sadness. Eventually, you notice that your ex begins hanging out with a new friend (not a new romantic partner, just a friend). Through higher-order conditioning, you may begin to develop feelings of resentment toward this person even though you have never met them. They could be the nicest person in the world, but because they are now associated with your ex, you begin to equate this person with your ex.

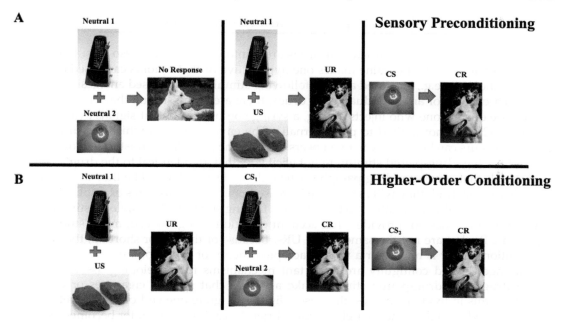

FIGURE 6.2 (A) An example of sensory preconditioning. In the left panel, two neutral stimuli are repeatedly paired in the absence of an unconditioned stimulus (US). The presentation of these two stimuli will not elicit a response from the animal. In the middle panel, only one neutral stimulus (in this case, a metronome) is paired with the US. Because the meat (US) is presented, the animal will salivate, which is an unconditioned response (UR). After repeated pairings of the metronome and the food, the researcher can present the second neutral stimulus, the blue light, by itself (depicted in the right panel). Even though the light was never explicitly paired with the meat, the light becomes an effective conditioned stimulus (CS), which elicits a conditioned response (CR). (B) An example of higher-order conditioning. In the left panel, a neutral stimulus is paired with a US as in a typical Pavlovian conditioning experiment. Over time, the metronome becomes an effective CS due to its repeated pairing with the US. In the middle panel, the CS is now paired with a second neutral stimulus. In the right panel, the second neutral stimulus acts as a second CS due to its repeated pairings with the first CS. Notice that sensory preconditioning and higher-order conditioning have almost identical steps. The key difference is the timing of when the two neutral stimuli are paired together. In sensory preconditioning, the neutral stimuli are paired together before one of them is paired with the US. In higher-order conditioning, one neutral stimulus (now a CS) is paired with a second neutral stimulus after the first one has been paired with the US. *The figure was created by the author using Microsoft PowerPoint. The image of the steaks was obtained through Wikimedia Commons and was posted by Venison Steaks (https://commons.wikimedia.org/wiki/File:Venison_Steaks.jpg). The image of the drooling dog was obtained through Wikimedia Commons and was posted by Ildar Sagdejev (https://commons.wikimedia.org/wiki/File:2008-08-12_White_German_Shepherd_drooling_in_the_sun.jpg). The image of the metronome was obtained through Wikimedia Commons and was posted by Vincent Quach (https://commons.wikimedia.org/wiki/File:Metronome_Nikko.jpg). The image of the dog playing with the ball was obtained through Pexels and was posted by Daniel Bendig (https://www.pexels.com/photo/photography-of-a-dog-biting-green-tennis-ball-931876/). The image of the blue light was obtained through Wikimedia Commons and was posted by Andrew Bossi (https://commons.wikimedia.org/wiki/File:IMG_6077_-_Bathroom_Light.JPG).*

In each of these examples, the CS appears to take on the same properties as the US, and the CR is strikingly similar to the UR, thus providing support for Pavlov's stimulus substitution theory. As you are about to learn, this theory has a major limitation that cannot account for all CRs.

Pavlovian processes involved in addiction

In the last chapter, you learned that drugs are reinforcers. They are also USs. Individuals do not need to learn that cocaine makes one hyperactive or that opioids cause sedation. The effects of the drug (e.g., increased activity following stimulants; increased analgesia following opioids) are URs. For an individual using drugs, there are many stimuli that can act as a CS. In the case of someone who injects a drug, a syringe become a CS that signals availability of the drug. CSs are not limited to paraphernalia. The environment in which the drug-taking behavior occurs can become a CS. Even peers can become a CS. In contrast to Pavlov's theory of conditioning, drug-paired stimuli do not elicit CRs that are identical to the drug's effects.[5] Individuals that encounter drug paraphernalia will not feel analgesia like they do when they inject heroin. Instead, they will experience increased pain sensitivity. This example highlights that CRs are not always identical to the UR. This finding suggests that Pavlov's stimulus substitution theory does not provide a full account of conditioning. Indeed, many studies have reported CRs that are not the same as the URs. Because of this major shortfall, the stimulus substitution theory is no longer a widely accepted theory of conditioning.

Contingency and contiguity are important mechanisms for developing associations between drugs and drug-paired stimuli. Take a person that smokes cigarettes or cannabis. Each time this individual smokes, they use a lighter to ignite one end of the cigarette/joint. The time needed to light the cigarette/joint is brief. Over time, the lighter becomes an effective CS. In addition to contingency and contiguity, sensory preconditioning and higher-order conditioning are relevant for SUDs. Before an individual begins smoking, lighters can become associated with other neutral stimuli, such as candles. If the individual uses the lighter frequently enough before smoking, something as innocuous as a candle may trigger a CR in the individual, which can lead to increased cravings. This process can occur after the individual begins smoking. If the individual begins using the lighter for other functions, new CSs can be created. The fact that numerous stimuli can become paired with drug use is one major reason why treating SUDs is difficult. An individual can abstain from drug use for years, but they may encounter a random stimulus that increases the urge to use the drug.

Pavlovian conditioned approach (PCA)

Although CSs elicit CRs, they can sometimes elicit more complex behaviors. One interesting finding in conditioning experiments is that some animals become almost fixated on the CS. In the case of Pavlov's dogs, a visiting researcher noted that one dog would approach the CS and then wag its tail before barking at and attempting to jump on the stimulus. An animal's interaction with and/or approach to a CS is known as **sign-tracking** and results from Pavlovian conditioning.[6] Not all animals engage in sign-tracking behavior. When the CS is presented, some animals will not approach it. Instead, they will wait for delivery of the US. These animals display **goal-tracking** behavior instead of sign-tracking behavior. Sign-tracking and goal-tracking are often measured in an operant chamber using **Pavlovian conditioned approach (PCA)**. When pigeons are used, a stimulus light is illuminated before food is delivered. Pigeons do not need to peck the light to receive food. The food is always delivered, regardless of the pigeons' actions. Sign-tracking pigeons will peck at the light

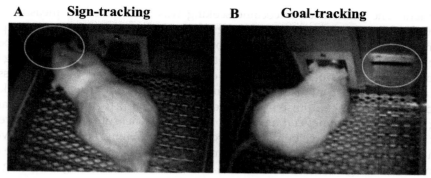

FIGURE 6.3 To measure sign-tracking/goal-tracking in rodents, a lever is inserted into the operant chamber for a short period of time (e.g., 8 s). Once the lever retracts, food is delivered to the center tray. In (A), the rat approaches the lever while it is extended and chews/presses on it. This is an example of sign-tracking behavior. In (B), the rat immediately goes to the food tray when the lever is extended into the operant chamber. Because obtaining food is the "goal" of the animal, going straight to the source of the food is goal tracking. In each panel, the extended lever is circled. *Image modified from Fig. 1 of Flagel SB, Akil H, Robinson TE. Individual differences in the attribution of incentive salience to reward-related cues: implications for addiction. Neuropharmacology. 2009;56(Supplement 1):139–148. https://doi. org/10.1016/j.neuropharm.2008.06.027. Copyright 2009, with permission from Elsevier.*

before receiving food. To measure sign-tracking in rats, a lever is extended for a certain amount of time before food is delivered. Sign-tracking rats will press/chew on the lever while it is extended into the chamber. If a pigeon or a rat is a goal-tracker, they will simply wait until the food is delivered. Animals that engage in sign-tracking are said to attribute *incentive salience* to the CS. Fig. 6.3 shows an image of a rat engaging in sign-tracking behavior and in goal-tracking behavior.

Animals often develop sign-tracking behavior when a CS is consistently paired with drug delivery. If a lever is extended into an operant chamber immediately before intravenous infusion of cocaine, rats will begin to approach the lever.[8] Remember, this is not an operant procedure; as such, the rat does not need to interact with the lever to receive drug infusions. Sign-tracking of a drug-paired CS has been observed for ethanol,[9,10] nicotine,[11] and the opioid remifentanil.[12] This research suggests that animals become "attracted" to drug-paired stimuli. In the last chapter, you learned about secondary reinforcers (e.g., money), which become reinforcing due to their repeated pairings with primary reinforcers. Secondary reinforcers are also known as conditioned reinforcers because this association represents Pavlovian conditioning. Drug-paired CSs become reinforcing as you learned when reading about the second-order schedule of reinforcement in the previous chapter. We will come back to this concept.

Understanding sign-tracking is important because research has shown that animals that attribute incentive salience to CSs are more likely to display certain addiction-like behaviors.[13] Sign-tracking rats self-administer more cocaine,[14] are more likely to treat cocaine as an inelastic good,[15] and choose cocaine over food more frequently[16] compared to goal-tracking rats. Increased sign-tracking is predictive of reinstatement of methamphetamine seeking[17] and nicotine seeking.[18] Pohořalá et al.[19] recently used the DSM-IV-based 3-criteria (3-CRIT) model to assess the role of sign-tracking on three addiction-like behaviors: (1) motivation to self-administer cocaine; (2) persistence of cocaine seeking; and (3) resistance to punishment. The persistence of cocaine seeking was determined by measuring rats' responses during alternating drug-ON and drug-OFF periods. During the drug-ON period, responses on a manipulandum led to drug reinforcement. During the drug-OFF period, cocaine was never delivered. Drug-seeking was characterized by increased response during the drug-OFF periods. Motivation to self-administer cocaine was assessed in a progressive ratio schedule, and resistance to punishment was measured in a compulsive drug-seeking paradigm. Recall that in the compulsive drug-seeking paradigm, foot shocks are paired with each drug infusion. If rats met certain criteria across each measure, they were given a score of one for each task, with three being the maximum score. Sign-tracking is positively correlated with resistance to punishment, but not motivation to self-administer cocaine or persistence of cocaine seeking. Overall, these results indicate that increased sign-tracking may be a predisposing factor for compulsive drug seeking.

Not only does sign-tracking predict addiction-like behaviors, but drugs can alter sign-tracking behavior. For example, cocaine and nicotine self-administration increase sign-tracking.[20-22] Interestingly, ethanol can either increase[23,24] or decrease[25] sign-tracking. The discrepancy across studies may be dependent on when animals are exposed to ethanol. McClory and Spear found that adolescent rats exposed to ethanol show increased sign-tracking behavior during adulthood; however, adult rats treated with ethanol do not show altered sign-tracking or goal-tracking behavior.[24] This finding suggests that adolescents are at heightened risk for attributing incentive salience to drug-paired stimuli. Collectively, these studies are important because they show that continued drug use can further increase the attraction to drug-paired stimuli, thus increasing the likelihood that these stimuli precipitate drug cravings when an individual tries to quit using the drug.

Attentional bias

Sign-tracking is not limited to pigeons or rats. Humans can become overly sensitive to drug-related stimuli, a phenomenon known as **attentional bias** or *cue reactivity*. In a typical cue reactivity experiment, participants are shown different images or videos, some of which include a drug or related paraphernalia and some of which include "neutral" objects (i.e., items not related to drug use). Participants then evaluate their subjective feelings of craving. Additionally, researchers can collect physiological measures like heart rate and skin conductance. Evidence has shown enhanced cue reactivity for ethanol,[26,27] cigarettes,[28] cocaine,[29] heroin,[30,31] and cannabis.[32] Higher-order conditioning has also been used to examine cue reactivity to ethanol-paired cues in individuals with low sensitivity to ethanol and in individuals with high sensitivity to ethanol.[33] The first CS was an olfactory cue isolated from either ethanol, sweet foods, or nonedible substances. This CS was paired with a neutral visual stimulus. Individuals with low sensitivity to ethanol, but not high sensitivity to ethanol, attributed greater incentive salience to the visual cues that had been paired with the ethanol-associated olfactory stimuli. These results suggest that individuals with low sensitivity to ethanol may be at higher risk for developing an alcohol use disorder because they attribute too much incentive salience to ethanol-paired cues.

Attentional bias can be measured with several tasks. Some of these tasks will be detailed below, as well as research examining the relationship between attentional bias and SUDs.

Eye-tracking experiments

Researchers can have individuals view words, single-object images, or complex scenes on a computer, and eye-tracking equipment can detect where an individual directs their gaze and for how long they look at a particular spot on the screen. To measure attentional bias, individuals look at drug-paired stimuli and neutral stimuli during the experiment. Given the ubiquity of tobacco and ethanol use, most eye-tracking experiments have focused on these two drugs. Increased cigarette cravings are positively correlated with an attentional bias toward smoking-related cues.[34] Similarly, individuals that score higher on ethanol craving and ethanol use problems spend more time looking at alcoholic beverages compared to nonalcoholic drinks.[35] Directly related to this finding, individuals that engage in heavy drinking spend more time looking at alcoholic beverages compared to light drinkers.[36] Even adolescent (14–16 years of age) social drinkers spend more time viewing ethanol-related stimuli compared to nondrinkers, although social drinkers spend a near equivalent amount of time viewing ethanol and neutral stimuli.[37] However, adolescent heavy drinkers spend significantly more time viewing ethanol stimuli compared to neutral stimuli.[36,37]

While the studies listed above had participants view drug-associated or neutral images on a computer screen, one study attempted to measure eye tracking in a more natural setting. Monem and Fillmore[38] had participants wear eye-tracking glasses and allowed them to explore a room containing ethanol-paired stimuli and neutral stimuli. During the first test session, participants spent a similar amount of time viewing both types of stimuli. However, during the second test session, participants spent less time viewing neutral stimuli. Monem and Fillmore also found that heavy drinkers spent more time viewing ethanol-paired stimuli.[38] Overall, these studies show increased attentional bias to drug-paired stimuli in individuals that use certain drugs. This increased attentional bias can promote individuals to continue using the drug as the drug-paired stimuli are more likely to capture the attention of these individuals.

Dot-probe task

In the *dot-probe task* (also known as the visual probe task), participants are asked to view a computer screen. A fixation cross (+) appears on the center of the screen for a short duration. Next, two images or words are quickly flashed on the screen for approximately 500 ms. After the images vanish, a dot appears in the location of one of the former stimuli. Individuals are asked to identify the location of the dot as quickly as possible. Faster reaction times indicate greater attentional bias. When the drug dot-probe task is used to measure attentional bias to drug-paired stimuli, individuals are presented with a neutral stimulus and a drug-paired stimulus (Fig. 6.4A). Increased attentional bias toward drug-paired stimuli has been observed in individuals that drink ethanol,[41,42] smoke cigarettes,[43,44] use opioids,[45–47] and use ketamine.[48] Increased attentional bias is also observed in those with cocaine dependence or concurrent cocaine/ethanol dependence.[49] Additionally, current smokers, but not former smokers or nonsmokers, spend more time viewing e-cigarette cues compared to neutral cues, suggesting that e-cigarette cues may contribute to tobacco cigarette cravings.[50] Attentional bias to one drug can be correlated to attentional bias to another drug. For example, individuals that show increased attentional bias to smoking-related cues are more likely to show attentional bias to ethanol-paired stimuli.[51] This finding makes sense considering the high concordance of ethanol and tobacco use.

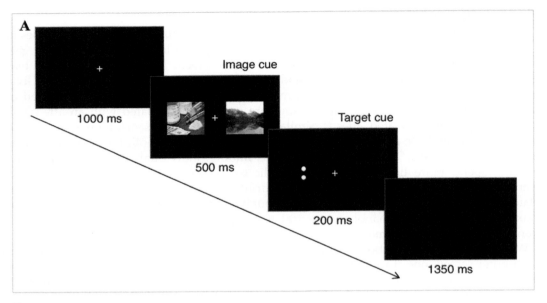

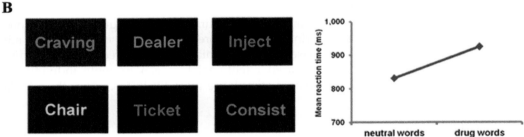

FIGURE 6.4 Two paradigms used to detect increased attentional bias to drug-paired stimuli: the dot-probe task (A) and the drug (emotional) Stroop task (B). In the dot-probe task, two stimuli are flashed on the screen. One image contains drug stimuli while the other image contains neutral stimuli (like nature scenes). After the images disappear, a dot (or dots in the illustration above) appears on one side of the screen, and participants must identify the location of the dot(s). If the dot(s) appear(s) on the same side as the drug-associated image, participants are much faster to respond. In the drug Stroop task, participants are presented drug-related words and neutral words in various colors. The object of the task is to name the color of the font as quickly as possible. Individuals with a SUD often take longer to respond when presented with drug-associated words, indicating that they become more fixated on the word as opposed to the color font. *Panel (A) comes from Fig. 1 of Zhao Q, Li H, Hu B, et al. Neural correlates of drug-related attentional bias in heroin dependence.* Front Hum Neurosci. *2018;1:646. https://doi.org/10.3389/fnhum.2017.00646. This is an open-access article distributed under the terms of the Creative Commons Attribution 4.0 International License (https://creativecommons.org/licenses/by/4.0/). Panel (B) comes from Fig. 2 of Murphy A, Taylor E, Elliott R. The detrimental effects of emotional process dysregulation on decision-making in substance dependence.* Front Integr Neurosci. *2012;6:101. https://doi.org/10.3389/fnint.2012.00101. This is an open-access article distributed under the terms of the Creative Commons Attribution 4.0 International License (https://creativecommons.org/licenses/by/4.0/).*

Drug cravings are often associated with increased attentional bias. Individuals with chronic pain that have become physically dependent on opioids are more likely to report opioid cravings if they show increased attentional bias to opioid-paired cues.[46] Not only is attentional bias associated with opioid cravings, but it is associated with an increased risk of opioid misuse in individuals with chronic pain.[47] Positive correlations between attentional bias and drug craving have also been observed for individuals with an alcohol use disorder[52] and in cannabis users.[53] Even social drinkers that show increased attentional bias are more likely to experience cue-induced craving.[54] However, the association between attentional bias and drug craving is not observed in individuals attempting to quit tobacco use.[55]

One interesting and somewhat paradoxical finding is that cocaine-dependent individuals who *avoid* ethanol-paired stimuli are *more* likely to drop out of treatment.[56] Why is this the case? One potential explanation is that those trying to abstain from drug use may develop an avoidance strategy; that is, they may try to avoid viewing drug-paired stimuli. In theory, this sounds like something one should do when going through therapy. However, actively avoiding these stimuli may make completing therapy more difficult. Think of it this way: there may be times in your life when you must encounter a stressful situation. Interviewing for a job is stressful. You may also have to have a difficult conversation with a loved one. Although these situations are stressful and unpleasant, they are important to face. The same thing applies to those with a SUD. Part of the therapy entails encountering drug-paired stimuli as you will learn at the end of this chapter. When individuals adopt an avoidance strategy, they may have increased difficulty during therapy because they are forced to confront difficult situations, which leads to dropout.[57] There is some evidence that supports this hypothesis. In one study, three groups of participants were asked to use one of three strategies to regulate emotions associated with smoking. Some individuals were tasked with reappraising their emotions, some were instructed to accept their emotions, and some were told to suppress (i.e., avoid) their emotions. Those that reappraised their emotions showed decreased cravings and attentional bias for cigarette-paired stimuli compared to those instructed to avoid their emotions.[58]

Drug Stroop Task

Another way to measure cue reactivity in individuals is by modifying the *Stroop task*. You will learn about the traditional Stroop task in the next chapter. In the drug-variant of the Stroop task (also known as the emotional Stroop task), words related to drugs (e.g., "cocktail", "syringe") and neutral words (e.g., "shirt", "coffee") are presented in various colors.[59] Participants are asked to name the color of the font, not the word itself (Fig. 6.4B). Research has shown that college students that engage in binge drinking (more than four alcoholic beverages during a single drinking episode for women/five drinks for men) show greater interference when presented with ethanol-associated words compared to neutral words, an effect that is absent in nonbinge drinkers.[60] Similar interference is observed in individuals that use cigarettes[61-63] or cannabis,[64-66] as well as in those with opioid use disorder/opioid dependence[67,68] and cocaine dependence.[67,69,70] Interestingly, while Carpenter et al.[69] did not observe differences in reaction time for cannabis-paired or heroin-paired words relative to neutral words in those with cannabis or heroin use disorder, they did find that these individuals were slower to respond to cocaine-associated words compared to neutral words.

Increased attentional bias in the drug Stroop task is predictive of poor treatment outcomes for those with a SUD. Relapse to ethanol is higher in individuals that show increased attentional bias.[71] Individuals that show increased cocaine Stroop interference provide more drug-positive urine samples and complete fewer weeks of therapy.[69] Individuals with increased attentional bias are more likely to use crack cocaine.[72] Attentional bias is also predictive of relapse to heroin[73] and to methamphetamine.[74] Similarly, in those with heroin use disorder, attentional bias increases immediately before relapse, and those that relapse show significantly higher attentional bias compared to those that do not relapse.[75] At least one exception has been noted in the literature; attentional bias to cannabis cues does not predict long-term (6 months) treatment progression for adolescents with a cannabis use disorder.[76]

Pavlovian-to-instrumental transfer (PIT)

Pavlovian conditioning and operant conditioning may measure distinct forms of learning, but they are closely intertwined with one another. In the previous chapter, you encountered an example of how Pavlovian conditioning influences operant conditioning: the second-order schedule of reinforcement. Even though animals earn few drug infusions in this schedule of reinforcement, they exhibit high rates of responding. This occurs because they can earn access to stimuli that have been previously associated with each drug infusion. These stimuli become conditioned reinforcers that can motivate the animal to continue responding. The fact that animals often approach drug-paired stimuli can be used to a researcher's advantage when first training an animal to self-administer a drug. An *autoshaping* procedure can be used to ensure animals acquire an operant response to earn drug infusions.[77] During an autoshaping session, animals earn drug reinforcement by responding on a manipulandum. This manipulandum will be made available to the animal for a certain amount of time. If the animal fails to respond on the manipulandum, the drug will be infused into the animal's catheter. Eventually, animals will develop a sign-tracking response in which they approach the manipulandum more frequently and emit a response to earn drug reinforcement. Ultimately, the researcher can stop using the autoshaping procedure after animals consistently respond on the manipulandum when it is available.

Given the influence of conditioned stimuli on behavior, one experimental design has been used to further examine how these stimuli affect operant conditioning: the **Pavlovian-to-instrumental transfer (PIT)** task. In a PIT task, animals are first conditioned to associate a neutral stimulus such as a light or a tone to the US, typically food. Following the Pavlovian conditioning phase, animals are trained to respond on a manipulandum to receive the US; the US is now a reinforcer as it maintains operant responding. Finally, the CS is presented to the animal, and responses on the manipulandum are recorded.[78] Importantly, reinforcement is not provided during this final part of testing. If the presentation of the CS increases operant responding, this means that the CS has become an effective secondary reinforcer that can increase motivation. Why have the PIT task? What does this tell us that we do not already know? Imagine someone that uses heroin by injecting it. They experience both Pavlovian conditioning and operant conditioning during the drug-taking experience. The individual will prepare for the heroin injection by heating the heroin, placing it in a syringe, and then inserting the needle into the skin. The operant response is to inject the heroin, which is reinforcing. Because these two learning phenomena occur at nearly the same time, there is some difficulty determining if continued drug taking purely results from operant conditioning or if it is influenced by Pavlovian conditioning. In the PIT task, if behavior is maintained purely by operant conditioning, presenting the CS should not increase responding on a manipulandum. If Pavlovian conditioning is important, one would expect to see an increase in responding when the CS is presented.

Most PIT studies involving drug self-administration have focused on the effects of a CS on ethanol consumption,[10,79–87] but other studies have used cocaine as the US.[88,89] Overall, these studies show that responding to drug-paired manipulanda, even when reinforcement is no longer made available, can be facilitated by exposure to a drug-paired CS. Additionally, animals that show enhanced PIT self-administer more cocaine.[90] Interestingly, Takahashi et al. argue that the positive correlation between PIT and cocaine self-administration does not necessarily mean that enhanced PIT is a risk factor for addiction.[90] Why? Takahashi et al. conducted an additional experiment in which they used the 3-CRIT model as described earlier in this chapter. Takahashi et al. did not observe any differences in PIT between addicted-like and nonaddicted-like rats. As such, they argue that the significant correlation observed between PIT and cocaine self-administration may reflect general differences in learning of the association between CSs and drug stimuli.[90] One issue with this interpretation is that relapse-like behavior is not explicitly measured in the 3-CRIT. As you will later learn, drug-paired stimuli are able to elicit intense drug cravings and can precipitate withdrawal symptoms, which can promote relapse. Overall, the PIT task shows that many behaviors are maintained by Pavlovian conditioning, including drug self-administration, providing further evidence of the influence of Pavlovian conditioning on addiction.

Pavlovian-based theories of addiction

As you learned earlier in this chapter, Pavlov developed the stimulus substitution theory to explain why Pavlovian conditioning occurs. One major weakness of Pavlov's theory is that it cannot explain why CRs to drug-paired CSs are opposite of the URs that result from exposure to the US. To understand why an individual with a SUD can experience intense physiological symptoms that mimic withdrawal when encountering a drug-paired CS, we need to cover Dr. Shepard Siegel's *compensatory response* theory.[91]

Compensatory response theory

Before we can dive into Siegel's theory, we need to back up a little and talk about Dr. Greg Kimble's preparatory response theory of conditioning.[92] Kimble argued that the function of the CR is to prepare an organism for the upcoming US. In the case of Pavlov's dogs, the dogs hear the metronome and begin to salivate (CR). According to Kimble, the dogs begin salivating to *prepare* themselves for the food that has been associated with the metronome previously. In humans, researchers can elicit a conditioned eyeblink response by pairing a tone with a puff of air to the eye. The US is the puff of air, and the CS is the tone. Both the puff of air and the tone elicit the same behavioral response: blinking of the eye. According to Pavlov's theory, this occurs because the tone takes on the same properties as the puff of air, thus causing the person to respond to it as they would to the tone. However, Kimble's theory states that this response occurs because the tone prepares the person for the arrival of the puff of air. By blinking early, the individual's eye is partially closed by the time the puff of air is delivered. Unlike Pavlov's theory, the preparatory response theory can account for CRs that are different from URs. For example, if a rat is given a foot shock, its response is to jump because getting shocked on the foot does not feel great. However, if a tone is repeatedly paired with the shock, the rat freezes when it hears the tone. Freezing most likely occurs because this is what rats do to avoid detection from a predator.

How does this relate to Siegel's theory? The compensatory response theory is a variant of the preparatory response theory. Siegel argued that the CR results from the organism preparing for delivery of the US by *compensating* for its effects. In one of his earlier studies, Siegel[91] gave rats injections of insulin. The US is the insulin, which decreases blood glucose levels (the UR). After repeated injections, the syringe becomes the CS. Eventually, Siegel injected rats with saline. The saline should not have affected blood glucose levels. However, Siegel found that blood glucose levels had increased! Now, imagine an individual using a syringe to inject heroin. Over time, the syringe becomes a CS that becomes paired with heroin (US). As I have already mentioned, the CR is not feeling analgesia; instead, individuals experience hyperalgesia (increased pain sensitivity). This occurs because the body is preparing itself to receive a substance that reduces pain. Recall from Chapter 5 that homeostasis refers to the body's attempts to maintain equilibrium. When an individual injects him/herself with a drug, homeostasis is lost. When the individual sees a CS like heroin, the body prepares itself for the incoming drug by producing symptoms that are opposite of the drug's effects like hyperalgesia. When the drug enters the body, its effects cancel out what the body has done before the drug enters the body; thus, homeostasis is maintained.

Conditioned tolerance

In one study, Siegel et al.[93] wanted to determine how Pavlovian conditioning controls tolerance to a drug's effects. Specifically, they wanted to determine if Pavlovian conditioning affects morphine's perceived analgesic effects. To test this possibility, Siegel et al. gave one group of rats morphine injections that occurred in the presence of distinct CSs (increasing brightness in the testing chamber and noise reduction). Following conditioning, rats were exposed to the CS before receiving an injection of morphine and being placed on a hot plate. Siegel et al. measured how much time was needed for rats to start licking their paws, a sign of pain/discomfort. If more time passed before a paw-lick response occurred, this indicated that the morphine's analgesic effects worked. To interpret the data, Siegel et al. had to test two additional groups of rats. One group of rats received injections of morphine but did not experience the CS. Another group of rats experienced the CS, but they received injections of saline instead of morphine. Siegel et al. found that rats given morphine injections in the presence of the CS took less time to lick their paws during the test session compared to the other two groups. This means that rats exposed to a CS during morphine injections developed tolerance to morphine's analgesic effects (i.e., the morphine became less effective at reducing pain). Importantly, the effect observed in this experiment does not merely reflect general tolerance, as rats injected with morphine in the absence of the CS took longer to lick their paws compared to the experimental group, and the time needed to lick their paws was similar to the group of rats that had previously received saline injections. This phenomenon is known as **conditioned tolerance** and highlights the role Pavlovian conditioning has on the development of drug tolerance. Siegel's work with rats has been replicated in humans using opioids.[94]

Conditioned tolerance is a potential explanation as to why some individuals overdose even when using the same dose over an extended period of time. If an individual uses a particular drug in the same location in the presence of the same stimuli (e.g., injecting heroin in one's living room), they may develop a conditioned tolerance to the drug's effects. It is important to emphasize that conditioned tolerance is specific to that one location. If the individual uses the drug in a novel environment (e.g., the bathroom of a club), their body will not be able to prepare itself for the drug. Even if the individual uses the same amount of drug they have used previously, this dose can now be potentially lethal as the body has not adequately compensated for the drug's effects. See Fig. 6.5 for a depiction of conditioned tolerance.

Conditioned withdrawal symptoms

Not only can conditioned stimuli increase tolerance to a drug's effects, but they can precipitate withdrawal symptoms, a phenomenon known as **conditioned withdrawal**. In one experiment, Krank and Perkins[95] tested three groups of rats. The first group of rats received injections of morphine in a distinct environment and saline injections in their home cage. The second group of rats received injections of saline in the distinct environment and morphine injections in their home cage. The final group of rats received saline injections in both the distinct environment and in the home cage. Eventually, each group of rats was placed in a distinct environment, and the researchers observed each rat for withdrawal symptoms such as wet dog shakes, jumping, twitches, ear wipes, head shakes, and paw tremors, to name a few. Krank and Perkins found that rats treated with morphine in a distinct environment showed more withdrawal symptoms compared to the other groups.[95] Because these rats showed more withdrawal symptoms compared to rats treated with morphine in the home cage, one can conclude that the environment, not just morphine exposure alone, can lead to the development of withdrawal symptoms. Fig. 6.6 provides an example of conditioned withdrawal.

The implications of conditioned withdrawal should be obvious. The ability of drug-paired stimuli to produce withdrawal symptoms in the absence of drugs may significantly contribute to increased drug use. As you learned in the previous chapter, individuals often continue using the drug to avoid withdrawal symptoms. If an individual goes into withdrawal by merely encountering a CS associated with the drug, they are going to be more likely to use the drug. Given that many stimuli become associated with drugs, whether directly or indirectly through sensory preconditioning/higher-order conditioning, this increases the difficulty of treating SUDs.

Incentive sensitization theory of addiction

Given that drugs increase sign-tracking and PIT and that drug-paired stimuli can become powerful conditioned reinforcers, Dr. Terry Robinson and Dr. Kent Berridge, both professors at the University of Michigan, developed the **incentive sensitization theory of addiction** to provide a framework for the neurobiological basis of addiction.[96] According to Robinson and Berridge, repeated drug use in susceptible individuals causes the neural circuits responsible for incentive salience attribution to become hypersensitive. To better understand this theory, we need to review a distinction that has already been discussed: "liking" something is not the same thing as "wanting" something. In the last chapter, we equated liking to the positive reinforcing effects of some experience (like eating cheesecake) and wanting as a form of negative reinforcement. However, wanting reflects more than negative reinforcement. In the incentive sensitization theory of addiction, wanting is equated more to a "desire" to use a drug that is independent of one's "liking" of the drug. When an individual uses a drug for the first time, dopamine levels increase in the mesocorticolimbic pathway. The individual may experience the positive reinforcing effects

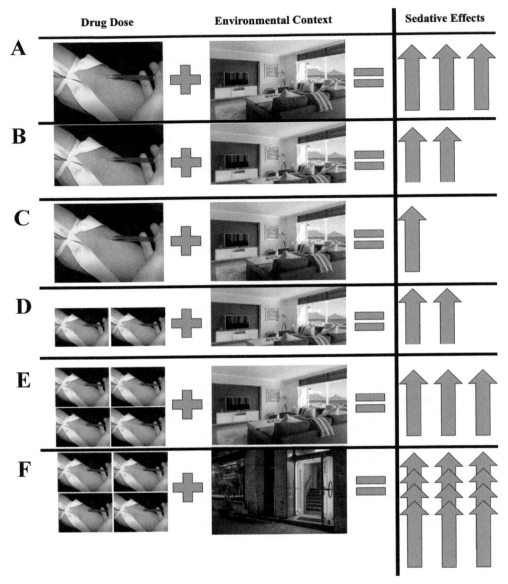

FIGURE 6.5 An example of conditioned tolerance to the sedative effects of heroin. With initial heroin use, the individual experiences a robust sedative effect (A). With repeated drug use, the individual develops tolerance to heroin's sedative effects (B and C). To compensate for tolerance, the individual increases the drug dose to experience the same "high" as they did initially (D and E). Notice that the environmental context in which the individual uses heroin is the same for panels A—E. In (F), the individual uses the same dose as they did in (E), but the drug use now occurs in a novel location. Because the body is not expecting an influx of heroin at this moment in time, there is no compensatory mechanism to protect the user from an overdose. Now, the same dose that the individual has been using produces a massive sedative effect, which can tragically lead to overdose and death. *The image was created by the author using Microsoft PowerPoint. The image of heroin injection comes from Wikimedia Commons and was posted by user Psychonaught (https://commons.wikimedia.org/wiki/File:Injecting_Heroin.JPG). The image of the living room and the bar come from Pexels and were posted by Jean van der Meulen (https://www.pexels.com/photo/photo-of-living-room-1457842/) and Mali Maeder (https://www.pexels.com/photo/white-motor-scooter-near-open-door-219095/), respectively.*

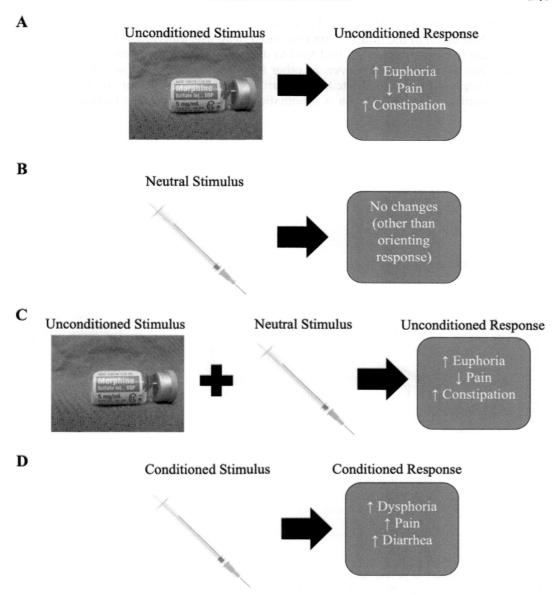

FIGURE 6.6 Illustration showing conditioned withdrawal. In (A), morphine (unconditioned stimulus [US]) produces euphoria while decreasing pain sensitivity and increasing constipation (unconditioned responses [URs]). In (B), the presentation of a neutral stimulus such as a syringe does not elicit much of a response from individuals. Panel (C) depicts conditioning. When an individual uses a syringe to inject morphine, they will experience the same effects as described in (A). As such, the individual will begin pairing the syringe with morphine. Over time, the presentation of the syringe, now a conditioned stimulus (CS), will elicit a conditioned response (CR) (D). However, the CRs are not the same as the URs. Individuals feel more dysphoric, experience increased pain sensitivity and experience diarrhea. The conditioned withdrawal symptoms can precipitate continued use of the drug. *The figure was created by the author of this textbook using Microsoft Powerpoint. The image of morphine comes from Wikimedia Commons and was posted by user Vaprotan (https://commons.wikimedia.org/wiki/File:Morphine_vial.JPG). The image of the syringe comes from https://openclipart.org/detail/1887/syringe.*

of the drug and may "like" those effects. At the same time, they may experience a "desire" to use the drug again in the future. With continued drug use, individuals begin to "like" the drug less due to neuroadaptations that lead to decreased dopamine release, as you learned about in Chapters 3 and 4. However, as other stimuli become paired with the drug (e.g., syringes, ashtrays, pint glasses, etc.), the neurocircuits responsible for incentive salience become sensitized. Fig. 6.7 shows a schematic of the incentive sensitization theory of addiction.

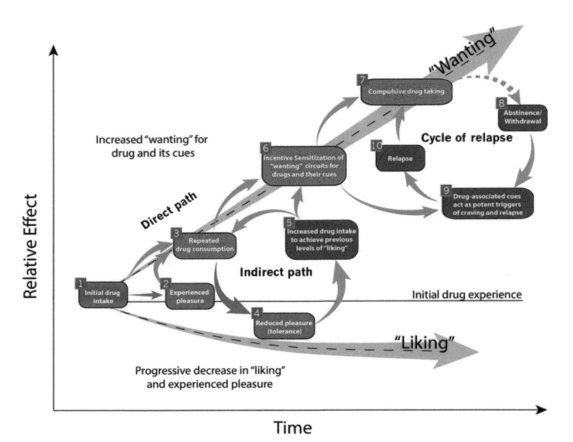

FIGURE 6.7 Schematic depicting the incentive sensitization theory of addiction. Initial drug use (1) leads to pleasure (2). This experience promotes the individual to continue using the drug (3), which then leads to tolerance to the reinforcing effects of the drug (4). To achieve the same "high" as initially experienced, the individual will increase their drug intake (5). Notice that this causes a loop to form between repeated drug consumption, increased tolerance, and increased drug intake. Eventually, the neural circuits responsible for attributing incentive salience to drugs and their cues become sensitized (6). Now, individuals have a strong desire to continue using drugs, which can promote compulsive drug seeking (7). If an individual stops using the drug (8), they may experience intense cravings after encountering drug-paired cues (9). These cravings can precipitate relapse (10), which then leads back to compulsive drug seeking. Over time, "liking" decreases, whereas "wanting" increases. *Image is from Fig. 2 of Robinson MJ, Fischer AM, Ahuja A, Lesser EN, Maniates H. Roles of "wanting" and "liking" in motivating behavior: gambling, food, and drug addictions. Curr Top Behav Neurosci. 2016;27:105–136. https://doi.org/10.1007/7854_2015_387.*

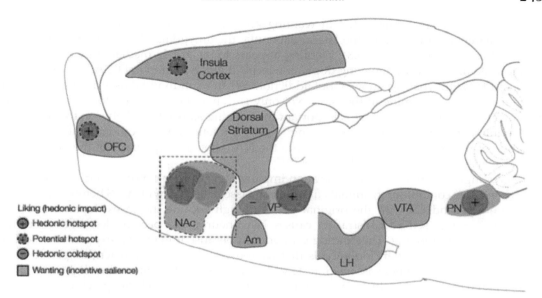

FIGURE 6.8 The neurobiology of incentive sensitization. Areas highlighted in gray indicate those that are associated with "wanting". Areas colored in red represent hedonic hotspots. A region surrounded by a dashed line is a potential hotspot, but more work is needed to verify if the region controls hedonic responses to drugs. Areas colored in blue are considered hedonic coldspots. Stimulating these areas decreases hedonic responses to sucrose. *Am*, amygdala; *LH*, lateral hypothalamus; *NAc*, nucleus accumbens; *OFC*, orbitofrontal cortex; *PN*, parabrachial nucleus (located in pons); *VP*, ventral pallidum; *VTA*, ventral tegmental area. *Image comes from Fig. 2 of Kringelbach ML, Berridge KC. Neuroscience of reward, motivation, and drive. In: Kim SI, Reeve J, Bong M. eds.* Recent Developments in Neuroscience Research on Human Motivation (Advances in Motivation and Achievement)*: Emerald Publishing Limited; 2016:23–35. https://doi.org/10.1108/S0749-742320160000019020. © Emerald Publishing Limited all rights reserved.*

The nucleus accumbens (NAc) shell is important for mediating the pleasure we experience when we receive a reward. As such, this region contains a *hedonic hot spot* that helps control liking for drug-paired stimuli. Importantly, this hot spot does not encompass the entire NAc shell; instead, it occupies a 1 cubic mm site in the rostrodorsal region of the shell.[98] The question becomes: "how does one determine the location of a hedonic hot spot?". Rodents will display certain hedonic facial expressions when given a sweet solution, such as sticking their tongue out or engaging in a licking response. Stimulating the rostrodorsal region of the NAc shell, but not other subregions of the shell increases these hedonic expressions. A hedonic hot spot has also been identified in the posterior region of the ventral pallidum.[99] Other brain regions have been proposed to contain hedonic hot spots, including the parabrachial nucleus of the pons, the insula, and the orbitofrontal cortex (OFC)[100] (Fig. 6.8). Interestingly, "liking" does not appear to be controlled by dopamine; instead, opioids and endocannabinoids are more important for this response.[98,99,101]

You previously learned that the NAc core and the ventral pallidum coordinate actions toward reward-paired cues. As such, these regions are involved in attributing incentive salience to drug-paired stimuli. In contrast to the "liking" system, the "wanting" system is composed of numerous brain regions, including those that contain hedonic hot spots (like the ventral pallidum). Other areas implicated in wanting are the NAc shell, the dorsal striatum, the ventral tegmental area (VTA), the lateral habenula, the amygdala, the insula, the OFC, and the parabrachial nucleus (Fig. 6.8). Wanting appears to be largely mediated by the dopaminergic system.[102] Indeed, depleting dopamine levels in the NAc impairs PCA.[103] This finding is consistent with what we know about addiction. When an individual uses a drug, there is a large increase of dopamine in the NAc. Not only is dopamine important for the direct reinforcing effects of the drug, but it is important for learning associations between the drug and environmental stimuli. However, "wanting" can be influenced by the opioid system as direct injection of mu-opioid receptor agonists into the NAc shell and into parts of the ventral pallidum increases consummatory behavior.[98,99]

Interestingly, cocaine self-administration impairs higher-order conditioning.[104] Recall that in higher-order conditioning, animals first learn to associate a CS with a US before learning to associate a second CS with the original CS. Although cocaine increases dopamine levels, long-term cocaine self-administration causes NAc neurons to become less active during higher-order conditioning. This finding is important because it may help explain why addiction is a relapsing disorder. Individuals that have used a drug for an extended period may have difficulty learning new associations (e.g., lighters are used for more than lighting a cigarette) upon drug cessation. Instead, the old associations persist, which can precipitate relapse.

Quiz 6.1

Read the following scenario before answering the questions:

An individual begins smoking cannabis daily in their basement. The individual places the cannabis in a pipe before lighting it. Initially, the individual feels relaxed when using cannabis. Over time, the individual notices that they no longer feel as relaxed when using cannabis, so they increase how much cannabis they use during the day.

1. What stimulus/stimuli can become paired with cannabis use?
2. What is the unconditioned stimulus?
   a. Basement
   b. Cannabis
   c. Lighter
   d. Pipe
3. What is most likely to happen if the individual encounters a drug-paired CS when not actively using cannabis?
   a. The individual will experience diarrhea
   b. The individual will feel irritable
   c. The individual will not feel anything
   d. The individual will feel relaxed
4. What could happen if the individual uses cannabis in a location other than his/her basement?
   a. Nothing will happen
   b. They can die
   c. They can experience anxiety/paranoia
   d. They will feel more relaxed than normal

5. According to the incentive sensitization theory of addiction, which brain region will become hypersensitive to stimuli paired with cannabis?
   a. Dorsal striatum
   b. NAc core
   c. NAc shell
   d. All of the above

Answers to quiz 6.1

1. The lighter, the pipe, the basement, and any items in the basement can become paired with cannabis use.
2. b — Cannabis.
3. b — The individual will feel irritable (remember, irritability is a sign of cannabis withdrawal).
4. c — They can experience anxiety/paranoia (these are symptoms of cannabis overdose).
5. d — All of the above.

Conditioned place preference

You are already familiar with the term conditioned place preference (CPP) as I briefly discussed this research paradigm in Chapter 3. CPP is used to measure the conditioned rewarding effects of a stimulus. CPP chambers are composed of separate compartments that have different colors and/or patterns on the walls of the compartments. The floors of each compartment can differ, as some compartments have smooth PVC floors, whereas others have steel rod floors or wire mesh floors. In some studies, researchers add scented items under the floor to provide different olfactory cues. A normal CPP experiment begins with a pretest. The animal, in a drug-free state, is allowed to explore each compartment. Then, in alternating sessions, the animal is given a drug of interest and is then isolated to one of the compartments. In sessions in which the animal does not receive the drug, they receive a vehicle injection and are isolated to a different compartment. In the case of a three-compartment CPP chamber, animals are never isolated to the center compartment. Animals are then given a posttest, in which they are allowed to explore all compartments in a drug-free state. Spending more time in the environment previously paired with the drug is considered CPP. Conversely, *conditioned place aversion (CPA)* occurs when the animal spends significantly less time in the compartment paired with the drug.

Not surprisingly, animals develop CPP to most drugs commonly used by humans,[105] but there are some inconsistencies in the literature. Historically, the barbiturate pentobarbital has failed to produce CPP in animals[106]; in fact, Lew and Parker found that pentobarbital produced CPA in animals.[106] However, Bossert and Franklin showed that rats could develop CPP to pentobarbital.[107] PCP is another drug that often produces CPA in animals.[108,109] Although PCP does not consistently produce CPP, there are newer synthetic dissociative drugs that reliably produce CPP in animals.[110,111]

Interestingly, if animals are pretreated with PCP (that is, they receive PCP injections before they are tested in the CPP experiment), they develop CPP.[112,113] This phenomenon has been observed with cocaine, *d*-amphetamine, nicotine, and morphine.[114,115] This finding somewhat defies what we know about Pavlovian conditioning. In a typical Pavlovian conditioning experiment, preexposing an animal to the US before pairing it with the CS weakens conditioning. This is known as the *US preexposure effect*.[116] In the case of drug CPP, why does preexposing an animal to the drug prior to conditioning enhance CPP instead of weakening it? According to Bardo et al.[117] there are a couple of reasons why preexposure to the drug can increase CPP. First, exposure to the drug can enhance its rewarding aspects, thus making it easier to associate them with a novel environment. Think of it this way: someone may not initially "enjoy" smoking a cigarette or drinking ethanol the first time they try it. Over time, the individual may start to enjoy the effects of the drug more. Second, preexposure to the drug may lead to tolerance to any aversive effects of the drug that may arise from acute withdrawal. By the time conditioning begins, the animal does not experience as many acute withdrawal symptoms but still experiences the rewarding effects of the drug.

One reason why drugs like PCP do not always produce CPP may be related to the methodology of the CPP experiment. When conducting a CPP experiment, the researcher must consider the dose(s) of the drug that will be paired with a specific environment. The researcher must also consider how much time passes between drug administration and being placed in the testing chamber. As mentioned, PCP often produces CPA instead of CPP. Marglin et al.[118] were able to establish CPP for acute injections of PCP. They were able to accomplish this by treating animals with a lower dose of PCP compared to other studies and by waiting less time to test the animals following drug administration. Another important methodological consideration will be discussed below.

Biased versus unbiased design

There are two major designs one can use in a CPP experiment: a biased design and an unbiased design. In a biased design, the researcher first measures preference for each compartment during the pretest. The researcher will then assign the initially nonpreferred chamber to be the drug-paired chamber. In an unbiased design, the researcher will counterbalance which compartment is paired with the drug. That is, half of the subjects will receive the drug in their initially preferred compartment, and half will receive the drug in the initially nonpreferred compartment. Is one design better than the other? The answer depends on what type of CPP apparatus one has. Just as there is a biased and unbiased design, there are biased and unbiased CPP chambers. A biased CPP chamber exists when animals spend most of their time in one compartment over the other during the pretest. For example, rats often spend more time in a compartment that has a wire mesh floor instead of a steel rod floor. If animals spend similar amounts of time in each compartment, the researcher has an unbiased CPP chamber. If one has an unbiased CPP chamber, using either a biased design or an unbiased design is acceptable. However, when using a biased apparatus, caution needs to be taken when using either design. Why? One concern with using a biased design with a biased apparatus is that any increased time spent in the initially nonpreferred compartment following conditioning may reflect decreased anxiety as opposed to increased reward.[119] Conversely, imagine some animals spend more time in the compartment that becomes paired with the drug. A researcher may not be able to observe an increase in the time spent in the drug-paired compartment because these animals already show a high preference for this compartment. This is known as a *ceiling effect*.

Cunningham et al.[120] showed how a biased/unbiased apparatus can alter CPP for ethanol. In experiment 1, an unbiased apparatus was used; in experiment 2, Cunningham et al. created a biased apparatus by changing the type of flooring in each compartment. In both experiments, an unbiased design was used. Results showed that when an unbiased apparatus was used, mice developed ethanol CPP, regardless of which floor type was used. However, when a biased apparatus was used, significant CPP was observed in animals that received ethanol in the compartment with the bar floor only (Fig. 6.9). When examining the data, what you will notice is that CPP was not observed in animals that had a high baseline preference for the compartment that was paired with ethanol. This made observing any further increases in preference following conditioning difficult. This is an example of a ceiling effect that I mentioned in the previous paragraph. Other studies have shown discrepant results when using a biased design compared to an unbiased design. For example, CB_1 knockout mice fail to develop MDMA CPP when an unbiased design is used, but they develop CPP when a biased design is used.[121]

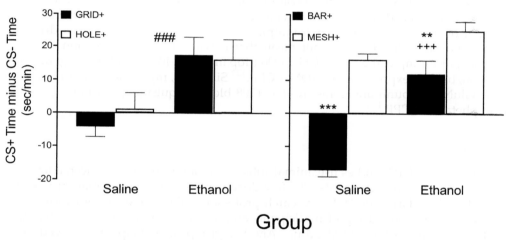

FIGURE 6.9 Cunningham et al. tested mice for ethanol CPP. There were two major experiments. In one experiment, mice were exposed to a CPP apparatus that had either a grid floor (2.3-mm steel rods mounted 6.4 mm apart) or a hole floor (perforated steel floor with 6.4-mm round holes on 9.5-mm staggered centers). Mice do not show a strong preference for one-floor type over the other. In the second experiment, mice were exposed to a CPP apparatus that had either a bar floor (6.4-mm steel rods mounted 13 mm apart) or a mesh floor (6.4-mm galvanized hardware cloth squares). In this experiment, mice show a strong bias toward the mesh floor. In both experiments, half of the mice were randomly assigned to receive ethanol in one compartment and saline in another compartment. The other half received saline in both compartments. The graphs show the time spent in the CS+ (ethanol) compartment subtracted by the time spent in the CS− (saline) compartment during the posttest. The left panel shows data for animals conditioned in an unbiased apparatus. The right panel shows data for animals conditioned in a biased apparatus. For the unbiased experiment, mice do not spend significantly more time in the compartment with the grid floor compared to the hole floor; this is why difference scores hover around 0 for each subgroup. Conversely, animals that received ethanol in either the grid-paired compartment or the hole-paired compartment developed significant CPP. For the biased experiment, mice treated with saline spent significantly more time in the mesh-paired compartment compared to the rod-paired compartment. Consequently, mice that received ethanol in the bar-paired compartment developed CPP, whereas mice that received ethanol in the mesh compartment did not. The null effect observed for this group most likely is the result of a ceiling effect. *Image recreated from Fig. 4 of Cunningham CL, Ferree NK, Howard MA. Apparatus bias and place conditioning with ethanol in mice. Psychopharmacology. 2003;170(4):409–422. https://doi.org/10.1007/s00213-003-1559-y. Copyright 2003.*

Acquisition versus expression of CPP

One major use of CPP is to screen potential pharmacotherapies for SUDs.[117] If a drug is effective in treating SUDs, it should be able to block the conditioned rewarding effects of a drug. One can test a potential pharmacotherapy for its ability to block the *acquisition* and/or the *expression* of drug-induced CPP. Acquisition is somewhat analogous to the initial learning of a drug-taking response. Think of an adolescent/young adult trying ethanol for the first time. If individuals enjoy ethanol, they will continue consuming it. They are not dependent on the drug, yet, but they have acquired the behavioral response of drinking. Expression is akin to someone that has been using a substance for an extended period. These descriptions of acquisition and expression are overly simplified. We will come back to the acquisition and the expression of CPP in the next chapter. For now, I will discuss how researchers test the effects of a potential pharmacotherapy on the acquisition and expression of CPP. To test acquisition, the potential pharmacotherapy is administered before the drug during conditioning sessions. To test expression, the potential pharmacotherapy is administered before the posttest only. Because individuals with a SUD have already acquired the drug-taking behavior, testing a potential pharmacotherapy on the expression of drug CPP may be more applicable to real life. Just because a drug blocks the acquisition of CPP, this does not mean that it will block the expression of CPP and vice-versa. For example, the drug SCH 23990 (dopamine D_1-like receptor antagonist) blocks acquisition, but not expression, of MDMA CPP.[122] Similarly, my students and I have shown that the GluN2B subunit antagonist Ro 63—1908 blocks acquisition, but not expression, of methamphetamine CPP.[123]

CPP in humans

Almost all drug CPP studies use animal subjects, although there have been a few studies that have used human participants. The first documented case of the human drug CPP was published by Dr. Harriet de Wit, a research professor at the University of Chicago. Childs and de Wit[124] gave one group of participants amphetamine in one room and placebo in a separate room; this is the paired group. Another group of participants received amphetamine and placebo in both rooms; this is the unpaired group. At the end of the experiment, participants rated their preference for each room. Participants in the paired group rated the amphetamine-paired room higher than the placebo-paired room, whereas participant ratings of each room did not differ in the unpaired group. In a follow-up study, Childs and de Wit found that participants in the paired group reported greater stimulation and drug craving the second time receiving amphetamine, an effect that was not observed in the unpaired group.[125] This finding suggests that environmental context enhances the stimulant effects and the rewarding effects of amphetamine. de Wit has also shown that humans develop CPP for ethanol. When social drinkers are allowed to explore one of two rooms, they spend more time in the room that had previously been associated with ethanol.[126] Overall, these studies show that humans can be conditioned to associate one environment with a specific drug. We can see examples of CPP in the real world. Individuals with alcohol use disorder prefer to stay in an environment that allows them to consume ethanol, such as at home or at a bar.

Advantages and disadvantages of CPP

A major advantage to CPP is that it can be used to study more than just drugs. Animals will develop CPP to natural rewards like food, water, social interaction, and novelty. This is important because if we want to determine if a potential pharmacotherapy blocks the conditioned rewarding effects of a drug, we want to ensure that the pharmacotherapy does not interfere with general reward learning. For example, Ma et al. found that the drug ifenprodil, an antagonist at GluN2B-containing NMDA receptors, blocks the acquisition and the expression of morphine CPP without affecting food CPP or social interaction CPP.[127] This finding provides greater credence to the idea of targeting the GluN2B subunit for the treatment of SUDs.

Another advantage of CPP is that it does not require catheter implantation surgery like drug self-administration. Catheter implantation surgeries can be laborious and can be difficult to train in some individuals. This poses a challenge for researchers that work primarily with undergraduate research assistants, although it is possible to conduct drug self-administration experiments at primarily undergraduate institutions. Researchers such as myself and Dr. Mark Smith of Davidson College conduct drug self-administration experiments while working with undergraduate students.[128,129] Regardless, because catheter implantation surgery is invasive, complications can arise during or following surgery. There is always the risk that an animal reacts poorly to anesthesia or has health complications following surgery. These issues can lead to *attrition*, thus decreasing one's sample size. Because CPP does not involve surgery, the risk of attrition is negligible.

One major disadvantage to CPP is that it requires more animal subjects compared to drug self-administration. In drug self-administration, multiple doses of a drug can be tested in the same animal to generate a *dose-response curve*. As discussed in the previous chapter, responding to a drug typically follows an inverted U-shape function; that is, animals respond maximally to intermediate doses of a drug. In CPP, because the same dose is used during conditioning, separate groups are needed to test the conditioned rewarding effects of various doses of the drug. To provide an illustration: in the previous chapter, you saw a graph of an inverted U-shaped dose-response curve for cocaine (see Fig. 5.6 of the previous chapter). Using six doses of cocaine (0.03, 0.06, 0.12, 0.25, 0.5, and 1.0 mg/kg/inf), Piazza et al. found that rats respond most for 0.12 mg/kg/inf of cocaine, followed by 0.25, 0.06, 0.5, 0.03, and 1.0 mg/kg/inf, respectively.[130] In a CPP experiment, generating a dose response curve would require six separate groups of animals, greatly increasing the number of animal subjects needed. This is cost prohibitive for some researchers.

Using Pavlovian concepts to treat addiction

Because Pavlovian conditioning appears to play a pivotal role in precipitating relapse, several therapeutic approaches have been used to mitigate the role of CSs on conditioned drug cravings. A few of the major techniques will be reviewed below.

Cue exposure therapy (extinction training)

In Pavlovian conditioning, repeatedly presenting the CS without presenting the US results in extinction; that is, over time, the animal will stop showing a CR when presented with the

CS. Given that individuals with a SUD experience cravings and physiological changes when exposed to a drug-paired stimulus, one therapeutic approach is to repeatedly expose an individual to drug-related stimuli without giving him/her the opportunity to use the drug. This is known as **cue exposure therapy**. Cue exposure therapy reduces cravings and drug use in a laboratory setting.[131–133] The major limitation to cue exposure therapy is that its effectiveness is limited to the environment in which one experiences the therapy. What I mean by this is that extinguishing craving in a lab setting does not mean that extinction has occurred in other types of settings. As you have learned, the environment can serve as a powerful stimulus that signals the availability of the drug. Because cue exposure therapy has occurred in a single environment, there is no guarantee that the individual will be able to avoid conditioned withdrawal and/or conditioned craving. How can one ensure that cue exposure therapy can be implemented outside of a laboratory setting?

As of February 2021, approximately 97% of Americans own a cell phone. Some research has attempted to take advantage of this fact by providing individuals *extinction reminders* during cue exposure therapy. During cue exposure therapy, a therapist can present a novel stimulus like a tone when an individual reports little to no drug craving. This tone becomes a CS that predicts the absence of drug craving. When the individual encounters an environment associated with drug use, the individual can call a number to hear the tone. Presentation of the tone can help reduce drug cravings.[134]

Virtual reality exposure therapy (VRET)

Virtual reality is a technology in which a user experiences computer-generated images while wearing a headset. In recent years, video games have incorporated virtual reality. You may be familiar with Oculus or PlayStation VR. However, virtual reality can be used to treat multiple psychiatric conditions, including SUDs, in the form of **virtual reality exposure therapy (VRET)**.[135] As the name suggests, VRET combines cue exposure therapy with virtual reality. The researcher generates images of environments in which drug-taking behaviors normally occur and can present the participant with images of drug-specific stimuli. For example, in one study, Hernández-Serrano et al. presented participants with images of four environments associated with ethanol use: bar, restaurant, pub, and home. Each environment included images of ethanol, and participants could interact with the alcohol bottles via the virtual reality setup.[136] So far, VRET has been effective at reducing craving for ethanol[136,137] and cigarettes.[138] One interesting finding is that the effectiveness of VRET has been shown to depend on the status of one's drug use. For example, using highly realistic scenarios decreases cravings for current smokers, but increases cravings in those that have recently quit smoking.[138]

Although VRET improves the generalizability of cue exposure therapy, there are several weaknesses to using this approach for the treatment of SUDs. As noted by Vinci et al., creating virtual environments is cost and time-prohibitive.[139] I can attest to the time commitment needed to develop a virtual environment. Although I am an animal researcher, I supervised an honors student who wanted to create a virtual environment to help those with social anxiety disorder. My student spent most of the semester creating a single environment. Now, imagine having to create multiple environments like the study described above. The second criticism of VRET is that the images projected to the user are not always realistic.

This can limit how much the individual immerses him or herself into the environment. Related to the first point, because creating multiple environments can take a considerable amount of time, studies often use a single environment. This is problematic because this limits the generalizability of the treatment to multiple contexts. One can use ethanol in multiple environments, such as at home, at a friend's house, at a bar/pub, at a sporting event, and so on.

Augmented reality

If you ever played *Pokémon Go*, you have experienced **augmented reality**, a technology that superimposes digital objects into the real world through a device like a smartphone or a headset. According to Vinci et al., addiction scientists should start using augmented reality to help treat individuals with a SUD.[139] Researchers can project drug-related stimuli such as glasses of ethanol, ashtrays, or syringes into the individual's environment. This approach is advantageous as it allows individuals to experience cue exposure therapy in their living space or in areas in which they would realistically use the drug. Another advantage is that VR equipment is not needed. Despite these advantages, augmented reality has not been extensively tested as a potential therapeutic for SUDs. However, one study used augmented reality to teach children about the health impacts of smoking.[140] Children from 5 to 13 years of age could point a tablet at a mannequin wearing a T-shirt. Images of the lungs or the heart appeared on the shirt. These images depicted the heart/lungs of a healthy individual or the heart/lungs of a chronic smoker (Fig. 6.10). Following the demonstration, children reported a decreased interest in tobacco products. Given that approximately 85% of individuals in the United States own a smartphone,[141] augmented reality is an easy way to engage individuals in cue exposure therapy.

Aversion therapy and counterconditioning

In **aversion therapy**, the goal is to pair something aversive with unwanted behavior. In the novel *A Clockwork Orange*, the main character Alex is the leader of a gang and engages in

FIGURE 6.10 An example of how augmented reality can be used to teach individuals about the risks of drug use. In the current application, individuals can use a tablet to project an image of internal organs on a T-shirt worn by a mannequin. Individuals can examine healthy lungs or lungs affected by chronic smoking. *Image comes from Fig. 4.31 Borovanska Z, Poyade M, Rea PM, Buksh ID. Engaging with children using augmented reality on clothing to prevent them from smoking. In: Rea PM, ed.* Advances in Experimental Medicine and Biology. *Springer; 2020;1262:59–94. https://doi.org/10.1007/978-3-030-43961-3_4. Copyright 2020.*

violent acts. Eventually, he is arrested and subjected to a form of aversion therapy known as the Ludovico Technique. Alex is injected with a drug that makes him sick and is forced to view violent videos. Following aversion therapy, Alex becomes physically ill when he thinks of violence. As related to SUDs, you have already learned about one form of aversion therapy: the use of disulfiram (Antabuse) for the treatment of alcohol use disorder. If an individual drinks ethanol after taking disulfiram, they will experience nausea. The goal is for the individual to associate nausea with ethanol use, thus decreasing their desire to use ethanol in the future.

Aversion therapy is a form of **counterconditioning**. The goal of counterconditioning, as the name suggests, is to reverse (counter) conditioning that has previously occurred. Because drug-paired stimuli can elicit strong cravings, the goal of aversion therapy is to make these stimuli less enticing. While disulfiram can decrease ethanol consumption, additional research has examined alternative forms of aversion therapy for other drugs. In an early study, individuals with a SUD received mild electric shocks to the wrist as they verbalized and imagined past drug-using experiences.[142] Participants reported decreased drug-related thoughts and showed low relapse rates 2 years following the experiment. One caveat to this experiment is that participants were also in group therapy; therefore, the low relapse rates may have been influenced more by this form of therapy instead of aversion therapy. In addition to cocaine, the use of electric shocks has been used to decrease ethanol,[143] nicotine,[144] and cannabis[145] use. While disulfiram is the only FDA-approved form of aversion therapy for alcohol use disorder, other chemicals have been used in counterconditioning experiments. In one experiment, Frawley and Smith[146] allowed patients to either snort a cocaine substitute (combination of 2% tetracaine and 1% quinine), smoke a cocaine substitute (either white candy or white soap that emits smoke when burned), or "inject" a white powder (D-xylose) dissolved in a liquid (note, patients did not actually inject anything into their arm; the liquid is pushed out of the syringe just above the forearm). Before patients engaged in these "drug-taking behaviors", they were given emetine. If you are not familiar with emetine, it is an alkaloid found in ipecac syrup, which causes extreme nausea and vomiting. The emetine was delivered such that its effects would occur around the same time as the patient "used" cocaine. After 18 months following the experiment, 38% of cocaine users reported complete abstinence.

One major drawback to aversion therapy is that compliance is often low. Individuals do not want to engage in activities that are uncomfortable. Counterconditioning does not have to incorporate aversive stimuli. Recently, virtual reality has been combined with counterconditioning to better decrease drug cravings. The Experiment Spotlight details such an experiment.

Experiment Spotlight: A virtual reality counterconditioning procedure to reduce methamphetamine cue-induced craving (Wang et al.[147])

Considering there are currently no approved pharmacotherapies for methamphetamine use disorder, Wang et al.[147] determined if a virtual reality counterconditioning procedure (VRCP) can be an effective treatment option for those with methamphetamine use disorder. Initially, 87 individuals were interviewed about their fears related to methamphetamine use. The six most common fears were: (1) being arrested by the police; (2) experiencing severe hallucinations; (3)

developing skin ulcers or an infection; (4) contracting a sexually transmitted illness; (5) tooth decay; and (6) death. The researchers used these concerns to design six VR videos. Each video consisted of two parts. The first part of each video was composed of three scenes: (1) a black screen with voices telling the participant to use methamphetamine; (2) a person using methamphetamine-related paraphernalia; and (3) individuals using methamphetamine in a room. The second part of the video depicted one of the six common fears described above (e.g., being arrested). The figure below shows example screenshots of one VR video.

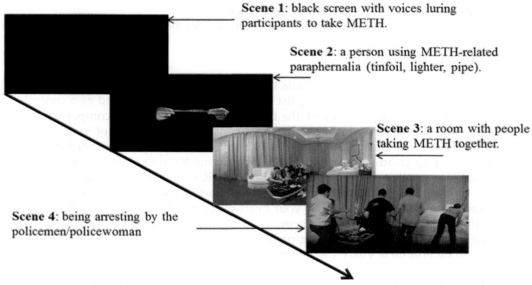

The image comes from Fig. 2 of Wang YG, Liu MH, Shen ZH. A virtual reality counterconditioning procedure to reduce methamphetamine cue-induced craving. J Psychiatr Res. 2019;116:88—94. https://doi.org/10.1016/j.jpsychires.2019.06.007.

There were two groups in this experiment. The first group experienced six VR videos and continued to receive their normal therapy. The second group was in a *waiting-list control group* in which they continued receiving their usual therapy only. Immediately before and during testing, individuals were given an electrocardiogram (ECG or EKG) to measure electrical signals in the heart. Participants were asked to rate their drug craving, drug liking, and drug use before and following the VRCP using a *visual analog scale*. The figure below shows the data for this experiment.

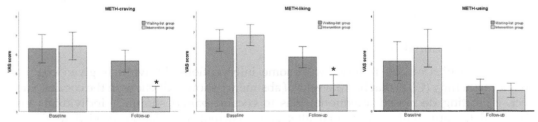

The image comes from Fig. 3 of Wang YG, Liu MH, Shen ZH. A virtual reality counterconditioning procedure to reduce methamphetamine cue-induced craving. J Psychiatr Res. 2019;116:88—94. https://doi.org/10.1016/j.jpsychires.2019.06.007. Copyright 2019, with permission from Elsevier.

Participants exposed to VRCP had decreased drug craving and drug liking scores compared to participants that received treatment as usual. Both groups showed decreased methamphetamine use at the follow-up compared to baseline.

Questions to consider:

(1) Why does the waiting-list control group receive treatment as well? Should the researchers have included a control group that does not receive any treatment?
(2) Why was an ECG used in the current experiment?
(3) Why do you think methamphetamine use decreased for both groups of participants?

Answers to questions:

(1) From an ethical standpoint, individuals that are currently seeking treatment for a SUD cannot be denied treatment. Waiting-list control groups are often used as a control group in these experiments because the treatment of interest can be compared to another group that does not receive that treatment. A control group of individuals that does not receive any treatment can be used. If an individual specifies that they are not currently seeking treatment for their SUD, they do not have to receive treatment.
(2) Because CRs can elicit physiological reactions, the ECG is used to measure something known as *heart rate variability (HRV)*. When our heart beats, it does not do so in a consistent fashion. For simplicity, if your heart beats 60 times per minute, this does not mean that your heart beats exactly every second. Instead, your heart may beat at 0.8 s and then at 1.1 s. This is HRV. Drug-paired cues can increase HRV. Thus, Wang et al. wanted to determine if VRCP can reduce this drug-induced increase in HRV. This is exactly what they found.
(3) If you look closely at the *y*-axis of each graph above, you will see that drug craving and drug liking scores go up to eight on the graph; however, the methamphetamine use scale ranges from 0 to 4. Considering that both groups were at 2–2.5, there is not much room to observe differential decreases in methamphetamine use across groups (i.e., this is a floor effect). Another thing to consider is that the control group received treatment as usual during the experiment. Thus, their normal treatment may have been effective at reducing methamphetamine use without altering cravings or liking for the drug.

One advantage to using virtual reality in conjunction with counterconditioning is that the therapist does not need to make the patient sick to create a negative association between the drug and drug-paired stimuli.

Reducing attentional bias

Attentional bias can be reduced with some interventions. Individuals given cognitive-behavioral therapy (CBT) that had 3 weeks of abstinence show reductions in the cocaine Stroop effect, indicating decreased attentional bias to cocaine-paired stimuli.[70] Individuals given behavioral therapy with transcranial alternating current stimulation applied to the PFC show reduced attentional bias toward cigarette-paired stimuli.[148] Certain pharmacological interventions can ameliorate attentional bias to drug-paired stimuli. For example, the ADHD medication atomoxetine (Strattera) reduces attentional bias to cocaine cues.[149]

Another way to improve attentional bias is by using the Alcohol/Drug Attention Control Training Program (AACTP/DACTP). Originally developed to reduce attentional bias to ethanol-paired cues,[150] the AACTP/DACTP exposes individuals to three "levels" of the drug Stroop task that increase in difficulty. During the first level, individuals are presented with a single image, either an alcoholic beverage/drug-paired stimulus or a neutral stimulus. These images are presented in random order and are surrounded by a colored background. The second level is similar to the first level, with the exception that each stimulus is surrounded by a colored outline as opposed to a background. During the third level, both stimuli are presented simultaneously, and each stimulus is surrounded by a colored outline (as in the second level). During the first two levels, individuals are asked to name the color of the background/border, and reaction times are recorded for each type of stimulus. During the third level, individuals name the color of the border for the neutral stimulus. This forces the individual to ignore the ethanol/drug-paired stimulus. To progress from one level to the next, individuals need to display a certain level of accuracy (e.g., 90%) while responding within a particular timeframe.[151] Within an individual level, the time needed to respond decreases. This manipulation is important because it forces individuals to decrease the amount of time needed to divert their attention away from drug-paired stimuli to neutral stimuli. During training, individuals receive feedback from the experimenter, which allows them to track their progress during the training program. So far, the AACTP/DACTP has been shown to reduce attentional bias to ethanol-, opioid-, and methamphetamine-paired cues.[150,151] Individuals given the DACTP also report less temptation to use drugs and are less likely to relapse following training.[151] Likewise, hazardous drinkers (i.e., drinking 22–50 units [men] or 15–35 units [women] of ethanol within a 1-week period) and harmful drinkers (drinking more than 50 units [men] or 35 units [women] of ethanol within a 1-week period) report less attentional bias following the AACTP, and harmful drinkers consume less ethanol following training.[150]

Quiz 6.2

1. Which of the following drugs is the most difficult to establish consistent CPP?
   a. Cocaine
   b. Ethanol
   c. Morphine
   d. PCP
2. What are two advantages of CPP over drug self-administration?
3. How does a CPP acquisition experiment differ from a CPP expression study?
4. Which of the following therapeutic approaches is most likely to cause compliance issues in patients?
   a. Aversion therapy
   b. Augmented reality
   c. Counterconditioning
   d. Cue exposure therapy

5. Which therapy enables the patient to use a smartphone to insert drug-paired stimuli into their environment?
   a. Aversion therapy
   b. Augmented reality
   c. Counterconditioning
   d. Virtual reality exposure therapy

Answers to quiz 6.2

1. d — PCP.
2. CPP can be tested with other types of USs like food, which allows one to determine if a potential pharmacotherapy specifically decreases CPP for drugs. CPP also does not require surgical interventions, which decreases the likelihood of attrition.
3. To determine if a potential pharmacotherapy alters the acquisition of CPP, the pharmacotherapy is administered before the drug during conditioning sessions. To determine if a potential pharmacotherapy alters the expression of CPP, the pharmacotherapy is administered before the posttest.
4. a — Aversion therapy.
5. b — Augmented reality.

Chapter summary

The finding that drug-paired stimuli can precipitate relapse poses a significant challenge for individuals trying to abstain from drug use. Numerous stimuli can become paired with the drug either directly (such as an ashtray or a syringe) or indirectly (objects in the environment in which drug use occurs). Individuals at risk of developing a SUD are drawn to these drug-paired stimuli (attentional bias), which can trigger intense physiological and emotional responses such as drug craving. This can then lead to relapse even if someone has been abstinent for years. Even if techniques like cue exposure therapy or counterconditioning can reduce the incentive salience to some drug-paired stimuli, there are many other stimuli that can still serve as a CS that can trigger withdrawal symptoms and/or drug cravings. Pavlovian conditioning may also be a major cause of overdose deaths. Take the story of Tom that you read about in the Introduction of the previous chapter. Tom's heroin overdose occurred in a motel room, which may not have been where Tom normally used heroin. Even if Tom used the same amount of heroin as normal, the change in environment, in conjunction with combining heroin with cocaine, could have significantly contributed to overdose. Overall, Pavlovian conditioning is a major contributing factor to SUDs and often works in conjunction with operant processes to influence an individual's continued drug use, even when they no longer enjoy the experience.

Glossary

Attentional bias — focusing on certain elements in the environment while ignoring other salient elements.
Augmented reality — a technology that allows one to project virtual images into the environment.

Aversion therapy — a form of conditioning in which an unpleasant stimulus is paired with another stimulus in an effort to decrease an organism's preference for that stimulus; e.g., disulfiram treatment is a form of aversion therapy aimed at reducing ethanol use.

Conditioned response (CR) — a type of behavior that only results after an organism has learned the association between two stimuli (e.g., salivating to the sound of a metronome that has been paired with food).

Conditioned stimulus (CS) — a type of stimulus that has been repeatedly paired with a stimulus that naturally elicits a response from an organism; over time, this stimulus can elicit a response from the organism (e.g., a metronome that has been repeatedly paired with food).

Conditioned tolerance — a phenomenon in which individuals develop tolerance to the effects of a drug in some environments, but not in others.

Conditioned withdrawal — a phenomenon in which the presentation of a drug-paired stimulus can elicit withdrawal symptoms.

Counterconditioning — a conditioning technique in which the pleasurable or aversive effects of a conditioned stimulus are reversed by pairing that stimulus with a new unconditioned stimulus.

Cue exposure therapy — a form of therapy in which individuals are repeatedly exposed to a conditioned stimulus without presenting the unconditioned stimulus. Eventually, individuals will learn that the conditioned stimulus no longer predicts the onset of the unconditioned stimulus (also known as extinction training).

Goal-tracking — a behavioral phenotype characterized by enhanced attention to a reinforcer instead of reinforcer-paired stimuli.

Higher-order (second-order) conditioning — a form of Pavlovian conditioning in which an already established conditioned stimulus can be paired with a new neutral stimulus, making the new neutral stimulus a second conditioned stimulus; note that the second conditioned stimulus is never paired with the unconditioned stimulus.

Incentive sensitization theory of addiction — emphasizes that continuous exposure to drug-paired cues causes the neural circuitry underlying incentive salience to become hypersensitive, causing these cues to elicit intense drug cravings.

Pavlovian (classical) conditioning — a form of learning in which an organism learns to associate two stimuli with one another.

Pavlovian conditioned approach (PCA) — an experimental procedure used to measure a type of behavioral response that occurs following the presentation of a conditioned stimulus that predicts the onset of an unconditioned stimulus.

Pavlovian-to-instrumental transfer (PIT) — the phenomenon in which conditioned stimuli alter operant conditioning.

Sensory preconditioning — a form of Pavlovian conditioning in which a neutral stimulus that has never been paired with an unconditioned stimulus can elicit a conditioned response; this occurs because the neutral stimulus was previously paired with a second neutral stimulus that was eventually paired with the unconditioned stimulus.

Sign-tracking — a behavioral phenotype characterized by excessive interaction with stimuli that become paired with a reinforcer.

Stimulus substitution theory of conditioning — Ivan Pavlov's theory of conditioning states that conditioned stimuli take on the same properties as unconditioned stimuli, thus eliciting conditioned responses that are identical to unconditioned responses.

Unconditioned response (UR) — behavior that results from the presentation of a stimulus that naturally elicits that behavior (e.g., salivation in response to food presentation).

Unconditioned stimulus (US) — a type of stimulus that naturally elicits a response from an organism (e.g., food).

Virtual reality exposure therapy (VRET) — a form of therapy in which individuals are exposed to problematic stimuli using a virtual reality headset; this allows the user to experience these stimuli without having to directly interact with them.

Supplementary data related to this chapter can be found online at https://educate.elsevier.com/9780323905787.

References

1. Chapel Hill Medical Detox. *Interview with a Heroin Addict*; May 27, 2020. https://www.chapelhilldetox.com/interview-with-a-heroin-addict/.

2. Pavlov IP. *Conditioned Reflexes: an Investigation of the Physiological Activity of the Cerebral Cortex*. Oxford University Press; 1927.
3. Brogden WJ. Sensory pre-conditioning. *J Exp Psychol*. 1939;25(4):323–332. https://doi.org/10.1037/h0058944.
4. Finch G, Culler E. Higher order conditioning with constant motivation. *Am J Psychol*. 1934;46(4):596–602.
5. Newlin DB. Conditioned compensatory response to alcohol placebo in humans. *Psychopharmacology*. 1986;88(2):247–251. https://doi.org/10.1007/BF00652249.
6. Hearst E, Jenkins H. *Sign-tracking: the Stimulus-Reinforcer Relation and Directed Action*. Austin: Monograph of the Psychonomic Society; 1974.
7. Flagel SB, Akil H, Robinson TE. Individual differences in the attribution of incentive salience to reward-related cues: implications for addiction. *Neuropharmacology*. 2009;56(Supplement 1):139–148. https://doi.org/10.1016/j.neuropharm.2008.06.027.
8. Uslaner JM, Acerbo MJ, Jones SA, Robinson TE. The attribution of incentive salience to a stimulus that signals an intravenous injection of cocaine. *Behav Brain Res*. 2006;169(2):320–324. https://doi.org/10.1016/j.bbr.2006.02.001.
9. Cunningham CL, Patel P. Rapid induction of Pavlovian approach to an ethanol-paired visual cue in mice. *Psychopharmacology*. 2007;192(2):231–241. https://doi.org/10.1007/s00213-007-0704-4.
10. Krank MD. Pavlovian conditioning with ethanol: sign-tracking (autoshaping), conditioned incentive, and ethanol self-administration. *Alcohol Clin Exp Res*. 2003;27(10):1592–1598. https://doi.org/10.1097/01.ALC.0000092060.09228.DE.
11. Yager LM, Robinson TE. Individual variation in the motivational properties of a nicotine cue: sign-trackers vs. goal-trackers. *Psychopharmacology*. 2015;232(17):3149–3160. https://doi.org/10.1007/s00213-015-3962-6.
12. Yager LM, Pitchers KK, Flagel SB, Robinson TE. Individual variation in the motivational and neurobiological effects of an opioid cue. *Neuropsychopharmacology*. 2015;40(5):1269–1277. https://doi.org/10.1038/npp.2014.314.
13. Tomie A, Grimes KL, Pohorecky LA. Behavioral characteristics and neurobiological substrates shared by Pavlovian sign-tracking and drug abuse. *Brain Res Rev*. 2008;58(1):121–135. https://doi.org/10.1016/j.brainresrev.2007.12.003.
14. Beckmann JS, Marusich JA, Gipson CD, Bardo MT. Novelty seeking, incentive salience and acquisition of cocaine self-administration in the rat. *Behav Brain Res*. 2011;216(1):159–165. https://doi.org/10.1016/j.bbr.2010.07.022.
15. Kawa AB, Bentzley BS, Robinson TE. Less is more: prolonged intermittent access cocaine self-administration produces incentive-sensitization and addiction-like behavior. *Psychopharmacology*. 2016;233(19–20):3587–3602. https://doi.org/10.1007/s00213-016-4393-8.
16. Tunstall BJ, Kearns DN. Sign-tracking predicts increased choice of cocaine over food in rats. *Behav Brain Res*. 2015;281:222–228. https://doi.org/10.1016/j.bbr.2014.12.034.
17. Everett NA, Carey HA, Cornish JL, Baracz SJ. Sign tracking predicts cue-induced but not drug-primed reinstatement to methamphetamine seeking in rats: effects of oxytocin treatment. *J Psychopharmacol*. 2020;34(11):1271–1279. https://doi.org/10.1177/0269881120954052.
18. Versaggi CL, King CP, Meyer PJ. The tendency to sign-track predicts cue-induced reinstatement during nicotine self-administration, and is enhanced by nicotine but not ethanol. *Psychopharmacology*. 2016;233(15–16):2985–x2997. https://doi.org/10.1007/s00213-016-4341-7.
19. Pohořalá V, Enkel T, Bartsch D, Spanagel R, Bernardi RE. Sign- and goal-tracking score does not correlate with addiction-like behavior following prolonged cocaine self-administration. *Psychopharmacology*. 2021;238(8):2335–2346. https://doi.org/10.1007/s00213-021-05858-z.
20. Overby PF, Daniels CW, Del Franco A, et al. Effects of nicotine self-administration on incentive salience in male Sprague Dawley rats. *Psychopharmacology*. 2018;235(4):1121–1130. https://doi.org/10.1007/s00213-018-4829-4.
21. Saddoris MP, Wang X, Sugam JA, Carelli RM. Cocaine self-administration experience induces pathological phasic accumbens dopamine signals and abnormal incentive behaviors in drug-abstinent rats. *J Neurosci*. 2016;36(1):235–250. https://doi.org/10.1523/JNEUROSCI.3468-15.2016.
22. Stringfield SJ, Madayag AC, Boettiger CA, Robinson DL. Sex differences in nicotine-enhanced Pavlovian conditioned approach in rats. *Biol Sex Differ*. 2019;10(1):37. https://doi.org/10.1186/s13293-019-0244-8.

References

23. Madayag AC, Stringfield SJ, Reissner KJ, Boettiger CA, Robinson DL. Sex and adolescent ethanol exposure influence Pavlovian conditioned approach. *Alcohol Clin Exp Res*. 2017;41(4):846–856. https://doi.org/10.1111/acer.13354.
24. McClory AJ, Spear LP. Effects of ethanol exposure during adolescence or in adulthood on Pavlovian conditioned approach in Sprague-Dawley rats. *Alcohol*. 2014;48(8):755–763. https://doi.org/10.1016/j.alcohol.2014.05.006.
25. Fiorenza AM, Shnitko TA, Sullivan KM, et al. Ethanol exposure history and alcoholic reward differentially alter dopamine release in the nucleus accumbens to a reward-predictive cue. *Alcohol Clin Exp Res*. 2018;42(6):1051–1061. https://doi.org/10.1111/acer.13636.
26. Walitzer KS, Sher KJ. Alcohol cue reactivity and ad lib drinking in young men at risk for alcoholism. *Addict Behav*. 1990;15(1):29–46. https://doi.org/10.1016/0306-4603(90)90005-i.
27. Witteman J, Post H, Tarvainen M, et al. Cue reactivity and its relation to craving and relapse in alcohol dependence: a combined laboratory and field study. *Psychopharmacology*. 2015;232(20):3685–3696. https://doi.org/10.1007/s00213-015-4027-6.
28. Payne TJ, Smith PO, Sturges LV, Holleran SA. Reactivity to smoking cues: mediating roles of nicotine dependence and duration of deprivation. *Addict Behav*. 1996;21(2):139–154. https://doi.org/10.1016/0306-4603(95)00043-7.
29. Robbins SJ, Ehrman RN, Childress AR, O'Brien CP. Using cue reactivity to screen medications for cocaine abuse: a test of amantadine hydrochloride. *Addict Behav*. 1992;17(5):491–499. https://doi.org/10.1016/0306-4603(92)90009-k.
30. Franken IH, de Haan HA, van der Meer CW, Haffmans PM, Hendriks VM. Cue reactivity and effects of cue exposure in abstinent posttreatment drug users. *J Subst Abuse Treat*. 1999;16(1):81–85. https://doi.org/10.1016/s0740-5472(98)00004-x.
31. Li Q, Li W, Wang H, et al. Predicting subsequent relapse by drug-related cue-induced brain activation in heroin addiction: an event-related functional magnetic resonance imaging study. *Addiction Biol*. 2015;20(5):968–978. https://doi.org/10.1111/adb.12182.
32. Gray KM, LaRowe SD, Upadhyaya P. Cue reactivity in young marijuana smokers: a preliminary investigation. *Psychol Addict Behav*. 2008;22(4):582–586. https://doi.org/10.1037/a0012985.
33. Fleming KA, Cofresí RU, Bartholow BD. Transfer of incentive salience from a first-order alcohol cue to a novel second-order alcohol cue among individuals at risk for alcohol use disorder: electrophysiological evidence. *Addiction*. 2021;116(7):1734–1746. https://doi.org/10.1111/add.15380.
34. Kang OS, Chang DS, Jahng GH, et al. Individual differences in smoking-related cue reactivity in smokers: an eye-tracking and fMRI study. *Prog Neuro Psychopharmacol Biol Psychiatr*. 2012;38(2):285–293. https://doi.org/10.1016/j.pnpbp.2012.04.013.
35. Soleymani A, Ivanov Y, Mathot S, de Jong PJ. Free-viewing multi-stimulus eye tracking task to index attention bias for alcohol versus soda cues: satisfactory reliability and criterion validity. *Addict Behav*. 2020;100:106117. https://doi.org/10.1016/j.addbeh.2019.106117.
36. McAteer AM, Hanna D, Curran D. Age-related differences in alcohol attention bias: a cross-sectional study. *Psychopharmacology*. 2018;235(8):2387–2393. https://doi.org/10.1007/s00213-018-4935-3.
37. McGivern C, Curran D, Hanna D. Alcohol attention bias in 14-16 year old adolescents: an eye tracking study. *Psychopharmacology*. 2021;238(3):655–664. https://doi.org/10.1007/s00213-020-05714-6.
38. Monem RG, Fillmore MT. Measuring heightened attention to alcohol in a naturalistic setting: a validation study. *Exp Clin Psychopharmacol*. 2017;25(6):496–502. https://doi.org/10.1037/pha0000157.
39. Zhao Q, Li H, Hu B, et al. Neural correlates of drug-related attentional bias in heroin dependence. *Front Hum Neurosci*. 2018;11:646. https://doi.org/10.3389/fnhum.2017.00646.
40. Murphy A, Taylor E, Elliott R. The detrimental effects of emotional process dysregulation on decision-making in substance dependence. *Front Integr Neurosci*. 2012;6:101. https://doi.org/10.3389/fnint.2012.00101.
41. Miller MA, Fillmore MT. The effect of image complexity on attentional bias towards alcohol-related images in adult drinkers. *Addiction*. 2010;105(5):883–890. https://doi.org/10.1111/j.1360-0443.2009.02860.x.
42. Townshend JM, Duka T. Attentional bias associated with alcohol cues: differences between heavy and occasional social drinkers. *Psychopharmacology*. 2001;157(1):67–74. https://doi.org/10.1007/s002130100764.

43. Ehrman RN, Robbins SJ, Bromwell MA, Lankford ME, Monterosso JR, O'Brien CP. Comparing attentional bias to smoking cues in current smokers, former smokers, and non-smokers using a dot-probe task. *Drug Alcohol Depend*. 2002;67(2):185–191. https://doi.org/10.1016/s0376-8716(02)00065-0.
44. Hogarth LC, Mogg K, Bradley BP, Duka T, Dickinson A. Attentional orienting towards smoking-related stimuli. *Behav Pharmacol*. 2003;14(2):153–160. https://doi.org/10.1097/00008877-200303000-00007.
45. Constantinou N, Morgan CJ, Battistella S, O'Ryan D, Davis P, Curran HV. Attentional bias, inhibitory control and acute stress in current and former opiate addicts. *Drug Alcohol Depend*. 2010;109(1–3):220–225. https://doi.org/10.1016/j.drugalcdep.2010.01.012.
46. Garland EL, Froeliger BE, Passik SD, Howard MO. Attentional bias for prescription opioid cues among opioid dependent chronic pain patients. *J Behav Med*. 2013;36(6):611–620. https://doi.org/10.1007/s10865-012-9455-8.
47. Garland EL, Howard MO. Opioid attentional bias and cue-elicited craving predict future risk of prescription opioid misuse among chronic pain patients. *Drug Alcohol Depend*. 2014;144:283–287. https://doi.org/10.1016/j.drugalcdep.2014.09.014.
48. Morgan CJ, Rees H, Curran HV. Attentional bias to incentive stimuli in frequent ketamine users. *Psychol Med*. 2008;38(9):1331–1340. https://doi.org/10.1017/S0033291707002450.
49. Marks KR, Pike E, Stoops WW, Rush CR. The magnitude of drug attentional bias is specific to substance use disorder. *Psychol Addict Behav*. 2015;29(3):690–695. https://doi.org/10.1037/adb0000084.
50. Lochbuehler K, Wileyto EP, Tang KZ, Mercincavage M, Cappella JN, Strasser AA. Do current and former cigarette smokers have an attentional bias for e-cigarette cues? *J Psychopharmacol*. 2018;32(3):316–323. https://doi.org/10.1177/0269881117728418.
51. MacLean RR, Sofuoglu M, Waters AJ. Naturalistic measurement of dual cue attentional bias in moderate to heavy-drinking smokers: a preliminary investigation. *Drug Alcohol Depend*. 2020;209:107892. https://doi.org/10.1016/j.drugalcdep.2020.107892.
52. Field M, Mogg K, Mann B, Bennett GA, Bradley BP. Attentional biases in abstinent alcoholics and their association with craving. *Psychol Addict Behav*. 2013;27(1):71–80. https://doi.org/10.1037/a0029626.
53. Field M, Mogg K, Bradley BP. Cognitive bias and drug craving in recreational cannabis users. *Drug Alcohol Depend*. 2004;74(1):105–111. https://doi.org/10.1016/j.drugalcdep.2003.12.005.
54. Manchery L, Yarmush DE, Luehring-Jones P, Erblich J. Attentional bias to alcohol stimuli predicts elevated cue-induced craving in young adult social drinkers. *Addict Behav*. 2017;70:14–17. https://doi.org/10.1016/j.addbeh.2017.01.035.
55. Begh R, Smith M, Ferguson SG, Shiffman S, Munafò MR, Aveyard P. Association between smoking-related attentional bias and craving measured in the clinic and in the natural environment. *Psychol Addict Behav*. 2016;30(8):868–875. https://doi.org/10.1037/adb0000231.
56. Díaz-Batanero C, Domínguez-Salas S, Moraleda E, Fernández-Calderón F, Lozano OM. Attentional bias toward alcohol stimuli as a predictor of treatment retention in cocaine dependence and alcohol user patients. *Drug Alcohol Depend*. 2018;182:40–47. https://doi.org/10.1016/j.drugalcdep.2017.10.005.
57. Spruyt A, De Houwer J, Tibboel H, et al. On the predictive validity of automatically activated approach/avoidance tendencies in abstaining alcohol-dependent patients. *Drug Alcohol Depend*. 2013;127(1–3):81–86. https://doi.org/10.1016/j.drugalcdep.2012.06.019.
58. Szasz PL, Szentagotai A, Hofmann SG. Effects of emotion regulation strategies on smoking craving, attentional bias, and task persistence. *Behav Res Ther*. 2012;50(5):333–340. https://doi.org/10.1016/j.brat.2012.02.010.
59. Cox WM, Fadardi JS, Pothos EM. The addiction-stroop test: theoretical considerations and procedural recommendations. *Psychol Bull*. 2006;132(3):443–476. https://doi.org/10.1037/0033-2909.132.3.443.
60. Hallgren KA, McCrady BS. Interference in the alcohol Stroop task with college student binge drinkers. *J Behav Health*. 2013;2(2):112–119. https://doi.org/10.5455/jbh.20130224082728.
61. Canamar CP, London E. Acute cigarette smoking reduces latencies on a Smoking Stroop test. *Addict Behav*. 2012;37(5):627–631. https://doi.org/10.1016/j.addbeh.2012.01.017.
62. Greenaway R, Mogg K, Bradley BP. Attentional bias for smoking-related information in pregnant women: relationships with smoking experience, smoking attitudes and perceived harm to foetus. *Addict Behav*. 2012;37(9):1025–1028. https://doi.org/10.1016/j.addbeh.2012.04.005.
63. Rusted JM, Caulfield D, King L, Goode A. Moving out of the laboratory: does nicotine improve everyday attention? *Behav Pharmacol*. 2000;11(7–8):621–629. https://doi.org/10.1097/00008877-200011000-00009.

64. Cousijn J, Watson P, Koenders L, Vingerhoets WA, Goudriaan AE, Wiers RW. Cannabis dependence, cognitive control and attentional bias for cannabis words. *Addict Behav.* 2013;38(12):2825–2832. https://doi.org/10.1016/j.addbeh.2013.08.011.
65. Field M. Cannabis 'dependence' and attentional bias for cannabis-related words. *Behav Pharmacol.* 2005;16(5–6):473–476. https://doi.org/10.1097/00008877-200509000-00021.
66. Metrik J, Aston ER, Kahler CW, et al. Cue-elicited increases in incentive salience for marijuana: craving, demand, and attentional bias. *Drug Alcohol Depend.* 2016;167:82–88. https://doi.org/10.1016/j.drugalcdep.2016.07.027.
67. Gardini S, Caffarra P, Venneri A. Decreased drug-cue-induced attentional bias in individuals with treated and untreated drug dependence. *Acta Neuropsychiatr.* 2009;21(4):179–185. https://doi.org/10.1111/j.1601-5215.2009.00389.x.
68. MacLean RR, Sofuoglu M, Brede E, Robinson C, Waters AJ. Attentional bias in opioid users: a systematic review and meta-analysis. *Drug Alcohol Depend.* 2018;191:270–278. https://doi.org/10.1016/j.drugalcdep.2018.07.012.
69. Carpenter KM, Schreiber E, Church S, McDowell D. Drug Stroop performance: relationships with primary substance of use and treatment outcome in a drug-dependent outpatient sample. *Addict Behav.* 2006;31(1):174–181. https://doi.org/10.1016/j.addbeh.2005.04.012.
70. DeVito EE, Kiluk BD, Nich C, Mouratidis M, Carroll KM. Drug Stroop: mechanisms of response to computerized cognitive behavioral therapy for cocaine dependence in a randomized clinical trial. *Drug Alcohol Depend.* 2018;183:162–168. https://doi.org/10.1016/j.drugalcdep.2017.10.022.
71. Cox WM, Hogan LM, Kristian MR, Race JH. Alcohol attentional bias as a predictor of alcohol abusers' treatment outcome. *Drug Alcohol Depend.* 2002;68(3):237–243. https://doi.org/10.1016/s0376-8716(02)00219-3.
72. Nuijten M, Blanken P, Van den Brink W, Goudriaan AE, Hendriks VM. Impulsivity and attentional bias as predictors of modafinil treatment outcome for retention and drug use in crack-cocaine dependent patients: results of a randomised controlled trial. *J Psychopharmacol.* 2016;30(7):616–626. https://doi.org/10.1177/0269881116645268.
73. Marissen MAE, Franken IHA, Waters AJ, Blanken P, van den Brink W, Hendriks VM. Attentional bias predicts heroin relapse following treatment. *Addiction.* 2006;101(9):1306–1312. https://doi.org/10.1111/j.1360-0443.2006.01498.x.
74. Hester R, Lee N, Pennay A, Nielsen S, Ferris J. The effects of modafinil treatment on neuropsychological and attentional bias performance during 7-day inpatient withdrawal from methamphetamine dependence. *Exp Clin Psychopharmacol.* 2010;18(6):489–497. https://doi.org/10.1037/a0021791.
75. Marhe R, Waters AJ, van de Wetering BJM, Franken IHA. Implicit and explicit drug-related cognitions during detoxification treatment are associated with drug relapse: an ecological momentary assessment study. *J Consult Clin Psychol.* 2013;81(1):1–12. https://doi.org/10.1037/a0030754.
76. Cousijn J, van Benthem P, van der Schee E, Spijkerman R. Motivational and control mechanisms underlying adolescent cannabis use disorders: a prospective study. *Develop Cogn Neurosci.* 2015;16:36–45. https://doi.org/10.1016/j.dcn.2015.04.001.
77. Carroll ME, Lac ST. Autoshaping i.v. cocaine self-administration in rats: effects of nondrug alternative reinforcers on acquisition. *Psychopharmacology.* 1993;110(1–2):5–12. https://doi.org/10.1007/BF02246944.
78. Cartoni E, Balleine B, Baldassarre G. Appetitive pavlovian-instrumental transfer: a review. *Neurosci Biobehav Rev.* 2016;71:829–848. https://doi.org/10.1016/j.neubiorev.2016.09.020.
79. Alarcón DE, Delamater AR. Outcome-specific Pavlovian-to-instrumental transfer (PIT) with alcohol cues and its extinction. *Alcohol.* 2019;76:131–146. https://doi.org/10.1016/j.alcohol.2018.09.003.
80. Corbit LH, Janak PH. Ethanol-associated cues produce general pavlovian-instrumental transfer. *Alcohol Clin Exp Res.* 2007;31(5):766–774. https://doi.org/10.1111/j.1530-0277.2007.00359.x.
81. Corbit LH, Janak PH. Changes in the influence of alcohol-paired stimuli on alcohol seeking across extended training. *Front Psychiatr.* 2016;7:169. https://doi.org/10.3389/fpsyt.2016.00169.
82. Glasner SV, Overmier JB, Balleine BW. The role of Pavlovian cues in alcohol seeking in dependent and nondependent rats. *J Stud Alcohol.* 2005;66(1):53–61. https://doi.org/10.15288/jsa.2005.66.53.
83. Krank MD, O'Neill S, Squarey K, Jacob J. Goal- and signal-directed incentive: conditioned approach, seeking, and consumption established with unsweetened alcohol in rats. *Psychopharmacology.* 2008;196(3):397–405. https://doi.org/10.1007/s00213-007-0971-0.

84. Lamb RJ, Ginsburg BC, Schindler CW. Effects of an ethanol-paired CS on responding for ethanol and food: comparisons with a stimulus in a Truly-Random-Control group and to a food-paired CS on responding for food. *Alcohol*. 2016;57:15–27. https://doi.org/10.1016/j.alcohol.2016.10.009.
85. Lamb RJ, Ginsburg BC, Greig A, Schindler CW. Effects of rat strain and method of inducing ethanol drinking on Pavlovian-Instrumental-Transfer with ethanol-paired conditioned stimuli. *Alcohol*. 2019;79:47–57. https://doi.org/10.1016/j.alcohol.2019.01.003.
86. Lamb RJ, Schindler CW, Ginsburg BC. Ethanol-paired stimuli can increase reinforced ethanol responding. *Alcohol*. 2020;85:27–34. https://doi.org/10.1016/j.alcohol.2019.10.007.
87. Milton AL, Schramm MJ, Wawrzynski JR, et al. Antagonism at NMDA receptors, but not β-adrenergic receptors, disrupts the reconsolidation of pavlovian conditioned approach and instrumental transfer for ethanol-associated conditioned stimuli. *Psychopharmacology*. 2012;219(3):751–761. https://doi.org/10.1007/s00213-011-2399-9.
88. Kruzich PJ, Congleton KM, See RE. Conditioned reinstatement of drug-seeking behavior with a discrete compound stimulus classically conditioned with intravenous cocaine. *Behav Neurosci*. 2001;115(5):1086–1092. https://doi.org/10.1037//0735-7044.115.5.1086.
89. LeBlanc KH, Ostlund SB, Maidment NT. Pavlovian-to-instrumental transfer in cocaine seeking rats. *Behav Neurosci*. 2012;126(5):681–689. https://doi.org/10.1037/a0029534.
90. Takahashi TT, Vengeliene V, Enkel T, Reithofer S, Spanagel R. Pavlovian to instrumental transfer responses do not correlate with addiction-like behavior in rats. *Front Behav Neurosci*. 2019;13:129. https://doi.org/10.3389/fnbeh.2019.00129.
91. Siegel S. Conditioning of insulin-induced glycemia. *J Comp Physiol Psychol*. 1972;78(2):233–241. https://doi.org/10.1037/h0032180.
92. Kimble GA. *Hilgard and Marquis' Conditioning and Learning*. 2nd ed. Appleton-Century-Crofts; 1961.
93. Siegel S, Hinson RE, Krank MD. The role of predrug signals in morphine analgesic tolerance: support for a Pavlovian conditioning model of tolerance. *J Exp Psychol Anim Behav Process*. 1978;4(2):188–196. https://doi.org/10.1037/0097-7403.4.2.188.
94. Ehrman R, Ternes J, O'Brien CP, McLellan AT. Conditioned tolerance in human opiate addicts. *Psychopharmacology*. 1992;108(1–2):218–224. https://doi.org/10.1007/BF02245311.
95. Krank MD, Perkins WL. Conditioned withdrawal signs elicited by contextual cues for morphine administration. *Psychobiology*. 1993;21(2):113–119.
96. Robinson TE, Berridge KC. The neural basis of drug craving: an incentive-sensitization theory of addiction. *Brain Res Rev*. 1993;18(3):247–291. https://doi.org/10.1016/0165-0173(93)90013-p.
97. Robinson MJ, Fischer AM, Ahuja A, Lesser EN, Maniates H. Roles of "wanting" and "liking" in motivating behavior: Gambling, food, and drug addictions. *Curr Top Behav Neurosci*. 2016;27:105–136. https://doi.org/10.1007/7854_2015_387.
98. Peciña S, Berridge KC. Hedonic hot spot in nucleus accumbens shell: where do mu-opioids cause increased hedonic impact of sweetness? *J Neurosci*. 2005;25(50):11777–11786. https://doi.org/10.1523/JNEUROSCI.2329-05.2005.
99. Smith KS, Berridge KC. The ventral pallidum and hedonic reward: neurochemical maps of sucrose "liking" and food intake. *J Neurosci*. 2005;25(38):8637–8649. https://doi.org/10.1523/JNEUROSCI.1902-05.2005.
100. Kringelbach ML, Berridge KC. Neuroscience of reward, motivation, and drive. In: Kim S-I, Reeve J, Bong M, eds. *Recent Developments in Neuroscience Research on Human Motivation (Advances in Motivation and Achievement)*. Emerald Publishing Limited; 2016:23–35. https://doi.org/10.1108/S0749-742320160000019020.
101. Mahler SV, Smith KS, Berridge KC. Endocannabinoid hedonic hotspot for sensory pleasure: nandamide in nucleus accumbens shell enhances 'liking' of a sweet reward. *Neuropsychopharmacology*. 2007;32(11):2267–2278. https://doi.org/10.1038/sj.npp.1301376.
102. Berridge KC, Robinson TE. Liking, wanting, and the incentive-sensitization theory of addiction. *Am Psychol*. 2016;71(8):670–679. https://doi.org/10.1037/amp0000059.
103. Parkinson JA, Dalley JW, Cardinal RN, et al. Nucleus accumbens dopamine depletion impairs both acquisition and performance of appetitive Pavlovian approach behaviour: implications for mesoaccumbens dopamine function. *Behav Brain Res*. 2002;137(1–2):149–163. https://doi.org/10.1016/s0166-4328(02)00291-7.

104. Saddoris MP, Carelli RM. Cocaine self-administration abolishes associative neural encoding in the nucleus accumbens necessary for higher-order learning. *Biol Psychiatr.* 2014;75(2):156–164. https://doi.org/10.1016/j.biopsych.2013.07.037.
105. Bardo MT, Bevins RA. Conditioned place preference: what does it add to our preclinical understanding of drug reward? *Psychopharmacology.* 2000;153:31–43. https://doi.org/10.1007/s002130000569.
106. Lew G, Parker LA. Pentobarbital-induced place aversion learning. *Anim Learn Behav.* 1998;26:219–224. https://doi.org/10.3758/BF03199214.
107. Bossert JM, Franklin KB. Pentobarbital-induced place preference in rats is blocked by GABA, dopamine, and opioid antagonists. *Psychopharmacology.* 2001;157(2):115–122. https://doi.org/10.1007/s002130100772.
108. Acquas E, Carboni E, Leone P, Di Chiara G. SCH 23390 blocks drug-conditioned place-preference and place-aversion: anhedonia (lack of reward) or apathy (lack of motivation) after dopamine-receptor blockade? *Psychopharmacology.* 1989;99(2):151–155. https://doi.org/10.1007/BF00442800.
109. Noda Y, Nabeshima T. Neuronal mechanisms of phencyclidine-induced place aversion and preference in the conditioned place preference task. *Methods Find Exp Clin Pharmacol.* 1998;20(7):607–611. https://doi.org/10.1358/mf.1998.20.7.485726.
110. Abiero A, Botanas CJ, Custodio RJ, et al. 4-Meo-PCP and 3-MeO-PCMo, new dissociative drugs, produce rewarding and reinforcing effects through activation of mesolimbic dopamine pathway and alteration of accumbal CREB, deltaFosB, and BDNF levels. *Psychopharmacology.* 2020;237(3):757–772. https://doi.org/10.1007/s00213-019-05412-y.
111. Ryu IS, Kim OH, Lee YE, et al. The abuse potential of novel synthetic phencyclidine derivative 1-(1-(4-fluorophenyl)cyclohexyl)piperidine (4′-F-PCP) in rodents. *Int J Mol Sci.* 2020;21(13):4631. https://doi.org/10.3390/ijms21134631.
112. Kitaichi K, Noda Y, Hasegawa T, Furukawa H, Nabeshima T. Acute phencyclidine induces aversion, but repeated phencyclidine induces preference in the place conditioning test in rats. *Eur J Pharmacol.* 1996;318(1):7–9. https://doi.org/10.1016/s0014-2999(96)00875-8.
113. Miyamoto Y, Noda Y, Komori Y, Sugihara H, Furukawa H, Nabeshima T. Involvement of nitric oxide in phencyclidine-induced place aversion and preference in mice. *Behav Brain Res.* 2000;116(2):187–196. https://doi.org/10.1016/s0166-4328(00)00274-6.
114. Lett BT. Repeated exposures intensify rather than diminish the rewarding effects of amphetamine, morphine, and cocaine. *Psychopharmacology.* 1989;98(3):357–362. https://doi.org/10.1007/BF00451687.
115. Shoaib M, Stolerman IP, Kumar RC. Nicotine-induced place preferences following prior nicotine exposure in rats. *Psychopharmacology.* 1994;113(3–4):445–452. https://doi.org/10.1007/BF02245221.
116. Randich A, LoLordo VM. Associative and nonassociative theories of the UCS preexposure phenomenon: implications for Pavlovian conditioning. *Psychol Bull.* 1979;86(3):523–548. https://doi.org/10.1037/0033-2909.86.3.523.
117. Bardo MT, Horton DB, Yates JR. Conditioned place preference as a preclinical model for screening pharmacotherapies for drug abuse. In: Markgraf CG, Hudzik TJ, Comptom DR, eds. *Nonclinical Assessment of Abuse Potential for New Pharmacotherapies.* London: Academic Press; 2015:151–196.
118. Marglin SH, Milano WC, Mattie ME, Reid LD. PCP and conditioned place preferences. *Pharmacol, Biochem Behav.* 1989;33(2):281–383. https://doi.org/10.1016/0091-3057(89)90500-5.
119. Carr GD, Fibiger HC, Phillips AG. Conditioned place preference as a measure of drug reward. In: Liebman JM, Cooper SJ, eds. *Neuropharmacological Basis of Reward.* Oxford; 1989:264–319.
120. Cunningham CL, Ferree NK, Howard MA. Apparatus bias and place conditioning with ethanol in mice. *Psychopharmacology.* 2003;170(4):409–422. https://doi.org/10.1007/s00213-003-1559-y.
121. Rodríguez-Arias M, Valverde O, Daza-Losada M, Blanco-Gandía MC, Aguilar MA, Miñarro J. Assessment of the abuse potential of MDMA in the conditioned place preference paradigm: role of CB1 receptors. *Prog Neuro Psychopharmacol Biol Psychiatr.* 2013;47:77–84. https://doi.org/10.1016/j.pnpbp.2013.07.013.
122. Vidal-Infer A, Roger-Sanchez C, Daza-Losada M, Aguilar MA, Minarro J, Rodriguez-Arias M. Role of the dopaminergic system in the acquisition, expression and reinstatement of MDMA-induced conditioned place preference in adolescent mice. *PLoS One.* 2012;7(8):e43107. https://doi.org/10.1371/journal.pone.0043107.
123. Yates JR, Campbell HL, Hawley LL, Horchar MJ, Kappesser JL, Wright MR. Effects of the GluN2B-selective antagonist Ro 63-1908 on acquisition and expression, of methamphetamine conditioned place preference in

male and female rats. *Drug Alcohol Depend.* 2021;225:108785. https://doi.org/10.1016/j.drugalcdep.2021.108785.
124. Childs E, de Wit H. Amphetamine-induced place preference in humans. *Biol Psychiatr.* 2009;65(10):900–904. https://doi.org/10.1016/j.biopsych.2008.11.016.
125. Childs E, de Wit H. Contextual conditioning enhances the psychostimulant and incentive properties of d-amphetamine in humans. *Addiction Biol.* 2013;18(6):985–992. https://doi.org/10.1111/j.1369-1600.2011.00416.x.
126. Childs E, de Wit H. Alcohol-induced place conditioning in moderate social drinkers. *Addiction.* 2016;111(12):2157–2165. https://doi.org/10.1111/add.13540.
127. Ma Y-Y, Guo C-Y, Yu P, Lee DY-W, Han J-S, Cui C-L. The role of NR2B containing NMDA receptor in place preference conditioned with morphine and natural reinforcers in rats. *Exp Neurol.* 2006;200:343–355. https://doi.org/10.1016/j.expneurol.2006.02.117.
128. Horchar MJ, Kappesser JL, Broderick MR, Wright MR, Yates JR. Effects of NMDA receptor antagonists on behavioral economic indices of cocaine self-administration. *Drug Alcohol Depend.* 2022;233:109348. https://doi.org/10.1016/j.drugalcdep.2022.109348.
129. Smith MA, Schmidt KT, Sharp JL, et al. Lack of evidence for positive reinforcing and prosocial effects of MDMA in pair-housed male and female rats. *Eur J Pharmacol.* 2021;913:174646. https://doi.org/10.1016/j.ejphar.2021.174646.
130. Piazza PV, Deroche-Gamonent V, Rouge-Pont F, Le Moal M. Vertical shifts in self-administration dose-response functions predict a drug-vulnerable phenotype predisposed to addiction. *J Neurosci.* 2000;20(11):4226–4232. https://doi.org/10.1523/JNEUROSCI.20-11-04226.2000.
131. Childress AR, McLellan AT, O'Brien CP. Abstinent opiate abusers exhibit conditioned craving, conditioned withdrawal and reductions in both through extinction. *Br J Addiction.* 1986;81(5):655–660. https://doi.org/10.1111/j.1360-0443.1986.tb00385.x.
132. Drummond DC, Glautier S. A controlled trial of cue exposure treatment in alcohol dependence. *J Consult Clin Psychol.* 1994;62(4):809–817. https://doi.org/10.1037//0022-006x.62.4.809.
133. Unrod M, Drobes DJ, Stasiewicz PR, et al. Decline in cue-provoked craving during cue exposure therapy for smoking cessation. *Nicotine Tob Res.* 2014;16(3):306–315. https://doi.org/10.1093/ntr/ntt145.
134. Rosenthal MZ, Kutlu MG. Translation of associative learning models into extinction reminders delivered via mobile phones during cue exposure interventions for substance use. *Psychol Addict Behav.* 2014;28(3):863–871. https://doi.org/10.1037/a0037082.
135. Maples-Keller JL, Bunnell BE, Kim SJ, Rothbaum BO. The use of virtual reality technology in the treatment of anxiety and other psychiatric disorders. *Harv Rev Psychiatr.* 2017;25(3):103–113. https://doi.org/10.1097/HRP.0000000000000138.
136. Hernández-Serrano O, Ghiţă A, Figueras-Puigderrajols N, et al. Predictors of changes in alcohol craving levels during virtual reality cue exposure treatment among patients with alcohol use disorder. *J Clin Med.* 2020;9(9):3018. https://doi.org/10.3390/jcm9093018.
137. Ghiţă A, Gutiérrez-Maldonado J. Applications of virtual reality in individuals with alcohol misuse: a systematic review. *Addict Behav.* 2018;81:1–11. https://doi.org/10.1016/j.addbeh.2018.01.036.
138. de Bruijn G-J, de Vries J, Bolman C, Wiers R. (No) escape from reality? Cigarette craving in virtual smoking environments. *J Behav Med.* 2021;44(1):138–143. https://doi.org/10.1007/s10865-020-00170-1.
139. Vinci C, Brandon KO, Kleinjan M, Brandon TH. The clinical potential of augmented reality. *Clin Psychol Sci Pract.* 2020;27(3):e12357. https://doi.org/10.1111/cpsp.12357.
140. Borovanska Z, Poyade M, Rea PM, Buksh ID. Engaging with children using augmented reality on clothing to prevent them from smoking. In: Rea PM, ed. *Advances in Experimental Medicine and Biology* Vol. 1262.
141. Pew Research Center. *Mobile Fact Sheet;* 2021. Accessed on May 29, 2021 https://www.pewresearch.org/internet/fact-sheet/mobile/.
142. Copemann CD. Drug addiction: II. An aversive counterconditioning technique for treatment. *Psychol Rep.* 1976;38(3 Pt 2):1271–1281. https://doi.org/10.2466/pr0.1976.38.3c.1271.
143. Jackson TR, Smith JW. A comparison of two aversion treatment methods for alcoholism. *J Stud Alcohol.* 1978;39(1):187–191. https://doi.org/10.15288/jsa.1978.39.187.
144. Smith JW. Long term outcome of clients treated in a commercial stop smoking program. *J Subst Abuse Treat.* 1988;5(1):33–36. https://doi.org/10.1016/0740-5472(88)90036-0.

145. Smith JW, Schmeling G, Knowles PL. A marijuana smoking cessation clinical trial utilizing THC-free marijuana, aversion therapy, and self-management counseling. *J Subst Abuse Treat*. 1988;5(2):89–98. https://doi.org/10.1016/0740-5472(88)90018-9.
146. Frawley PJ, Smith JW. Chemical aversion therapy in the treatment of cocaine dependence as part of a multimodal treatment program: treatment outcome. *J Subst Abuse Treat*. 1990;7(1):21–29. https://doi.org/10.1016/0740-5472(90)90033-m.
147. Wang YG, Liu MH, Shen ZH. A virtual reality counterconditioning procedure to reduce methamphetamine cue-induced craving. *J Psychiatr Res*. 2019;116:88–94. https://doi.org/10.1016/j.jpsychires.2019.06.007.
148. Mondino M, Lenglos C, Cinti A, Renauld E, Fecteau S. Eye tracking of smoking-related stimuli in tobacco use disorder: a proof-of-concept study combining attention bias modification with alpha-transcranial alternating current stimulation. *Drug Alcohol Depend*. 2020;214:108152. https://doi.org/10.1016/j.drugalcdep.2020.108152.
149. Passamonti L, Luijten M, Ziauddeen H, et al. Atomoxetine effects on attentional bias to drug-related cues in cocaine dependent individuals. *Psychopharmacology*. 2017;234(15):2289–2297. https://doi.org/10.1007/s00213-017-4643-4.
150. Fadardi JS, Cox WM. Reversing the sequence: reducing alcohol consumption by overcoming alcohol attentional bias. *Drug Alcohol Depend*. 2009;101(3):137–145. https://doi.org/10.1016/j.drugalcdep.2008.11.015.
151. Ziaee SS, Fadardi JS, Cox WM, Yazdi SA. Effects of attention control training on drug abusers' attentional bias and treatment outcome. *J Consult Clin Psychol*. 2016;84(10):861–873. https://doi.org/10.1037/a0040290.

CHAPTER 7

Attentional and memory processes underlying addiction

Introduction

"I mean every time I went to treatment I had some good clean time behind me, but I don't know, I always went back to using again" (p. 721).[1]

This quote highlights one of the most challenging aspects of addiction: relapse. Even if someone has been abstinent for years, relapse can occur, which can lead to overdose and death. The actor Phillip Seymour Hoffman is just one example of an individual who relapsed after years of sobriety. Within 2 years of his relapse, Hoffman died from a drug overdose. You have already learned that environmental stimuli that become paired with drugs can elicit intense cravings, thus precipitating relapse. Another difficulty that individuals face when trying to abstain from drug use is that these drug-paired stimuli are highly effective in directing one's attention away from other stimuli. To further understand why this is the case, you need to understand some of the cognitive processes involved in addiction. This chapter will introduce the cognitive model of addiction and will discuss two major cognitive processes implicated in the addiction process: attention and memory. In particular, memory dysfunction has been proposed to underlie relapse. By understanding the cognitive deficits of individuals with a substance use disorder (SUD), we can develop better strategies for treating someone with an addiction.

Learning objectives

By the end of this chapter, you should be able to ...

(1) Identify cognitive distortions commonly observed in individuals with a SUD.
(2) Discuss how drugs alter attentional processing.
(3) Compare and contrast the various methods used to measure relapse-like behavior.

(4) Explain how the neurobiological and cellular mechanisms of memory overlap with the neuromechanisms associated with addiction.
(5) Describe how cognitive-behavioral therapy is used to treat those with a SUD.

Cognitive model of addiction

Whereas traditional disease models view SUDs as a biological disease that can be worsened by environmental stimuli, the cognitive model proposes that individuals possess negative psychological beliefs/attitudes that prevent them from abstaining from drug use. Specifically, Dr. Aaron Beck (1921–2021) proposed that SUDs result from **cognitive distortions**, or irrational thought patterns.[2] Examples of cognitive distortions include:

1. All-or-nothing thinking—also known as *splitting* or "black-and-white thinking", this type of distortion is characterized by thinking in extremes. For example, an individual with a SUD may think that they are a total failure because they have recently relapsed.
2. Magnification and minimization—giving greater weight to perceived failure or weaknesses and less weight to perceived success or strengths. Related to the example above, the individual places greater weight on relapsing than on the progress they have made to stop using a substance.
3. Catastrophizing—a form of magnification in which the individual gives greater weight to the worst possible outcome or claims that a mildly uncomfortable situation as being unbearable or impossible to handle. An individual that has relapsed may begin thinking that they have "ruined" their life and that there is no path to recovery.
4. Overgeneralization—assumption that a single event or occurrence applies to all situations. For example, following relapse, an individual may think "I can never do anything right". Words like "always" or "never" are commonly used when overgeneralizations are made. *Labeling* is a form of overgeneralization in which an individual associates him or herself with a single attribute (label). If an individual relapses, they may label themselves as a "loser" or as a "failure" as you saw in the example for all-or-nothing thinking.
5. Disqualifying the positive—even when a positive experience occurs, an individual may downplay it. Imagine an individual has abstained from drug use for a 1-week period. Others may congratulate this individual for remaining abstinent for 1 week. The individual may think "these people don't actually care that I've been abstinent for a week. They are just acting nice".
6. Filtering — related to disqualifying the positive, filtering occurs when an individual ruminates on one negative event while ignoring other positive events. Even if 50 people praise the individual for attempting to quit drug use, the individual may focus on the single person that expressed doubts about their ability to achieve abstinence.[3,4]

The cognitive model of addiction focuses heavily on the concept of **self-efficacy**, a term first proposed by psychologist Dr. Albert Bandura (1925–2021).[5] Self-efficacy is defined as one's *perceived* ability to accomplish certain tasks or to cope with certain situations. The key word in this definition is "perceived". An individual with high self-efficacy has high confidence in his or her ability to complete tasks or to respond in an appropriate manner when faced with a hardship. If an individual with a SUD has low self-efficacy, their actions can lead to a *self-fulfilling prophecy*. They assume that they are a "loser" or that there is no way for them to "beat" their addiction. When they relapse, the individual can say "see, I knew I couldn't overcome my addiction. I guess I'm destined to be an addict".[6] Not surprisingly, self-efficacy is a major predictor of treatment outcomes for those with a SUD.[6] Individuals with low self-efficacy are more likely to relapse following treatment.[7] These cognitive distortions severely interfere with one's ability to make meaningful changes when seeking treatment for a SUD.

Now that you have a general understanding of the cognitive model of addiction, I need to discuss some of the specific cognitive processes involved in addiction. Cognition is a broad construct that encompasses multiple processes. As noted in the Introduction, this chapter will focus on just two specific cognitive processes: attention and memory. These are terms that you are most likely familiar with as these are processes that we constantly use. We need to take notice of stimuli in our environment (attention) to make sense of what is happening around us, and we need to be able to recall information that we have learned previously (memory). While these processes are critically important, they can become dysregulated in individuals with SUDs, as you saw in the previous chapter when I discussed *attentional bias*. Because the previous chapter discussed the contribution of attentional bias to the addiction process, I will focus on how drugs affect attentional control before discussing the contributions of memory to addiction.

Drug effects on attention and automaticity

Before discussing how drugs affect attention and **automaticity**, I need to briefly discuss what these cognitive processes are. I imagine each individual reading this book is familiar with attention. There are several major types of attention. One type of attention is *sustained attention*, which involves focusing on a stimulus or an event for an extended period of time. Another type of attention is *selective attention*, in which individuals must sustain their attention on a specific stimulus/event while ignoring other stimuli/events. For example, when having a conversation with another individual, you must engage in both sustained attention and selective attention. You want to focus on what the other individual has to say (sustained attention) while focusing less on other things going on around you (e.g., sounds from a TV, people walking by, the vibration of a cell phone, etc.). The opposite of selective attention is *divided attention*, also known as *multitasking*. Here, attention is split between two or more stimuli/activities. Driving and talking on a cell phone/texting is a dangerous form of divided attention.

Although attention is important for many activities, like taking notes during class or following instructions at work, there are numerous things that we do that require little effortful attention. Think back to the concept of habit formation. When we perform an action repeatedly, we often perform these behaviors without thinking too much about them. In cognitive terms, this is known as automaticity. When you were a child, tying your shoes may have been difficult at first. Now, you most likely do not even think when tying your shoes. Even complex actions can become automatized, such as driving. In the previous chapter, you learned about the emotional Stroop task. In the traditional Stroop task, participants view words for colors that are either printed in the same color as the word or in a different color as the word. For example, during a congruent trial, the word "blue" is presented in blue font. During incongruent trials, the word "blue" is presented in a different color font like red or green. The participant is asked to name the color of the font, not the word that is presented. When the color is incongruent with the word presented, individuals take longer to respond. Because reading becomes an automatized process, individuals have to focus their attention on the color of the font, not the word itself.

Given that drugs often produce intoxication, one would expect to observe altered attention after drug use. If you have ever seen someone inebriated after consuming ethanol, you have experienced some of the deleterious effects that this drug has on attention. Individuals may have increased difficulty paying attention to what you are saying, and they may be more easily distracted by other stimuli. Unsurprisingly, acute administration of depressant-like drugs impairs performance on measures of attention and automaticity. Participants given ethanol have difficulty maintaining their gaze on a fixation point on a screen, indicating impaired selective attention.[8] Ethanol administration also impairs sustained attention in humans[9] and in rats,[10] worsens divided attention,[11] and impairs performance on the Stroop task.[12] Similarly, cannabis use decreases accuracy and slows reaction time in a selective attention task,[13] impairs divided attention,[14] and negatively affects Stroop performance.[15] The opioid fentanyl leads to decreased sustained attention[16] while the dissociative hallucinogen ketamine impairs performance on the Stroop task.[17]

Not all drugs impair attention when given acutely. Certain drugs can *improve* aspects of attention. Acute administration of *d*-amphetamine or cocaine enhances sustained attention in animals, but only at lower doses; higher doses of *d*-amphetamine impair selective attention.[18] Amphetamine's ability to improve sustained attention has been replicated in humans.[19] However, recreational cocaine use decreases sustained attention[20]; interpreting this finding is difficult because the recreational cocaine users in this study smoked significantly higher levels of cannabis compared to the control group.[20] Thus, the increased cannabis use observed in recreational cocaine users may account for the decreased sustained attention reported in that study, not recreational cocaine use. Regardless, the ability of amphetamine to improve aspects of attention makes sense as stimulants are often used to treat attention deficit/hyperactivity disorder (ADHD). In fact, during World War II, Japanese pilots were often given methamphetamine to stay alert during flights.[21] Interestingly, although methamphetamine has been used to sustain attention and can be used to treat ADHD (as Desoxyn), it worsens performance in a Stroop task,[22] suggesting that this drug increases habitual-like responses.

Nicotine is another drug that can improve performance in attention-based tasks. Smoking improves performance in a *rapid visual information processing task*[23]; that is, individuals are better able to detect specific sequences of numbers that are flashed quickly on a screen. Even nicotine gum can improve sustained attention in smokers,[24] and nicotine gum increases reaction time on the Stroop task, indicating improved performance.[25] Related to this finding, when individuals are forced to abstain from cigarettes, they perform worse on the Stroop task.[26] Allowing individuals to smoke cigarettes before completing the Stroop task improves performance.[26] In animals, nicotine improves attentional processing.[27]

Why do cigarettes/nicotine improve attention? Remember that nicotine acts on nicotinic acetylcholine receptors. These receptors are important for attentional processes. Because nicotine stimulates these receptors, one hypothesis is that nicotine is a "cognitive enhancer".[28] Directly related to this finding is the high prevalence of tobacco use in individuals with schizophrenia (as you learned in Chapter 1). Schizophrenia is a debilitating cognitive disorder characterized by multiple symptoms, such as delusions, hallucinations, blunted affect, and disorganized thinking. Individuals with schizophrenia may smoke as a form of self-medication to ameliorate some of these cognitive disruptions. There is support to this hypothesis, as nicotine administration enhances attention in individuals with schizophrenia.[29] In rodents, researchers can model symptoms of schizophrenia by injecting a pregnant rodent with a bacterium that causes a specific immune response to occur. This can lead to cognitive deficits in rat pups that mirror schizophrenia. "Schizophrenia-like" rats that self-administer nicotine show improvements in attentional deficits,[30] further corroborating studies performed on individuals with schizophrenia.

Although acute administration of psychostimulants can improve attention, long-term drug use often negatively impacts attention[31] and Stroop performance.[32] Adults with comorbid ADHD and cocaine use disorder have difficulty sustaining their attention.[33] Additionally, increased use of ethanol, cigarettes, cannabis, amphetamine, or MDMA is linked to decrements in selective attention,[34-36] while chronic use of cannabis, cocaine, methamphetamine, or heroin is associated with increased errors in the Stroop task.[37-39] Individuals that began using cigarettes, cannabis, or cocaine at an earlier age have more difficulty performing a sustained attention task.[40-42] Early onset of cannabis use is also associated with greater decrements in Stroop performance.[43] In animals, chronic administration of *d*-amphetamine leads to increased difficulty ignoring a stimulus that is irrelevant, mirroring selective attention deficits.[44] Withdrawal from drug self-administration also impairs performance in an animal model of attention.[45]

If long-term stimulant use impairs attention, how do drugs like amphetamine (Adderall) and methylphenidate (Ritalin) improve attention in individuals with ADHD? Because individuals with ADHD have lower levels of dopamine and norepinephrine,[46] psychostimulant drugs normalize the amount of dopamine and norepinephrine in the brain. Misuse of psychostimulants leads to increased levels of dopamine and norepinephrine. In Chapter 3, I discussed the dopamine depletion hypothesis of addiction. This hypothesis posits that increased drug use causes neurons to release too much dopamine; over time, drugs become less effective at releasing further dopamine, leading to a hypoactive dopaminergic system. Similarly, overuse of psychostimulants can expend too much dopamine/norepinephrine. As dopamine and norepinephrine are important for attentional processes,[47] depleted dopamine and/or norepinephrine levels may lead to attentional deficits in individuals with or without ADHD that misuse psychostimulants over time.

One assumption an individual could make is that stopping long-term drug use should improve attentional deficits, right? The answer, unfortunately, is no. Withdrawal from drugs can further impair attentional control. Both acute and long-term withdrawal from ethanol decreases selective attention.[48] Individuals experiencing a "hangover" have impaired selective attention.[49] This finding makes sense if you have ever experienced a hangover (I was young once and experienced this feeling more times than I want to admit). Because hangovers are unpleasant, individuals can have difficulty focusing, especially when completing boring, tedious computer tasks. Even when individuals are not hungover, drinking ethanol the night before testing impairs sustained attention and selective attention, although divided attention is unaltered.[50] Withdrawal effects on attention are not isolated to ethanol. As already covered, nicotine withdrawal impairs Stroop performance. Nicotine withdrawal also negatively affects sustained attention.[51] Withdrawal from nicotine, as well as other drugs such as ethanol, negatively affects automaticity.[48,52] When animals go into acute withdrawal from nicotine, cocaine, or heroin, they make more errors in an attention-based task.[45,53]

Related to withdrawal effects on attention, one shocking finding is that impaired attention is observed in individuals that have been abstinent from drug use for an extended period. For example, although former drinkers have normal selective attention, they struggle in a divided attention task.[52] Individuals receiving methadone replacement therapy or that have abstained from opioid use perform worse on measures of sustained attention compared to nonusers.[54] This effect is not observed for all drugs, as 1-week abstinence from cannabis does not impair selective attention.[55] Collectively, these results suggest that certain drugs like ethanol and opioids can cause long-lasting effects on the brain that can lead to attentional deficits.

In Chapter 1, you learned about fetal alcohol syndrome, a condition marked by physical and cognitive alterations that occur in children exposed to ethanol during development. Prenatal exposure to ethanol, as well as prenatal exposure to other drugs, impairs sustained/selective attention.[56–59] More specifically, individuals prenatally exposed to ethanol or cocaine show impairments in selective attention when given visual stimuli, but not auditory stimuli.[60,61] These results suggest that prenatal drug exposure may have detrimental effects on visual processing but not necessarily on auditory processing. For those prenatally exposed to cocaine, the time needed to respond to stimuli decreases across the session, indicating a loss of sustained effort.[61] Prenatal exposure to ethanol and methamphetamine leads to impaired performance on the Stroop task.[62,63]

The relevance of the research findings highlighted above should be obvious. If you drive a vehicle, you (hopefully) know that driving under the influence of ethanol and/or drugs is a crime. This is because many substances impair our attention and increase the amount of time we need to respond to changes in our environment. In 2019, there were approximately 33,244 traffic-related fatalities in the United States, of which 9,236 (28%) involved an ethanol-impaired driver.[64] Experiments have been conducted to measure how drug use affects attention while driving. Obviously, for ethical and legal reasons, researchers cannot give participants ethanol and then let them drive a vehicle on the street. Instead, participants use a driving simulator following exposure to a substance. The researcher can have the participant engage in a secondary task while driving. For example, in the divided attention steering simulator (DASS) task, participants attempt to keep a car in the center of a road outline while scanning their periphery for a specific number (e.g., 2).[65] This mimics the divided attention individuals experience when driving (e.g., changing the radio station, talking with passengers, etc.). This may come as a shock to you, but drugs, particularly ethanol, severely impair one's ability to multitask while driving (sarcasm intended).[65,66] Even if someone has not met the legal requirement of being intoxicated (blood alcohol content of 0.08%), ethanol administration can impair simulated driving performance when individuals are given a divided attention task, an effect that is magnified if they have been instructed to stay awake for an extended period.[67] Furthermore, acute withdrawal from ethanol can impair simulated driving performance.[68] In other words, even driving with a hangover can be dangerous.

Ethanol is not the only substance that impairs driving. Cannabis also severely impacts one's ability to drive. One review found that driving under the influence of cannabis can double or triple the risk of being in a vehicular accident; this most likely occurs because of cannabis' deleterious effects on reaction time and divided attention.[69] Experiments have also shown that individuals under the influence of cannabis are more likely to depart from their lane during a simulated driving task.[14] With cannabis becoming legal or decriminalized across the United States, there are concerns that fatal car crashes will increase as a result of more individuals driving while under the influence of cannabis.[70]

Attention to drug warnings

Given that ethanol and tobacco products contain warning labels about the health consequences associated with their use, and advertisements for ethanol often encourage individuals to drink responsibly, eye-tracking experiments have been conducted to measure how much time individuals spend looking at these warnings/responsibility messages. Overall, warning labels often fail to capture the attention of individuals, particularly adolescents. In one study, adolescents failed to look at a warning label on a tobacco advertisement approximately 44% of the time.[71] Another study reported that adolescents spend the least amount of time viewing ethanol responsibility messages in magazine advertisements; instead, they spend most of their time viewing beverage bottles, product logos, and cartoon illustrations.[72] Furthermore, college-aged individuals spend little time reading antibinge drinking messages.[73] Evidence has even shown that smokers will actively avoid viewing warning labels on cigarette packages.[74]

Because individuals do not spend much time examining warning labels, efforts have been made to adjust these warnings to make them more salient. Something as simple as moving warning labels from the bottom of a cigarette package to the top can increase visual attention to these warnings.[75] In one study, individuals viewed alcoholic beverage containers that had either a moderately severe graphic health warning or a highly severe graphic health warning. For example, when highlighting the association between ethanol consumption and cirrhosis of the liver, the moderately severe image depicted a computer-generated image of the torso with the liver highlighted in red. The highly severe image depicted a real liver of an individual with cirrhosis. Individuals presented with highly severe images are more likely to report less desire to drink ethanol compared to those presented with moderately severe images.[76] Including graphic images in advertisements for tobacco products increases the time adolescents spend looking at a warning message[77] and increases attention to and recall of these types of messages.[78] At first glance, these results suggest that alcoholic beverages/tobacco products should contain graphic images to dissuade individuals from drinking/using tobacco. However, the use of graphic images can be counterproductive and can lead to a reduced desire to stop drinking.[79] This effect may occur because overly graphic images may pull attention away from health warnings or antidrinking messages. Somewhat similarly, the use of "disgusting" images on cigarette packages draws attention away from the warning text to the image itself.[80]

An alternative to using graphic images is to require standardized packaging that does not contain logos or branding. This practice has been adopted in some countries to curb tobacco use. In 2012, Australia became the first country to enforce standardized packaging for tobacco packaging. Eye-tracking studies have shown that individuals spend more time viewing health warning labels on standardized packages compared to packages that have brands/logos.[81] Another alternative is to use a warning that emphasizes the negative social impacts of drug use. For example, a cigarette package can contain an image depicting a child being exposed to cigarette smoke. Smokers and nonsmokers report increased motivation to quit/refrain from smoking when exposed to the socially framed graphic warning; however, these messages are less effective compared to graphic health warnings (e.g., depicting someone with lung cancer).[82]

Quiz 7.1

(1) Which drug typically improves attention when administered acutely?
   a. Cannabis
   b. Ethanol
   c. Heroin
   d. Nicotine
(2) Which cognitive procedure is used to measure automaticity?
   a. Dichotic listening task
   b. Dot-probe task
   c. Stroop task
   d. None of the above
(3) Which of the following is an example of disqualifying the positive?
   a. An individual calls themself a loser after relapsing.
   b. An individual experiences strong drug cravings following 6 months of sobriety; their first thought is "I'm never going to beat these urges. I may as start using again".
   c. An individual thinks they only received a job offer because the manager felt sorry for them.
   d. All of the above
(4) Which type of warning label appears to be the most effective at capturing one's attention?
   a. One in which the warning is placed at the bottom of the package
   b. One located on a package with no brands or logos
   c. One that contains highly graphic images
   d. One that contains moderately graphic images
(5) How is self-efficacy related to addiction?

Answers to quiz 7.1

(1) d — Nicotine
(2) c — Stroop task

(3) c — An individual thinks they only received a job offer because the manager felt sorry for them.
(4) b — One located on a package with no brands or logos
(5) Individuals with low self-efficacy have increased difficulty achieving abstinence after experiencing a relapse. They are more likely to adopt a view in which they believe they are destined to be "addicted" to the substance.

Drug effects on memory

As you have already learned, drug-paired stimuli can elicit intense cravings due to Pavlovian conditioning, thus precipitating relapse. Pavlovian conditioning requires the activation of information stored in memory (i.e., the association between the drug and other stimuli). This implies that memory plays a critical role in SUDs. Memory is not a unitary construct, as it can be fractioned into short-term (STM)/working memory and long-term memory (LTM), with each of these systems being further fractioned. In particular, LTM is divided into explicit and implicit memory. Explicit memory (also known as declarative memory) involves the conscious recollection of personal memories (*episodic memory*) and facts/concepts that one has learned (*semantic memory*). Implicit memory (also known as nondeclarative memory) stores processes that occur unconsciously. Pavlovian and operant conditioning are examples of implicit memory. Another type of implicit memory is *procedural memory*, which encodes processes in which we engage on a routine basis. Habits are a form of procedural memory.

Drug use, whether acute or chronic, is associated with memory deficits. Increased adolescent ethanol consumption is somewhat associated with impaired working memory performance measured 3 years later.[83] Acute cannabis use impairs working memory,[84] although the effects of chronic cannabis on working memory are mixed.[85,86] Recreational cannabis users perform worse on a measure of STM compared to controls.[87] Additionally, synthetic cannabinoid users perform worse on a working memory task compared to recreational cannabis users and noncannabis users, and they have impaired LTM.[88] Impaired working memory performance has been observed in individuals that use cocaine,[89] crack cocaine,[90] and ketamine,[91] and individuals that began using cocaine at an earlier age perform worse on measures of working memory.[41]

Those with a SUD experience memory deficits. Working memory is impaired in individuals with a cannabis use disorder[92] and a methamphetamine use disorder.[93] Individuals with a cocaine use disorder that also use cannabis have impaired working memory.[94] Even after abstinence, impaired working memory and LTM are observed.[95,96] For example, individuals on methadone maintenance therapy have impaired working memory,[97] although it is unclear if the methadone itself leads to memory deficits or if this is the result of long-term opioid use.

Adolescents that are prenatally exposed to ethanol exhibit impaired STM and LTM.[98] Importantly, these adolescents do not meet the diagnostic criteria for fetal alcohol syndrome. Young adults prenatally exposed to cannabis are more likely to begin using cannabis at an earlier age, which leads to memory deficits.[99] Prenatal exposure to opioids is associated with impaired working memory.[100] Rats given cocaine during adolescence show impairments in working memory and in LTM during early adulthood, but this impairment improves over time,[101] suggesting that drug-induced decrements in memory can recover with prolonged abstinence. While prenatal exposure to cocaine does not alter working memory in rats,[102] prenatal exposure to crack cocaine leads to impaired LTM.[103] One question you may be asking yourself is "how are rats exposed to crack?". Read the Experiment Spotlight to find out more about prenatal crack cocaine exposure.

Experiment Spotlight: Maternal crack cocaine use in rats leads to depressive- and anxiety-like behavior, memory impairment, and increased seizure susceptibility in the offspring (Pacheco et al.[103])

The number of infants exposed to drugs during gestation has increased since the year 2000. As described above, prenatal exposure to drugs like ethanol and cannabis leads to memory deficits later in life. The goal of the highlighted study was to determine if prenatal crack cocaine exposure negatively affects cognition (as well as other behaviors) in rats. Using rodents to answer this question is beneficial because other factors like maternal diet, potential exposure to toxins like lead, and genetics can be finely controlled. Because crack cocaine is primarily smoked, Pacheco et al.[103] wanted to expose pregnant rats to crack cocaine via the inhalation method. Obviously, the researchers cannot hold a pipe up to the rat and have them smoke it. Instead, the researchers placed the rats in an enclosed box connected to one end of a hose. The other end of the hose was connected to a contraption that transmits the smoke from heated crack cocaine to the enclosed box (see the figure on the next page). Pregnant rats were exposed to 200 mg of crack cocaine for 10 min per day for 16 days (fifth gestational day to 21st gestational day).

Following birth, pups lived with the mother until postnatal day (PND) 21. At PND 30, corresponding to adolescence, rats were given a battery of tests to determine how prenatal crack cocaine exposure influences anxiety-like behavior, locomotor activity, depressive-like behavior, and LTM. LTM was measured using a *step-down avoidance test*. In the step-down avoidance test, rats were placed on a platform. Rats could step off the platform to a grid floor, but the floor would deliver a 3-s foot shock. The researchers measured the latency for the rats to step off the platform. One day later, rats were tested to determine how long they would stay on the platform before stepping off. At PND 60 (corresponding to adulthood), rats were tested in the step-down avoidance test again.

The figure on the next page shows a schematic of the step-down avoidance test, as well as the effects of prenatal crack cocaine exposure on LTM. At PND 30, only male rats prenatally exposed to crack cocaine showed LTM impairments, whereas, at PND 60, both males and females experienced LTM deficits following prenatal crack cocaine.

Questions to consider:
(1) How is the step-down avoidance test a measure of LTM?
(2) Although I did not explain the rest of the results of the experiment, do you think that prenatal exposure to crack cocaine affects anxiety-like behavior, hyperactivity, and/or depressive-like behavior? Why or why not?
(3) Why do you think prenatal crack cocaine exposure affected LTM in male rats only at PND 30?

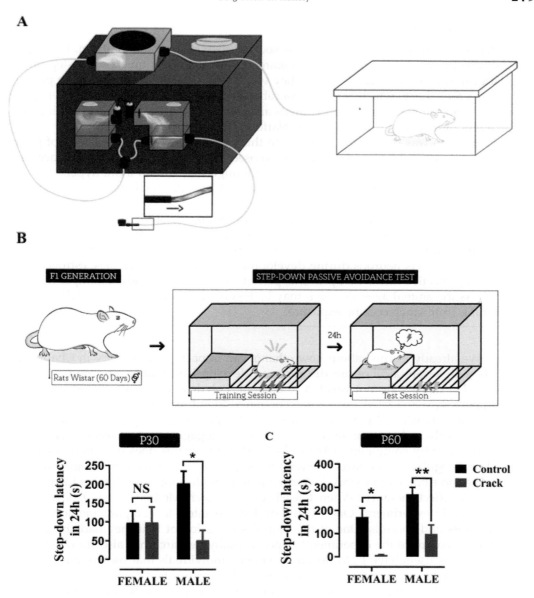

Panel (A) comes from panel (A) Fig. 1, and Panel (B) comes from Fig. 6 of Pacheco ALD, de Melo IS, de Souza FMA, et al. *Maternal crack cocaine use in rats leads to depressive- and anxiety-like behavior, memory impairment, and increased seizure susceptibility in the offspring.* Eur Neuropsychopharmacol. 2021;44:34–50. https://doi.org/10.1016/j.euroneuro.2020.12.011. Copyright 2021, with permission from Elsevier.

Answers to questions:

(1) In this test, rats receive a foot shock as soon as they step off the platform. During the test session, which is 24 h later, the researcher measures how quickly rats step off the platform. If rats have "good" LTM, they will remember that stepping off the platform led to a foot shock, which is not pleasurable. As such, they should take longer to step down from the platform. When animals are prenatally exposed to crack cocaine, they take less time to step down from the platform, indicating impaired LTM.

(2) This question was designed to get you to think about other potential effects of prenatal exposure to crack cocaine on rats. Rats exposed to crack cocaine experience increased anxiety-like and depressive-like behavior, although they are no more hyperactive compared to control rats. These results show that prenatal crack cocaine exposure has multiple deleterious effects in animals.

(3) As you will learn in Chapter 12, sex has a large influence on drug effects. Specifically, you will learn how hormones affect the reinforcing and the subjective effects of drugs. One important hormone you will read about is estrogen, which is found in females. Estrogen also affects neurons during development. In adolescent females, estrogen could act as a protective factor that prevents prenatal crack cocaine-induced impairments in LTM. As the animal ages, estrogen may be less effective in ameliorating LTM deficits resulting from crack cocaine exposure.

You have already learned about how ethanol can impact LTM memory in Chapter 1. Individuals with severe alcohol use disorder can develop Korsakoff's syndrome, marked by amnesia. LTM deficits are not isolated to ethanol. Polydrug use is associated with impairments in both STM and LTM, even when compared to individuals with alcohol use disorder or cocaine use disorder.[104] Drug use is associated with impaired episodic memory.[34] More specifically, methamphetamine users have difficulty engaging in free recall,[22] and MDMA users have difficulty completing LTM tasks.[105] While cannabis does not affect one's ability to recall or recognize certain facts,[106] individuals given cannabis have more difficulty recalling information from a short story.[15] Allow me to provide some anecdotal evidence of how cannabis can affect memory. When I teach our Research Methods class, I have students design a simple experiment. One of my students wanted to compare memory between cannabis users and controls. One of the cannabis users forgot that they had shown up the day before to complete the experiment! Experimentally, research has also shown that recreational cannabis users have LTM deficits.[87] Finally, individuals that use cocaine or MDMA have deficits in prospective memory (memory for future events).[107]

In contrast to the drugs listed above, nicotine tends to improve working memory.[95] Importantly, this finding is specific to individuals that regularly use nicotine products as those that do not use nicotine show impaired performance on a working memory task following acute nicotine exposure.[95] Nicotine administration improves immediate recall of words and tends to improve delayed recall in individuals with a stimulant use disorder, but not in individuals with alcohol use disorder.[108] Nicotine withdrawal worsens working memory,[109] but this impaired performance is not ameliorated if participants are allowed to smoke.[110] Nicotine withdrawal also negatively impacts episodic memory.[111] It is important to note that some inconsistencies have been reported concerning nicotine's effects on memory. Some work has shown impairments in working/STM and LTM in chronic cigarette users.[36,112]

Animal models have largely corroborated what clinical studies have shown: drug administration can have serious effects on working memory and on LTM. In monkeys, acute administration of cocaine impairs performance in a working memory task,[113] and ketamine negatively affects working memory.[114] Prenatal exposure to ethanol causes deficits in the radial arm maze in mice.[115] Chronic administration of methamphetamine impairs both working memory and LTM in rodents.[116,117] Just four injections of methamphetamine can lead to deficits in LTM.[118] Like humans, prenatal exposure to methamphetamine impairs both STM and LTM in mice.[119] In rodent models, nicotine improves performance in a radial arm maze, but nicotine withdrawal worsens performance.[120] One study reported an interesting effect of prenatal THC exposure on LTM in rats. THC-treated rats show impaired performance on an LTM task 1 day after learning the task; however, their LTM capabilities were comparable to controls 1 week after learning the task.[121] These results suggest that memory impairments resulting from prenatal drug exposure can improve over time, similar to what has been observed for drug-induced attentional deficits.

Cognitive impairments as a predictor of drug use problems

You have learned how acute and chronic drug administration can affect attention and memory. While drugs have large effects on these cognitive domains, research has also shown that impaired cognitive functioning can act as a predisposing factor for the development of a SUD.[122] Adolescents that have low working memory are at higher risk of developing a SUD, especially if they begin using a substance at an earlier age.[123] The correlation between the urge to smoke and nicotine dependence is observed in those with low to average working memory only.[124] Although working memory has not been directly linked to the development of cannabis use disorder, lower working memory is associated with problems resulting from cannabis use, such as procrastination, decreased energy, and decreased productivity.[125]

Impairments in cognition are also associated with poor treatment outcomes for those with a SUD.[126] Individuals with lower working memory are more likely to use methamphetamine during treatment[127] and resume cigarette smoking more rapidly following a brief period of abstinence.[128] Using the Montreal Cognitive Assessment to measure attention, memory, executive function, and language capabilities, Copersino et al. found that individuals with impaired cognitive functioning were less likely to attend each group therapy session.[129] In another study, individuals with a cocaine use disorder were given the Stroop task. Performance on the Stroop task was a significant predictor of treatment compliance. That is, individuals that performed worse on this task were more likely to stop treatment.[130] What does automaticity have to do with compliance? In the Stroop task, individuals must refrain from blurting out the word that they see printed on the screen. Instead, they must state the color in which the word is printed. This requires executive functioning. Someone with poor executive control may have difficulty completing therapy as treatment requires a long-term focus.

Modeling relapse-like behavior

In Chapters 3 and 4 I described how pharmacological and genetic manipulations can alter relapse-like behavior in animals. In Chapter 3, you were introduced to reinstatement and incubation of craving. The purpose of this section is to go into further detail as to how researchers measure relapse-like behavior. In addition to discussing reinstatement and incubation of craving in more detail, you will also learn about other methods researchers can use to elucidate the neuromechanisms of relapse (Fig. 7.1). Paradigms measuring relapse-like behavior are often performed following drug self-administration, although most can be measured with conditioned place preference (CPP). Before measuring relapse-like behavior in animals, researchers need to first decrease the reinforcing or conditioned rewarding effects of the drug. As such, I need to describe extinction in more detail.

Extinction

During extinction sessions, the drug reinforcer or the unconditioned stimulus are no longer presented to the animal. In a drug self-administration experiment, there are several ways to go about extinction. First, responses on the manipulandum no longer lead to a drug infusion.[131] Second, responses lead to an infusion of saline, which is physiologically inert.[132] Third, drug infusions can be paired with an aversive stimulus like foot shock.[131] Like drug self-administration, there are a couple of ways to extinguish an established CPP. The researcher can allow the animal to explore all compartments of the CPP chamber in a drug-free state for several days.[133] Alternatively, the researcher can inject the animal with saline before isolating them to one compartment on alternating sessions (i.e., both compartments become paired with saline).[134] One interesting finding observed in self-administration experiments is an **extinction burst**. At the beginning of extinction, animals will respond more on the manipulandum that was previously paired with a drug (or a natural) reinforcer.[135]

Importantly, during extinction, animals are *not forgetting* the original association between an operant response and drug delivery or the association between a drug and an environmental context. Instead, animals are learning a new association, specifically that the operant response no longer leads to reinforcement or that an environmental context now predicts the absence of drug.[136] We know that extinction and forgetting are not isomorphic concepts because of phenomena such as conditioned craving and relapse. If an individual has "forgotten" that a bathroom is associated with drug use, exposure to this environmental context should not be able to elicit cravings. Yet, environmental contexts can induce robust drug cravings. Even after years of abstinence, exposure to drug-associated cues or environments can elicit strong cravings that can precipitate relapse.

Because extinction is analogous to abstinence (i.e., cue exposure therapy is also known as extinction training), preclinical work has sought to determine if potential pharmacotherapies for SUDs can increase the rate at which animals extinguish drug self-administration or drug CPP. I will highlight just a few examples here. Baclofen, a $GABA_B$ receptor agonist, facilitates the extinction of methamphetamine CPP[137] and ethanol consumption in primates, but it also decreases consumption of a nonalcoholic beverage, suggesting that baclofen may cause a

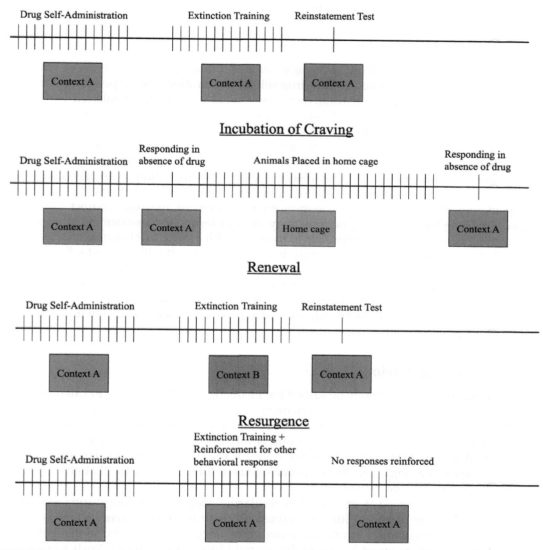

FIGURE 7.1 Representative timelines for different tasks that measure relapse-like behavior. In the traditional reinstatement model, animals self-administer a drug for a certain number of sessions (14 in the current example). Animals then receive extinction training (14 days in the example). Following extinction, animals are given a reinstatement test. In the incubation of drug craving model, animals do not receive extinction training like in the reinstatement model. Following drug self-administration, animals receive one extinction session, but then they are placed in their home cage for approximately 30 days. Finally, animals are placed back in the operant conditioning chamber and are allowed to respond on the manipulandum previously paired with the drug. No drug reinforcement is delivered during this session. If responses are higher on the extinction session following the 30-day abstinence period compared to the first day of abstinence, incubation of craving has occurred. The renewal model is similar to the reinstatement model, with one important difference. Extinction training occurs in a different environmental context compared to drug self-administration. In renewal, the reinstatement test can occur in a third environmental context (Context C) as opposed to the original context. In resurgence, animals are trained to emit a different response during the extinction of drug self-administration. There is no reinstatement test. Instead, animals receive several sessions in which responses for either reinforcer are not reinforced. Resurgence occurs when responding on the drug-paired manipulandum increases during these sessions. *Image created by the author of the textbook using Microsoft PowerPoint.*

general suppression in consummatory behavior.[138] The mu-opioid receptor antagonist naltrexone increases the rate of extinction of ethanol self-administration.[139] THC potentiates extinction of amphetamine and cocaine CPP.[140]

Reinstatement of drug-seeking behavior

You have already learned about two major forms of reinstatement: drug-induced reinstatement and cue-induced reinstatement. In drug self-administration, each type of reinstatement can be tested. In drug-induced reinstatement, the animal is systemically injected with a low dose of the drug before being placed in the operant conditioning chamber. This is analogous to an individual with an alcohol use disorder saying "I will have just one drink tonight" before they relapse. In cue-induced reinstatement, the animal is placed in the operant chamber and is then exposed to stimuli previously paired with each drug infusion, such as lights and/or tones. As discussed in the previous chapter, drug-paired stimuli can elicit drug cravings, thus precipitating relapse. In CPP, drug-induced reinstatement is primarily used.

Considering drug-induced and cue-induced reinstatement measure distinct aspects of relapse, studying the underlying neurobiology of both forms of reinstatement is important for designing better treatment options for individuals with a SUD. Just because one treatment can block drug-induced reinstatement, this does not mean that it will be able to diminish the impact of drug-paired stimuli on precipitating withdrawal symptoms and/or drug cravings. In Chapter 4 I highlighted a study that examined chemogenetic manipulations on reinstatement. Recall that Mahler et al. found that chemogenetic inhibition of intra-ventral tegmental area (VTA) dopaminergic neurons attenuate drug-induced reinstatement without significantly altering cue-induced reinstatement.[141]

Incubation of drug craving

As with reinstatement, I briefly described the incubation of drug craving in Chapter 3. The difference between incubation and reinstatement is what happens between the final drug self-administration session and the test for relapse-like behavior. In the reinstatement paradigm, animals experience extinction training. If one is interested in measuring the incubation of craving, animals are placed in their home cage for an extended period (i.e., they do not receive extinction training). This model is analogous to forced abstinence that an individual may experience if they are incarcerated.

Are differences observed if an animal "voluntarily" stops responding during extinction training or if they are placed in forced abstinence? There is some evidence to suggest that reinstatement and incubation of craving measure distinct aspects of relapse-like behavior. In one study, animals received lesions to the ventral hippocampus, which leads to a phenotype that mimics schizophrenia. While "schizophrenia-like" rats showed greater reinstatement of cocaine-seeking behavior compared to controls, they showed a similar level of incubation of craving as controls.[142] In another study, rats were allowed to self-administer cocaine either during adolescence or during adulthood. Cocaine-primed reinstatement was similar across groups, but self-administration during adolescence attenuated the incubation of craving effect.[143] Thus, reinstatement and incubation of craving do not necessarily model the same aspects of relapse. If treatment interventions can decrease one type of relapse-like behavior but not the other, this can have serious ramifications for those that have been incarcerated. That is, if behavioral interventions are more effective in individuals that voluntarily decide to abstain from drug use, individuals in prison may not benefit from this intervention.

Renewal

Like reinstatement, **renewal** is a phenomenon in which responding for drug reinforcement reemerges following extinction.[144] The difference between renewal and reinstatement is the environment in which extinction training occurs. Envision a rat that has been trained to self-administer heroin in one operant conditioning chamber. If one is interested in reinstatement, extinction training and reinstatement will occur in the same operant chamber where self-administration occurred. In renewal, the rat will self-administer the drug in one environmental context (context A). Extinction will occur in a separate environmental context (e.g., a distinct operant chamber) (context B). Finally, the rat will be placed back in context A. Placing the animal in the original environment in which drug self-administration occurred can lead to reinstatement of drug-seeking behavior. This type of renewal is known as ABA renewal. Other forms of renewal exist, such as AAB renewal and ABC renewal. In AAB renewal, drug self-administration and extinction occur in the same environmental context, whereas reinstatement is tested in a separate environmental context. In ABC renewal, drug self-administration, extinction, and reinstatement occur in separate environmental contexts. Renewal of drug seeking has been observed across multiple drugs.[145–147] Research has also shown that extinction of morphine and cocaine CPP only occurs if animals receive extinction training in the environment in which they originally received either drug.[148]

In the previous chapter, you learned about sign tracking and goal tracking. Typically, sign-tracking animals self-administer more drug compared to goal-tracking animals. Interestingly, goal-tracking animals show greater renewal of cocaine-seeking behavior.[149] Perhaps this occurs because goal-tracking animals are better at learning associations between environmental stimuli and reinforcers. What I mean by this: goal-tracking animals learn to go to the food receptacle once they see a conditioned stimulus that predicts the upcoming food delivery whereas sign trackers interact directly with the conditioned stimulus before retrieving the food. In renewal, goal trackers may be better at learning to associate the entire environmental context with drug reinforcement whereas sign trackers become fixated with a specific stimulus that becomes paired with drug delivery (such as a stimulus light). When animals are brought back to the original environment in which they learned to self-administer the drug, the environmental context is better remembered by the goal-tracking rats, which leads to greater renewal of drug-seeking behavior.

Renewal is highly applicable to real life. Individuals with a SUD often experience therapy in an environmental context that is distinct from where drug use normally occurs. Even if the individual responds well to therapy, relapse can occur when the individual is placed back in the environment in which drug use occurred. As you learned in the previous chapter, renewal effects are a major reason why newer therapeutic approaches utilize virtual reality or augmented reality. The goal of these approaches is to expose individuals to drug-paired stimuli in numerous environmental settings. The findings with sign-tracking and goal-tracking animals suggest that individuals that do not show enhanced attentional bias to drug-paired stimuli may be at greater risk of relapse when exposed to the environment in which they originally learned to use drugs.

Resurgence

One interesting phenomenon that can occur is **resurgence**, the reappearance of a previously extinguished behavior during extinction training of subsequent behavior. Dr. Tim Shahan of Utah State University has extensively studied resurgence in animals using a three-phase procedure. During the first phase, animals earn drug infusions by completing a response requirement (e.g., responding on a lever five times). During the second phase, responses on the drug-paired manipulandum are extinguished while responses on a separate manipulandum are now reinforced with a nondrug stimulus such as food. During the third phase, animals no longer receive reinforcement for responding on either manipulandum. When the nondrug reinforcer is no longer made available, animals begin responding on the manipulandum that was previously associated with drug reinforcement. So far, resurgence has been observed for ethanol- and cocaine-seeking behavior.[150] As noted by Nall and Shahan,[151] traditional models of resurgence use extinction training to reduce responding for drug reinforcement during phase two. One limitation to this approach is that extinction is not the main reason motivating individuals to abstain from drug use. Instead, aversive consequences of drug use (e.g., loss of employment, health issues, etc.) are often reported as major motivators to seek abstinence. To better model resurgence, Nall and Shahan[151] modified phase two such that rats continued to receive cocaine infusions but also received a mild foot shock. Responses on a separate manipulandum led to food delivery that was never paired with shock. Resurgence of cocaine seeking was observed in this study, corroborating the results of previous studies using the traditional three-phase resurgence model.

How does resurgence apply to real life? During contingency management, individuals receive alternative reinforcement, such as food vouchers or money, for abstaining from drug use. The goal of alternative reinforcement is to decrease the drug use. This is effectively a form of extinction training. At the end of contingency management, individuals no longer receive the alternative reinforcer. Therefore, the behavior that maintains abstinence is extinguished. This leads to resurgence of behaviors related to drug use.

Conditioned place preference revisited

In the previous chapter, CPP was a major topic as this paradigm is built on Pavlovian conditioning concepts. To briefly recap CPP: (1) animals explore all compartments of the CPP apparatus in a drug-free state; (2) animals receive alternating injections of the drug of interest and vehicle and are isolated to one compartment—this allows animals to associate one compartment with drug and one compartment with vehicle, and (3) animals are allowed to explore all compartments of the CPP apparatus in a drug-free state. During each conditioning session, **consolidation** occurs.[152] That is, animals begin transferring information about the environmental context and the drug effects from STM to LTM. During each subsequent conditioning session, animals retrieve the memory from the previous session(s). This memory becomes *destabilized*, which allows the animal to update the information they already have stored in memory. Not to anthropomorphize animals too much, but imagine this scenario: a rat is being tested for morphine CPP. During the initial pretest, the rat explores all three compartments of the CPP chamber. Here, the animal does not receive any injections. The animal just learns that it is encountering new environmental contexts and begins to consolidate this information into LTM. During the first conditioning session, the animal receives an injection of morphine and is placed in one CPP compartment that is black. When the animal is placed in the black box, this may activate the rat's memory from the previous session. The rat may remember seeing the black box the previous day, but now it experiences the sensations of being on an opioid. The animal now needs to modify the memory it has of the black box. The modification of an existing memory is called **reconsolidation**.[152] During the next conditioning session, additional reconsolidation occurs as the animal learns that a distinct environmental context (e.g., a white compartment) is not associated with the effects of morphine. This process will repeat itself with each subsequent conditioning session.

Because the posttest occurs during a drug-free state, the rat must "remember" that the drug was explicitly paired with one environmental context (the black box). If the drug is rewarding, the animal will spend more time in the environment paired with that drug. In memory terms, the rat can retrieve this information from LTM.[152] As Pavlovian conditioning is a form of implicit learning, the animal is not consciously aware that it is learning to associate an environmental context with the drug's effects.

In the last chapter, I discussed the acquisition and the expression of CPP. When testing a potential intervention (e.g., pharmacotherapy) on CPP, a researcher can administer the intervention before each conditioning session (acquisition) or before the posttest (expression). Applying the intervention before each conditioning session determines if the intervention disrupts the ability of the animal to learn the association between the drug and the environmental context. Providing the intervention before the posttest allows one to determine if it affects memory retrieval. How are consolidation and reconsolidation tested in the CPP paradigm? To determine if an intervention affects consolidation, it is presented *after* each conditioning session. To test reconsolidation, the intervention is provided *after* the posttest.[153] In this case, animals need to be given an additional posttest to determine if the intervention altered reconsolidation.

As with renewal and resurgence, these principles are applicable to real life. An individual that begins using a substance will implicitly start associating environmental stimuli and drug paraphernalia with the drug itself and will consolidate this information into LTM (similar to the acquisition of CPP). As the individual continues using the drug, reconsolidation occurs as they begin associating even more environmental stimuli with drug use. Over time, exposure to drug-paired paraphernalia and/or environmental stimuli can retrieve drug-associated memories, precipitating withdrawal symptoms and/or intense drug cravings. Even when someone tries to abstain from using the drug, these cues can retrieve drug-associated memories, thus potentially leading to relapse.

Quiz 7.2

(1) What is the difference between renewal and reinstatement? What is the difference between incubation of craving and reinstatement?

(2) Which of the following is an example of procedural memory?
   a. An individual knows that heroin is a semisynthetic drug derived from the opium poppy plant.
   b. An individual recalls the first time that they used heroin.
   c. An individual remembers to attend a therapy session that they made 2 months prior.
   d. An individual that has been using heroin is able to prepare the drug and use it without consciously thinking about what they are doing.

(3) Which of the following statements about drug effects on memory is true?
   a. Acute cocaine administration improves working memory.
   b. Nicotine improves working memory, but this effect is typically isolated to those that regularly use tobacco products.
   c. Prenatal exposure to drugs negatively affects working memory but not long-term memory.
   d. Withdrawal from drugs has no effect on memory.

(4) If a researcher wants to measure the effects of a potential pharmacotherapy on the reconsolidation of methamphetamine CPP, when should the researcher expose the animals to the treatment of interest?
   a. After each methamphetamine injection
   b. After the posttest
   c. Before each methamphetamine injection
   d. Before the posttest

Answers to quiz 7.2

(1) In a reinstatement experiment, extinction training occurs in the same testing chamber as drug self-administration. In a renewal experiment, extinction training occurs in a different testing chamber as drug self-administration. The difference between incubation of craving and reinstatement is the absence of extinction training in the former.

(2) d — An individual that has been using heroin is able to prepare the drug and use it without consciously thinking about what they are doing.

(3) b — Nicotine improves working memory, but this effect is typically isolated to those that regularly use tobacco products.

(4) b — After the posttest

Neurobiology of attention and memory: relevance to addiction

The multiple memory systems view of addiction primarily focuses on the contribution of three brain regions to SUDs: the hippocampus, the amygdala, and the dorsal striatum.[154] The hippocampus is primarily responsible for consolidating STM into LTM, but it also plays a role in spatial memory. As such, the hippocampus is important for encoding information about drug-related situations. For example, if an individual uses heroin and experiences euphoria, the hippocampus will consolidate information about the environment in which the drug use occurred, as well as the internal state (e.g., euphoric feeling) the individual experiences during drug use. In adult rodents, lesions to the hippocampus suppress ethanol consumption.[155] Interestingly, lesioning the hippocampus of young rodents *increases* the reinforcing effects of ethanol and psychostimulants and increases relapse-like behavior.[142,156,157] These results suggest that damaging the hippocampus at an early age may lead to neural adaptations that increase susceptibility to drug dependence and addiction, which is consistent with what has already been discussed in this chapter: memory impairments are associated with addiction-like behaviors. We will continue discussing the hippocampus and addiction in the next section.

While the hippocampus encodes explicit information about the drug-taking experience, the amygdala consolidates emotionally driven memories. The amygdala is important for establishing Pavlovian associations between neutral stimuli and the drug. Over time, drug-paired stimuli can elicit strong emotional responses like craving, as you have already learned. To support this claim, blocking glutamate NMDA receptors in the basolateral amygdala (BLA) attenuates amphetamine CPP while inactivating the BLA prevents the extinction of an established amphetamine CPP.[158] The BLA is also involved in relapse-like behavior; specifically, increased BLA c-Fos expression is observed in animals that show renewal of drug-seeking behavior.[145]

Although not specifically identified in the multiple memory systems view, the nucleus accumbens (NAc) is involved in forming associations between neutral stimuli and reinforcers like drugs.[159] NAc dopaminergic activity is also involved in reinstatement. Across extinction sessions, dopamine levels decrease, but they drastically increase during drug-induced reinstatement.[160] This finding is consistent with what has already been covered concerning the role of dopamine in the mesocorticolimbic pathway. Dopamine is involved in reinforcement learning. Recall the seminal work of Dr. Wolfram Schultz who showed that dopaminergic neurons in the ventral tegmental area (VTA) increase their firing in response to conditioned stimuli and show decreased responding when an expected reinforcer is not delivered (i.e., extinction).

The basal ganglia, including the dorsal striatum, are primarily involved in procedural memory. As covered previously, the dorsal striatum contributes to habit formation. As an individual uses a drug repeatedly, drug-paired stimuli can elicit repetitive, automatic behaviors. For example, exposure to tobacco-associated cues can trigger someone to reach for a cigarette. Importantly, the dorsal striatum is involved in the transition from controlled drug use to compulsive drug use (see Chapter 3 for research examining the role of the dorsal striatum on addiction-like behaviors).

While the hippocampus consolidates STM to LTM, this does not mean that memories are stored there. Where exactly are memories stored? Evidence has indicated that the PFC acts as a sort of storage bin for our memories. When individuals engage in recall or recognition, prefrontal cortical regions are activated.[161,162] If LTM is involved in addiction, one would expect the PFC to play an important role in relapse-like behavior. Indeed, evidence has implicated the PFC in extinction and reinstatement. Like the BLA, blocking NMDA receptors in the medial PFC (mPFC) prevents the extinction of amphetamine CPP,[163] and c-Fos levels increase in the infralimbic cortex during the renewal of cocaine seeking.[145] Monoaminergic activity in the prelimbic region plays an important role in extinction learning as concomitant stimulation of dopamine D_1-like and D_2-like receptors or norepinephrine depletion increases the rate of extinction of amphetamine CPP in mice.[164,165]

When you learned about the neurobiology of addiction, you learned about the decreased dopamine D_2-like receptor expression in the PFC of those with a SUD. The altered neurochemical signaling in the PFC observed in individuals with a SUD could also explain why these individuals experience attentional and memory deficits. In the next chapter, we will come back to this idea when we discuss the relationship between maladaptive decision-making and addiction.

Like the NAc, the cerebellum is not included in the multiple memory systems view, but it plays an important role in conditioning. Specifically, recent evidence has indicated that neuronal activity in the cerebellum helps drive cocaine CPP.[166] Long-term drug use can also alter cerebellar activity, as cocaine-exposed monkeys performing a working memory task show greater metabolic activity in the cerebellum compared to controls.[167] The increased cerebellar activity may occur as a compensatory mechanism for the decreased activity observed in the PFC of these animals. More work is needed to fully understand the role of the cerebellum in the development of SUDs. For now, let's discuss some of the cellular mechanisms of learning and how they are associated with addiction.

Cellular mechanisms of learning and addiction

To better understand the memory processes underlying addiction, you need to learn more about the cellular mechanisms of learning. In Chapter 2, I mentioned that glutamate is involved in learning and memory. To be more specific, glutamate plays an important role in the cellular mechanisms of learning: **long-term potentiation (LTP)** and **long-term depression (LTD)**.[168] LTP and LTD are forms of *synaptic plasticity*. Synaptic plasticity refers to the ability of synapses to strengthen (as in the case of LTP) or weaken (as in the case of LTD) over time. The ionotropic AMPA and NMDA glutamate receptors are important for LTP.[168] When AMPA receptors are stimulated, they depolarize the neuron, which causes a magnesium (Mg^{++}) ion to become displaced from a nearby NMDA receptor. This then allows calcium (Ca^{++}) to enter the postsynaptic neuron. When Ca^{++} enters the neuron, it activates protein kinases like CaMKII, leading to phosphorylation. This phosphorylation leads to the insertion of additional AMPA receptors at the postsynaptic membrane. By increasing the number of AMPA receptors, the postsynaptic neuron is now more sensitive to glutamate that is released from the presynaptic neuron (i.e., the synapse has become strengthened). This process is known as *early phase LTP*.[168] In addition to activating CaMKII, increased Ca^{++} levels lead to protein synthesis and gene transcription via the activation of other protein kinases (e.g., protein kinase A) (Fig. 7.2A). This is known as *late phase LTP*.[168] If this sounds familiar to you, it is because you have already read about CaMKII and other protein kinases in Chapter 4 when I discussed the cellular pathways of addiction.

LTD also involves AMPA and NMDA receptors.[168] The difference is that in LTD, AMPA receptors are only moderately stimulated. The depolarizations that occur are not strong enough to displace the Mg^{++} ions from each NMDA receptor located on the postsynaptic

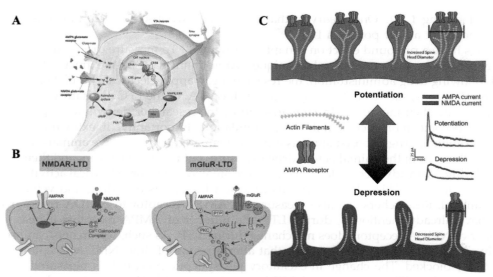

FIGURE 7.2 (A) Example of long-term potentiation (LTP) in a postsynaptic neuron. When AMPA receptors receive enough stimulation from glutamate, a Mg^{++} ion is displaced from neighboring NMDA receptors, thus further depolarizing the postsynaptic neuron. Stimulation of NMDA receptors allows Ca^{++} into the neuron, which activates signaling pathways. Additional AMPA receptors can be inserted into the cell membrane (not shown here). Eventually, gene transcription occurs in the cell nucleus. This can cause morphological changes to occur to the neuron, such as the creation of additional synapses. This causes the neuron to become more sensitive to subsequent excitation. (B) Long-term depression (LTD) is mediated either by the NMDA receptor or the mGluR. The result is the same: AMPA receptors are removed from the postsynaptic membrane. (C) Morphological and electrophysiological changes that result from LTP and LTD. During LTP, dendritic spines become enlarged, which allows the insertion of additional AMPA receptors. Actin filaments, which provide structural support to the dendrite, increase in number. Because there are increased AMPA receptors, this increases the AMPA receptor to NMDA receptor ratio. Notice that there is more activity at AMPA receptors compared to NMDA receptors. During LTD, opposite changes in morphology and electrophysiology are observed (e.g., decreased dendritic spine size and decreased AMPA receptor to NDMA receptor ratio). *Panel (A) comes from Fig. 1 of Gould TJ. Addiction and cognition. Addiction Sci Clin Pract. 2010;5(2):4–14. This is an open-access article distributed under the terms of the Creative Commons Attribution 4.0 International License (https://creativecommons.org/licenses/by/4.0/). Panel B comes from Fig. 2 of Pinar C, Fontaine CJ, Triviño-Paredes J, Lottenberg CP, Gil-Mohapel J, Christie BR. Revisiting the flip side: long-term depression of synaptic efficacy in the hippocampus. Neurosci Biobehav Rev. 2017;80:394–413. https://doi.org/10.1016/j.neubiorev.2017.06.001. Copyright 2017, with permission from Elsevier. Panel (C) comes from Fig. 3 of Scofield MD, Heinsbroek JA, Gipson CD, et al. The nucleus accumbens: mechanisms of addiction across drug classes reflect the importance of glutamate homeostasis. Pharmacol Rev. 2016;68(3):816–871. https://doi.org/10.1124/pr.116.012484. Permission conveyed through Copyright Clearance Center, Inc.*

neuron; therefore, not as much Ca^{++} enters the neuron. The reduced Ca^{++} levels are not sufficient to activate the same protein kinases that lead to the insertion of additional AMPA receptors. Instead, this results in a different cascade of events that leads to a reduction of AMPA receptors on the postsynaptic membrane. Metabotropic glutamate receptors (mGluRs) and the endocannabinoid system have also been implicated in LTD.[168] Low-frequency stimulation of postsynaptic mGluRs has a result similar to that of low-frequency stimulation of AMPA receptors: reduced cell surface expression of AMPA receptors (Fig. 7.2B). Stimulation of cannabinoid CB_1 receptors leads to reduced glutamate release, leading to LTD. What is the point of having LTD? One theory is that LTD resets synapses, which allows them to be strengthened at a later point via LTP.

Drugs have a profound effect on synaptic plasticity.[169] To measure drug-induced changes in LTP, researchers can use several techniques. After brains have been extracted from the animal following drug administration, the researcher can apply electrical stimulation to a select group of neurons, such as in the hippocampus, the VTA, or the PFC. Excitatory postsynaptic potentials are then measured. In layman's terms, the researcher determines if postsynaptic neurons become more active following stimulation. Importantly, if extracting the brain from the animal, drug-exposed animals must be compared to drug-naïve animals. We cannot take the brain from the animal before drug exposure, put the brain back in, and then give the animal drug. With advances in technology, there are ways to record electrical activity in alive, conscious animals. This allows a researcher to test the effects of a drug and vehicle in the same animal. Researchers can also measure the **AMPA receptor to NMDA receptor ratio**. As I have already mentioned, during LTP, the number of AMPA receptors increases. The number of NMDA receptors does not change during LTP. As such, researchers can use electrophysiology as just described, except that after a period of time, NMDA receptors can be selectively blocked. The change in excitatory postsynaptic potentials is measured, yielding a measure of AMPA receptor to NMDA receptor ratio.[170] Higher ratio values indicate increased AMPA receptor expression, meaning that LTP has occurred (Fig. 7.2C). Researchers can also determine if physical changes have occurred to the neuron, such as altered dendritic spine morphology or density. During LTP, dendritic spine density increases, and the head of the dendritic spine increases to make more room for the increased AMPA receptor expression.

Stimulant administration tends to enhance LTP in the hippocampus,[171] the VTA,[170] the PFC,[172,173] and the hypothalamus[174] (Fig. 7.3). Although methamphetamine increases LTP in rats, it reduces LTP in the mouse hippocampus.[175] The discrepancy in methamphetamine-induced alterations in LTP observed between rats and mice could reflect a species difference, or it could be related to differences in the dosing regimen (four injections of 5 mg/kg for rats vs. 10 injections of 10 mg/kg for mice). Amazingly, a single exposure to drugs can induce LTP in dopaminergic VTA neurons.[170] Not only do stimulants enhance LTP, but they impair LTP/increase LTD at inhibitory synapses in the VTA.[176]

In contrast to stimulants, depressant-like drugs decrease LTP in the hippocampus.[177] In humans, binge drinking leads to impaired LTP.[178] Ethanol prevents LTP in the NAc,[179] and cannabinoids prevent LTP in the VTA.[180] Ethanol and opioids prevent LTP of GABAergic neurons in the VTA.[176,181] While ethanol impairs LTP in the hippocampus, ethanol enhances LTP at intra-VTA synapses.[182] Ethanol also enhances LTD in the hippocampus[183] while endocannabinoids induce LTD in the VTA.[184] However, research has shown LTP induction in rats following expression of morphine CPP.[185] Animals do not have to be directly exposed to morphine to experience reduced hippocampal LTP. Offspring of male and female rats treated with morphine have reduced LTP.[186] One interesting result is that a single injection of morphine can prevent LTD of GABAergic neurons in the VTA.[187]

Withdrawal from drugs induces LTP in several regions, including the hippocampus,[188] the lateral amygdala,[189] and the central nucleus of the amygdala.[190] Methamphetamine withdrawal is associated with decreased synaptic plasticity.[191] One important consideration is that LTP resulting from withdrawal is time-dependent. For example, LTP is enhanced in

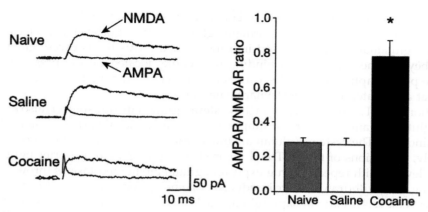

FIGURE 7.3 A single injection of cocaine increases the AMPA receptor to NMDA receptor ratio in the ventral tegmental area compared to naïve animals that have never received any injections and to animals given an injection of saline. This finding shows that long-term potentiation can occur after a single exposure to psychostimulants. *Image modified from Fig. 1 of Ungless MA, Whistler JL, Malenka RC, Bonci A. Single cocaine exposure in vivo induces long-term potentiation in dopamine neurons. Nature. 2001;411(6837):583—587. https://doi.org/10.1038/35079077. Copyright 2001.*

animals that have experienced 3 days of cocaine withdrawal but is suppressed following 100 days of withdrawal.[192] These results indicate that long-term withdrawal reduces LTP. Another important finding is that drug withdrawal prevents LTP and LTD in the NAc following stimulation of the PFC, an example of **metaplasticity**.[193] Similar to withdrawal, relapse-like behavior restores LTP in the hippocampus,[177] as well as in the NAc.[194,195] Furthermore, exposing heroin-dependent rats to morphine restores LTP in the hippocampus.[196]

What does this all mean? In the VTA, many drugs, including some depressant-like drugs, decrease LTP of inhibitory synapses. Normally, the VTA projects GABAergic neurons to the NAc to inhibit the release of dopamine. By reducing LTP or inducing LTD in this region, these GABAergic neurons are no longer as effective at inhibiting dopamine release. Simultaneously, stimulants increase LTP of non-GABAergic neurons in the VTA. Now, the system that regulates dopamine release is impaired. These findings provide a cellular mechanism by which drug use leads to dependence. Because LTP is often impaired in the hippocampus, this can account for many of the memory deficits observed in individuals with a SUD.

Glutamate homeostasis hypothesis of addiction

In Chapter 3, you learned about the dopamine depletion hypothesis of addiction. As you have learned by now, other neurotransmitter systems play an integral role in addiction, particularly glutamate. Dr. Peter Kalivas, a professor at the Medical University of South Carolina, proposed the **glutamate homeostasis hypothesis of addiction**.[197] We first need to discuss what glutamate homeostasis is before diving into this hypothesis. You previously learned about homeostasis when I discussed the hedonic allostasis model of addiction. Homeostasis can be observed at the neuronal level. In Chapter 2, you learned that glial cells play an important role in the synthesis and breakdown of glutamate. They are also major contributors of glutamate homeostasis. Normally, glutamate transporters (GLTs) located on glial cells uptake glutamate that is released into the synapse. This limits excess glutamate from traveling beyond the synapse and binding to mGluRs located on the presynaptic neuron and on the postsynaptic neuron. This causes the leftover glutamate to bind to ionotropic receptors that are located on the postsynaptic neuron.

In addition to GLT, astrocytes contain the **cysteine/glutamate transporter** (also known as the cysteine/glutamate antiporter or just simply the glutamate exchanger).[198] Cysteine is a nonessential amino acid. The cysteine/glutamate transporter trades one cysteine for one glutamate. Collectively, the actions of the GLT and the cysteine/glutamate transporter tightly regulate glutamate levels. With repeated drug exposure, glutamate homeostasis is thrown out of balance. During drug withdrawal, there is reduced cysteine/glutamate transporter activity, which results in lower levels of extra-synaptic glutamate levels. At the same time, presynaptic mGluRs are not as active. Remember that presynaptic mGluRs act as autoreceptors. Because these autoreceptors are not able to effectively regulate glutamate release, increased glutamate is released into the synapse. Another change that is observed is increased ionotropic glutamate receptor expression on the postsynaptic neuron. Specifically, the AMPA receptor to NMDA receptor ratio increases. Fig. 7.4 shows how drugs can alter glutamate homeostasis.

Research from Dr. Kalivas' lab, as well as from other labs, has provided support for the glutamate homeostasis hypothesis of addiction. In humans, basal glutamate levels are lower in the NAc of individuals with cocaine use disorder, but glutamate concentrations increase during cue-induced drug cravings.[199] In rats, sequential self-administration of cocaine and ethanol leads to decreased GLT expression.[200] Similar downregulation of GLT is observed following nicotine self-administration.[201] A single cocaine injection reduces the number of dendritic spines and postsynaptic density 95 (PSD95) (a protein found in postsynaptic dendrites) in the mPFC of adolescent rats.[202] Loss of glutamate homeostasis is also observed during relapse-like behavior in animals, as stimulants potentiate LTP, as evidenced by increased AMPA receptor to NMDA receptor ratio[194] while depressants induce LTD, as evidenced by decreased AMPA receptor to NMDA receptor ratio.[195]

In Chapter 2, I introduced dendritic spines. Dendritic spines can be classified into different categories: long and thin with no head (*filopodia*), thin with a head (either short or long), head with no neck (*stubby*), and short with a large head (*mushroom*). As reviewed by Pchitskaya and Bezprozvanny,[203] the various dendritic spines have different functional roles. Filopodia spines are primarily associated with immature neurons. Thin spines are involved in learning. During LTP, the heads of thin spines enlarge, leading to mushroom spines. Stubby spines are posited to form as mushroom spines disappear. In Chapter 4, I described some of the ways in which drugs can alter the number of dendritic spines and the morphology of dendritic spines. While stimulants increase the number of dendritic spines and increase the size of dendritic spines, depressant-like drugs have the opposite effect. Those results mirror the effects these drugs have on LTP and LTD. Drug withdrawal also increases dendritic spine head diameter, which is further increased following a priming injection of the drug.[197] These results are consistent with the effects of drug withdrawal and relapse on LTP. One interesting finding is that chronic ethanol administration increases the proportion of mushroom spines while decreasing the total number of dendritic spines in the NAc.[204]

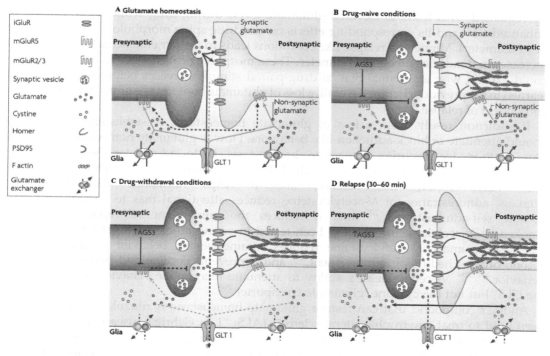

FIGURE 7.4 The glutamate homeostasis hypothesis of addiction. Panel (A) shows how glutamate homeostasis works. When glutamate is released from the presynaptic membrane, it can bind to ionotropic receptors (iGluR; e.g., AMPA/NMDA) or to metabotropic glutamate receptors (mGluRs; e.g., mGluR5). When glutamate binds to these receptors, excitation occurs at the postsynaptic neuron. Glutamate can also bind to the mGluR$_{2/3}$; these receptors are inhibitory and act as a negative feedback loop to suppress further glutamate release. Excess glutamate is removed from the synapse via the glutamate transporter (GLT-1) located on surrounding astrocytes. Astrocytes contain the cysteine/glutamate transporter (glutamate exchanger). Cysteine is taken back up into the astrocyte while glutamate is released into the extrasynaptic space. These mechanisms ensure a consistent level of glutamate. Panel (B) shows glutamate homeostasis in drug-naïve animals. This image shows some proteins found in the dendrite of the postsynaptic neuron (Homer, postsynaptic density 95 [PSD95], and F actin). Panel (C) shows how drug withdrawal disrupts glutamate homeostasis. Not as much glutamate is released from the cysteine/glutamate transporter. This, in turn, does not allow glutamate to stimulate mGluR$_{2/3}$, resulting in increased glutamate binding to AMPA/NMDA receptors. There is also an upregulation of AMPA receptors on the postsynaptic neuron, as well as an upregulation of PSD95 and F actin. Conversely, there is a downregulation of the Homer protein. Panel (D) shows further alterations in glutamate homeostasis during relapse-like behavior. More glutamate is released from the presynaptic neuron and binds to ionotropic receptors, which have been further upregulated. Due to the increased AMPA receptor expression, dendritic spine heads increase in size. *Image comes from Fig. 2 of Kalivas, PW. The glutamate homeostasis hypothesis of addiction. Nat Rev Neurosci. 2009;10(8):561–572. https://doi.org/10.1038/nrn2515. Copyright 2009.*

Let's quickly recap what this all means. We know that dopamine levels increase following drug administration, which is important for reinforcement learning. While dopamine mediates reinforcement learning, glutamate controls general learning via LTP and LTD. With prolonged drug use, dopamine levels decrease; this is the basis of the dopamine depletion hypothesis. As dopamine levels decrease with prolonged drug use, glutamate homeostasis becomes dysregulated. This leads to elevated glutamate levels during drug withdrawal, leading to enhanced LTP and its corresponding effects on dendritic spine morphology (e.g., increased head diameter). Mimicking relapse-like conditions leads to further enhancements in LTP and increases in dendritic spine head diameter. These morphological changes to dendritic spines could explain why attentional bias to drug-paired stimuli increases before relapse, as you learned in the previous chapter. Overall, a loss of glutamate homeostasis appears to be a critical component of relapse.

If addiction results from a loss of glutamate homeostasis, reestablishing homeostasis should attenuate or block addiction-like behaviors. How is glutamate homeostasis restored? One method for restoring glutamate homeostasis is to administer *N*-acetylcysteine. Remember that drug exposure disrupts the balance of glutamate and cysteine. *N*-acetylcysteine helps restore this balance, and it helps normalize GLT levels. In humans, administration of *N*-acetylcysteine reduces attentional bias to cocaine cues and cocaine self-administration,[205] and it reduces the number of self-reported cigarettes smoked.[201] *N*-acetylcysteine has also been tested in young adults with cannabis use disorder.[206] In animals, *N*-acetylcysteine reduces cocaine-seeking behavior as assessed in a second-order schedule of reinforcement,[207] and it prevents cocaine and heroin relapse-like behavior.[208,209] *N*-acetylcysteine's ability to attenuate relapse-like behavior is mediated by the GLT, but not the cysteine/glutamate transporter.[209]

Another compound that can restore glutamate homeostasis is riluzole, a drug that has been used to treat amytrophic lateral sclerosis (ALS; "Lou Gehrig's disease"). Riluzole prevents drug- and cue-induced reinstatement of cocaine-seeking behavior.[210] The ability of riluzole to prevent relapse-like behavior may result from its effects on the GLT, as it reverses drug-induced decreases in expression of this transporter.[210] Interestingly, antibiotic drugs like ceftriaxone and Augmentin (amoxicillin/clavulanate) can reduce ethanol consumption and attenuate relapse-like behavior as they increase GLT levels.[211,212] Finally, restoring glutamate homeostasis in the NAc by infusing a cannabinoid prevents drug-induced and cue-induced reinstatement.[213] This last finding shows that the endocannabinoid system is important for maintaining glutamate homeostasis. This finding should not be too surprising given that the endocannabinoid system can regulate LTD, which offsets the LTP that results during drug withdrawal and during reinstatement of drug-seeking behavior.

Cognitive-based treatments for addiction

Because individuals with a SUD experience distorted thoughts and deficits in attentional control and in memory, several therapies aim to reverse these deficits. This section will highlight one major therapeutic approach, as well as some other strategies, that have been used to help improve cognition, thus improving treatment outcomes for those with a SUD.

Cognitive-behavioral therapy (CBT)

Originally developed by Dr. Beck[214] (who you read about at the beginning of this chapter) to treat major depressive disorder, **cognitive-behavioral therapy (CBT)** has been expanded to treat various psychiatric conditions, including SUDs. CBT aims to help individuals change distorted thoughts, beliefs, and/or attitudes, as well as to enable them to improve emotion regulation and to develop healthy coping strategies. Even if an individual receives pharmacotherapy and/or contingency management, cognitive distortions can render these interventions ineffective. CBT emphasizes the importance of modifying one's thoughts and beliefs to eliminate cognitive distortions. Not only do therapists work with individuals to alter their thought processes, but they help individuals modify or adapt certain behaviors (hence the name *cognitive-behavioral* therapy).

One fundamental step in CBT is identifying cognitive distortions. Individuals need to be able to recognize these distortions to modify them. Additionally, understanding what triggers drug craving is crucial as this provides an early alert system for those at risk of relapse. If an individual learns to recognize that their cravings increase when they go to a specific place in town, they can learn to modify their behavior to avoid or severely limit how many times they go to that location. The therapist will then work with the individual to develop strategies to address cognitive distortions. Individuals with a SUD need to learn how to distract themselves when cravings emerge. This can include many activities, such as spending time with those that are supportive or engaging in a hobby. This is similar to what you learned in Chapter 5 concerning alternative reinforcement. Individuals in CBT also learn how to better solve problems without resorting to ethanol/drug use to cope with those problems. The goal is to have individuals make lifestyle changes that allow the individual to achieve sobriety.[215]

Multiple studies have provided support for the use of CBT in the treatment of SUDs. CBT delivered in 17 sessions across an 8-month period is effective in reducing ethanol consumption in individuals with alcohol use disorder.[216] CBT also reduces the use of the following substances: cigarettes,[217] cocaine/crack cocaine,[218,219] methamphetamine,[220] prescription opioids,[221] and cannabis.[222] CBT reduces drug use in individuals with a comorbid condition, such as ethanol/problem substance use in individuals with depression[223] and cannabis use disorder in individuals with anxiety.[224] CBT, whether administered alone[225] or in conjunction with a support group[226] improves self-efficacy in individuals with a SUD.[6] At the beginning of this chapter, I mentioned that self-efficacy is a predictor of treatment outcomes. The ability of CBT to increase self-efficacy may explain its effectiveness in treating SUDs.

One important aspect of CBT is the completion of "homework", exercises individuals complete on their own time. The exercises help reinforce the concepts covered during a therapy session. Completion of these homework assignments is predictive of cocaine use during treatment. That is, individuals who complete more homework assignments use less cocaine[227]; however, this finding is only observed in individuals that report a greater desire to change their drug use habits. For individuals that do not report a high readiness to change, the exercises do not have an impact on cocaine use.[228]

To make CBT more accessible to individuals, computerized forms of CBT can be administered. For example, Carroll et al.[229] had 77 individuals seeking treatment for a SUD complete the computer-based training in CBT (CBT4CBT). This program consists of six lessons and covers the following topics:

(1) Understanding and changing patterns of substance use
(2) Coping with craving
(3) Refusing offers of drugs and alcohol
(4) Problem-solving skills

(5) Identifying and changing thoughts about drugs and alcohol, and
(6) Improving decision-making skills (p. 882)[229]

Individuals that completed the CBT4CBT program submitted more drug-free urine samples; also, completion of homework assignments was associated with how well individuals did during treatment. Computerized forms of CBT have been shown to reduce cocaine cravings.[230] In addition to computerized CBT, CBT can be implemented via text messaging. So far, text messaging-delivered CBT appears to be a promising intervention for alcohol use disorder.[231]

Another way to incorporate technology into CBT is through video games. I want to highlight one study that utilized Microsoft's Kinect (for the Xbox). Although the Kinect is now discontinued, Metcalf et al.[232] developed a game individuals could play in which they countered drug-paired cues by performing certain actions. For example, participants could swipe away alcohol bottles (Fig. 7.5). One important finding from this study is that individuals reported greater self-efficacy after spending several weeks playing the game. Now that virtual reality and augmented reality are becoming more popular, I anticipate seeing more clinicians incorporate these technologies into CBT.

One advantage of CBT is that it can be administered in a group setting, allowing for the simultaneous treatment of multiple individuals.[216,223] Another advantage of CBT is that it can easily be used in conjunction with other behavioral treatments like contingency management, as well as with pharmacotherapy. Recall from Chapter 5 that contingency management involves providing individuals an alternative reinforcer (e.g., money, voucher) for each drug-free urine sample. When compared to CBT, contingency management leads to greater drug-free urine samples during treatment, but CBT leads to comparable long-term abstinence when compared to contingency management.[233] Somewhat related, CBT combined with contingency management leads to fewer cocaine-negative urine samples compared to contingency management or CBT administered alone while patients go through treatment. Yet, once treatment ends, individuals given the combined CBT/contingency management treatment are more likely to abstain from cocaine use compared to the individual

FIGURE 7.5 An individual playing a game on Microsoft's Kinect, in which they can swipe away drug-paired stimuli. *Image comes from Fig. 1 of Metcalf M, Rossie K, Stokes K, Tallman C, Tanner B. Virtual reality cue refusal video game for alcohol and cigarette recovery support: summative study. JMIR Serious Games. 2018;6(2):e7. https://doi.org/10.2196/games.9231. This is an open-access article distributed under the terms of the Creative Commons Attribution License (https://creativecommons.org/licenses/by/4.0/).*

treatment groups.[234] Collectively, these results suggest that contingency management may be more effective early in treatment, but CBT may be more beneficial for maintaining long-term abstinence. In addition to contingency management, a recent meta-analysis indicated the utility of combining CBT with pharmacotherapy for the treatment of alcohol use disorder and other SUDs.[235] To provide a specific example, individuals given naltrexone while receiving CBT drink less ethanol, take longer to relapse, and exert more control over ethanol cravings compared to those given placebo and CBT.[236]

Improving cognitive functions in individuals with a SUD

One hypothesis is that improving memory functions in individuals will help reduce drug use and will help promote long-term abstinence. Working memory training helps increase feelings of self-control in individuals with a methamphetamine use disorder[237] and tends to decrease the number of drinks an individual with an alcohol use disorder consumes during a single drinking session.[238] However, working memory training does not affect drug craving, drug use, or attentional bias to drug-paired stimuli.[239] Working memory training also fails to reduce ethanol cravings and fails to increase abstinence rates 2 weeks following discharge from treatment.[240] Overall, targeting working memory on its own may not be enough to help improve treatment outcomes in individuals with a SUD.

In addition to cognitive-based treatments, several potential pharmacotherapies have been tested to determine if they are able to improve cognitive abnormalities in those with a SUD. I will not detail each study that has been published. Instead, I will highlight some findings that have been reported in the literature. The anti-Alzheimer's disease drug galantamine improves sustained attention in abstinent cocaine users[241] and improves working memory in individuals with cocaine use disorder.[242] The antiinflammatory agent ibudilast improves sustained attention during early abstinence from methamphetamine.[243]

The drug guanfacine ameliorates cocaine-induced deficits and THC-induced deficits in working memory.[113,244] While individuals with poor working memory initiate cigarette use more rapidly following abstinence, this is not true for those treated with varenicline.[128] Impaired working memory performance is observed in "relapsers" treated with placebo compared to "abstainers", but impaired working memory performance is not observed in those treated with varenicline.[128] In individuals with alcohol use disorder, larger varenicline-induced improvements in working memory lead to reduced drinking.[245] Given that withdrawal from cigarettes can negatively impact attention/memory, one potential treatment approach is to give individuals a drug that can relieve withdrawal symptoms. Bupropion is one such drug, and it reduces withdrawal-induced working memory and attentional deficits.[246] Modafinil improves working memory, as well as sustained attention, in individuals with cocaine use disorder[247] and in those with alcohol use disorder that have poor working memory.[248] In those with a methamphetamine use disorder, varenicline enhances reaction time in a working memory task but does not improve how much information is stored in working memory.[249] Finally, in mice, treatment with memantine (used to treat Alzheimer's disease) can ameliorate methamphetamine-induced memory deficits.[118]

Quiz 7.3

(1) Which of the following statements about cognitive-behavioral therapy is true?
   a. CBT can be effectively administered in person or online.
   b. CBT is only effective when administered to a single individual.
   c. Combining CBT with other therapies is ineffective for reducing drug cravings.
   d. Individuals with a comorbid condition cannot be treated with CBT.
(2) Which brain structure is involved in habitual or ritualistic actions related to drug use?
   a. Amygdala
   b. Cerebellum
   c. Dorsal striatum
   d. Hippocampus
(3) After an individual uses heroin, what happens to the number of AMPA receptors compared to NMDA receptors?
   a. They decrease
   b. They increase
   c. They stay the same
(4) In addition to glutamate, which neurotransmitter system is implicated in long-term depression?
   a. Dopamine
   b. Endocannabinoid
   c. GABA
   d. Opioid
(5) Which drug can restore glutamate homeostasis?
   a. Galantamine
   b. Guanfacine
   c. Modafinil
   d. N-acetylcysteine

Answers to quiz 7.3

(1) a — CBT can be effectively administered in person or online.
(2) c — Dorsal striatum
(3) a — They decrease
(4) b — Endocannabinoid
(5) d — N-acetylcysteine

Chapter summary

In this chapter, you learned that drug use negatively affects attentional and memory processes. Not only do drugs affect attention and memory, but attentional and memory deficits are linked to SUDs. At the molecular level, the glutamatergic system plays a critical role in relapse, as glutamate is highly involved in LTP and LTD, the cellular mechanisms of learning.

Drugs cause widespread changes to the glutamatergic system, which ultimately causes morphological and functional changes to occur to neurons. These changes are implicated in relapse. Because cognitive distortions are observed in those with an addiction, several therapeutic approaches have been adopted to reduce these distortions. CBT has received considerable attention as evidence indicates this approach can help individuals maintain long-term abstinence. CBT can be easily combined with other behavioral therapies and/or pharmacotherapy. Given the role of glutamate homeostasis in addiction, efforts have been made to determine if restoring this homeostasis with drugs like N-acetylcysteine can reduce drug use and prevent relapse. While attention and memory are important mediators of addiction, they are not the only cognitive factors involved in SUDs. In the next chapter, you will learn about the role of decision making, specifically impulsive and risky decision making, in the development and maintenance of a SUD.

Glossary

AMPA receptor to NMDA receptor ratio — a measure of the number of AMPA receptors compared to NMDA receptors; during long-term potentiation, the number of AMPA receptors increases, meaning this ratio can be used to determine if long-term potentiation has occurred.
Automaticity — a phenomenon in which we perform certain actions without the need for effortful, conscious control.
Cognitive-behavioral therapy (CBT) — a therapeutic approach that aims to help individuals change distorted thoughts, beliefs, and/or attitudes, as well as to enable them to improve emotion regulation and develop healthy coping strategies.
Cognitive distortions — exaggerated or irrational thought patterns that are characteristic of individuals with various psychiatric conditions.
Consolidation — process of transferring learned information from short-term memory to long-term memory.
Cysteine/glutamate transporter — a protein that is responsible for exchanging glutamate with cysteine.
Extinction burst — a pattern of responding that occurs at the beginning of extinction training in which animals respond more on a manipulandum that is now no longer reinforced.
Glutamate homeostasis hypothesis of addiction — a proposed explanation of how addiction may stem from dysregulation of glutamate homeostasis in the brain.
Long-term depression — a form of synaptic plasticity that results from moderate stimulation of glutamate AMPA receptors, leading to decreased AMPA receptor expression on neurons.
Long-term potentiation — a form of synaptic plasticity that results from high frequency stimulation of glutamate AMPA receptors, leading to increased AMPA receptor expression on neurons.
Metaplasticity — the plasticity of synaptic plasticity; the alteration of a synapse's susceptibility for subsequent long-term potentiation or long-term depression.
Reconsolidation — process of recalling memories to modify them before consolidating them back into long-term memory.
Renewal — a phenomenon in which a behavior that has been extinguished in one environment reemerges when the organism is exposed to the environment in which the behavior was originally learned.
Resurgence — a phenomenon in which a previously extinguished behavior reappears during extinction training for another behavior.
Self-efficacy — refers to a personal assessment of how well an individual can perform a task or to cope with a specific situation.

Supplementary data related to this chapter can be found online at https://educate.elsevier.com/9780323905787.

References

1. Hammer RR, Dingel MJ, Ostergren JE, Nowakowski KE, Koenig BA. The experience of addiction as told by the addicted: incorporating biological understandings into self-story. *Cult Med Psychiatr*. 2012;36(4):712–734. https://doi.org/10.1007/s11013-012-9283-x.
2. Beck AT, Wright FD, Newman CF, Liese BS. *Cognitive Therapy of Substance Abuse*. New York: Guilford Press; 1993.
3. Beck AT, Rush A, Shaw B, Emery G. *Cognitive Therapy of Depression*. Guilford; 1979.
4. Burns DD. *Feeling Good: The New Mood Therapy*. Signet; 1980.
5. Bandura A. Self-efficacy: toward a unifying theory of behavioral change. *Psychol Rev*. 1977;84(2):191–215. https://doi.org/10.1037/0033-295X.84.2.191.
6. Kadden RM, Litt MD. The role of self-efficacy in the treatment of substance use disorders. *Addict Behav*. 2011;36(12):1120–1126. https://doi.org/10.1016/j.addbeh.2011.07.032.
7. Burleson JA, Kaminer Y. Self-efficacy as a predictor of treatment outcome in adolescent substance use disorders. *Addict Behav*. 2005;30(9):1751–1764. https://doi.org/10.1016/j.addbeh.2005.07.006.
8. Abroms BD, Gottlob LR, Fillmore MT. Alcohol effects on inhibitory control of attention: distinguishing between intentional and automatic mechanisms. *Psychopharmacology*. 2006;188(3):324–334. https://doi.org/10.1007/s00213-006-0524-y.
9. Rohrbaugh JW, Stapleton JM, Parasuraman R, et al. Alcohol intoxication reduces visual sustained attention. *Psychopharmacology*. 1988;96(4):442–446. https://doi.org/10.1007/BF02180021.
10. Givens B. Effect of ethanol on sustained attention in rats. *Psychopharmacology*. 1997;129(2):135–140. https://doi.org/10.1007/s002130050173.
11. Schulte T, Müller-Oehring EM, Strasburger H, Warzel H, Sabel BA. Acute effects of alcohol on divided and covert attention in men. *Psychopharmacology*. 2001;154(1):61–69. https://doi.org/10.1007/s002130000603.
12. Marinkovic K, Rickenbacher E, Azma S, Artsy E. Acute alcohol intoxication impairs top-down regulation of Stroop incongruity as revealed by blood oxygen level-dependent functional magnetic resonance imaging. *Hum Brain Mapp*. 2012;33(2):319–333. https://doi.org/10.1002/hbm.21213.
13. Böcker KB, Gerritsen J, Hunault CC, Kruidenier M, Mensinga TT, Kenemans JL. Cannabis with high δ9-THC contents affects perception and visual selective attention acutely: an event-related potential study. *Pharmacol Biochem Behav*. 2010;96(1):67–74. https://doi.org/10.1016/j.pbb.2010.04.008.
14. Arkell TR, Lintzeris N, Kevin RC, et al. Cannabidiol (CBD) content in vaporized cannabis does not prevent tetrahydrocannabinol (THC)-induced impairment of driving and cognition. *Psychopharmacology*. 2019;236(9):2713–2724. https://doi.org/10.1007/s00213-019-05246-8.
15. Hooker WD, Jones RT. Increased susceptibility to memory intrusions and the Stroop interference effect during acute marijuana intoxication. *Psychopharmacology*. 1987;91(1):20–24. https://doi.org/10.1007/BF00690920.
16. Schneider U, Bevilacqua C, Jacobs R, et al. Effects of fentanyl and low doses of alcohol on neuropsychological performance in healthy subjects. *Neuropsychobiology*. 1999;39(1):38–43. https://doi.org/10.1159/000026558.
17. Zeng H, Su D, Jiang X, Zhu L, Ye H. The similarities and differences in impulsivity and cognitive ability among ketamine, methadone, and non-drug users. *Psychiatr Res*. 2016;243:109–114. https://doi.org/10.1016/j.psychres.2016.04.095.
18. Grilly DM, Gowans GC, McCann DS, Grogan TW. Effects of cocaine and d-amphetamine on sustained and selective attention in rats. *Pharmacol Biochem Behav*. 1989;33(4):733–739. https://doi.org/10.1016/0091-3057(89)90463-2.
19. Servan-Schreiber D, Carter CS, Bruno RM, Cohen JD. Dopamine and the mechanisms of cognition: Part II. D-amphetamine effects in human subjects performing a selective attention task. *Biol Psychiatr*. 1998;43(10):723–729. https://doi.org/10.1016/s0006-3223(97)00449-6.
20. Soar K, Mason C, Potton A, Dawkins L. Neuropsychological effects associated with recreational cocaine use. *Psychopharmacology*. 2012;222(4):633–643. https://doi.org/10.1007/s00213-012-2666-4.
21. Sato A. Methamphetamine use in Japan after the second world war: transformation of narratives. *Contemp Drug Probl*. 2008;35(4):717–746. https://doi.org/10.1177/009145090803500410.
22. Simon SL, Domier C, Carnell J, Brethen P, Rawson R, Ling W. Cognitive impairment in individuals currently using methamphetamine. *Am J Addict*. 2000;9(3):222–231. https://doi.org/10.1080/10550490050148053.
23. Parrott AC, Craig D. Cigarette smoking and nicotine gum (0, 2 and 4 mg): effects upon four visual attention tasks. *Neuropsychobiology*. 1992;25(1):34–43. https://doi.org/10.1159/000118807.

24. Kelemen WL, Fulton EK. Cigarette abstinence impairs memory and metacognition despite administration of 2 mg nicotine gum. *Exp Clin Psychopharmacol.* 2008;16(6):521–531. https://doi.org/10.1037/a0014246.
25. Provost SC, Woodward R. Effects of nicotine gum on repeated administration of the Stroop test. *Psychopharmacology.* 1991;104(4):536–540. https://doi.org/10.1007/BF02245662.
26. Domier CP, Monterosso JR, Brody AL, et al. Effects of cigarette smoking and abstinence on Stroop task performance. *Psychopharmacology.* 2007;195(1):1–9. https://doi.org/10.1007/s00213-007-0869-x.
27. Young JW, Finlayson K, Spratt C, et al. Nicotine improves sustained attention in mice: evidence for involvement of the alpha7 nicotinic acetylcholine receptor. *Neuropsychopharmacology.* 2004;29(5):891–900. https://doi.org/10.1038/sj.npp.1300393.
28. Valentine G, Sofuoglu M. Cognitive effects of nicotine: recent progress. *Curr Neuropharmacol.* 2018;16(4):403–414. https://doi.org/10.2174/1570159X15666171103152136.
29. Dondé C, Brunelin J, Mondino M, Cellard C, Rolland B, Haesebaert F. The effects of acute nicotine administration on cognitive and early sensory processes in schizophrenia: a systematic review. *Neurosci Biobehav Rev.* 2020;118:121–133. https://doi.org/10.1016/j.neubiorev.2020.07.035.
30. Waterhouse U, Brennan KA, Ellenbroek BA. Nicotine self-administration reverses cognitive deficits in a rat model for schizophrenia. *Addiction Biol.* 2018;23(2):620–630. https://doi.org/10.1111/adb.12517.
31. Slobodin O, Blankers M, Kapitány-Fövény M, et al. Differential diagnosis in patients with substance use disorder and/or attention-deficit/hyperactivity disorder using continuous performance test. *Eur Addiction Res.* 2020;26(3):151–162. https://doi.org/10.1159/000506334.
32. Saraswat N, Ranjan S, Ram D. Set-shifting and selective attentional impairment in alcoholism and its relation with drinking variables. *Indian J Psychiatr.* 2006;48(1):47–51. https://doi.org/10.4103/0019-5545.31619.
33. Miguel CS, Martins PA, Moleda N, et al. Cognition and impulsivity in adults with attention deficit hyperactivity disorder with and without cocaine and/or crack dependence. *Drug Alcohol Depend.* 2016;160:97–104. https://doi.org/10.1016/j.drugalcdep.2015.12.040.
34. Indlekofer F, Piechatzek M, Daamen M, et al. Reduced memory and attention performance in a population-based sample of young adults with a moderate lifetime use of cannabis, ecstasy and alcohol. *J Psychopharmacol.* 2009;23(5):495–509. https://doi.org/10.1177/0269881108091076.
35. McKetin R, Solowij N. Event-related potential indices of auditory selective attention in dependent amphetamine users. *Biol Psychiatr.* 1999;45(11):1488–1497. https://doi.org/10.1016/s0006-3223(98)00200-5.
36. Nadar MS, Hasan AM, Alsaleh M. The negative impact of chronic tobacco smoking on adult neuropsychological function: a cross-sectional study. *BMC Publ Health.* 2021;21(1):1278. https://doi.org/10.1186/s12889-021-11287-6.
37. Battisti RA, Roodenrys S, Johnstone SJ, Pesa N, Hermens DF, Solowij N. Chronic cannabis users show altered neurophysiological functioning on Stroop task conflict resolution. *Psychopharmacology.* 2010;212(4):613–624. https://doi.org/10.1007/s00213-010-1988-3.
38. Salo R, Nordahl TE, Natsuaki Y, et al. Attentional control and brain metabolite levels in methamphetamine abusers. *Biol Psychiatr.* 2007;61(11):1272–1280. https://doi.org/10.1016/j.biopsych.2006.07.031.
39. Verdejo-García AJ, Perales JC, Pérez-García M. Cognitive impulsivity in cocaine and heroin polysubstance abusers. *Addict Behav.* 2007;32(5):950–966. https://doi.org/10.1016/j.addbeh.2006.06.032.
40. Fontes MA, Bolla KI, Cunha PJ, et al. Cannabis use before age 15 and subsequent executive functioning. *Br J Psychiatr.* 2011;198(6):442–447. https://doi.org/10.1192/bjp.bp.110.077479.
41. Lopes BM, Gonçalves PD, Ometto M, et al. Distinct cognitive performance and patterns of drug use among early and late onset cocaine users. *Addict Behav.* 2017;73:41–47. https://doi.org/10.1016/j.addbeh.2017.04.013.
42. Mashhoon Y, Betts J, Farmer SL, Lukas SE. Early onset tobacco cigarette smokers exhibit deficits in response inhibition and sustained attention. *Drug Alcohol Depend.* 2018;184:48–56. https://doi.org/10.1016/j.drugalcdep.2017.11.020.
43. Sagar KA, Dahlgren MK, Gönenç A, Racine MT, Dreman MW, Gruber SA. The impact of initiation: early onset marijuana smokers demonstrate altered Stroop performance and brain activation. *Develop Cogn Neurosci.* 2015;16:84–92. https://doi.org/10.1016/j.dcn.2015.03.003.
44. Crider A, Solomon PR, McMahon MA. Disruption of selective attention in the rat following chronic d-amphetamine administration: relationship to schizophrenic attention disorder. *Biol Psychiatr.* 1982;17(3):351–361.

45. Dalley JW, Lääne K, Pena Y, Theobald DE, Everitt BJ, Robbins TW. Attentional and motivational deficits in rats withdrawn from intravenous self-administration of cocaine or heroin. *Psychopharmacology*. 2005;182(4):579−587. https://doi.org/10.1007/s00213-005-0107-3.
46. Biederman J, Spencer T. Attention-deficit/hyperactivity disorder (ADHD) as a noradrenergic disorder. *Biol Psychiatr*. 1999;46(9):1234−1242. https://doi.org/10.1016/s0006-3223(99)00192-4.
47. Mueller A, Hong DS, Shepard S, Moore T. Linking ADHD to the neural circuitry of attention. *Trends Cognit Sci*. 2017;21(6):474−488. https://doi.org/10.1016/j.tics.2017.03.009.
48. Cordovil De Sousa Uva M, Luminet O, Cortesi M, Constant E, Derely M, De Timary P. Distinct effects of protracted withdrawal on affect, craving, selective attention and executive functions among alcohol-dependent patients. *Alcohol Alcohol*. 2010;45(3):241−246. https://doi.org/10.1093/alcalc/agq012.
49. Devenney LE, Coyle KB, Verster JC. Memory and attention during an alcohol hangover. *Hum Psychopharmacol*. 2019;34(4):e2701. https://doi.org/10.1002/hup.2701.
50. McKinney A, Coyle K, Penning R, Verster JC. Next day effects of naturalistic alcohol consumption on tasks of attention. *Hum Psychopharmacol*. 2012;27(6):587−594. https://doi.org/10.1002/hup.2268.
51. Hughes JR, Keenan RM, Yellin A. Effect of tobacco withdrawal on sustained attention. *Addict Behav*. 1989;14(5):577−580. https://doi.org/10.1016/0306-4603(89)90079-8.
52. Tedstone D, Coyle K. Cognitive impairments in sober alcoholics: performance on selective and divided attention tasks. *Drug Alcohol Depend*. 2004;75(3):277−286. https://doi.org/10.1016/j.drugalcdep.2004.03.005.
53. Shoaib M, Bizarro L. Deficits in a sustained attention task following nicotine withdrawal in rats. *Psychopharmacology*. 2005;178(2−3):211−222. https://doi.org/10.1007/s00213-004-2004-6.
54. Prosser J, London ED, Galynker II. Sustained attention in patients receiving and abstinent following methadone maintenance treatment for opiate dependence: performance and neuroimaging results. *Drug Alcohol Depend*. 2009;104(3):228−240. https://doi.org/10.1016/j.drugalcdep.2009.04.022.
55. Jager G, Kahn RS, Van Den Brink W, Van Ree JM, Ramsey NF. Long-term effects of frequent cannabis use on working memory and attention: an fMRI study. *Psychopharmacology*. 2006;185(3):358−368. https://doi.org/10.1007/s00213-005-0298-7.
56. Bandstra ES, Morrow CE, Anthony JC, Accornero VH, Fried PA. Longitudinal investigation of task persistence and sustained attention in children with prenatal cocaine exposure. *Neurotoxicol Teratol*. 2001;23(6):545−559. https://doi.org/10.1016/s0892-0362(01)00181-7.
57. Fried PA. Adolescents prenatally exposed to marijuana: examination of facets of complex behaviors and comparisons with the influence of in utero cigarettes. *J Clin Pharmacol*. 2002;42(S1):97S−102S. https://doi.org/10.1002/j.1552-4604.2002.tb06009.x.
58. Konijnenberg C, Melinder A. Visual selective attention is impaired in children prenatally exposed to opioid agonist medication. *Eur Addiction Res*. 2015;21(2):63−70. https://doi.org/10.1159/000366018.
59. Kooistra L, Crawford S, Gibbard B, Ramage B, Kaplan BJ. Differentiating attention deficits in children with fetal alcohol spectrum disorder or attention-deficit-hyperactivity disorder. *Dev Med Child Neurol*. 2010;52(2):205−211. https://doi.org/10.1111/j.1469-8749.2009.03352.x.
60. Coles CD, Platzman KA, Lynch ME, Freides D. Auditory and visual sustained attention in adolescents prenatally exposed to alcohol. *Alcohol Clin Exp Res*. 2002;26(2):263−271.
61. Singer LT, Min MO, Minnes S, et al. Prenatal and concurrent cocaine, alcohol, marijuana, and tobacco effects on adolescent cognition and attention. *Drug Alcohol Depend*. 2018;191:37−44. https://doi.org/10.1016/j.drugalcdep.2018.06.022.
62. Derauf C, Lagasse LL, Smith LM, et al. Prenatal methamphetamine exposure and inhibitory control among young school-age children. *J Pediatr*. 2012;161(3):452−459. https://doi.org/10.1016/j.jpeds.2012.02.002.
63. Connor PD, Sampson PD, Bookstein FL, Barr HM, Streissguth AP. Direct and indirect effects of prenatal alcohol damage on executive function. *Dev Neuropsychol*. 2000;18(3):331−354. https://doi.org/10.1207/S1532694204Connor.
64. National Highway Traffic Safety Administration. *Table 31. Fatal Crashes and Percentage Alcohol-Impaired Driving by Time of Day and Crash Type*. State: USA; 2019. https://cdan.nhtsa.gov/SASStoredProcess/guest.
65. Wester AE, Verster JC, Volkerts ER, Böcker KB, Kenemans JL. Effects of alcohol on attention orienting and dual-task performance during simulated driving: an event-related potential study. *J Psychopharmacol*. 2010;24(9):1333−1348. https://doi.org/10.1177/0269881109348168.

66. Simons R, Martens M, Ramaekers J, Krul A, Klöpping-Ketelaars I, Skopp G. Effects of dexamphetamine with and without alcohol on simulated driving. *Psychopharmacology*. 2012;222(3):391–399. https://doi.org/10.1007/s00213-011-2549-0.
67. Iudice A, Bonanni E, Gelli A, et al. Effects of prolonged wakefulness combined with alcohol and hands-free cell phone divided attention tasks on simulated driving. *Hum Psychopharmacol*. 2005;20(2):125–132. https://doi.org/10.1002/hup.664.
68. Alford C, Broom C, Carver H, et al. The impact of alcohol hangover on simulated driving performance during a 'commute to work'-zero and residual alcohol effects compared. *J Clin Med*. 2020;9(5):1435. https://doi.org/10.3390/jcm9051435.
69. Busardò FP, Pellegrini M, Klein J, di Luca NM. Neurocognitive correlates in driving under the influence of cannabis. *CNS Neurol Disord - Drug Targets*. 2017;16(5):534–540. https://doi.org/10.2174/1871527316666170424115455.
70. National Conference of State Legislatures. *Drugged Driving, Marijuana-Impaired Driving*; September 23, 2021. https://www.ncsl.org/research/transportation/drugged-driving-overview.aspx.
71. Fischer PM, Richards Jr JW, Berman EJ, Krugman DM. Recall and eye tracking study of adolescents viewing tobacco advertisements. *JAMA*. 1989;261(1):84–89.
72. Thomsen SR, Fulton K. Adolescents' attention to responsibility messages in magazine alcohol advertisements: an eye-tracking approach. *J Adolesc Health*. 2007;41(1):27–34. https://doi.org/10.1016/j.jadohealth.2007.02.014.
73. Yzer M, Han J, Choi K. Eye movement patterns in response to anti-binge drinking messages. *Health Commun*. 2018;33(12):1454–1461. https://doi.org/10.1080/10410236.2017.1359032.
74. Maynard OM, Attwood A, O'Brien L, et al. Avoidance of cigarette pack health warnings among regular cigarette smokers. *Drug Alcohol Depend*. 2014;136:170–174. https://doi.org/10.1016/j.drugalcdep.2014.01.001.
75. Sillero-Rejon C, Leonards U, Munafò MR, et al. Avoidance of tobacco health warnings? An eye-tracking approach. *Addiction*. 2021;116(1):126–138. https://doi.org/10.1111/add.15148.
76. Sillero-Rejon C, Attwood AS, Blackwell A, Ibáñez-Zapata JA, Munafò MR, Maynard OM. Alcohol pictorial health warning labels: the impact of self-affirmation and health warning severity. *BMC Publ Health*. 2018;18(1):1403. https://doi.org/10.1186/s12889-018-6243-6.
77. Peterson EB, Thomsen S, Lindsay G, John K. Adolescents' attention to traditional and graphic tobacco warning labels: an eye-tracking approach. *J Drug Educ*. 2010;40(3):227–244. https://doi.org/10.2190/DE.40.3.b.
78. Klein EG, Quisenberry AJ, Shoben AB, et al. Health warning labels for smokeless tobacco: the impact of graphic images on attention, recall, and craving. *Nicotine Tob Res*. 2017;19(10):1172–1177. https://doi.org/10.1093/ntr/ntx021.
79. Brown SL, Richardson M. The effect of distressing imagery on attention to and persuasiveness of an antialcohol message: a gaze-tracking approach. *Health Educ Behav*. 2012;39(1):8–17. https://doi.org/10.1177/1090198111404411.
80. Kemp D, Niederdeppe J, Byrne S. Adolescent attention to disgust visuals in cigarette graphic warning labels. *J Adolesc Health*. 2019;65(6):769–775. https://doi.org/10.1016/j.jadohealth.2019.07.007.
81. McNeill A, Gravely S, Hitchman SC, Bauld L, Hammond D, Hartmann-Boyce J. Tobacco packaging design for reducing tobacco use. *Cochrane Database Syst Rev*. 2017;4(4):CD011244. https://doi.org/10.1002/14651858.CD011244.pub2.
82. Park H, Hong MY, Lee IS, Chae Y. Effects of different graphic health warning types on the intention to quit smoking. *Int J Environ Res Publ Health*. 2020;17(9):3267. https://doi.org/10.3390/ijerph17093267.
83. Mahedy L, Field M, Gage S, et al. Alcohol use in adolescence and later working memory: findings from a large population-based birth cohort. *Alcohol Alcohol*. 2018;53(3):251–258. https://doi.org/10.1093/alcalc/agx113.
84. Ilan AB, Smith ME, Gevins A. Effects of marijuana on neurophysiological signals of working and episodic memory. *Psychopharmacology*. 2004;176(2):214–222. https://doi.org/10.1007/s00213-004-1868-9.
85. Cousijn J, Vingerhoets WA, Koenders L, et al. Relationship between working-memory network function and substance use: a 3-year longitudinal fMRI study in heavy cannabis users and controls. *Addiction Biol*. 2014;19(2):282–293. https://doi.org/10.1111/adb.12111.
86. Wittemann M, Brielmaier J, Rubly M, et al. Cognition and cortical thickness in heavy cannabis users. *Eur Addiction Res*. 2021;27(2):115–122. https://doi.org/10.1159/000509987.

87. Nestor L, Roberts G, Garavan H, Hester R. Deficits in learning and memory: parahippocampal hyperactivity and frontocortical hypoactivity in cannabis users. *Neuroimage*. 2008;40(3):1328−1339. https://doi.org/10.1016/j.neuroimage.2007.12.059.
88. Cohen K, Kapitány-Fövény M, Mama Y, et al. The effects of synthetic cannabinoids on executive function. *Psychopharmacology*. 2017;234(7):1121−1134. https://doi.org/10.1007/s00213-017-4546-4.
89. Albein-Urios N, Martinez-González JM, Lozano O, Clark L, Verdejo-García A. Comparison of impulsivity and working memory in cocaine addiction and pathological gambling: implications for cocaine-induced neurotoxicity. *Drug Alcohol Depend*. 2012;126(1−2):1−6. https://doi.org/10.1016/j.drugalcdep.2012.03.008.
90. Sanvicente-Vieira B, Kommers-Molina J, De Nardi T, Francke I, Grassi-Oliveira R. Crack-cocaine dependence and aging: effects on working memory. *Braz J Psychiatr*. 2016;38(1):58−60. https://doi.org/10.1590/1516-4446-2015-1708.
91. Morgan CJ, Muetzelfeldt L, Curran HV. Consequences of chronic ketamine self-administration upon neurocognitive function and psychological wellbeing: a 1-year longitudinal study. *Addiction*. 2010;105(1):121−133. https://doi.org/10.1111/j.1360-0443.2009.02761.x.
92. Smith MJ, Cobia DJ, Wang L, et al. Cannabis-related working memory deficits and associated subcortical morphological differences in healthy individuals and schizophrenia subjects. *Schizophr Bull*. 2014;40(2):287−299. https://doi.org/10.1093/schbul/sbt176.
93. Potvin S, Pelletier J, Grot S, Hébert C, Barr AM, Lecomte T. Cognitive deficits in individuals with methamphetamine use disorder: a meta-analysis. *Addict Behav*. 2018;80:154−160. https://doi.org/10.1016/j.addbeh.2018.01.021.
94. De Oliveira Jr HP, Gonçalves PD, Ometto M, et al. Distinct effects of cocaine and cocaine + cannabis on neurocognitive functioning and abstinence: a six-month follow-up study. *Drug Alcohol Depend*. 2019;205:107642. https://doi.org/10.1016/j.drugalcdep.2019.107642.
95. Grundey J, Amu R, Ambrus GG, Batsikadze G, Paulus W, Nitsche MA. Double dissociation of working memory and attentional processes in smokers and non-smokers with and without nicotine. *Psychopharmacology*. 2015;232(14):2491−2501. https://doi.org/10.1007/s00213-015-3880-7.
96. Wang ZX, Xiao ZW, Zhang DR, Liang CY, Zhang JX. Verbal working memory deficits in abstinent heroin abusers. *Acta Neuropsychiatr*. 2008;20(5):265−268. https://doi.org/10.1111/j.1601-5215.2008.00293.x.
97. Mintzer MZ, Stitzer ML. Cognitive impairment in methadone maintenance patients. *Drug Alcohol Depend*. 2002;67(1):41−51. https://doi.org/10.1016/s0376-8716(02)00013-3.
98. Willford JA, Richardson GA, Leech SL, Day NL. Verbal and visuospatial learning and memory function in children with moderate prenatal alcohol exposure. *Alcohol Clin Exp Res*. 2004;28(3):497−507. https://doi.org/10.1097/01.alc.0000117868.97486.2d.
99. Willford JA, Goldschmidt L, De Genna NM, Day NL, Richardson GA. A longitudinal study of the impact of marijuana on adult memory function: prenatal, adolescent, and young adult exposures. *Neurotoxicol Teratol*. 2021;84:106958. https://doi.org/10.1016/j.ntt.2021.106958.
100. Sirnes E, Griffiths ST, Aukland SM, Eide GE, Elgen IB, Gundersen H. Functional MRI in prenatally opioid-exposed children during a working memory-selective attention task. *Neurotoxicol Teratol*. 2018;66:46−54. https://doi.org/10.1016/j.ntt.2018.01.010.
101. Santucci AC, Capodilupo S, Bernstein J, Gomez-Ramirez M, Milefsky R, Mitchell H. Cocaine in adolescent rats produces residual memory impairments that are reversible with time. *Neurotoxicol Teratol*. 2004;26(5):651−661. https://doi.org/10.1016/j.ntt.2004.06.002.
102. Gendle MH, Strawderman MS, Mactutus CF, Booze RM, Levitsky DA, Strupp BJ. Prenatal cocaine exposure does not alter working memory in adult rats. *Neurotoxicol Teratol*. 2004;26(2):319−329. https://doi.org/10.1016/j.ntt.2003.12.001.
103. Pacheco ALD, de Melo IS, de Souza FMA, et al. Maternal crack cocaine use in rats leads to depressive- and anxiety-like behavior, memory impairment, and increased seizure susceptibility in the offspring. *Eur Neuropsychopharmacol*. 2021;44:34−50. https://doi.org/10.1016/j.euroneuro.2020.12.011.
104. Selby MJ, Azrin RL. Neuropsychological functioning in drug abusers. *Drug Alcohol Depend*. 1998;50(1):39−45. https://doi.org/10.1016/s0376-8716(98)00002-7.
105. Brown J, McKone E, Ward J. Deficits of long-term memory in ecstasy users are related to cognitive complexity of the task. *Psychopharmacology*. 2010;209(1):51−67. https://doi.org/10.1007/s00213-009-1766-2.

106. Darley CF, Tinklenberg JR, Roth WT, Vernon S, Kopell BS. Marijuana effects on long-term memory assessment and retrieval. *Psychopharmacology*. 1977;52(3):239—241. https://doi.org/10.1007/BF00426706.
107. Hadjiefthyvoulou F, Fisk JE, Montgomery C, Bridges N. Prospective memory functioning among ecstasy/polydrug users: evidence from the Cambridge Prospective Memory Test (CAMPROMPT). *Psychopharmacology*. 2011;215(4):761—774. https://doi.org/10.1007/s00213-011-2174-y.
108. Gilbertson R, Boissoneault J, Prather R, Nixon SJ. Nicotine effects on immediate and delayed verbal memory after substance use detoxification. *J Clin Exp Neuropsychol*. 2011;33(6):609—618. https://doi.org/10.1080/13803395.2010.543887.
109. Falcone M, Wileyto EP, Ruparel K, et al. Age-related differences in working memory deficits during nicotine withdrawal. *Addiction Biol*. 2014;19(5):907—917. https://doi.org/10.1111/adb.12051.
110. Mendrek A, Monterosso J, Simon SL, et al. Working memory in cigarette smokers: comparison to non-smokers and effects of abstinence. *Addict Behav*. 2006;31(5):833—844. https://doi.org/10.1016/j.addbeh.2005.06.009.
111. Hirshman E, Rhodes DK, Zinser M, Merritt P. The effect of tobacco abstinence on recognition memory, digit span recall, and attentional vigilance. *Exp Clin Psychopharmacol*. 2004;12(1):76—83. https://doi.org/10.1037/1064-1297.12.1.76.
112. Jacobsen LK, Krystal JH, Mencl WE, Westerveld M, Frost SJ, Pugh KR. Effects of smoking and smoking abstinence on cognition in adolescent tobacco smokers. *Biol Psychiatr*. 2005;57(1):56—66. https://doi.org/10.1016/j.biopsych.2004.10.022.
113. Terry Jr AV, Callahan PM, Schade R, Kille NJ, Plagenhoef M. Alpha 2A adrenergic receptor agonist, guanfacine, attenuates cocaine-related impairments of inhibitory response control and working memory in animal models. *Pharmacol Biochem Behav*. 2014;126:63—72. https://doi.org/10.1016/j.pbb.2014.09.010.
114. Taffe MA, Davis SA, Gutierrez T, Gold LH. Ketamine impairs multiple cognitive domains in rhesus monkeys. *Drug Alcohol Depend*. 2002;68(2):175—187. https://doi.org/10.1016/s0376-8716(02)00194-1.
115. Schambra UB, Lewis CN, Harrison TA. Deficits in spatial learning and memory in adult mice following acute, low or moderate levels of prenatal ethanol exposure during gastrulation or neurulation. *Neurotoxicol Teratol*. 2017;62:42—54. https://doi.org/10.1016/j.ntt.2017.05.001.
116. Lee K-W, Kim H-C, Lee S-Y, Jang C-G. Methamphetamine-sensitized mice are accompanied by memory impairment and reduction of N-methyl-d-aspartate receptor ligand binding in the prefrontal cortex and hippocampus. *Neuroscience*. 2011;178:101—107. https://doi.org/10.1016/j.neuroscience.2011.01.025.
117. Seyedhosseini Tamijani SM, Beirami E, Ahmadiani A, Dargahi L. Effect of three different regimens of repeated methamphetamine on rats' cognitive performance. *Cognit Process*. 2018;19(1):107—115. https://doi.org/10.1007/s10339-017-0839-0.
118. Long J-D, Liu Y, Jiao D-L, et al. The neuroprotective effect of memantine on methamphetamine-induced cognitive deficits. *Behav Brain Res*. 2017;323:133—140. https://doi.org/10.1016/j.bbr.2017.01.042.
119. Dong N, Zhu J, Han W, et al. Maternal methamphetamine exposure causes cognitive impairment and alteration of neurodevelopment-related genes in adult offspring mice. *Neuropharmacology*. 2018;140:25—34. https://doi.org/10.1016/j.neuropharm.2018.07.024.
120. Levin ED, Lee C, Rose JE, et al. Chronic nicotine and withdrawal effects on radial-arm maze performance in rats. *Behav Neural Biol*. 1990;53(2):269—276. https://doi.org/10.1016/0163-1047(90)90509-5.
121. Silva L, Zhao N, Popp S, Dow-Edwards D. Prenatal tetrahydrocannabinol (THC) alters cognitive function and amphetamine response from weaning to adulthood in the rat. *Neurotoxicol Teratol*. 2012;34(1):63—71. https://doi.org/10.1016/j.ntt.2011.10.006.
122. Melugin PR, Nolan SO, Siciliano CA. Bidirectional causality between addiction and cognitive deficits. *Int Rev Neurobiol*. 2021;157:371—407. https://doi.org/10.1016/bs.irn.2020.11.001.
123. Khurana A, Romer D, Betancourt LM, Hurt H. Working memory ability and early drug use progression as predictors of adolescent substance use disorders. *Addiction*. 2017;112(7):1220—1228. https://doi.org/10.1111/add.13792.
124. Lechner WV, L Gunn R, Minto A, et al. Effects of negative affect, urge to smoke, and working memory performance (n-back) on nicotine dependence. *Subst Use Misuse*. 2018;53(7):1177—1183. https://doi.org/10.1080/10826084.2017.1400569.
125. Day AM, Metrik J, Spillane NS, Kahler CW. Working memory and impulsivity predict marijuana-related problems among frequent users. *Drug Alcohol Depend*. 2013;131(1—2):171—174. https://doi.org/10.1016/j.drugalcdep.2012.12.016.

126. Aharonovich E, Nunes E, Hasin D. Cognitive impairment, retention and abstinence among cocaine abusers in cognitive-behavioral treatment. *Drug Alcohol Depend*. 2003;71(2):207–211. https://doi.org/10.1016/s0376-8716(03)00092-9.
127. Rubenis AJ, Fitzpatrick RE, Lubman DI, Verdejo-Garcia A. Working memory predicts methamphetamine hair concentration over the course of treatment: moderating effect of impulsivity and implications for dual-systems model. *Addiction Biol*. 2019;24(1):145–153. https://doi.org/10.1111/adb.12575.
128. Patterson F, Jepson C, Loughead J, et al. Working memory deficits predict short-term smoking resumption following brief abstinence. *Drug Alcohol Depend*. 2010;106(1):61–64. https://doi.org/10.1016/j.drugalcdep.2009.07.020.
129. Copersino ML, Schretlen DJ, Fitzmaurice GM, et al. Effects of cognitive impairment on substance abuse treatment attendance: predictive validation of a brief cognitive screening measure. *Am J Drug Alcohol Abuse*. 2012;38(3):246–250. https://doi.org/10.3109/00952990.2012.670866.
130. Streeter CC, Terhune DB, Whitfield TH, et al. Performance on the Stroop predicts treatment compliance in cocaine-dependent individuals. *Neuropsychopharmacology*. 2008;33(4):827–836. https://doi.org/10.1038/sj.npp.1301465.
131. Grove RN, Schuster CR. Suppression of cocaine self-administration by extinction and punishment. *Pharmacol Biochem Behav*. 1974;2(2):199–208. https://doi.org/10.1016/0091-3057(74)90053-7.
132. Davis WM, Smith SG. Role of conditioned reinforcers in the initiation, maintenance and extinction of drug-seeking behavior. *Pavlovian J Biol Sci*. 1976;11(4):222–236. https://doi.org/10.1007/BF03000316.
133. Brenhouse HC, Thompson BS, Sonntag KC, Andersen SL. Extinction and reinstatement to cocaine-associated cues in male and female juvenile rats and the role of D1 dopamine receptor. *Neuropharmacology*. 2015;95:22–28. https://doi.org/10.1016/j.neuropharm.2015.02.017.
134. Font L, Houck CA, Cunningham CL. Naloxone effects on extinction of ethanol- and cocaine-induced conditioned place preference in mice. *Psychopharmacology*. 2017;234(18):2747–2759. https://doi.org/10.1007/s00213-017-4672-z.
135. Pushparaj A, Pryslawsky Y, Forget B, Yan Y, Le Foll B. Extinction bursts in rats trained to self-administer nicotine or food in 1-h daily sessions. *Am J Tourism Res*. 2012;4(4):422–431.
136. Bouton ME. Context, ambiguity, and unlearning: sources of relapse after behavioral extinction. *Biol Psychiatr*. 2002;52(10):976–986. https://doi.org/10.1016/s0006-3223(02)01546-9.
137. Voigt RM, Herrold AA, Napier TC. Baclofen facilitates the extinction of methamphetamine-induced conditioned place preference in rats. *Behav Neurosci*. 2011;125(2):261–267. https://doi.org/10.1037/a0022893.
138. Duke AN, Kaminski BJ, Weerts EM. Baclofen effects on alcohol seeking, self-administration and extinction of seeking responses in a within-session design in baboons. *Addiction Biol*. 2014;19(1):16–26. https://doi.org/10.1111/j.1369-1600.2012.00448.x.
139. Kaminski BJ, Duke AN, Weerts EM. Effects of naltrexone on alcohol drinking patterns and extinction of alcohol seeking in baboons. *Psychopharmacology*. 2012;223(1):55–66. https://doi.org/10.1007/s00213-012-2688-y.
140. Parker LA, Burton P, Sorge RE, Yakiwchuk C, Mechoulam R. Effect of low doses of delta9-tetrahydrocannabinol and cannabidiol on the extinction of cocaine-induced and amphetamine-induced conditioned place preference learning in rats. *Psychopharmacology*. 2004;175(3):360–366. https://doi.org/10.1007/s00213-004-1825-7.
141. Mahler SV, Brodnik ZD, Cox BM, et al. Chemogenetic manipulations of ventral tegmental area dopamine neurons reveal multifaceted roles in cocaine abuse. *J Neurosci*. 2019;39(3):503–518. https://doi.org/10.1523/JNEUROSCI.0537-18.2018.
142. Karlsson RM, Kircher DM, Shaham Y, O'Donnell P. Exaggerated cue-induced reinstatement of cocaine seeking but not incubation of cocaine craving in a developmental rat model of schizophrenia. *Psychopharmacology*. 2013;226(1):45–51. https://doi.org/10.1007/s00213-012-2882-y.
143. Li C, Frantz KJ. Attenuated incubation of cocaine seeking in male rats trained to self-administer cocaine during periadolescence. *Psychopharmacology*. 2009;204(4):725–733. https://doi.org/10.1007/s00213-009-1502-y.
144. Crombag HS, Shaham Y. Renewal of drug seeking by contextual cues after prolonged extinction in rats. *Behav Neurosci*. 2002;116(1):169–173. https://doi.org/10.1037//0735-7044.116.1.169.
145. Hamlin AS, Clemens KJ, McNally GP. Renewal of extinguished cocaine-seeking. *Neuroscience*. 2008;151(3):659–670. https://doi.org/10.1016/j.neuroscience.2007.11.018.

146. Rubio FJ, Liu QR, Li X, et al. Context-induced reinstatement of methamphetamine seeking is associated with unique molecular alterations in Fos-expressing dorsolateral striatum neurons. *J Neurosci.* 2015;35(14):5625–5639. https://doi.org/10.1523/JNEUROSCI.4997-14.2015.
147. Willcocks AL, McNally GP. The role of medial prefrontal cortex in extinction and reinstatement of alcohol-seeking in rats. *Eur J Neurosci.* 2013;37(2):259–268. https://doi.org/10.1111/ejn.12031.
148. Parker LA, Limebeer CL, Slomke J. Renewal effect: context-dependent extinction of a cocaine- and a morphine-induced conditioned floor preference. *Psychopharmacology.* 2006;187(2):133–137. https://doi.org/10.1007/s00213-006-0422-3.
149. Saunders BT, O'Donnell EG, Aurbach EL, Robinson TE. A cocaine context renews drug seeking preferentially in a subset of individuals. *Neuropsychopharmacology.* 2014;39(12):2816–2823. https://doi.org/10.1038/npp.2014.131.
150. Nall RW, Craig AR, Browning KO, Shahan TA. Longer treatment with alternative non-drug reinforcement fails to reduce resurgence of cocaine or alcohol seeking in rats. *Behav Brain Res.* 2018;341:54–62. https://doi.org/10.1016/j.bbr.2017.12.020.
151. Nall RW, Shahan TA. Resurgence of punishment-suppressed cocaine seeking in rats. *Exp Clin Psychopharmacol.* 2020;28(3):365–374. https://doi.org/10.1037/pha0000317.
152. McKendrick G, Graziane NM. Drug-induced conditioned place preference and its practical use in substance use disorder research. *Front Behav Neurosci.* 2020;14:582147. https://doi.org/10.3389/fnbeh.2020.582147.
153. Robinson MJF, Franklin KB. Effects of anisomycin on consolidation and reconsolidation of a morphine-conditioned place preference. *Behav Brain Res.* 2007;178(1):146–153. https://doi.org/10.1016/j.bbr.2006.12.013.
154. White NM. Addictive drugs as reinforcers: multiple partial actions on memory systems. *Addiction.* 1996;91(7):921–965.
155. Myers RD, Swartzwelder HS, Holahan W. Effect of hippocampal lesions produced by intracerebroventricular kainic acid on alcohol drinking in the rat. *Brain Res Bull.* 1983;10(3):333–338. https://doi.org/10.1016/0361-9230(83)90100-4.
156. Chambers RA, Self DW. Motivational responses to natural and drug rewards in rats with neonatal ventral hippocampal lesions: an animal model of dual diagnosis schizophrenia. *Neuropsychopharmacology.* 2002;27(6):889–905. https://doi.org/10.1016/S0893-133X(02)00365-2.
157. Sentir AM, Bell RL, Engleman EA, Chambers RA. Polysubstance addiction vulnerability in mental illness: concurrent alcohol and nicotine self-administration in the neurodevelopmental hippocampal lesion rat model of schizophrenia. *Addiction Biol.* 2020;25(1):e12704. https://doi.org/10.1111/adb.12704.
158. Goodman J, Hsu E, Packard MG. NMDA receptors in the basolateral amygdala mediate acquisition and extinction of an amphetamine conditioned place preference. *Behav Neurosci.* 2019;133(4):428–436. https://doi.org/10.1037/bne0000323.
159. Day JJ, Carelli RM. The nucleus accumbens and Pavlovian reward learning. *Neuroscientist.* 2007;13(2):148–159. https://doi.org/10.1177/1073858406295854.
160. Ranaldi R, Pocock D, Zereik R, Wise RA. Dopamine fluctuations in the nucleus accumbens during maintenance, extinction, and reinstatement of intravenous D-amphetamine self-administration. *J Neurosci.* 1999;19(10):4102–4109. https://doi.org/10.1523/JNEUROSCI.19-10-04102.1999.
161. Milton F, Muhlert N, Butler CR, Smith A, Benattayallah A, Zeman AZ. An fMRI study of long-term everyday memory using SenseCam. *Memory.* 2011;19(7):733–744. https://doi.org/10.1080/09658211.2011.552185.
162. Ranganath C, Johnson MK, D'Esposito M. Prefrontal activity associated with working memory and episodic long-term memory. *Neuropsychologia.* 2003;41(3):378–389. https://doi.org/10.1016/s0028-3932(02)00169-0.
163. Hsu E, Packard MG. Medial prefrontal cortex infusions of bupivacaine or AP-5 block extinction of amphetamine conditioned place preference. *Neurobiol Learn Mem.* 2008;89(4):504–512. https://doi.org/10.1016/j.nlm.2007.08.006.
164. Latagliata EC, Saccoccio P, Milia C, Puglisi-Allegra S. Norepinephrine in prelimbic cortex delays extinction of amphetamine-induced conditioned place preference. *Psychopharmacology.* 2016;233(6):973–982. https://doi.org/10.1007/s00213-015-4177-6.
165. Latagliata EC, Coccia G, Chiacchierini G, Milia C, Puglisi-Allegra S. Concomitant D1 and D2 dopamine receptor agonist infusion in prelimbic cortex is required to foster extinction of amphetamine-induced conditioned place preference. *Behav Brain Res.* 2020;392:112716. https://doi.org/10.1016/j.bbr.2020.112716.

166. Gil-Miravet I, Guarque-Chabrera J, Carbo-Gas M, Olucha-Bordonau F, Miquel M. The role of the cerebellum in drug-cue associative memory: functional interactions with the medial prefrontal cortex. *Eur J Neurosci.* 2019;50(3):2613–2622. https://doi.org/10.1111/ejn.14187.
167. Porter JN, Minhas D, Lopresti BJ, Price JC, Bradberry CW. Altered cerebellar and prefrontal cortex function in rhesus monkeys that previously self-administered cocaine. *Psychopharmacology.* 2014;231(21):4211–4218. https://doi.org/10.1007/s00213-014-3560-z.
168. Malenka RC, Bear MF. LTP and LTD: an embarrassment of riches. *Neuron.* 2004;44(1):5–21. https://doi.org/10.1016/j.neuron.2004.09.012.
169. Kauer JA, Malenka RC. Synaptic plasticity and addiction. *Nat Rev Neurosci.* 2007;8(11):844–858. https://doi.org/10.1038/nrn2234.
170. Ungless MA, Whistler JL, Malenka RC, Bonci A. Single cocaine exposure in vivo induces long-term potentiation in dopamine neurons. *Nature.* 2001;411(6837):583–587. https://doi.org/10.1038/35079077.
171. Heysieattalab S, Naghdi N, Hosseinmardi N, Zarrindast MR, Haghparast A, Khoshbouei H. Methamphetamine-induced enhancement of hippocampal long-term potentiation is modulated by NMDA and GABA receptors in the shell-accumbens. *Synapse.* 2016;70(8):325–335. https://doi.org/10.1002/syn.21905.
172. Huang CC, Lin HJ, Hsu KS. Repeated cocaine administration promotes long-term potentiation induction in rat medial prefrontal cortex. *Cerebr Cortex.* 2007;17(8):1877–1888. https://doi.org/10.1093/cercor/bhl096.
173. Ruan H, Yao WD. Cocaine promotes coincidence detection and lowers induction threshold during hebbian associative synaptic potentiation in prefrontal cortex. *J Neurosci.* 2017;37(4):986–997. https://doi.org/10.1523/JNEUROSCI.2257-16.2016.
174. Rao Y, Mineur YS, Gan G, et al. Repeated in vivo exposure of cocaine induces long-lasting synaptic plasticity in hypocretin/orexin-producing neurons in the lateral hypothalamus in mice. *J Physiol.* 2013;591(7):1951–1966. https://doi.org/10.1113/jphysiol.2012.246983.
175. Swant J, Chirwa S, Stanwood G, Khoshbouei H. Methamphetamine reduces LTP and increases baseline synaptic transmission in the CA1 region of mouse hippocampus. *PLoS One.* 2010;5(6):e11382. https://doi.org/10.1371/journal.pone.0011382.
176. Niehaus JL, Murali M, Kauer JA. Drugs of abuse and stress impair LTP at inhibitory synapses in the ventral tegmental area. *Eur J Neurosci.* 2010;32(1):108–117. https://doi.org/10.1111/j.1460-9568.2010.07256.x.
177. Portugal GS, Al-Hasani R, Fakira AK, et al. Hippocampal long-term potentiation is disrupted during expression and extinction but is restored after reinstatement of morphine place preference. *J Neurosci.* 2014;34(2):527–538. https://doi.org/10.1523/JNEUROSCI.2838-13.2014.
178. Loheswaran G, Barr MS, Rajji TK, Blumberger DM, Le Foll B, Daskalakis ZJ. Alcohol intoxication by binge drinking impairs neuroplasticity. *Brain Stimul.* 2016;9(1):27–32. https://doi.org/10.1016/j.brs.2015.08.011.
179. Mishra D, Zhang X, Chergui K. Ethanol disrupts the mechanisms of induction of long-term potentiation in the mouse nucleus accumbens. *Alcohol Clin Exp Res.* 2012;36(12):2117–2125. https://doi.org/10.1111/j.1530-0277.2012.01824.x.
180. Kortleven C, Fasano C, Thibault D, Lacaille JC, Trudeau LE. The endocannabinoid 2-arachidonoylglycerol inhibits long-term potentiation of glutamatergic synapses onto ventral tegmental area dopamine neurons in mice. *Eur J Neurosci.* 2011;33(10):1751–1760. https://doi.org/10.1111/j.1460-9568.2011.07648.x.
181. Guan Y-Z, Ye J-H. Ethanol blocks long-term potentiation of GABAergic synapses in the ventral tegmental area involving mu-opioid receptors. *Neuropsychopharmacology.* 2010;35(9):1841–1849. https://doi.org/10.1038/npp.2010.51.
182. Bernier BE, Whitaker LR, Morikawa H. Previous ethanol experience enhances synaptic plasticity of NMDA receptors in the ventral tegmental area. *J Neurosci.* 2011;31(14):5205–5212. https://doi.org/10.1523/JNEUROSCI.5282-10.2011.
183. Hendricson AW, Miao CL, Lippmann MJ, Morrisett RA. Ifenprodil and ethanol enhance NMDA receptor-dependent long-term depression. *J Pharmacol Exp Therapeut.* 2002;301(3):938–944. https://doi.org/10.1124/jpet.301.3.938.
184. Liu Z, Han J, Jia L, et al. Synaptic neurotransmission depression in ventral tegmental dopamine neurons and cannabinoid-associated addictive learning. *PLoS One.* 2010;5(12):e15634. https://doi.org/10.1371/journal.pone.0015634.

185. Li Y-J, Ping X-J, Qi C, et al. Re-exposure to morphine-associated context facilitated long-term potentiation in the vSUB-NAc glutamatergic pathway via GluN2B-containing receptor activation. *Addiction Biol.* 2017;22(2):435–445. https://doi.org/10.1111/adb.12343.
186. Sarkaki A, Assaei R, Motamedi F, Badavi M, Pajouhi N. Effect of parental morphine addiction on hippocampal long-term potentiation in rats offspring. *Behav Brain Res.* 2008;186(1):72–77. https://doi.org/10.1016/j.bbr.2007.07.041.
187. Dacher M, Nugent FS. Morphine-induced modulation of LTD at GABAergic synapses in the ventral tegmental area. *Neuropharmacology.* 2011;61(7):1166–1171. https://doi.org/10.1016/j.neuropharm.2010.11.012.
188. Keralapurath MM, Briggs SB, Wagner JJ. Cocaine self-administration induces changes in synaptic transmission and plasticity in ventral hippocampus. *Addiction Biol.* 2017;22(2):446–456. https://doi.org/10.1111/adb.12345.
189. Goussakov I, Chartoff EH, Tsvetkov E, et al. LTP in the lateral amygdala during cocaine withdrawal. *Eur J Neurosci.* 2006;23(1):239–250. https://doi.org/10.1111/j.1460-9568.2005.04538.x.
190. Pollandt S, Liu J, Orozco-Cabal L, et al. Cocaine withdrawal enhances long-term potentiation induced by corticotropin-releasing factor at central amygdala glutamatergic synapses via CRF, NMDA receptors and PKA. *Eur J Neurosci.* 2006;24(6):1733–1743. https://doi.org/10.1111/j.1460-9568.2006.05049.x.
191. North A, Swant J, Salvatore MF, et al. Chronic methamphetamine exposure produces a delayed, long-lasting memory deficit. *Synapse.* 2013;67(5):245–257. https://doi.org/10.1002/syn.21635.
192. Thompson AM, Swant J, Gosnell BA, Wagner JJ. Modulation of long-term potentiation in the rat hippocampus following cocaine self-administration. *Neuroscience.* 2004;127(1):177–185. https://doi.org/10.1016/j.neuroscience.2004.05.001.
193. Shen H, Kalivas PW. Reduced LTP and LTD in prefrontal cortex synapses in the nucleus accumbens after heroin self-administration. *Int J Neuropsychopharmacol.* 2013;16(5):1165–1167. https://doi.org/10.1017/S1461145712001071.
194. Gipson CD, Kupchik YM, Shen H, Reissner KJ, Thomas CA, Kalivas PW. Relapse induced by cues predicting cocaine depends on rapid, transient synaptic potentiation. *Neuron.* 2013;77(5):867–872. https://doi.org/10.1016/j.neuron.2013.01.005.
195. Shen H, Moussawi K, Zhou W, Toda S, Kalivas PW. Heroin relapse requires long-term potentiation-like plasticity mediated by NMDA2b-containing receptors. *Proc Natl Acad Sci USA.* 2011;108(48):19407–19412. https://doi.org/10.1073/pnas.1112052108.
196. Bao G, Kang L, Li H, et al. Morphine and heroin differentially modulate in vivo hippocampal LTP in opiate-dependent rat. *Neuropsychopharmacology.* 2007;32(8):1738–1749. https://doi.org/10.1038/sj.npp.1301308.
197. Kalivas PW. The glutamate homeostasis hypothesis of addiction. *Nat Rev Neurosci.* 2009;10(8):561–572. https://doi.org/10.1038/nrn2515.
198. Bridges RJ, Natale NR, Patel SA. System xc⁻ cystine/glutamate antiporter: an update on molecular pharmacology and roles within the CNS. *Br J Pharmacol.* 2012;165(1):20–34. https://doi.org/10.1111/j.1476-5381.2011.01480.x.
199. Engeli E, Zoelch N, Hock A, et al. Impaired glutamate homeostasis in the nucleus accumbens in human cocaine addiction. *Mol Psychiatry.* 2021;26(9):5277–5285. https://doi.org/10.1038/s41380-020-0828-z.
200. Stennett BA, Padovan-Hernandez Y, Knackstedt LA. Sequential cocaine-alcohol self-administration produces adaptations in rat nucleus accumbens core glutamate homeostasis that are distinct from those produced by cocaine self-administration alone. *Neuropsychopharmacology.* 2020;45(3):441–450. https://doi.org/10.1038/s41386-019-0452-2.
201. Knackstedt LA, LaRowe S, Mardikian P, et al. The role of cystine-glutamate exchange in nicotine dependence in rats and humans. *Biol Psychiatr.* 2009;65(10):841–845. https://doi.org/10.1016/j.biopsych.2008.10.040.
202. Caffino L, Messa G, Fumagalli F. A single cocaine administration alters dendritic spine morphology and impairs glutamate receptor synaptic retention in the medial prefrontal cortex of adolescent rats. *Neuropharmacology.* 2018;140:209–216. https://doi.org/10.1016/j.neuropharm.2018.08.006.
203. Pchitskaya E, Bezprozvanny I. Dendritic spines shape analysis-classification or clusterization? Perspective. *Front Synaptic Neurosci.* 2020;12:31. https://doi.org/10.3389/fnsyn.2020.00031.
204. Zhou FC, Anthony B, Dunn KW, Lindquist WB, Xu ZC, Deng P. Chronic alcohol drinking alters neuronal dendritic spines in the brain reward center nucleus accumbens. *Brain Res.* 2007;1134(1):148–161. https://doi.org/10.1016/j.brainres.2006.11.046.

205. Bolin BL, Alcorn 3rd JL, Lile JA, et al. N-Acetylcysteine reduces cocaine-cue attentional bias and differentially alters cocaine self-administration based on dosing order. *Drug Alcohol Depend*. 2017;178:452–460. https://doi.org/10.1016/j.drugalcdep.2017.05.039.
206. Gray KM, Watson NL, Carpenter MJ, Larowe SD. N-acetylcysteine (NAC) in young marijuana users: an open-label pilot study. *Am J Addict*. 2010;19(2):187–189. https://doi.org/10.1111/j.1521-0391.2009.00027.x.
207. Murray JE, Everitt BJ, Belin D. N-Acetylcysteine reduces early- and late-stage cocaine seeking without affecting cocaine taking in rats. *Addiction Biol*. 2012;17(2):437–440. https://doi.org/10.1111/j.1369-1600.2011.00330.x.
208. Hodebourg R, Murray JE, Fouyssac M, Puaud M, Everitt BJ, Belin D. Heroin seeking becomes dependent on dorsal striatal dopaminergic mechanisms and can be decreased by N-acetylcysteine. *Eur J Neurosci*. 2019;50(3):2036–2044. https://doi.org/10.1111/ejn.13894.
209. Reissner KJ, Gipson CD, Tran PK, Knackstedt LA, Scofield MD, Kalivas PW. Glutamate transporter GLT-1 mediates N-acetylcysteine inhibition of cocaine reinstatement. *Addiction Biol*. 2015;20(2):316–323. https://doi.org/10.1111/adb.12127.
210. Sepulveda-Orengo MT, Healey KL, Kim R, et al. Riluzole impairs cocaine reinstatement and restores adaptations in intrinsic excitability and GLT-1 expression. *Neuropsychopharmacology*. 2018;43(6):1212–1223. https://doi.org/10.1038/npp.2017.244.
211. Alhaddad H, Das SC, Sari Y. Effects of ceftriaxone on ethanol intake: a possible role for xCT and GLT-1 isoforms modulation of glutamate levels in P rats. *Psychopharmacology*. 2014;231(20):4049–4057. https://doi.org/10.1007/s00213-014-3545-y.
212. Hakami AY, Alshehri FS, Althobaiti YS, Sari Y. Effects of orally administered Augmentin on glutamate transporter 1, cystine-glutamate exchanger expression and ethanol intake in alcohol-preferring rats. *Behav Brain Res*. 2017;320:316–322. https://doi.org/10.1016/j.bbr.2016.12.016.
213. Zhang L-Y, Zhou Y-Q, Yu Z-P, Zhang X-Q, Shi J, Shen H-W. Restoring glutamate homeostasis in the nucleus accumbens via endocannabinoid-mimetic drug prevents relapse to cocaine seeking behavior in rats. *Neuropsychopharmacology*. 2021;46(5):970–981. https://doi.org/10.1038/s41386-021-00955-1.
214. Beck AT. *Cognitive Therapy and the Emotional Disorders*. International Universities Press; 1976.
215. Newman C. *Treating Substance Misuse Disorders with CBT*. Beck Institute Cognitive Behavior Therapy; March 28, 2016. https://beckinstitute.org/treating-substance-misuse-disorders-cbt/.
216. Marques AC, Formigoni ML. Comparison of individual and group cognitive-behavioral therapy for alcohol and/or drug-dependent patients. *Addiction*. 2001;96(6):835–846. https://doi.org/10.1046/j.1360-0443.2001.9668355.x.
217. Laude JR, Bailey SR, Crew E, et al. Extended treatment for cigarette smoking cessation: a randomized control trial. *Addiction*. 2017;112(8):1451–1459. https://doi.org/10.1111/add.13806.
218. Maude-Griffin PM, Hohenstein JM, Humfleet GL, Reilly PM, Tusel DJ, Hall SM. Superior efficacy of cognitive-behavioral therapy for urban crack cocaine abusers: main and matching effects. *J Consult Clin Psychol*. 1998;66(5):832–837. https://doi.org/10.1037//0022-006x.66.5.832.
219. Petitjean SA, Dürsteler-MacFarland KM, Krokar MC, et al. A randomized, controlled trial of combined cognitive-behavioral therapy plus prize-based contingency management for cocaine dependence. *Drug Alcohol Depend*. 2014;145:94–100. https://doi.org/10.1016/j.drugalcdep.2014.09.785.
220. Alammehrjerdi Z, Briggs NE, Biglarian A, Mokri A, Dolan K. A randomized controlled trial of brief cognitive behavioral therapy for regular methamphetamine use in methadone treatment. *J Psychoact Drugs*. 2019;51(3):280–289. https://doi.org/10.1080/02791072.2019.1578445.
221. Moore BA, Fiellin DA, Cutter CJ, et al. Cognitive behavioral therapy improves treatment outcomes for prescription opioid users in primary care buprenorphine treatment. *J Subst Abuse Treat*. 2016;71:54–57. https://doi.org/10.1016/j.jsat.2016.08.016.
222. Budney AJ, Stanger C, Tilford JM, et al. Computer-assisted behavioral therapy and contingency management for cannabis use disorder. *Psychol Addict Behav*. 2015;29(3):501–511. https://doi.org/10.1037/adb0000078.
223. Watkins KE, Hunter SB, Hepner KA, et al. An effectiveness trial of group cognitive behavioral therapy for patients with persistent depressive symptoms in substance abuse treatment. *Arch Gen Psychiatr*. 2011;68(6):577–584. https://doi.org/10.1001/archgenpsychiatry.2011.53.
224. Buckner JD, Zvolensky MJ, Ecker AH, et al. Integrated cognitive behavioral therapy for comorbid cannabis use and anxiety disorders: a pilot randomized controlled trial. *Behav Res Ther*. 2019;115:38–45. https://doi.org/10.1016/j.brat.2018.10.014.

225. Stephens RS, Wertz JS, Roffman RA. Self-efficacy and marijuana cessation: a construct validity analysis. *J Consult Clin Psychol*. 1995;63(6):1022–1031. https://doi.org/10.1037//0022-006x.63.6.1022.
226. DeMarce JM, Stephens RS, Roffman RA. Psychological distress and marijuana use before and after treatment: testing cognitive-behavioral matching hypotheses. *Addict Behav*. 2005;30(5):1055–1059. https://doi.org/10.1016/j.addbeh.2004.09.009.
227. Decker SE, Kiluk BD, Frankforter T, Babuscio T, Nich C, Carroll KM. Just showing up is not enough: homework adherence and outcome in cognitive-behavioral therapy for cocaine dependence. *J Consult Clin Psychol*. 2016;84(10):907–912. https://doi.org/10.1037/ccp0000126.
228. Gonzalez VM, Schmitz JM, DeLaune KA. The role of homework in cognitive-behavioral therapy for cocaine dependence. *J Consult Clin Psychol*. 2006;74(3):633–637. https://doi.org/10.1037/0022-006X.74.3.633.
229. Carroll KM, Ball SA, Martino S, et al. Computer-assisted delivery of cognitive-behavioral therapy for addiction: a randomized trial of CBT4CBT. *Am J Psychiatr*. 2008;165(7):881–888. https://doi.org/10.1176/appi.ajp.2008.07111835.
230. Strickland JC, Reynolds AR, Stoops WW. Regulation of cocaine craving by cognitive strategies in an online sample of cocaine users. *Psychol Addict Behav*. 2016;30(5):607–612. https://doi.org/10.1037/adb0000180.
231. Glasner S, Chokron Garneau H, Ang A, et al. Preliminary efficacy of a cognitive behavioral therapy text messaging intervention targeting alcohol use and antiretroviral therapy adherence: a randomized clinical trial. *PLoS One*. 2020;15(3):e0229557. https://doi.org/10.1371/journal.pone.0229557.
232. Metcalf M, Rossie K, Stokes K, Tallman C, Tanner B. Virtual reality cue refusal video game for alcohol and cigarette recovery support: Summative study. *JMIR Serious Games*. 2018;6(2):e7. https://doi.org/10.2196/games.9231.
233. Rawson RA, McCann MJ, Flammino F, et al. A comparison of contingency management and cognitive-behavioral approaches for stimulant-dependent individuals. *Addiction*. 2006;101(2):267–274. https://doi.org/10.1111/j.1360-0443.2006.01312.x.
234. Epstein DH, Hawkins WE, Covi L, Umbricht A, Preston KL. Cognitive-behavioral therapy plus contingency management for cocaine use: findings during treatment and across 12-month follow-up. *Psychol Addict Behav*. 2003;17(1):73–82. https://doi.org/10.1037/0893-164x.17.1.73.
235. Ray LA, Meredith LR, Kiluk BD, Walthers J, Carroll KM, Magill M. Combined pharmacotherapy and cognitive behavioral therapy for adults with alcohol or substance use disorders: a systematic review and meta-analysis. *JAMA Netw Open*. 2020;3(6):e208279. https://doi.org/10.1001/jamanetworkopen.2020.8279.
236. Anton RF, Moak DH, Waid LR, Latham PK, Malcolm RJ, Dias JK. Naltrexone and cognitive behavioral therapy for the treatment of outpatient alcoholics: results of a placebo-controlled trial. *Am J Psychiatr*. 1999;156(11):1758–1764. https://doi.org/10.1176/ajp.156.11.1758.
237. Brooks SJ, Wiemerslage L, Burch KH, et al. The impact of cognitive training in substance use disorder: the effect of working memory training on impulse control in methamphetamine users. *Psychopharmacology*. 2017;234(12):1911–1921. https://doi.org/10.1007/s00213-017-4597-6.
238. Khemiri L, Brynte C, Stunkel A, Klingberg T, Jayaram-Lindström N. Working memory training in alcohol use disorder: a randomized controlled trial. *Alcohol Clin Exp Res*. 2019;43(1):135–146. https://doi.org/10.1111/acer.13910.
239. Wanmaker S, Leijdesdorff S, Geraerts E, van de Wetering B, Renkema PJ, Franken I. The efficacy of a working memory training in substance use patients: a randomized double-blind placebo-controlled clinical trial. *J Clin Exp Neuropsychol*. 2018;40(5):473–486. https://doi.org/10.1080/13803395.2017.1372367.
240. Manning V, Mroz K, Garfield J, et al. Combining approach bias modification with working memory training during inpatient alcohol withdrawal: an open-label pilot trial of feasibility and acceptability. *Subst Abuse Treat Prev Pol*. 2019;14(1):24. https://doi.org/10.1186/s13011-019-0209-2.
241. Sofuoglu M, Waters AJ, Poling J, Carroll KM. Galantamine improves sustained attention in chronic cocaine users. *Exp Clin Psychopharmacol*. 2011;19(1):11–19. https://doi.org/10.1037/a0022213.
242. DeVito EE, Carroll KM, Babuscio T, Nich C, Sofuoglu M. Randomized placebo-controlled trial of galantamine in individuals with cocaine use disorder. *J Subst Abuse Treat*. 2019;107:29–37. https://doi.org/10.1016/j.jsat.2019.08.009.
243. Birath JB, Briones M, Amaya S, et al. Ibudilast may improve attention during early abstinence from methamphetamine. *Drug Alcohol Depend*. 2017;178:386–390. https://doi.org/10.1016/j.drugalcdep.2017.05.016.

244. Mathai DS, Holst M, Rodgman C, et al. Guanfacine attenuates adverse effects of dronabinol (THC) on working memory in adolescent-onset heavy cannabis users: a pilot study. *J Neuropsychiatry*. 2018;30(1):66–76. https://doi.org/10.1176/appi.neuropsych.16120328.
245. Roberts W, McKee SA. Effects of varenicline on cognitive performance in heavy drinkers: dose-response effects and associations with drinking outcomes. *Exp Clin Psychopharmacol*. 2018;26(1):49–57. https://doi.org/10.1037/pha0000161.
246. Perkins KA, Karelitz JL, Jao NC, Gur RC, Lerman C. Effects of bupropion on cognitive performance during initial tobacco abstinence. *Drug Alcohol Depend*. 2013;133(1):283–286. https://doi.org/10.1016/j.drugalcdep.2013.05.003.
247. Kalechstein AD, Mahoney 3rd JJ, Yoon JH, Bennett R, De la Garza 2nd R. Modafinil, but not escitalopram, improves working memory and sustained attention in long-term, high-dose cocaine users. *Neuropharmacology*. 2013;64:472–478. https://doi.org/10.1016/j.neuropharm.2012.06.064.
248. Joos L, Goudriaan AE, Schmaal L, van den Brink W, Sabbe BG, Dom G. Effect of modafinil on cognitive functions in alcohol dependent patients: a randomized, placebo-controlled trial. *J Psychopharmacol*. 2013;27(11):998–1006. https://doi.org/10.1177/0269881113503505.
249. Kalechstein AD, Mahoney 3rd JJ, Verrico CD, De La Garza 2nd R. Short-term, low-dose varenicline administration enhances information processing speed in methamphetamine-dependent users. *Neuropharmacology*. 2014;85:493–498. https://doi.org/10.1016/j.neuropharm.2014.05.045.

CHAPTER 8

Maladaptive decision making and addiction

Introduction

Each day begins with a decision we must face: do we get out of bed or not? There are multiple reasons one may want to get out of bed instead of sleeping for a few extra hours. We may need to go to class or to work. We may want to eat some breakfast, or we may just need to use the bathroom. Even after deliberating staying in bed, we make decisions throughout the rest of the day. What do we eat for lunch and for dinner? What activities do we do when we are finished with school and/or work? Do we watch a movie, read a book, listen to music, spend time with friends/family, or sit and relax? If watching a movie at home, one must decide what genre of movie they want to watch and then decide on multiple films within that genre. When I am at work, I must make certain decisions. Do I spend my day working on this textbook, working on my grant application, or working on lectures? Decision making goes beyond activities we perform in an individual day. We make decisions that can greatly impact our future. Some decide they want to pursue college and then determine which college to attend. Others may choose to pursue a career following high school graduation. Although daunting, most of us can make long-term plans. Even if those plans ultimately change, we still think about our future. As you are about to learn, individuals with a substance use disorder (SUD) do not think far into the future. In fact, SUDs are characterized by maladaptive decision making, the focus of the current chapter.

Learning objectives

By the end of this chapter, you should be able to ...

(1) Describe the various ways decision making can be measured in a laboratory.
(2) Explain how acute and long-term drug use alters general decision making, impulsive decision making, and risky decision making.
(3) Identify the ways in which maladaptive decision making acts as a risk factor for future drug use problems.

(4) Identify the shared neuromechanisms of maladaptive decision making and addiction.
(5) Define metacognition and explain the relationship between metacognition, mindfulness-based training, and addiction.
(6) Recognize some of the interventions used to ameliorate drug-induced impairments in decision making.

Overview of decision making

Decision making is the process of choosing between available options and involves a *judgment* that we make about these alternatives. In the Introduction of the chapter, I highlighted just some of the decisions we make during the day. Ideally, when making a decision, we use all of the available information we have to make a cost-benefit analysis between options. However, our decision making is not always rational. For now, I want to highlight how researchers measure decision-making capabilities in an individual.

Measuring decision making

There are numerous ways one can measure decision making in a laboratory. I will highlight two prominent tasks that have been and are currently used to gauge decision making. In the Iowa gambling task (IGT), participants are presented with four decks of cards on a computer screen. Choosing one card results in delivery or removal of game money from the participant's "bank". The object of the IGT is to earn as much game money as possible. The four decks differ from one another in the distribution of rewarded cards and penalized cards, as well as in the amount of money gained or lost during a trial. In the original IGT,[1] Decks A and B led to delivery of $100 while Decks C and D led to delivery of $50. However, consistently choosing Decks A and B is disadvantageous because the probability of losing money is highest in Deck A, and the amounts lost during a penalized trial are greatest for Deck B. Across the session, participants earn more money if they consistently select cards from Decks C and D. Individuals that consistently choose cards from Decks A and B are considered to exhibit impaired decision making. Fig. 8.1A shows a schematic of the IGT.

An animal analog of the IGT has been developed to measure decision making in rodents, aptly named the rat gambling task (rGT).[2] In the rGT, rats are tested in an operant conditioning chamber in which they can respond on one of four nosepoke apertures. The apertures are associated with different contingencies of reinforcement. The magnitude of reinforcement, the probability of receiving reinforcement, and the time-out period (i.e., the time the animal must wait before they can respond again) associated with each aperture differs (see Fig. 8.1B for more details). Like the IGT, consistently choosing the options associated with smaller reinforcer magnitudes is more advantageous as the rat can earn more food during the session.

The Wason selection task,[3] like the IGT, involves the use of four cards; however, this task measures deductive reasoning (one's ability to reach a *logical* conclusion from a series of statements). Here is an example of the Wason selection task:

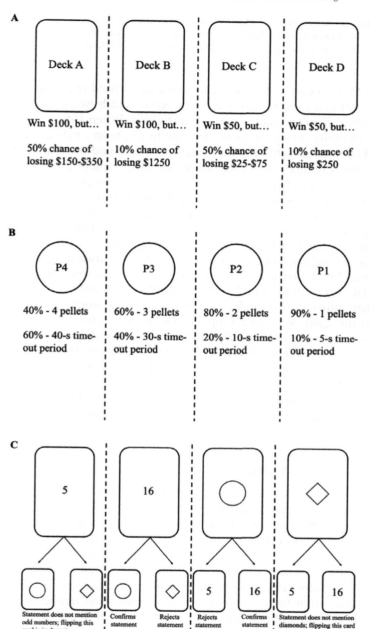

FIGURE 8.1 Schematic representations of the Iowa gambling task (IGT) (A), the rat gambling task (rGT) (B), and the Wason selection task (C). In the IGT, consistently choosing cards from Decks C and D is advantageous as one will earn more money in the long run. Similarly, in the rGT, rats will earn more food during the experimental session if they consistently choose the P1 and P2 options. In the Wason selection task, individuals are given statements like "When looking at the four cards pictured here, you notice that two cards have a number on them (5 and 16) and that two cards have a symbol on them (a *circle* and a *diamond*). Which card(s) must you turn over to test the following statement, 'If a card shows an even number on its face, then its opposite face is a circle'?". In this case, the individual must flip the cards with the 16 and the circle. *Figure was created by the author of this textbook using Microsoft PowerPoint.*

When looking at the four cards pictured here, you notice that two cards have a number on them (5 and 16) and that two cards have a symbol on them (a circle and a diamond). Which card(s) must you turn over to test the following statement, "If a card shows an even number on its face, then its opposite face is a circle"? (Fig. 8.1C).

Which card(s) would you flip? First, you must decide how many cards need to be flipped. To help you out, I will tell you that you need to flip two cards. Now, which two cards will you flip? If you find this problem difficult to solve, you are not alone. When I discuss the Wason selection task with my students, many of them struggle to get both cards correct. In the current example, the cards you need to flip are the ones with the 16 and the diamond. So, why these two? Let's break the statement "If a card shows an even number on its face, then its opposite face is a circle" down into two parts. We want to first focus on the expression "If a card shows an even number on its face". Notice that the focus is on even-number cards. Flipping the odd-numbered card is completely irrelevant to what this expression states. The odd-numbered card could have a circle or a diamond or any other shape, but this does not matter for the question at hand. Therefore, we can ignore the card labeled with the number 5. Now, let's focus on the second half of the expression "… then its opposite face is a circle". To test the statement that a circle is on the opposite side of an even-numbered card, we can first flip the card labeled with the number 16. If one flips the card with the 16 and sees a diamond or any shape other than a circle, we know that the statement is false. Most individuals recognize that this card needs to be flipped. Deciding on the second card is more difficult for individuals. Intuitively, individuals think that the second card to flip is the one labeled with the circle. However, this is incorrect. The expression only states that if an even-numbered card is flipped, a circle will be on the other side. Even if flipping the circle card reveals an odd-numbered card, this does not violate the *given* expression. Circles could be found on both even- and odd-numbered cards. Instead, one needs to flip the diamond card. If one flips the diamond card and sees an even number, this violates the given expression.

Impaired decision making resulting from drug use

Impaired decision making is often observed in individuals that use drugs.[4–6] Specifically, individuals that use drugs make fewer advantageous choices in the IGT.[4,6–8] One interesting finding is that individuals who have been convicted of driving under the influence of ethanol twice make more disadvantageous choices in the IGT.[9] These results suggest that impaired decision making may partially account for why individuals are willing to drive while intoxicated, despite the dangers associated with this behavior. Also, individuals that have had to receive detoxification treatment for ethanol dependence on two or more occasions perform worse on the IGT compared to those that have received fewer detoxification treatments.[10] Collectively, these results indicate that suboptimal decision making in the IGT is associated with problematic drug use (e.g., driving under the influence of ethanol) and with relapse.

Cocaine-dependent individuals select the disadvantageous decks more frequently compared to recreational cocaine users in the IGT,[11] suggesting that decision making worsens with continued drug use. One important consideration is that individuals completing the IGT typically earn hypothetical (i.e., fake) money. While long-term cocaine users show impaired performance in the IGT when hypothetical monetary rewards are used, they perform just as well as nondrug users when real money is delivered.[12] These results are important as they indicate that impaired decision making often observed in the IGT may reflect motivational differences between cocaine users and nonusers.

Considering individuals with a SUD often use more than one substance, some research has examined if polysubstance use can further exacerbate impairments in decision making. While opioid-dependent individuals that smoke cigarettes perform worse on a gambling task compared to opioid-dependent individuals that do not smoke,[13] cannabis use does not further impair task performance in opioid-dependent individuals.[14] Additionally, concomitant use of methamphetamine and ketamine does not further affect IGT performance.[15] More work is clearly needed to further examine how polysubstance use impacts general decision making.

While many studies have used the IGT to measure decision making in substance users, Kornreich et al.[16] tested deductive reasoning in those with an alcohol use disorder with several variants of the Wason selection task. Individuals were given statements regarding social contracts ("If you borrow a car, then you must fill up the tank with gas"), precautions ("If you work with [tuberculosis] patients, then you must wear a surgical mask"), and general descriptions ("If a person becomes a biologist, then that person enjoys camping") (p. 954).[16] Compared to controls, individuals with an alcohol use disorder perform significantly worse on all three statement types, indicating impaired deductive reasoning.

Following short- and long-term abstinence, individuals make poor decisions in the IGT,[17] suggesting that drug-induced impairments in decision making do not immediately return to normal following drug use cessation. Fortunately, decision making following drug use can improve during abstinence. Specifically, individuals that have been abstinent from heroin for 3–24 months perform better on the IGT compared to those that have been abstinent for 3–30 days.[18] Abstinent-induced improvements in decision making have also been observed in those with methamphetamine dependence[19] and cannabis dependence.[20]

So far, few preclinical studies have examined the acute or chronic effects of commonly used drugs on decision making as assessed in the rGT. Acute administration of d-amphetamine generally impairs performance in the rGT.[2] Acute ethanol modestly reduces responding for the optimal option in the rGT while repeated ethanol administration substantially increases preference for the high risk/high reward option,[21] meaning that animals make worse decisions on this task following long-term ethanol exposure. Like humans, rats that self-administer cocaine make worse decisions in the rGT, an effect that persists during withdrawal.[22] Cocaine-induced impairments in the rGT are prominently observed in rats that prefer the high-risk/high-reward options.[23]

Impulsive decision making

Sometimes we have to make a decision between immediate satisfaction or a more beneficial reward delivered at a later point in time. For example, a student can decide to go out with friends or can spend the night studying for an exam. Although going out provides immediate reinforcement, studying and obtaining a good grade on the exam is more beneficial. To study delayed gratification, Dr. Walter Mischel (1930–2018) performed the seminal "Stanford Marshmallow Experiment".[24] In this experiment, preschool children, tested one at a time, were given a choice between one marshmallow or one pretzel. If the child selected the marshmallow, the experimenter told the child that they could have it after the experimenter returned to the room. If they could not wait for the experimenter to return, they could have the nonpreferred treat. Mischel et al.[24] wanted to determine if giving children overt and covert activities would increase their ability to wait to receive the preferred reward. In subsequent follow-up studies, Mischel and his colleagues found that children that were unwilling to delay gratification tended to have poorer life outcomes, such as lower SAT scores, higher body mass index, and lower educational attainment.[25,26] However, the results of this research have been questioned in recent years.[27,28] While doubts have been raised about the original findings obtained by Mischel and his colleagues, this work was highly influential in promoting research on **impulsive decision making**.

Although the marshmallow test is not as great of a predictor of life outcomes as originally reported, this research was and is still important because it led to the development of a procedure that is commonly used to measure impulsive decision making: delay discounting. The term delay discounting refers to the decrease in subjective value of a reinforcer as a function of the delay to its delivery. This is a fancy way of saying that rewards lose their appeal the longer one must wait for them. Delay discounting is often measured in humans by asking participants to choose between two hypothetical monetary rewards differing in magnitude.[29] The value of the larger, delayed monetary reward is held constant, whereas the value of the smaller, immediate reward is decreased systematically. The indifference point is the point at which a person switches their preference from the large, delayed reinforcer to the small, immediate reinforcer (Fig. 8.2).

Acute drug effects on impulsive decision making

The ADHD medications amphetamine and methylphenidate decrease impulsive decision making in healthy humans[30] and in rodents.[31–33] Methamphetamine and cocaine also decrease impulsive decision making.[34,35] These findings make sense as psychostimulants are the primary treatment option for those with ADHD. In contrast to most psychostimulants, acute nicotine fails to alter impulsive decision making in humans,[36] although acute nicotine increases delay discounting in rats.[37,38]

In contrast to psychostimulants, acute administration of depressants often fails to alter delay discounting, including ethanol,[39–42] the prescription opioid oxycodone,[43] and the benzodiazepine diazepam.[44] Even more surprising is the finding that acute ethanol does not affect impulsive decision making in late adolescents.[45] Given that adolescence is marked by increased impulsive decision making, one would expect to see this population become even more impulsive following administration of a substance that is known to lead to disinhibition of behavior. You may also be asking yourself "how does one test the acute effects of ethanol on adolescents? Isn't that illegal?". Read the Experiment Spotlight to find out more.

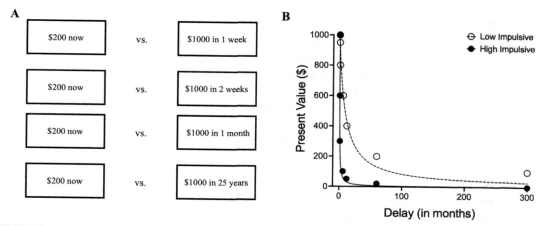

FIGURE 8.2 (A) Representation of a delay-discounting procedure in humans. Participants indicate their preference for one monetary reward delivered immediately and a larger monetary reward delivered after a certain delay. For example, participants can rate their preference for one of 27 monetary amounts ($1,000 down to $1) compared to $1,000 delivered after a delay. The delay to the large monetary reward systematically increases throughout the experiment (1 week, 2 weeks, 1 month, 6 months, 1 year, 5 years, 25 years). The example above shows four choices between $200 delivered immediately and $1,000 delivered after one of four different delays. (B) Hypothetical discounting data showing the differences between an individual that displays lower impulsive decision making and an individual that displays higher impulsive decision making. At the fifth delay (1 year), the indifference point for the low impulsive individual is $400 while the indifference point for the high impulsive individual is $50. *Image created by the author of this textbook using Microsoft PowerPoint and GraphPad Prism 9.0.*

Experiment Spotlight: Acute alcohol effects on impulsive choice in adolescents (Bernhardt et al.[45])

This experiment occurred in Germany, where the legal drinking age is lower (16 for beer and wine/18 for liquor) compared to the United States (21). Bernhardt et al.[45] tested 54 males aged 18–19 years (late adolescence/early adulthood). Before testing the participants, the researchers measured liver function. Participants were excluded from the study if they had elevated liver enzymes or if they used certain medications, submitted a positive drug screen on testing days or the day prior to testing, or if they had consumed ethanol on or 1 day before the test day. Instead of allowing participants to drink ethanol at their own pace, they received an intravenous infusion of ethanol (6.0%) or saline on two separate test days. Following ethanol/saline infusion, participants were given several self-report questionnaires that asked about impulsiveness, ethanol use, and problems related to drinking before being tested in a battery of cognitive tasks, including delay discounting.

Intravenous infusion of ethanol did not alter impulsive decision making. Ethanol also had no effect on how quickly participants responded during delay discounting. While ethanol administration did not affect impulsive decision making, Bernhardt et al. found that participants that showed increased ethanol-induced delay discounting were more likely to report feeling more stimulated following ethanol administration and were more likely to report increased ethanol cravings and wanting to drink because of its stimulating properties.

Questions to consider:

(1) Why do you think participants were excluded from the study if they had elevated liver enzymes or took certain medications?
(2) Why did participants receive an infusion of ethanol instead of drinking ethanol?
(3) Why was the ethanol concentration set to 6.0%?
(4) Although ethanol did not increase delay discounting, why are the current results important?
(5) What is one limitation about the current experiment?

Answers to questions:

(1) The presence of elevated liver enzymes can signal chronic ethanol use (i.e., alcohol use disorder). From an ethical standpoint, providing ethanol to individuals with an alcohol use disorder can be problematic if these individuals have cirrhosis of the liver. Certain medications, when combined with ethanol, can become dangerous. For example, combining opioids with ethanol can lead to overdose. This exclusion criterion was included as a safety measure.
(2) When participants are allowed to drink ethanol at their own pace, this can lead to differences in blood alcohol content (BAC) across participants. For example, if someone chugs their beer in 1 min, their BAC levels will peak at a faster rate compared to someone that "nurses" their beer. By infusing the drug, researchers can ensure that BACs are equivalent in participants as they complete each cognitive task.
(3) This ethanol concentration corresponds to a typical beer.
(4) These results suggest that ethanol-induced alterations in delay discounting are more sensitive to the physical effects of ethanol and desire to drink ethanol more so than those that are unaffected by ethanol. This could serve as a risk factor for the development of an alcohol use disorder. You will learn about how impulsive decision making can serve as a predictor for addiction-like behaviors later in this chapter.
(5) This study only examined male participants. Theoretically, females could become more impulsive compared to males following ethanol administration.

Animal models have provided mixed results concerning the acute effects of depressants on impulsive decision making. In rodents, some studies have found no effect of ethanol on impulsive decision making[46] while others have found increases in delay discounting following ethanol administration.[47] One study even found that ethanol intoxication *decreases* delay discounting.[48] The effects of opioids on delay discounting appear to be species dependent, as morphine increases impulsive decision making in rats[49] and in pigeons,[50] but has no effect in monkeys.[51] Interestingly, the CB_1/CB_2 receptor agonist CP55940 fails to alter delay discounting in monkeys; however, when CP55940 is combined with morphine, monkeys choose the large, delayed reinforcer more frequently.[51] Thus, concomitant stimulation of mu opioid and cannabinoid receptors appears to be needed to decrease impulsive decision making in primates.

In rats, benzodiazepines largely increase impulsive decision making,[52] although there is one notable exception. The drug alprazolam (Xanax) decreases delay discounting in a T-maze.[52] To complicate matters more, the effects of benzodiazepines on impulsive decision making in rodents appears to be strain dependent as well. Diazepam has no effect on delay discounting in Lewis rats (a type of inbred rat that shows high sensitivity to the reinforcing effects of drugs), mirroring the results obtained in humans; however, diazepam decreases impulsive decision making in Fischer F344 rats (an inbred rat strain that shows lows sensitivity to the reinforcing effects of drugs).[53] Finally, flunitrazepam (Rohypnol) decreases delay discounting in pigeons.[54]

Long-term consequences of drug use on impulsive decision making

Earlier you learned that individuals with a SUD display impaired decision making. One remarkable finding is that individuals with a SUD often make impulsive decisions.[55] Increased delay discounting is observed across numerous SUDs, including those for ethanol,[56] nicotine[36,57] (but see[58]), cocaine,[11,57,58] methamphetamine,[59] and opioids.[60] College-aged dependent smokers are more impulsive compared to nondependent cigarette users.[61] Individuals that become dependent on ethanol at an earlier age are more impulsive compared to those that develop an alcohol use disorder at a later age.[62] Cocaine users, in particular, have difficulty delaying gratification. In most delay-discounting studies, participants are asked if they prefer a smaller reward compared to a larger reward delivered after 1 week, 2 weeks, 2 months, 6 months, 1 year, 5 years, or 25 years. Because most cocaine-dependent individuals could not tolerate waiting just 1 week for the larger reward, Coffey et al. had to include much shorter delays (5 min–5 days)![57]

Impulsive decision making is exacerbated in individuals that use certain drugs relative to other drugs. Individuals that use crack cocaine discount reinforcers to a greater extent compared to those that use heroin,[63] and individuals that use cocaine or heroin display greater impulsive decision making compared to those that consume ethanol.[64] Individuals that use heroin are more impulsive compared to those that use prescription opioids.[65] Another factor that may be related to impulsive decision making is the route of drug administration. Those that smoke cocaine are more impulsive than those that snort cocaine.[66] Withdrawal from drugs can also exacerbate delay discounting, an effect that has been observed in smokers.[67]

Even if an individual is not diagnosed with a SUD, long-term or heavy drug use is linked to impulsive decision making. Cigarette smokers and e-cigarette smokers are more impulsive in delay discounting compared to never/nonsmokers,[36,55,68–71] and e-cigarette users display greater impulsive decision making compared to cigarette smokers.[71] Adolescents that experiment with tobacco use are more impulsive in delay discounting than those that have never smoked.[69] Similarly, adults that are light smokers (fewer than 10 cigarettes a day) discount monetary rewards more than nonsmokers.[72]

Adolescents and adults that engage in heavy ethanol consumption display greater impulsive decision making.[39,41,73] Delay discounting can even be used to predict how intoxicated an individual will become while drinking.[74] Individuals that report more lifetime problems with ethanol use are more impulsive.[75] In Chapter 1, you learned that mixing energy drinks with ethanol is a dangerous practice. One interesting study examined the association between weekly energy drink consumption and delay discounting. Those that consume energy drinks on a weekly basis show greater impulsive decision making.[76] This finding, in conjunction with other studies showing increased delay discounting in ethanol users, suggests that individuals that prefer to mix energy drinks with ethanol may be especially prone to impulsive decision making.

One drug that has yielded mixed results in the literature is cannabis.[77] Some reports have indicated that cannabis users display greater impulsive decision making compared to nonusers,[8,78] but other studies have failed to observe such differences.[79] Given that cannabis use is often correlated with tobacco use, increased impulsive decision making observed in cannabis users may be affected by cooccurring nicotine use. Indeed, a previous study showed just this: delay discounting is more closely related to tobacco use compared to cannabis use.[80] Furthermore, a more recent study found that the association between cannabis use and delay discounting disappears after controlling for other drug use.[81] Overall, these results collectively show that the effects of cannabis on impulsive decision making are smaller in magnitude compared to other substances.

Other studies have investigated how polysubstance use impacts impulsive decision making. In contrast to what has been reported for polydrug use and impaired decision making, using two or more substances is consistently linked to delay discounting. Cigarette smokers that use another substance are more impulsive compared to individuals that use cigarettes only.[82] Those that consume ethanol and smoke are more impulsive than those that only drink or only smoke cigarettes.[83] Finally, those that engage in polydrug use show greater delay discounting during abstinence compared to controls and compared to ethanol-abstinent individuals.[84]

Even after individuals have stopped using drugs, they display impairments in impulsive decision making. Former heroin users are more impulsive in a delay-discounting task.[17] Similarly, former opioid users maintained on methadone replacement therapy are more impulsive in delay discounting compared to those that have never used opioids.[60] Because these individuals are maintained on methadone, there are two potential interpretations to these results. One, opioid use causes long-term changes in delay discounting that can persist for 2 years, as previously reported.[60] Two, methadone treatment itself increases impulsivity in abstinent individuals. Remember that methadone is a full agonist at mu opioid receptors just like heroin. If impulsive decision making is caused by long-term opioid use, replacing heroin with methadone would not ameliorate heroin-induced impairments in impulsive decision making.

Animal models have largely corroborated clinical findings that drug use and impulsive decision making are positively associated. Drug self-administration of various substances increases impulsive decision making in rodents, including cocaine,[85] d-amphetamine,[86] synthetic cathinones,[87] and heroin.[88] Similarly, chronic experimenter-administered cocaine or WIN 55,212-2 (cannabinoid CB_1 agonist) increase impulsive decision making in rodents.[89,90] High ethanol-preferring rodents display greater impulsive decision making.[91] However, neither neonatal/adolescent exposure to nicotine nor chronic intermittent ethanol exposure during adolescence or during adulthood alters impulsive decision making later in life.[92,93] Also, repeated experimenter-delivered heroin injections fail to alter delay discounting,[94] which contrasts with research showing that self-administration of heroin increases impulsive decision making. In animals, acute withdrawal from nicotine increases impulsive decision making,[37] but recent work has shown that exposure to nicotine vapor does not produce long-lasting changes in delay discounting.[38] One potential explanation for the discrepant results is that baseline differences in impulsive decision making moderate the effects of nicotine withdrawal on impulsive decision making. Not only does acute administration increase impulsive decision making in low impulsive animals, but nicotine withdrawal further increases impulsive decision making; conversely, neither nicotine administration nor acute withdrawal alter delay discounting in high impulsive animals.[95] Additionally, after animals have stopped self-administering *d*-amphetamine, drug-induced increases in impulsive decision making return to normal,[86] showing that impulsive decision making can return to normal after cessation of drug use.

Quiz 8.1

(1) Which of the following statements about acute drug effects on impulsive decision making is true?
   a. Benzodiazepines increase impulsive decision making in both humans and in rodents.
   b. Ethanol consistently increases impulsive decision making.
   c. Nicotine increases impulsive decision making in animals, but not in humans.
   d. Psychostimulants increase impulsive decision making.

(2) Which group of drug-dependent individuals shows the greatest sensitivity to delayed reinforcement in delay discounting?
   a. Ethanol
   b. Cocaine
   c. Nicotine
   d. Opioids

(3) How does long-term drug use affect performance in the IGT?
   a. Drug-dependent individuals are more likely to select the high-risk/high-reward option than controls.
   b. Drug-dependent individuals are more likely to select the low-risk/low-reward option than controls.
   c. Drug-dependent individuals perform similarly as controls.
   d. Only stimulant-dependent individuals perform worse on the IGT compared to controls.

(4) Which of the following statements about long-term drug effects on impulsive decision making is true?
   a. Impulsive decision making is characteristic of most SUDs.
   b. Individuals that use nicotine long-term become less impulsive over time.
   c. Long-term drug users are only more impulsive than controls when hypothetical monetary rewards are used.
   d. Snorting cocaine is associated with greater impulsive decision making than smoking cocaine.

(5) Here is a problem from the Wason selection task: You are presented with four cards labeled "16", "21", "beer", and "soda". Which card(s) must you turn over to test the following statement, "If a card shows beer, then its opposite face is 21"?

Answers to quiz 8.1

(1) c — Nicotine increases impulsive decision making in animals, but not in humans.
(2) b — Cocaine

(3) a — Drug-dependent individuals are more likely to select the high-risk/high-reward option than controls.

(4) a — Impulsive decision making is characteristic of most SUDs.

(5) You should flip the cards labeled "beer" and "16". If you found this question easier to answer compared to the example used earlier in the chapter, you are not alone. Using concrete examples like the one here makes identifying the correct cards easier in this task.

Risky decision making

Risky decision making is often defined as making a choice in the face of uncertainty.[96] This type of decision making is not inherently maladaptive or "bad". Sometimes, we need to take risks. Think of an entrepreneur who takes a risk by developing a new product or a filmmaker that takes a risk by creating a movie unlike any other film. There is no guarantee that consumers will buy the new product or will go to see the novel film. We can take a risk by moving somewhere far from home to attend college or to start a new career. We are not sure if this decision will pay off. Unlike risky decisions that can pay dividends in the future, this section will focus on risky decision making that is disadvantageous. Substance misuse is a risky behavior in itself. Individuals can become sick from drug use, such as contracting hepatitis or HIV from sharing needles, developing cancer from tobacco use, or developing cirrhosis of the liver from excessive ethanol consumption, just to name a few examples. Individuals can lose their jobs or can become incarcerated for illicit drug use. The biggest risk someone takes when using many substances is overdose, which can lead to death.

In humans, several techniques can be used to measure risky decision making. First, the balloon analog risk task (BART) allows participants to inflate a virtual balloon presented on a computer screen to earn money.[97] Each time an individual inflates the balloon, they earn more money. However, the balloon may pop. If the balloon pops, the individual loses the money they have accumulated during the trial. Risky decision making can also be measured with probability discounting. Probability-discounting sessions are structured similarly to delay discounting. The difference is that delivery of the large magnitude reinforcer is probabilistic.[29] Risky decision making can also be examined in driving simulator tasks. The experimenter can determine how likely a participant will run a red light or drive over the speed limit during simulated driving.

In animals, probability discounting is often used to measure risky decision making.[98] Earlier, you learned about the rGT, an animal analog of the IGT. Although I covered the rGT when discussing general decision making, the rGT is often labeled as a risky decision-making task.[99] Probability discounting and the rGT rely on negative punishment to assess risk; that is, rats may or may not receive food following a response. Another way to measure risky decision making is to provide probabilistic delivery of positive punishment. In the risky decision task (RDT),[100] rats are trained to respond for one of two reinforcer alternatives. One reinforcer consists of a single food pellet, and the second reinforcer consists of two or more food pellets that are paired with probabilistic delivery of a mild foot shock. Across the session, the probability of receiving the foot shock increases.

Acute drug effects on risky decision making

While acute psychostimulant administration consistently improves impulsive decision making, psychostimulant effects on risky decision making are mixed. Amphetamine increases the likelihood of failing to stop at a red traffic light in a driving simulator task,[101] but methamphetamine does not alter driving performance in recreational illicit stimulant users.[102] Individuals that use MDMA show *decreased* risky decision making in a simulated driving task.[103] This does not mean that MDMA improves driving as Brookhuis et al. found that multiple exposures to MDMA severely impair driving performance.[103] The decreased risk taking observed in this population may have occurred as a compensatory mechanism. That is, MDMA users may have acknowledged that they were intoxicated and tried to adjust their driving to compensate. Despite this compensatory response, repeated MDMA use leads to impaired driving.

Not surprisingly, acute ethanol administration increases risky driving.[104] Furthermore, participants that consume ethanol at a faster rate are more likely to engage in risky driving.[105] However, the effects of ethanol administration on other forms of risky decision making are mixed, with some studies reporting increased risky decision making[40] and others reporting no changes in risky decision making[42] following ethanol consumption. One area that has received attention is the role of ethanol in risky sexual behavior. Binge drinking is associated with unplanned sexual activity,[106] and acute ethanol administration increases the likelihood that individuals forgo delayed protected sex to have immediate unprotected sex.[40] In HIV-serodiscordant couples (i.e., only one individual in the relationship is HIV+), heavy drinking is associated with increased sex outside of the relationship and condomless sex with one's partner.[107] The ramifications of this research should be obvious. Ethanol-induced increases in risky decision making can increase the likelihood of a vehicular accident or can increase unprotected sex, leading to unwanted pregnancies or the spread of sexually transmitted infections.

While cannabis does not consistently affect impulsive decision making, acute administration of cannabis increases risky decision making.[108] As covered in the previous chapter, cannabis use negatively impacts one's ability to drive. Specific to risky driving behavior, driving under the influence of cannabis is correlated with increased risky driving.[109] Giving participants cannabis also increases risky driving in a simulator task.[110] Similar to cannabis, benzodiazepines increase risky decision making in humans.[111] One shocking finding is that administration of benzodiazepines the night before being tested in a driving simulator task increases driving speed during the simulation.[112] This finding shows that benzodiazepines can negatively affect driving even if the individual is not currently intoxicated. To date, little research has examined the acute effects of opioids on risky decision making. Similar to impulsive decision making, oxycodone administration does not alter risky decision making as measured with the BART or with probability discounting.[43] Overall, depressant-like drugs increase risky decision making, although more work is needed with opioids. Given that depressants lead to disinhibition, increased risky decision making following acute administration of depressant-like drugs makes sense as individuals are less likely to inhibit risky behaviors.

In animals, *d*-amphetamine increases risky decision making in probability discounting,[33,98] but *decreases* risky decision making as assessed in the RDT.[100,113] Nicotine, like amphetamine, differentially alters risky decision making in probability discounting and in the RDT. Nicotine increases risky choice in probability discounting[85] but decreases risky choice in the RDT.[113] In one procedure, Kaminski and Ator[114] had rats choose between one alternative that always delivered three food pellets and one alternative that delivered either 15 food pellets (probability of 0.33) or no food pellets (probability of 0.67). In addition to testing several drugs on risky decision making, Kaminski and Ator manipulated the intertrial interval (ITI) of each trial; that is, in one condition, rats had to wait 20 s before initiating another trial, or they had to wait 80 s. If animals had to wait 20 s, baseline risky choice was low. Conversely, if they had to wait 80 s, baseline risky decision making was high. Why does this matter? The effects of amphetamine on risky decision making are dependent on which ITI is used. When baseline risky decision making is low, amphetamine increases risky decision making and vice versa. In contrast to amphetamine, ethanol decreases risky decision making, regardless of ITI length.

In contrast to what has been observed with amphetamine, ethanol administration increases risky decision making when reinforcement delivery is probabilistic, regardless of ITI length.[114] However, ethanol does not affect performance in the RDT.[113] More work is needed to determine how depressants like opioids, cannabis, and benzodiazepines alter risky decision making.

Long-term consequences of drug use on risky decision making

While those with a SUD consistently display greater impulsive decision making in delay discounting compared to those without a SUD, inconsistencies have been reported in the literature concerning the relationship between drug dependence and probability discounting. Heavy cigarette users have been shown to be more risk taking[115] but have also been shown to be more risk averse[70] compared to nonsmokers. Some studies have also found that probability discounting does not differ between current smokers and nonsmokers,[68,116] but individuals that have recently tried cigarettes are more risk averse.[116] Abstinence from cigarette smoking does not alter probability discounting.[117] As noted by Yi et al., inconsistencies in the literature may be due to floor effects.[118] That is, when examining just the high probabilities of obtaining reinforcement, smokers are more likely to choose the probabilistic reinforcer.

One study reported an interesting finding concerning long-term ethanol use and performance on the BART. Across all trials, ethanol users were less willing to pump the balloon.[119] At first glance, this finding suggests that ethanol use is associated with decreased risky decision making. Campbell et al. argue that these results indicate a general impairment in decision making as pumping the balloon is advantageous on trials in which the probability of popping the balloon is low.[119] Similar findings have been reported for cigarette smokers[120] and cocaine users.[121] Therefore, these findings are more similar to those reporting impaired decision in the IGT in those that use these substances. Like the finding that weekly energy drink consumption is associated with higher levels of impulsive decision making, individuals that consume energy drinks are more likely to engage in risky behaviors, such as having unprotected sex, using more illicit drugs, and smoking more cigarettes.[76]

One study found that stimulant users tend to make riskier decisions.[122] More specifically, cocaine users are more willing to take more risks in a randomized lottery task.[123] In a gambling-like task, cocaine users take more risks when the probability of receiving a reward is low, but they are risk averse when the probability of winning increases.[124] However, those with a cocaine use disorder show similar probability discounting as controls.[58,125] MDMA and methamphetamine users are more likely to engage in risky driving behaviors.[126,127] More interestingly, methamphetamine users show greater probability discounting compared to controls,[59] meaning they make more risk-averse decisions in this task. Does this mean that chronic methamphetamine use decreases risky decision making? Yi et al. argue that delay and probability discounting measure the same underlying construct: *psychological horizon*.[59] In other words, delay discounting and probability discounting may measure the same thing. Think of it this way: in probability discounting, participants do not always receive a reward. They may have to go through several trials before receiving the reward. Thus, a de facto delay is imposed on this reward. There have been cases in which certain populations show greater discounting of both delayed and probabilistic reinforcers.[91] Somewhat tangentially related, adolescents that smoke cigarettes are more likely to rate delayed reinforcement as being uncertain.[128]

Opiate-dependent individuals display greater risky decision making compared to controls.[129] To be more specific, when gains are involved (e.g., $750 guaranteed vs. $1,000 with a probability of 0.5), opioid-dependent individuals are risk averse; however, when losses are involved (e.g., guaranteed loss of $750 vs. 50% chance of losing $1,000 or losing no money), opioid-dependent individuals exhibit increased risky decision making.[130] A recent study took an interesting approach to measuring probability discounting in individuals currently using heroin. Instead of manipulating the probability of receiving monetary rewards, Dolan et al.[131] asked participants to evaluate the likelihood of using a sample of heroin based on the probability it contained impurities (e.g., fentanyl added to the heroin) or the probability of overdose. Less discounting (i.e., greater risky decision making) was observed in those with more severe opioid dependence.

Like those with a cocaine use disorder, individuals with a cannabis use disorder show similar probability discounting as controls.[125] Additionally, cannabis use is not significantly correlated with probability discounting.[78] Adolescent cannabis users show increased risky decision making in the BART,[132] but adult cannabis users are either more risk averse in the BART[133] or perform similarly to controls on the BART.[134] Cannabis users that expect to experience greater impairment are less willing to take risks in the BART.[135] Cannabis users may not engage in risk-taking behavior when money is involved, but they are more willing to take risks in other avenues; for example, cannabis users are more likely to take socials risks (e.g., wearing unconventional clothing), ethical risks (e.g., cheating on an exam), and health risks (e.g., unprotected sex).[136] This study highlights the need to explore other reinforcers than money. Just because risk-taking for monetary rewards is unaffected by drug use, this does not mean that drug use fails to alter risky decision making. As reported earlier, individuals may be more likely to engage in risky sexual activities, which are independent of monetary gains or losses, or they may be more willing to use a drug that has been adulterated.

Research on long-term drug effects on animal models of risky decision making is scarce. Thus far, animal models have shown that adolescent exposure to ethanol increases risky decision making later in life.[137] Concerning nicotine, neonatal exposure to nicotine does not alter RDT performance during adulthood,[93] which contrasts with findings showing that acute nicotine administration decreases risky decision making in this task.

Maladaptive decision making as a risk factor for substance use disorders

So far, I have discussed how drug use alters decision making. One important question is if maladaptive decision making results from drug use or if this form of decision making predisposes an individual to begin using drugs. Research suggests that the relationship between drug use and maladaptive decision making is bidirectional. Just as drugs impair decision making or exacerbate impulsive decision making, and to a lesser extent risky decision making, maladaptive decision making is predictive of drug use. For example, adolescents that show greater delay discounting are more likely to use tobacco products, ethanol, and cannabis, and they are more likely to binge drink.[138,139] Cannabis-dependent military veterans with high levels of impulsive decision making are more likely to experience increased cravings for cannabis and are more likely to have initiated cannabis use at an earlier age.[140] Some evidence has suggested that performance on the IGT can be used to make predictions about one's future drug use. Individuals with comorbid bipolar disorder and psychostimulant use disorder (either cocaine or methamphetamine) that perform poorly on the IGT are more likely to use drugs in the future.[141] However, performance on the IGT does not predict adolescent substance use.[142] Increased risky decision making, but not impulsive decision making, is positively associated with ethanol consumption in social drinkers.[143] High impulsive individuals with an alcohol use disorder report higher levels of subjective stimulation and positive mood after consuming ethanol.[144] The increased positive mood following ethanol consumption can act as a positive reinforcer, encouraging further ethanol consumption.

Concerning tobacco use, one experiment allowed individuals to earn money for each 30-second period that they refrained from smoking. Individuals with greater delay discounting were more likely to smoke during these abstinence reinforcement sessions.[145] In a separate experiment, participants were asked to abstain from smoking for 48 h. Individuals with greater delay discounting were more likely to smoke during those 48 h.[146] Greater delay discounting is also associated with problematic cannabis use.[147] Individuals with higher levels of impulsive decision making are less willing to change their cannabis use; furthermore, those with cannabis use disorder are more likely to smoke cigarettes if they display greater levels of impulsivity,[148] and they are more likely to use cannabis during treatment for a cannabis use disorder or for cannabis dependence.[149]

Clinical studies have also shown that delay discounting is a risk factor for smoking relapse[150] (but see[151]). In those enrolled in a smoking cessation treatment, higher delay discounting is associated with near immediate smoking following the cessation treatment.[152] Delay discounting is also associated with one's willingness to seek treatment for a SUD.[153] Furthermore, increased delay discounting is associated with length of recovery from a SUD, and individuals that are more impulsive state that they feel they will relapse.[154] However, one study found that greater impulsive decision making during acute nicotine withdrawal leads to less cigarette smoking during a 1-week period in which participants are reinforced for each drug-free urine sample.[155]

Increased risky decision making in the BART is negatively associated with treatment seeking in opioid-dependent individuals.[156] Risky decision making is also a risk factor for dropout from a relapse prevention program in methamphetamine users.[157] Impulsive decision making has been implicated in treatment outcomes for those with a SUD. Delay discounting is associated with abstinence during contingency management; that is, individuals with lower impulsive decision making achieve more consecutive weeks of cocaine abstinence,[158] and they are less likely to smoke cigarettes during treatment.[159] Individuals that show a bias for the disadvantageous decks in the IGT are more likely to drop out of therapy prematurely[160] and are more likely to relapse 3 months following treatment.[161] While Grosskopf et al. did not find that delay discounting predicts treatment outcomes for those seeking treatment for nicotine dependence, they observed that individuals with high variability in their impulsive decision making are more likely to begin smoking during treatment.[151] Finally, increased impulsive decision making is predictive of lower improvement in psychological and social quality of life during early treatment for methamphetamine use disorder.[162]

Animal models have been particularly useful in parsing the association between maladaptive decision making (particularly impulsive decision making) and addiction. Animals that are classified as high impulsive in delay discounting acquire cocaine self-administration at a faster rate,[163] self-administer more cocaine and methylphenidate,[32,164] show greater escalation of cocaine self-administration,[165] consume more ethanol,[166] develop greater d-amphetamine CPP,[167] treat nicotine and cocaine as inelastic goods,[168,169] emit more responses during extinction of cocaine self-administration,[170] and show greater reinstatement of nicotine- and cocaine-seeking behavior[170,171] compared to low impulsive rats. Mice that are more impulsive in a delay-discounting task are less active following ethanol administration, but they develop greater behavioral sensitization to the locomotor-stimulant effects of ethanol over time,[172] suggesting that these animals are at higher risk of engaging in addiction-like behaviors. Interestingly, high impulsive rats are *less* sensitive to the suppressant effects of ethanol.[46] While ethanol normally leads to sedation with high enough concentrations, this finding is important because it suggests that high impulsive individuals may not become as sedated with continued ethanol consumption. This, in turn, may cause these individuals to misjudge how much ethanol they have consumed, which can increase the likelihood of overdose.

Rats that make more suboptimal choices in the rGT show greater incubation of craving for cocaine-paired cues (i.e., relapse-like behavior).[22] In one study, Ferland et al.[23] wanted to determine if pairing highly salient cues with reinforcement delivery for the high risk/high reward options would impair decision making in the rGT and increase cocaine self-administration. As hypothesized, rats trained on a cued version of the rGT made more suboptimal choices compared to rats trained on an uncued variant. More importantly, rats trained on the cued version of the rGT self-administered more cocaine. In animal models, high risk-taking rats as assessed in the RDT self-administer more cocaine,[173] develop greater escalation of cocaine self-administration,[174] and show greater demand inelasticity for cocaine[175] compared to risk averse rats. However, increased risky decision making in the RDT is associated with decreased cocaine CPP.[176] Increased risk-taking behavior as measured in a radial arm variant of a gambling task predicts methamphetamine CPP.[177] In addition to drug self-administration and CPP, risky decision making in the RDT is correlated with locomotor activity following nicotine exposure.[178] In sum, both human studies and animal studies have largely shown that maladaptive decision making is a predictor of addiction-like behaviors.

Quiz 8.2

(1) Which drug has been shown to be self-administered at greater levels in high impulsive, high risk-taking, and suboptimal decision-making animals?
   a. Cannabis
   b. Cocaine
   c. Ethanol
   d. Nicotine
(2) Which of the following statements about acute/long-term drug effects on risky decision making is true?
   a. Acute amphetamine administration consistently increases risky decision making in animals.
   b. Long-term drug use is consistently linked to increased risky decision making.
   c. Opioids administered acutely do not impact risky decision making, but long-term opioid use is associated with increased risky decision making.
   d. Stimulants administered acutely do not impact risky decision, but long-term stimulant use is associated with increased risky decision making.
(3) Is the following statement true or false: Maladaptive decision making is a risk factor for relapse.

Answers to quiz 8.2

(1) b — Cocaine
(2) c — Opioids administered acutely do not impact risky decision making, but long-term opioid use is associated with increased risky decision making.
(3) True

Shared neuromechanisms of maladaptive decision making and addiction

Given that individuals with a SUD often display impaired decision making, this section will highlight some of the major brain regions that have been implicated in both addiction and in maladaptive decision making. As you have already learned about these structures in Chapter 3, this section will focus more on research that has examined the contribution of each of these regions on decision making.

Prefrontal cortex

In addition to mediating addictive-like behaviors and attentional/memory processes, the prefrontal cortex (PFC) is heavily involved in decision making. Individuals with damage to the ventromedial PFC perform poorly on the IGT.[1] Bechara et al.[5] compared decision making between healthy controls, individuals with bilateral lesions to the ventromedial PFC, and individuals with a SUD. Those with a SUD were much more likely to perform on the same level as those with lesions to the PFC. These results suggest that individuals with a SUD have impaired PFC functioning. Indeed, one startling finding is that individuals with severe ethanol use disorder have decreased gray matter volume in the ventromedial PFC.[7] Gray matter is composed of cell bodies of neurons. Less gray matter means there are fewer neuronal cell bodies, which may account for some of the cognitive impairments observed in this population. In cocaine users, impaired decision making following abstinence may be linked to altered activity of the frontal cortex; specifically, abstinent cocaine users show increased activity in the right orbitofrontal cortex (OFC) but decreased activity in the right dorsolateral PFC (dlPFC) and in the left medial PFC (mPFC) while performing the IGT.[179] In rats, temporarily inactivating the mPFC increases suboptimal decision making in the rGT.[180]

Prefrontal cortical structures also mediate impulsive/risky decision making. Individuals with more severe ethanol use problems show greater activation of the OFC and the dlPFC when completing a delay-discounting task.[181,182] Activating the mPFC with transcranial magnetic stimulation reduces delay discounting.[183] When individuals choose a risky option compared to a safe option, the mPFC is activated.[184] Animal models have also shown that structures in the PFC mediate aspects of decision making, although mixed results have been observed concerning the role of the PFC to impulsive decision making. Lesions to the OFC or the mPFC either increase impulsive decision making[185,186] or decrease impulsive decision making.[187–189] The contribution of the OFC to impulsive decision making is especially nuanced. Depending on which subregion of the OFC is damaged, impulsive decision making can increase or decrease. Lesions to the medial OFC (mOFC) increase delay discounting while lesions to the lateral OFC (lOFC) decrease delay discounting.[190] In contrast to impulsive decision making, lesions to the OFC decrease risky decision making.[186,187,191] While lesioning the mOFC increases delay discounting, temporary inactivation of this area with GABAergic agonists has no effect on delay discounting.[192] Stopper et al. also found that inactivating the OFC increases risky decision making,[192] which contrasts with the effects of OFC lesions. Collectively, these results highlight the complex ways in which the PFC regulates decision making, as subregions of the OFC appear to differentially regulate impulsive decision making, and techniques such as lesions and temporary inactivation can differentially alter decision making.

Recall that individuals with a SUD have decreased dopamine D_2-like receptors in the PFC, which may partially account for the impaired decision making observed in individuals with a SUD. In animals, blockade of PFC D_2-like receptors increases impulsive decision making.[193] While downregulation of D_2-like receptors is associated with addiction, overexpression of D_1-like receptors in the PFC leads to increased impulsive decision making, increased cocaine CPP, and increased cocaine self-administration in rats.[194] Other research has shown that high impulsive animals have increased dopamine D_1-like receptor gene expression in the mPFC and that stimulating intra-mPFC D_1-like receptors increases impulsive decision making.[195] It is important to note that dopaminergic activity in the PFC does not fully account for the maladaptive decision making observed in those with a SUD. For example, not only do smokers engage in maladaptive decision making, but they have decreased glutamate levels in the dlPFC.[196] This finding is not surprising given the role of glutamate in learning and memory. In animal models, administration of ketamine (NMDA receptor antagonist) increases impulsive decision making.[31]

Limbic structures

Beyond the PFC, structures in the limbic system significantly contribute to decision making. This section will focus exclusively on two brain structures: the nucleus accumbens (NAc) and the amygdala. In animals, lesions to or inactivation of the NAc consistently increase impulsive decision making[197] and suboptimal decision making in delay discounting.[198] Disconnecting part of the PFC from the NAc core also increases impulsive decision making.[199] As with impaired decision making, research has focused extensively on the contribution of NAc dopamine receptors on impulsive decision making. Interestingly, both D_1-like receptor binding and D_2-like receptor binding are decreased in the NAc core of high impulsive animals.[200,201] At first glance, these results seem to contradict each other as D_1-like receptors and D_2-like receptors have opposing functions (i.e., D_1-like receptors are excitatory while D_2-like receptors are inhibitory). The view that these receptor subclasses merely oppose one another is perhaps too simplistic. Other research has shown that both D_1-like receptors and D_2-like receptors are involved in motivation. More specifically, animals exposed to food-predictive cues show greater activation of both receptor types when responding following presentation of these cues.[202] Therefore, high impulsive animals may not be as motivated to wait for delayed reinforcement, which corresponds to the decreased D_1-like/D_2-like receptor binding observed in these animals.

Like the mPFC, choosing a risky option activates the NAc.[184] The finding that both mPFC and NAc are activated during decision making is not surprising considering these two regions share important connections with one another. When performing a risky gains task, noncocaine users show increased activity in the ventral striatum (including the NAc) when choosing high-risk/high-reward options; however, cocaine users do not have altered activity in this region, suggesting that extended drug use causes neural adaptations in response to risky decision making. This neural adaptation may account for why cocaine-dependent individuals are more likely to choose the risky option following a trial in which they lose money.[203] In adolescents, activation of the ventral striatum, as well as increased risky decision making, is predictive of future binge drinking.[204]

Dopamine levels in the NAc contribute significantly to risky decision making. In probability discounting, dopamine levels in the NAc increase as uncertainty of reinforcement increases,[205] and high risk-taking rats (as assessed in the RDT) show greater release of dopamine in the NAc.[206] Adolescent risky decision making is associated with decreased dopamine D_2-like mRNA expression in the NAc shell[173] while blocking NAc dopamine D_1-like receptors increases risk aversion in rats.[207] When NAc D_2-like receptors are pharmacologically or optogenetically stimulated, risk-taking decreases.[173,208] Using probability discounting, Jenni et al.[209] showed that simultaneous inactivation of the NAc and intra-PFC blockade of dopamine D_1-like receptors decreases risky decision making. In short, increasing dopaminergic activity in the NAc potentiates risky decision making while decreasing such activity attenuates risky decision making.

In the rGT, chemogenetic inhibition of NAc-projecting dopamine neurons improves decision making,[210] which mirrors what is observed in risky decision making. As you have already learned, presenting highly salient cues with disadvantageous alternatives in the rGT increases maladaptive decision making and increases cocaine self-administration. Ferland et al. also showed that rats exposed to these highly salient cues have lower basal dopamine release in the NAc but have greater cocaine-induced dopamine release.[23] These results suggest that exposure to salient cues while gambling causes a downregulation of the dopaminergic system, increasing drug misuse vulnerability. Think about gambling in the real world. Casinos are inundated with flashing lights and sounds from slot machines. These salient cues encourage an individual to keep "playing", which can drive an addiction phenotype. We will discuss gambling disorder in more detail in Chapter 13.

Although the amygdala is most often associated with emotional memories and fear conditioning, this brain region is also important for decision making. Patients with lesions to the amygdala have difficulty performing the IGT.[211] Similarly, lesions to the basolateral nucleus of the amygdala (BLA) impair decision making in the rGT,[212] and they increase impulsive decision making[185,189] and risky decision making.[191] Inactivating the BLA increases suboptimal decision making; that is, rats are less likely to choose a risky option when the probability of receiving reinforcement is high.[213] Functional disconnection of the PFC from the BLA (i.e., pharmacological inactivation of the PFC in one hemisphere of the brain and inactivation of the BLA in the opposite side of the brain) substantially increases risky decision making in probability discounting. More specifically, PFC-BLA disconnection decreases the propensity of rats to shift their responses to the small, certain reinforcer following a loss.[214] Optogenetic research has been able to provide a clearer role of the amygdala in decision making. In the RDT, inhibiting the BLA before a rat makes a choice between a small, safe reward and a large, risky reward decreases risky decision making; however, when the BLA is inhibited while the animal receives punishment for selecting the risky option, they are more likely to show increased risky decision making.[215] Notice that the NAc and the amygdala have similar roles in decision making. Damaging either brain structure increases impulsive decision making and risky decision making. The BLA sends neurons to the NAc and can excite this region. Thus, when the BLA is stimulated, it will stimulate the NAc.

Anterior cingulate cortex

While you have already learned about the PFC and limbic structures such as the NAc and the BLA, you now need to learn about another important structure: the anterior cingulate cortex (ACC). The ACC acts as a bridge between the PFC and the limbic system; as such, this region is also an important mediator of decision making. Specifically, this region is important for signaling outcome-relevant information and for resolving conflict.[216] In healthy individuals, ACC activation is potentiated when individuals choose a high-risk alternative compared to a low-risk alternative,[217] engage in risk-taking behaviors,[218] or make suboptimal choices in a gambling task.[219] Also, increased connectivity between the ACC and the OFC is associated with gambling-like behavior.[220] Long-term drug use appears to decrease the activity of the ACC. Methamphetamine users performing a risky gains task have decreased ACC activity compared to healthy controls; additionally, lower activation of the

ACC in methamphetamine users is associated with an increased propensity for choosing a risky option following a loss.[221] Abstinent drug users also show decreased activity in the ACC while performing a risky decision-making task compared to nonusers,[222] and ethanol-dependent individuals have less gray matter in the ACC.[7] In smokers, craving during abstinence is positively associated with ACC activation following exposure to a smoking-related cue,[223] suggesting that activation of this region may precipitate relapse. Indeed, increased connectivity between the NAc and the ACC is associated with impulsive decision making and cocaine relapse.[224]

At the neurotransmitter level, work has shown that glutamate concentrations in the ACC are negatively correlated with delay discounting in healthy individuals. That is, higher impulsive decision making is associated with increased glutamate levels in this region.[225] This finding contrasts with what has been observed in the PFC of smokers, providing further support that the PFC and the ACC mediate different aspects of decision making. Beyond glutamate, smokers have decreased ACC GABA levels compared to nonsmokers, and GABA levels in the ACC are positively correlated with decision making in smokers.[226] This finding is consistent with preclinical work showing that GABA receptor binding is lower in high impulsive animals.[201]

Metacognition, mindfulness, and addiction

One final cognitive construct I want to cover is **metacognition**. In short, metacognition refers to thinking about one's own mental processes or activities ("thinking about thinking"). Given that individuals with a SUD often display impaired decision making, one important question to address is if these individuals have difficulty gauging their own thoughts and judgments. Wasmuth et al.[227] measured four subscales of metacognition in those with a SUD: self-reflectivity (ability to understand one's own mental states), understanding others' thoughts, decentration (ability to see the world as viewable from multiple perspectives), and mastery (ability to use metacognitive knowledge to address social and psychological dilemmas). Individuals with a SUD have low metacognitive mastery; they can identify psychological problems but have difficulty addressing them effectively. In another study, Balconi et al.[4] had cocaine-dependent individuals complete the IGT. Following the experiment, participants were asked about self-knowledge of the strategy used during the IGT. These questions included "Were you able to apply a strategic plan during the game?", "Were you aware of using a strategy during the game?", "Did you change your strategy during the game?", and "Do you think you used an efficacious strategy?" (p. 402).[227] Cocaine-dependent individuals were less likely to report using a strategy during the IGT, and they reported lower awareness of any strategies that they used. However, they were more likely to report that they were more flexible and made more efficacious choices compared to the control group. These findings indicate that individuals with cocaine dependence have impaired metacognitive ability as they have an unrealistic representation of IGT performance. That is, they feel that they performed better than they did. This unrealistic representation may partially account for the decision-making impairments observed in SUDs.

As related to SUDs, Spada et al.[228] state that metacognition can be divided into positive metacognition and negative metacognition. Positive metacognition refers to a belief that substance use will improve cognition and/or mood (i.e., self-medication). Negative metacognition can refer to one of several beliefs: (1) drug use continues to occur due to a lack of executive function (e.g., "No matter what I do, I cannot quit drinking"); (2) uncontrollable thoughts centered around drug use (e.g., "I can't stop thinking about drinking alcohol"); (3) *thought-action fusion*, the idea that simply thinking about an action is the same as performing the action (e.g., "Thinking about alcohol makes me drink alcohol"); and (4) substance use will negatively impact cognition (e.g., "Drinking will make me more forgetful"). Both positive and negative metacognitive beliefs are associated with addiction-like behaviors, such as ethanol consumption,[229] cigarette use,[230] nicotine dependence,[231] and cannabis use.[232] While problem drinkers score higher on positive metacognitive beliefs compared to nonproblem drinkers, ethanol-dependent individuals score higher on both positive and negative metacognitive beliefs compared to problem and nonproblem drinkers.[233]

According to Spada and Wells, individuals engage in ethanol use because of positive metacognitive beliefs and negative affect (i.e., depressed mood).[234] Think of an individual that is dealing with a stressful situation, such as struggling in school, working with a difficult colleague, or going through a breakup. This individual will experience negative affect and may think that drinking ethanol will help improve their mood. However, once the individual begins drinking, they experience a reduction in *metacognitive monitoring*, the ability to monitor one's thought processes. This loss of metacognitive monitoring leads to continued drinking. Because excessive drinking is associated with aversive side effects like hangover, negative metacognitive beliefs related to ethanol consumption increase. This causes an increase in negative affect, which propels the individual to continue drinking. Consistent with this hypothesis is the finding that individuals with high maladaptive metacognition are more likely to develop symptoms of ethanol dependence as negative affect increases whereas individuals with low maladaptive metacognition are more likely to avoid ethanol as negative affect increases.[235]

Mindfulness is a meditative-like state in which individuals focus their awareness on the present moment while acknowledging and accepting one's feeling, thoughts, and bodily sensations without judgment. Mindfulness requires focused attention as the individual must concentrate on a sensory process, such as breathing, and it requires open monitoring. Open monitoring is a form of metacognition in which the individual reflects on their state of consciousness. Mindfulness-based interventions have been used to help those with a SUD. In a typical mindfulness-based intervention, individuals are given approximately 8 weeks of training in which a clinician teaches them mindfulness practices like mindful breathing and body scan meditations.[236] Overall, mindfulness-based interventions are effective in reducing drug use.[237] Garland et al. argue that the beneficial effects of mindfulness-based interventions for SUDs result from enhanced functional connectivity within cortical structures involved in executive functioning (e.g., PFC) and between structures like the PFC and those that control automaticity/habit formation (basal ganglia), memory consolidation (hippocampus), reactivity (amygdala), and reward (NAc)[238] (Fig. 8.3). In other words, mindfulness-induced increases in PFC functioning can regulate habitual actions associated with drug use and can control cravings associated with drug-paired stimuli.

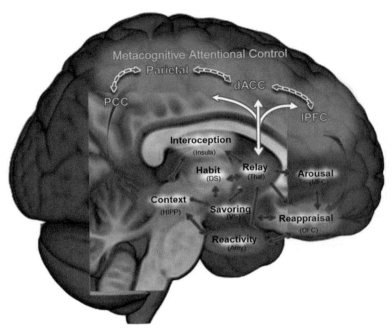

FIGURE 8.3 A proposed explanation of the neuromechanisms associated with mindfulness-based interventions for SUDs. In this framework, mindfulness training improves connectivity within cortical structures involved in decision making such as the prefrontal cortex (PFC) and the anterior cingulate cortex (ACC). Mindfulness training also improves connectivity between these cortical areas and subcortical structures like the hippocampus (HIPP), the dorsal striatum (DS), the ventral striatum (VS), and the amygdala (Amy). When individuals encounter drug-paired stimuli, areas like the PFC can exert greater control over areas involved in compulsive drug seeking (DS), reactivity (Amy), and wanting/savoring (VS). *Image comes from Fig. 2 of Garland EL, Froeliger B, Howard MO. Mindfulness training targets neurocognitive mechanisms of addiction at the attention-appraisal-emotion interface.* Front Psychiatr. *2014;4:173. https://doi.org/10.3389/fpsyt.2013.00173, This is an open-access article distributed under the terms of the Creative Commons Attribution 4.0 International License (https://creativecommons.org/licenses/by/4.0/).*

Treatments for improving drug-induced impairments in decision making

As impaired decision making is associated with treatment dropout and relapse, an important goal during treatment is to improve decision-making processes in those with a SUD. Studies have examined the efficacy of behavioral, cognitive, and pharmacological treatments for improving drug-induced deficits in decision making. You just read that mindfulness-based interventions are effective in combating SUDs. Enrollment in mindfulness training also improves performance in the IGT in abstinent polysubstance-dependent individuals.[239] Like mindfulness, cognitive-behavioral therapy (CBT) has been shown to improve decision making in the IGT in polysubstance-dependent individuals,[240] but CBT does not appear to be effective in attenuating impulsive decision making in those with a SUD.[240] Individuals given CBT for smoking dependence are less impulsive at 1 year following therapy.[241] Thus, CBT may not produce immediate reductions in impulsive decision making but can lead to alterations in other behaviors that may lead to decreased impulsivity in the future. One paradoxical finding is that individuals with cannabis use disorder become *more* impulsive during CBT,[148] indicating that this therapy may not be beneficial in high impulsive cannabis users.

In Chapter 5 you learned about the use of contingency management to decrease short-term drug use. Some mixed results have been observed concerning the ability of contingency management to decrease delay discounting in smokers.[242,243] Although Yoon et al. did not see a decrease in delay discounting for hypothetical money, they found that smokers were more likely to choose delayed monetary rewards over cigarettes following contingency management.[243] Additionally, individuals with depression given contingency management for smoking cessation are less impulsive when tested at follow-up.[244] Contingency management, when combined with CBT, decreases delay discounting in female, but not in male, smokers.[245] While contingency management does not decrease impulsive decision making in those with cannabis use disorder, it prevents further increases in impulsivity over time.[148]

Although not a treatment outcome *per se*, Bickel et al.[246] investigated if impulsive decision making predicts the frequency at which an individual receiving contingency management redeems vouchers. In this experiment, opioid-dependent individuals receiving buprenorphine treatment could immediately redeem obtained vouchers or save them in a "bank" for later use. Those with higher impulsive decision making were much more likely to redeem their vouchers immediately after receiving them. Somewhat similarly, methamphetamine users and/or opioid users redeem vouchers at a faster rate compared to those that have abstained from those substances.[247] The increased propensity to redeem vouchers is consistent with work showing increased impulsive decision making in those with a SUD.

Beyond cognitive and behavioral interventions, some studies have tested the effects of pharmacotherapies on decision making in individuals with a SUD. Recall that bupropion (as Zyban) and varenicline (Chantix) are used as treatments for nicotine dependence. Bupropion and varenicline decrease delay discounting, but only in men.[248] Cannabidiol, a nonintoxicating component of cannabis, has also been tested in cigarette users. Cannabidiol fails to ameliorate nicotine withdrawal-induced increases in impulsive decision making.[249]

As covered in Chapter 2, one major treatment for opioid use disorder is agonist treatment (i.e., replacing heroin with methadone or buprenorphine). Individuals given agonist treatment show improvements in the IGT, indicating enhanced decision making.[250] However, as discussed earlier in this chapter, individuals on methadone maintenance therapy are more impulsive in delay discounting compared to those that have never used opioids.[60] Additionally, methadone-maintained patients are more likely to make risky decisions when given punishment.[251] If given a choice between a small, but more likely reward and a large, but less likely reward, methadone-maintained individuals are more likely to "gamble" if they failed to win during the previous trial. In contrast to methadone, buprenorphine treatment is associated with reduced impulsive decision making in opioid-dependent individuals,[252] indicating that buprenorphine may be a better treatment for those with high levels of impulsive decision making.

A few other drugs have been tested in decision making tasks in individuals with a SUD. The stimulant-like drug modafinil decreases impulsive decision making in those with an alcohol use disorder[253] and decreases impulsive decision making in rats prenatally exposed to ethanol.[254] Modafinil also improves suboptimal decision making observed in cocaine users performing the BART.[121] The catechol-*O*-methyltransferase inhibitor tolcapone decreases self-reported ethanol consumption in those with an alcohol use disorder and decreases impulsive decision making.[255] The anxiolytic buspirone decreases sexual risk-taking intent, but it fails to reduce cocaine self-administration.[256] Buspirone also fails to alter delay discounting in cocaine users.[257] Clearly, more work is needed to determine which pharmacotherapies can be administered in treatment-seeking substance users to alleviate maladaptive decision making. Care is needed when prescribing certain pharmacotherapies, and even cognitive-behavioral treatments, as certain interventions can exacerbate poor decision making, which could then lead someone to relapse.

Quiz 8.3

(1) Which statement best summarizes the effects of OFC lesions on decision making?
    a. OFC lesions do not affect decision making in animals.
    b. OFC lesions impair general decision making but decrease impulsive decision making.
    c. OFC lesions impair general decision making but decrease risky decision making.
    d. OFC lesions improve maladaptive decision making.
(2) Which of the following drugs has been shown to improve impulsive decision making in those with a SUD?
    a. Buspirone
    b. Methadone
    c. Modafinil
    d. All of the above
(3) Which of the following exemplifies the metacognitive domain mastery?
    a. "I know that my families and friends are upset with me."
    b. "I notice that I drink more when I am upset."
    c. "I realize I drink more when I get stressed. Next time I get stressed, I need to practice the deep breathing exercises I learned earlier and need to find some other activity to do."
    d. "I understand why my partner is upset with me."
(4) Which brain region is largely responsible for conflict monitoring?
    a. Amygdala
    b. ACC
    c. Nucleus accumbens
    d. PFC

Answers to quiz 8.3

(1) c — OFC lesions impair general decision making but decrease risky decision making
(2) c — Modafinil
(3) c — "I realize I drink more when I get stressed. Next time I get stressed, I need to practice the deep breathing exercises I learned earlier and need to find some other activity to do."
(4) b — ACC

Chapter summary

Like the relationship between attentional/memory deficits and addiction, the relationship between decision making and SUDs is bidirectional. Long-term drug use severely impairs decision making, and maladaptive decision making is a risk factor for relapse, particularly impaired decision making and impulsive decision making. Several brain regions involved in addiction are important mediators of decision making (e.g., PFC, NAc). To better treat SUDs, mindfulness-based interventions have been employed to alter impaired decision making observed in individuals with a SUD. Other interventions have been used to ameliorate drug-induced deficits in decision making, including contingency management, CBT, and pharmacotherapies. While this chapter focused heavily on discounting-type procedures such as delay discounting, probability discounting, and the RDT, you will learn more about impulsivity and risky decision making and their relationship with addiction in the next chapter.

Glossary

Decision making — an aspect of cognition in which we make choices between two or more alternatives.
Impulsive decision making — a form of decision making in which individuals are unable to delay gratification even if the delayed alternative is more advantageous.
Metacognition — the process by which we reflect on our own decision making and cognitive processing ("thinking about thinking").
Mindfulness — a meditative-like state in which individuals focus on the sensations they are currently experiencing and reflect on their thoughts and feelings without judgment.
Risky decision making — a form of decision making in which individuals are more likely to choose options that are more uncertain or are linked to harmful outcomes.

Supplementary data related to this chapter can be found online at https://educate.elsevier.com/9780323905787.

References

1. Bechara A, Damasio AR, Damasio H, Anderson SW. Insensitivity to future consequences following damage to human prefrontal cortex. *Cognition*. 1994;50(1–3):7–15. https://doi.org/10.1016/0010-0277(94)90018-3.
2. Zeeb FD, Robbins TW, Winstanley CA. Serotonergic and dopaminergic modulation of gambling behavior as assessed using a novel rat gambling task. *Neuropsychopharmacology*. 2009;34(10):2329–2343. https://doi.org/10.1038/npp.2009.62.
3. Wason PC. Reasoning about a rule. *Q J Exp Psychol*. 1968;20(3):273–281. https://doi.org/10.1080/14640746808400161.
4. Balconi M, Finocchiaro R, Campanella S. Reward sensitivity, decisional bias, and metacognitive deficits in cocaine drug addiction. *J Addiction Med*. 2014;8(6):399–406. https://doi.org/10.1097/ADM.0000000000000065.
5. Bechara A, Dolan S, Denburg N, Hindes A, Anderson SW, Nathan PE. Decision-making deficits, linked to a dysfunctional ventromedial prefrontal cortex, revealed in alcohol and stimulant abusers. *Neuropsychologia*. 2001;39(4):376–389. https://doi.org/10.1016/s0028-3932(00)00136-6.
6. Hanson KL, Luciana M, Sullwold K. Reward-related decision-making deficits and elevated impulsivity among MDMA and other drug users. *Drug Alcohol Depend*. 2008;96(1–2):99–110. https://doi.org/10.1016/j.drugalcdep.2008.02.003.
7. Le Berre AP, Rauchs G, La Joie R, et al. Impaired decision-making and brain shrinkage in alcoholism. *Eur Psychiatr*. 2014;29(3):125–133. https://doi.org/10.1016/j.eurpsy.2012.10.002.

8. O'Donnell BF, Skosnik PD, Hetrick WP, Fridberg DJ. Decision making and impulsivity in young adult cannabis users. *Front Psychol.* 2021;12:679904. https://doi.org/10.3389/fpsyg.2021.679904.
9. Kasar M, Gleichgerrcht E, Keskinkilic C, Tabo A, Manes FF. Decision-making in people who relapsed to driving under the influence of alcohol. *Alcohol Clin Exp Res.* 2010;34(12):2162−2168. https://doi.org/10.1111/j.1530-0277.2010.01313.x.
10. Loeber S, Duka T, Welzel H, et al. Impairment of cognitive abilities and decision making after chronic use of alcohol: the impact of multiple detoxifications. *Alcohol Alcohol.* 2009;44(4):372−381. https://doi.org/10.1093/alcalc/agp030.
11. Hulka LM, Eisenegger C, Preller KH, et al. Altered social and non-social decision-making in recreational and dependent cocaine users. *Psychol Med.* 2014;44(5):1015−1028. https://doi.org/10.1017/S0033291713001839.
12. Vadhan NP, Hart CL, Haney M, van Gorp WG, Foltin RW. Decision-making in long-term cocaine users: effects of a cash monetary contingency on Gambling task performance. *Drug Alcohol Depend.* 2009;102(1−3):95−101. https://doi.org/10.1016/j.drugalcdep.2009.02.003.
13. Rotheram-Fuller E, Shoptaw S, Berman SM, London ED. Impaired performance in a test of decision-making by opiate-dependent tobacco smokers. *Drug Alcohol Depend.* 2004;73(1):79−86. https://doi.org/10.1016/j.drugalcdep.2003.10.003.
14. Ghosh A, Basu D, Mattoo SK, Kumar Rana D, Roub F. Does cannabis dependence add on to the neurocognitive impairment among patients with opioid dependence? A cross-sectional comparative study. *Am J Addict.* 2020;29(2):120−128. https://doi.org/10.1111/ajad.12986.
15. Chen YC, Wang LJ, Lin SK, Chen CK. Neurocognitive profiles of methamphetamine users: comparison of those with or without concomitant ketamine use. *Subst Use Misuse.* 2015;50(14):1778−1785. https://doi.org/10.3109/10826084.2015.1050110.
16. Kornreich C, Delle-Vigne D, Knittel J, et al. Impaired conditional reasoning in alcoholics: a negative impact on social interactions and risky behaviors? *Addiction.* 2011;106(5):951−959. https://doi.org/10.1111/j.1360-0443.2010.03346.x.
17. Li X, Zhang F, Zhou Y, Zhang M, Wang X, Shen M. Decision-making deficits are still present in heroin abusers after short- to long-term abstinence. *Drug Alcohol Depend.* 2013;130(1−3):61−67. https://doi.org/10.1016/j.drugalcdep.2012.10.012.
18. Zhang XL, Shi J, Zhao LY, et al. Effects of stress on decision-making deficits in formerly heroin-dependent patients after different durations of abstinence. *Am J Psychiatr.* 2011;168(6):610−616. https://doi.org/10.1176/appi.ajp.2010.10040499.
19. Wang G, Shi J, Chen N, et al. Effects of length of abstinence on decision-making and craving in methamphetamine abusers. *PLoS One.* 2013;8(7):e68791. https://doi.org/10.1371/journal.pone.0068791.
20. Delibas DH, Akseki HS, Erdoğan E, Zorlu N, Gülseren S. Impulsivity, sensation seeking, and decision-making in long-term abstinent cannabis dependent patients. *Archiv Neuropsychiatr.* 2018;55(4):315−319. https://doi.org/10.5152/npa.2017.19304.
21. Spoelder M, Lesscher HM, Hesseling P, et al. Altered performance in a rat gambling task after acute and repeated alcohol exposure. *Psychopharmacology.* 2015;232(19):3649−3662. https://doi.org/10.1007/s00213-015-4020-0.
22. Ferland J-N, Winstanley CA. Risk-preferring rats make worse decisions and show increased incubation of craving after cocaine self-administration. *Addiction Biol.* 2017;22(4):991−1001. https://doi.org/10.1111/adb.12388.
23. Ferland J-N, Hynes TJ, Hounjet CD, et al. Prior exposure to salient win-paired cues in a rat gambling task increases sensitivity to cocaine self-administration and suppresses dopamine efflux in nucleus accumbens: support for the reward deficiency hypothesis of addiction. *J Neurosci.* 2019;39(10):1842−1854. https://doi.org/10.1523/JNEUROSCI.3477-17.2018.
24. Mischel W, Ebbesen EB, Zeiss AR. Cognitive and attentional mechanisms in delay of gratification. *J Pers Soc Psychol.* 1972;21(2):204−218. https://doi.org/10.1037/h0032198.
25. Schlam TR, Wilson NL, Shoda Y, Mischel W, Ayduk O. Preschoolers' delay of gratification predicts their body mass 30 years later. *J Pediatr.* 2013;162(1):90−93. https://doi.org/10.1016/j.jpeds.2012.06.049.
26. Shoda Y, Mischel W, Peake PK. Predicting adolescent cognitive and self-regulatory competencies from preschool delay of gratification: identifying diagnostic conditions. *Dev Psychol.* 1990;26(6):978−986. https://doi.org/10.1037/0012-1649.26.6.978.

27. Benjamin DJ, Laibson D, Mischel W, et al. Predicting mid-life capital formation with pre-school delay of gratification and life-course measures of self-regulation. *J Econ Behav Organ.* 2020;179:743–756. https://doi.org/10.1016/j.jebo.2019.08.016.
28. Watts TW, Duncan GJ, Quan H. Revisiting the marshmallow test: a conceptual replication investigating links between early delay of gratification and later outcomes. *Psychol Sci.* 2018;29(7):1159–1177. https://doi.org/10.1177/0956797618761661.
29. Rachlin H, Raineri A, Cross D. Subjective probability and delay. *J Exp Anal Behav.* 1991;55(2):233–244. https://doi.org/10.1901/jeab.1991.55-233.
30. de Wit H, Enggasser JL, Richards JB. Acute administration of d-amphetamine decreases impulsivity in healthy volunteers. *Neuropsychopharmacology.* 2002;27(5):813–825. https://doi.org/10.1016/S0893-133X(02)00343-3.
31. Floresco SB, Tse MT, Ghods-Sharifi S. Dopaminergic and glutamatergic regulation of effort- and delay-based decision making. *Neuropsychopharmacology.* 2008;33(8):1966–1979. https://doi.org/10.1038/sj.npp.1301565.
32. Freund N, Jordan CJ, Lukkes JL, Norman KJ, Andersen SL. Juvenile exposure to methylphenidate and guanfacine in rats: effects on early delay discounting and later cocaine-taking behavior. *Psychopharmacology.* 2019;236(2):685–698. https://doi.org/10.1007/s00213-018-5096-0.
33. Yates JR, Day HA, Evans KE, et al. Effects of d-amphetamine and MK-801 on impulsive choice: modulation by schedule of reinforcement and delay length. *Behav Brain Res.* 2019;376:112228. https://doi.org/10.1016/j.bbr.2019.112228.
34. Li Y, Zuo Y, Yu P, Ping X, Cui C. Role of basolateral amygdala dopamine D2 receptors in impulsive choice in acute cocaine-treated rats. *Behav Brain Res.* 2015;287:187–195. https://doi.org/10.1016/j.bbr.2015.03.039.
35. Xue Z, Siemian JN, Johnson BN, Zhang Y, Li JX. Methamphetamine-induced impulsivity during chronic methamphetamine treatment in rats: effects of the TAAR 1 agonist RO5263397. *Neuropharmacology.* 2018;129:36–46. https://doi.org/10.1016/j.neuropharm.2017.11.012.
36. Kobiella A, Ripke S, Kroemer NB, et al. Acute and chronic nicotine effects on behaviour and brain activation during intertemporal decision making. *Addiction Biol.* 2014;19(5):918–930. https://doi.org/10.1111/adb.12057.
37. Dallery J, Locey ML. Effects of acute and chronic nicotine on impulsive choice in rats. *Behav Pharmacol.* 2005;16(1):15–23. https://doi.org/10.1097/00008877-200502000-00002.
38. Flores RJ, Alshbool FZ, Giner P, O'Dell LE, Mendez IA. Exposure to nicotine vapor produced by an electronic nicotine delivery system causes short-term increases in impulsive choice in adult male rats. *Nicotine Tob Res.* 2022;24(3):358–365. https://doi.org/10.1093/ntr/ntab141.
39. Adams S, Attwood AS, Munafò MR. Drinking status but not acute alcohol consumption influences delay discounting. *Hum Psychopharmacol.* 2017;32(5):e2617. https://doi.org/10.1002/hup.2617.
40. Johnson PS, Sweeney MM, Herrmann ES, Johnson MW. Alcohol increases delay and probability discounting of condom-protected sex: a novel vector for alcohol-related HIV transmission. *Alcohol Clin Exp Res.* 2016;40(6):1339–1350. https://doi.org/10.1111/acer.13079.
41. Reed SC, Levin FR, Evans SM. Alcohol increases impulsivity and abuse liability in heavy drinking women. *Exp Clin Psychopharmacol.* 2012;20(6):454–465. https://doi.org/10.1037/a0029087.
42. Richards JB, Zhang L, Mitchell SH, de Wit H. Delay or probability discounting in a model of impulsive behavior: effect of alcohol. *J Exp Anal Behav.* 1999;71(2):121–143. https://doi.org/10.1901/jeab.1999.71-121.
43. Zacny JP, de Wit H. The prescription opioid, oxycodone, does not alter behavioral measures of impulsivity in healthy volunteers. *Pharmacol Biochem Behav.* 2009;94(1):108–113. https://doi.org/10.1016/j.pbb.2009.07.010.
44. Acheson A, Reynolds B, Richards JB, de Wit H. Diazepam impairs behavioral inhibition but not delay discounting or risk taking in healthy adults. *Exp Clin Psychopharmacol.* 2006;14(2):190–198. https://doi.org/10.1037/1064-1297.14.2.190.
45. Bernhardt N, Obst E, Nebe S, et al. Acute alcohol effects on impulsive choice in adolescents. *J Psychopharmacol.* 2019;33(3):316–325. https://doi.org/10.1177/0269881118822063.
46. Moschak TM, Mitchell SH. Sensitivity to reinforcer delay predicts ethanol's suppressant effects, but itself is unaffected by ethanol. *Drug Alcohol Depend.* 2013;132(1–2):22–28. https://doi.org/10.1016/j.drugalcdep.2013.07.009.
47. Olmstead MC, Hellemans KG, Paine TA. Alcohol-induced impulsivity in rats: an effect of cue salience? *Psychopharmacology.* 2006;184(2):221–228. https://doi.org/10.1007/s00213-005-0215-0.
48. Ortner CN, MacDonald TK, Olmstead MC. Alcohol intoxication reduces impulsivity in the delay-discounting paradigm. *Alcohol Alcohol.* 2003;38(2):151–156. https://doi.org/10.1093/alcalc/agg041.

49. Harvey-Lewis C, Franklin KB. The effect of acute morphine on delay discounting in dependent and non-dependent rats. *Psychopharmacology*. 2015;232(5):885–895. https://doi.org/10.1007/s00213-014-3724-x.
50. Eppolito AK, France CP, Gerak LR. Effects of acute and chronic morphine on delay discounting in pigeons. *J Exp Anal Behav*. 2013;99(3):277–289. https://doi.org/10.1002/jeab.25.
51. Minervini V, France CP. Effects of opioid/cannabinoid mixtures on impulsivity and memory in rhesus monkeys. *Behav Pharmacol*. 2020;31(2&3):233–248. https://doi.org/10.1097/FBP.0000000000000551.
52. Bizot J, Le Bihan C, Puech AJ, Hamon M, Thiébot M. Serotonin and tolerance to delay of reward in rats. *Psychopharmacology*. 1999;146(4):400–412. https://doi.org/10.1007/pl00005485.
53. Huskinson SL, Anderson KG. Effects of acute and chronic administration of diazepam on delay discounting in Lewis and Fischer 344 rats. *Behav Pharmacol*. 2012;23(4):315–330. https://doi.org/10.1097/FBP.0b013e3283564da4.
54. Eppolito AK, France CP, Gerak LR. Effects of acute and chronic flunitrazepam on delay discounting in pigeons. *J Exp Anal Behav*. 2011;95(2):163–174. https://doi.org/10.1901/jeab.2011.95-163.
55. Businelle MS, McVay MA, Kendzor D, Copeland A. A comparison of delay discounting among smokers, substance abusers, and non-dependent controls. *Drug Alcohol Depend*. 2010;112(3):247–250. https://doi.org/10.1016/j.drugalcdep.2010.06.010.
56. Petry NM. Delay discounting of money and alcohol in actively using alcoholics, currently abstinent alcoholics, and controls. *Psychopharmacology*. 2001;154(3):243–250. https://doi.org/10.1007/s002130000638.
57. Coffey SF, Gudleski GD, Saladin ME, Brady KT. Impulsivity and rapid discounting of delayed hypothetical rewards in cocaine-dependent individuals. *Exp Clin Psychopharmacol*. 2003;11(1):18–25. https://doi.org/10.1037//1064-1297.11.1.18.
58. Cox DJ, Dolan SB, Johnson P, Johnson MW. Delay and probability discounting in cocaine use disorder: comprehensive examination of money, cocaine, and health outcomes using gains and losses at multiple magnitudes. *Exp Clin Psychopharmacol*. 2020;28(6):724–738. https://doi.org/10.1037/pha0000341.
59. Yi R, Carter AE, Landes RD. Restricted psychological horizon in active methamphetamine users: future, past, probability, and social discounting. *Behav Pharmacol*. 2012;23(4):358–366. https://doi.org/10.1097/FBP.0b013e3283564e11.
60. Robles E, Huang BE, Simpson PM, McMillan DE. Delay discounting, impulsiveness, and addiction severity in opioid-dependent patients. *J Subst Abuse Treat*. 2011;41(4):354–362. https://doi.org/10.1016/j.jsat.2011.05.003.
61. Heyman GM, Gibb SP. Delay discounting in college cigarette chippers. *Behav Pharmacol*. 2006;17(8):669–679. https://doi.org/10.1097/FBP.0b013e3280116cfe.
62. Dom G, D'haene P, Hulstijn W, Sabbe B. Impulsivity in abstinent early- and late-onset alcoholics: differences in self-report measures and a discounting task. *Addiction*. 2006;101(1):50–59. https://doi.org/10.1111/j.1360-0443.2005.01270.x.
63. Bornovalova MA, Daughters SB, Hernandez GD, Richards JB, Lejuez CW. Differences in impulsivity and risk-taking propensity between primary users of crack cocaine and primary users of heroin in a residential substance-use program. *Exp Clin Psychopharmacol*. 2005;13(4):311–318. https://doi.org/10.1037/1064-1297.13.4.311.
64. Kirby KN, Petry NM. Heroin and cocaine abusers have higher discount rates for delayed rewards than alcoholics or non-drug-using controls. *Addiction*. 2004;99(4):461–471. https://doi.org/10.1111/j.1360-0443.2003.00669.x.
65. Karakula SL, Weiss RD, Griffin ML, Borges AM, Bailey AJ, McHugh RK. Delay discounting in opioid use disorder: differences between heroin and prescription opioid users. *Drug Alcohol Depend*. 2016;169:68–72. https://doi.org/10.1016/j.drugalcdep.2016.10.009.
66. Reed SC, Evans SM. The effects of oral d-amphetamine on impulsivity in smoked and intranasal cocaine users. *Drug Alcohol Depend*. 2016;163:141–152. https://doi.org/10.1016/j.drugalcdep.2016.04.013.
67. Heckman BW, MacQueen DA, Marquinez NS, MacKillop J, Bickel WK, Brandon TH. Self-control depletion and nicotine deprivation as precipitants of smoking cessation failure: a human laboratory model. *J Consult Clin Psychol*. 2017;85(4):381–396. https://doi.org/10.1037/ccp0000197.
68. Ohmura Y, Takahashi T, Kitamura N. Discounting delayed and probabilistic monetary gains and losses by smokers of cigarettes. *Psychopharmacology*. 2005;182(4):508–515. https://doi.org/10.1007/s00213-005-0110-8.
69. Reynolds B, Fields S. Delay discounting by adolescents experimenting with cigarette smoking. *Addiction*. 2012;107(2):417–424. https://doi.org/10.1111/j.1360-0443.2011.03644.x.

70. Reynolds B, Richards JB, Horn K, Karraker K. Delay discounting and probability discounting as related to cigarette smoking status in adults. *Behav Process.* 2004;65(1):35–42. https://doi.org/10.1016/s0376-6357(03)00109-8.
71. Weidberg S, González-Roz A, Secades-Villa R. Delay discounting in e-cigarette users, current and former smokers. *Int J Clin Health Psychol.* 2017;17(1):20–27. https://doi.org/10.1016/j.ijchp.2016.07.004.
72. Johnson MW, Bickel WK, Baker F. Moderate drug use and delay discounting: a comparison of heavy, light, and never smokers. *Exp Clin Psychopharmacol.* 2007;15(2):187–194. https://doi.org/10.1037/1064-1297.15.2.187.
73. Field M, Christiansen P, Cole J, Goudie A. Delay discounting and the alcohol Stroop in heavy drinking adolescents. *Addiction.* 2007;102(4):579–586. https://doi.org/10.1111/j.1360-0443.2007.01743.x.
74. Moore SC, Cusens B. Delay discounting predicts increase in blood alcohol level in social drinkers. *Psychiatr Res.* 2010;179(3):324–327. https://doi.org/10.1016/j.psychres.2008.07.024.
75. Bailey AJ, Gerst K, Finn PR. Intelligence moderates the relationship between delay discounting rate and problematic alcohol use. *Psychol Addict Behav.* 2020;34(1):175–181. https://doi.org/10.1037/adb0000471.
76. Meredith SE, Sweeney MM, Johnson PS, Johnson MW, Griffiths RR. Weekly energy drink use is positively associated with delay discounting and risk behavior in a nationwide sample of young adults. *J Caffeine Res.* 2016;6(1):10–19. https://doi.org/10.1089/jcr.2015.0024.
77. Strickland JC, Lee DC, Vandrey R, Johnson MW. A systematic review and meta-analysis of delay discounting and cannabis use. *Exp Clin Psychopharmacol.* 2020. https://doi.org/10.1037/pha0000378. Advance online publication.
78. Parlar M, MacKillop E, Petker T, Murphy J, MacKillop J. Cannabis use, age of initiation, and neurocognitive performance: findings from a large sample of heavy drinking emerging adults. *J Int Neuropsychol Soc.* 2021;27(6):533–545. https://doi.org/10.1017/S1355617721000618.
79. Jarmolowicz DP, Reed DD, Stancato SS, et al. On the discounting of cannabis and money: sensitivity to magnitude vs. delay. *Drug Alcohol Depend.* 2020;212:107996. https://doi.org/10.1016/j.drugalcdep.2020.107996.
80. Johnson MW, Bickel WK, Baker F, Moore BA, Badger GJ, Budney AJ. Delay discounting in current and former marijuana-dependent individuals. *Exp Clin Psychopharmacol.* 2010;18(1):99–107. https://doi.org/10.1037/a0018333.
81. Patel H, Naish KR, Amlung M. Discounting of delayed monetary and cannabis rewards in a crowdsourced sample of adults. *Exp Clin Psychopharmacol.* 2020;28(4):462–470. https://doi.org/10.1037/pha0000327.
82. Moody L, Franck C, Hatz L, Bickel WK. Impulsivity and polysubstance use: a systematic comparison of delay discounting in mono-, dual-, and trisubstance use. *Exp Clin Psychopharmacol.* 2016;24(1):30–37. https://doi.org/10.1037/pha0000059.
83. Moallem NR, Ray LA. Dimensions of impulsivity among heavy drinkers, smokers, and heavy drinking smokers: singular and combined effects. *Addict Behav.* 2012;37(7):871–874. https://doi.org/10.1016/j.addbeh.2012.03.002.
84. Taylor EM, Murphy A, Boyapati V, et al, ICCAM Platform. Impulsivity in abstinent alcohol and polydrug dependence: a multidimensional approach. *Psychopharmacology.* 2016;233(8):1487–1499. https://doi.org/10.1007/s00213-016-4245-6.
85. Mendez IA, Simon NW, Hart N, et al. Self-administered cocaine causes long-lasting increases in impulsive choice in a delay discounting task. *Behav Neurosci.* 2010;124(4):470–477. https://doi.org/10.1037/a0020458.
86. Gipson CD, Bardo MT. Extended access to amphetamine self-administration increases impulsive choice in a delay discounting task in rats. *Psychopharmacology.* 2009;207(3):391–400. https://doi.org/10.1007/s00213-009-1667-4.
87. Hyatt WS, Berquist MD, Chitre NM, et al. Repeated administration of synthetic cathinone 3,4-methylenedioxypyrovalerone persistently increases impulsive choice in rats. *Behav Pharmacol.* 2019;30(7):555–565. https://doi.org/10.1097/FBP.0000000000000492.
88. Schippers MC, Binnekade R, Schoffelmeer AN, Pattij T, De Vries TJ. Unidirectional relationship between heroin self-administration and impulsive decision-making in rats. *Psychopharmacology.* 2012;219(2):443–452. https://doi.org/10.1007/s00213-011-2444-8.
89. Dandy KL, Gatch MB. The effects of chronic cocaine exposure on impulsivity in rats. *Behav Pharmacol.* 2009;20(5–6):400–405. https://doi.org/10.1097/FBP.0b013e328330ad89.
90. Johnson KR, Boomhower SR, Newland MC. Behavioral effects of chronic WIN 55,212-2 administration during adolescence and adulthood in mice. *Exp Clin Psychopharmacol.* 2019;27(4):348–358. https://doi.org/10.1037/pha0000271.

91. Wilhelm CJ, Mitchell SH. Rats bred for high alcohol drinking are more sensitive to delayed and probabilistic outcomes. *Gene Brain Behav.* 2008;7(7):705–713. https://doi.org/10.1111/j.1601-183X.2008.00406.x.
92. Mejia-Toiber J, Boutros N, Markou A, Semenova S. Impulsive choice and anxiety-like behavior in adult rats exposed to chronic intermittent ethanol during adolescence and adulthood. *Behav Brain Res.* 2014;266:19–28. https://doi.org/10.1016/j.bbr.2014.02.019.
93. Mitchell MR, Mendez IA, Vokes CM, Damborsky JC, Winzer-Serhan UH, Setlow B. Effects of developmental nicotine exposure in rats on decision-making in adulthood. *Behav Pharmacol.* 2012;23(1):34–42. https://doi.org/10.1097/FBP.0b013e32834eb04a.
94. Harty SC, Whaley JE, Halperin JM, Ranaldi R. Impulsive choice, as measured in a delay discounting paradigm, remains stable after chronic heroin administration. *Pharmacol Biochem Behav.* 2011;98(3):337–340. https://doi.org/10.1016/j.pbb.2011.02.004.
95. Kayir H, Semenova S, Markou A. Baseline impulsive choice predicts the effects of nicotine and nicotine withdrawal on impulsivity in rats. *Prog Neuro Psychopharmacol Biol Psychiatr.* 2014;48:6–13. https://doi.org/10.1016/j.pnpbp.2013.09.007.
96. Platt ML, Huettel SA. Risky business: the neuroeconomics of decision making under uncertainty. *Nat Neurosci.* 2008;11(4):398–403. https://doi.org/10.1038/nn2062.
97. Lejuez CW, Read JP, Kahler CW, et al. Evaluation of a behavioral measure of risk taking: the balloon analogue risk task (BART). *J Exp Psychol Appl.* 2002;8(2):75–84. https://doi.org/10.1037//1076-898x.8.2.75.
98. St Onge JR, Floresco SB. Dopaminergic modulation of risk-based decision making. *Neuropsychopharmacology.* 2009;34(3):681–697. https://doi.org/10.1038/npp.2008.121.
99. Zeeb FD, Wong AC, Winstanley CA. Differential effects of environmental enrichment, social-housing, and isolation-rearing on a rat gambling task: dissociations between impulsive action and risky decision-making. *Psychopharmacology.* 2013;225(2):381–395. https://doi.org/10.1007/s00213-012-2822-x.
100. Simon NW, Gilbert RJ, Mayse JD, Bizon JL, Setlow B. Balancing risk and reward: a rat model of risky decision making. *Neuropsychopharmacology.* 2009;34(10):2208–2217. https://doi.org/10.1038/npp.2009.48.
101. Silber BY, Papafotiou K, Croft RJ, Ogden E, Swann P, Stough C. The effects of dexamphetamine on simulated driving performance. *Psychopharmacology.* 2005;179(3):536–543. https://doi.org/10.1007/s00213-004-2061-x.
102. Silber BY, Croft RJ, Downey LA, et al. The effect of d,l-methamphetamine on simulated driving performance. *Psychopharmacology.* 2012;219(4):1081–1087. https://doi.org/10.1007/s00213-011-2437-7.
103. Brookhuis KA, de Waard D, Samyn N. Effects of MDMA (ecstasy), and multiple drugs use on (simulated) driving performance and traffic safety. *Psychopharmacology.* 2004;173(3–4):440–445. https://doi.org/10.1007/s00213-003-1714-5.
104. Van Dyke NA, Fillmore MT. Laboratory analysis of risky driving at 0.05% and 0.08% blood alcohol concentration. *Drug Alcohol Depend.* 2017;175:127–132. https://doi.org/10.1016/j.drugalcdep.2017.02.005.
105. Bernosky-Smith KA, Aston ER, Liguori A. Rapid drinking is associated with increases in driving-related risk-taking. *Hum Psychopharmacol.* 2012;27(6):622–625. https://doi.org/10.1002/hup.2260.
106. Townshend JM, Kambouropoulos N, Griffin A, Hunt FJ, Milani RM. Binge drinking, reflection impulsivity, and unplanned sexual behavior: impaired decision-making in young social drinkers. *Alcohol Clin Exp Res.* 2014;38(4):1143–1150. https://doi.org/10.1111/acer.12333.
107. Joseph Davey D, Kilembe W, Wall KM, et al. Risky sex and HIV acquisition among HIV serodiscordant couples in Zambia, 2002–2012: what does alcohol have to do with it? *AIDS Behav.* 2017;21(7):1892–1903. https://doi.org/10.1007/s10461-017-1733-6.
108. Lane SD, Cherek DR, Tcheremissine OV, Lieving LM, Pietras CJ. Acute marijuana effects on human risk taking. *Neuropsychopharmacology.* 2005;30(4):800–809. https://doi.org/10.1038/sj.npp.1300620.
109. Bergeron J, Paquette M. Relationships between frequency of driving under the influence of cannabis, self-reported reckless driving and risk-taking behavior observed in a driving simulator. *J Saf Res.* 2014;49:19–24. https://doi.org/10.1016/j.jsr.2014.02.002.
110. Ortiz-Peregrina S, Ortiz C, Anera RG. Aggressive driving behaviours in cannabis users. The influence of consumer characteristics. *Int J Environ Res Publ Health.* 2021;18(8):3911. https://doi.org/10.3390/ijerph18083911.
111. Lane SD, Cherek DR, Nouvion SO. Modulation of human risky decision making by flunitrazepam. *Psychopharmacology.* 2008;196(2):177–188. https://doi.org/10.1007/s00213-007-0951-4.
112. Meskali M, Berthelon C, Marie S, Denise P, Bocca ML. Residual effects of hypnotic drugs in aging drivers submitted to simulated accident scenarios: an exploratory study. *Psychopharmacology.* 2009;207(3):461–467. https://doi.org/10.1007/s00213-009-1677-2.

113. Mitchell MR, Vokes CM, Blankenship AL, Simon NW, Setlow B. Effects of acute administration of nicotine, amphetamine, diazepam, morphine, and ethanol on risky decision-making in rats. *Psychopharmacology.* 2011;218(4):703—712. https://doi.org/10.1007/s00213-011-2363-8.
114. Kaminski BJ, Ator NA. Behavioral and pharmacological variables affecting risky choice in rats. *J Exp Anal Behav.* 2001;75(3):275—297. https://doi.org/10.1901/jeab.2001.75-275.
115. Yan W-S, Chen R-T, Liu M-M, Zheng D-H. Monetary reward discounting, inhibitory control, and trait impulsivity in young adults with internet gaming disorder and nicotine dependence. *Front Psychiatr.* 2021;12:628933. https://doi.org/10.3389/fpsyt.2021.628933.
116. Reynolds B, Karraker K, Horn K, Richards JB. Delay and probability discounting as related to different stages of adolescent smoking and non-smoking. *Behav Process.* 2003;64(3):333—344. https://doi.org/10.1016/s0376-6357(03)00168-2.
117. Yi R, Landes RD. Temporal and probability discounting by cigarette smokers following acute smoking abstinence. *Nicotine Tob Res.* 2012;14(5):547—558. https://doi.org/10.1093/ntr/ntr252.
118. Yi R, Chase WD, Bickel WK. Probability discounting among cigarette smokers and nonsmokers: molecular analysis discerns group differences. *Behav Pharmacol.* 2007;18(7):633—639. https://doi.org/10.1097/FBP.0b013e3282effbd3.
119. Campbell JA, Samartgis JR, Crowe SF. Impaired decision making on the Balloon Analogue Risk Task as a result of long-term alcohol use. *J Clin Exp Neuropsychol.* 2013;35(10):1071—1081. https://doi.org/10.1080/13803395.2013.856382.
120. Dean AC, Sugar CA, Hellemann G, London ED. Is all risk bad? Young adult cigarette smokers fail to take adaptive risk in a laboratory decision-making test. *Psychopharmacology.* 2011;215(4):801—811. https://doi.org/10.1007/s00213-011-2182-y.
121. Canavan SV, Forselius EL, Bessette AJ, Morgan PT. Preliminary evidence for normalization of risk taking by modafinil in chronic cocaine users. *Addict Behav.* 2014;39(6):1057—1061. https://doi.org/10.1016/j.addbeh.2014.02.015.
122. Leland DS, Paulus MP. Increased risk-taking decision-making but not altered response to punishment in stimulant-using young adults. *Drug Alcohol Depend.* 2005;78(1):83—90. https://doi.org/10.1016/j.drugalcdep.2004.10.001.
123. Wittwer A, Hulka LM, Heinimann HR, Vonmoos M, Quednow BB. Risky decisions in a lottery task are associated with an increase of cocaine use. *Front Psychol.* 2016;7:640. https://doi.org/10.3389/fpsyg.2016.00640.
124. Kluwe-Schiavon B, Kexel A, Manenti G, et al. Sensitivity to gains during risky decision-making differentiates chronic cocaine users from stimulant-naïve controls. *Behav Brain Res.* 2020;379:112386. https://doi.org/10.1016/j.bbr.2019.112386.
125. Mejía-Cruz D, Green L, Myerson J, Morales-Chainé S, Nieto J. Delay and probability discounting by drug-dependent cocaine and marijuana users. *Psychopharmacology.* 2016;233(14):2705—2714. https://doi.org/10.1007/s00213-016-4316-8.
126. Bosanquet D, Macdougall HG, Rogers SJ, et al. Driving on ice: impaired driving skills in current methamphetamine users. *Psychopharmacology.* 2013;225(1):161—172. https://doi.org/10.1007/s00213-012-2805-y.
127. Dastrup E, Lees MN, Bechara A, Dawson JD, Rizzo M. Risky car following in abstinent users of MDMA. *Accid Anal Prev.* 2010;42(3):867—873. https://doi.org/10.1016/j.aap.2009.04.015.
128. Reynolds B, Patak M, Shroff P. Adolescent smokers rate delayed rewards as less certain than adolescent non-smokers. *Drug Alcohol Depend.* 2007;90(2—3):301—303. https://doi.org/10.1016/j.drugalcdep.2007.04.008.
129. Brand M, Roth-Bauer M, Driessen M, Markowitsch HJ. Executive functions and risky decision-making in patients with opiate dependence. *Drug Alcohol Depend.* 2008;97(1—2):64—72. https://doi.org/10.1016/j.drugalcdep.2008.03.017.
130. Garami J, Moustafa AA. Probability discounting of monetary gains and losses in opioid-dependent adults. *Behav Brain Res.* 2019;364:334—339. https://doi.org/10.1016/j.bbr.2019.02.017.
131. Dolan SB, Johnson MW, Dunn KE, Huhn AS. The discounting of death: probability discounting of heroin use by fatal overdose likelihood and drug purity. *Exp Clin Psychopharmacol.* 2021;29(3):219—228. https://doi.org/10.1037/pha0000486.
132. Hanson KL, Thayer RE, Tapert SF. Adolescent marijuana users have elevated risk-taking on the balloon analog risk task. *J Psychopharmacol.* 2014;28(11):1080—1087. https://doi.org/10.1177/0269881114550352.

133. Raymond DR, Paneto A, Yoder KK, et al. Does chronic cannabis use impact risky decision-making: an examination of fMRI activation and effective connectivity? *Front Psychiatr.* 2020;11:599256. https://doi.org/10.3389/fpsyt.2020.599256.
134. Fischer BA, McMahon RP, Kelly DL, et al. Risk-taking in schizophrenia and controls with and without cannabis dependence. *Schizophr Res.* 2015;161(2−3):471−477. https://doi.org/10.1016/j.schres.2014.11.009.
135. Gunn RL, Skalski L, Metrik J. Expectancy of impairment attenuates marijuana-induced risk taking. *Drug Alcohol Depend.* 2017;178:39−42. https://doi.org/10.1016/j.drugalcdep.2017.04.027.
136. Gilman JM, Calderon V, Curran MT, Evins AE. Young adult cannabis users report greater propensity for risk-taking only in non-monetary domains. *Drug Alcohol Depend.* 2015;147:26−31. https://doi.org/10.1016/j.drugalcdep.2014.12.020.
137. McMurray MS, Amodeo LR, Roitman JD. Consequences of adolescent ethanol consumption on risk preference and orbitofrontal cortex encoding of reward. *Neuropsychopharmacology.* 2016;41(5):1366−1375. https://doi.org/10.1038/npp.2015.288.
138. Richardson CG, Edalati H. Application of a brief measure of delay discounting to examine the relationship between delay discounting and the initiation of substance use among adolescents. *Subst Use Misuse.* 2016;51(4):540−544. https://doi.org/10.3109/10826084.2015.1126740.
139. Wulfert E, Block JA, Santa Ana E, Rodriguez ML, Colsman M. Delay of gratification: impulsive choices and problem behaviors in early and late adolescence. *J Pers.* 2002;70(4):533−552. https://doi.org/10.1111/1467-6494.05013.
140. Heinz AJ, Peters EN, Boden MT, Bonn-Miller MO. A comprehensive examination of delay discounting in a clinical sample of cannabis-dependent military veterans making a self-guided quit attempt. *Exp Clin Psychopharmacol.* 2013;21(1):55−65. https://doi.org/10.1037/a0031192.
141. Nejtek VA, Kaiser KA, Zhang B, Djokovic M. Iowa Gambling Task scores predict future drug use in bipolar disorder outpatients with stimulant dependence. *Psychiatr Res.* 2013;210(3):871−879. https://doi.org/10.1016/j.psychres.2013.08.021.
142. Ernst M, Luckenbaugh DA, Moolchan ET, et al. Decision-making and facial emotion recognition as predictors of substance-use initiation among adolescents. *Addict Behav.* 2010;35(3):286−289. https://doi.org/10.1016/j.addbeh.2009.10.014.
143. Fernie G, Cole JC, Goudie AJ, Field M. Risk-taking but not response inhibition or delay discounting predict alcohol consumption in social drinkers. *Drug Alcohol Depend.* 2010;112(1−2):54−61. https://doi.org/10.1016/j.drugalcdep.2010.05.011.
144. Westman JG, Bujarski S, Ray LA. Impulsivity moderates subjective responses to alcohol in alcohol-dependent individuals. *Alcohol Alcohol.* 2017;52(2):249−255. https://doi.org/10.1093/alcalc/agw096.
145. Dallery J, Raiff BR. Delay discounting predicts cigarette smoking in a laboratory model of abstinence reinforcement. *Psychopharmacology.* 2007;190(4):485−496. https://doi.org/10.1007/s00213-006-0627-5.
146. Muench C, Juliano LM. Predictors of smoking lapse during a 48-hour laboratory analogue smoking cessation attempt. *Psychol Addict Behav.* 2017;31(4):415−422. https://doi.org/10.1037/adb0000246.
147. Lopez-Vergara HI, Jackson KM, Meshesha LZ, Metrik J. Dysregulation as a correlate of cannabis use and problem use. *Addict Behav.* 2019;95:138−144. https://doi.org/10.1016/j.addbeh.2019.03.010.
148. Peters EN, Petry NM, Lapaglia DM, Reynolds B, Carroll KM. Delay discounting in adults receiving treatment for marijuana dependence. *Exp Clin Psychopharmacol.* 2013;21(1):46−54. https://doi.org/10.1037/a0030943.
149. Stanger C, Ryan SR, Fu H, et al. Delay discounting predicts adolescent substance abuse treatment outcome. *Exp Clin Psychopharmacol.* 2012;20(3):205−212. https://doi.org/10.1037/a0026543.
150. Sheffer CE, Christensen DR, Landes R, Carter LP, Jackson L, Bickel WK. Delay discounting rates: a strong prognostic indicator of smoking relapse. *Addict Behav.* 2014;39(11):1682−1689. https://doi.org/10.1016/j.addbeh.2014.04.019.
151. Grosskopf CM, Kroemer NB, Pooseh S, Böhme F, Smolka MN. Temporal discounting and smoking cessation: choice consistency predicts nicotine abstinence in treatment-seeking smokers. *Psychopharmacology.* 2021;238(2):399−410. https://doi.org/10.1007/s00213-020-05688-5.
152. MacKillop J, Kahler CW. Delayed reward discounting predicts treatment response for heavy drinkers receiving smoking cessation treatment. *Drug Alcohol Depend.* 2009;104(3):197−203. https://doi.org/10.1016/j.drugalcdep.2009.04.020.
153. Athamneh LN, Stein JS, Bickel WK. Will delay discounting predict intention to quit smoking? *Exp Clin Psychopharmacol.* 2017;25(4):273−280. https://doi.org/10.1037/pha0000129.

154. Turner JK, Athamneh LN, Basso JC, Bickel WK. The phenotype of recovery V: does delay discounting predict the perceived risk of relapse among individuals in recovery from alcohol and drug use disorders. *Alcohol Clin Exp Res*. 2021;45(5):1100−1108. https://doi.org/10.1111/acer.14600.
155. Miglin R, Kable JW, Bowers ME, Ashare RL. Withdrawal-related changes in delay discounting predict short-term smoking abstinence. *Nicotine Tob Res*. 2017;19(6):694−702. https://doi.org/10.1093/ntr/ntw246.
156. Aklin WM, Severtson SG, Umbricht A, et al. Risk-taking propensity as a predictor of induction onto naltrexone treatment for opioid dependence. *J Clin Psychiatr*. 2012;73(8):e1056−e1061. https://doi.org/10.4088/JCP.09m05807.
157. Chen YC, Chen CK, Wang LJ. Predictors of relapse and dropout during a 12-week relapse prevention program for methamphetamine users. *J Psychoact Drugs*. 2015;47(4):317−324. https://doi.org/10.1080/02791072.2015.1071447.
158. Washio Y, Higgins ST, Heil SH, et al. Delay discounting is associated with treatment response among cocaine-dependent outpatients. *Exp Clin Psychopharmacol*. 2011;19(3):243−248. https://doi.org/10.1037/a0023617.
159. Harvanko AM, Strickland JC, Slone SA, Shelton BJ, Reynolds BA. Dimensions of impulsive behavior: predicting contingency management treatment outcomes for adolescent smokers. *Addict Behav*. 2019;90:334−340. https://doi.org/10.1016/j.addbeh.2018.11.031.
160. Barreno EM, Domínguez-Salas S, Díaz-Batanero C, Lozano ÓM, Marín J, Verdejo-García A. Specific aspects of cognitive impulsivity are longitudinally associated with lower treatment retention and greater relapse in therapeutic community treatment. *J Subst Abuse Treat*. 2019;96:33−38. https://doi.org/10.1016/j.jsat.2018.10.004.
161. Verdejo-García A, Albein-Urios N, Martinez-Gonzalez JM, Civit E, de la Torre R, Lozano O. Decision-making impairment predicts 3-month hair-indexed cocaine relapse. *Psychopharmacology*. 2014;231(21):4179−4187. https://doi.org/10.1007/s00213-014-3563-9.
162. Rubenis AJ, Fitzpatrick RE, Lubman DI, Verdejo-Garcia A. Impulsivity predicts poorer improvement in quality of life during early treatment for people with methamphetamine dependence. *Addiction*. 2018;113(4):668−676. https://doi.org/10.1111/add.14058.
163. Perry JL, Nelson SE, Carroll ME. Impulsive choice as a predictor of acquisition of IV cocaine self- administration and reinstatement of cocaine-seeking behavior in male and female rats. *Exp Clin Psychopharmacol*. 2008;16(2):165−177. https://doi.org/10.1037/1064-1297.16.2.165.
164. Marusich JA, Bardo MT. Differences in impulsivity on a delay-discounting task predict self-administration of a low unit dose of methylphenidate in rats. *Behav Pharmacol*. 2009;20(5−6):447−454. https://doi.org/10.1097/FBP.0b013e328330ad6d.
165. Anker JJ, Perry JL, Gliddon LA, Carroll ME. Impulsivity predicts the escalation of cocaine self-administration in rats. *Pharmacol Biochem Behav*. 2009;93(3):343−348. https://doi.org/10.1016/j.pbb.2009.05.013.
166. Poulos CX, Le AD, Parker JL. Impulsivity predicts individual susceptibility to high levels of alcohol self-administration. *Behav Pharmacol*. 1995;6(8):810−814.
167. Yates JR, Marusich JA, Gipson CD, Beckmann JS, Bardo MT. High impulsivity in rats predicts amphetamine conditioned place preference. *Pharmacol Biochem Behav*. 2012;100(3):370−376. https://doi.org/10.1016/j.pbb.2011.07.012.
168. Diergaarde L, van Mourik Y, Pattij T, Schoffelmeer AN, De Vries TJ. Poor impulse control predicts inelastic demand for nicotine but not alcohol in rats. *Addiction Biol*. 2012;17(3):576−587. https://doi.org/10.1111/j.1369-1600.2011.00376.x.
169. Koffarnus MN, Woods JH. Individual differences in discount rate are associated with demand for self-administered cocaine, but not sucrose. *Addiction Biol*. 2013;18(1):8−18. https://doi.org/10.1111/j.1369-1600.2011.00361.x.
170. Broos N, Diergaarde L, Schoffelmeer AN, Pattij T, De Vries TJ. Trait impulsive choice predicts resistance to extinction and propensity to relapse to cocaine seeking: a bidirectional investigation. *Neuropsychopharmacology*. 2012;37(6):1377−1386. https://doi.org/10.1038/npp.2011.323.
171. Diergaarde L, Pattij T, Poortvliet I, et al. Impulsive choice and impulsive action predict vulnerability to distinct stages of nicotine seeking in rats. *Biol Psychiatr*. 2008;63(3):301−308. https://doi.org/10.1016/j.biopsych.2007.07.011.
172. Mitchell SH, Reeves JM, Li N, Phillips TJ. Delay discounting predicts behavioral sensitization to ethanol in outbred WSC mice. *Alcohol Clin Exp Res*. 2006;30(3):429−437. https://doi.org/10.1111/j.1530-0277.2006.00047.x.

173. Mitchell MR, Weiss VG, Beas BS, Morgan D, Bizon JL, Setlow B. Adolescent risk taking, cocaine self-administration, and striatal dopamine signaling. *Neuropsychopharmacology*. 2014;39(4):955—962. https://doi.org/10.1038/npp.2013.295.
174. Orsini CA, Blaes SL, Dragone RJ, et al. Distinct relationships between risky decision making and cocaine self-administration under short- and long-access conditions. *Prog Neuro Psychopharmacol Biol Psychiatr*. 2020;98:109791. https://doi.org/10.1016/j.pnpbp.2019.109791.
175. Horchar MJ, Kappesser JL, Broderick MR, Wright MR, Yates JR. Effects of NMDA receptor antagonists on behavioral economic indices of cocaine self-administration. *Drug Alcohol Depend*. 2022;233:109348. https://doi.org/10.1016/j.drugalcdep.2022.109348.
176. Yates JR, Horchar MJ, Kappesser JL, Broderick MR, Ellis AL, Wright MR. The association between risky decision making and cocaine conditioned place preference is moderated by sex. *Drug Alcohol Depend*. 2021;228:109079. https://doi.org/10.1016/j.drugalcdep.2021.109079.
177. Takahashi K, Toyoshima M, Ichitani Y, Yamada K. Enhanced methamphetamine-induced conditioned place preference in risk-taking rats. *Behav Brain Res*. 2020;378:112299. https://doi.org/10.1016/j.bbr.2019.112299.
178. Gabriel D, Freels TG, Setlow B, Simon NW. Risky decision-making is associated with impulsive action and sensitivity to first-time nicotine exposure. *Behav Brain Res*. 2019;359:579—588. https://doi.org/10.1016/j.bbr.2018.10.008.
179. Bolla KI, Eldreth DA, London ED, et al. Orbitofrontal cortex dysfunction in abstinent cocaine abusers performing a decision-making task. *Neuroimage*. 2003;19(3):1085—1094. https://doi.org/10.1016/s1053-8119(03)00113-7.
180. Zeeb FD, Baarendse PJ, Vanderschuren LJ, Winstanley CA. Inactivation of the prelimbic or infralimbic cortex impairs decision-making in the rat gambling task. *Psychopharmacology*. 2015;232(24):4481—4491. https://doi.org/10.1007/s00213-015-4075-y.
181. Amlung M, Sweet LH, Acker J, Brown CL, MacKillop J. Dissociable brain signatures of choice conflict and immediate reward preferences in alcohol use disorders. *Addiction Biol*. 2014;19(4):743—753. https://doi.org/10.1111/adb.12017.
182. Claus ED, Kiehl KA, Hutchison KE. Neural and behavioral mechanisms of impulsive choice in alcohol use disorder. *Alcohol Clin Exp Res*. 2011;35(7):1209—1219. https://doi.org/10.1111/j.1530-0277.2011.01455.x.
183. Cho SS, Koshimori Y, Aminian K, et al. Investing in the future: stimulation of the medial prefrontal cortex reduces discounting of delayed rewards. *Neuropsychopharmacology*. 2015;40(3):546—553. https://doi.org/10.1038/npp.2014.211.
184. Matthews SC, Simmons AN, Lane SD, Paulus MP. Selective activation of the nucleus accumbens during risk-taking decision making. *Neuroreport*. 2004;15(13):2123—2127. https://doi.org/10.1097/00001756-200409150-00025.
185. Churchwell JC, Morris AM, Heurtelou NM, Kesner RP. Interactions between the prefrontal cortex and amygdala during delay discounting and reversal. *Behav Neurosci*. 2009;123(6):1185—1196. https://doi.org/10.1037/a0017734.
186. Mobini S, Body S, Ho MY, et al. Effects of lesions of the orbitofrontal cortex on sensitivity to delayed and probabilistic reinforcement. *Psychopharmacology*. 2002;160(3):290—298. https://doi.org/10.1007/s00213-001-0983-0.
187. Abela AR, Chudasama Y. Dissociable contributions of the ventral hippocampus and orbitofrontal cortex to decision-making with a delayed or uncertain outcome. *Eur J Neurosci*. 2013;37(4):640—647. https://doi.org/10.1111/ejn.12071.
188. Feja M, Koch M. Ventral medial prefrontal cortex inactivation impairs impulse control but does not affect delay-discounting in rats. *Behav Brain Res*. 2014;264:230—239. https://doi.org/10.1016/j.bbr.2014.02.013.
189. Winstanley CA, Theobald DE, Cardinal RN, Robbins TW. Contrasting roles of basolateral amygdala and orbitofrontal cortex in impulsive choice. *J Neurosci*. 2004;24(20):4718—4722. https://doi.org/10.1523/JNEUROSCI.5606-03.2004.
190. Mar AC, Walker AL, Theobald DE, Eagle DM, Robbins TW. Dissociable effects of lesions to orbitofrontal cortex subregions on impulsive choice in the rat. *J Neurosci*. 2011;31(17):6398—6404. https://doi.org/10.1523/JNEUROSCI.6620-10.2011.
191. Orsini CA, Trotta RT, Bizon JL, Setlow B. Dissociable roles for the basolateral amygdala and orbitofrontal cortex in decision-making under risk of punishment. *J Neurosci*. 2015;35(4):1368—1379. https://doi.org/10.1523/JNEUROSCI.3586-14.2015.
192. Stopper CM, Green EB, Floresco SB. Selective involvement by the medial orbitofrontal cortex in biasing risky, but not impulsive, choice. *Cerebr Cortex*. 2014;24(1):154—162. https://doi.org/10.1093/cercor/bhs297.

193. Yates JR, Perry JL, Meyer AC, Gipson CD, Charnigo R, Bardo MT. Role of medial prefrontal and orbitofrontal monoamine transporters and receptors in performance in an adjusting delay discounting procedure. *Brain Res.* 2014;1574:26–36. https://doi.org/10.1016/j.brainres.2014.06.004.
194. Sonntag KC, Brenhouse HC, Freund N, Thompson BS, Puhl M, Andersen SL. Viral over-expression of D1 dopamine receptors in the prefrontal cortex increase high-risk behaviors in adults: comparison with adolescents. *Psychopharmacology.* 2014;231(8):1615–1626. https://doi.org/10.1007/s00213-013-3399-8.
195. Loos M, Pattij T, Janssen MC, et al. Dopamine receptor D1/D5 gene expression in the medial prefrontal cortex predicts impulsive choice in rats. *Cerebr Cortex.* 2010;20(5):1064–1070. https://doi.org/10.1093/cercor/bhp167.
196. Durazzo TC, Meyerhoff DJ, Mon A, Abé C, Gazdzinski S, Murray DE. Chronic cigarette smoking in healthy middle-aged individuals is associated with decreased regional brain N-acetylaspartate and glutamate levels. *Biol Psychiatr.* 2016;79(6):481–488. https://doi.org/10.1016/j.biopsych.2015.03.029.
197. Valencia-Torres L, Olarte-Sánchez CM, da Costa Araújo S, Body S, Bradshaw CM, Szabadi E. Nucleus accumbens and delay discounting in rats: evidence from a new quantitative protocol for analysing inter-temporal choice. *Psychopharmacology.* 2012;219(2):271–283. https://doi.org/10.1007/s00213-011-2459-1.
198. Steele CC, Peterson JR, Marshall AT, Stuebing SL, Kirkpatrick K. Nucleus accumbens core lesions induce suboptimal choice and reduce sensitivity to magnitude and delay in impulsive choice tasks. *Behav Brain Res.* 2018;339:28–38. https://doi.org/10.1016/j.bbr.2017.11.013.
199. Bezzina G, Body S, Cheung TH, et al. Effect of disconnecting the orbital prefrontal cortex from the nucleus accumbens core on inter-temporal choice behaviour: a quantitative analysis. *Behav Brain Res.* 2008;191(2):272–279. https://doi.org/10.1016/j.bbr.2008.03.041.
200. Barlow RL, Gorges M, Wearn A, et al. Ventral striatal D2/3 receptor availability is associated with impulsive choice behavior as well as limbic corticostriatal connectivity. *Int J Neuropsychopharmacol.* 2018;21(7):705–715. https://doi.org/10.1093/ijnp/pyy030.
201. Jupp B, Caprioli D, Saigal N, et al. Dopaminergic and GABA-ergic markers of impulsivity in rats: evidence for anatomical localisation in ventral striatum and prefrontal cortex. *Eur J Neurosci.* 2013;37(9):1519–1528. https://doi.org/10.1111/ejn.12146.
202. Soares-Cunha C, Coimbra B, David-Pereira A, et al. Activation of D2 dopamine receptor-expressing neurons in the nucleus accumbens increases motivation. *Nat Commun.* 2016;7:11829. https://doi.org/10.1038/ncomms11829.
203. Gowin JL, May AC, Wittmann M, Tapert SF, Paulus MP. Doubling down: increased risk-taking behavior following a loss by individuals with cocaine use disorder is associated with striatal and anterior cingulate dysfunction. *Biol Psychiatr Cogn Neurosci & Neuroimag.* 2017;2(1):94–103. https://doi.org/10.1016/j.bpsc.2016.02.002.
204. Morales AM, Jones SA, Ehlers A, Lavine JB, Nagel BJ. Ventral striatal response during decision making involving risk and reward is associated with future binge drinking in adolescents. *Neuropsychopharmacology.* 2018;43(9):1884–1890. https://doi.org/10.1038/s41386-018-0087-8.
205. St Onge JR, Ahn S, Phillips AG, Floresco SB. Dynamic fluctuations in dopamine efflux in the prefrontal cortex and nucleus accumbens during risk-based decision making. *J Neurosci.* 2012;32(47):16880–16891. https://doi.org/10.1523/JNEUROSCI.3807-12.2012.
206. Freels TG, Gabriel D, Lester DB, Simon NW. Risky decision-making predicts dopamine release dynamics in nucleus accumbens shell. *Neuropsychopharmacology.* 2020;45(2):266–275. https://doi.org/10.1038/s41386-019-0527-0.
207. Stopper CM, Khayambashi S, Floresco SB. Receptor-specific modulation of risk-based decision making by nucleus accumbens dopamine. *Neuropsychopharmacology.* 2013;38(5):715–728. https://doi.org/10.1038/npp.2012.240.
208. Zalocusky KA, Ramakrishnan C, Lerner TN, Davidson TJ, Knutson B, Deisseroth K. Nucleus accumbens D2R cells signal prior outcomes and control risky decision-making. *Nature.* 2016;531(7596):642–646. https://doi.org/10.1038/nature17400.
209. Jenni NL, Larkin JD, Floresco SB. Prefrontal dopamine D_1 and D_2 receptors regulate dissociable aspects of decision making via distinct ventral striatal and amygdalar circuits. *J Neurosci.* 2017;37(26):6200–6213. https://doi.org/10.1523/JNEUROSCI.0030-17.2017.
210. Hynes TJ, Ferland JM, Feng TL, et al. Chemogenetic inhibition of dopaminergic projections to the nucleus accumbens has sexually dimorphic effects in the rat gambling task. *Behav Neurosci.* 2020;134(4):309–322. https://doi.org/10.1037/bne0000372.

211. Bechara A, Damasio H, Damasio AR, Lee GP. Different contributions of the human amygdala and ventromedial prefrontal cortex to decision-making. *J Neurosci*. 1999;19(13):5473−5481. https://doi.org/10.1523/JNEUROSCI.19-13-05473.1999.
212. Zeeb FD, Winstanley CA. Lesions of the basolateral amygdala and orbitofrontal cortex differentially affect acquisition and performance of a rodent gambling task. *J Neurosci*. 2011;31(6):2197−2204. https://doi.org/10.1523/JNEUROSCI.5597-10.2011.
213. van Holstein M, MacLeod PE, Floresco SB. Basolateral amygdala - nucleus accumbens circuitry regulates optimal cue-guided risk/reward decision making. *Prog Neuro Psychopharmacol Biol Psychiatr*. 2020;98:109830. https://doi.org/10.1016/j.pnpbp.2019.109830.
214. St Onge JR, Stopper CM, Zahm DS, Floresco SB. Separate prefrontal-subcortical circuits mediate different components of risk-based decision making. *J Neurosci*. 2012;32(8):2886−2899. https://doi.org/10.1523/JNEUROSCI.5625-11.2012.
215. Orsini CA, Hernandez CM, Singhal S, et al. Optogenetic inhibition reveals distinct roles for basolateral amygdala activity at discrete time points during risky decision making. *J Neurosci*. 2017;37(48):11537−11548. https://doi.org/10.1523/JNEUROSCI.2344-17.2017.
216. Krawczyk DC. Contributions of the prefrontal cortex to the neural basis of human decision making. *Neurosci Biobehav Rev*. 2002;26(6):631−664. https://doi.org/10.1016/s0149-7634(02)00021-0.
217. Christopoulos GI, Tobler PN, Bossaerts P, Dolan RJ, Schultz W. Neural correlates of value, risk, and risk aversion contributing to decision making under risk. *J Neurosci*. 2009;29(40):12574−12583. https://doi.org/10.1523/JNEUROSCI.2614-09.2009.
218. Vorobyev V, Kwon MS, Moe D, Parkkola R, Hämäläinen H. Risk-taking behavior in a computerized driving task: brain activation correlates of decision-making, outcome, and peer influence in male adolescents. *PLoS One*. 2015;10(6):e0129516. https://doi.org/10.1371/journal.pone.0129516.
219. Hewig J, Straube T, Trippe RH, et al. Decision-making under risk: an fMRI study. *J Cognit Neurosci*. 2009;21(8):1642−1652. https://doi.org/10.1162/jocn.2009.21112.
220. Wang M, Chen Z, Zhang S, et al. High self-control reduces risk preference: the role of connectivity between right orbitofrontal cortex and right anterior cingulate cortex. *Front Neurosci*. 2019;13:194. https://doi.org/10.3389/fnins.2019.00194.
221. Gowin JL, Stewart JL, May AC, et al. Altered cingulate and insular cortex activation during risk-taking in methamphetamine dependence: losses lose impact. *Addiction*. 2014;109(2):237−247. https://doi.org/10.1111/add.12354.
222. Fishbein DH, Eldreth DL, Hyde C, et al. Risky decision making and the anterior cingulate cortex in abstinent drug abusers and nonusers. *Brain Res Cogn Brain Res*. 2005;23(1):119−136. https://doi.org/10.1016/j.cogbrainres.2004.12.010.
223. McClernon FJ, Kozink RV, Lutz AM, Rose JE. 24-h smoking abstinence potentiates fMRI-BOLD activation to smoking cues in cerebral cortex and dorsal striatum. *Psychopharmacology*. 2009;204(1):25−35. https://doi.org/10.1007/s00213-008-1436-9.
224. Contreras-Rodríguez O, Albein-Urios N, Perales JC, et al. Cocaine-specific neuroplasticity in the ventral striatum network is linked to delay discounting and drug relapse. *Addiction*. 2015;110(12):1953−1962. https://doi.org/10.1111/add.13076.
225. Schmaal L, Goudriaan AE, van der Meer J, van den Brink W, Veltman DJ. The association between cingulate cortex glutamate concentration and delay discounting is mediated by resting state functional connectivity. *Brain and Behav*. 2012;2(5):553−562. https://doi.org/10.1002/brb3.74.
226. Durazzo TC, Meyerhoff DJ. GABA concentrations in the anterior cingulate and dorsolateral prefrontal cortices: associations with chronic cigarette smoking, neurocognition, and decision making. *Addiction Biol*. 2021;26(3):e12948. https://doi.org/10.1111/adb.12948.
227. Wasmuth SL, Outcalt J, Buck K, Leonhardt BL, Vohs J, Lysaker PH. Metacognition in persons with substance abuse: findings and implications for occupational therapists. *Can J Occup Ther*. 2015;82(3):150−159. https://doi.org/10.1177/0008417414564865.
228. Spada MM, Caselli G, Nikčević AV, Wells A. Metacognition in addictive behaviors. *Addict Behav*. 2015;44:9−15. https://doi.org/10.1016/j.addbeh.2014.08.002.
229. Janssen AG, Kennair L, Hagen R, Hjemdal O, Havnen A, Solem S. Positive and negative metacognitions about alcohol: validity of the Norwegian PAMS and NAMS. *Addict Behav*. 2020;108:106466. https://doi.org/10.1016/j.addbeh.2020.106466.

230. Spada MM, Nikcević AV, Moneta GB, Wells A. Metacognition as a mediator of the relationship between emotion and smoking dependence. *Addict Behav*. 2007;32(10):2120–2129. https://doi.org/10.1016/j.addbeh.2007.01.012.
231. Nikčević AV, Alma L, Marino C, et al. Modelling the contribution of negative affect, outcome expectancies and metacognitions to cigarette use and nicotine dependence. *Addict Behav*. 2017;74:82–89. https://doi.org/10.1016/j.addbeh.2017.06.002.
232. Brosnan T, Kolubinski DC, Spada MM. Parenting styles and metacognitions as predictors of cannabis use. *Addic Behav Rep*. 2020;11:100259. https://doi.org/10.1016/j.abrep.2020.100259.
233. Spada MM, Wells A. Metacognitions across the continuum of drinking behavior. *Pers Indiv Differ*. 2010;49(5):425–429. https://doi.org/10.1016/j.paid.2010.04.011.
234. Spada MM, Wells A. A metacognitive model of problem drinking. *Clin Psychol Psychother*. 2009;16(5):383–393. https://doi.org/10.1002/cpp.620.
235. Moneta GB. Metacognition, emotion, and alcohol dependence in college students: a moderated mediation model. *Addict Behav*. 2011;36(7):781–784. https://doi.org/10.1016/j.addbeh.2011.02.010.
236. Garland EL, Howard MO. Mindfulness-based treatment of addiction: current state of the field and envisioning the next wave of research. *Addiction Sci Clin Pract*. 2018;13(1):14. https://doi.org/10.1186/s13722-018-0115-3.
237. Goldberg SB, Tucker RP, Greene PA, et al. Mindfulness-based interventions for psychiatric disorders: a systematic review and meta-analysis. *Clin Psychol Rev*. 2018;59:52–60. https://doi.org/10.1016/j.cpr.2017.10.011.
238. Garland EL, Froeliger B, Howard MO. Mindfulness training targets neurocognitive mechanisms of addiction at the attention-appraisal-emotion interface. *Front Psychiatr*. 2014;4:173. https://doi.org/10.3389/fpsyt.2013.00173.
239. Alfonso JP, Caracuel A, Delgado-Pastor LC, Verdejo-García A. Combined Goal Management Training and Mindfulness meditation improve executive functions and decision-making performance in abstinent polysubstance abusers. *Drug Alcohol Depend*. 2011;117(1):78–81. https://doi.org/10.1016/j.drugalcdep.2010.12.025.
240. De Wilde B, Bechara A, Sabbe B, Hulstijn W, Dom G. Risky decision-making but not delay discounting improves during inpatient treatment of polysubstance dependent alcoholics. *Front Psychiatr*. 2013;4:91. https://doi.org/10.3389/fpsyt.2013.00091.
241. Secades-Villa R, Weidberg S, García-Rodríguez O, Fernández-Hermida JR, Yoon JH. Decreased delay discounting in former cigarette smokers at one year after treatment. *Addict Behav*. 2014;39(6):1087–1093. https://doi.org/10.1016/j.addbeh.2014.03.015.
242. Yi R, Johnson MW, Giordano LA, Landes RD, Badger GJ, Bickel WK. The effects of reduced cigarette smoking on discounting future rewards: an initial evaluation. *Psychol Rec*. 2008;58(2):163–174. https://doi.org/10.1007/BF03395609.
243. Yoon JH, Higgins ST, Bradstreet MP, Badger GJ, Thomas CS. Changes in the relative reinforcing effects of cigarette smoking as a function of initial abstinence. *Psychopharmacology*. 2009;205(2):305–318. https://doi.org/10.1007/s00213-009-1541-4.
244. García-Pérez Á, Vallejo-Seco G, Weidberg S, González-Roz A, Secades-Villa R. Long-term changes in delay discounting following a smoking cessation treatment for patients with depression. *Drug Alcohol Depend*. 2020;212:108007. https://doi.org/10.1016/j.drugalcdep.2020.108007.
245. Weidberg S, Landes RD, García-Rodríguez O, Yoon JH, Secades-Villa R. Interaction effect of contingency management and sex on delay-discounting changes among treatment-seeking smokers. *Exp Clin Psychopharmacol*. 2015;23(5):361–368. https://doi.org/10.1037/pha0000043.
246. Bickel WK, Jones BA, Landes RD, Christensen DR, Jackson L, Mancino M. Hypothetical intertemporal choice and real economic behavior: delay discounting predicts voucher redemptions during contingency-management procedures. *Exp Clin Psychopharmacol*. 2010;18(6):546–552. https://doi.org/10.1037/a0021739.
247. Fletcher JB, Dierst-Davies R, Reback CJ. Contingency management voucher redemption as an indicator of delayed gratification. *J Subst Abuse Treat*. 2014;47(1):73–77. https://doi.org/10.1016/j.jsat.2014.03.003.
248. Ashare RL, McKee SA. Effects of varenicline and bupropion on cognitive processes among nicotine-deprived smokers. *Exp Clin Psychopharmacol*. 2012;20(1):63–70. https://doi.org/10.1037/a0025594.
249. Hindocha C, Freeman TP, Grabski M, et al. The effects of cannabidiol on impulsivity and memory during abstinence in cigarette dependent smokers. *Sci Rep*. 2018;8(1):7568. https://doi.org/10.1038/s41598-018-25846-2.
250. Kriegler J, Wegener S, Richter F, Scherbaum N, Brand M, Wegmann E. Decision making of individuals with heroin addiction receiving opioid maintenance treatment compared to early abstinent users. *Drug Alcohol Depend*. 2019;205:107593. https://doi.org/10.1016/j.drugalcdep.2019.107593.

251. Ersche KD, Roiser JP, Clark L, London M, Robbins TW, Sahakian BJ. Punishment induces risky decision-making in methadone-maintained opiate users but not in heroin users or healthy volunteers. *Neuropsychopharmacology*. 2005;30(11):2115−2124. https://doi.org/10.1038/sj.npp.1300812.
252. Landes RD, Christensen DR, Bickel WK. Delay discounting decreases in those completing treatment for opioid dependence. *Exp Clin Psychopharmacol*. 2012;20(4):302−309. https://doi.org/10.1037/a0027391.
253. Schmaal L, Goudriaan AE, Joos L, et al. Neural substrates of impulsive decision making modulated by modafinil in alcohol-dependent patients. *Psychol Med*. 2014;44(13):2787−2798. https://doi.org/10.1017/S0033291714000312.
254. Heyer-Osorno R, Juárez J. Modafinil reduces choice impulsivity while increasing motor activity in preadolescent rats treated prenatally with alcohol. *Pharmacol, Biochem Behav*. 2020;194:172936. https://doi.org/10.1016/j.pbb.2020.172936.
255. Coker AR, Weinstein DN, Vega TA, Miller CS, Kayser AS, Mitchell JM. The catechol-O-methyltransferase inhibitor tolcapone modulates alcohol consumption and impulsive choice in alcohol use disorder. *Psychopharmacology*. 2020;237(10):3139−3148. https://doi.org/10.1007/s00213-020-05599-5.
256. Bolin BL, Lile JA, Marks KR, Beckmann JS, Rush CR, Stoops WW. Buspirone reduces sexual risk-taking intent but not cocaine self-administration. *Exp Clin Psychopharmacol*. 2016;24(3):162−173. https://doi.org/10.1037/pha0000076.
257. Strickland JC, Bolin BL, Romanelli MR, Rush CR, Stoops WW. Effects of acute buspirone administration on inhibitory control and sexual discounting in cocaine users. *Hum Psychopharmacol*. 2017;32(1). https://doi.org/10.1002/hup.2567.

SECTION IV

Individual and sociocultural factors linked to addiction

CHAPTER 9

Individual differences in addiction: focus on personality traits

Introduction

Why are some individuals more susceptible to developing an addiction compared to others? Are there common traits that predispose someone to addiction? These are important questions that have received considerable attention in addiction science, and rightfully so. In the previous chapter, you learned about the relationship between maladaptive decision making and addiction. Constructs like impulsive decision making and suboptimal decision making vary across individuals. Some individuals are capable of delaying gratification for extended periods of time whereas others have no desire to wait to receive something. In other words, there are **individual differences** in impulsive decision making. As you learned in the previous chapter, these individual differences are associated with substance use disorders (SUDs). Yet, decision making is not the only area in which individual differences are observed. You have also learned previously that individuals that attribute greater incentive salience to drug-paired stimuli are more susceptible to developing a SUD. The purpose of this chapter is to highlight additional individual differences that are predictive of addiction-like behaviors. In particular, this chapter will focus on *personality traits* associated with addiction. **Personality** refers to "*individual differences* in characteristic patterns of thinking, feeling and behaving".[1] Our personalities are what make us unique. As you will learn, certain personality traits can predispose someone to use drugs and to develop a SUD. We will return to impulsivity as impulsive decision making is just one facet of this construct. Like impulsive decision making, other facets of impulsivity are important determinants of addiction. This chapter will conclude by discussing the association between attention-deficit/hyperactivity disorder (ADHD), a neurocognitive disorder characterized by excessive impulsivity, and addiction.

Learning objectives

By the end of this chapter, you should be able to …

(1) Compare and contrast the various scales that have been used to measure personality traits.
(2) Discuss which personality traits have been implicated in SUDs.
(3) Describe the various facets of impulsivity and how they are measured in humans and in animals.
(4) Explain the contribution of impulsivity to addiction-like behaviors.
(5) Discuss the relationship between ADHD and addiction, and explain how ADHD symptoms are modeled in rodents.

Measuring personality traits

Think about all the people you know and how they differ. You most likely know individuals that are outgoing, able to strike up a conversation with anyone, and you most likely know individuals that would rather do anything other than interact with others. Some individuals may engage in activities without stopping to think through their actions while others meticulously plan out their days. Given the complex nature of personality, several theories and scales (questionnaires) have been developed to explain how personality can be described. Some, but certainly not all, of the major theories/scales will be detailed below.

Eysenck personality questionnaire

Dr. Hans Eysenck (1916–1997) believed that personality is largely influenced by our nervous system.[2] After working with soldiers admitted to a psychiatric hospital, Eysenck proposed that personality can be explained by two dimensions measured on a continuum: extraversion/introversion and neuroticism/stability. Extraversion relates to how comfortable someone is around others. Extraverts can easily start conversations with others, and they enjoy being the center of attention. Conversely, introverts prefer to spend time alone or with a small group of close friends/family members. Neuroticism is characterized by emotional instability, anxiety, and irritability while individuals high in stability are calm and even-tempered. According to Eysenck, individuals can be placed in one of four different categories: *sanguine* (high extraversion and high stability), *phlegmatic* (high introversion and high stability), *choleric* (high extraversion and high neuroticism), and *melancholic* (high introversion and high neuroticism) (see Fig. 9.1 for words used to describe these types of individuals).

To measure extraversion/introversion and neuroticism/stability, Eysenck and his wife Dr. Sybil Eysenck (1927–2020) originally used the Eysenck Personality Inventory (EPI).[3] The EPI consisted of 57 yes/no questions, 24 of which measured extraversion/introversion (e.g., "Do you stop to think things over before doing anything?") and 24 of which measured neuroticism/stability (e.g., "Do your moods go up and down?"). Nine questions measured *social desirability*; that is, some questions were used to determine if participants responded in such a way to make themselves look better (i.e., participants are not being truthful in their responses). Eventually, Eysenck and Eysenck modified the EPI by adding another dimension: psychoticism/socialization. Psychoticism is used to describe individuals that act in a hostile, aggressive manner. The questionnaire developed by Eysenck and Eysenck is now known as the **Eysenck Personality Questionnaire (EPQ)**.[4] The latest version of the EPQ asks a series of 100 yes/no questions, although a shorter version containing 48 questions exists (EPQ-RS).

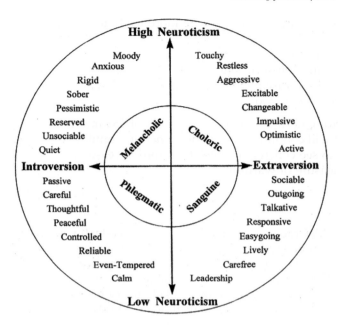

FIGURE 9.1 Eysenck's model of personality. The horizontal line depicts extraversion-introversion, and the vertical line depicts neuroticism-stability (labeled as low neuroticism here). Notice that individuals can fall under one of four dimensions in the original model proposed by Eysenck: choleric, melancholic, sanguine, and phlegmatic. The chart shows words used to describe individuals that fall into each dimension. *Image comes from Fig. 1 of Rybanská J. Selected personality characteristics as predictors of emotional consumer behaviour. Eur J Business Sci and Technol. 2015;1(2):128–136. https://doi.org/10.11118/ejobsat.v1i2.26. This is an open-access article distributed under the terms of the Creative Commons Attribution 4.0 International License (https://creativecommons.org/licenses/by/4.0/).*

Sixteen personality factor questionnaire

The Sixteen Personality Factor Questionnaire's origins began in the 1930s when Dr. Gordon Allport (1897–1967) and Henry Odbert methodically identified words in the dictionary that described personality. Allport and Odbert created a list of 4,500 words that described observable traits that were unrelated to one's physical appearance.[5] Eventually, Dr. Raymond Cattell (1905–1998) narrowed down the list of 4,500 words to create 16 personality factors.[6] These factors are as follows:

1. Warmth—describes how outgoing an individual is and how attentive they are to others (somewhat similar to Eysenck's extraversion)
2. Reasoning—refers to one's intelligence and type of thinking (e.g., more abstract or more concrete)
3. Emotional stability—this is similar to Eysenck's neuroticism/stability dimension
4. Dominance—describes how competitive, assertive, and/or stubborn someone is
5. Liveliness—measures an individual's enthusiasm, expressiveness, and impulsiveness
6. Rule-consciousness—refers to one's willingness to conform to expected norms
7. Social boldness—used to gauge how venturesome and thick-skinned an individual is
8. Sensitivity—used to describe individuals that are sentimental and tender-minded
9. Vigilance—refers to how distrustful/skeptical an individual is
10. Abstractedness—individuals that score high in this area are more imaginative, but more absentminded
11. Privateness—describes one's willingness to disclose information to others

12. Apprehension—individuals that score high in this area are more likely to doubt themselves and are more insecure
13. Openness to change—more experimental and freethinking
14. Self-reliance—more individualistic and self-sufficient
15. Perfectionism—more organized and self-disciplined; also more compulsive
16. Tension—more impatient and easily frustrated

Today, the Sixteen Personality Factor Questionnaire (16pf) consists of 185 multiple-choice questions. Some questions include: "I will continue to work on something until it is perfect" (measure of perfectionism); "I prefer to see concrete results" (measure of reasoning); and "I enjoy being around other people" (measure of warmth).

Five-factor model of personality

Largely associated with Dr. Robert McCrae and Dr. Paul Costa, the **five-factor model of personality** suggests that personality can be separated into five main categories (known as the "Big Five" personality traits): *openness to experience, conscientiousness, extraversion, agreeableness*, and *neuroticism*.[7] The acronym OCEAN can be used to remember the big five personality traits. Individuals that score higher on openness are more likely to try new things and are more imaginative. Conscientiousness is synonymous with thoughtfulness; that is, individuals that score high in this area are more goal directed and are less impulsive. High agreeableness is associated with prosocial behaviors, like helping those in need, showing kindness to others, and displaying empathy. Extraversion and neuroticism are conceptualized in a similar manner as proposed by Eysenck.

McCrae and Costa developed the NEO Personality Inventory in the late 1970s and have revised this scale several times, with the most recent revision (NEO PI-3) being published in 2005.[8] The NEO PI-3 consists of 240 Likert-scale questions and measures individual *facets* of the "Big Five" factors. For example, a facet of extraversion is warmth, one of the factors described above when discussing the 16pf. Table 9.1 shows some of the individual facets associated with each trait.

Alternative five factor model of personality

The alternative five-factor model of personality was proposed by Dr. Marvin Zuckerman (1928–2018) and includes the following factors: **sensation seeking**, neuroticism-anxiety, aggression-hostility, sociability, and activity.[9] Sensation seeking is characterized by a desire to participate in experiences that are intense (e.g., skydiving) and is often associated with risk-taking behaviors, although risk is not required for sensation seeking. Sensation seeking is largely associated with impulsive behaviors. As such, sensation seeking will be described in more detail later in this chapter. There is overlap between the traditional five factor model of personality and the alternative model proposed by Zuckerman. Sensation seeking is similar to openness to experience and is the opposite of conscientiousness. Neuroticism-anxiety is similar to the neuroticism scale, aggression-hostility is analogous to the agreeableness scale, and sociability is akin to the extraversion scale. Activity refers to an individual's energy levels and persistence. The alternative five model of personality is assessed with the Zukerman-Kuhlman Personality Questionnaire (ZKPQ). The ZKPQ (currently known as the ZKPQ–III–R) asks participants 99 true/false questions.[10]

TABLE 9.1 The "Big Five" personality factors and their individual facets.

Big five traits	Personality facets, adjectives	Respondent see himself herself as someone who
Openness to experience (openness)	Ideas (curious) Fantasy (imaginative) Aesthetics (artistic) Actions (wide interests) Feelings (excitable) Values (unconventional)	01 is original, comes up with ideas 02 values artistic, aesthetic experiences 03 has an active imagination
Conscientiousness	Competence (efficient) Order (organized) Dutifulness (not careless) Achievement striving (thorough) Self-discipline (not lazy) Deliberation (not impulsive)	C1 does a thorough job C2 tends to be lazy (reversed score) C3 does things efficiently
Extraversion	Gregariousness (sociable) Assertiveness (forceful) Activity (energetic) Excitement-seeking (adventurous) Positive emotions (enthusiastic) Warmth (outgoing)	E1 is talkative E2 is outgoing, sociable E3 is reserved (reversed score)
Agreeableness	Trust (forgiving) Straightforwardness (not demanding) Altruism (warm) Compliance (not stubborn) Modesty (not show-off) Tender-mindedness (sympathetic)	A1 is sometime rude to others (reversed score) A2 has a forgiving nature A3 is considerate and kind
Neuroticism	Anxiety (tense) Angry hostility (irritable) Depression (not contented) Self-consciousness (shy) Impulsiveness (moody) Vulnerability (not self-confident)	N1 worries a lot N2 gets nervous easily N3 is relaxed, handles stress well (reversed score)

The right column shows example questions related to each factor.
Table comes from Table 4 of Nandi A, Nicoletti C. Explaining personality pay gaps in the UK. Appl Econ. 2014;46(26):3131−3150, https://doi.org/10.1080/00036846.2014.922670. This is an open-access article distributed under the terms of the Creative Commons Attribution 4.0 International License (https://creativecommons.org/licenses/by/4.0/).

Temperament and character inventory

Like Eysenck, Dr. Robert Cloninger was interested in the role of biology on personality. Using genetic, neurobiological, and neuropharmacological data, Cloninger developed the Tridimensional Personality Questionnaire (TPQ),[11] now known as the Temperament and Character Inventory (TCI).[12] The TCI focuses on seven personality traits:

1. **Novelty seeking**—increased exploratory activity in response to new stimulation; also characterized by increased impulsivity
2. Harm avoidance—similar to the construct of neuroticism

3. Reward dependence—marked by increased sensitivity to rewards
4. Persistence—perseverance despite frustration or fatigue
5. Self-directedness—similar to the construct of conscientiousness
6. Cooperativeness—similar to the construct of agreeableness
7. Self-transcendence—relating spirituality to oneself

The first four traits are referred to as *temperaments* and are more biological in nature, whereas the last three traits are known as *characters* and result from learning. Each temperament has four subscales while each character has three to five subscales. Table 9.2 list each subscale.

Notice that sensation seeking as proposed by Zuckerman and novelty seeking as proposed by Cloninger sound very similar. Indeed, previous research has shown that sensation seeking and novelty seeking are highly correlated with one another.[13] Due to the similarity between these constructs, studies assessing the relationship between sensation seeking/novelty seeking will be discussed together in the impulsivity section of this chapter.

Minnesota multiphasic personality inventory

The Minnesota Multiphasic Personality Inventory (MMPI) is used to measure psychopathology. The MMPI was developed by two University of Minnesota faculty, Dr. Starke Hathaway (1903–1984) and Dr. John McKinley (1891–1950).[14] The newest version of the MMPI (MMPI-3, released in 2020[15]) contains 335 self-report items, and individuals are evaluated based on eight clinical scales:

1. Demoralization—general unhappiness and dissatisfaction
2. Somatic complaints—diffuse physical health complaints
3. Low positive emotions—lack of positive emotional responsiveness
4. Antisocial behavior—rule breaking and irresponsible behavior
5. Ideas of persecution—self-referential beliefs that others pose a threat
6. Dysfunctional negative emotions—maladaptive anxiety, anger, irritability
7. Aberrant experiences—unusual perceptions or dysfunctional thoughts
8. Hypomanic activation—overactivation, aggression, impulsivity, and grandiosity

Because the MMPI was recently revised, an overview of the MMPI-2 is warranted as studies examining psychopathology in those with SUDs have used this scale. This version of the MMPI[16] evaluates individuals based on 10 clinical scales:

1. Hypochondriasis—neurotic concern over one's physical functioning
2. Depression—similar to the Demoralization scale of the MMPI-3
3. Hysteria—similar to the Somatic complaints scale of the MMPI-3
4. Psychopathic deviate—similar to the Antisocial behavior scale of the MMPI-3
5. Masculinity-femininity—refers to how much an individual identifies with stereotypical male and female gender roles

TABLE 9.2 Subscales of each temperament and character in Cloninger's Temperament and Character Inventory (TCI), as well as example questions for each subscale.

	Temperament/character	Subscale	Example question[a]
Temperament	Novelty seeking	Exploratory excitability Impulsiveness Extravagance Disorderliness	I prefer variety to routine. I jump into things without thinking I spend more money than I have. I know how to get around the rules.
	Harm avoidance	Anticipatory worry Fear of uncertainty Shyness Fatigability	I fear for the worst I would never make a high risk investment. I often fed uncomfortable around others. I become overwhelmed by Events.
	Reward dependence	Sentimentality Openness to warm communication Attachment Dependence	I am easily moved to tears. I am open about myself to others. I enjoy bringing people together. I do what others want me to do.
	Persistence	Eagerness of effort Work hardened Ambitious Perfectionist	I get chores done right away. I accept challenging tasks. I want to be the very best I am exacting in my work.
Character	Self-directedness	Responsibility Purposeful Resourcefulness Self-acceptance Englightened second nature	I know how to enjoy myself. I work on improving myself. I face problems directly. I take things as they come. I easily resist temptations.
	Cooperativeness	Social Acceptance Empathy Helpfulness Compassion Pure-hearted conscience	I accept people as they are. I anticipate the needs of others. I like to be of service to others. I try to forgive and forget I return extra change when a cashier makes a mistake.
	Self-transcendence	Self-forgetful Transpersonal Identification Spiritual Acceptance	I get so involved with things that I forget the time. I see beauty in things that others might not notice. I believe that there are universal truths.

[a] Note, the example questions listed here do not come directly from the TCI, but instead come from the International Personality Item Pool (IPIP).
Table made by the author of this textbook using Microsoft Excel.

6. Paranoia—similar to the Ideas of persecution scale of the MMPI-3
7. Psychasthenia—used to describe individuals with high anxiety, depression, and obsessive-compulsive disorder; somewhat similar to the Dysfunctional negative emotions scale of the MMPI-3
8. Schizophrenia—encompasses several scales of the MMPI-3, including Low positive emotions, Ideas of persecution, and Aberrant experiences

9. Hypomania—similar to the Hypomanic activation scale of the MMPI-3
10. Social introversion—assesses one's shyness and tendency to withdraw from social interactions

Personality factors associated with drug use and addiction

As most of the studies assessing the link between personality traits and addiction have relied on the "Big Five" traits, this section will focus primarily on these traits. Due to the overlap between the "Big Five" traits and traits discussed by Eysenck, Cattell, and Zuckerman, these scales will be discussed together.

"Big Five" traits associated with addiction

One common finding is that drug use is positively associated with extraversion, neuroticism, and openness to experience,[17-33] although cocaine and heroin users have been shown to score lower in extraversion[34,35] while methamphetamine using women score lower in openness to experience[25] Additionally, medical students that score high on the aggressive-hostility scale are more likely to use psychostimulants.[36] Agreeableness and conscientiousness are negatively associated with drug use.[17,20,21,24,26,27,32,33,37,38] Personality differences have even been observed between natural cannabis users and synthetic cannabinoid users. Read the Experiment Spotlight below to find out more.

Experiment Spotlight: Personality traits and psychotic proneness among chronic synthetic cannabinoid users (Cohen et al.[19])

One concerning finding that has emerged in the literature is that chronic cannabis use is linked to the development of schizophrenia. Additionally, natural cannabis users score higher in measures of schizotypal personality traits. Given the rise in synthetic cannabis use, Cohen et al.[19] wanted to determine if synthetic cannabis users differ from natural cannabis users and noncannabis users in personality, substance use history, depression, and anxiety. The researchers recruited 42 synthetic cannabis users, 39 natural cannabis users, and 47 nonusing individuals. Participants completed several questionnaires, including the Big-Five Factors Inventory (BFI) (measure of the "Big Five" traits), the Schizotypal Personality Questionnaire-Brief (SPQ-B) (measure of psychotic proneness), and the Beck Depression Inventory (BDI) (measure of depression and anxiety). Participants also completed demographic information, including their education level, age, gender, current/past neurological and/or psychiatric conditions, and drug use histories.

Results showed that synthetic cannabis users scored higher on neuroticism and lower on agreeableness, conscientiousness, and extraversion compared to natural cannabis users and nonusers (see figure on next page). Furthermore, synthetic users scored higher on the SPQ-B compared to the other two groups, indicating increased risk of psychosis, and they scored higher in the BDI, indicating elevated levels of depression/anxiety. While ethanol consumption did not differ across groups, synthetic cannabis users reported smoking more cigarettes and reported beginning cannabis use at an earlier age (15 ± 6.56 years) compared to natural cannabis users (18.41 ± 4.72 years).

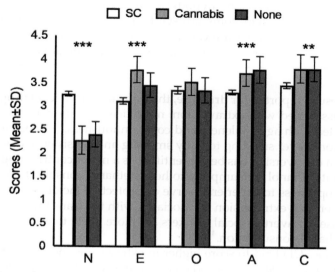

Scores from the BFI for synthetic cannabis users (SC), natural cannabis users (Cannabis), and noncannabis users (None). A, agreeableness; C, conscientiousness; E, extraversion; N, neuroticism and O, openness to experience. Asterisks (*) indicate significant differences across each group of participants. Synthetic cannabis users score higher in neuroticism compared to cannabis and noncannabis users. However, they score lower in extraversion, agreeableness, and conscientiousness. *Image recreated from Fig. 1 of Cohen K, Rosenzweig S, Rosca P, Pinhasov A, Weizman A, Weinstein A. Personality traits and psychotic proneness among chronic synthetic cannabinoid users. Front Psychiatr. 2020;11:355. https://doi.org/10.3389/fpsyt.2020.00355. This is an open-access article distributed under the terms of the Creative Commons Attribution 4.0 International License (https://creativecommons.org/licenses/by/4.0/).*

Questions to consider:

(1) What could account for the decreased extraversion observed in synthetic cannabis users, but increased extraversion observed in natural cannabis users?
(2) In your opinion, which of the following seems to provide a better account of the present data: using synthetic cannabinoids increases the risk of developing psychosis or individuals with elevated psychosis are more likely to use synthetic cannabinoids to "self-medicate"?

Answers to questions:

(1) Natural cannabis is often used in social settings while synthetic cannabis users prefer to be alone when using the drug. This difference could largely explain the differential extraversion scores observed in natural cannabis users and synthetic cannabis users.
(2) The results of this study do not provide a clear answer to this question. However, Cohen et al. argue that both explanations could contribute to the results of the study. Synthetic cannabinoids are full agonists at CB receptors. Remember that cannabis is a

partial agonist at these receptors. Thus, this pharmacological difference could contribute to the increased psychotic proneness observed in synthetic cannabis users. Conversely, the "self-medication" hypothesis can be used to explain why individuals with increased psychotic symptoms use synthetic cannabinoids. Due to the pharmacological differences between natural cannabis and synthetic cannabinoids, these individuals may receive greater benefits from using synthetic cannabinoids.

Similar to the results reported for drug use, the transition from legal drug use to illicit drug use is positively associated with extraversion, neuroticism, and openness to experience, but negatively associated with agreeableness and conscientiousness.[39] Neuroticism is also linked to the transition from never smoking to daily smoking and persistence of smoking over a 10-year period.[33] High extraversion has been identified as a major contributing factor in the transition from moderate ethanol consumption to heavy ethanol consumption, while high agreeableness and low openness to experience serve as protective factors against this transition.[37] One potential reason that extraversion is associated with the transition to heavy ethanol consumption is because extraverted social drinkers are more likely to experience mood enhancement following ethanol consumption.[40]

Individuals with a SUD tend to score higher in extraversion, neuroticism, and openness to experience, but lower in agreeableness and conscientiousness,[34,41-45] but there are some exceptions. For example, individuals with an alcohol use disorder score lower in extraversion and openness to experience.[44] Another paradoxical finding is that severity of nicotine dependence is worse in high conscientious individuals, but lower in high neurotic individuals.[18] Why are extraversion, neuroticism, and openness to experience associated with addiction? The link between openness to experience and addiction is somewhat self-explanatory. These individuals are more willing to try new things, such as using a drug. Extraverted individuals like to spend time with others. As discussed in the Experimental Spotlight, certain drugs (e.g., natural cannabis, ethanol, and cigarettes) are often used in a social setting. If individuals spend time with others that use these substances, they may be more likely to use the substance as well. We will return to this concept in the next chapter. Given that neuroticism is characterized by anxiety and emotional disturbances, drug use may serve as a coping mechanism for these individuals, as was observed with synthetic cannabinoid users above. Not only is neuroticism linked to drug use, but it has been identified as a critical mediator of relapse.[24,46,47]

Personality factors have also been associated with problematic drug-taking behaviors. Increased risky ethanol use is linked to increased levels of extraversion and neuroticism, as well as lower levels of agreeableness and conscientiousness.[48] Women that score high in extraversion and openness to experience, but low in conscientiousness, are more likely to consume ethanol while pregnant.[49-51] Interestingly, women that score high in neuroticism are *less* likely to smoke while pregnant.[51] Cocaine-dependent individuals are more likely to experience cocaine-induced psychosis if they score high on neuroticism-anxiety of the ZKPQ.[52] One study determined if certain traits predict opioid overdose deaths. Tacheva and Ivanov[53] did not measure personality traits in individual participants. Instead, they examined social media posts of residents in 2,891 U S. counties between 2014 and 2016 and calculated the average county-level scores for each personality trait. Three of the "Big Five" traits were significantly associated with opioid overdose deaths: extraversion, neuroticism, and conscientiousness.

Temperaments and characters associated with addiction

Not as many studies have examined the associations between personality traits as measured with the TCI and addiction-like behaviors. Overall, individuals that use drugs score higher in harm avoidance (similar to what is observed for individuals that score high in neuroticism), but lower in persistence and self-directedness.[54–56] One exception is that drinking frequency is associated with low levels of harm avoidance.[57] Pomerleau et al.[58] argue that reward dependence is an important personality factor that leads to the initiation of smoking. While one study found lower reward dependence in opioid users,[56] another study found that opioid users that do not become dependent score higher in harm avoidance, as well as in self-transcendence, compared to never users.[59] Nondependent opioid users score lower in self-directedness and cooperativeness compared to never users; however, they score higher in reward dependence and self-directedness (as well as in self-transcendence) compared to dependent opioid users. Zaaijer et al.[59] propose that reward dependence and self-directedness may act as protective factors against the development of opioid dependence because these individuals are more sensitive to social approval (part of reward dependence) and have higher self-efficacy (component of self-directedness; recall self-efficacy from Chapter 7).

Like neuroticism, harm avoidance has been largely implicated in SUDs.[58,60] Individuals admitted to the hospital for ethanol overdose score higher in harm avoidance and self-transcendence, but score lower in reward dependence, cooperativeness, and self-directedness compared to others admitted to the hospital for nonalcohol-related events.[61] While harm avoidance has been proposed to be a personality trait critically involved in SUDs, relapse is associated with lower levels of persistence, reward dependence, and cooperativeness.[62] Harm avoidance is conceptually similar to neuroticism; therefore, the finding that harm avoidance is positively associated with addiction should not be surprising. Because maintaining abstinence is a life-long process, individuals that score low in persistence are more likely to "give up" and start using drugs following difficult times in their life. The association between cooperativeness and relapse is not surprising either. Because cooperativeness is similar to the construct of agreeableness, individuals that score low in this area are less concerned about how others feel about their actions. Individuals that score high in cooperativeness may decide to remain abstinent because they want to be able to provide for their family, or they may realize they need to take care of aging parents. The one somewhat surprising finding is that low reward dependence is linked to relapse. At first glance, one may assume that high reward dependence would be associated with relapse as these individuals are more sensitive to rewards, such as drugs. Remember, reward dependence entails more than the rewarding properties of specific stimuli. One important facet of reward dependence is the degree to which individuals are willing to help others. In a sense, reward dependence is similar to cooperativeness/agreeableness. When viewed in this manner, the lower reward dependence observed in those that relapse makes sense.

Psychopathology and addiction

The primary trait from the MMPI implicated in drug use is psychopathic deviance. Individuals that score high in this trait are more likely to use tobacco, cannabis, and amphetamines,[63] and they are more likely to engage in heavy drug use.[64] Furthermore, individuals diagnosed with alcohol use disorder by 25 years of age score higher on the psychopathic deviate clinical scale.[65] Related to these findings is that psychoticism/antisocial personality traits have been linked to heroin use disorder[34] and methadone use.[66] In addition to psychopathic deviance, higher scores in hysteria, hypomania, paranoia, psychasthenia, and schizophrenia have been associated with drug use.[63,64]

The MMPI-2 has also been used to assess psychopathological traits associated with driving while intoxicated (DWI). Individuals that engage in this behavior score higher in hypochondriasis, psychopathic deviance, paranoia, psychasthenia, schizophrenia, and hypomania while scoring lower in social introversion.[67] DWI recidivists score higher in the psychopathic deviate and hypomania scales.[68]

While depression is a scale of the MMPI-2 (now low emotional experiences in the MMPI-3), depressive symptoms are most often measured using the Beck Depression Inventory, as you saw in the Experiment Spotlight. Elevated depression scores are observed in individuals that are dependent on various substances, including nicotine,[69] ethanol,[70] cannabis,[71] cocaine,[72] methamphetamine,[73] and heroin.[74] Individuals that show increased scores on the BDI report elevated subjective responses to cocaine,[75] and they are less sensitive to increasing prices in cigarettes.[76] Depression has also been implicated in relapse.[77-81] Collectively, these findings suggest that treating depressive symptoms may be an important target for improving treatment outcomes in those with a SUD. However, this does not appear to be a foolproof method for treating SUDs. Giving opioid-dependent individuals an antidepressant in conjunction with buprenorphine does not increase treatment retention rates, nor does it decrease illicit drug use.[82] Additional work has shown that depression *per se* is not always an important predictor of relapse. Anhedonia, inability to feel pleasure, has been shown to predict treatment outcomes for cocaine-dependent individuals instead of depressive symptoms.[83]

Quiz 9.1

(1) Which of the following personality factors acts as a protective factor against drug use?
    a. Agreeableness
    b. Extraversion
    c. Neuroticism
    d. Openness to experience

(2) Which of the following statements best describes someone with high harm avoidance?
    a. I avoid helping others.
    b. I do not believe that there is a higher power.
    c. I would rather spend my weekend at home than out with friends.
    d. When I meet someone new, I am willing to disclose a lot about myself in a short period of time.

(3) Which of the following drug-related behaviors is not associated with harm avoidance?
    a. Cigarette smoking
    b. Drinking frequency
    c. Methamphetamine use during pregnancy
    d. Opioid overdose

(4) Which MMPI personality trait is largely associated with addiction-like behaviors?
   a. Hypomania
   b. Psychopathic deviant
   c. Schizophrenia
   d. Social introversion
(5) According to Eysenck, an individual that displays high leadership qualities would be classified as _____.
   a. choleric
   b. melancholic
   c. phlegmatic
   d. sanguine

Answers to quiz 9.1

(1) a — Agreeableness
(2) c — I would rather spend my weekend at home than out with friends.
(3) b — Drinking frequency
(4) b — Psychopathic deviant
(5) d — sanguine

Impulsivity

In the last chapter, you learned about impulsive decision making, which is characterized by an inability to delay gratification. Impulsivity is not a unitary construct, meaning it can be subdivided into different subtypes. Before discussing the relationship between impulsivity and addiction, a discussion of the impulsivity subtypes and how they are measured is merited.

Behavioral measures of impulsivity

From a behavioral perspective, impulsivity is often separated into two broad categories. **Impulsive choice** is often used synonymously with impulsive decision making. Because we have already discussed this type of impulsivity, it will not be discussed in this chapter. **Impulsive action** is also known as motor impulsivity. Some individuals have difficulty withholding a *prepotent response*, that is, an ongoing response that is no longer relevant or a response that conflicts with a current goal. For example, if someone is attempting to diet, eating junk food is a prepotent response that interferes with the plan of dieting. In rodents, continuously responding on a manipulandum during a "time-out" period in which reinforcement is unavailable is considered a prepotent response (see below for more details).

The primary behavioral paradigms used to measure impulsive action in humans are the stop signal reaction time task (SSRTT) and the go/no-go task. In the SSRTT, subjects are required to inhibit responses they have already initiated when presented with a cue.[84] In go/no-go tasks, subjects are required to either initiate a response (go) or inhibit a response (no-go) when presented with different cues[85] (see Fig. 9.2 for an example of a go/no-go task). Like impulsive decision making, impulsive action can be modeled in animals, such

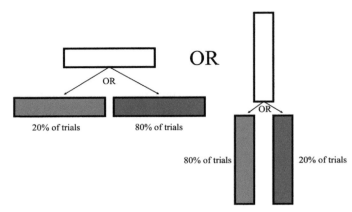

FIGURE 9.2 A schematic of a go/no-go task. Participants are first presented with one type of stimulus (either a horizontal bar or a vertical bar). Next, participants are presented with a colored bar. If a green bar is presented, the participant should respond on the keyboard. If a blue bar is presented, the participant needs to withhold the response. Notice that when the horizontal bar is presented, "go" trials are presented on 20% of trials. Conversely, when the vertical bar is presented, "go" trials are presented on 80% of trials. *Image created by the author of this textbook using Microsoft PowerPoint using a description detailed in Marczinski CA, Fillmore MT. Preresponse cues reduce the impairing effects of alcohol on the execution and suppression of responses. Exp Clin Psychopharmacol. 2003;11(1):110−117. https://doi.org/10.1037//1064-1297.11.1.110.*

as the SSRTT and the go/no-go task. In the animal analog of the SSRTT, animals are required to stop responding on a manipulandum when a stimulus light is illuminated. In the go/no-go task, rats receive reinforcement when a stimulus light is presented, but they do not receive reinforcement when the light is extinguished. The five-choice serial reaction time task (5CSRTT) measures both attention and motor impulsivity. In the 5CSRTT, rats must respond in one of several apertures depending on which stimulus light is illuminated. Additionally, rodents must wait a certain amount of time before they respond. Premature responses are used as an index of impulsive action.[86]

Trait measures of impulsivity

In addition to measuring impulsivity with behavioral tasks, personality questionnaires can be used to gauge one's trait impulsivity. Two major impulsivity inventories will be highlighted here. The Barratt Impulsiveness Scale (BIS-11)[87] is one of the most widely used self-report questionnaires for measuring impulsivity. The BIS-11 contains 30 items that measure three main domains: attention impulsivity, motor impulsivity, and nonplanning impulsivity. The three main domains can be fractioned into six dimensions of impulsivity:

1. Attention—having difficulty concentrating
2. Motor—doing things without thinking about them first
3. Self-control—having difficulty refraining from certain actions, like spending money
4. Cognitive complexity—having difficulty focusing on a problem
5. Perseverance—giving up on things quickly
6. Cognitive instability—having extraneous thoughts when thinking

The Impulsivity Scale 12 (IS-12) is a simplified version of the BIS-11 that focuses on two dimensions of impulsivity: cognitive impulsivity and behavioral impulsivity.[88]

Developed by Dr. Stephen Whiteside and Dr. Donald Lynam, the UPPS Scale focuses on four major impulsive traits: **urgency,** (lack of) premeditation, (lack of) perseverance, and sensation seeking.[89] Urgency is the tendency to act rashly when in a heightened emotional state. There are two forms of urgency: negative urgency (acting rashly in a distressed state) and positive urgency (acting rashly in a euphoric state). Imagine an individual is watching their favorite sports team in the championship game (at the time I wrote this chapter, my beloved Atlanta Braves won the World Series!). If the team loses, the individual may become irrationally angry, breaking their television. This is negative urgency. Conversely, imagine this team wins the championship. This individual may engage in reckless behavior, like excessive ethanol consumption. This is positive urgency.

Impulsive action and addiction

In the last chapter, you learned that long-term drug use increases impulsive decision making. Similarly, long-term substance use increases impulsive action.[90,91] Specifically, increased impulsive action is observed in binge/heavy drinkers[92–94] and smokers,[95] as well as in individuals that use cannabis,[96,97] cocaine,[98–103] crack cocaine,[104] methamphetamine,[105] and ketamine.[66] Ten hours of abstinence from nicotine leads to impaired performance in the SSRTT.[106] One notable exception has been reported in the literature as little evidence has linked long-term opioid use to motor impulsivity.[90] This discrepancy may be due to the large inhibitory effects opioids have on the nervous system. Although they do not show increased motor impulsivity, individuals with opioid use disorder show increased impulsivity (see below and see Chapter 8).

Not only do drugs influence impulsive action, but impulsive action is predictive of addiction. Impulsive action predicts the age of onset of cocaine dependence,[107] and poor response inhibition predicts problem ethanol consumption and illicit drug use in adolescents.[108] Decreased inhibitory control in the go/no-go task predicts ethanol consumption in young adults.[109] To determine if impulsive action is a risk factor for future drug dependence, Ersche et al.[110,111] measured impulsivity in siblings. One sibling met diagnostic criteria for stimulant dependence, whereas the other sibling did not use drugs. Ersche et al. found that impulsive action was significantly correlated between siblings, suggesting that impulsivity has a genetic component. Ersche et al. used these findings to argue that impulsive action may represent an *endophenotype* associated with cocaine/amphetamine dependence. To better measure the direct relationship between impulsive action and addiction, animal models can be used, as discussed in the last chapter. In animals, high impulsive action is predictive of escalation of cocaine self-administration,[112] compulsive cocaine self-administration,[113] and reinstatement of nicotine-, cocaine-, and MDMA-seeking behavior.[114–116] Impulsive action also predicts ethanol consumption in adolescent rats,[116] but not in adult rats.[117,118]

Increased impulsive action, like increased impulsive decision making, predicts lower improvement in psychological and social quality of life during early treatment for methamphetamine use disorder.[119] High impulsive individuals, as assessed with a go/no-go task, are less likely to adhere to naltrexone treatment for opioid use disorder.[120] These results suggest that impulsive action can interfere with one's ability to successfully complete a treatment regimen for their SUD.

Trait impulsivity and addiction

Individuals that smoke are more impulsive according to the BIS-11,[121] particularly showing greater disinhibition.[37] Impulsivity is also associated with habitual tobacco use,[122] cannabis use,[123] cocaine use,[124] ketamine use,[66] and methadone use.[66] Ethanol-induced disinhibition (e.g., driving while intoxicated or getting into fights while inebriated) is more pronounced in high impulsive individuals.[125] Self-reported impulsivity fluctuates with cocaine use; as cocaine use increases, self-reported impulsivity increases accordingly.[126] Impulsivity, as measured with the BIS-11, is positively associated with nicotine dependence,[127] alcohol use disorder,[128,129] opioid use disorder,[129,130] and cannabis use disorder.[131] Both behavioral and cognitive impulsivity dimensions of the IS-12 are associated with substance misuse[88] while behavioral impulsivity is a strong predictor of alcohol use disorder severity.[132] Increased trait impulsivity is also associated with relapse to ethanol.[133]

Urgency has largely been studied in context of ethanol dependence. Individuals that score high in positive urgency are more likely to consume ethanol[134] and are more likely to increase their ethanol consumption over time.[135] Children in fifth grade are at increased risk of initiating ethanol consumption by the end of sixth grade if they score high in positive urgency.[136] Positive and negative urgency are associated with ethanol consumption and problematic drinking in adolescents.[137] Individuals celebrating their 21st birthday consume more alcoholic beverages and engage in more problematic drinking if they score higher in positive urgency.[138] Positive urgency is also associated with ethanol cravings.[139] Urgency, both negative and positive, is a significant predictor of problematic ethanol consumption.[140,141] In particular, negative urgency acts as an important mediator between neuroticism and problematic ethanol consumption.[142] In addition to urgency, other impulsivity traits have been associated with ethanol use. In adolescents, lack of premeditation and lack of perseverance are positively associated with ethanol use and problematic ethanol consumption.[137]

Like those with ethanol dependence, smokers are more likely to score higher in urgency (both positive and negative), lack of premeditation, and lack of perseverance compared to nonsmokers.[143,144] Both positive and negative urgency are associated with nicotine dependence.[143,145,146] The increased nicotine dependence observed in those with high negative urgency may be explained by the finding that these individuals experience greater withdrawal symptoms, negative affect, and tobacco cravings during abstinence.[147]

Lack of premeditation and negative urgency are associated with poorer treatment outcomes for SUDs.[148] Conversely, treatment for SUDs can decrease negative urgency and lack of premeditation.[148,149] Furthermore, total BIS-11 scores are negatively associated with quality of life in those with methamphetamine use disorder.[150] During early withdrawal of methamphetamine, high impulsive individuals are more likely to experience depressive symptoms compared to low impulsive individuals.[151] Even after abstaining from methamphetamine, former methamphetamine users report higher levels of impulsivity compared to never users.[152]

Overall, impulsivity is a major risk factor for drug dependence, but some studies have shown that increased impulsivity is associated with better treatment outcomes. In one study, individuals with comorbid cocaine use disorder and ADHD received amphetamine as a potential treatment for both conditions. Individuals with *higher* self-reported impulsivity were more likely to continuing to achieve 1-week cocaine abstinence over time.[153] Additionally, high impulsive individuals have 11.2% more cocaine-free days following topiramate treatment compared to low impulsive individuals.[154] These results highlight the need for clinicians to tailor treatment plans for each individual as opposed to trying to give each patient the same plan. One's baseline impulsivity can greatly influence how they respond to pharmacological interventions.

Novelty seeking/sensation seeking and addiction

High novelty/sensation seekers are more likely to initiate drug use[22] and are more likely to smoke,[38,143,146] consume ethanol,[57] use cannabis,[56,122] and use cocaine.[56,124] One factor that may contribute to the increased drug use in high novelty seekers is that they experience greater positive reinforcement when using the drug. Indeed, this has been observed in high novelty seeking individuals that use tobacco[155] and ethanol.[156] However, novelty seekers also experience greater *negative* subjective effects of ethanol.[156] Because they are more likely to experience symptoms like hangover, high novelty seekers may be more likely to consume ethanol to escape the aversive symptoms of ethanol (i.e., negative reinforcement).

Novelty seeking is also associated with heavy or daily smoking,[54,157] heavy ethanol consumption,[158] and escalation of cannabis use from adolescence to adulthood.[159] Sensation seeking is a predictor of increased ethanol consumption frequency[135] and is associated with problematic ethanol consumption.[137] Not surprisingly, novelty/sensation seeking is associated with several SUDs, including those for ethanol, nicotine, cannabis, and other illicit substances.[160,161] Sensation seeking is higher in those that develop alcohol use disorder by 25 years of age.[65] Similarly, individuals high in novelty seeking develop cocaine dependence at a younger age.[107] Increased novelty/sensation seeking is linked to problems associated with drug use. Sensation seeking is associated with severity of nicotine dependence.[143] University students are more likely to experience alcohol-related problems if they score higher in novelty seeking.[162] High novelty seeking women are more likely to consume ethanol during pregnancy.[163] Individuals admitted to the hospital for ethanol overdose score higher in novelty seeking.[61]

Novelty seeking appears to increase the difficulty of abstaining from drug use. For example, smokers high in novelty seeking experience greater negative affect and cigarette cravings during abstinence,[164] and individuals that relapse to ethanol score higher in novelty seeking.[62,133] While novelty seeking is a risk factor for relapse, receiving treatment for a SUD can decrease sensation seeking.[147]

Novelty seeking/sensation making can be measured in animals using a variety of methods.[165] I will detail two common approaches here. One approach to measuring novelty seeking is to place a rodent in a novel environment and record its activity levels during a specified period of time. This is known as *inescapable novelty*. Some rodents will show increased exploration (i.e., will have greater locomotor activity), whereas some rodents will explore the novel area less. Novelty seeking can also be assessed with a *novelty place preference* task. This task is similar to the conditioned place preference (CPP) paradigm you learned about in Chapters 3 and 6. Instead of pairing a drug with a specific environment, the researcher can isolate an animal to the same environmental context repeatedly. Eventually, the researcher will allow the animal to explore the familiar environment and a novel environment (i.e., one that the animal has never seen before). If the animal spends more time in the novel environment compared to the familiar environment, they are classified as high novelty seekers.

In animals, increased novelty/sensation seeking predicts consumption of a nicotine solution,[166] consumption of a morphine solution,[167] ethanol self-administration,[168] *d*-amphetamine self-administration,[169] compulsive cocaine self-administration,[170] and CPP of stimulants and opioids.[167,171–173] Additionally, high novelty seeking rats are more likely to treat cocaine as an inelastic good, but only when cocaine is made available on an intermittent basis.[174] Rats that are selectively bred to exhibit increased sensation seeking show greater preference for ethanol,[175] acquire cocaine self-administration at a faster rate,[176] and are more motivated to respond for cocaine in a progressive ratio task.[177] In Chapter 6, you learned about the association between incentive salience and addiction. Preclinical research has shown that novelty seeking is an important mediator of this association.[178] That is, animals that are more likely to attribute incentive salience to food-paired stimuli show greater novelty seeking and self-administer more cocaine.

As detailed here and in Chapter 8, impulsivity is a major predictor of SUDs. One neurocognitive disorder associated with elevated impulsivity is ADHD. ADHD was mentioned briefly in Chapter 1 when discussing comorbidities in individuals with a SUD. The following section will discuss the relationship between ADHD and SUDs in more detail.

Attention deficit/hyperactivity disorder

Like impulsivity, there are different types of ADHD, depending on what impulsive behaviors individuals show. There is a hyperactivity-impulsive subtype. As the name suggests, individuals diagnosed with this ADHD subtype have difficulty sitting still, and they talk excessively. These individuals also have difficulty waiting their turn and interrupt others during conversation. Another facet of ADHD is inattention, in which individuals have difficulty sustaining attention or focusing on a particular task. Finally, some individuals are diagnosed with a combination subtype in which they display both hyperactivity and inattentiveness.

ADHD and addiction

SUDs and ADHD often co-occur,[179] and those with ADHD have a higher chance of developing a SUD later in life.[180–183] Approximately 11%—23% of drug treatment seekers meet diagnostic criteria for ADHD.[184–187] Those with ADHD have a longer history of cocaine use compared to non-ADHD individuals,[188] smoke more cigarettes,[189,190] and are more likely to be dependent on ethanol, nicotine, or cannabis.[191,192] Research has also shown that ADHD individuals report earlier onset of nicotine and illicit drug use[179,190,193] while the presence of ADHD symptoms at age 20 is associated with the late initiation (between 20 and 25 years of age) of several substances, including psychostimulants, MDMA, and hallucinogens.[194] An ADHD diagnosis is a significant predictor of daily tobacco smoking[195,196] and number of cigarettes smoked.[193] In particular, women with ADHD are 5 times more likely to develop tobacco use disorder compared to women without ADHD.[197] Individuals with ADHD are also more likely to engage in risky ethanol and nicotine use[198] and experience more severe withdrawal symptoms upon nicotine cessation.[195,199,200]

Why are individuals with ADHD more likely to use substances? Overall, there appears to be genetic overlap between ADHD and SUDs that may account for the increased rates of SUDs in this population.[201] One potential genetic link between ADHD and SUDs is a variant of the *SYT1* gene.[202] This gene encodes for synaptotagmin-1, which regulates the release of neurotransmitters from presynaptic vesicles into the synapse. Because individuals with ADHD are proposed to have too little dopamine/norepinephrine in the mesocorticolimbic pathway, drug use may act as a form of self-medication. Approximately 36% of individuals diagnosed with ADHD report that they use tobacco and other substances to self-medicate while 25% reported using substances for their intoxicating effects.[203]

One important question is if SUD severity differs across ADHD subtypes. Individuals with the combined ADHD subtype tend to display more severe SUDs compared to individuals diagnosed with the inattentive subtype[204,205] or the hyperactive-impulsive subtype,[204] but the inattentive subtype is a significant predictor of alcohol use disorder.[206] The number of inattentive and hyperactive symptoms is associated with ethanol, tobacco, and cannabis use during the past month,[207] and both ADHD subtypes significantly predict lifetime smoking status[208] and nicotine dependence.[209] However, a different study reported that hyperactivity/impulsivity subtype poses greater risk of developing SUDs for ethanol, tobacco, and other illicit drugs.[210] Additional research has shown that individuals diagnosed with the inattentive subtype are less likely to have cocaine dependence compared to individuals with the hyperactive-impulsive subtype.[211,212] Thus, there is evidence that shows that ADHD subtypes can differentially predict later substance use problems.

Does psychostimulant treatment for ADHD increase the risk of developing a SUD?

Given that ADHD is commonly treated with psychostimulant medications like amphetamine and methylphenidate and that ADHD individuals are at higher risk of developing a SUD, there have been some concerns about prescribing these medications. Although adults with ADHD treated with methylphenidate smoke more cigarettes during a 4-h period[213] and following 3 months of treatment,[189] there is not much support to the claim that treating children and adolescents with ADHD medications increases the risk of developing a SUD later in life.[214,215] For example, individuals that use stimulant medications long term are *less* likely to smoke cigarettes or use cocaine.[216,217] Other studies have shown that ADHD medication use decreases the odds of concurrent substance use,[218,219] with one study even showing that treating ADHD individuals with psychostimulant medications reduces the risk of developing a SUD by 85%[214]!

Modeling ADHD in rodents

During the 1960s, Japanese researchers selected Wistar-Kyoto (WKY) rats that had high hypertension and bred them. The goal was to develop a rodent model of hypertension.[220] Selectively inbreeding WKY rats with high blood pressure resulted in the spontaneously hypertensive rat (SHR). Over time, researchers noticed something odd about SHRs. They were hyperactive and appeared to have attentional deficits. As an animal researcher, I have witnessed firsthand the hyperactivity of SHRs. Because of these ADHD-like symptoms, the

SHR has become a popular preclinical model of ADHD. Compared to other strains such as the WKY, SHRs are more impulsive in delay discounting[221–227] (but see[228,229]) and display greater impulsive action[230] (but see[231]). It is worth mentioning that there are two major substrains of SHRs: the SHR/NCrl and the SHR/NHsd. The SHR/NCrl comes from a company called Charles River Laboratories while the SHR/NHsd comes from Harlan Industries (now known as Envigo). Amazingly, SHRs from these two companies display strikingly different behavioral phenotypes. SHRs coming from Charles River display greater motor impulsivity compared to those coming from Envigo.[232]

If SHRs are a model of ADHD, one would expect ADHD medications to improve impulsivity in these rats. Surprisingly, d-amphetamine does not appear to improve impulsive choice as measured in delay discounting,[224,227] nor improve attention.[233] However, a different isomer of amphetamine (l-amphetamine) reduces prepotent responding[234] and improves impulsivity and attention in a visual discrimination task.[235] SHRs treated with methylphenidate during adolescence become less hyperactive, but they display greater impulsive action during adulthood.[236] Because these rats were no longer on methylphenidate, the increased impulsivity could be due to drug withdrawal. Regardless, acute methylphenidate administration, whether during adolescence or adulthood, fails to decrease impulsive choice[222] or impulsive action[231] in SHRs.

Given that individuals with ADHD are more likely to develop a SUD, SHRs should demonstrate an increased propensity to self-administer drugs. Indeed, SHRs acquire methylphenidate self-administration at a faster rate compared to WKY and Sprague Dawley (an outbred strain) rats,[237] and they self-administer more methylphenidate compared to Wistar rats (another outbred strain).[238] Similarly, SHRs acquire cocaine self-administration at a faster rate and self-administer more cocaine compared to WKY and Wistar rats.[239–242] Not only are SHR/NCrl more impulsive than SHR/NHsd, but they self-administer more cocaine.[232]

Although psychostimulant treatment during adolescence is not necessarily a risk factor for development of a SUD in humans, preclinical studies have shown that SHRs treated with methylphenidate during adolescence acquire cocaine self-administration at a faster rate, self-administer more cocaine, and show greater reinstatement of cocaine seeking compared to those treated with vehicle and compared to Wistar/WKY rats.[240,243,244] However, repeated treatment of methylphenidate reduces self-administration of methylphenidate in a progressive ratio schedule in SHRs and prevents relapse-like behavior.[245] Also, early-life exposure to methylphenidate decreases cocaine CPP later in life.[246] While methylphenidate treatment potentiates cocaine self-administration in SHRs, atomoxetine (nonstimulant ADHD medication) administration during adolescence does not increase the reinforcing effects of cocaine.[242] In fact, atomoxetine treatment can protect against reinstatement of cocaine-seeking behavior.[240] Furthermore, amphetamine administration during adolescence protects SHRs from cocaine self-administration[241] and relapse-like behavior,[241] mirroring what has been reported in clinical studies.

Is the SHR a valid model of ADHD? There is some debate concerning the validity of the SHR as an animal model of ADHD. Some have argued that the SHR is not an appropriate model of ADHD.[231,247,248] Perhaps the biggest proponent of using SHRs to model ADHD-like symptoms was Dr. Terje Sagvolden (1945–2011), who spent a large portion of his career studying SHRs. He was the first individual to publish the finding that this strain is hyperactive.[249] Throughout his career he argued that the SHR, particularly the SHR/NCrl substrain, closely mimics ADHD.[250] Sagvolden also argued that some of the discrepant findings reported in the literature may be due to the use of inappropriate control strains in addition to using the SHR/NHsd substrain.[251] For example, studies often compare the SHR to WKY rats and either Wistar rats[231] or Sprague Dawley rats.[248] Like SHRs, there are two substrains of the WKY rat: one from Charles River (WKY/NCrl) and one from Harlan/Envigo (WKY/NHsd). Because WKY/NCrl rats display behaviors that are similar to the inattentive subtype of ADHD, Sagvolden et al. argue that the most appropriate control strain is the WKY/NHsd.[251]

Quiz 9.2

(1) Which of the following statements about long-term ADHD treatment is true?
   a. Acute administration of methylphenidate decreases cigarette smoking.
   b. Individuals treated with ADHD medications during adolescence are more likely to develop a SUD later in life.
   c. Long-term treatment with ADHD medications can act as a protective factor against the development of a SUD.
   d. Preclinical models show that long-term methylphenidate administration decreases cocaine self-administration.
(2) According to Terje Sagvolden, which rat strain is considered the best animal model of ADHD?
   a. Sprague Dawley
   b. Spontaneously Hypertensive Rat from Charles River
   c. Spontaneously Hypertensive Rat from Envigo
   d. Wistar Kyoto rat
(3) What is the difference between novelty seeking and sensation seeking?
(4) Which of the following statements best summarizes the relationship between ADHD and SUDs?
   a. Individuals with ADHD are less likely to develop a SUD compared to non-ADHD individuals.
   b. Individuals with ADHD are more likely develop a SUD compared to non-ADHD individuals.
   c. Individuals with ADHD are more likely to develop psychostimulant dependence compared to non-ADHD individuals, but are not at increased risk of developing a SUD to depressants.
   d. SUD rates are similar between ADHD individuals and non-ADHD individuals.
(5) Which impulsivity construct is associated with alcohol use disorder?
   a. Impulsive action
   b. Novelty seeking
   c. Urgency
   d. All of the above

Answers to quiz 9.2

(1) c — Long-term treatment with ADHD medications can act a protective factor against the development of a SUD.
(2) b — Spontaneously Hypertensive Rat from Charles River

(3) They are nearly identical constructs. Both describe impulsive behaviors characterized by novelty exploration and are both associated with SUDs.
(4) b — Individuals with ADHD are more likely develop a SUD compared to non-ADHD individuals.
(5) d — All of the above

Chapter summary

Individual differences are an important predictor of drug use and the development of SUDs. In particular, personality traits such as extraversion, neuroticism, openness to experience, harm avoidance, and psychopathic deviance have consistently been linked to addiction-like behaviors. Individual differences have also been shown to predict how individuals respond to interventions for SUDs and have been shown to predict problems associated with substance use (e.g., psychostimulant-induced psychosis). Preclinical models have been able to assess individual differences in addiction, with a focus on impulsive behaviors. These preclinical models have largely corroborated clinical findings: increased impulsivity is a predisposing factor for drug use. Not surprisingly, individuals with ADHD, which is characterized by increased impulsivity, are at increased risk of developing a SUD. However, contrary to some beliefs, there is little evidence to support the claim that long-term psychostimulant treatment for ADHD increases drug use/misuse vulnerability. While certain traits are associated with addiction, it is important to emphasize there are numerous people that score high in extraversion or in neuroticism that do not use drugs. The focus of the next chapter is to discuss important sociocultural factors that influence substance use and can moderate the associations between personality traits and addiction.

Glossary

Five-factor model of personality — theory that states that personality can be fractioned into five categories: openness to experience, conscientiousness, extraversion, agreeableness, and neuroticism. Also referred to as the "Big Five" factors of personality.
Eysenck Personality Questionnaire (EPQ) — questionnaire used to measure personality on four continuums: extraversion/introversion, neuroticism/stability, psychoticism/socialization, and lie/social desirability.
Impulsive action — form of impulsivity in which individuals have difficulty withholding a response (also known as motor impulsivity).
Impulsive choice — form of impulsivity in which individuals have difficulty delaying gratification (often used interchangeably with impulsive decision making).
Individual differences — stable traits or characteristics that vary across individuals.
Novelty seeking — increased exploratory activity in response to new stimulation; also characterized by increased impulsivity.
Personality — the combination of characteristics or qualities that form an individual's distinctive character.
Sensation seeking — desire to participate in intense experiences (see novelty seeking).
Urgency — tendency to act rashly when in a heightened emotional state.

Supplementary data related to this chapter can be found online at https://educate.elsevier.com/9780323905787.

References

1. American Psychological Association. Personality; n.d. https://www.apa.org/topics/personality.
2. Eysenck H. *Dimensions of Personality*. Kegan Paul, Trench, Trubner & Co., Ltd; 1947.
3. Eysenck HJ, Eysenck SBG. *Manual of the Eysenck Personality Inventory*. Hodder and Stoughton; 1964.
4. Eysenck HJ, Eysenck SBG. *Manual of the Eysenck Personality Questionnaire*. Hodder and Stoughton; 1975.
5. Allport GW, Odbert HS. Trait-names: a psycho-lexical study. *Psychol Monogr*. 1936;47(1):i–171. https://doi.org/10.1037/h0093360.
6. Cattell RB, Eber HW, Tatsuoka MM. *Handbook for the Sixteen Personality Factor Questionnaire (16 PF)*. Institute for Personality and Ability Testing; 1970.
7. Costa Jr PT, McCrae RR. *The NEO Personality Inventory Manual*. Psychological Assessment Resources; 1985.
8. McCrae RR, Costa Jr PT, Martin TA. The NEO-PI-3: a more readable revised NEO personality inventory. *J Pers Assess*. 2005;84(3):261–270. https://doi.org/10.1207/s15327752jpa8403_05.
9. Zuckerman M, Kuhlman DM, Thornquist M, Kiers H. Five (or three) robust questionnaire scale factors of personality without culture. *Pers Indiv Differ*. 1991;12(9):929–941. https://doi.org/10.1016/0191-8869(91)90182-B.
10. Zuckerman M, Kuhlman DM, Joireman J, Teta P, Kraft M. A comparison of three structural models for personality: the big three, the big five, and the alternative five. *J Pers Soc Psychol*. 1993;65(4):757–768. https://doi.org/10.1037/0022-3514.65.4.757.
11. Cloninger CR, Przybeck TR, Svrakic DM. The tridimensional personality questionnaire: U.S. Normative data. *Psychol Rep*. 1991;69(3 Pt 1):1047–1057. https://doi.org/10.2466/pr0.1991.69.3.1047.
12. Cloninger CR. *The Temperament and Character Inventory (TCI): A Guide to its Development and Use*. Center for Psychobiology of Personality, Washington University; 1994.
13. McCourt WF, Gurrera RJ, Cutter HS. Sensation seeking and novelty seeking. Are they the same? *J Nerv Ment Dis*. 1993;181(5):309–312. https://doi.org/10.1097/00005053-199305000-00006.
14. Hathaway SR, McKinley JC. A multiphasic personality schedule (Minnesota): I. Construction of the schedule. *J Psychol*. 1940;10:249–254. https://doi.org/10.1080/00223980.1940.9917000.
15. Ben-Porath YS, Tellegen A. *Minnesota Multiphasic Personality Inventory-3 (MMPI-3): Manual for Administration, Scoring, and Interpretation*. University of Minnesota Press; 2020.
16. Butcher JN, Graham JR, Tellegen A, Kaemmer B. *Manual for the Restandardized Minnesota Multiphasic Personality Inventory: MMPI-2*. University of Minnesota Press; 1989.
17. Adan A, Forero DA, Navarro JF. Personality traits related to binge drinking: a systematic review. *Front Psychiatr*. 2017;8:134. https://doi.org/10.3389/fpsyt.2017.00134.
18. Choi J-S, Payne TJ, Ma JZ, Li MD. Relationship between personality traits and nicotine dependence in male and female smokers of African-American and European-American samples. *Front Psychiatr*. 2017;8:122. https://doi.org/10.3389/fpsyt.2017.00122.
19. Cohen K, Rosenzweig S, Rosca P, Pinhasov A, Weizman A, Weinstein A. Personality traits and psychotic proneness among chronic synthetic cannabinoid users. *Front Psychiatr*. 2020;11:355. https://doi.org/10.3389/fpsyt.2020.00355.
20. Flory K, Lynam D, Milich R, Leukefeld C, Clayton R. The relations among personality, symptoms of alcohol and marijuana abuse, and symptoms of comorbid psychopathology: results from a community sample. *Exp Clin Psychopharmacol*. 2002;10(4):425–434. https://doi.org/10.1037//1064-1297.10.4.425.
21. Fridberg DJ, Vollmer JM, O'Donnell BF, Skosnik PD. Cannabis users differ from non-users on measures of personality and schizotypy. *Psychiatr Res*. 2011;186(1):46–52. https://doi.org/10.1016/j.psychres.2010.07.035.
22. Greenbaum L, Kanyas K, Karni O, et al. Why do young women smoke? I. Direct and interactive effects of environment, psychological characteristics and nicotinic cholinergic receptor genes. *Mol Psychiatr*. 2006;11(3). https://doi.org/10.1038/sj.mp.4001774, 312–223.
23. Grzywacz A, Suchanecka A, Chmielowiec J, et al. Personality traits or genetic determinants-which strongly influences E-cigarette users? *Int J Environ Res Publ Health*. 2020;17(1):365. https://doi.org/10.3390/ijerph17010365.
24. Hakulinen C, Hintsanen M, Munafò MR, et al. Personality and smoking: individual-participant meta-analysis of nine cohort studies. *Addiction*. 2015;110(11):1844–1852. https://doi.org/10.1111/add.13079.
25. Hojjat SK, Golmakani E, Bayazi MH, Mortazavi R, Norozi Khalili M, Akaberi A. Personality traits and identity styles in methamphetamine-dependent women: a comparative study. *Global J Health Sci*. 2015;8(1):14–20. https://doi.org/10.5539/gjhs.v8n1p14.

26. Listabarth S, Vyssoki B, Waldhoer T, et al. Hazardous alcohol consumption among older adults: a comprehensive and multi-national analysis of predictive factors in 13,351 individuals. *Eur Psychiatr*. 2020;64(1):e4. https://doi.org/10.1192/j.eurpsy.2020.112.
27. Lui PP, Chmielewski M, Trujillo M, Morris J, Pigott TD. Linking big five personality domains and facets to alcohol (mis)use: a systematic review and meta-analysis. *Alcohol Alcohol*. 2022;57(1):58–73. https://doi.org/10.1093/alcalc/agab030.
28. Munafò MR, Zetteler JI, Clark TG. Personality and smoking status: a meta-analysis. *Nicotine Tob Res*. 2007;9(3):405–413. https://doi.org/10.1080/14622200701188851.
29. Pashapour H, Musavi S, Dadashzadeh H, Mohammadpoorasl A. Relationship between extraversion and tobacco smoking among high school students. *Int J Prev Med*. 2020;11:134. https://doi.org/10.4103/ijpvm.IJPVM_177_19.
30. Phillips KT, Phillips MM, Duck KD. Factors associated with marijuana use and problems among college students in Colorado. *Subst Use Misuse*. 2018;53(3):477–483. https://doi.org/10.1080/10826084.2017.1341923.
31. Tate DL, Charette L. Personality, alcohol consumption, and menstrual distress in young women. *Alcohol Clin Exp Res*. 1991;15(4):647–652. https://doi.org/10.1111/j.1530-0277.1991.tb00573.x.
32. Terracciano A, Löckenhoff CE, Crum RM, Bienvenu OJ, Costa Jr PT. Five-Factor Model personality profiles of drug users. *BMC Psychiatr*. 2008;8:22. https://doi.org/10.1186/1471-244X-8-22.
33. Zvolensky MJ, Taha F, Bono A, Goodwin RD. Big five personality factors and cigarette smoking: a 10-year study among US adults. *J Psychiatr Res*. 2015;63:91–96. https://doi.org/10.1016/j.jpsychires.2015.02.008.
34. Sahasi G, Chawla HM, Bhushan B, Kacker C. Eysenck's personality questionnaire scores of heroin addicts in India. *Indian J Psychiatr*. 1990;32(1):25–29.
35. Spotts JV, Shontz FC. Drugs and personality: extraversion-introversion. *J Clin Psychol*. 1984;40(2):624–628. https://doi.org/10.1002/1097-4679(198403)40:2<624::aid-jclp2270400243>3.0.co;2-s.
36. Bucher JT, Vu DM, Hojat M. Psychostimulant drug abuse and personality factors in medical students. *Med Teach*. 2013;35(1):53–57. https://doi.org/10.3109/0142159X.2012.731099.
37. Hakulinen C, Elovainio M, Batty GD, Virtanen M, Kivimäki M, Jokela M. Personality and alcohol consumption: pooled analysis of 72,949 adults from eight cohort studies. *Drug Alcohol Depend*. 2015;151:110–114. https://doi.org/10.1016/j.drugalcdep.2015.03.008.
38. Rass O, Ahn WY, O'Donnell BF. Resting-state EEG, impulsiveness, and personality in daily and nondaily smokers. *Clin Neurophysiol*. 2016;127(1):409–418. https://doi.org/10.1016/j.clinph.2015.05.007.
39. Ozuna Esprinosa MS, Candia Arredondo JS, Alonso Castillo MM, López García KS, Guzmán Facundo FR. Factors in the transition from legal to illicit drug use in young adults from Northern Mexico. *Invest Educ Enfermería*. 2019;37(3):e11. https://doi.org/10.17533/udea.iee.v37n3e11.
40. Fairbairn CE, Sayette MA, Wright AG, Levine JM, Cohn JF, Creswell KG. Extraversion and the rewarding effects of alcohol in a social context. *J Abnorm Psychol*. 2015;124(3):660–673. https://doi.org/10.1037/abn0000024.
41. Kawakami N, Takai A, Takatsuka N, Shimizu H. Eysenck's personality and tobacco/nicotine dependence in male ever-smokers in Japan. *Addict Behav*. 2000;25(4):585–591. https://doi.org/10.1016/s0306-4603(99)00019-2.
42. Kleinjan M, Vitaro F, Wanner B, Brug J, Van den Eijnden RJ, Engels RC. Predicting nicotine dependence profiles among adolescent smokers: the roles of personal and social-environmental factors in a longitudinal framework. *BMC Publ Health*. 2012;12:196. https://doi.org/10.1186/1471-2458-12-196.
43. Rajapaksha RMDS, Hammonds R, Filbey F, Choudhary PK, Biswas S. A preliminary risk prediction model for cannabis use disorder. *Prevent Med Rep*. 2020;20:101228. https://doi.org/10.1016/j.pmedr.2020.101228.
44. Zilberman N, Yadid G, Efrati Y, Neumark Y, Rassovsky Y. Personality profiles of substance and behavioral addictions. *Addict Behav*. 2018;82:174–181. https://doi.org/10.1016/j.addbeh.2018.03.007.
45. Zilberman N, Yadid G, Efrati Y, Rassovsky Y. Who becomes addicted and to what? psychosocial predictors of substance and behavioral addictive disorders. *Psychiatr Res*. 2020;291:113221. https://doi.org/10.1016/j.psychres.2020.113221.
46. Gilbert DG, Crauthers DM, Mooney DK, McClernon FJ, Jensen RA. Effects of monetary contingencies on smoking relapse: influences of trait depression, personality, and habitual nicotine intake. *Exp Clin Psychopharmacol*. 1999;7(2):174–181. https://doi.org/10.1037//1064-1297.7.2.174.
47. Powell J, Dawe S, Richards D, et al. Can opiate addicts tell us about their relapse risk? Subjective predictors of clinical prognosis. *Addict Behav*. 1993;18(4):473–490. https://doi.org/10.1016/0306-4603(93)90065-h.

48. Hakulinen C, Jokela M. Alcohol use and personality trait change: pooled analysis of six cohort studies. *Psychol Med.* 2019;49(2):224–231. https://doi.org/10.1017/S0033291718000636.
49. Beijers C, Burger H, Verbeek T, Bockting CL, Ormel J. Continued smoking and continued alcohol consumption during early pregnancy distinctively associated with personality. *Addict Behav.* 2014;39(5):980–986. https://doi.org/10.1016/j.addbeh.2014.01.022.
50. Leszko M, Keenan-Devlin L, Adam EK, et al. Are personality traits associated with smoking and alcohol use prior to and during pregnancy? *PLoS One.* 2020;15(5):e0232668. https://doi.org/10.1371/journal.pone.0232668.
51. Ystrom E, Vollrath ME, Nordeng H. Effects of personality on use of medications, alcohol, and cigarettes during pregnancy. *Eur J Clin Pharmacol.* 2012;68(5):845–851. https://doi.org/10.1007/s00228-011-1197-y.
52. Roncero C, Daigre C, Barral C, et al. Neuroticism associated with cocaine-induced psychosis in cocaine-dependent patients: a cross-sectional observational study. *PLoS One.* 2014;9(9):e106111. https://doi.org/10.1371/journal.pone.0106111.
53. Tacheva Z, Ivanov A. Exploring the association between the "Big Five" personality traits and fatal opioid overdose: county-level empirical analysis. *JMIR Mental Health.* 2021;8(3):e24939. https://doi.org/10.2196/24939.
54. Etter J-F. Smoking and Cloninger's temperament and character inventory. *Nicotine Tob Res.* 2010;12(9):919–926. https://doi.org/10.1093/ntr/ntq116.
55. Schneider Jr R, Ottoni GL, de Carvalho HW, Elisabetsky E, Lara DR. Temperament and character traits associated with the use of alcohol, cannabis, cocaine, benzodiazepines, and hallucinogens: evidence from a large Brazilian web survey. *Rev Bras Psiquiatr.* 2015;37(1):31–39. https://doi.org/10.1590/1516-4446-2014-1352.
56. Teh LK, Izuddin AF, Fazkeen HMH, Zakaria ZA, Salleh MZ. Tridimensional personalities and polymorphism of dopamine D2 receptor among heroin addicts. *Biol Res Nurs.* 2012;14(2):188–196. https://doi.org/10.1177/1099800411405030.
57. Galen LW, Henderson MJ, Whitman RD. The utility of novelty seeking, harm avoidance, and expectancy in the prediction of drinking. *Addict Behav.* 1997;22(1):93–106. https://doi.org/10.1016/s0306-4603(96)00018-4.
58. Pomerleau CS, Pomerleau OF, Flessland KA, Basson SM. Relationship of Tridimensional Personality Questionnaire scores and smoking variables in female and male smokers. *J Subst Abuse.* 1992;4(2):143–154. https://doi.org/10.1016/0899-3289(92)90014-o.
59. Zaaijer ER, Bruijel J, Blanken P, et al. Personality as a risk factor for illicit opioid use and a protective factor for illicit opioid dependence. *Drug Alcohol Depend.* 2014;145:101–105. https://doi.org/10.1016/j.drugalcdep.2014.09.783.
60. Lee YS, Son JH, Park JH, Kim SM, Kee BS, Han DH. The comparison of temperament and character between patients with internet gaming disorder and those with alcohol dependence. *J Ment Health.* 2017;26(3):242–247. https://doi.org/10.1080/09638237.2016.1276530.
61. Estedlal AR, Mani A, Vardanjani HM, et al. Temperament and character of patients with alcohol toxicity during COVID-19 pandemic. *BMC Psychiatr.* 2021;21(1):49. https://doi.org/10.1186/s12888-021-03052-1.
62. Foulds J, Newton-Howes G, Guy NH, Boden JM, Mulder RT. Dimensional personality traits and alcohol treatment outcome: a systematic review and meta-analysis. *Addiction.* 2017;112(8):1345–1357. https://doi.org/10.1111/add.13810.
63. Andrucci GL, Archer RP, Pancoast DL, Gordon RA. The relationship of MMPI and Sensation Seeking Scales to adolescent drug use. *J Pers Assess.* 1989;53(2):253–266. https://doi.org/10.1207/s15327752jpa5302_4.
64. Goldstein JW, Sappington JT. Personality characteristics of students who became heavy drug users: an MMPI study of an avant-garde. *Am J Drug Alcohol Abuse.* 1977;4(3):401–412. https://doi.org/10.3109/00952997709002774.
65. Varma VK, Basu D, Malhotra A, Sharma A, Mattoo SK. Correlates of early- and late-onset alcohol dependence. *Addict Behav.* 1994;19(6):609–619. https://doi.org/10.1016/0306-4603(94)90016-7.
66. Zeng H, Su D, Jiang X, Zhu L, Ye H. The similarities and differences in impulsivity and cognitive ability among ketamine, methadone, and non-drug users. *Psychiatr Res.* 2016;243:109–114. https://doi.org/10.1016/j.psychres.2016.04.095.
67. Zhao RJ, Sun W, Zhang LL, et al. Psychopathology and personality traits associated with driving while intoxicated in Beijing, China: implications for interventions. *Am J Addict.* 2017;26(4):374–378. https://doi.org/10.1111/ajad.12536.
68. Roma P, Mazza C, Ferracuti G, Cinti ME, Ferracuti S, Burla F. Drinking and driving relapse: data from BAC and MMPI-2. *PLoS One.* 2019;14(1):e0209116. https://doi.org/10.1371/journal.pone.0209116.

69. Rezvanfard M, Ekhtiari H, Mokri A, Djavid G, Kaviani H. Psychological and behavioral traits in smokers and their relationship with nicotine dependence level. *Arch Iran Med*. 2010;13(5):395–405.
70. Petit G, Deschietere G, Loas G, Luminet O, de Timary P. Link between anhedonia and depression during early alcohol abstinence: gender matters. *Alcohol Alcohol*. 2020;55(1):71–77. https://doi.org/10.1093/alcalc/agz090.
71. Braidwood R, Mansell S, Waldron J, Rendell PG, Kamboj SK, Curran HV. Non-dependent and dependent daily cannabis users differ in mental health but not prospective memory ability. *Front Psychiatr*. 2018;9:97. https://doi.org/10.3389/fpsyt.2018.00097.
72. Mahoney 3rd JJ, Thompson-Lake DG, Cooper K, Verrico CD, Newton TF, De La Garza 2nd R. A comparison of impulsivity, depressive symptoms, lifetime stress and sensation seeking in healthy controls versus participants with cocaine or methamphetamine use disorders. *J Psychopharmacol*. 2015;29(1):50–56. https://doi.org/10.1177/0269881114560182.
73. Li S-X, Yan S-Y, Bao Y-P, et al. Depression and alterations in hypothalamic-pituitary-adrenal and hypothalamic-pituitary-thyroid axis function in male abstinent methamphetamine abusers. *Hum Psychopharmacol*. 2013;28(5):477–483. https://doi.org/10.1002/hup.2335.
74. Seifert CL, Magon S, Sprenger T, et al. Reduced volume of the nucleus accumbens in heroin addiction. *Eur Arch Psychiatr Clin Neurosci*. 2015;265(8):637–645. https://doi.org/10.1007/s00406-014-0564-y.
75. Sofuoglu M, Brown S, Dudish-Poulsen S, Hatsukami DK. Individual differences in the subjective response to smoked cocaine in humans. *Am J Drug Alcohol Abuse*. 2000;26(4):591–602. https://doi.org/10.1081/ada-100101897.
76. Secades-Villa R, Weidberg S, González-Roz A, Reed DD, Fernández-Hermida JR. Cigarette demand among smokers with elevated depressive symptoms: an experimental comparison with low depressive symptoms. *Psychopharmacology*. 2018;235(3):719–728. https://doi.org/10.1007/s00213-017-4788-1.
77. Curran GM, Flynn HA, Kirchner J, Booth BM. Depression after alcohol treatment as a risk factor for relapse among male veterans. *J Subst Abuse Treat*. 2000;19(3):259–265. https://doi.org/10.1016/s0740-5472(00)00107-0.
78. Greenfield SF, Weiss RD, Muenz LR, et al. The effect of depression on return to drinking: a prospective study. *Arch Gen Psychiatr*. 1998;55(3):259–265. https://doi.org/10.1001/archpsyc.55.3.259.
79. Keizer I, Gex-Fabry M, Croquette P, Humair JP, Khan AN. Tobacco craving and withdrawal symptoms in psychiatric patients during a motivational enhancement intervention based on a 26-hour smoking abstinence period. *Tobac Prevent & Cessat*. 2019;5:22. https://doi.org/10.18332/tpc/109785.
80. Ranjit A, Latvala A, Kinnunen TH, Kaprio J, Korhonen T. Depressive symptoms predict smoking cessation in a 20-year longitudinal study of adult twins. *Addict Behav*. 2020;108:106427. https://doi.org/10.1016/j.addbeh.2020.106427.
81. Sonne SC, Nunes EV, Jiang H, Tyson C, Rotrosen J, Reid MS. The relationship between depression and smoking cessation outcomes in treatment-seeking substance abusers. *Am J Addict*. 2010;19(2):111–118. https://doi.org/10.1111/j.1521-0391.2009.00015.x.
82. Stein MD, Herman DS, Kettavong M, et al. Antidepressant treatment does not improve buprenorphine retention among opioid-dependent persons. *J Subst Abuse Treat*. 2010;39(2):157–166. https://doi.org/10.1016/j.jsat.2010.05.014.
83. Crits-Christoph P, Wadden S, Gaines A, et al. Symptoms of anhedonia, not depression, predict the outcome of treatment of cocaine dependence. *J Subst Abuse Treat*. 2018;92:46–50. https://doi.org/10.1016/j.jsat.2018.06.010.
84. Logan GD, Cowan WB, Davis KA. On the ability to inhibit simple and choice reaction time responses: a model and a method. *J Exp Psychol Hum Percept Perform*. 1984;10(2):276–291. https://doi.org/10.1037//0096-1523.10.2.276.
85. Newman JP, Widom CS, Nathan S. Passive avoidance in syndromes of disinhibition: psychopathy and extraversion. *J Pers Soc Psychol*. 1985;48(5):1316–1327. https://doi.org/10.1037//0022-3514.48.5.1316.
86. Winstanley CA, Olausson P, Taylor JR, Jentsch JD. Insight into the relationship between impulsivity and substance abuse from studies using animal models. *Alcohol Clin Exp Res*. 2010;34(8):1306–1318. https://doi.org/10.1111/j.1530-0277.2010.01215.x.
87. Patton JH, Stanford MS, Barratt ES. Factor structure of the Barratt impulsiveness scale. *J Clin Psychol*. 1995;51(6):768–774. https://doi.org/10.1002/1097-4679(199511)51:6<768::aid-jclp2270510607>3.0.co;2-1.
88. Kahn J-P, Cohen RF, Etain B, et al. Reconsideration of the factorial structure of the Barratt Impulsiveness Scale (BIS-11): assessment of impulsivity in a large population of euthymic bipolar patients. *J Affect Disord*. 2019;253:203–209. https://doi.org/10.1016/j.jad.2019.04.060.

89. Whiteside SP, Lynam DR. The Five Factor Model and impulsivity: using a structural model of personality to understand impulsivity. *Pers Indiv Differ*. 2001;30(4):669–689. https://doi.org/10.1016/S0191-8869(00)00064-7.
90. Smith JL, Mattick RP, Jamadar SD, Iredale JM. Deficits in behavioural inhibition in substance abuse and addiction: a meta-analysis. *Drug Alcohol Depend*. 2014;145:1–33. https://doi.org/10.1016/j.drugalcdep.2014.08.009.
91. Verdejo-García A, Bechara A, Recknor EC, Pérez-García M. Executive dysfunction in substance dependent individuals during drug use and abstinence: an examination of the behavioral, cognitive and emotional correlates of addiction. *J Int Neuropsychol Soc*. 2006;12(3):405–415. https://doi.org/10.1017/s1355617706060486.
92. Kreusch F, Quertemont E, Vilenne A, Hansenne M. Alcohol abuse and ERP components in Go/No-go tasks using alcohol-related stimuli: impact of alcohol avoidance. *Int J Psychophysiol*. 2014;94(1):92–99. https://doi.org/10.1016/j.ijpsycho.2014.08.001.
93. Sanchez-Roige S, Peña-Oliver Y, Ripley TL, Stephens DN. Repeated ethanol exposure during early and late adolescence: double dissociation of effects on waiting and choice impulsivity. *Alcohol Clin Exp Res*. 2014;38(10):2579–2589. https://doi.org/10.1111/acer.12535.
94. Zhao X, Qian W, Fu L, Maes J. Deficits in go/no-go task performance in male undergraduate high-risk alcohol users are driven by speeded responding to go stimuli. *Am J Drug Alcohol Abuse*. 2017;43(6):656–663. https://doi.org/10.1080/00952990.2017.1282502.
95. Zhao X, Liu X, Zan X, Jin G, Maes JH. Male smokers' and non-smokers' response inhibition in go/no-go tasks: effect of three task parameters. *PLoS One*. 2016;11(8):e0160595. https://doi.org/10.1371/journal.pone.0160595.
96. Behan B, Connolly CG, Datwani S, et al. Response inhibition and elevated parietal-cerebellar correlations in chronic adolescent cannabis users. *Neuropharmacology*. 2014;84:131–137. https://doi.org/10.1016/j.neuropharm.2013.05.027.
97. Moreno M, Estevez AF, Zaldivar F, et al. Impulsivity differences in recreational cannabis users and binge drinkers in a university population. *Drug Alcohol Depend*. 2012;124(3):355–362. https://doi.org/10.1016/j.drugalcdep.2012.02.011.
98. Colzato LS, van den Wildenberg WP, Hommel B. Impaired inhibitory control in recreational cocaine users. *PLoS One*. 2007;2(11):e1143. https://doi.org/10.1371/journal.pone.0001143.
99. Fillmore MT, Rush CR. Impaired inhibitory control of behavior in chronic cocaine users. *Drug Alcohol Depend*. 2002;66(3):265–273. https://doi.org/10.1016/s0376-8716(01)00206-x.
100. Lane SD, Moeller FG, Steinberg JL, Buzby M, Kosten TR. Performance of cocaine dependent individuals and controls on a response inhibition task with varying levels of difficulty. *Am J Drug Alcohol Abuse*. 2007;33(5):717–726. https://doi.org/10.1080/00952990701522724.
101. Li CS, Milivojevic V, Kemp K, Hong K, Sinha R. Performance monitoring and stop signal inhibition in abstinent patients with cocaine dependence. *Drug Alcohol Depend*. 2006;85(3):205–212. https://doi.org/10.1016/j.drugalcdep.2006.04.008.
102. Pike E, Marks KR, Stoops WW, Rush CR. Cocaine-related stimuli impair inhibitory control in cocaine users following short stimulus onset asynchronies. *Addiction*. 2015;110(8):1281–1286. https://doi.org/10.1111/add.12947.
103. Verdejo-García AJ, Perales JC, Pérez-García M. Cognitive impulsivity in cocaine and heroin polysubstance abusers. *Addict Behav*. 2007;32(5):950–966. https://doi.org/10.1016/j.addbeh.2006.06.032.
104. Hess A, Menezes CB, de Almeida R. Inhibitory control and impulsivity levels in women crack users. *Subst Use Misuse*. 2018;53(6):972–979. https://doi.org/10.1080/10826084.2017.1387568.
105. Monterosso JR, Aron AR, Cordova X, Xu J, London ED. Deficits in response inhibition associated with chronic methamphetamine abuse. *Drug Alcohol Depend*. 2005;79(2):273–277. https://doi.org/10.1016/j.drugalcdep.2005.02.002.
106. Charles-Walsh K, Furlong L, Munro DG, Hester R. Inhibitory control dysfunction in nicotine dependence and the influence of short-term abstinence. *Drug Alcohol Depend*. 2014;143:81–86. https://doi.org/10.1016/j.drugalcdep.2014.07.008.
107. Prisciandaro JJ, Korte JE, McRae-Clark AL, Brady KT. Associations between behavioral disinhibition and cocaine use history in individuals with cocaine dependence. *Addict Behav*. 2012;37(10):1185–1188. https://doi.org/10.1016/j.addbeh.2012.05.015.
108. Nigg JT, Wong MM, Martel MM, et al. Poor response inhibition as a predictor of problem drinking and illicit drug use in adolescents at risk for alcoholism and other substance use disorders. *J Am Acad Child Adolesc Psychiatr*. 2006;45(4):468–475. https://doi.org/10.1097/01.chi.0000199028.76452.a9.

109. Henges AL, Marczinski CA. Impulsivity and alcohol consumption in young social drinkers. *Addict Behav.* 2012;37(2):217–220. https://doi.org/10.1016/j.addbeh.2011.09.013.
110. Ersche KD, Jones PS, Williams GB, Turton AJ, Robbins TW, Bullmore ET. Abnormal brain structure implicated in stimulant drug addiction. *Science*. 2012;335(6068):601–604. https://doi.org/10.1126/science.1214463.
111. Ersche KD, Turton AJ, Chamberlain SR, Müller U, Bullmore ET, Robbins TW. Cognitive dysfunction and anxious-impulsive personality traits are endophenotypes for drug dependence. *Am J Psychiatr.* 2012;169(9):926–936. https://doi.org/10.1176/appi.ajp.2012.11091421.
112. Dalley JW, Fryer TD, Brichard L, et al. Nucleus accumbens D2/3 receptors predict trait impulsivity and cocaine reinforcement. *Science*. 2007;315(5816):1267–1270. https://doi.org/10.1126/science.1137073.
113. Belin D, Mar AC, Dalley JW, Robbins TW, Everitt BJ. High impulsivity predicts the switch to compulsive cocaine-taking. *Science*. 2008;320(5881):1352–1355. https://doi.org/10.1126/science.1158136.
114. Bird J, Schenk S. Contribution of impulsivity and novelty-seeking to the acquisition and maintenance of MDMA self-administration. *Addiction Biol.* 2013;18(4):654–664. https://doi.org/10.1111/j.1369-1600.2012.00477.x.
115. Diergaarde L, Pattij T, Poortvliet I, et al. Impulsive choice and impulsive action predict vulnerability to distinct stages of nicotine seeking in rats. *Biol Psychiatr.* 2008;63(3):301–308. https://doi.org/10.1016/j.biopsych.2007.07.011.
116. Economidou D, Pelloux Y, Robbins TW, Dalley JW, Everitt BJ. High impulsivity predicts relapse to cocaine-seeking after punishment-induced abstinence. *Biol Psychiatr.* 2009;65(10):851–856. https://doi.org/10.1016/j.biopsych.2008.12.008.
117. Hammerslag LR, Belagodu AP, Aladesuyi Arogundade OA, et al. Adolescent impulsivity as a sex-dependent and subtype-dependent predictor of impulsivity, alcohol drinking and dopamine D2 receptor expression in adult rats. *Addiction Biol.* 2019;24(2):193–205. https://doi.org/10.1111/adb.12586.
118. Pattij T, van Mourik Y, Diergaarde L, de Vries TJ. The role of impulsivity as predisposing behavioural trait in different aspects of alcohol self-administration in rats. *Drug Alcohol Depend.* 2020;212:107984. https://doi.org/10.1016/j.drugalcdep.2020.107984.
119. Rubenis AJ, Fitzpatrick RE, Lubman DI, Verdejo-Garcia A. Impulsivity predicts poorer improvement in quality of life during early treatment for people with methamphetamine dependence. *Addiction*. 2018;113(4):668–676. https://doi.org/10.1111/add.14058.
120. Shi Z, Jagannathan K, Wang AL, et al. Behavioral and accumbal responses during an affective go/no-go task predict adherence to injectable naltrexone treatment in opioid use disorder. *Int J Neuropsychopharmacol.* 2019;22(3):180–185. https://doi.org/10.1093/ijnp/pyz002.
121. Round JT, Fozard TE, Harrison AA, Kolokotroni KZ. Disentangling the effects of cannabis and cigarette smoking on impulsivity. *J Psychopharmacol.* 2020;34(9):955–968. https://doi.org/10.1177/0269881120926674.
122. Chase HW, Hogarth L. Impulsivity and symptoms of nicotine dependence in a young adult population. *Nicotine Tob Res.* 2011;13(12):1321–1325. https://doi.org/10.1093/ntr/ntr114.
123. Dugas EN, Sylvestre MP, Ewusi-Boisvert E, Chaiton M, Montreuil A, O'Loughlin J. Early risk factors for daily cannabis use in young adults. *Can J Psychiatr.* 2019;64(5):329–337. https://doi.org/10.1177/0706743718804541.
124. Vonmoos M, Hulka LM, Preller KH, et al. Differences in self-reported and behavioral measures of impulsivity in recreational and dependent cocaine users. *Drug Alcohol Depend.* 2013;133(1):61–70. https://doi.org/10.1016/j.drugalcdep.2013.05.032.
125. Choi KW, Na EJ, Hong JP, et al. Alcohol-induced disinhibition is associated with impulsivity, depression, and suicide attempt: a nationwide community sample of Korean adults. *J Affect Disord.* 2018;227:323–329. https://doi.org/10.1016/j.jad.2017.11.001.
126. Hulka LM, Vonmoos M, Preller KH, et al. Changes in cocaine consumption are associated with fluctuations in self-reported impulsivity and gambling decision-making. *Psychol Med.* 2015;45(14):3097–3110. https://doi.org/10.1017/S0033291715001063.
127. Ryan KK, Mackillop J, Carpenter MJ. The relationship between impulsivity, risk-taking propensity and nicotine dependence among older adolescent smokers. *Addict Behav.* 2013;38(1):1431–1434. https://doi.org/10.1016/j.addbeh.2012.08.013.
128. Jakubczyk A, Trucco EM, Kopera M, et al. The association between impulsivity, emotion regulation, and symptoms of alcohol use disorder. *J Subst Abuse Treat.* 2018;91:49–56. https://doi.org/10.1016/j.jsat.2018.05.004.

129. Sübay B, Sönmez MB. Interoceptive awareness, decision-making and impulsiveness in male patients with alcohol or opioid use disorder. *Subst Use Misuse*. 2021;56(9):1275–1283. https://doi.org/10.1080/10826084.2021.1914108.
130. Peters L, Soyka M. Interrelationship of opioid dependence, impaired impulse control, and depressive symptoms: an open-label cross-sectional study of patients in maintenance therapy. *Neuropsychobiology*. 2019;77(2):73–82. https://doi.org/10.1159/000494697.
131. Akbari M, Bahadori MH, Mohammadkhani S, Kolubinski DC, Nikčević AV, Spada MM. A discriminant analysis model of psychosocial predictors of problematic Internet use and cannabis use disorder in university students. *Addict Behav Rep*. 2021;14:100354. https://doi.org/10.1016/j.abrep.2021.100354.
132. Szczypiński J, Jakubczyk A, Kopera M, Trucco E, Wojnar M. Impulsivity Scale-12 and its utilization in alcohol use disorder. *Drug Alcohol Depend*. 2021;225:108809. https://doi.org/10.1016/j.drugalcdep.2021.108809.
133. Evren C, Durkaya M, Evren B, Dalbudak E, Cetin R. Relationship of relapse with impulsivity, novelty seeking and craving in male alcohol-dependent inpatients. *Drug Alcohol Rev*. 2012;31(1):81–90. https://doi.org/10.1111/j.1465-3362.2011.00303.x.
134. Dinc L, Cooper AJ. Positive affective states and alcohol consumption: the moderating role of trait positive urgency. *Addict Behav*. 2015;47:17–21. https://doi.org/10.1016/j.addbeh.2015.03.014.
135. Cyders MA, Flory K, Rainer S, Smith GT. The role of personality dispositions to risky behavior in predicting first-year college drinking. *Addiction*. 2009;104(2):193–202. https://doi.org/10.1111/j.1360-0443.2008.02434.x.
136. Settles RE, Zapolski TC, Smith GT. Longitudinal test of a developmental model of the transition to early drinking. *J Abnorm Psychol*. 2014;123(1):141–151. https://doi.org/10.1037/a0035670.
137. Stautz K, Cooper A. Impulsivity-related personality traits and adolescent alcohol use: a meta-analytic review. *Clin Psychol Rev*. 2013;33(4):574–592. https://doi.org/10.1016/j.cpr.2013.03.003.
138. Whitt ZT, Bernstein M, Spillane N, et al. Positive urgency worsens the impact of normative feedback on 21st birthday drinking. *Drug Alcohol Depend*. 2019;204:107559. https://doi.org/10.1016/j.drugalcdep.2019.107559.
139. Waddell JT, Corbin WR, Leeman RF. Differential effects of UPPS-P impulsivity on subjective alcohol response and craving: an experimental test of acquired preparedness. *Exp Clin Psychopharmacol*. 2021. https://doi.org/10.1037/pha0000524.
140. Coskunpinar A, Dir AL, Cyders MA. Multidimensionality in impulsivity and alcohol use: a meta-analysis using the UPPS model of impulsivity. *Alcohol Clin Exp Res*. 2013;37(9):1441–1450. https://doi.org/10.1111/acer.12131.
141. Fischer S, Smith GT, Annus A, Hendricks M. The relationship of neuroticism and urgency to negative consequences of alcohol use in women with bulimic symptoms. *Pers Indiv Differ*. 2007;43(5):1199–1209. https://doi.org/10.1016/j.paid.2007.03.011.
142. Papachristou H, Nederkoorn C, Jansen A. Neuroticism and negative urgency in problematic alcohol use: a pilot study. *Subst Use Misuse*. 2016;51(11):1529–1533. https://doi.org/10.1080/10826084.2016.1178294.
143. Kale D, Stautz K, Cooper A. Impulsivity related personality traits and cigarette smoking in adults: a meta-analysis using the UPPS-P model of impulsivity and reward sensitivity. *Drug Alcohol Depend*. 2018;185:149–167. https://doi.org/10.1016/j.drugalcdep.2018.01.003.
144. Kale D, Pickering A, Cooper A. Examining the relationship between impulsivity-related personality traits and e-cigarette use in adults. *Addict Behav*. 2020;106:106348. https://doi.org/10.1016/j.addbeh.2020.106348.
145. Pang RD, Hom MS, Geary BA, et al. Relationships between trait urgency, smoking reinforcement expectancies, and nicotine dependence. *J Addict Dis*. 2014;33(2):83–93. https://doi.org/10.1080/10550887.2014.909695.
146. Spillane NS, Smith GT, Kahler CW. Impulsivity-like traits and smoking behavior in college students. *Addict Behav*. 2010;35(7):700–705. https://doi.org/10.1016/j.addbeh.2010.03.008.
147. Park AD, Farrahi LN, Pang RD, Guillot CR, Aguirre CG, Leventhal AM. Negative urgency is associated with heightened negative affect and urge during tobacco abstinence in regular smokers. *J Stud Alcohol Drugs*. 2016;77(5):766–773. https://doi.org/10.15288/jsad.2016.77.766.
148. Hershberger AR, Um M, Cyders MA. The relationship between the UPPS-P impulsive personality traits and substance use psychotherapy outcomes: a meta-analysis. *Drug Alcohol Depend*. 2017;178:408–416. https://doi.org/10.1016/j.drugalcdep.2017.05.032.
149. Mulhauser K, Weinstock J, Van Patten R, McGrath AB, Merz ZC, White CN. Examining the stability of the UPPS-P and MCQ-27 during residential treatment for substance use disorder. *Exp Clin Psychopharmacol*. 2019;27(5):474–481. https://doi.org/10.1037/pha0000255.

150. Wang Y, Zuo J, Hao W, et al. Quality of life in patients with methamphetamine use disorder: relationship to impulsivity and drug use characteristics. *Front Psychiatr*. 2020;11:579302. https://doi.org/10.3389/fpsyt.2020.579302.
151. Zhang J, Su H, Tao J, et al. Relationship of impulsivity and depression during early methamphetamine withdrawal in Han Chinese population. *Addict Behav*. 2015;43:7–10. https://doi.org/10.1016/j.addbeh.2014.10.032.
152. Huang S, Dai Y, Zhang C, et al. Higher impulsivity and lower grey matter volume in the bilateral prefrontal cortex in long-term abstinent individuals with severe methamphetamine use disorder. *Drug Alcohol Depend*. 2020;212:108040. https://doi.org/10.1016/j.drugalcdep.2020.108040.
153. Blevins D, Choi CJ, Pavlicova M, et al. Impulsiveness as a moderator of amphetamine treatment response for cocaine use disorder among ADHD patients. *Drug Alcohol Depend*. 2020;213:108082. https://doi.org/10.1016/j.drugalcdep.2020.108082.
154. Blevins D, Wang XQ, Sharma S, Ait-Daoud N. Impulsiveness as a predictor of topiramate response for cocaine use disorder. *Am J Addict*. 2019;28(2):71–76. https://doi.org/10.1111/ajad.12858.
155. Perkins KA, Lerman C, Coddington SB, et al. Initial nicotine sensitivity in humans as a function of impulsivity. *Psychopharmacology*. 2008;200(4):529–544. https://doi.org/10.1007/s00213-008-1231-7.
156. Bidwell LC, Knopik VS, Audrain-McGovern J, et al. Novelty seeking as a phenotypic marker of adolescent substance use. *Subst Abuse*. 2015;9(Suppl 1):1–10. https://doi.org/10.4137/SART.S22440.
157. Gurpegui M, Jurado D, Luna JD, Fernández-Molina C, Moreno-Abril O, Gálvez R. Personality traits associated with caffeine intake and smoking. *Prog Neuro Psychopharmacol Biol Psychiatr*. 2007;31(5):997–1005. https://doi.org/10.1016/j.pnpbp.2007.02.006.
158. Vladimirov D, Niemelä S, Keinänen-Kiukaanniemi S, et al. Cloninger's Temperament Dimensions and longitudinal alcohol use in early midlife: a Northern Finland birth cohort 1966 study. *Alcohol Clin Exp Res*. 2018;42(10):1924–1932. https://doi.org/10.1111/acer.13857.
159. Passarotti AM, Crane NA, Hedeker D, Mermelstein RJ. Longitudinal trajectories of marijuana use from adolescence to young adulthood. *Addict Behav*. 2015;45:301–308. https://doi.org/10.1016/j.addbeh.2015.02.008.
160. Chen F, Yang H, Bulut O, Cui Y, Xin T. Examining the relation of personality factors to substance use disorder by explanatory item response modeling of DSM-5 symptoms. *PLoS One*. 2019;14(6):e0217630. https://doi.org/10.1371/journal.pone.0217630.
161. Foulds JA, Boden JM, Newton-Howes GM, Mulder RT, Horwood LJ. The role of novelty seeking as a predictor of substance use disorder outcomes in early adulthood. *Addiction*. 2017;112(9):1629–1637. https://doi.org/10.1111/add.13838.
162. Hosier SG, Cox WM. Personality and motivational correlates of alcohol consumption and alcohol-related problems among excessive drinking university students. *Addict Behav*. 2011;36(1–2):87–94. https://doi.org/10.1016/j.addbeh.2010.08.029.
163. Magnusson A, Göransson M, Heilig M. Hazardous alcohol users during pregnancy: psychiatric health and personality traits. *Drug Alcohol Depend*. 2007;89(2–3):275–281. https://doi.org/10.1016/j.drugalcdep.2007.01.015.
164. Leventhal AM, Waters AJ, Boyd S, et al. Associations between Cloninger's temperament dimensions and acute tobacco withdrawal. *Addict Behav*. 2007;32(12):2976–2989. https://doi.org/10.1016/j.addbeh.2007.06.014.
165. Arenas MC, Aguilar MA, Montagud-Romero S, et al. Influence of the novelty-seeking endophenotype on the rewarding effects of psychostimulant drugs in animal models. *Curr Neuropharmacol*. 2016;14(1):87–100. https://doi.org/10.2174/1570159x13666150921112841.
166. Abreu-Villaça Y, Queiroz-Gomes F, Dal Monte AP, Filgueiras CC, Manhães AC. Individual differences in novelty-seeking behavior but not in anxiety response to a new environment can predict nicotine consumption in adolescent C57BL/6 mice. *Behav Brain Res*. 2006;167(1):175–182. https://doi.org/10.1016/j.bbr.2005.09.003.
167. Pelloux Y, Costentin J, Duterte-Boucher D. Novelty preference predicts place preference conditioning to morphine and its oral consumption in rats. *Pharmacol Biochem Behav*. 2006;84(1):43–50. https://doi.org/10.1016/j.pbb.2006.04.004.
168. Nadal R, Armario A, Janak PH. Positive relationship between activity in a novel environment and operant ethanol self-administration in rats. *Psychopharmacology*. 2002;162(3):333–338. https://doi.org/10.1007/s00213-002-1091-5.
169. Cain ME, Saucier DA, Bardo MT. Novelty seeking and drug use: contribution of an animal model. *Exp Clin Psychopharmacol*. 2005;13(4):367–375. https://doi.org/10.1037/1064-1297.13.4.367.

170. Belin D, Berson N, Balado E, Piazza PV, Deroche-Gamonet V. High-novelty-preference rats are predisposed to compulsive cocaine self-administration. *Neuropsychopharmacology*. 2011;36(3):569–579. https://doi.org/10.1038/npp.2010.188.
171. Klebaur JE, Bardo MT. Individual differences in novelty seeking on the playground maze predict amphetamine conditioned place preference. *Pharmacol, Biochem Behav*. 1999;63(1):131–136. https://doi.org/10.1016/s0091-3057(98)00258-5.
172. Arenas MC, Daza-Losada M, Vidal-Infer A, Aguilar MA, Miñarro J, Rodríguez-Arias M. Capacity of novelty-induced locomotor activity and the hole-board test to predict sensitivity to the conditioned rewarding effects of cocaine. *Physiol Behav*. 2014;133:152–160. https://doi.org/10.1016/j.physbeh.2014.05.028.
173. Vidal-Infer A, Arenas MC, Daza-Losada M, Aguilar MA, Miñarro J, Rodríguez-Arias M. High novelty-seeking predicts greater sensitivity to the conditioned rewarding effects of cocaine. *Pharmacol Biochem Behav*. 2012;102(1):124–132. https://doi.org/10.1016/j.pbb.2012.03.031.
174. O'Connor SL, Aston-Jones G, James MH. The sensation seeking trait confers a dormant susceptibility to addiction that is revealed by intermittent cocaine self-administration in rats. *Neuropharmacology*. 2021;195:108566. https://doi.org/10.1016/j.neuropharm.2021.108566.
175. Manzo L, Gómez MJ, Callejas-Aguilera JE, et al. Relationship between ethanol preference and sensation/novelty seeking. *Physiol Behav*. 2014;133:53–60. https://doi.org/10.1016/j.physbeh.2014.05.003.
176. Davis BA, Clinton SM, Akil H, Becker JB. The effects of novelty-seeking phenotypes and sex differences on acquisition of cocaine self-administration in selectively bred high-responder and low-responder rats. *Pharmacol Biochem Behav*. 2008;90(3):331–338. https://doi.org/10.1016/j.pbb.2008.03.008.
177. Cummings JA, Gowl BA, Westenbroek C, Clinton SM, Akil H, Becker JB. Effects of a selectively bred novelty-seeking phenotype on the motivation to take cocaine in male and female rats. *Biol Sex Differ*. 2011;2:3. https://doi.org/10.1186/2042-6410-2-3.
178. Beckmann JS, Marusich JA, Gipson CD, Bardo MT. Novelty seeking, incentive salience and acquisition of cocaine self-administration in the rat. *Behav Brain Res*. 2011;216(1):159–165. https://doi.org/10.1016/j.bbr.2010.07.022.
179. Kaye S, Gilsenan J, Young JT, et al. Risk behaviours among substance use disorder treatment seekers with and without adult ADHD symptoms. *Drug Alcohol Depend*. 2014;144:70–77. https://doi.org/10.1016/j.drugalcdep.2014.08.008.
180. Brook JS, Balka EB, Zhang C, Brook DW. ADHD, conduct disorder, substance use disorder, and nonprescription stimulant use. *J Atten Disord*. 2017;21(9):776–782. https://doi.org/10.1177/1087054714528535.
181. Capusan AJ, Bendtsen P, Marteinsdottir I, Larsson H. Comorbidity of adult adhd and its subtypes with substance use disorder in a large population-based epidemiological study. *J Atten Disord*. 2019;23(12):1416–1426. https://doi.org/10.1177/1087054715626511.
182. Groenman AP, Oosterlaan J, Rommelse N, et al. Substance use disorders in adolescents with attention deficit hyperactivity disorder: a 4-year follow-up study. *Addiction*. 2013;108(8):1503–1511. https://doi.org/10.1111/add.12188.
183. Knop J, Penick EC, Nickel EJ, et al. Childhood ADHD and conduct disorder as independent predictors of male alcohol dependence at age 40. *J Stud Alcohol Drugs*. 2009;70(2):169–177. https://doi.org/10.15288/jsad.2009.70.169.
184. Kaye S, Ramos-Quiroga JA, van de Glind G, et al. Persistence and subtype stability of ADHD among substance use disorder treatment seekers. *J Atten Disord*. 2019;23(12):1438–1453. https://doi.org/10.1177/1087054716629217.
185. Lugoboni F, Levin FR, Pieri MC, et al, Gruppo InterSert Collaborazione Scientifica Gics. Co-occurring Attention Deficit Hyperactivity Disorder symptoms in adults affected by heroin dependence: patients characteristics and treatment needs. *Psychiatr Res*. 2017;250:210–216. https://doi.org/10.1016/j.psychres.2017.01.052.
186. Simon N, Rolland B, Karila L. Methylphenidate in adults with attention deficit hyperactivity disorder and substance use disorders. *Curr Pharmaceut Des*. 2015;21(23):3359–3366. https://doi.org/10.2174/1381612821666150619093254.
187. van der Burg D, Crunelle CL, Matthys F, van den Brink W. Diagnosis and treatment of patients with comorbid substance use disorder and adult attention-deficit and hyperactivity disorder: a review of recent publications. *Curr Opin Psychiatr*. 2019;32(4):300–306. https://doi.org/10.1097/YCO.0000000000000513.

188. Congia P, Mannarino S, Deiana S, Maulu M, Muscas E. Association between adult ADHD, self-report, and behavioral measures of impulsivity and treatment outcome in cocaine use disorder. *J Subst Abuse Treat*. 2020;118:108120. https://doi.org/10.1016/j.jsat.2020.108120.
189. Bron TI, Bijlenga D, Kasander MV, Spuijbroek AT, Beekman AT, Kooij JJ. Long-term relationship between methylphenidate and tobacco consumption and nicotine craving in adults with ADHD in a prospective cohort study. *Eur Neuropsychopharmacol*. 2013;23(6):542−554. https://doi.org/10.1016/j.euroneuro.2012.06.004.
190. Matthies S, Holzner S, Feige B, et al. ADHD as a serious risk factor for early smoking and nicotine dependence in adulthood. *J Atten Disord*. 2013;17(3):176−186. https://doi.org/10.1177/1087054711428739.
191. Breyer JL, Lee S, Winters KC, August GJ, Realmuto GM. A longitudinal study of childhood ADHD and substance dependence disorders in early adulthood. *Psychol Addict Behav*. 2014;28(1):238−246. https://doi.org/10.1037/a0035664.
192. Ohlmeier MD, Peters K, Kordon A, et al. Nicotine and alcohol dependence in patients with comorbid attention-deficit/hyperactivity disorder (ADHD). *Alcohol Alcohol*. 2007;42(6):539−543. https://doi.org/10.1093/alcalc/agm069.
193. Sánchez-García NC, González RA, Ramos-Quiroga JA, et al. Attention deficit hyperactivity disorder increases nicotine addiction severity in adults seeking treatment for substance use disorders: the role of personality disorders. *Eur Addiction Res*. 2020;26(4−5):191−200. https://doi.org/10.1159/000508545.
194. Moggi F, Schorno D, Soravia LM, et al. Screened attention deficit/hyperactivity disorder as a predictor of substance use initiation and escalation in early adulthood and the role of self-reported conduct disorder and sensation seeking: a 5-year longitudinal study with young adult Swiss men. *Eur Addiction Res*. 2020;26(4−5):233−244. https://doi.org/10.1159/000508304.
195. Mitchell JT, Howard AL, Belendiuk KA, et al. Cigarette smoking progression among young adults diagnosed with ADHD in childhood: a 16-year longitudinal study of children with and without ADHD. *Nicotine Tob Res*. 2019;21(5):638−647. https://doi.org/10.1093/ntr/nty045.
196. Rhodes JD, Pelham WE, Gnagy EM, Shiffman S, Derefinko KJ, Molina BS. Cigarette smoking and ADHD: an examination of prognostically relevant smoking behaviors among adolescents and young adults. *Psychol Addict Behav*. 2016;30(5):588−600. https://doi.org/10.1037/adb0000188.
197. Elkins IJ, Saunders G, Malone SM, Wilson S, McGue M, Iacono WG. Differential implications of persistent, remitted, and late-onset ADHD symptoms for substance abuse in women and men: a twin study from ages 11 to 24. *Drug Alcohol Depend*. 2020;212:107947. https://doi.org/10.1016/j.drugalcdep.2020.107947.
198. Estévez-Lamorte N, Foster S, Eich-Höchli D, Moggi F, Gmel G, Mohler-Kuo M. Adult attention-deficit/hyperactivity disorder, risky substance use and substance use disorders: a follow-up study among young men. *Eur Arch Psychiatr Clin Neurosci*. 2019;269(6):667−679. https://doi.org/10.1007/s00406-018-0958-3.
199. McClernon FJ, Van Voorhees EE, English J, Hallyburton M, Holdaway A, Kollins SH. Smoking withdrawal symptoms are more severe among smokers with ADHD and independent of ADHD symptom change: results from a 12-day contingency-managed abstinence trial. *Nicotine Tob Res*. 2011;13(9):784−792. https://doi.org/10.1093/ntr/ntr073.
200. Pomerleau CS, Downey KK, Snedecor SM, Mehringer AM, Marks JL, Pomerleau OF. Smoking patterns and abstinence effects in smokers with no ADHD, childhood ADHD, and adult ADHD symptomatology. *Addict Behav*. 2003;28(6):1149−1157. https://doi.org/10.1016/s0306-4603(02)00223-x.
201. Vilar-Ribó L, Sánchez-Mora C, Rovira P, et al. Genetic overlap and causality between substance use disorder and attention-deficit and hyperactivity disorder. *Am J Med Genet Part B Neuropsychiatr Genet*. 2021;186(3):140−150. https://doi.org/10.1002/ajmg.b.32827.
202. da Silva BS, Cupertino RB, Schuch JB, et al. The association between SYT1-rs2251214 and cocaine use disorder further supports its role in psychiatry. *Prog Neuro Psychopharmacol Biol Psychiatr*. 2019;94:109642. https://doi.org/10.1016/j.pnpbp.2019.109642.
203. Wilens TE, Adamson J, Sgambati S, et al. Do individuals with ADHD self-medicate with cigarettes and substances of abuse? Results from a controlled family study of ADHD. *Am J Addict*. 2007;16(Suppl 1):14−23. https://doi.org/10.1080/10550490601082742.
204. Cumyn L, French L, Hechtman L. Comorbidity in adults with attention-deficit hyperactivity disorder. *Can J Psychiatr*. 2009;54(10):673−683. https://doi.org/10.1177/070674370905401004.

205. Tamm L, Adinoff B, Nakonezny PA, Winhusen T, Riggs P. Attention-deficit/hyperactivity disorder subtypes in adolescents with comorbid substance-use disorder. *Am J Drug Alcohol Abuse*. 2012;38(1):93–100. https://doi.org/10.3109/00952990.2011.600395.
206. Bozkurt M, Evren C, Umut G, Evren B. Relationship of attention-deficit/hyperactivity disorder symptom severity with severity of alcohol-related problems in a sample of inpatients with alcohol use disorder. *Neuropsychiatric Dis Treat*. 2016;12:1661–1667. https://doi.org/10.2147/NDT.S105190.
207. Upadhyaya HP, Carpenter MJ. Is attention deficit hyperactivity disorder (ADHD) symptom severity associated with tobacco use? *Am J Addict*. 2008;17(3):195–198. https://doi.org/10.1080/10550490802021937.
208. Kollins SH, McClernon FJ, Fuemmeler BF. Association between smoking and attention-deficit/hyperactivity disorder symptoms in a population-based sample of young adults. *Arch Gen Psychiatr*. 2005;62(10):1142–1147. https://doi.org/10.1001/archpsyc.62.10.1142.
209. Fuemmeler BF, Kollins SH, McClernon FJ. Attention deficit hyperactivity disorder symptoms predict nicotine dependence and progression to regular smoking from adolescence to young adulthood. *J Pediatr Psychol*. 2007;32(10):1203–1213. https://doi.org/10.1093/jpepsy/jsm051.
210. Elkins IJ, McGue M, Iacono WG. Prospective effects of attention-deficit/hyperactivity disorder, conduct disorder, and sex on adolescent substance use and abuse. *Arch Gen Psychiatr*. 2007;64(10):1145–1152. https://doi.org/10.1001/archpsyc.64.10.1145.
211. Liebrenz M, Gamma A, Ivanov I, Buadze A, Eich D. Adult attention-deficit/hyperactivity disorder: associations between subtype and lifetime substance use - a clinical study. *F1000Research*. 2015;4:407. https://doi.org/10.12688/f1000research.6780.2.
212. Saules KK, Pomerleau CS, Schubiner H. Patterns of inattentive and hyperactive symptomatology in cocaine-addicted and non-cocaine-addicted smokers diagnosed with adult attention deficit hyperactivity disorder. *J Addict Dis*. 2003;22(2):71–78. https://doi.org/10.1300/J069v22n02_06.
213. Vansickel AR, Stoops WW, Glaser PE, Poole MM, Rush CR. Methylphenidate increases cigarette smoking in participants with ADHD. *Psychopharmacology*. 2011;218(2):381–390. https://doi.org/10.1007/s00213-011-2328-y.
214. Biederman J, Wilens T, Mick E, Spencer T, Faraone SV. Pharmacotherapy of attention-deficit/hyperactivity disorder reduces risk for substance use disorder. *Pediatrics*. 1999;104(2):e20. https://doi.org/10.1542/peds.104.2.e20.
215. Wilens TE, Adamson J, Monuteaux MC, et al. Effect of prior stimulant treatment for attention-deficit/hyperactivity disorder on subsequent risk for cigarette smoking and alcohol and drug use disorders in adolescents. *Arch Pediatr Adolesc Med*. 2008;162(10):916–921. https://doi.org/10.1001/archpedi.162.10.916.
216. Manni C, Cipollone G, Pallucchini A, Maremmani A, Perugi G, Maremmani I. Remarkable reduction of cocaine use in dual disorder (adult attention deficit hyperactive disorder/cocaine use disorder) patients treated with medications for ADHD. *Int J Environ Res Publ Health*. 2019;16(20):3911. https://doi.org/10.3390/ijerph16203911.
217. Schoenfelder EN, Faraone SV, Kollins SH. Stimulant treatment of ADHD and cigarette smoking: a meta-analysis. *Pediatrics*. 2014;133(6):1070–1080. https://doi.org/10.1542/peds.2014-0179.
218. Groenman AP, Oosterlaan J, Rommelse NN, et al. Stimulant treatment for attention-deficit hyperactivity disorder and risk of developing substance use disorder. *Br J Psychiatr*. 2013;203(2):112–119. https://doi.org/10.1192/bjp.bp.112.124784.
219. Quinn PD, Chang Z, Hur K, et al. ADHD medication and substance-related problems. *Am J Psychiatr*. 2017;174(9):877–885. https://doi.org/10.1176/appi.ajp.2017.16060686.
220. Okamoto K, Aoki K. Development of a strain of spontaneously hypertensive rats. *Jpn Circ J*. 1963;27:282–293. https://doi.org/10.1253/jcj.27.282.
221. Aparicio CF, Hennigan PJ, Mulligan LJ, Alonso-Alvarez B. Spontaneously hypertensive (SHR) rats choose more impulsively than Wistar-Kyoto (WKY) rats on a delay discounting task. *Behav Brain Res*. 2019;364:480–493. https://doi.org/10.1016/j.bbr.2017.09.040.
222. Bizot J-C, Chenault N, Houzé B, et al. Methylphenidate reduces impulsive behaviour in juvenile Wistar rats, but not in adult Wistar, SHR and WKY rats. *Psychopharmacology*. 2007;193(2):215–223. https://doi.org/10.1007/s00213-007-0781-4.
223. Fox AT, Hand DJ, Reilly MP. Impulsive choice in a rodent model of attention-deficit/hyperactivity disorder. *Behav Brain Res*. 2008;187(1):146–152. https://doi.org/10.1016/j.bbr.2007.09.008.

224. Hand DJ, Fox AT, Reilly MP. Differential effects of d-amphetamine on impulsive choice in spontaneously hypertensive and Wistar-Kyoto rats. *Behav Pharmacol.* 2009;20(5–6):549–553. https://doi.org/10.1097/FBP.0b013e3283305ee1.
225. Leffa DT, Ferreira SG, Machado NJ, et al. Caffeine and cannabinoid receptors modulate impulsive behavior in an animal model of attentional deficit and hyperactivity disorder. *Eur J Neurosci.* 2019;49(12):1673–1683. https://doi.org/10.1111/ejn.14348.
226. Orduña V. Impulsivity and sensitivity to amount and delay of reinforcement in an animal model of ADHD. *Behav Brain Res.* 2015;294:62–71. https://doi.org/10.1016/j.bbr.2015.07.046.
227. Wooters TE, Bardo MT. Methylphenidate and fluphenazine, but not amphetamine, differentially affect impulsive choice in spontaneously hypertensive, Wistar-Kyoto and Sprague-Dawley rats. *Brain Res.* 2011;1396:45–53. https://doi.org/10.1016/j.brainres.2011.04.040.
228. Garcia A, Kirkpatrick K. Impulsive choice behavior in four strains of rats: evaluation of possible models of Attention-Deficit/Hyperactivity Disorder. *Behav Brain Res.* 2013;238:10–22. https://doi.org/10.1016/j.bbr.2012.10.017.
229. Turner M, Wilding E, Cassidy E, Dommett EJ. Effects of atomoxetine on locomotor activity and impulsivity in the spontaneously hypertensive rat. *Behav Brain Res.* 2013;243:28–37. https://doi.org/10.1016/j.bbr.2012.12.025.
230. Sanabria F, Killeen PR. Evidence for impulsivity in the Spontaneously Hypertensive Rat drawn from complementary response-withholding tasks. *Behav Brain Funct.* 2008;4:7. https://doi.org/10.1186/1744-9081-4-7.
231. van den Bergh FS, Bloemarts E, Chan JS, Groenink L, Olivier B, Oosting RS. Spontaneously hypertensive rats do not predict symptoms of attention-deficit hyperactivity disorder. *Pharmacol Biochem Behav.* 2006;83(3):380–390. https://doi.org/10.1016/j.pbb.2006.02.018.
232. Kantak KM, Stots C, Mathieson E, Bryant CD. Spontaneously Hypertensive Rat substrains show differences in model traits for addiction risk and cocaine self-administration: implications for a novel rat reduced complexity cross. *Behav Brain Res.* 2021;411:113406. https://doi.org/10.1016/j.bbr.2021.113406.
233. Bizot J-C, Cogrel N, Massé F, et al. D-amphetamine improves attention performance in adolescent Wistar, but not in SHR rats, in a two-choice visual discrimination task. *Psychopharmacology.* 2015;232(17):3269–3286. https://doi.org/10.1007/s00213-015-3974-2.
234. Sagvolden T, Xu T. l-Amphetamine improves poor sustained attention while d-amphetamine reduces overactivity and impulsiveness as well as improves sustained attention in an animal model of Attention-Deficit/Hyperactivity Disorder (ADHD). *Behav Brain Funct.* 2008;4:3. https://doi.org/10.1186/1744-9081-4-3.
235. Sagvolden T. Impulsiveness, overactivity, and poorer sustained attention improve by chronic treatment with low doses of l-amphetamine in an animal model of Attention-Deficit/Hyperactivity Disorder (ADHD). *Behav Brain Funct.* 2011;7:6. https://doi.org/10.1186/1744-9081-7-6.
236. Somkuwar SS, Kantak KM, Bardo MT, Dwoskin LP. Adolescent methylphenidate treatment differentially alters adult impulsivity and hyperactivity in the Spontaneously Hypertensive Rat model of ADHD. *Pharmacol Biochem Behav.* 2016;141:66–77. https://doi.org/10.1016/j.pbb.2015.12.002.
237. Marusich JA, McCuddy WT, Beckmann JS, Gipson CD, Bardo MT. Strain differences in self-administration of methylphenidate and sucrose pellets in a rat model of attention-deficit hyperactivity disorder. *Behav Pharmacol.* 2011;22(8):794–804. https://doi.org/10.1097/FBP.0b013e32834d623e.
238. dela Peña IC, Ahn HS, Choi JY, Shin CY, Ryu JH, Cheong JH. Methylphenidate self-administration and conditioned place preference in an animal model of attention-deficit hyperactivity disorder: the spontaneously hypertensive rat. *Behav Pharmacol.* 2011;22(1):31–39. https://doi.org/10.1097/FBP.0b013e328342503a.
239. Baskin BM, Nic Dhonnchadha BÁ, Dwoskin LP, Kantak KM. Blockade of α2-adrenergic receptors in prelimbic cortex: impact on cocaine self-administration in adult spontaneously hypertensive rats following adolescent atomoxetine treatment. *Psychopharmacology.* 2017;234(19):2897–2909. https://doi.org/10.1007/s00213-017-4681-y.
240. Jordan CJ, Harvey RC, Baskin BB, Dwoskin LP, Kantak KM. Cocaine-seeking behavior in a genetic model of attention-deficit/hyperactivity disorder following adolescent methylphenidate or atomoxetine treatments. *Drug Alcohol Depend.* 2014;140:25–32. https://doi.org/10.1016/j.drugalcdep.2014.04.020.
241. Jordan CJ, Lemay C, Dwoskin LP, Kantak KM. Adolescent d-amphetamine treatment in a rodent model of attention deficit/hyperactivity disorder: impact on cocaine abuse vulnerability in adulthood. *Psychopharmacology.* 2016;233(23–24):3891–3903. https://doi.org/10.1007/s00213-016-4419-2.

242. Somkuwar SS, Jordan CJ, Kantak KM, Dwoskin LP. Adolescent atomoxetine treatment in a rodent model of ADHD: effects on cocaine self-administration and dopamine transporters in frontostriatal regions. *Neuropsychopharmacology*. 2013;38(13):2588–2597. https://doi.org/10.1038/npp.2013.163.
243. Baskin BM, Dwoskin LP, Kantak KM. Methylphenidate treatment beyond adolescence maintains increased cocaine self-administration in the spontaneously hypertensive rat model of attention deficit/hyperactivity disorder. *Pharmacol, Biochem Behav*. 2015;131:51–56. https://doi.org/10.1016/j.pbb.2015.01.019.
244. Harvey RC, Sen S, Deaciuc A, Dwoskin LP, Kantak KM. Methylphenidate treatment in adolescent rats with an attention deficit/hyperactivity disorder phenotype: cocaine addiction vulnerability and dopamine transporter function. *Neuropsychopharmacology*. 2011;36(4):837–847. https://doi.org/10.1038/npp.2010.223.
245. dela Peña I, Yoon SY, Lee JC, et al. Methylphenidate treatment in the spontaneously hypertensive rat: influence on methylphenidate self-administration and reinstatement in comparison with Wistar rats. *Psychopharmacology*. 2012;221(2):217–226. https://doi.org/10.1007/s00213-011-2564-1.
246. Augustyniak PN, Kourrich S, Rezazadeh SM, Stewart J, Arvanitogiannis A. Differential behavioral and neurochemical effects of cocaine after early exposure to methylphenidate in an animal model of attention deficit hyperactivity disorder. *Behav Brain Res*. 2006;167(2):379–382. https://doi.org/10.1016/j.bbr.2005.09.014.
247. Alsop B. Problems with spontaneously hypertensive rats (SHR) as a model of attention-deficit/hyperactivity disorder (AD/HD). *J Neurosci Methods*. 2007;162(1–2):42–48. https://doi.org/10.1016/j.jneumeth.2006.12.002.
248. Bull E, Reavill C, Hagan JJ, Overend P, Jones DN. Evaluation of the spontaneously hypertensive rat as a model of attention deficit hyperactivity disorder: acquisition and performance of the DRL-60s test. *Behav Brain Res*. 2000;109(1):27–35. https://doi.org/10.1016/s0166-4328(99)00156-4.
249. Knardahl S, Sagvolden T. Open-field behavior of spontaneously hypertensive rats. *Behav Neural Biol*. 1979;27(2):187–200. https://doi.org/10.1016/s0163-1047(79)91801-6.
250. Sagvolden T, Russell VA, Aase H, Johansen EB, Farshbaf M. Rodent models of attention-deficit/hyperactivity disorder. *Biol Psychiatr*. 2005;57(11):1239–1247. https://doi.org/10.1016/j.biopsych.2005.02.002.
251. Sagvolden T, Johansen EB, Wøien G, et al. The spontaneously hypertensive rat model of ADHD–The importance of selecting the appropriate reference strain. *Neuropharmacology*. 2009;57(7–8):619–626. https://doi.org/10.1016/j.neuropharm.2009.08.004.

CHAPTER 10

Social and sociocultural factors associated with addiction

Introduction

My father-in-law used to smoke cigarettes. When he decided he wanted to quit smoking, he would meet with a friend to discuss strategies to quit smoking. The problem was that each time my father-in-law met with his friend, he would crave cigarettes. He told me he was able to finally quit smoking after he stopped spending time with his friend. This example highlights the role that peers have on drug use and substance use disorders (SUDs). In addition to peer influences, family dynamics can contribute to the initiation and the continuation of drug use. On a broader level, certain sociocultural factors are highly associated with SUDs. This chapter will detail the social, racial, cultural, and religious influences on SUDs. Just as social factors influence continued drug use, they can be used to help combat SUDs. The last part of this chapter will focus on treatments that incorporate family members and friends. You have already learned about one such method in Chapter 5 in the form of community reinforcement and family training (CRAFT). In addition to providing alternative reinforcement to the individual seeking drug abstinence, CRAFT involves this individual's family members by teaching them how to watch for potential triggers that could precipitate relapse.

Learning objectives

By the end of this chapter, you should be able to ...

(1) Explain the contribution of peers and family to the initiation and maintenance of drug-taking behaviors.
(2) Differentiate social learning theory and social control theory, and discuss how these theories account for peer/familial influences on drug use.
(3) Discuss the ways in which social influences are measured in animals.
(4) Explain how sociocultural factors such as socioeconomic status, race/ethnicity, and religiosity/spirituality influence substance use disorders
(5) Compare and contrast treatment options incorporating social components for SUDs.

Peer and familial influences on drug use

During development, children spend a large portion of their day with peers. Most children attend schools, where they interact with their peers 5 days a week. As children age, they begin spending more time with friends outside of the classroom, particularly during adolescence. Because our peers can greatly influence our decision-making and can bias our choices, researchers have been interested in studying the contribution of social influences on the initiation and the maintenance of drug-taking behaviors. As expected, peers influence adolescent smoking and tobacco use, ethanol consumption, cannabis use, and cocaine use.[1–9] College-aged students are more likely to consume multiple alcoholic beverages when with friends as opposed to by themselves, with significant others, or with family members.[10] Similarly, cannabis use occurs more frequently in social settings, particularly if others are using cannabis as well.[11] One idea that has been proposed to explain these findings is the concept of *peer pressure*. However, Reed and Rountree have suggested that peers do not influence drug use by overt peer pressure. Instead, adolescents seek out peers who share similar behavioral patterns, such as positive attitudes toward drug use.[12] More recent work has also shown that *suggestibility* (tendency to accept and internalize messages), but not peer conformity, is related to ethanol consumption in a laboratory setting.[13]

Not only do peers influence drug use, particularly during adolescence, but affiliating with those that use drugs is predictive of SUDs, including those for ethanol and cannabis.[6,14] In a large cross-sectional study, 12% of participants noted that social influences were a contributing factor to their inability to quit tobacco smoking.[15] This percentage may not seem high when viewed in isolation. Consider the numerous risk factors that we have already covered in this book (e.g., maladaptive decision making, personality factors, conditioned stimuli) and have yet to cover (e.g., education level, race, gender, and stress). A single factor accounting for over 10% of behavior is substantial.

In addition to peers, family dynamics affect drug use in adolescents. Families that are less cohesive and lack strong intellectual, religious, and political beliefs are more likely to have adolescents who use ethanol.[16] Familial dysfunction increases problematic ethanol use in parents, which correlates with increased problematic ethanol use in children.[17] Adolescents are more inclined to use ethanol if they have a sibling close in age that uses ethanol, and they are more susceptible to developing a SUD if they have a sibling that smokes cigarettes.[18] Research has identified protective family behaviors that decrease the risk of adolescent drug use, including parental involvement, family dinners, positive parent-child relationships, and **parental monitoring**. Parental monitoring encompasses several practices like tracking and supervising a child's behaviors (i.e., where the child is going, with whom they are spending time, what activities they will be doing, etc.). A protective family context decreases adolescent substance use across various drugs.[19]

Concerning familial influences on drug use, the specific contribution of parents to this behavior has received considerable attention. Parental influences are highly associated with adolescent drug use.[2] Before discussing these parental influences, I first need to discuss the concept of **parenting styles**. Four major parenting styles have been identified: uninvolved, permissive (sometimes referred to as indulgent), authoritarian, and authoritative.[20] The uninvolved parenting style is what it sounds like; parents pay little attention to their child/children and offer little guidance or nurturing. Simply, these parents do not appear to enjoy their children. Permissive parents nurture their children, but they rarely enforce rules. Children of these parents can seemingly do whatever they want with little consequence. Conversely, authoritarian parents are overly strict, with an emphasis on obedience and punishment. Think of the expression "it's my way or the highway". Lastly, authoritative parents are nurturing, but they enforce rules. These parents have discussions with their children as to why the enforced rules are important. As such, the authoritative parenting style is ideal.

Parenting style is an important determinant of drug use. Parents who show more hostility and harsh discipline (i.e., authoritarian parenting style) are more likely to have children who use tobacco while parental warmth protects against adolescent smoking,[21] and this effect is more pronounced when neither parent has a history of cigarette use.[22] Related to these findings, the authoritative parenting style is associated with decreased cigarette use in adolescents.[23] Children raised by permissive, uninvolved, or authoritarian parents are more likely to consume ethanol, including heavy drinking and binge drinking, compared to children of authoritative parents.[24–28] University students that have an authoritarian mother drink more, experience more ethanol-related problems, and show impaired control over drinking.[29] Cannabis use is also positively associated with authoritarian and permissive parenting styles.[30] Finally, compared to those with neglectful parents, college students raised by authoritative parents are less likely to use MDMA or engage in polydrug use.[31] Overall, children of authoritative parents are less likely to use ethanol and/or drugs.

I mentioned parental monitoring earlier in this section. This form of monitoring differs across parenting styles. Uninvolved and permissive parents have low parental monitoring, whereas authoritative and authoritarian parents have higher parental monitoring. Low parental monitoring is associated with ethanol use,[7,32,33] cigarette use,[21,23,32] cannabis use,[7,32] and other illicit drug use.[34,35] There is some dispute concerning the direct role of parental monitoring on ethanol/drug use. Given that parental monitoring is high in the authoritarian parenting style, one would expect to observe decreased drug use in children of these parents. However, as noted above, this is not the case. Bahr et al.[3] argue that parental monitoring does not have direct effects on adolescent drug taking. Instead, monitoring is negatively related to the number of drug-taking peers with whom the adolescent affiliates. That is, the more parents monitor their children, the less likely they are to associate with adolescents who use drugs.

Although both peer and parental influences contribute to adolescent drug use, peer influences appear to have a greater impact on drug use in this population. Using monozygotic and dizygotic twins, Walden et al. found that drug use is shaped more by the environment compared to heredity, and they found that peer influences contribute more to adolescent drug use compared to parental influences.[36] Bahr et al. also found similar results regarding family drug use.[3] Family drug use did not directly affect adolescent drug use, but it was positively related to the number of drug-taking peers with whom an adolescent affiliated. Bahr et al. found that parental influences, such as monitoring and attachment to their child, were negatively associated with adolescent drug use, but these associations were suppressed when peer influences were added to the analysis.[4] In a Swedish sample, the neglectful parenting style is associated with worse substance use outcomes across a variety of drugs; however, when other factors, such as having deviant peers, were added to the analysis, the associations between neglectful parenting and substance use outcomes disappeared.[37]

Theories of social influence on drug use

In the previous section, you learned that social factors influence one's drug use. However, *how* these factors affect drug use was not fully discussed. The purpose of this section is to discuss the theoretical underpinnings of social influence on drug use.

Social learning theory

Dr. Albert Bandura (1925—2021) was a social-cognitive psychologist best known for developing **social learning theory**. According to Bandura, an individual's learning can be divided into four processes: attention, retention, motor reproduction, and motivation.[38] First, individuals must attend to the behaviors of a particular person for learning to occur. Often, attention is directed toward those that have the greatest impact on one's life (e.g., parents and friends). Retention is also important for learning to occur. Someone who wants to perform a behavior when a social demonstrator is absent must be able to recall this behavior from memory. Bandura states that there are two representational systems involved in retention: an imaginal system that stores images of modeled behavior that can be retrieved later and a verbal system that converts modeled behaviors into verbal codes.[38] The verbal system is responsible for the ability of humans to quickly learn and retain modeled behavior. Another process involved in learning is motor reproduction. One must take symbolic representations and convert them to the correct action. This simply means that individuals need to mimic the actions observed in others. Finally, a motivational process must be present for learning to occur. For someone to model a particular behavior, there must be some incentive to perform it. For example, if an adolescent observes a friend having fun following ethanol consumption, they will be more likely to consume ethanol as well. This is known as **vicarious reinforcement**. Conversely, if the individual observes their friend get extremely sick following ethanol consumption, they may be less motivated to try ethanol. This is an example of **vicarious punishment**. Vicarious reinforcement and vicarious punishment are critical components of social learning, just as reinforcement and punishment are critically important for nonsocial learning (see Chapter 5).

Another important contribution of Bandura is the concept of **reciprocal determinism**. Bandura argued that our behavior simultaneously influences and is influenced by our environment and by personal factors (e.g., cognition, beliefs, physical characteristics).[39] Hence, all three of these traits are *determined reciprocally* by one another. Reciprocal determinism can be used to explain how the environment and personal factors influence drug-taking behavior.[40] Imagine an individual with alcohol use disorder. The major behavior of interest is ethanol consumption. How does ethanol consumption influence one's environment? This individual may elect to spend more time in environments that allow drinking to occur, such as in a bar or at home. The individual can also choose to spend more time with individuals that enjoy drinking. Because these environments allow ethanol consumption, this increases drinking behavior. As an individual drinks, they may feel more relaxed, reflecting the influence of behavior on personal factors. If the individual experiences withdrawal symptoms, this will promote the individual to resume drinking to alleviate these symptoms. Fig. 10.1 shows how these three factors can interact with one another.

There is modest support that reciprocal determinism is an important mechanism of addiction. Wardell and Read[41] determined if norms for drinking behavior and positive alcohol expectancies (PAEs) (e.g., reduced tension, increased social "lubrication", and performance enhancement) were bidirectionally related to ethanol use. Norms were measured by asking participants to state how often they believed students at their college consume ethanol. While norms and quantity of ethanol consumption were reciprocally influenced by one another, PAEs influenced frequency and quantity of ethanol use, but ethanol use did not affect future PAEs. These results are consistent with a longitudinal study that examined drug use from adolescence into adulthood.[42] Overall, views on drug use are a strong predictor of future drug use, but drug use does not appear to predict one's views on drug use. Therefore, some aspects of drug use may not be reciprocally related.

I think it is important to mention that Bandura was not an addiction scientist. He was more interested in determining how social learning influences children to engage in

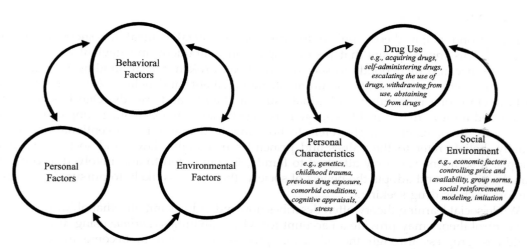

FIGURE 10.1 Albert Bandura's reciprocal determinism model is presented in panel (A). Panel (B) illustrates a model of addiction adopted from Bandura's theory of reciprocal determinism. *The figure is recreated from Fig. 1 of Smith MA, Social learning and addiction. Behav Brain Res. 2021;398:112954. https://doi.org/10.1016/j.bbr.2020.112954, Copyright 2021, with permission from Elsevier.*

aggressive behaviors. This means that Bandura was not interested in developing a theory to specifically account for why individuals use drugs. Dr. Ronald Akers applied social learning theory to deviant behaviors, including drug use. According to Akers,[43] learning involves operant conditioning and observational learning, as Bandura argued. Akers argued that there are four mechanisms by which drug-taking behavior is learned. First, individuals can acquire drug-taking behaviors through rewards and punishments for that behavior and through rewards and punishments for alternative behaviors. This is known as differential reinforcement. Next, individuals develop definitions that include the norms and opinions of individuals, and they identify with those that share similar definitions. Third, groups can reinforce one's behaviors and can expose one to definitions (differential association). The most important groups are peers and family, which makes sense given what we have already covered in this chapter. Finally, imitation is important for the initiation of drug use, but this becomes less important after one has had experience using the drug.

Akers' social learning theory has received empirical support. Drinking in both adolescents and older individuals is influenced by one's own norms, norms of one's peer groups, and differential reinforcement.[44,45] These variables, as well as imitation, explain 68% of the variation in adolescent cannabis use.[44] Specifically, ethanol use in adolescents is largely driven by peer beliefs on ethanol use; when peers show greater approval of ethanol use and have greater PAEs, individuals begin to share the same views, which then predicts ethanol consumption.[46] One or more elements of Akers' social learning theory account for cigarette smoking and cocaine use in adolescents,[1,47] nonmedical use of stimulants in college students,[48] and use of prescription drugs and ethanol.[49] At least with cigarette smoking, Akers' social learning theory better accounts for the maintenance of smoking in adolescents as opposed to the initiation of smoking.[50] Collectively, the social learning theories proposed by Bandura and Akers adequately explain how peers and family members influence one to use drugs.

Social control theory

The second major theory linked to drug use is **social control** (or **social bonding**) **theory**.[51] This theory posits that people engage in delinquent and deviant acts, such as drug use, because their bonds to society have been broken. Social bonding is composed of four elements: attachment, commitment, involvement, and beliefs. Attachment refers to how close one is to other individuals such as family and friends. One is said to be committed if they are inclined to engage in conventional activities (e.g., getting a job, going to school). Involvement simply refers to how involved one is in conventional activities. Finally, beliefs refer to the values and principles of parents, laws, and society that one adopts. If someone is closely attached to family, is committed to and involved in conventional activities, and adopts the beliefs of society, they are less likely to pursue deviant behaviors such as drug seeking.

While social learning theory often describes how individuals initiate ethanol/drug use, social control theory may provide an account for why individuals *continue* using drugs. As individuals progress through the addiction process, they tend to become more socially withdrawn.[52] An individual with a SUD may avoid social interaction because they want to be alone to use ethanol/drugs, or they want to avoid confrontation with others that are concerned about their substance use. Social control theory is somewhat supported. However, this support is stronger when cross-sectional studies have been used[53–55] as longitudinal studies have historically found weaker support for social control on drug use.[56,57] However, a more recent longitudinal-based study showed that variables of the social control theory are protective against adolescent drug use while social learning theory factors are predictive of drug use.[58] That is, peers are highly influential in promoting drug use; however, attachment to nondrug-using family members and high involvement with other activities can counteract the influence of peers.

Social influences on drug use in animal models

Although social influences are involved in the initiation and the maintenance of drug use in humans, the underlying neuromechanisms linking social influences and drug use in humans are not clear. By using animal models, these neuromechanisms can be studied, which can further clarify the relationship between social context and drug dependence.

Social influences on drug preference

The mere presence of a **conspecific** can potentiate preference for a drug reward. Using the conditioned place preference (CPP) paradigm, animals can be exposed to a conspecific in one compartment following administration of a *subthreshold dose* of a drug (i.e., often does not produce CPP on its own). When rats are exposed to a conspecific following administration of a subthreshold dose of cocaine[59] or nicotine,[60] they develop significant CPP; however, exposure to either social interaction or subthreshold doses of cocaine/nicotine alone fails to produce CPP. In a somewhat related study, adolescent rats were allowed to interact with either an intoxicated rat exposed to ethanol or an ethanol-free conspecific. Following this interaction, rats were tested to determine if they preferred a cotton ball scented with ethanol or vanilla. Rats that had interacted with nonintoxicated conspecifics showed a preference for the vanilla scented cotton ball, but rats paired with an intoxicated partner showed no preference for either cotton ball. These rats were more willing to explore both cotton balls, unlike rats that were paired with a nonintoxicated partner.[61] Pairing an environmental context with cocaine or methamphetamine elicits robust CPP when the context is also associated with a conspecific that receives an injection of cocaine/methamphetamine.[62,63] Interestingly, mice develop greater morphine CPP if they are exposed to a nonintoxicated conspecific; if exposed to a conspecific treated with morphine, mice show less CPP.[64] The effect of morphine may be due to its sedative effects as mice may not have interacted directly with each other in the CPP compartment (i.e., both mice spent time resting on opposite ends of the compartment).

CPP can also be used to compare the conditioned rewarding effects of social interaction to drugs. In one experiment I conducted as a graduate student, I found that adolescent male rats prefer spending time in a compartment paired with social interaction over a compartment paired with *d*-amphetamine whereas adult males prefer the amphetamine-paired compartment.[65] Similar results have been observed for cocaine CPP,[66] although it is important to note that these effects appear to be more specific to rats as Kummer et al. failed to observe a protective effect of social interaction in mice.[66] Subregions of the nucleus accumbens (NAc) appear to differentially mediate the conditioned rewarding effects of social interaction and cocaine. Lesions to the NAc core increase preference for social interaction, but lesions to the NAc shell increase preference for cocaine.[67]

In certain contexts, social interaction can act as a protective factor against drug preference. Giving rodents four 15-minute sessions in which they interact with a conspecific blocks reacquisition and reexpression of cocaine CPP,[68–71] and prevents reinstatement of cocaine CPP.[72] Similarly, access to a conspecific placed in the vehicle-paired compartment during the CPP posttest blocks cocaine CPP in mice.[73] These interactions may act as an alternative reinforcer that competes with the conditioned rewarding effects of drug-paired stimuli. At a neurobiological level, social interaction introduced during extinction training decreases the expression of immediate early genes like FosB and early growth response protein 1 (EGR1) in NAc core and shell regions during a CPP reinstatement test.[72,74] The ability of social interaction to attenuate cocaine-induced alterations in ERG1 levels in the NAc is primarily localized to dopamine D_1-containing medium spiny neurons (MSNs).[74]

Drug self-administration paradigms can also be used to measure social influences on drug-taking behaviors. Rats will respond more to cocaine if they can earn contingent access to a social partner, and if rats are raised in isolation and then presented with a social partner during self-administration sessions, they will respond more for cocaine.[75] Preclinical research has recently shown how social interaction can facilitate relapse-like behavior. Read the Experiment Spotlight to find out more.

Experiment Spotlight: Social interaction with relapsed partner facilitates cocaine relapse in rats (Meng et al.[76])

As you know by now, social factors contribute to the initiation of drug use and to relapse. While preclinical studies have assessed how social influences contribute to drug preference, very few studies have examined the contributions of social peers to relapse-like behavior. The goal of the present study was to develop "an animal model of social transfer of relapse" (p. 2). To accomplish this goal, Meng et al.[76] designed several experiments.

In Experiment 1, pair-housed rats were trained to self-administer cocaine (0.75 mg/kg/infusion) during three 1-h sessions for 10 days. During each session, the maximum number of infusions that each rat could earn was set to 20. The rats did not self-administer cocaine in the presence of their cage mate, but they were trained at the same time in different chambers. Rats were then given extinction training until the number of active responses decreased to below 20% of the mean number of responses during the last three sessions of self-administration training. Like cocaine self-administration, rats were tested at the same time as their cage mate, but in different chambers. After extinction, one rat (designated as "relapsed") was given a priming injection of cocaine (10 mg/kg) before being placed in the operant chamber. Drug-induced reinstatement was measured in this rat. The relapsed rat was placed in its homecage, where it could interact with its cage mate (referred to as the "targeted" rat). Following social interaction, the targeted rat was placed in the operant chamber, and drug-seeking behavior was measured. Unlike the relapsed rat, the targeted rat was never given a priming dose of cocaine before the reinstatement test session.

Experiment 2 was similar to Experiment 1, except that one rat received a saline injection before the reinstatement test ("unrelapsed" rat). The targeted rat interacted with the unrelapsed rat before being tested for reinstatement of cocaine seeking. Experiment 5 was also similar to Experiment 1, except that the targeted rat interacted with a relapsed rat that was unfamiliar to the rat (i.e., these rats did not live together). Experiments 3 and 4 determined some of the parameters of social influences on relapse-like behavior. In Experiment 3, Meng et al. manipulated how much time the targeted rat could spend with its relapsed cage mate (either 0, 10, 30, or 60 min). Experiment 4 determined the time course of socially mediated reinstatement. That is, rats interacted with their relapsed cage mate for 30 min and then were tested for reinstatement either immediately or 1−7 days following social interaction.

Experiment 6 was similar to Experiments 1, 2, and 5, except that rats were tested in CPP instead of drug self-administration. Rats received a baseline session in which they were allowed to explore all three compartments of the CPP apparatus for 15 min. Rats received eight conditioning sessions; rats received injections of cocaine (10 mg/kg) or saline (1 mL/kg) and were isolated to one of the compartments on alternating days (i.e., cocaine on days 2, 4, 6, and 8 and saline on days 3, 5, 7, and 9) for 45 min. Rats received a test session to assess CPP on day 10, followed by eight sessions of extinction training. During extinction sessions, rats were confined to one of the compartments, but they did not receive any injections. Rats were isolated to each compartment in alternating sessions. Extinction was verified by giving the rats another test session in which they explored all three compartments for 15 min. Some rats were then exposed to a relapsed cage mate, an unrelapsed cage mate, and a relapsed stranger before being tested in drug-induced reinstatement. Another control group was used for this experiment; some rats received no social interaction or a priming injection of cocaine before being tested for reinstatement of cocaine-seeking behavior.

Instead of describing the results for each experiment, I want you to look at the graph on the next page. Panel (A) provides a schematic showing the experimental timeline. Panels (B) and (C) present infusions earned during self-administration sessions and active responses during extinction sessions, respectively, for the relapsed rat and the targeted rat. Panels (D) and (E) present active and inactive responses during the reinstatement test for the relapsed rat and the targeted rat, respectively. Panels (F−I) are similar to Panels (B-E), except that these data are for the unrelapsed rat and their cage mate.

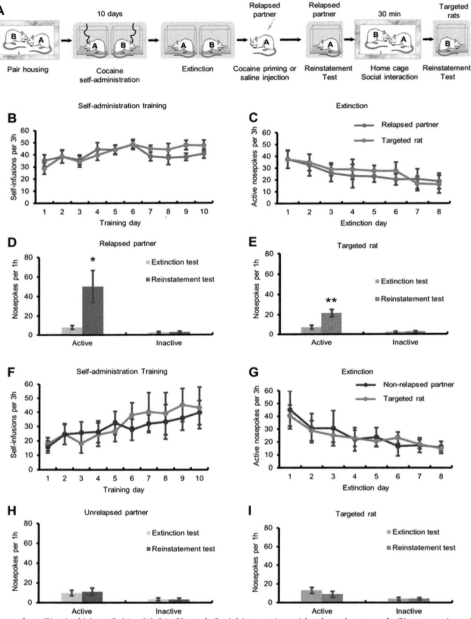

Figure comes from Fig. 1 of Meng S, Yan W, Liu X, et al. Social interaction with relapsed partner facilitates cocaine relapse in rats. Front Pharmacol. 2021;12:750397. https://doi.org/10.3389/fphar.2021.750397. This is an open-access article distributed under the terms of the Creative Commons Attribution 4.0 International License (https://creativecommons.org/licenses/by/4.0/).

Questions to consider:

(1) Did exposure to a relapsed rat increase relapse-like behavior in targeted rats? What evidence are you using to support your answer?
(2) I did not present the data for unfamiliar relapsed rats. Do you think that targeted rats exposed to an unfamiliar partner showed increased cocaine-seeking behavior following interaction with unfamiliar relapsed rats?
(3) Why were rats limited to 20 cocaine infusions during self-administration sessions?

Answers to questions:

(1) Yes. Look at Panel (E) of the figure. The active responses of the targeted rat increased during the reinstatement test (second bar on the left of Panel (E)) compared to active responses during extinction sessions. This effect is not as robust as in relapsed rats given a priming injection of cocaine (Panel (D)). If you look at Panel (I), you will see that targeted rats exposed to an unrelapsed rat do not show an increase in active responses during the reinstatement test. These data provide support for the claim that social interaction with a relapsed partner can facilitate relapse-like behavior. These results were observed in the CPP experiment as well.
(2) The results of this experiment are interesting as targeted rats exposed to an unfamiliar relapsed rat do not show an increase in responding during the reinstatement test. These results suggest that familiarity with the conspecific is important for social-induced reinstatement of cocaine seeking.
(3) Cocaine infusions were limited to prevent overdose deaths. Because rats were trained in three separate 1-h sessions each day, the researchers had to ensure that rats did not earn too much cocaine during the first session, which could have increased the risk of overdose during the second or the third sessions. Capping the amount of cocaine that rats could earn also helped ensure that rats in each pair earned approximately the same amount of cocaine in each session (see Panel (B) of the figure).

Not only can interaction with a "relapsed" rat facilitate reinstatement of drug seeking, but social peers can serve as discriminative stimuli that signal the availability of the drug, thus increasing the probability of relapse-like behavior. As a graduate student, I helped conduct an experiment in which we trained rats to self-administer cocaine in the presence of one conspecific. The conspecific was in an adjacent compartment separated by a wire mesh partition. Rats were then given extinction training in the absence of their partner. During the reinstatement test, rats were reexposed to their partner. We found that the presence of the social partner increased cocaine-seeking responses, although this effect was observed during the first reinstatement session only. In subsequent experiments, rats were trained to self-administer cocaine in the presence of one conspecific and saline in the presence of a different conspecific. Reinstatement of cocaine seeking occurred in the presence of the cocaine-paired rat only.[77] These results provide a mechanism explaining why being around certain individuals can make abstaining from drugs difficult. If one spends time with others that use drugs, these individuals can act as discriminative stimuli signaling the availability of drug. Even if trying to abstain from drugs, being in the presence of these other individuals can precipitate cravings and relapse. Think of my father-in-law wanting to smoke cigarettes each time he met with his friend.

Collectively, the results of preclinical studies suggest that social factors influence individuals to use drugs even if they do not particularly like to do so. This is analogous to individuals referring to themselves as social drinkers and/or smokers. On their own, these individuals may not be highly motivated to drink ethanol or to smoke cigarettes. However, social situations involving drinking/smoking can act as a powerful motivator to use these substances. To provide some anecdotal evidence: during my first year in college, I did not consume ethanol; however, almost every single one of my friends drank at parties. Over time, I grew tired of being the proverbial "odd man out". During my second year, I decided to drink while hanging out with friends. Similar to the findings reported by Reed and Rountree,[12] my friends did not overtly pressure me to drink at these parties. During my remaining time in college, I only drank if I was with others. I never desired a beer with my dinner if I was eating alone in my apartment. For me, as well as many individuals, drug-taking experiences are socially bound.

Observational learning

Given that imitation is an important variable in Akers' social learning theory of deviant behaviors, some preclinical research has determined if ethanol consumption in rodents can be influenced by observing a conspecific receiving ethanol as well. Consistent with social learning theory, periadolescent rats that observe a sibling consume ethanol, but not water or coffee, consume more ethanol than water or coffee.[78–80] Additionally, adolescents that observe an adult female ingesting ethanol show a preference for ethanol over water.[81] These results have been extended to show that partner familiarity is a critical determinant in ethanol consumption following social observation. While both males and females consume more ethanol following exposure to a familiar cagemate, adolescent males exposed to a novel conspecific show decreased ethanol consumption while females consume more ethanol in the presence of an unfamiliar conspecific.[82] The increased ethanol consumption in observer rats is specific to watching another rat receive ethanol. If rats observe another rat receive cocaine injections, subsequent ethanol consumption is unaffected.[83]

Social facilitation

The mere presence of another person can motivate individuals to perform a task. This phenomenon is known as **social facilitation** and was first formally described by Dr. Floyd Allport (1890–1979).[84] Social facilitation of drug self-administration can be measured in rats with a special operant conditioning chamber composed of two compartments separated by either Plexiglas or meshed wire. One rat is placed in one compartment and is trained to self-administer a drug. A conspecific can be placed in the adjacent compartment during training or following acquisition of self-administration (Fig. 10.2). To date, only a few studies have observed the social facilitation of drug self-administration in rodents. Introducing a conspecific following acquisition enhances d-amphetamine self-administration, but this effect is observed following the first exposure only.[86] A similar effect has been observed in rats trained to self-administer nicotine.[87]

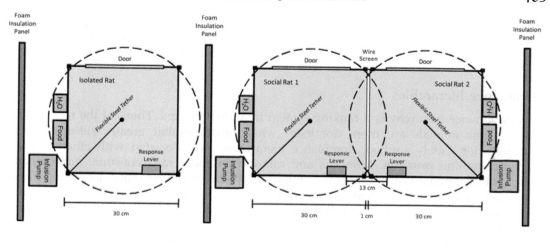

FIGURE 10.2 A schematic of an operant conditioning chamber equipped for rats performing self-administration training in isolation (top left image) or in the presence of another rat (top right image). The bottom left image shows the operant conditioning chamber used to measure social facilitation of drug self-administration, and the bottom right image shows two rats performing drug self-administration in each compartment of the operant chamber. Notice that each compartment is separated by a wire mesh partition. *Image comes from Fig. 1 of Smith MA Peer influences on drug self-administration: social facilitation and social inhibition of cocaine intake in male rats.* Psychopharmacology. *2012;224(1):81–90. https://doi.org/10.1007/s00213-012-2737-6, Copyright 2012.*

In one study, Smith[85] tested the social facilitation of cocaine self-administration in individually and pair-housed rats. Pair-housed rats were separated into two additional groups. One group of rats learned to self-administer cocaine at the same time, one in each compartment separated by a wire mesh partition. For the second group, only one rat learned to self-administer cocaine while the other rat had no access to reinforcement in the operant chamber. Interestingly, social facilitation of cocaine self-administration was observed in pair-housed rats that were concurrently trained, whereas social *inhibition* was observed in pair-housed rats whose partner did not receive self-administration training. These results suggest that self-administration is enhanced when both peers learn to self-administer the drug at the same time. However, this does not always appear to be the case. Rats show increased remifentanil self-administration when exposed to a conspecific in an adjacent chamber, regardless if the peer is self-administering remifentanil or saline.[88] More work is needed to elucidate the situations in which social facilitation or social inhibition are observed.

Dominance hierarchies

A dominance hierarchy is a ranking system in animal groups. Think of the term "alpha male". Some animals are more dominant while others exhibit greater submissiveness. Although not widely studied, dominance hierarchies are associated with ethanol use in both rodents and primates. Monkeys and rats low in social rank consume more ethanol compared to dominant conspecifics.[89–91] Interestingly, rats given ethanol before being placed in group housing fail to develop dominance hierarchies and show less aggression compared to untreated rats.[92] Perhaps repeated ethanol exposure produces alterations to neural systems associated with aggression.

Beyond ethanol, some work has examined cocaine self-administration in dominant and submissive monkeys. Submissive monkeys self-administer more cocaine than dominant monkeys.[93] The neurobiological mechanisms explaining why submissive animals self-administer more drugs may be partially related to alterations in the dopaminergic system as submissive monkeys have lower levels of monoamine oxidase activity,[89] and dominant monkeys have higher D_2-like receptor distribution volume ratios and reduced levels of extracellular dopamine.[93] Dominant monkeys also have increased D_2-like receptor availability in the caudate nucleus during cocaine abstinence compared to subordinates.[94] Placing subordinate monkeys into new groups such that they become dominant increases availability of D_2-like receptors.[95] These studies demonstrate a relationship between monoamine activity, especially dopamine, and dominance in nonhuman primates.

Quiz 10.1

(1) Which parenting style is most protective against drug-taking behaviors in adolescents?
   a. Authoritarian
   b. Authoritative
   c. Permissive
   d. Uninvolved
(2) Which of the following statements about preclinical research on social interaction and drug preference is true?
   a. Access to social peers always increases the rewarding effects of a drug.
   b. Providing access to a social partner with each infusion of the drug decreases the reinforcing effects of the drug.
   c. Rats are incapable of learning to self-administer a drug by merely observing a conspecific self-administer the drug.
   d. Social interaction can increase the rewarding effects of a drug that normally does not produce a reward on its own.

(3) According to Akers' social learning theory, which of the following does NOT influence drug use?
   a. Differential association
   b. Differential reinforcement
   c. Imitation
   d. All of the above influence drug use
(4) According to Bandura, what will most likely occur if an adolescent sees his/her friend get arrested for possessing cannabis?
   a. They will be less likely to use cannabis in the future.
   b. They will be more likely to use cannabis in the future.
   c. This event will have no effect on future cannabis use.
(5) True/false: adolescents often use drugs because they are pressured by their friends to do so.

Answers to quiz 10.1

(1) b — Authoritative
(2) d — Social interaction can increase the rewarding effects of a drug that normally does not produce a reward on its own.
(3) d — All of the above influence drug use.
(4) a — They will be less likely to use cannabis in the future.
(5) False

Socioeconomic status (SES)

Dominance hierarchies are not limited to lower-order species like nonhuman primates and rodents. We have our own dominance hierarchy in the form of wealth and prestige. This is known as **socioeconomic status (SES)**, an individual's social or class standing, and is often associated with one's income, education level, and occupation. Individual differences in SES have long been associated with addiction-like behaviors. However, results across studies have revealed a complex relationship between SES and drug use. One consistent finding is that low SES is associated with cigarette smoking.[96-100] According to the Centers for Disease Control and Prevention, over 30% of adults living below the poverty line smoke cigarettes compared to approximately 16% of those living more than twice above the poverty line.[101] Similarly, over 30% of adults with less than a high school degree use cigarettes, but only 10% with a college degree smoke.[101] This finding seems counterintuitive considering the high costs of cigarettes (see Chapter 1). However, cigarette prices have not always been as high as they are now. Because low SES individuals have already developed a nicotine use disorder, they have a difficult time abandoning cigarettes as prices increase (real-life example of an inelastic good you learned about in Chapter 5). Another factor influencing cigarette use in low SES areas is the increased density of tobacco retailers.[102] With greater access to tobacco products, individuals in low SES neighborhoods have more opportunities to smoke cigarettes. Unsurprisingly, poverty at the neighborhood level, beyond the individual level, is predictive of tobacco use.[103]

In addition to cigarette use, low SES is linked to problematic use of several substances, including ethanol, cannabis, and cocaine,[104,105] including crack cocaine.[106] Like cigarettes, poverty at the neighborhood level is predictive of cocaine and heroin use.[107] Across 28 European countries, low SES students (~15–16 years of age) are more likely to engage in frequent cannabis use (i.e., 20+ instances during the past month), episodic cannabis use (i.e., more than twice during their lifetime but less than 20 times during the past month), experimental cocaine use (i.e., 1–2 times during their lifetime), and episodic cocaine use.[108]

The SES level of one's parents can influence drug use as well. Adolescents with mothers with a high school degree only are at increased risk of using inhalants, cocaine (including crack cocaine), psychedelic hallucinogens, and "club" drugs like MDMA.[109] Also, individuals that grew up in a low SES household drink more heavily during times of unemployment.[110] However, young adults with high SES parents are *more* likely to binge drink, use cannabis, and use cocaine.[111,112] Other studies have also found increased ethanol consumption and cannabis use in high SES adolescents and young adults.[97,100,113,114] There appears to be contradictory findings concerning the relationship between SES and cannabis use. Whereas one study found increased episodic cannabis use in low SES students, other work has observed the opposite association. One potential reason for this discrepancy can be how cannabis use is quantified. For example, adolescents from high SES families are more likely to experiment with cannabis, but they are less likely to develop heavy or problematic cannabis use.[115] The association between SES and drug use becomes more complex when other factors are examined, such as race. We will come back to this later in the chapter.

Modeling the association between SES and addiction in animals

In addition to being able to spend more money on leisurely activities, high SES individuals are more easily able to live in areas that provide them with better access to healthcare and education. Obviously, animals do not use money as we do. However, researchers can model aspects of SES in animals by manipulating the environment in which the animal lives. In environmental enrichment studies, some animals live in isolation/pairs while others live in social cohorts with access to novel objects (see Fig. 10.3 for an example of environmental enrichment for rats).

While early reports showed that environmentally enriched rats self-administer more cocaine[117] or consume more ethanol[118] compared to isolated rats, many studies have since shown that environmental enrichment is a protective factor against addiction-like behaviors. Environmentally enriched rats acquire self-administration at a slower rate[119,120] and are less likely to acquire self-administration compared to isolated rats.[121,122] Environmentally enriched rats self-administer/consume less ethanol,[123–125] d-amphetamine,[126–130] methylphenidate,[131] nicotine,[132] morphine,[133] and remifentanil[134] compared to isolated rats. Environmental enrichment also decreases the reinforcing efficacy of drugs as assessed in progressive ratio schedules[126,127,129,131,135] and blunts reinstatement/renewal of drug-seeking behavior.[121,135–146] Behavioral economic analyses have shown that environmentally enriched animals treat drug reinforcers as more elastic compared to isolated animals.[134,147] When combined with extinction training, environmental enrichment is effective at preventing reacquisition of cocaine self-administration[148] and reinstatement of drug-seeking behavior.[149] Environmental enrichment has also been shown to prevent the incubation of cocaine and heroin seeking.[150,151]

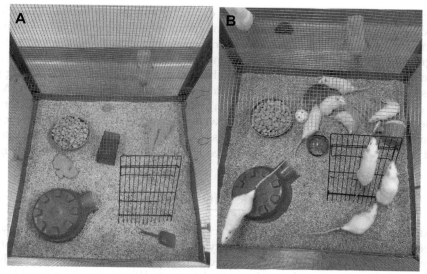

FIGURE 10.3 One example of environmental enrichment for rats. In this experiment, rats are given access to toys that are changed weekly, and the walls are made of wire mesh to allow the rats to climb. Panel (A) shows the enriched environment without rats while panel (B) shows the rats in the enriched environment. *Image comes from Fig. 1 of Smail MA, Smith BL, Nawreen N, Herman JP. Differential impact of stress and environmental enrichment on corticolimbic circuits. Pharmacol Biochem Behav. 2020;197:172993. https://doi.org/10.1016/j.pbb.2020.172993, Copyright 2020, with permission from Elsevier.*

One consideration needs to be taken into account when comparing enriched rats to isolated rats: baseline rates of responding. For example, Smith et al. found that isolated rats self-administer more cocaine compared to enriched rats, but they also respond more for saline.[152] When responses were expressed as a percentage change relative to saline, the results completely changed. Enriched rats showed a greater percentage increase in cocaine self-administration compared to isolated rats. These results better align with earlier studies examining environmental enrichment effects on drug self-administration.[117]

The effects of environmental enrichment on conditioned place preference (CPP) are more mixed compared to drug self-administration. Although environmental enrichment consistently attenuates the conditioned rewarding effects of opioids[138,153–155] (but see[156]) and diazepam,[157] it potentiates CPP for ethanol[158,159] (but see[160]), nicotine,[161] and d-amphetamine[162,163] (but see[164]). The findings for enrichment effects on cocaine CPP have been largely mixed, with some studies reporting beneficial effects of enrichment on cocaine reward[165–169] and others reporting detrimental effects of enrichment on CPP.[152,170,171] One study has even shown that environmental enrichment fails to decrease cocaine CPP until *after* animals have already developed CPP.[172] So far, no differences in methamphetamine CPP have been observed between isolated and enriched animals.[140,173] Similar to studies examining enrichment effects on self-administration, environment enrichment decreases reinstatement of ethanol[174,175] and methamphetamine CPP.[176]

Because environmental enrichment entails exposing animals to social interaction and to novel objects, some research has examined if one component is more important than the other. In addition to testing isolated animals and animals raised in social cohorts given access to novelty, one group of animals can have access to novelty while living in isolation, and a separate group of rats can live in social cohorts without having access to novel objects. Using these control groups, Gipson et al.[120] measured the escalation of cocaine self-administration. When given a low unit dose of cocaine (0.1 mg/kg/infusion), isolated rats, regardless of whether they have access to novelty, escalate their cocaine intake, whereas group-housed rats do not increase their self-administration when given extended access to cocaine. These results show that social interaction, not novelty, is critically important in reducing the reinforcing effects of drugs.

Racial and ethnic differences in addiction

Before discussing the findings regarding racial and ethnic differences in drug use patterns, we first need to differentiate race from ethnicity. **Race** refers to physical characteristics shared across a group of individuals that are inherited across generations. In contrast, **ethnicity** encompasses one's shared national or *cultural* traditions.[177] While race describes biological features, ethnicity is more concerned about social-behavioral facets. The United States recognizes five races: White, African American (or Black), Asian American, Indian/Alaska Native, and Native Hawaiian/Pacific Islander. The U.S. divides individuals into one of two ethnicities: Hispanic/Latino and non-Hispanic/Latino.

In 2019, African Americans made up approximately 14% of the U.S. population, and Hispanics/Latinos made up slightly over 18% of the population. However, approximately 75% of all federally incarcerated individuals for drug offenses are racial minorities[178]! Does this mean that minorities are at greater risk of using drugs or developing a SUD? Absolutely not. Although African Americans have higher rates of past year cocaine use compared to other races,[179] they are *less* likely to use cocaine with cannabis.[180] Furthermore, adolescent African Americans are less likely to use ethanol, cigarettes, inhalants, and methamphetamine[181,182] and are more likely to abstain from ethanol, cigarettes, and cannabis compared to Whites.[183] Similarly, African Americans, along with Asians, are less likely to report using drugs before and during college compared to White and Hispanic students.[184,185] African Americans are even less likely to use ethanol and cigarettes compared to Hispanics.[182] When comparing African American and White drug offenders, African Americans report greater preference for cannabis, but they report fewer severe drug problems compared to Whites.[186]

Why is there a disconnect between empirical findings supporting decreased drug use in African Americans and incarceration rates for drug-related offenses in this group? To answer this, we need to have a brief history lesson. During the 1970s, former U.S. president Richard Nixon initiated the "War on Drugs" to decrease drug use in Americans. With the introduction of minimum sentencing, a disproportionate number of African Americans and Hispanics have received long prison sentences for drug violations. Why? Minorities tend to live in more urban areas, which are more densely populated, and urban areas receive more scrutiny from the police compared to rural areas. Others have argued that the war on drugs was a way to suppress the voting rights of minorities.[187]

Because minorities are more likely to earn less compared to Whites[188] and SES is an important determinant of drug use, controlling for SES is vitally important when examining racial/ethnic differences in SUDs. For example, African Americans are reported to have an increased risk for both lifetime and recent crack cocaine use; however, this effect disappears when other SES factors are statistically controlled.[106] Similarly, African Americans have been shown to accelerate their use of ethanol, cannabis, and cocaine to a greater extent compared to Whites, an effect that disappears when SES is controlled.[104] Indeed, when controlling for SES, Whites initiate ethanol use, cannabis use, and cocaine use earlier than African Americans, and they are more likely to enter treatment following problematic drinking, cannabis, or cocaine use at an earlier age compared to African Americans.[104] Racial minorities are less likely to use powdered cocaine, and Hispanics are less likely to use crack cocaine compared to Whites.[106]

Race can also act as a moderating variable when examining the relationship between SES and drug use. In adolescents, decreased SES is associated with increased cigarette smoking in Whites,[113] which is expected; however, the association between SES and cigarette smoking is reversed in non-Whites, meaning that high SES minorities are more likely to smoke.[113,189] Another study found that low SES is associated with cannabis use in African Americans only.[190] These findings further highlight the complex relationship between SES and drug use.

Religion/spirituality and addiction

In Chapter 1, I mentioned that the prevalence of alcohol use disorder is lower in Muslim-majority countries. This finding is not isolated to Muslims as different religious beliefs act as a protective factor against drug use. Both Christians and Muslims are less likely to smoke cigarettes compared to others that do not follow either religion,[99] although one study found that Muslims living in South Africa are more likely to smoke waterpipes,[191] demonstrating an interaction between religiosity and cultural factors. Compared to individuals that are agnostic/atheist, individuals that belong to a religious group (e.g., Christian, Hindu, Jewish, Muslim, Sikh, etc.) are less likely to use tobacco products[192,193] or to engage in risky drinking/ethanol misuse.[192,194] Frequent church attenders are less likely to use cigarettes, ethanol, cocaine, and opioids[195–204] and are less likely to be diagnosed with an alcohol use disorder.[205,206] Even if individuals believe in God, they are more likely to smoke cigarettes if they do not belong to a religion.[207] Overall, individuals with high levels of religiosity are less likely to use cigarettes,[96,208] ethanol,[200] cocaine,[179,180,209] either alone or in combination with cannabis,[180] and are less likely to misuse opioids with another substance.[210] Paradoxically, religiosity is positively associated with relapse to crack cocaine.[211]

Even in adolescents, theism is negatively associated with ethanol/drug use.[24,26,181,212–219] Not surprisingly, adolescents that view drug use as sinful are less likely to use drugs and report decreased sensation seeking.[213] Importantly, attending church is not necessarily protective against drug use. In a sample of Mexican adolescents, frequent church attendance is associated with an elevated risk of ethanol and cigarette use if individuals intrinsically do not believe religion to be important.[217] Additionally, the protective effects of religiousness on the initiation of drug use are mediated by delay discounting.[220] That is, strong religious beliefs are associated with decreased impulsive decision making, which is then associated with decreased drug use initiation.

There are individuals that do not prescribe to a specific religion but still believe in a connection to something larger than themselves. This is known as *spirituality*. The relationship between spirituality and drug use is interesting as some individuals use drugs as a spiritual practice.[221] In the previous chapter, you learned about self-transcendence, which taps into spirituality. Recall that opioid users often score higher in self-transcendence compared to those that score lower in this dimension. If individuals identify using drugs as part of their spirituality, they are more likely to engage in cigarette smoking and hallucinogen use compared to nondrug-endorsing spiritual individuals.[221] Related to this finding, spirituality often acts as a protective factor against drug use, like religiosity.[222-225] Indeed, one study found that high levels of spirituality, but not high levels of religiosity, predict decreased self-reported heroin and cocaine/crack cocaine use.[226] Conner et al. also found that individuals high in spirituality submitted fewer heroin-positive urine samples compared to those low in spirituality.[226]

Increased spirituality/religiosity has also been shown to predict outcomes for those seeking treatment for a SUD. Those that report spending more time on religious/spiritual activities have higher retention rates when seeking treatment for cocaine or heroin dependence.[227,228] Additional studies have shown that individuals with high religiosity have lower relapse rates for drugs such as ethanol, heroin, cocaine, and cannabis.[211] Religious activities can be successfully combined with traditional treatments. For example, Muslim heroin users receiving methadone maintenance in mosques abstain from drug use for at least 12 months following treatment.[229]

While religiosity/spirituality appears to protect against ethanol/drug use, there are some variables that have been shown to moderate the benefits of religion. One such variable is peer influences. Palamar et al. found that religious attendance protects against recent cannabis and cocaine use; however, this effect disappears when controlling for exposure to those that use these substances.[230] Race/ethnicity appears to be a moderating factor between religiosity and substance use, as the protective effects of religion are more pronounced in Whites compared to African Americans and to Hispanic Americans.[231] Specifically, religiosity is an important protector against ethanol use and binge drinking in Whites, but not in non-Whites.[232] While adolescent African Americans are more likely to abstain from drug use compared to White adolescents, Wallace Jr. et al. found that highly religious Whites are more likely to abstain from drug use compared to highly religious African Americans.[183] Concerning spirituality, while Conner et al. found beneficial effects of spirituality in a sample of African Americans and Hispanic Americans,[226] a separate study found no benefits of spirituality for Hispanic Americans enrolled in maintenance replacement therapy.[233] Furthermore, increased spirituality is linked to increased cigarette and cannabis use in American Indian/Alaska Native Adolescents.[234]

Socially based treatments for SUDs

If peers and families can influence drug experiences, treatment programs should be able to utilize social groups to help facilitate abstinence. Similar to what you learned when reading preclinical studies about social influences on addiction, social interaction can be used to our advantage to curb drug-taking behaviors. This section will highlight some of the ways social interaction can be used for the treatment of SUDs.

Family therapy

At the beginning of this chapter, I brought up CRAFT, which was discussed in Chapter 5. CRAFT incorporates elements of **family therapy**.[235] As the name suggests, therapists work with families and couples in intimate relationships to foster changes in behavior (in this case, drug/ethanol cessation). There are several variations of family therapy, including multidimensional family therapy, multisystemic family therapy, family systems therapy, and marriage and family therapy. A major advantage of family therapy is that sessions can be completed at home with the therapist and with the individual's family. As such, dropout rates tend to be low. For example, in one study, 98% of families assigned to family therapy completed treatment while only 15% of families assigned to treatment as usual received any sort of substance use treatment.[236]

Overall, family therapy, when combined with usual care, is more effective than nonfamily treatments for reducing drug use in adolescents.[237] Also, family therapy is just as effective as cognitive-behavioral therapy (CBT) in reducing ethanol and cannabis use in adolescents.[238,239] However, age is an important moderator when comparing the effectiveness of family therapy to CBT for the reduction of cannabis use, as 17-18-year-olds benefit more from CBT while younger adolescents benefit more from family therapy.[240] While family therapy and CBT are equally effective in reducing cannabis use, family therapy is more effective than individual psychotherapy in decreasing cannabis dependence.[241] Although family therapy is largely used to treat adolescent drug use,[242,243] this approach can be used with adults.[244-246] Family therapy decreases ethanol consumption in parents and decreases substance use in adolescents of drug-using parents.[247]

Social-network therapy utilizes more than family members or intimate partners. Friends, neighbors, and coworkers can be used as an emotional support system as the individual goes through therapy.[248] Social-network therapy appears to be beneficial in reducing cocaine use[249,250] and heroin use.[251] In addition to observing a decrease in heroin use, Copello et al. found that social-network therapy decreases self-reported dependence-like behaviors and increases family satisfaction.[251] The increased family satisfaction may be an important mechanism to help individuals abstain from drugs. Recall the social control theory, which states that individuals engage in deviant behaviors because they have low commitment to their families.

Residential treatment centers

For severe SUDs, individuals may need to go to a **residential treatment center**, often colloquially referred to as "rehab". A residential treatment center allows an individual to stay in a home-like facility that is fully staffed. One of the most famous residential treatment centers specific to SUDs is the Betty Ford Center in Rancho Mirage, California. The name comes from former U.S. First Lady Betty Ford (1918—2011), wife of Gerald Ford, who cofounded the residential treatment center in 1982. Ford was interested in opening a treatment center for SUDs as she battled alcohol use disorder and was dependent on opioids. In 2014, the Betty Ford Center merged with the Hazelden Foundation to form the Hazelden Betty Ford Foundation, headquartered in Minnesota.

Staying in a residential treatment center is effective in decreasing drug use. Individuals needing long-term care for cocaine use disorder provide fewer cocaine-positive urine samples if enrolled in a residential treatment center compared to individuals enrolled in a short-term inpatient program and to individuals enrolled in an outpatient program.[252] Staying in a residential treatment center can decrease drug cravings, in addition to decreasing psychiatric symptoms and increasing cognitive flexibility.[253]

In a residential treatment center, individuals can receive different types of therapy, including family therapy, contingency management, CBT, and mindfulness training.[254] Receiving therapy while residing in a residential treatment center is important for improving treatment outcomes. For example, participation in mindfulness sessions decreases dropout rates.[254] Another important consideration is one's education level. When comparing those that successfully completed treatment in a residential treatment center to those that did not, Federer et al. found higher education levels in the former group.[255] Length of time spent in a residential treatment center also predicts the effectiveness of this treatment option.[256] Specifically, individuals that spend at least 6 months in a residential treatment center report decreased cocaine use, decreased engagement in illegal activities, and increased employment 1 year following treatment.[257]

One disadvantage of residential treatment centers is that they typically provide short-term treatment, with 30–90 days being common lengths of stay. Another major deterrent to staying in a residential treatment center is cost. In 2006, the average cost for a 12-week program was approximately $10,000.[258] Accounting for inflation, this would be over $13,000 in 2021. If individuals seeking treatment are uninsured, this cost makes staying at a residential treatment center extremely difficult. This can exacerbate drop-out rates. In fact, one study reported a drop-out of over 40%.[253]

Halfway houses

While residential treatment centers can help individuals with a SUD, these individuals cannot permanently live there. How do we help these individuals transition from living in a residential treatment center to living on their own? Individuals can live in a **halfway house**, a type of treatment facility that is staffed to provide support services and to monitor the individual. Historically, halfway houses have been used to help those returning to society after serving time in prison. However, halfway houses have been shown to help individuals with a SUD reintegrate into society. Van Ryswyk et al.[259] measured the number of admissions to a detoxification center, dependence on public assistance, involvement with the criminal justice system, employment, and abstinence. Compared to a 14-month period before being placed in a halfway house, individuals showed improvements in each area 14 months after living in a halfway house. Those living in a halfway house stay in treatment for a SUD longer compared to those living on their own or with their friends/family.[260] Although men living in a halfway house report a similar frequency of ethanol-induced intoxications compared to controls, they are less likely to be arrested for ethanol intoxication.[261]

Because halfway houses are primarily concerned with reintegrating the individual back into society, individuals receive training in "life skills" like money management, job coaching, and coping skills. These skills do not necessarily help individuals abstain from drug use. Today, there are alternatives to halfway houses that are geared toward promoting abstinence. Keep reading to find out more.

Therapeutic communities

An alternative to the halfway house is the **therapeutic community**. Therapeutic communities are similar to halfway houses, except that they do not always provide support staff. These communities can also focus on a specific demographic, such as adolescents,[262–264] pregnant women/women with children,[265] prisoners,[266] and homeless veterans.[267] Like residential treatment centers, other therapies can be used within the therapeutic community, like methadone maintenance,[268] mindfulness training,[269,270] and even art therapy.[262]

One example of a therapeutic community is the *Oxford house*. Individuals staying in an Oxford house independently operate the home and are responsible for maintaining their house. Additionally, residents of an Oxford house can stay as long as needed. However, if someone uses ethanol or drugs, they can be asked to leave the house by the other residents. A major benefit of the Oxford house is that individuals feel like they are in a home and not in an institution; by providing a sense of "home", individuals living in the Oxford house may be more inclined to spend time caring for the home instead of using drugs (form of alternative reinforcement).[271]

Multiple studies have shown the utility of therapeutic communities for the treatment of SUDs. Spending 1 year in a therapeutic community leads to decreased drug use, as well as decreased arrests and increased employment.[272] The benefits of therapeutic communities are more pronounced in those who "graduate" from the community relative to those that drop out prematurely.[273] Adolescents examined 1 year after spending time in a therapeutic community report reductions in drug use and criminal activity.[263] Ethanol and drug dependence have been observed to decrease by as much as 60% after living in a therapeutic community,[274] and therapeutic communities are more effective than other treatment strategies such as methadone maintenance and hospitalization.[275] Compared to treatment as usual, staying in an Oxford house modestly increases 2-year abstinence rates (31.4% vs. 21.2%).[276] However, the number of days spent in an Oxford house is related to abstinence.[277,278] Given the success of therapeutic communities like the Oxford house, the University of North Carolina (UNC) established the first Oxford house for college students in 2016.[279] Therapeutic communities have also been established all over the world, with such communities located in Australia,[262] New Zealand,[280] Italy,[281] Spain,[282] Croatia,[275] Thailand,[283] Peru,[284] and Saudi Arabia,[285] just to name a few.

There are several factors that influence the effectiveness of therapeutic communities. An obvious determinant of success is the length of stay.[286] Specifically, living in an Oxford house for at least 6 months leads to improved outcomes related to substance use.[277,287] Another factor that affects the utility of therapeutic communities is education level.[286] As you learned in Chapter 8, impaired decision-making is a risk factor for SUDs. Poor decision-making as assessed in the Iowa gambling task (IGT) is linked to premature drop out from therapeutic communities.[288] Similarly, impaired response inhibition accounts for treatment dropout,[282] and individuals with attention-deficit/hyperactivity disorder (ADHD) are less likely to complete treatment.[289] However, individuals who complete 9 months in a therapeutic community become less impulsive.[290]

12-Step programs

While treatment options such as contingency management and CBT may not be easily recognizable to those with little knowledge of SUDs, I suspect many of you reading this book have heard the term *Alcoholics Anonymous (AA)* at some point. AA was founded in 1935 by William (Bill) Wilson (1895–1971) and Dr. Robert (Bob) Smith (1879–1950). Wilson, who had been abstinent from ethanol use for 6 months, met Smith, a surgeon who was detoxing from ethanol use in an Ohio hospital. Four years later, Wilson and Smith helped write *Alcoholics Anonymous: The Story of How More Than 100 Men Have Recovered from Alcoholism*.[291] AA's guiding principles are largely influenced by religion as Wilson was a "born-again" Christian, having joined the evangelical Oxford Group (the name Oxford house is derived from this group). AA is considered a **twelve (12)-step program** because individuals move through 12 stages (steps) during the recovery process. The steps are listed below[291]:

1. Individuals first admit that they are powerless over ethanol and that their lives have become unmanageable.
2. Individuals accept that there is a higher power that can help them overcome their alcohol use disorder.
3. Individuals turn their lives over to God.
4. Individuals make a moral inventory of themselves.
5. Individuals admit their mistakes to themselves, to others, and to God.
6. Individuals ready themselves to have God remove their "defects".
7. Individuals ask God to remove these defects.
8. Individuals make a list of all persons that they have harmed, and they ready themselves to make amends with these persons.
9. Individuals attempt to make amends with the persons identified in the previous step.
10. Individuals continue taking personal inventory, quickly admitting when they are wrong.
11. Individuals engage in prayer/meditation to strengthen their relationship with God.
12. Individuals, after having a "spiritual awakening", attempt to spread the tenets of AA to others.

One important feature of AA is that individuals remain somewhat anonymous by stating their first names only. Also, those enrolled in AA are given a *sponsor*, someone that is further along in the recovery process. The sponsor guides the individual as they complete the steps of AA. As individuals reach milestones (i.e., achieving so many months of sobriety), members of AA receive a medal celebrating this achievement (Fig. 10.4). Twelve-step programs are not isolated to those with an alcohol use disorder. Narcotics Anonymous (NA) and Marijuana Anonymous (MA) use the same principles outlined above to help those with nonalcohol use disorders. Although 12-step programs are deeply rooted in religion, organizations like AA and NA allow nonreligious individuals to go through treatment.[292] For example, one study found that while 94% of participants enrolled in NA identified as spiritual, only

29.6% designated themselves as religious.[293] Another study found that only 43% of those attending AA meetings reported a sense of spirituality.[294]

How well do 12-step programs work? Anecdotal evidence supports the use of programs like AA/NA. As told by one 37-year-old with methamphetamine use disorder:

> The best way to get off of meth, is to get into a 12-step program, CMA, crystal meth anonymous ... Just show up. Secondly, when you show up at these meetings these people will embrace you and accept you, and then all you got to do is cut the ties with those old friends because they're not friends. They're not people that care about you. Cut those ties with them. Leave them alone. (p. 296)[295]

There is also empirical evidence to suggest the effectiveness of 12-step programs to promote short-term and long-term abstinence.[296–301] Over 80% of adolescents that participate in more than one AA/NA meeting per week were abstinent 3 months following treatment compared to fewer than 50% of adolescents enrolled in standard outpatient therapy.[302] Somewhat similar findings have been observed for women returning to society following incarceration: attending at least one AA meeting weekly decreases the number of drinking days and reduces negative drinking consequences.[303] In one study, almost 400 members of NA answered questions about their previous drug dependence and how long they had been abstinent from drug use; on average, respondents had been abstinent for 5.7 years.[293] An important determinant of long-term abstinence is *involvement* in 12-step programs, not just merely attending meetings. Majer et al.[278] found that 1-year abstinence rates were higher for those that engaged in activities like having a sponsor, reading 12-step literature, performing service work, and calling other members for help. Conversely, the number of AA/NA meetings attended was not a significant predictor of long-term abstinence. Majer et al. extended these findings when examining 2-year abstinence rates.[304]

One advantage of 12-step programs is that they can be implemented with other treatments, like pharmacotherapy and CBT.[299,305–307] Attending AA/NA meetings is common in those living in a halfway/Oxford house.[294] Additionally, individuals highly involved in a 12-step program are more likely to achieve abstinence when living in an Oxford house compared to those assigned to treatment as usual.[276] Twelve-step programs also appear to be just as effective in individuals with comorbid SUDs and psychiatric conditions (e.g., mood or anxiety disorders) as those with a SUD only.[291]

FIGURE 10.4 Two examples of medals given to AA members achieving 6 months of sobriety (green medal) and 9 months of sobriety (purple medal). The front of each medal indicates how long the individual has been abstinent. The back of each medal contains the *Serenity Prayer*. Image comes from Wikimedia Commons and was posted by Jonn Leffmann (https://commons.wikimedia.org/wiki/File:AA_-_Medalj.jpg).

Due to the effectiveness of 12-step programs, there are efforts to incorporate 12-step practices into treatment plans for those with a SUD.[308] This is known as 12-step facilitation (TSF) treatment. In one pilot clinical study,[309] adolescents were assigned to a TSF treatment that was integrated with motivational enhancement therapy (MET) and CBT (integrated TSF or iTSF). The iTSF group was compared to another group that received MET/CBT only. The primary outcome variable was the percent of days abstinent, with secondary outcomes including 12-step attendance and longest period of abstinence. While the percentage days of abstinence did not differ between the iTSF and the MET/CBT groups, adolescents assigned to iTSF attended more 12-step meetings. Unfortunately, increased 12-step attendance was observed during the implementation of the iTSF treatment only, meaning that iTSF did not produce long-lasting changes in adolescents' attendance of 12-step meetings.

Quiz 10.2

(1) What are some differences between a halfway house and a therapeutic community?
(2) Which factor can act as a third variable explaining the relationship between race and drug use?
   a. Impulsivity
   b. Psychopathy
   c. Religiosity
   d. SES
(3) How is SES modeled in animals?
(4) Which of the following statements regarding 12-step programs is true?
   a. Individuals can receive other treatments while enrolled in a 12-step program.
   b. One must be a member of the church to join a program like AA.
   c. The most important factor determining the effectiveness of 12-step programs is how many sessions one attends.
   d. Twelve-step programs are ineffective at treating adolescents.
(5) Which treatment strategy recruits family members, neighbors, and coworkers to help the individual recover from a SUD?
   a. CRAFT
   b. Multidimensional therapy
   c. Multisystemic therapy
   d. Social-network therapy

Answers to quiz 10.2

(1) Halfway houses tend to provide support staff like what would be seen in a residential treatment center, whereas therapeutic communities are more autonomously run by the individuals living there. Halfway houses also focus on reintegrating the individual back into society by offering training in money management, job interviewing, and coping skills while therapeutic communities are specifically focused on promoting drug abstinence.

(2) d — SES
(3) Environmental enrichment is used to model SES in animals. Animals raised in an enriched environment (i.e., access to social partners and novel objects) are compared to animals raised in isolation or in pairs.
(4) a — Individuals can receive other treatments while enrolled in a 12-step program.
(5) d — Social-network therapy.

Chapter summary

Peers and family members can greatly influence one's drug use, simultaneously acting as risk factors for and protective factors against drug use. According to social learning theory, individuals, particularly adolescents, are more likely to use drugs if they observe those closest to them receive reinforcement for their drug use (vicarious reinforcement). If parents do not monitor their children and utilize a neglectful parenting style, they are more likely to have children that use drugs. Therapists can utilize social influences to their advantage when treating someone with a SUD. Multiple treatment options have included family members and social peers to decrease drug use, such as family therapy, therapeutic communities, and 12-step programs. Beyond immediate family members and peers, sociocultural factors are known to affect drug use. Living in poverty or receiving lower levels of education are significant risk factors for the development of a SUD. Conversely, belonging to a religious group or having high levels of spirituality can decrease the likelihood that one uses drugs.

Glossary

Conspecific — animal of the same species.
Ethnicity — refers to the shared national or cultural traditions of a group of individuals.
Family therapy — a form of treatment in which a therapist works with families or couples in intimate relationships to foster changes in behavior and to strengthen interpersonal relations between the individual seeking treatment and his/her family and/or partner.
Halfway house — a treatment facility that allows one to transition from staying in a residential treatment center or serving time in prison to living on their own.
Parental monitoring — the extent to which a parent tracks and supervises a child's behaviors.
Parenting style — the way in which a parent raises and disciplines their child.
Race — physical characteristics shared across a group of individuals that are inherited across generations.
Reciprocal determinism — a theory proposed by Albert Bandura suggesting that our behaviors influence and are affected by the environment and by personal factors such as cognition.
Residential treatment center — a live-in health care facility that provides different types of therapy for various psychiatric illnesses, including SUDs.
Social control theory — the idea that people engage in delinquent and deviant acts, such as drug use, because their bonds to society have been broken. Also known as social bonding theory.
Social facilitation — the increase in behavior due to the mere presence of a conspecific.
Social learning theory — states that individuals learn to engage in specific behaviors by observing how peers/family members are reinforced and punished by performing the same behaviors.
Socioeconomic status (SES) — an individual's social or class standing; is often measured by education level and wealth.
Therapeutic community — a nontreatment facility characterized by self-help supportive housing in which small groups of individuals recovering from a SUD live together.

Twelve (12)-step program — a form of therapy rooted in religion in which individuals progress through 12 stages of recovery by attending meetings with other individuals battling a SUD.

Vicarious punishment — a decrease in the likelihood of engaging in a behavior if witnessing another receive a negative consequence for performing the behavior.

Vicarious reinforcement — an increase in the likelihood of engaging in a behavior if witnessing another receive a positive consequence for performing the behavior.

Supplementary data related to this chapter can be found online at https://educate.elsevier.com/9780323905787.

References

1. Akers R, Lee G. A longitudinal test of social learning theory: adolescent smoking. *J Drug Issues*. 1996;26(2):317—343. https://doi.org/10.1177/002204269602600203.
2. Allen M, Donohue WA, Griffin A, Ryan D, Turner MM. Comparing the influence of parents and peers on the choice to use drugs: a meta-analytic summary of the literature. *Crim Justice Behav*. 2003;30(2):163—186. https://doi.org/10.1177/0093854802251002.
3. Bahr SJ, Hawks RD, Wang G. Family and religious influences on adolescent substance abuse. *Youth Soc*. 1993;24(4):443—465. https://doi.org/10.1177/0044118X93024004007.
4. Bahr SJ, Hoffmann JP, Yang X. Parental and peer influences on the risk of adolescent drug use. *J Primar Prevent*. 2005;26(6):529—551. https://doi.org/10.1007/s10935-005-0014-8.
5. Boyle RG, Claxton AJ, Forster JL. The role of social influences and tobacco availability on adolescent smokeless tobacco use. *J Adolesc Health*. 1997;20(4):279—285. https://doi.org/10.1016/S1054-139X(96)00272-8.
6. Defoe IN, Khurana A, Betancourt LM, Hurt H, Romer D. Disentangling longitudinal relations between youth cannabis use, peer cannabis use, and conduct problems: developmental cascading links to cannabis use disorder. *Addiction*. 2019;114(3):485—493. https://doi.org/10.1111/add.14456.
7. Dishion TJ, Loeber R. Adolescent marijuana and alcohol use: the role of parents and peers revisited. *Am J Drug Alcohol Abuse*. 1985;11(1—2):11—25. https://doi.org/10.3109/00952998509016846.
8. Hoffmann J. Exploring the direct and indirect family effects on adolescent drug use. *J Drug Issues*. 1993;23(3):535—557. https://doi.org/10.1177/002204269302300312.
9. Kandel DB. On processes of peer influence in adolescent drug use: a developmental perspective. In: Brook JS, Lettieri DJ, Brook DW, Stimmel B, eds. *Alcohol and Substance Abuse in Adolescence*. 1985:139—164 (Haworth).
10. Varela A, Pritchard ME. Peer influence: use of alcohol, tobacco, and prescription medications. *J Am Coll Health*. 2011;59(8):751—756. https://doi.org/10.1080/07448481.2010.544346.
11. Buckner JD, Zvolensky MJ, Ecker AH. Cannabis use during a voluntary quit attempt: an analysis from ecological momentary assessment. *Drug Alcohol Depend*. 2013;132(3):610—616. https://doi.org/10.1016/j.drugalcdep.2013.04.013.
12. Reed MD, Rountree P. Peer pressure and adolescent substance use. *J Quant Criminol*. 1997;13(2):143—180. https://doi.org/10.1007/BF02221306.
13. Stangl BL, Schuster RM, Schneider A, et al. Suggestibility is associated with alcohol self-administration, subjective alcohol effects, and self-reported drinking behavior. *J Psychopharmacol*. 2019;33(7):769—778. https://doi.org/10.1177/0269881119827813.
14. Goldstick JE, Walton MA, Bohnert A, Heinze JE, Cunningham RM. Predictors of alcohol use transitions among drug-using youth presenting to an urban emergency department. *PLoS One*. 2019;14(12):e0227140. https://doi.org/10.1371/journal.pone.0227140.
15. Irfan M, Haque AS, Shahzad H, Samani ZA, Awan S, Khan JA. Reasons for failure to quit: a cross-sectional survey of tobacco use in major cities in Pakistan. *Int J Tubercul Lung Dis*. 2016;20(5):673—678. https://doi.org/10.5588/ijtld.15.0271.
16. McGue M, Sharma A, Benson P. Parent and sibling influences on adolescent alcohol use and misuse: evidence from a U.S. adoption cohort. *J Stud Alcohol*. 1995;57(1):8—18. https://doi.org/10.15288/jsa.1996.57.8.
17. Kluck AS, Carriere L, Dallesasse S, et al. Pathways of family influence: alcohol use and disordered eating in daughters. *Addict Behav*. 2014;39(10):1404—1407. https://doi.org/10.1016/j.addbeh.2014.05.015.

18. Gau SS, Chong MY, Yang P, Yen CF, Liang KY, Cheng AT. Psychiatric and psychosocial predictors of substance use disorders among adolescents: longitudinal study. *Br J Psychiatr*. 2007;190:42–48. https://doi.org/10.1192/bjp.bp.106.022871.
19. Constante K, Huntley ED, Si Y, Schillinger E, Wagner C, Keating DP. Conceptualizing protective family context and its effect on substance use: comparisons across diverse ethnic-racial youth. *Subst Abuse*. 2021;42(4):796–805. https://doi.org/10.1080/08897077.2020.1856289.
20. Sanvictores T, Mendez MD. *Types of Parenting Styles and Effects on Children*. StatPearls Publishing; 2021.
21. Melby JN, Conger RD, Conger KJ, Lorenz FO. Effects of parental behavior on tobacco use by young male adolescents. *J Marriage Fam*. 1993;55(2):439–454. https://doi.org/10.2307/352814.
22. Foster SE, Jones DJ, Olson AL, et al. Family socialization of adolescent's self-reported cigarette use: the role of parents' history of regular smoking and parenting style. *J Pediatr Psychol*. 2007;32(4):481–493. https://doi.org/10.1093/jpepsy/jsl030.
23. Miller BA, Byrnes HF, Cupp PK, et al. Thai parenting practices, family rituals and risky adolescent behaviors: alcohol use, cigarette use and delinquency. *Int J Child & Adolesc Health*. 2011;4(4):367–378.
24. Bahr SJ, Hoffmann JP. Parenting style, religiosity, peers, and adolescent heavy drinking. *J Stud Alcohol Drugs*. 2010;71(4):539–543. https://doi.org/10.15288/jsad.2010.71.539.
25. Garcia OF, Serra E, Zacares JJ, Calafat A, Garcia F. Alcohol use and abuse and motivations for drinking and non-drinking among Spanish adolescents: do we know enough when we know parenting style? *Psychol Health*. 2020;35(6):645–664. https://doi.org/10.1080/08870446.2019.1675660.
26. Hoffmann JP, Bahr SJ. Parenting style, religiosity, peer alcohol use, and adolescent heavy drinking. *J Stud Alcohol Drugs*. 2014;75(2):222–227. https://doi.org/10.15288/jsad.2014.75.222.
27. Stafström M. Influence of parental alcohol-related attitudes, behavior and parenting styles on alcohol use in late and very late adolescence. *Eur Addiction Res*. 2014;20(5):233–240. https://doi.org/10.1159/000357319.
28. Zuquetto CR, Opaleye ES, Feijó MR, Amato TC, Ferri CP, Noto AR. Contributions of parenting styles and parental drunkenness to adolescent drinking. *Braz J Psychiatr*. 2019;41(6):511–517. https://doi.org/10.1590/1516-4446-2018-0041.
29. Hartman JD, Patock-Peckham JA, Corbin WR, et al. Direct and indirect links between parenting styles, self-concealment (secrets), impaired control over drinking and alcohol-related outcomes. *Addict Behav*. 2015;40:102–108. https://doi.org/10.1016/j.addbeh.2014.08.009.
30. Brosnan T, Kolubinski DC, Spada MM. Parenting styles and metacognitions as predictors of cannabis use. *Addic Behav Rep*. 2020;11:100259. https://doi.org/10.1016/j.abrep.2020.100259.
31. Montgomery C, Fisk JE, Craig L. The effects of perceived parenting style on the propensity for illicit drug use: the importance of parental warmth and control. *Drug Alcohol Rev*. 2008;27(6):640–649. https://doi.org/10.1080/09595230802392790.
32. Martins SS, Storr CL, Alexandre PK, Chilcoat HD. Adolescent ecstasy and other drug use in the National Survey of Parents and Youth: the role of sensation-seeking, parental monitoring and peer's drug use. *Addict Behav*. 2008;33(7):919–933. https://doi.org/10.1016/j.addbeh.2008.02.010.
33. Šumskas L, Zaborskis A. Family social environment and parenting predictors of alcohol use among adolescents in Lithuania. *Int J Environ Res Publ Health*. 2017;14(9):1037. https://doi.org/10.3390/ijerph14091037.
34. Clark HK, Shamblen SR, Ringwalt CL, Hanley S. Predicting high risk adolescents' substance use over time: the role of parental monitoring. *J Primary Prevent*. 2012;33(2–3):67–77. https://doi.org/10.1007/s10935-012-0266-z.
35. Wu P, Liu X, Fan B. Factors associated with initiation of ecstasy use among US adolescents: findings from a national survey. *Drug Alcohol Depend*. 2010;106(2–3):193–198. https://doi.org/10.1016/j.drugalcdep.2009.08.020.
36. Walden B, McGue M, Lacono WG, Burt SA, Elkins I. Identifying shared environmental contributions to early substance use: the respective roles of peers and parents. *J Abnorm Psychol*. 2004;113(3):440–450. https://doi.org/10.1037/0021-843X.113.3.440.
37. Berge J, Sundell K, Öjehagen A, Håkansson A. Role of parenting styles in adolescent substance use: results from a Swedish longitudinal cohort study. *BMJ Open*. 2016;6(1):e008979. https://doi.org/10.1136/bmjopen-2015-008979.
38. Bandura A. *Social Learning Theory*. Prentice-Hall; 1977.
39. Bandura A. *Social Foundations of Thought and Action: A Social Cognitive Theory*. Prentice-Hall; 1986.

40. Smith MA. Social learning and addiction. *Behav Brain Res*. 2021;398:112954. https://doi.org/10.1016/j.bbr.2020.112954.
41. Wardell JD, Read JP. Alcohol expectancies, perceived norms, and drinking behavior among college students: examining the reciprocal determinism hypothesis. *Psychol Addict Behav*. 2013;27(1):191–196. https://doi.org/10.1037/a0030653.
42. Stacy AW, Newcomb MD, Bentler PM. Cognitive motivation and drug use: a 9-year longitudinal study. *J Abnorm Psychol*. 1991;100(4):502–515. https://doi.org/10.1037//0021-843x.100.4.502.
43. Akers RL. *Deviant Behavior: A Social Learning Approach*. 2nd ed. Wadsworth; 1977.
44. Akers RL, Krohn MD, Lanza-Kaduce L, Radosevich M. Social learning and deviant behavior: a specific test of a general theory. *Am Socio Rev*. 1979;44(4):636–655. https://doi.org/10.2307/2094592.
45. Akers RL, La Grecam AJ, Cochran J, Sellers C. Social learning theory and alcohol behavior among the elderly. *Socio Q*. 1989;30(4):625–638. https://doi.org/10.1111/j.1533-8525.1989.tb01539.x.
46. Ragan DT. Revisiting "what they think": adolescent drinking and the importance of peer beliefs. *Criminology*. 2014;52(3):488–513. https://doi.org/10.1111/1745-9125.12044.
47. Schaefer BP, Vito AG, Marcum CD, Higgins GE, Ricketts ML. Examining adolescent cocaine use with social learning and self-control theories. *Deviant Behav*. 2015;36(10):823–833. https://doi.org/10.1080/01639625.2014.977178.
48. Ford JA, Ong J. Non-medical use of prescription stimulants for academic purposes among college students: a test of social learning theory. *Drug Alcohol Depend*. 2014;144:279–282. https://doi.org/10.1016/j.drugalcdep.2014.09.011.
49. Steele JL, Peralta RL, Elman C. The co-ingestion of nonmedical prescription drugs and alcohol: a partial test of social learning theory. *J Drug Issues*. 2011;41(4):561–585. https://doi.org/10.1177/002204261104100406.
50. Krohn MD, Skinner WF, Massey JL, Akers RL. Social learning theory and adolescent smoking: a longitudinal study. *Soc Probl*. 1985;32(5):455–473. https://doi.org/10.2307/800775.
51. Hirschi T. *Causes of Delinquency*. University of California Press; 1969.
52. Christie NC. The role of social isolation in opioid addiction. *Soc Cognit Affect Neurosci*. 2021;16(7):645–656. https://doi.org/10.1093/scan/nsab029.
53. Krohn MD, Massey JL. Social control and delinquent behavior: an examination of the elements of the social bond. *Socio Q*. 1980;21(4):529–544.
54. Marcos AC, Bahr SJ. Control theory and adolescent drug use. *Youth Soc*. 1988;19(4):395–425. https://doi.org/10.1177/0044118X88019004003.
55. Wiatrowski MD, Griswold DB, Roberts MK. Social control theory and delinquency. *Am Psychol Rev*. 1981;46(5):525–541.
56. Agnew R. Social control theory and delinquency: a longitudinal test. *Criminology*. 1985;23(1):47–61. https://doi.org/10.1111/j.1745-9125.1985.tb00325.x.
57. Paternoster R, Iovanni L. The deterrent effect of perceived severity: a reexamination. *Soc Forces*. 1986;64(3):751–777. https://doi.org/10.2307/2578823.
58. Krohn MD, Loughran TA, Thornberry TP, Jang DW, Freeman-Gallant A, Castro ED. Explaining adolescent drug use in adjacent generations: testing the generality of theoretical explanations. *J Drug Issues*. 2016;46(4):373–395. https://doi.org/10.1177/0022042616659758.
59. Thiel KJ, Okun AC, Neisewander JL. Social reward-conditioned place preference: a model revealing an interaction between cocaine and social context rewards in rats. *Drug Alcohol Depend*. 2008;96(3):202–212. https://doi.org/10.1016/j.drugalcdep.2008.02.013.
60. Thiel KJ, Sanabria F, Neisewander JL. Synergistic interaction between nicotine and social rewards in adolescent male rats. *Psychopharmacology*. 2009;204(3):391–402. https://doi.org/10.1007/s00213-009-1470-2.
61. Fernández-Vidal JM, Molina JC. Socially mediated alcohol preferences in adolescent rats following interactions with an intoxicated peer. *Pharmacol, Biochem Behav*. 2004;79(2):229–241. https://doi.org/10.1016/j.pbb.2004.07.010.
62. Smith MA, Strickland JC, Bills SE, Lacy RT. The effects of a shared history of drug exposure on social choice. *Behav Pharmacol*. 2015;26(7 Spec No):631–635. https://doi.org/10.1097/FBP.0000000000000139.
63. Watanabe S. Drug-social interactions in the reinforcing property of methamphetamine in mice. *Behav Pharmacol*. 2011;22(3):203–206. https://doi.org/10.1097/FBP.0b013e328345c815.

64. Watanabe S. Social factors in conditioned place preference with morphine in mice. *Pharmacol Biochem Behav.* 2013;103(3):440–443. https://doi.org/10.1016/j.pbb.2012.10.001.
65. Yates JR, Beckmann JS, Meyer AC, Bardo MT. Concurrent choice for social interaction and amphetamine using conditioned place preference in rats: effects of age and housing condition. *Drug Alcohol Depend.* 2013;129(3):240–246. https://doi.org/10.1016/j.drugalcdep.2013.02.024.
66. Kummer KK, Hofhansel L, Barwitz CM, et al. Differences in social interaction- vs. cocaine reward in mouse vs. rat. *Front Behav Neurosci.* 2014;8:363. https://doi.org/10.3389/fnbeh.2014.00363.
67. Fritz M, El Rawas R, Klement S, et al. Differential effects of accumbens core vs. shell lesions in a rat concurrent conditioned place preference paradigm for cocaine vs. social interaction. *PLoS One.* 2011;6(10):e26761. https://doi.org/10.1371/journal.pone.0026761.
68. Bregolin T, Pinheiro BS, El Rawas R, Zernig G. Preventive strength of dyadic social interaction against reacquisition/reexpression of cocaine conditioned place preference. *Front Behav Neurosci.* 2017;11:225. https://doi.org/10.3389/fnbeh.2017.00225.
69. Fritz M, El Rawas R, Salti A, et al. Reversal of cocaine-conditioned place preference and mesocorticolimbic Zif268 expression by social interaction in rats. *Addiction Biol.* 2011;16(2):273–284. https://doi.org/10.1111/j.1369-1600.2010.00285.x.
70. Zernig G, Pinheiro BS. Dyadic social interaction inhibits cocaine-conditioned place preference and the associated activation of the accumbens corridor. *Behav Pharmacol.* 2015;26(6):580–594. https://doi.org/10.1097/FBP.0000000000000167.
71. Zernig G, Kummer KK, Prast JM. Dyadic social interaction as an alternative reward to cocaine. *Front Psychiatr.* 2013;4:100. https://doi.org/10.3389/fpsyt.2013.00100.
72. El Rawas R, Klement S, Salti A, et al. Preventive role of social interaction for cocaine conditioned place preference: correlation with FosB/DeltaFosB and pCREB expression in rat mesocorticolimbic areas. *Front Behav Neurosci.* 2012;6:8. https://doi.org/10.3389/fnbeh.2012.00008.
73. Sampedro-Piquero P, Ávila-Gámiz F, Moreno Fernández RD, Castilla-Ortega E, Santín LJ. The presence of a social stimulus reduces cocaine-seeking in a place preference conditioning paradigm. *J Psychopharmacol.* 2019;33(12):1501–1511. https://doi.org/10.1177/0269881119874414.
74. Prast JM, Schardl A, Schwarzer C, Dechant G, Saria A, Zernig G. Reacquisition of cocaine conditioned place preference and its inhibition by previous social interaction preferentially affect D1-medium spiny neurons in the accumbens corridor. *Front Behav Neurosci.* 2014;8:317. https://doi.org/10.3389/fnbeh.2014.00317.
75. Smith MA, Cha HS, Griffith AK, Sharp JL. Social contact reinforces cocaine self-administration in young adult male rats: the role of social reinforcement in vulnerability to drug use. *Front Behav Neurosci.* 2021;15:771114. https://doi.org/10.3389/fnbeh.2021.771114.
76. Meng S, Yan W, Liu X, et al. Social interaction with relapsed partner facilitates cocaine relapse in rats. *Front Pharmacol.* 2021;12:750397. https://doi.org/10.3389/fphar.2021.750397.
77. Weiss VG, Yates JR, Beckmann JS, Hammerslag LR, Bardo MT. Social reinstatement: a rat model of peer-induced relapse. *Psychopharmacology.* 2018;235(12):3391–3400. https://doi.org/10.1007/s00213-018-5048-8.
78. Hallmark RA, Hunt PS. Social learning about ethanol in preweanling rats: role of endogenous opioids. *Dev Psychobiol.* 2004;44(2):132–139. https://doi.org/10.1002/dev.10163.
79. Hunt PS, Lant GM, Carroll CA. Enhanced intake of ethanol in preweanling rats following interactions with intoxicated siblings. *Dev Psychobiol.* 2000;37(2):90–99.
80. Hunt PS, Holloway JL, Scordalakes EM. Social interaction with an intoxicated sibling can result in increased intake of ethanol by periadolescent rats. *Dev Psychobiol.* 2001;38(2):101–109. https://doi.org/10.1002/1098-23020338:2<101::aid-dev1002>3.0.co;2-4.
81. Honey PL, Varley KR, Galef Jr BG. Effects of ethanol consumption by adult female rats on subsequent consumption by adolescents. *Appetite.* 2004;42(3):299–306. https://doi.org/10.1016/j.appet.2004.01.002.
82. Maldonado AM, Finkbeiner LM, Kirstein CL. Social interaction and partner familiarity differentially alter voluntary ethanol intake in adolescent male and female rats. *Alcohol.* 2008;42(8):641–648. https://doi.org/10.1016/j.alcohol.2008.08.003.
83. Gamble DN, Josefson CC, Hennessey MK, et al. Social interaction with an alcohol-intoxicated or cocaine-injected peer selectively alters social behaviors and drinking in adolescent male and female rats. *Alcohol Clin Exp Res.* 2019;43(12):2525–2535. https://doi.org/10.1111/acer.14208.

84. Allport FH. The influence of the group upon association and thought. *J Exp Psychol.* 1920;3(3):159–182. https://doi.org/10.1037/h0067891.
85. Smith MA. Peer influences on drug self-administration: social facilitation and social inhibition of cocaine intake in male rats. *Psychopharmacology.* 2012;224(1):81–90. https://doi.org/10.1007/s00213-012-2737-6.
86. Gipson CD, Yates JR, Beckmann JS, Marusich JA, Zentall TR, Bardo MT. Social facilitation of d-amphetamine self-administration in rats. *Exp Clin Psychopharmacol.* 2011;19(6):409–419. https://doi.org/10.1037/a0024682.
87. Peartree NA, Hatch KN, Goenaga JG, et al. Social context has differential effects on acquisition of nicotine self-administration in male and female rats. *Psychopharmacology.* 2017;234(12):1815–1828. https://doi.org/10.1007/s00213-017-4590-0.
88. Hofford RS, Bond PN, Chow JJ, Bardo MT. Presence of a social peer enhances acquisition of remifentanil self-administration in male rats. *Drug Alcohol Depend.* 2020;213:108125. https://doi.org/10.1016/j.drugalcdep.2020.108125.
89. Fahlke C, Garpenstrand H, Oreland L, Suomi SJ, Higley JD. Platelet monoamine oxidase activity in a nonhuman primate model of type 2 excessive alcohol consumption. *Am J Psychiatr.* 2002;159(12):2107. https://doi.org/10.1176/appi.ajp.159.12.2107.
90. McKenzie-Quirk SD, Miczek KA. Social rank and social separation as determinants of alcohol drinking in squirrel monkeys. *Psychopharmacology.* 2008;201(1):137–145. https://doi.org/10.1007/s00213-008-1256-y.
91. Wolffgramm J, Heyne A. Social behavior, dominance, and social deprivation of rats determine drug choice. *Pharmacol Biochem Behav.* 1991;38(2):389–399. https://doi.org/10.1016/0091-3057(91)90297-f.
92. Duncan EA, Tamashiro KL, Nguyen MM, Gardner SR, Woods SC, Sakai RR. The impact of moderate daily alcohol consumption on aggression and the formation of dominance hierarchies in rats. *Psychopharmacology.* 2006;189(1):83–94. https://doi.org/10.1007/s00213-006-0536-7.
93. Morgan D, Grant KA, Gage HD, et al. Social dominance in monkeys: dopamine D2 receptors and cocaine self-administration. *Nat Neurosci.* 2002;5(2):169–174. https://doi.org/10.1038/nn798.
94. Czoty PW, Gage HD, Nader MA. Differences in D2 dopamine receptor availability and reaction to novelty in socially housed male monkeys during abstinence from cocaine. *Psychopharmacology.* 2010;208(4):585–592. https://doi.org/10.1007/s00213-009-1756-4.
95. Czoty PW, Gould RW, Gage HD, Nader MA. Effects of social reorganization on dopamine D2/D3 receptor availability and cocaine self-administration in male cynomolgus monkeys. *Psychopharmacology.* 2017;234(18):2673–2682. https://doi.org/10.1007/s00213-017-4658-x.
96. Boyas JF, Valera P, Ruiz E. Subjective well-being among Latino day laborers: examining the role of religiosity, social networks, and cigarette use. *Health Promot Perspect.* 2018;8(1):46–53. https://doi.org/10.15171/hpp.2018.06.
97. Charitonidi E, Studer J, Gaume J, Gmel G, Daeppen JB, Bertholet N. Socioeconomic status and substance use among Swiss young men: a population-based cross-sectional study. *BMC Publ Health.* 2016;16:333. https://doi.org/10.1186/s12889-016-2949-5.
98. Doku D, Darteh EK, Kumi-Kyereme A. Socioeconomic inequalities in cigarette smoking among men: evidence from the 2003 and 2008 Ghana demographic and health surveys. *Arch Publ Health.* 2013;71(1):9. https://doi.org/10.1186/0778-7367-71-9.
99. Nketiah-Amponsah E, Afful-Mensah G, Ampaw S. Determinants of cigarette smoking and smoking intensity among adult males in Ghana. *BMC Publ Health.* 2018;18(1):941. https://doi.org/10.1186/s12889-018-5872-0.
100. Patrick ME, Wightman P, Schoeni RF, Schulenberg JE. Socioeconomic status and substance use among young adults: a comparison across constructs and drugs. *J Stud Alcohol Drugs.* 2012;73(5):772–782. https://doi.org/10.15288/jsad.2012.73.772.
101. Centers for Disease Control and Prevention. Cigarette smoking and tobacco use among people of low socioeconomic status. Retrieved from: https://www.cdc.gov/tobacco/disparities/low-ses/index.htm; 2019.
102. Yu D, Peterson NA, Sheffer MA, Reid RJ, Schnieder JE. Tobacco outlet density and demographics: analysing the relationships with a spatial regression approach. *Publ Health.* 2010;124(7):412–416. https://doi.org/10.1016/j.puhe.2010.03.024.
103. Karriker-Jaffe KJ. Neighborhood socioeconomic status and substance use by U.S. adults. *Drug Alcohol Depend.* 2013;133(1):212–221. https://doi.org/10.1016/j.drugalcdep.2013.04.033.
104. Lewis B, Hoffman L, Garcia CC, Nixon SJ. Race and socioeconomic status in substance use progression and treatment entry. *J Ethn Subst Abuse.* 2018;17(2):150–166. https://doi.org/10.1080/15332640.2017.1336959.

105. Wiers CE, Shokri-Kojori E, Cabrera E, et al. Socioeconomic status is associated with striatal dopamine D2/D3 receptors in healthy volunteers but not in cocaine abusers. *Neurosci Lett.* 2016;617:27−31. https://doi.org/10.1016/j.neulet.2016.01.056.
106. Palamar JJ, Davies S, Ompad DC, Cleland CM, Weitzman M. Powder cocaine and crack use in the United States: an examination of risk for arrest and socioeconomic disparities in use. *Drug Alcohol Depend.* 2015;149:108−116. https://doi.org/10.1016/j.drugalcdep.2015.01.029.
107. Williams CT, Latkin CA. Neighborhood socioeconomic status, personal network attributes, and use of heroin and cocaine. *Am J Prev Med.* 2007;32(6 Suppl):S203−S210. https://doi.org/10.1016/j.amepre.2007.02.006.
108. Gerra G, Benedetti E, Resce G, Potente R, Cutilli A, Molinaro S. Socioeconomic status, parental education, school connectedness and individual socio-cultural resources in vulnerability for drug use among students. *Int J Environ Res Publ Health.* 2020;17(4):1306. https://doi.org/10.3390/ijerph17041306.
109. Aschengrau A, Grippo A, Winter MR. Influence of family and community socioeconomic status on the risk of adolescent drug use. *Subst Use Misuse.* 2021;56(5):577−587. https://doi.org/10.1080/10826084.2021.1883660.
110. Lee JO, Hill KG, Hartigan LA, et al. Unemployment and substance use problems among young adults: does childhood low socioeconomic status exacerbate the effect? *Soc Sci Med (1982).* 2015;143:36−44. https://doi.org/10.1016/j.socscimed.2015.08.016.
111. Humensky JL. Are adolescents with high socioeconomic status more likely to engage in alcohol and illicit drug use in early adulthood? *Subst Abuse Treat Prev Pol.* 2010;5:19. https://doi.org/10.1186/1747-597X-5-19.
112. Martin CC. High socioeconomic status predicts substance use and alcohol consumption in U.S. undergraduates. *Subst Use Misuse.* 2019;54(6):1035−1043. https://doi.org/10.1080/10826084.2018.1559193.
113. Goodman E, Huang B. Socioeconomic status, depressive symptoms, and adolescent substance use. *Arch Pediatr Adolesc Med.* 2002;156(5):448−453. https://doi.org/10.1001/archpedi.156.5.448.
114. Petruzelka B, Vacek J, Gavurova B, et al. Interaction of socioeconomic status with risky internet use, gambling and substance use in adolescents from a structurally disadvantaged region in Central Europe. *Int J Environ Res Publ Health.* 2020;17(13):4803. https://doi.org/10.3390/ijerph17134803.
115. Legleye S, Beck F, Khlat M, Peretti-Watel P, Chau N. The influence of socioeconomic status on cannabis use among French adolescents. *J Adolesc Health.* 2012;50(4):395−402. https://doi.org/10.1016/j.jadohealth.2011.08.004.
116. Smail MA, Smith BL, Nawreen N, Herman JP. Differential impact of stress and environmental enrichment on corticolimbic circuits. *Pharmacol Biochem Behav.* 2020;197:172993. https://doi.org/10.1016/j.pbb.2020.172993.
117. Hill SY, Powell BJ. Cocaine and morphine self-administration: effects of differential rearing. *Pharmacol Biochem Behav.* 1976;5(6):701−704. https://doi.org/10.1016/0091-3057(76)90315-4.
118. Rockman GE, Gibson JE, Benarroch A. Effects of environmental enrichment on voluntary ethanol intake in rats. *Pharmacol Biochem Behav.* 1989;34(3):487−490. https://doi.org/10.1016/0091-3057(89)90545-5.
119. Bellés L, Dimiziani A, Herrmann FR, Ginovart N. Early environmental enrichment and impoverishment differentially affect addiction-related behavioral traits, cocaine-taking, and dopamine $D_{2/3}$ receptor signaling in a rat model of vulnerability to drug abuse. *Psychopharmacology.* 2021. https://doi.org/10.1007/s00213-021-05971-z.
120. Gipson CD, Beckmann JS, El-Maraghi S, Marusich JA, Bardo MT. Effect of environmental enrichment on escalation of cocaine self-administration in rats. *Psychopharmacology.* 2011;214(2):557−566. https://doi.org/10.1007/s00213-010-2060-z.
121. Lü X, Zhao C, Zhang L, et al. The effects of rearing condition on methamphetamine self-administration and cue-induced drug seeking. *Drug Alcohol Depend.* 2012;124(3):288−298. https://doi.org/10.1016/j.drugalcdep.2012.01.022.
122. Puhl MD, Blum JS, Acosta-Torres S, Grigson PS. Environmental enrichment protects against the acquisition of cocaine self-administration in adult male rats, but does not eliminate avoidance of a drug-associated saccharin cue. *Behav Pharmacol.* 2012;23(1):43−53. https://doi.org/10.1097/FBP.0b013e32834eb060.
123. Deehan Jr GA, Cain ME, Kiefer SW. Differential rearing conditions alter operant responding for ethanol in outbred rats. *Alcohol Clin Exp Res.* 2007;31(10):1692−1698. https://doi.org/10.1111/j.1530-0277.2007.00466.x.
124. Deehan Jr GA, Palmatier MI, Cain ME, Kiefer SW. Differential rearing conditions and alcohol-preferring rats: consumption of and operant responding for ethanol. *Behav Neurosci.* 2011;125(2):184−193. https://doi.org/10.1037/a0022627.

125. Lopez MF, Laber K. Impact of social isolation and enriched environment during adolescence on voluntary ethanol intake and anxiety in C57BL/6J mice. *Physiol Behav.* 2015;148:151−156. https://doi.org/10.1016/j.physbeh.2014.11.012.
126. Arndt DL, Johns KC, Dietz ZK, Cain ME. Environmental condition alters amphetamine self-administration: role of the MGluR$_5$ receptor and schedule of reinforcement. *Psychopharmacology.* 2015;232(20):3741−3752. https://doi.org/10.1007/s00213-015-4031-x.
127. Bardo MT, Klebaur JE, Valone JM, Deaton C. Environmental enrichment decreases intravenous self-administration of amphetamine in female and male rats. *Psychopharmacology.* 2001;155(3):278−284. https://doi.org/10.1007/s002130100720.
128. Garcia EJ, Cain ME. Environmental enrichment and a selective metabotropic glutamate receptor$_{2/3}$ (mGluR$_{2/3}$) agonist suppress amphetamine self-administration: characterizing baseline differences. *Pharmacol Biochem Behav.* 2020;192:172907. https://doi.org/10.1016/j.pbb.2020.172907.
129. Green TA, Gehrke BJ, Bardo MT. Environmental enrichment decreases intravenous amphetamine self-administration in rats: dose-response functions for fixed- and progressive-ratio schedules. *Psychopharmacology.* 2002;162(4):373−378. https://doi.org/10.1007/s00213-002-1134-y.
130. Stairs DJ, Ewin SE, Kangiser MM, Pfaff MN. Effects of environmental enrichment on d-amphetamine self-administration following nicotine exposure. *Exp Clin Psychopharmacol.* 2017;25(5):393−401. https://doi.org/10.1037/pha0000137.
131. Alvers KM, Marusich JA, Gipson CD, Beckmann JS, Bardo MT. Environmental enrichment during development decreases intravenous self-administration of methylphenidate at low unit doses in rats. *Behav Pharmacol.* 2012;23(7):650−657. https://doi.org/10.1097/FBP.0b013e3283584765.
132. Venebra-Muñoz A, Corona-Morales A, Santiago-García J, Melgarejo-Gutiérrez M, Caba M, García-García F. Enriched environment attenuates nicotine self-administration and induces changes in ΔFosB expression in the rat prefrontal cortex and nucleus accumbens. *Neuroreport.* 2014;25(9):688−692. https://doi.org/10.1097/WNR.0000000000000157.
133. Mohammadian J, Najafi M, Miladi-Gorji H. Effect of enriched environment during adolescence on spatial learning and memory, and voluntary consumption of morphine in maternally separated rats in adulthood. *Dev Psychobiol.* 2019;61(4):615−625. https://doi.org/10.1002/dev.21808.
134. Hofford RS, Chow JJ, Beckmann JS, Bardo MT. Effects of environmental enrichment on self-administration of the short-acting opioid remifentanil in male rats. *Psychopharmacology.* 2017;234(23−24):3499−3506. https://doi.org/10.1007/s00213-017-4734-2.
135. Imperio CG, McFalls AJ, Hadad N, et al. Exposure to environmental enrichment attenuates addiction-like behavior and alters molecular effects of heroin self-administration in rats. *Neuropharmacology.* 2018;139:26−40. https://doi.org/10.1016/j.neuropharm.2018.06.037.
136. Chauvet C, Lardeux V, Goldberg SR, Jaber M, Solinas M. Environmental enrichment reduces cocaine seeking and reinstatement induced by cues and stress but not by cocaine. *Neuropsychopharmacology.* 2009;34(13):2767−2778. https://doi.org/10.1038/npp.2009.127.
137. Frankowska M, Miszkiel J, Pomierny-Chamioło L, et al. Alternation in dopamine D$_2$-like and metabotropic glutamate type 5 receptor density caused by differing housing conditions during abstinence from cocaine self-administration in rats. *J Psychopharmacol.* 2019;33(3):372−382. https://doi.org/10.1177/0269881118821113.
138. Galaj E, Manuszak M, Ranaldi R. Environmental enrichment as a potential intervention for heroin seeking. *Drug Alcohol Depend.* 2016;163:195−201. https://doi.org/10.1016/j.drugalcdep.2016.04.016.
139. Garcia EJ, Cain ME. Isolation housing elevates amphetamine seeking independent of nucleus accumbens glutamate receptor adaptations. *Eur J Neurosci.* 2021;54(7):6382−6396. https://doi.org/10.1111/ejn.15441.
140. Hofford RS, Darna M, Wilmouth CE, Dwoskin LP, Bardo MT. Environmental enrichment reduces methamphetamine cue-induced reinstatement but does not alter methamphetamine reward or VMAT2 function. *Behav Brain Res.* 2014;270:151−158. https://doi.org/10.1016/j.bbr.2014.05.007.
141. Nicolas C, Hofford RS, Dugast E, et al. Prevention of relapse to methamphetamine self-administration by environmental enrichment: involvement of glucocorticoid receptors. *Psychopharmacology.* 2022;239(4):1009−1018. https://doi.org/10.1007/s00213-021-05770-6.
142. Powell GL, Vannan A, Bastle RM, et al. Environmental enrichment during forced abstinence from cocaine self-administration opposes gene network expression changes associated with the incubation effect. *Sci Rep.* 2020;10(1):11291. https://doi.org/10.1038/s41598-020-67966-8.

143. Ranaldi R, Kest K, Zellner M, Hachimine-Semprebom P. Environmental enrichment, administered after establishment of cocaine self-administration, reduces lever pressing in extinction and during a cocaine context renewal test. *Behav Pharmacol*. 2011;22(4):347−353. https://doi.org/10.1097/FBP.0b013e3283487365.
144. Sikora M, Nicolas C, Istin M, Jaafari N, Thiriet N, Solinas M. Generalization of effects of environmental enrichment on seeking for different classes of drugs of abuse. *Behav Brain Res*. 2018;341:109−113. https://doi.org/10.1016/j.bbr.2017.12.027.
145. Stairs DJ, Klein ED, Bardo MT. Effects of environmental enrichment on extinction and reinstatement of amphetamine self-administration and sucrose-maintained responding. *Behav Pharmacol*. 2006;17(7):597−604. https://doi.org/10.1097/01.fbp.0000236271.72300.0e.
146. Thiel KJ, Sanabria F, Pentkowski NS, Neisewander JL. Anti-craving effects of environmental enrichment. *Int J Neuropsychopharmacol*. 2009;12(9):1151−1156. https://doi.org/10.1017/S1461145709990472.
147. Yates JR, Bardo MT, Beckmann JS. Environmental enrichment and drug value: a behavioral economic analysis in male rats. *Addiction Biol*. 2019;24(1):65−75. https://doi.org/10.1111/adb.12581.
148. Gauthier JM, Lin A, Nic Dhonnchadha BÁ, Spealman RD, Man HY, Kantak KM. Environmental enrichment facilitates cocaine-cue extinction, deters reacquisition of cocaine self-administration and alters AMPAR GluA1 expression and phosphorylation. *Addiction Biol*. 2017;22(1):152−162. https://doi.org/10.1111/adb.12313.
149. Thiel KJ, Engelhardt B, Hood LE, Peartree NA, Neisewander JL. The interactive effects of environmental enrichment and extinction interventions in attenuating cue-elicited cocaine-seeking behavior in rats. *Pharmacol Biochem Behav*. 2011;97(3):595−602. https://doi.org/10.1016/j.pbb.2010.09.014.
150. Barrera ED, Loughlin L, Greenberger S, Ewing S, Hachimine P, Ranaldi R. Environmental enrichment reduces heroin seeking following incubation of craving in both male and female rats. *Drug Alcohol Depend*. 2021;226:108852. https://doi.org/10.1016/j.drugalcdep.2021.108852.
151. Chauvet C, Goldberg SR, Jaber M, Solinas M. Effects of environmental enrichment on the incubation of cocaine craving. *Neuropharmacology*. 2012;63(4):635−641. https://doi.org/10.1016/j.neuropharm.2012.05.014.
152. Smith MA, Iordanou JC, Cohen MB, et al. Effects of environmental enrichment on sensitivity to cocaine in female rats: importance of control rates of behavior. *Behav Pharmacol*. 2009;20(4):312−321. https://doi.org/10.1097/FBP.0b013e32832ec568.
153. El Rawas R, Thiriet N, Lardeux V, Jaber M, Solinas M. Environmental enrichment decreases the rewarding but not the activating effects of heroin. *Psychopharmacology*. 2009;203(3):561−570. https://doi.org/10.1007/s00213-008-1402-6.
154. Khalaji S, Bigdeli I, Ghorbani R, Miladi-Gorji H. Environmental enrichment attenuates morphine-induced conditioned place preference and locomotor sensitization in maternally separated rat pups. *Basic Clin Neurosci*. 2018;9(4):241−250. https://doi.org/10.32598/bcn.9.4.241.
155. Xu Z, Hou B, Gao Y, He F, Zhang C. Effects of enriched environment on morphine-induced reward in mice. *Exp Neurol*. 2007;204(2):714−719. https://doi.org/10.1016/j.expneurol.2006.12.027.
156. Smith MA, Chisholm KA, Bryant PA, et al. Social and environmental influences on opioid sensitivity in rats: importance of an opioid's relative efficacy at the mu-receptor. *Psychopharmacology*. 2005;181(1):27−37. https://doi.org/10.1007/s00213-005-2218-2.
157. Haider S, Nawaz A, Batool Z, Tabassum S, Perveen T. Alleviation of diazepam-induced conditioned place preference and its withdrawal-associated neurobehavioral deficits following pre-exposure to enriched environment in rats. *Physiol Behav*. 2019;208:112564. https://doi.org/10.1016/j.physbeh.2019.112564.
158. Pautassi RM, Suárez AB, Hoffmann LB, et al. Effects of environmental enrichment upon ethanol-induced conditioned place preference and pre-frontal BDNF levels in adolescent and adult mice. *Sci Rep*. 2017;7(1):8574. https://doi.org/10.1038/s41598-017-08795-0.
159. Rae M, Zanos P, Georgiou P, Chivers P, Bailey A, Camarini R. Environmental enrichment enhances conditioned place preference to ethanol via an oxytocinergic-dependent mechanism in male mice. *Neuropharmacology*. 2018;138:267−274. https://doi.org/10.1016/j.neuropharm.2018.06.013.
160. Bahi A. Environmental enrichment reduces chronic psychosocial stress-induced anxiety and ethanol-related behaviors in mice. *Prog Neuro Psychopharmacol Biol Psychiatr*. 2017;77:65−74. https://doi.org/10.1016/j.pnpbp.2017.04.001.
161. Ewin SE, Kangiser MM, Stairs DJ. The effects of environmental enrichment on nicotine condition place preference in male rats. *Exp Clin Psychopharmacol*. 2015;23(5):387−394. https://doi.org/10.1037/pha0000024.

162. Bardo MT, Bowling SL, Rowlett JK, Manderscheid P, Buxton ST, Dwoskin LP. Environmental enrichment attenuates locomotor sensitization, but not in vitro dopamine release, induced by amphetamine. *Pharmacol Biochem Behav*. 1995;51(2−3):397−405. https://doi.org/10.1016/0091-3057(94)00413-d.
163. Bowling SL, Bardo MT. Locomotor and rewarding effects of amphetamine in enriched, social, and isolate reared rats. *Pharmacol, Biochem Behav*. 1994;48(2):459−464. https://doi.org/10.1016/0091-3057(94)90553-3.
164. Wongwitdecha N, Marsden CA. Isolation rearing prevents the reinforcing properties of amphetamine in a conditioned place preference paradigm. *Eur J Pharmacol*. 1995;279(1):99−103. https://doi.org/10.1016/0014-2999(95)00212-4.
165. Chauvet C, Lardeux V, Jaber M, Solinas M. Brain regions associated with the reversal of cocaine conditioned place preference by environmental enrichment. *Neuroscience*. 2011;184:88−96. https://doi.org/10.1016/j.neuroscience.2011.03.068.
166. Freese L, Almeida FB, Heidrich N, et al. Environmental enrichment reduces cocaine neurotoxicity during cocaine-conditioned place preference in male rats. *Pharmacol Biochem Behav*. 2018;169:10−15. https://doi.org/10.1016/j.pbb.2018.04.001.
167. Mustroph ML, Pinardo H, Merritt JR, Rhodes JS. Parameters for abolishing conditioned place preference for cocaine from running and environmental enrichment in male C57BL/6J mice. *Behav Brain Res*. 2016;312:366−373. https://doi.org/10.1016/j.bbr.2016.06.049.
168. Solinas M, Chauvet C, Thiriet N, El Rawas R, Jaber M. Reversal of cocaine addiction by environmental enrichment. *Proc Natl Acad Sci U S A*. 2008;105(44):17145−17150. https://doi.org/10.1073/pnas.0806889105.
169. Zakharova E, Miller J, Unterwald E, Wade D, Izenwasser S. Social and physical environment alter cocaine conditioned place preference and dopaminergic markers in adolescent male rats. *Neuroscience*. 2009;163(3):890−897. https://doi.org/10.1016/j.neuroscience.2009.06.068.
170. Dow-Edwards D, Iijima M, Stephenson S, Jackson A, Weedon J. The effects of prenatal cocaine, post-weaning housing and sex on conditioned place preference in adolescent rats. *Psychopharmacology*. 2014;231(8):1543−1555. https://doi.org/10.1007/s00213-013-3418-9.
171. Green TA, Alibhai IN, Roybal CN, et al. Environmental enrichment produces a behavioral phenotype mediated by low cyclic adenosine monophosphate response element binding (CREB) activity in the nucleus accumbens. *Biol Psychiatr*. 2010;67(1):28−35. https://doi.org/10.1016/j.biopsych.2009.06.022.
172. Galaj E, Shukur A, Manuszak M, Newman K, Ranaldi R. No evidence that environmental enrichment during rearing protects against cocaine behavioral effects but as an intervention reduces an already established cocaine conditioned place preference. *Pharmacol, Biochem Behav*. 2017;156:56−62. https://doi.org/10.1016/j.pbb.2017.04.005.
173. Thiriet N, Gennequin B, Lardeux V, et al. Environmental enrichment does not reduce the rewarding and neurotoxic effects of methamphetamine. *Neurotox Res*. 2011;19(1):172−182. https://doi.org/10.1007/s12640-010-9158-2.
174. Bahi A, Dreyer JL. Environmental enrichment decreases chronic psychosocial stress-impaired extinction and reinstatement of ethanol conditioned place preference in C57BL/6 male mice. *Psychopharmacology*. 2020;237(3):707−721. https://doi.org/10.1007/s00213-019-05408-8.
175. Li X, Meng L, Huang K, Wang H, Li D. Environmental enrichment blocks reinstatement of ethanol-induced conditioned place preference in mice. *Neurosci Lett*. 2015;599:92−96. https://doi.org/10.1016/j.neulet.2015.05.035.
176. Althobaiti YS, Almalki AH. Effects of environmental enrichment on reinstatement of methamphetamine-induced conditioned place preference. *Behav Brain Res*. 2020;379:112372. https://doi.org/10.1016/j.bbr.2019.112372.
177. Ford ME, Kelly PA. Conceptualizing and categorizing race and ethnicity in health services research. *Health Serv Res*. 2005;40(5 Pt 2):1658−1675. https://doi.org/10.1111/j.1475-6773.2005.00449.x.
178. Motivans M. *Federal Justice Statistics, 2019*. U.S. Department of Justice Office of Justice Programs, Bureau of Justice Statistics. Report NCJ 301158; 2021. https://bjs.ojp.gov/content/pub/pdf/fjs19.pdf.
179. Nicholson Jr HL, Ford JA. Sociodemographic, neighborhood, psychosocial, and substance use correlates of cocaine use among Black adults: findings from a pooled analysis of national data. *Addict Behav*. 2019;88:182−186. https://doi.org/10.1016/j.addbeh.2018.08.042.

180. Jones AA, Webb FJ, Lasopa SO, Striley CW, Cottler LB. The association between religiosity and substance use patterns among women involved in the criminal justice system. *J Drug Issues*. 2018;48(3):327−336. https://doi.org/10.1177/0022042618757208.
181. Herman-Stahl MA, Krebs CP, Kroutil LA, Heller DC. Risk and protective factors for nonmedical use of prescription stimulants and methamphetamine among adolescents. *J Adolesc Health*. 2006;39(3):374−380. https://doi.org/10.1016/j.jadohealth.2006.01.006.
182. Vega WA, Zimmerman RS, Warheit GJ, Apospori E, Gil AG. Risk factors for early adolescent drug use in four ethnic and racial groups. *Am J Publ Health*. 1993;83(2):185−189. https://doi.org/10.2105/ajph.83.2.185.
183. Wallace Jr JM, Brown TN, Bachman JG, LaVeist TA. The influence of race and religion on abstinence from alcohol, cigarettes and marijuana among adolescents. *J Stud Alcohol*. 2003;64(6):843−848. https://doi.org/10.15288/jsa.2003.64.843.
184. McCabe SE, Morales M, Cranford JA, Delva J, McPherson MD, Boyd CJ. Race/ethnicity and gender differences in drug use and abuse among college students. *J Ethn Subst Abuse*. 2007;6(2):75−95. https://doi.org/10.1300/J233v06n02_06.
185. Wallace Jr JM, Bachman JG, O'Malley PM, Johnston LD, Schulenberg JE, Cooper SM. Tobacco, alcohol, and illicit drug use: racial and ethnic differences among U.S. high school seniors, 1976-2000. *Publ Health Rep*. 2002;117(Suppl 1):S67−S75.
186. Rosenberg A, Groves AK, Blankenship KM. Comparing black and white drug offenders: implications for racial disparities in criminal justice and reentry policy and programming. *J Drug Issues*. 2017;47(1):132−142. https://doi.org/10.1177/0022042616678614.
187. Alexander M. *The New Jim Crow: Mass Incarceration in the Age of Colorblindness*. The New Press; 2010.
188. Chetty R, Hendren N, Jones MR, Porter SR. Race and economic opportunity in the United States: an intergenerational perspective. *Q J Econ*. 2020;135(2):711−783. https://doi.org/10.1093/qje/qjz042.
189. Epperson AE, Gonzalez M, Skorek M, Song AV. Challenging assumptions about race/ethnicity, socioeconomic status, and cigarette smoking among adolescents. *J Rac & Ethn Health Disparit*. 2022;9(2):436−443. https://doi.org/10.1007/s40615-021-00974-0.
190. White HR, Bechtold J, Loeber R, Pardini D. Divergent marijuana trajectories among men: socioeconomic, relationship, and life satisfaction outcomes in the mid-30s. *Drug Alcohol Depend*. 2015;156:62−69. https://doi.org/10.1016/j.drugalcdep.2015.08.031.
191. Kruger L, van Walbeek C, Vellios N. Waterpipe and cigarette smoking among university students in the Western Cape, South Africa. *Am J Health Behav*. 2016;40(4):416−426. https://doi.org/10.5993/AJHB.40.4.3.
192. Blay SL, Batista AD, Andreoli SB, Gastal FL. The relationship between religiosity and tobacco, alcohol use, and depression in an elderly community population. *Am J Geriatr Psychiatr*. 2008;16(11):934−943. https://doi.org/10.1097/JGP.0b013e3181871392.
193. Primack BA, Mah J, Shensa A, Rosen D, Yonas MA, Fine MJ. Associations between race, ethnicity, religion, and waterpipe tobacco smoking. *J Ethn Subst Abuse*. 2014;13(1):58−71. https://doi.org/10.1080/15332640.2013.850462.
194. Tuck A, Robinson M, Agic B, Ialomiteanu AR, Mann RE. Religion, alcohol use and risk drinking among Canadian adults living in Ontario. *J Relig Health*. 2017;56(6):2023−2038. https://doi.org/10.1007/s10943-016-0339-z.
195. Bowie JV, Ensminger ME, Robertson JA. Alcohol-use problems in young black adults: effects of religiosity, social resources, and mental health. *J Stud Alcohol*. 2006;67(1):44−53. https://doi.org/10.15288/jsa.2006.67.44.
196. Bowie JV, Parker LJ, Beadle-Holder M, Ezema A, Bruce MA, Thorpe Jr RJ. The Influence of religious attendance on smoking among black men. *Subst Use Misuse*. 2017;52(5):581−586. https://doi.org/10.1080/10826084.2016.1245342.
197. Gillum RF. Frequency of attendance at religious services and cigarette smoking in American women and men: the Third National Health and Nutrition Examination Survey. *Prev Med*. 2005;41(2):607−613. https://doi.org/10.1016/j.ypmed.2004.12.006.
198. Gillum RF, Sullins DP. Cigarette smoking during pregnancy: independent associations with religious participation. *South Med J*. 2008;101(7):686−692. https://doi.org/10.1097/SMJ.0b013e31817a76cc.
199. Koenig HG, George LK, Cohen HJ, Hays JC, Larson DB, Blazer DG. The relationship between religious activities and cigarette smoking in older adults. *J. Gerontol. Series A, Biol Sci & Med Sci*. 1998;53(6):M426−M434. https://doi.org/10.1093/gerona/53a.6.m426.

200. Lin H-C, Hu Y-H, Barry AE, Russell A. Assessing the associations between religiosity and alcohol use stages in a representative U.S. sample. *Subst Use Misuse*. 2020;55(10):1618−1624. https://doi.org/10.1080/10826084.2020.1756331.
201. Lucchetti G, Peres MF, Lucchetti AL, Koenig HG. Religiosity and tobacco and alcohol use in a Brazilian shantytown. *Subst Use Misuse*. 2012;47(7):837−846. https://doi.org/10.3109/10826084.2012.673142.
202. Torchalla I, Li K, Strehlau V, Linden IA, Krausz M. Religious participation and substance use behaviors in a Canadian sample of homeless people. *Community Ment Health J*. 2014;50(7):862−869. https://doi.org/10.1007/s10597-014-9705-z.
203. Wang Z, Koenig HG, Al Shohaib S. Religious involvement and tobacco use in mainland China: a preliminary study. *BMC Publ Health*. 2015;15:155. https://doi.org/10.1186/s12889-015-1478-y.
204. Whooley MA, Boyd AL, Gardin JM, Williams DR. Religious involvement and cigarette smoking in young adults: the CARDIA study (Coronary Artery Risk Development in Young Adults) study. *Arch Intern Med*. 2002;162(14):1604−1610. https://doi.org/10.1001/archinte.162.14.1604.
205. Borders TF, Booth BM. Stimulant use, religiosity, and the odds of developing or maintaining an alcohol use disorder over time. *J Stud Alcohol Drugs*. 2013;74(3):369−377.
206. Meyers JL, Brown Q, Grant BF, Hasin D. Religiosity, race/ethnicity, and alcohol use behaviors in the United States. *Psychol Med*. 2017;47(1):103−114. https://doi.org/10.1017/S0033291716001975.
207. Martinez EZ, Bueno-Silva CC, Bartolomeu IM, Ribeiro-Pizzo LB, Zucoloto ML. Relationship between religiosity and smoking among undergraduate health sciences students. *Trends Psychiatr & Psychother*. 2021;43(1):17−22. https://doi.org/10.47626/2237-6089-2019-0031.
208. Klassen BJ, Smith KZ, Grekin ER. Differential relationships between religiosity, cigarette smoking, and waterpipe use: implications for college student health. *J Am Coll Health*. 2013;61(7):381−385. https://doi.org/10.1080/07448481.2013.819806.
209. Cucciare MA, Han X, Curran GM, Booth BM. Associations between religiosity, perceived social support, and stimulant use in an untreated rural sample in the U.S.A. *Subst Use Misuse*. 2016;51(7):823−834. https://doi.org/10.3109/10826084.2016.1155611.
210. Acheampong AB, Lasopa S, Striley CW, Cottler LB. Gender differences in the association between religion/spirituality and simultaneous polysubstance use (SPU). *J Relig Health*. 2016;55(5):1574−1584. https://doi.org/10.1007/s10943-015-0168-5.
211. Schoenthaler SJ, Blum K, Braverman ER, et al. NIDA-Drug Addiction Treatment Outcome Study (DATOS) relapse as a function of spirituality/religiosity. *J Reward Defici Syndr*. 2015;1(1):36−45. https://doi.org/10.17756/jrds.2015-007.
212. Charro Baena B, Meneses C, Caperos JM, Prieto M, Uroz J. The role of religion and religiosity in alcohol consumption in adolescents in Spain. *J Relig Health*. 2019;58(5):1477−1487. https://doi.org/10.1007/s10943-018-0694-z.
213. D'Onofrio BM, Murrelle L, Eaves LJ, McCullough ME, Landis JL, Maes HH. Adolescent religiousness and its influence on substance use: Preliminary findings from the Mid-Atlantic School Age Twin Study. *Twin Res*. 1999;2(2):156−168. https://doi.org/10.1375/136905299320566022.
214. Gryczynski J, Ward BW. Social norms and the relationship between cigarette use and religiosity among adolescents in the United States. *Health Educ Behav*. 2011;38(1):39−48. https://doi.org/10.1177/1090198110372331.
215. Heath AC, Madden PA, Grant JD, McLaughlin TL, Todorov AA, Bucholz KK. Resiliency factors protecting against teenage alcohol use and smoking: Influences of religion, religious involvement and values, and ethnicity in the Missouri Adolescent Female Twin Study. *Twin Res*. 1999;2(2):145−155. https://doi.org/10.1375/136905299320566013.
216. Jeynes WH. Adolescent religious commitment and their consumption of marijuana, cocaine, and alcohol. *J Health Soc Pol*. 2006;21(4):1−20. https://doi.org/10.1300/J045v21n04_01.
217. Marsiglia FF, Ayers SL, Hoffman S. Religiosity and adolescent substance use in central Mexico: Exploring the influence of internal and external religiosity on cigarette and alcohol use. *Am J Community Psychol*. 2012;49(1−2):87−97. https://doi.org/10.1007/s10464-011-9439-9.
218. Miller L, Davies M, Greenwald S. Religiosity and substance use and abuse among adolescents in the National Comorbidity Survey. *J Am Acad Child Adolesc Psychiatr*. 2000;39(9):1190−1197. https://doi.org/10.1097/00004583-200009000-00020.

219. Russell AM, Yu B, Thompson CG, Sussman SY, Barry AE. Assessing the relationship between youth religiosity and their alcohol use: a meta-analysis from 2008 to 2018. *Addict Behav.* 2020;106:106361. https://doi.org/10.1016/j.addbeh.2020.106361.
220. Kim-Spoon J, McCullough ME, Bickel WK, Farley JP, Longo GS. Longitudinal associations among religiousness, delay discounting, and substance use initiation in early adolescence. *J Res Adolesc.* 2015;25(1):36–43. https://doi.org/10.1111/jora.12104.
221. Sussman S, Skara S, Rodriguez Y, Pokhrel P. Non drug use- and drug use-specific spirituality as one-year predictors of drug use among high-risk youth. *Subst Use Misuse.* 2006;41(13):1801–1816. https://doi.org/10.1080/10826080601006508.
222. Bakken NW, DeCamp W, Visher CA. Spirituality and desistance from substance use among reentering offenders. *Int J Offender Ther Comp Criminol.* 2014;58(11):1321–1339. https://doi.org/10.1177/0306624X13494076.
223. Leigh J, Bowen S, Marlatt GA. Spirituality, mindfulness and substance abuse. *Addict Behav.* 2005;30(7):1335–1341. https://doi.org/10.1016/j.addbeh.2005.01.010.
224. Stanton M, Webster JM, Hiller ML, Rostosky S, Leukefeld C. An exploratory examination of spiritual well-being, religiosity, and drug use among incarcerated men. *J Soc Work Pract Addict.* 2003;3(3):87–103. https://doi.org/10.1300/J160v03n03_06.
225. Stewart C. The influence of spirituality on substance use of college students. *J Drug Educ.* 2001;31(4):343–351. https://doi.org/10.2190/HEPQ-CR08-MGYF-YYLW.
226. Conner BT, Anglin MD, Annon J, Longshore D. Effect of religiosity and spirituality on drug treatment outcomes. *J Behav Health Serv Res.* 2009;36(2):189–198. https://doi.org/10.1007/s11414-008-9145-z.
227. Heinz A, Epstein DH, Preston KL. Spiritual/Religious experiences and in-treatment outcome in an inner-city program for heroin and cocaine dependence. *J Psychoact Drugs.* 2007;39(1):41–49. https://doi.org/10.1080/02791072.2007.10399863.
228. Petry NM, Lewis MW, Ostvik-White EM. Participation in religious activities during contingency management interventions is associated with substance use treatment outcomes. *Am J Addict.* 2008;17(5):408–413. https://doi.org/10.1080/10550490802268512.
229. Rashid RA, Kamali K, Habil MH, Shaharom MH, Seghatoleslam T, Looyeh MY. A mosque-based methadone maintenance treatment strategy: Implementation and pilot results. *Int J Drug Pol.* 2014;25(6):1071–1075. https://doi.org/10.1016/j.drugpo.2014.07.003.
230. Palamar JJ, Kiang MV, Halkitis PN. Religiosity and exposure to users in explaining illicit drug use among emerging adults. *J Relig Health.* 2014;53(3):658–674. https://doi.org/10.1007/s10943-012-9660-3.
231. Wallace Jr JM, Delva J, O'Malley PM, et al. Race/ethnicity, religiosity and adolescent alcohol, cigarette and marijuana use. *Soc Work Publ Health.* 2007;23(2–3):193–213. https://doi.org/10.1080/19371910802152059.
232. Hai AH. Are there gender, racial, or religious denominational differences in religiosity's effect on alcohol use and binge drinking among youth in the United States? A propensity score weighting approach. *Subst Use Misuse.* 2019;54(7):1096–1105. https://doi.org/10.1080/10826084.2018.1555598.
233. Wong EC, Longshore D. Ethnic identity, spirituality, and self-efficacy influences on treatment outcomes among Hispanic American methadone maintenance clients. *J Ethn Subst Abuse.* 2008;7(3):328–340. https://doi.org/10.1080/15332640802313478.
234. Unger JB, Sussman S, Begay C, Moerner L, Soto C. Spirituality, ethnic identity, and substance use among American Indian/Alaska Native adolescents in California. *Subst Use Misuse.* 2020;55(7):1194–1198. https://doi.org/10.1080/10826084.2020.1720248.
235. Stanton MD, Todd TC. *The Family Therapy of Drug Abuse and Addiction.* Guilford Press; 1982.
236. Henggeler SW, Pickrel SG, Brondino MJ, Crouch JL. Eliminating (almost) treatment dropout of substance abusing or dependent delinquents through home-based multisystemic therapy. *Am J Psychiatr.* 1996;153(3):427–428. https://doi.org/10.1176/ajp.153.3.427.
237. Hogue A, Dauber S, Henderson CE, et al. Randomized trial of family therapy versus nonfamily treatment for adolescent behavior problems in usual care. *J Clin Child Adolesc Psychol.* 2015;44(6):954–969. https://doi.org/10.1080/15374416.2014.963857.
238. Hendriks V, van der Schee E, Blanken P. Treatment of adolescents with a cannabis use disorder: main findings of a randomized controlled trial comparing multidimensional family therapy and cognitive behavioral therapy in The Netherlands. *Drug Alcohol Depend.* 2011;119(1–2):64–71. https://doi.org/10.1016/j.drugalcdep.2011.05.021.

239. Liddle HA, Dakof GA, Turner RM, Henderson CE, Greenbaum PE. Treating adolescent drug abuse: a randomized trial comparing multidimensional family therapy and cognitive behavior therapy. *Addiction.* 2008;103(10):1660–1670. https://doi.org/10.1111/j.1360-0443.2008.02274.x.
240. Hendriks V, van der Schee E, Blanken P. Matching adolescents with a cannabis use disorder to multidimensional family therapy or cognitive behavioral therapy: treatment effect moderators in a randomized controlled trial. *Drug Alcohol Depend.* 2012;125(1–2):119–126. https://doi.org/10.1016/j.drugalcdep.2012.03.023.
241. Rigter H, Henderson CE, Pelc I, et al. Multidimensional family therapy lowers the rate of cannabis dependence in adolescents: A randomised controlled trial in Western European outpatient settings. *Drug Alcohol Depend.* 2013;130(1–3):85–93. https://doi.org/10.1016/j.drugalcdep.2012.10.013.
242. Smit E, Verdurmen J, Monshouwer K, Smit F. Family interventions and their effect on adolescent alcohol use in general populations; a meta-analysis of randomized controlled trials. *Drug Alcohol Depend.* 2008;97(3):195–206. https://doi.org/10.1016/j.drugalcdep.2008.03.032.
243. Szapocznik J, Perez-Vidal A, Brickman AL, et al. Engaging adolescent drug abusers and their families in treatment: a strategic structural systems approach. *J Consult Clin Psychol.* 1988;56(4):552–557. https://doi.org/10.1037//0022-006x.56.4.552.
244. Garrido-Fernández M, Marcos-Sierra JA, López-Jiménez A, Ochoa de Alda I. Multi-family therapy with a reflecting team: a preliminary study on efficacy among opiate addicts in methadone maintenance treatment. *J Marital Fam Ther.* 2017;43(2):338–351. https://doi.org/10.1111/jmft.12195.
245. Plant CP, Holland JM. Family behavior therapy for alcohol and drug problems in later-life. *Clin Gerontol.* 2018;41(5):508–515. https://doi.org/10.1080/07317115.2017.1349701.
246. Slesnick N, Zhang J. Family systems therapy for substance-using mothers and their 8- to 16-year-old children. *Psychol Addict Behav.* 2016;30(6):619–629. https://doi.org/10.1037/adb0000199.
247. Horigian VE, Feaster DJ, Brincks A, Robbins MS, Perez MA, Szapocznik J. The effects of Brief Strategic Family Therapy (BSFT) on parent substance use and the association between parent and adolescent substance use. *Addict Behav.* 2015;42:44–50. https://doi.org/10.1016/j.addbeh.2014.10.024.
248. Galanter M. Network therapy for addiction: a model for office practice. *Am J Psychiatr.* 1993;150(1):28–36. https://doi.org/10.1176/ajp.150.1.28.
249. Galanter M, Dermatis H, Keller D, Trujillo M. Network therapy for cocaine abuse: use of family and peer supports. *Am J Addict.* 2002;11(2):161–166. https://doi.org/10.1080/10550490290087938.
250. Glazer SS, Galanter M, Megwinoff O, Dermatis H, Keller DS. The role of therapeutic alliance in network therapy: a family and peer support-based treatment for cocaine abuse. *Subst Abuse.* 2003;24(2):93–100. https://doi.org/10.1080/08897070309511537.
251. Copello A, Williamson E, Orford J, Day E. Implementing and evaluating Social Behaviour and Network Therapy in drug treatment practice in the UK: a feasibility study. *Addict Behav.* 2006;31(5):802–810. https://doi.org/10.1016/j.addbeh.2005.06.005.
252. Simpson DD, Joe GW, Fletcher BW, Hubbard RL, Anglin MD. A national evaluation of treatment outcomes for cocaine dependence. *Arch Gen Psychiatr.* 1999;56(6):507–514. https://doi.org/10.1001/archpsyc.56.6.507.
253. Kamp F, Proebstl L, Hager L, et al. Effectiveness of methamphetamine abuse treatment: predictors of treatment completion and comparison of two residential treatment programs. *Drug Alcohol Depend.* 2019;201:8–15. https://doi.org/10.1016/j.drugalcdep.2019.04.010.
254. Black DS, Amaro H. Moment-by-Moment in Women's Recovery (MMWR): mindfulness-based intervention effects on residential substance use disorder treatment retention in a randomized controlled trial. *Behav Res Ther.* 2019;120:103437. https://doi.org/10.1016/j.brat.2019.103437.
255. Federer MB, McKenry PC, Howard L. Factors related to the treatment success of drug addicts enrolled in a residential rehabilitation facility. *Adv Alcohol Subst Abuse.* 1986;5(4):85–97. https://doi.org/10.1300/J251v05n04_07.
256. Charuvastra VC, Dalali ID, Cassuci M, Ling W. Outcome study: comparison of short-term vs long-term treatment in a residential community. *Int J Addict.* 1992;27(1):15–23. https://doi.org/10.3109/10826089109063459.
257. Hubbard RL, Craddock SG, Anderson J. Overview of 5-year followup outcomes in the drug abuse treatment outcome studies (DATOS). *J Subst Abuse Treat.* 2003;25(3):125–134. https://doi.org/10.1016/s0740-5472(03)00130-2.

258. French MT, Popovici I, Tapsell L. The economic costs of substance abuse treatment: updated estimates and cost bands for program assessment and reimbursement. *J Subst Abuse Treat*. 2008;35(4):462−469. https://doi.org/10.1016/j.jsat.2007.12.008.
259. Van Ryswyk C, Churchill M, Velasquez J, McGuire R. Effectiveness of halfway house placement for alcohol and drug abusers. *Am J Drug Alcohol Abuse*. 1981;8(4):499−512. https://doi.org/10.3109/00952998109016932.
260. Hitchcock HC, Stainback RD, Roque GM. Effects of halfway house placement on retention of patients in substance abuse aftercare. *Am J Drug Alcohol Abuse*. 1995;21(3):379−390. https://doi.org/10.3109/00952999509002704.
261. Annis HM, Liban CB. A follow-up study of male halfway-house residents and matched nonresident controls. *J Stud Alcohol*. 1979;40(1):63−69. https://doi.org/10.15288/jsa.1979.40.63.
262. Foster M, Nathan S, Ferry M. The experience of drug-dependent adolescents in a therapeutic community. *Drug Alcohol Rev*. 2010;29(5):531−539. https://doi.org/10.1111/j.1465-3362.2010.00169.x.
263. Jainchill N, Hawke J, De Leon G, Yagelka J. Adolescents in therapeutic communities: one-year posttreatment outcomes. *J Psychoact Drugs*. 2000;32(1):81−94. https://doi.org/10.1080/02791072.2000.10400214.
264. Jainchill N, Hawke J, Messina M. Post-treatment outcomes among adjudicated adolescent males and females in modified therapeutic community treatment. *Subst Use Misuse*. 2005;40(7):975−996. https://doi.org/10.1081/ja-200058857.
265. Stevens SJ, Arbiter N. A therapeutic community for substance-abusing pregnant women and women with children: process and outcome. *J Psychoact Drugs*. 1995;27(1):49−56. https://doi.org/10.1080/02791072.1995.10471672.
266. Kreager DA, Schaefer DR, Davidson KM, Zajac G, Haynie DL, De Leon G. Evaluating peer-influence processes in a prison-based therapeutic community: a dynamic network approach. *Drug Alcohol Depend*. 2019;203:13−18. https://doi.org/10.1016/j.drugalcdep.2019.05.018.
267. Burling TA, Seidner AL, Salvio MA, Marshall GD. A cognitive-behavioral therapeutic community for substance dependent and homeless veterans: treatment outcome. *Addict Behav*. 1994;19(6):621−629. https://doi.org/10.1016/0306-4603(94)90017-5.
268. Sorensen JL, Deitch DA, Acampora A. Treatment collaboration of methadone maintenance programs and therapeutic communities. *Am J Drug Alcohol Abuse*. 1984;10(3):347−359. https://doi.org/10.3109/00952998409001476.
269. Garland EL, Gaylord SA, Boettiger CA, Howard MO. Mindfulness training modifies cognitive, affective, and physiological mechanisms implicated in alcohol dependence: results of a randomized controlled pilot trial. *J Psychoact Drugs*. 2010;42(2):177−192. https://doi.org/10.1080/02791072.2010.10400690.
270. Garland EL, Schwarz NM, Kelly A, Whitt A, Howard MO. Mindfulness-oriented recovery enhancement for alcohol dependence: therapeutic mechanisms and intervention acceptability. *J Soc Work Pract Addict*. 2012;12(3):242−263. https://doi.org/10.1080/1533256X.2012.702638.
271. Ferrari JR, Jason LA, Sasser KC, Davis MI, Olson BD. Creating a home to promote recovery: the physical environments of Oxford House. *J Prev Interv Community*. 2006;31(1−2):27−39. https://doi.org/10.1300/J005v31n01_03.
272. Messina N, Wish E, Nemes S. Predictors of treatment outcomes in men and women admitted to a therapeutic community. *Am J Drug Alcohol Abuse*. 2000;26(2):207−227. https://doi.org/10.1081/ada-100100601.
273. De Leon G. Alcohol use among drug abusers: treatment outcome in a therapeutic community. *Alcohol Clin Exp Res*. 1987;11(5):430−436. https://doi.org/10.1111/j.1530-0277.1987.tb01917.x.
274. Staiger PK, Liknaitzky P, Lake AJ, Gruenert S. Longitudinal substance use and biopsychosocial outcomes following therapeutic community treatment for substance dependence. *J Clin Med*. 2020;9(1):118. https://doi.org/10.3390/jcm9010118.
275. Vidjak N. Treating heroin addiction: comparison of methadone therapy, hospital therapy without methadone, and therapeutic community. *Croat Med J*. 2003;44(1):59−64.
276. Groh DR, Jason LA, Ferrari JR, Davis MI. Oxford House and Alcoholics Anonymous: the impact of two mutual-help models on abstinence. *J Groups Addict Recov*. 2009;4(1−2):23−31. https://doi.org/10.1080/15560350802712363.
277. Jason LA, Salina D, Ram D. Oxford recovery housing: length of stay correlated with improved outcomes for women previously involved with the criminal justice system. *Subst Abuse*. 2016;37(1):248−254. https://doi.org/10.1080/08897077.2015.1037946.
278. Majer JM, Jason LA, Ferrari JR, Miller SA. 12-Step involvement among a U.S. national sample of Oxford House residents. *J Subst Abuse Treat*. 2011;41(1):37−44. https://doi.org/10.1016/j.jsat.2011.01.010.

279. Nguyen RL, Cope CE, Wiedbusch EK, Guerrero M, Jason LA. "This program helped save our lives so we all can bond over that": a preliminary study of the first oxford house collegiate recovery home. *Alcohol Treat Q.* 2021;39(4):489−504. https://doi.org/10.1080/07347324.2021.1898295.
280. Mulder RT, Frampton CM, Peka H, Hampton G, Marsters T. Predictors of 3-month retention in a drug treatment therapeutic community. *Drug Alcohol Rev.* 2009;28(4):366−371. https://doi.org/10.1111/j.1465-3362.2009.00050.x.
281. Vigna-Taglianti F, Mathis F, Diecidue R, et al. Factors predicting patient's allocation to short- and long-term therapeutic community treatments in the Italian VOECT Cohort Study. *Community Ment Health J.* 2017;53(8):972−983. https://doi.org/10.1007/s10597-017-0105-z.
282. Vergara-Moragues E, Verdejo-García A, Lozano OM, et al. Association between executive function and outcome measure of treatment in therapeutic community among cocaine dependent individuals. *J Subst Abuse Treat.* 2017;78:48−55. https://doi.org/10.1016/j.jsat.2017.04.014.
283. Johnson KW, Young L, Shamblen S, Suresh G, Browne T, Chookhare KW. Evaluation of the therapeutic community treatment model in Thailand: policy implications for compulsory and prison-based treatment. *Subst Use Misuse.* 2012;47(8−9):889−909. https://doi.org/10.3109/10826084.2012.663279.
284. Johnson K, Pan Z, Young L, et al. Therapeutic community drug treatment success in Peru: a follow-up outcome study. *Subst Abuse Treat Prev Pol.* 2008;3:26. https://doi.org/10.1186/1747-597X-3-26.
285. Alshomrani AT, Khoja AT, Alseraihah SF, Mahmoud MA. Drug use patterns and demographic correlations of residents of Saudi therapeutic communities for addiction. *J Taibah Univ Med Sci.* 2017;12(4):304−312. https://doi.org/10.1016/j.jtumed.2017.02.006.
286. Cutter HS, Samaraweera A, Price B, Haskell D, Schaeffer C. Prediction of treatment effectiveness in a drug-free therapeutic community. *Int J Addict.* 1977;12(2−3):301−321. https://doi.org/10.3109/10826087709027226.
287. Jason LA, Davis MI, Ferrari JR. The need for substance abuse after-care: longitudinal analysis of Oxford House. *Addict Behav.* 2007;32(4):803−818. https://doi.org/10.1016/j.addbeh.2006.06.014.
288. Barreno EM, Domínguez-Salas S, Díaz-Batanero C, Lozano ÓM, Marín J, Verdejo-García A. Specific aspects of cognitive impulsivity are longitudinally associated with lower treatment retention and greater relapse in therapeutic community treatment. *J Subst Abuse Treat.* 2019;96:33−38. https://doi.org/10.1016/j.jsat.2018.10.004.
289. Levin FR, Evans SM, Vosburg SK, Horton T, Brooks D, Ng J. Impact of attention-deficit hyperactivity disorder and other psychopathology on treatment retention among cocaine abusers in a therapeutic community. *Addict Behav.* 2004;29(9):1875−1882. https://doi.org/10.1016/j.addbeh.2004.03.041.
290. Bankston SM, Carroll DD, Cron SG, et al. Substance abuser impulsivity decreases with a nine-month stay in a therapeutic community. *Am J Drug Alcohol Abuse.* 2009;35(6):417−420. https://doi.org/10.3109/00952990903410707.
291. Bergman BG, Greene MC, Hoeppner BB, Slaymaker V, Kelly JF. Psychiatric comorbidity and 12-step participation: a longitudinal investigation of treated young adults. *Alcohol Clin Exp Res.* 2014;38(2):501−510. https://doi.org/10.1111/acer.12249.
292. Christo G, Franey C. Drug users' spiritual beliefs, locus of control and the disease concept in relation to Narcotics Anonymous attendance and six-month outcomes. *Drug Alcohol Depend.* 1995;38(1):51−56. https://doi.org/10.1016/0376-8716(95)01103-6.
293. Galanter M, Dermatis H, Post S, Santucci C. Abstinence from drugs of abuse in community-based members of Narcotics Anonymous. *J Stud Alcohol Drugs.* 2013;74(2):349−352. https://doi.org/10.15288/jsad.2013.74.349.
294. Nealon-Woods MA, Ferrari JR, Jason LA. Twelve-step program use among Oxford House residents: spirituality or social support in sobriety? *J Subst Abuse.* 1995;7(3):311−318. https://doi.org/10.1016/0899-3289(95)90024-1.
295. Boshears P, Boeri M, Harbry L. Addiction and sociality: perspectives from methamphetamine users in suburban USA. *Addiction Res Theor.* 2011;19(4):289−301. https://doi.org/10.3109/16066359.2011.566654.
296. Hoeppner BB, Hoeppner SS, Kelly JF. Do young people benefit from AA as much, and in the same ways, as adult aged 30+? A moderated multiple mediation analysis. *Drug Alcohol Depend.* 2014;143:181−188. https://doi.org/10.1016/j.drugalcdep.2014.07.023.
297. Humphreys K, Blodgett JC, Wagner TH. Estimating the efficacy of Alcoholics Anonymous without self-selection bias: an instrumental variables re-analysis of randomized clinical trials. *Alcohol Clin Exp Res.* 2014;38(11):2688−2694. https://doi.org/10.1111/acer.12557.

298. Morgenstern J, Labouvie E, McCrady BS, Kahler CW, Frey RM. Affiliation with alcoholics anonymous after treatment: a study of its therapeutic effects and mechanisms of action. *J Consult Clin Psychol.* 1997;65(5):768–777. https://doi.org/10.1037//0022-006x.65.5.768.
299. Pisani VD, Fawcett J, Clark DC, McGuire M. The relative contributions of medication adherence and AA meeting attendance to abstinent outcome for chronic alcoholics. *J Stud Alcohol.* 1993;54(1):115–119. https://doi.org/10.15288/jsa.1993.54.115.
300. Thurstin AH, Alfano AM, Nerviano VJ. The efficacy of AA attendance for aftercare of inpatient alcoholics: some follow-up data. *Int J Addict.* 1987;22(11):1083–1090. https://doi.org/10.3109/10826088709027471.
301. Tonigan JS, Beatty GK. Twelve-step program attendance and polysubstance use: interplay of alcohol and illicit drug use. *J Stud Alcohol Drugs.* 2011;72(5):864–871. https://doi.org/10.15288/jsad.2011.72.864.
302. Kelly JF, Dow SJ, Yeterian JD, Kahler CW. Can 12-step group participation strengthen and extend the benefits of adolescent addiction treatment? A prospective analysis. *Drug Alcohol Depend.* 2010;110(1–2):117–125. https://doi.org/10.1016/j.drugalcdep.2010.02.019.
303. Schonbrun YC, Strong DR, Anderson BJ, Caviness CM, Brown RA, Stein MD. Alcoholics Anonymous and hazardously drinking women returning to the community after incarceration: predictors of attendance and outcome. *Alcohol Clin Exp Res.* 2011;35(3):532–539. https://doi.org/10.1111/j.1530-0277.2010.01370.x.
304. Majer JM, Jason LA, Aase DM, Droege JR, Ferrari JR. Categorical 12-step involvement and continuous abstinence at 2 years. *J Subst Abuse Treat.* 2013;44(1):46–51. https://doi.org/10.1016/j.jsat.2012.03.001.
305. Ronel N, Gueta K, Abramsohn Y, Caspi N, Adelson M. Can a 12-step program work in methadone maintenance treatment? *Int J Offender Ther Comp Criminol.* 2011;55(7):1135–1153. https://doi.org/10.1177/0306624X10382570.
306. Stewart SH, Walitzer KS, Blanco J, et al. Medication-enhanced behavior therapy for alcohol use disorder: naltrexone, Alcoholics Anonymous Facilitation, and OPRM1 genetic variation. *J Subst Abuse Treat.* 2019;104:7–14. https://doi.org/10.1016/j.jsat.2019.05.004.
307. Winner NA. Bridging the gap: the compatibility of CBT-based approaches with twelve-step programs in the treatment of substance use disorders. *Subst Use Misuse.* 2021;56(11):1662–1669. https://doi.org/10.1080/10826084.2021.1949606.
308. Labbe AK, Slaymaker V, Kelly JF. Toward enhancing 12-step facilitation among young people: a systematic qualitative investigation of young adults' 12-step experiences. *Subst Abuse.* 2014;35(4):399–407. https://doi.org/10.1080/08897077.2014.950001.
309. Kelly JF, Kaminer Y, Kahler CW, et al. A pilot randomized clinical trial testing integrated 12-Step facilitation (iTSF) treatment for adolescent substance use disorder. *Addiction.* 2017;112(12):2155–2166. https://doi.org/10.1111/add.13920.

CHAPTER 11

Stress and addiction

Introduction

Imagine how you feel when taking an exam, asking someone to go on a date, delivering a presentation, going to a job interview, or having an unexpected cost. You may notice that your heart beats faster and that you sweat more than usual. Your stomach may feel upset, and you may have difficulty sleeping before engaging in one of these activities or throughout the duration of the event if prolonged like dealing with financial problems. These "symptoms" are signs of **stress**, physical and mental responses to an external cause (known as a *stressor*). There are numerous stressors one can experience. The examples above are just some of the stressors we can encounter. Some stressors are acute, like taking an exam, while other stressors are chronic (e.g., long-term financial problems). There are large individual differences in what we consider to be stressful and how we deal with these stressors. Some individuals may meditate when they feel stressed, or they may exercise. However, others may turn to ethanol and/or drugs as a way to cope with stress. As you will learn in this chapter, stress and addiction are closely linked. Individuals are more likely to use a substance when under stress, and stress is a major trigger for relapse. In the words of one individual dealing with stress at work:

> ... I enjoy the job, but the increase in work duties just keep piling up where the stress was built up against me ... and the stress just got worse and that is why I started [using ethanol] again. It just kept back and forth, back and forth (p. 720).[1]

Because stress is a major antecedent to relapse, methods for reducing stress need to be incorporated into treatment interventions. Such methods will be discussed at the end of this chapter.

Learning objectives

By the end of this chapter, you should be able to ...

(1) Describe how the hypothalamic-pituitary-adrenal (HPA) axis regulates physiological responses to stress.

(2) Discuss the ways in which stress and anxiety can exacerbate addiction-like behaviors.
(3) Identify the methods researchers use to induce stress in animals.
(4) Explain the neuromechanisms beyond the HPA axis that mediate the relationship between stress and addiction.
(5) Identify strategies that can be used to reduce stress-induced increases in drug craving and/or seeking.

Neurobiology of stress

As a medical student, Dr. Hans Selye (1907–1982) noticed that patients with chronic illnesses often displayed similar physiological symptoms. This led him to focus his research on what he would ultimately call stress. To understand why individuals experience physical symptoms when encountering stressors, I need to discuss the **neuroendocrine system**, the intersection between the nervous system and the *endocrine system*. Whereas the nervous system uses neurotransmitters/peptides to control neuronal actions, the endocrine system relies on the transmission of **hormones**, signaling molecules secreted by glands/organs that travel through the bloodstream and bind to receptors located on other glands. Notice that hormones are very similar to neurotransmitters in that they travel from one source (organ/gland instead of axon terminal) and bind to a receptor to exert an effect. In fact, some neurotransmitters act as hormones in the endocrine system. Epinephrine is a neurotransmitter, but many of you know it better as the hormone adrenaline. Fig. 11.1 shows the major structures of the endocrine system.

The neuroendocrine system is composed of several subsystems. This chapter will focus on one such system: the **hypothalamic-pituitary-adrenal (HPA) axis** (Fig. 11.2). The hypothalamus is a complex brain structure composed of distinct nuclei that help us maintain homeostasis (see Chapter 5). More specifically, the hypothalamus controls feeding and drinking behavior, metabolism, blood pressure, reproductive behaviors, and our "fight or flight" response, which is what we will be discussing in detail. When faced with a stressor, the hypothalamus, specifically the paraventricular nucleus (PVN), secretes *corticotropin-releasing hormone* (CRH) (commonly known as a corticotropin-releasing factor [CRF]), which stimulates the pituitary gland, a small structure in the brain at the base of the hypothalamus. The pituitary gland then secretes *adrenocorticotropic hormone* (ACTH). ACTH stimulates the adrenal glands, located above the kidneys. The adrenal glands release adrenaline and noradrenaline, as well as *cortisol*, a major **corticosteroid**. Corticosteroids are divided into two major types of steroid hormones: mineralocorticoids and **glucocorticoids**. Glucocorticoids are our "stress hormones", such as cortisol. After cortisol is released, it travels to both the hypothalamus and the pituitary gland to inhibit the further synthesis and release of CRH and ACTH. This is important because too much cortisol can lead to health complications. For example, Cushing's syndrome is caused by persistently elevated cortisol levels. Individuals with Cushing's syndrome experience weight gain, leading to a rounded face ("moon face"), and they bruise easily. The increased weight gain can lead to other complications like high blood pressure, diabetes, heart attack, and stroke.

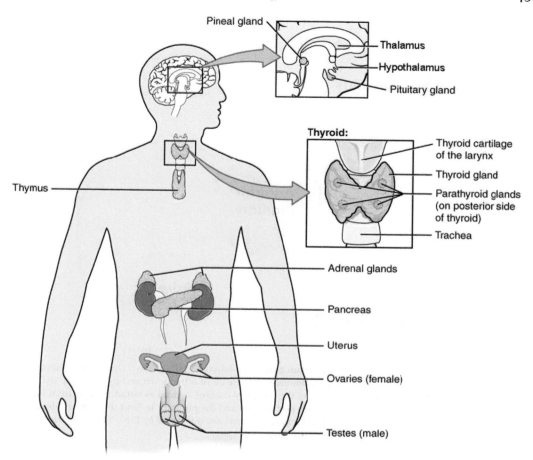

FIGURE 11.1 Schematic showing some of the major organs/glands that make up the endocrine system. Both male and female reproductive organs are shown in this schematic. *Image modified from Wikimedia Commons user OpenStax College (https://commons.wikimedia.org/wiki/File:1801_The_Endocrine_System.jpg).*

Norepinephrine/noradrenaline is the major neurotransmitter/hormone used by our *sympathetic nervous system*. Pretend that you are taking a hike in the mountains when you encounter a mountain lion. This is a stressful situation as there is the possibility that the mountain lion will attack. At this point, ACTH stimulates your adrenal glands to produce noradrenaline. Noradrenaline binds to noradrenergic receptors on various organs. Recall from Chapter 2 that noradrenergic receptors are metabotropic and can be either excitatory or inhibitory. When noradrenaline binds to excitatory receptors on the heart, your heart begins to pump more blood through your body. This is what produces the "racing heart" you experience during a stressful situation. Conversely, noradrenaline binds to inhibitory receptors in the digestive system. When faced with something like a mountain lion, the last thing your body is going to worry about is digesting

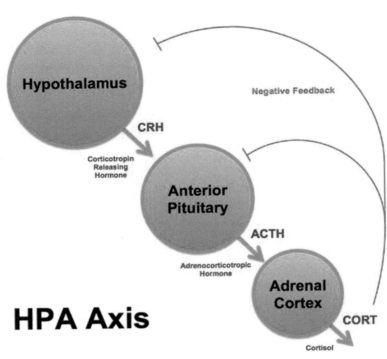

FIGURE 11.2 A schematic of the hypothalamic-pituitary-adrenal (HPA) axis. The hypothalamus releases corticotropin-releasing hormone (CRH), which stimulates the release of adrenocorticotropic hormone (ACTH) from the anterior pituitary gland. ACTH then stimulates the release of cortisol, as well as noradrenaline, from the adrenal gland (cortex). Cortisol binds to receptors in the hypothalamus and the pituitary to limit the further release of CRH and ACTH, respectively. *The image comes from Wikimedia Commons and was posted by Brian M. Sweis (https://commons.wikimedia.org/wiki/File:HPA_Axis_Diagram_(Brian_M_Sweis_2012).png).*

what you ate for breakfast. Instead, your body wants to prepare you to fight the mountain lion or to get away from the mountain lion, hence the term "fight or flight". We will return to the HPA axis when we discuss the neuromechanisms mediating the relationship between stress and addiction. First, I need to discuss the relationship between stress and drug use.

Stress and drug use patterns in humans

Multiple stressors are predictive of substance use and addiction, including losing a loved one, being abandoned, being homeless, being a victim of a violent act, living through a terrorist attack, and being abused (physical, sexual, and/or emotional).[2-4] Individuals with a substance use disorder (SUD) often report experiencing more negative life events compared to positive life events[5] and experiencing more distress compared to nondependent individuals.[6] One study found that family-related stressors are a better predictor of substance use compared to external stressors.[7] Similarly, former prison inmates that experience problems with family, friends, and/or partners are at risk of using drugs or engaging in hazardous drinking.[8] Chronic stress is associated with ethanol-related problems.[9] Individuals that have experienced more lifetime stress score higher for cocaine dependence,[10] and stress is associated with longer durations of cocaine use.[11] Lifetime trauma, particularly childhood trauma, is significantly associated with SUDs.[12,13]

Adolescence appears to be a critical period in which stress potentiates substance use. Overall, experiencing negative events (e.g., having a seriously ill family member, facing economic hardships) is a risk factor for substance use and heavy ethanol drinking.[14] Younger adolescents who experience stress are especially at risk of developing ethanol-related problems.[15] Research has also shown that family stress is positively associated with cannabis use.[16] Adolescents and young adults that are victims of violent acts or sexual abuse are more likely to use illicit drugs.[17] During a 5-year period following a traumatic injury, 14% of adolescents receive an opioid use disorder diagnosis.[18] Even before reaching adolescence, stress can have indirect effects on substance use. Specifically, experiencing stress between the ages of 2 and 5 is negatively associated with inhibitory control, which is then predictive of problematic behaviors later in childhood; these behaviors ultimately serve as risk factors for early adolescent substance use.[19]

Stress is a major risk factor for relapse.[20,21] Individuals experiencing stress report greater drug cravings,[22] and these cravings are exacerbated when drug-paired cues are present.[23,24] Premature treatment dropout is associated with abnormal cortisol levels,[25] and treatment dropout is associated with increased stress and a recent history of emotional abuse.[26] Indeed, veterans that experience severe chronic stress are 2.5 times more likely to use substances following treatment.[27] Additionally, those that drop out of methadone maintenance therapy report high levels of sexual abuse (36%), physical abuse (60%), emotional abuse (57%), and child physical neglect (66%), with 25% experiencing all four forms of abuse.[28] Following treatment, active substance use is significantly correlated with mental distress.[29] On the flip side, individuals that maintain long-term abstinence (1 year following treatment) report greater improvements in psychological distress.[30]

While research has implicated stress as an important determinant of substance use and relapse, there have been some questions about the direct role of stress on the addiction process. Think of it this way: we all experience stress in some capacity, but not all of us have a SUD. There are most likely other factors that moderate the association between stress and substance use. Indeed, one study found that ethanol-related problems were similar in individuals who experienced stress before the emergence of ethanol problems and those who experienced stress *after* the emergence of problems.[15] Furthermore, we also must consider the fact that some individuals handle stress better than others. Individuals that use more coping strategies are less likely to experience ethanol-related problems and are more likely to remain abstinent from drugs following treatment, even if they experience high levels of stress.[15,31] As you will read at the end of this chapter, several approaches can be used to better control stress, thus attenuating the effects of stress on drug craving and relapse.

Anxiety disorders and addiction

In Chapter 1, I mentioned that SUDs often cooccur with other psychiatric disorders. Because stress is a major component of anxiety disorders, discussing the relationship between anxiety disorders and SUDs is merited. There are multiple anxiety disorders, and many of them are associated with increased odds of substance use and dependence. I will cover some of the major anxiety disorders here. Generalized anxiety disorder (GAD) is marked by persistent worrying/anxiety even when no stressor is present. Individuals with GAD often have difficulty relaxing and feel "on edge". Approximately 49% of those with GAD also have a SUD,[32] and they are more likely to engage in chronic cigarette smoking, ethanol consumption, and cannabis use.[33] In contrast to GAD, panic disorder is characterized by repeated episodes of intense fear that appear suddenly and unexpectedly. Panic attacks are accompanied by physical symptoms like chest pain, heart palpitations, and shortness of breath. Individuals with panic disorder often state they feel like they are having a heart attack during a panic attack. Like GAD, panic disorder is highly associated with SUDs, as one study found that 60.4% of individuals that have panic attacks have a SUD compared to just 27.5% of individuals without panic attacks.[34]

Phobic disorders represent a broad class of anxiety disorders, but each phobic disorder has the same symptomatology. Individuals with a phobia have a persistent, exaggerated fear of a specific stimulus, event, or circumstance. Common phobic disorders include acrophobia (heights), agoraphobia (open spaces), claustrophobia (enclosed spaces), cynophobia (dogs), ophidiophobia (snakes), and trypanophobia (needles). Closely related to phobic disorder is social anxiety disorder (SAD), which is an intense fear of social situations. Social phobias, especially SAD, are highly linked to SUDs. Approximately one-third of individuals with a SUD and a mood/anxiety disorder experience social phobia.[35] In particular, SAD is predictive of alcohol use disorder and tobacco and cannabis use.[36,37] While SAD is predictive of an alcohol use disorder, there is evidence that SAD is more prevalent in poly-substance-dependent individuals (51% of the sample) compared to ethanol-dependent individuals (34% of the sample).[38]

Obsessive-compulsive disorder (OCD) is characterized by intrusive thoughts that cause discomfort in the individual (*obsessions*) and *compulsions*, which are actions that are performed to alleviate the discomfort associated with the obsessions. The actor and former TV game show host Howie Mandel has OCD that is exacerbated by his mysophobia (fear of dirt/germs). Due to his OCD, Mandel does not shake others' hands, instead opting to give them his trademark fist bump. He also shaved his head as he felt cleaner bald compared to having hair. As already covered in this book, SUDs involve compulsive drug taking, which is somewhat similar to the compulsions an individual with OCD performs to alleviate stressful obsessions. To alleviate withdrawal symptoms, an individual will use the drug. Because OCD and SUDs are both marked by compulsions, it should be of no surprise that there is a high comorbidity between the two conditions. In one sample, 27% of OCD patients have a SUD, with 70% of these individuals commenting that their OCD symptoms preceded their drug use.[39]

The final anxiety disorder I want to discuss is posttraumatic stress disorder (PTSD). As the name suggests, PTSD is associated with a traumatic experience. From a historical perspective, the term "shell shocked" was used to describe the behavioral/psychological disturbances observed in soldiers during World War I. We now know that those suffering from shell shock were most likely experiencing symptoms of PTSD. These symptoms include an increased fight-or-flight response, increased mental and physical distress, memory/cognitive impairments, and a "reliving" of the traumatic event. PTSD is not isolated to veterans; individuals that have experienced domestic violence, sexual assault, child abuse, or severe injury can develop PTSD. Individuals with PTSD are more likely to use drugs, are more likely to initiate drug use at an earlier age, and are more likely to be diagnosed with a SUD.[37,40–43] Specifically, 21.9% of veterans with PTSD have a SUD.[44] Adolescents who self-report having PTSD are more likely to use MDMA compared to those that have not experienced a traumatic event[45] (see Experiment Spotlight below). PTSD symptoms are also predictive of substance use severity,[46] future ethanol use,[47] and craving during early abstinence.[48] Individuals with a dual diagnosis of PTSD and a SUD have worse addiction-related psychiatric problems, family and social problems, and medical problems compared to individuals with a SUD only.[49] The relationship between PTSD and substance use appears to be bidirectional, as drug use can exacerbate PTSD symptoms[47] while detoxification from drugs can decrease PTSD symptoms.[48]

Experiment Spotlight: Self-reported PTSD is associated with increased use of MDMA in adolescents with substance use disorders (Basedow et al.[45])

As I mentioned above, individuals with anxiety disorders are more likely to have a SUD compared to the general population. I cited one study that showed that adolescents who self-report having PTSD are more likely to use MDMA. The Experiment Spotlight discusses this study in more detail.

Basedow et al.[45] recruited 121 German adolescents diagnosed with a SUD, with 42% of the sample consisting of females. Participants completed the University of California at Los Angeles Posttraumatic Stress Disorder Reaction Index for DSM-IV (UCLA RI-IV), a questionnaire that screens for traumatic experiences and PTSD symptoms. Participants select a traumatic event that affects them the most from a list of potential events. They then identify features of the traumatic event. Participants are then asked to indicate the frequency of PTSD symptoms during the past month. Based on participants' responses, they were grouped into one of three groups: No traumatic event experience (NoTE), traumatic event experience without PTSD symptoms (TE), and probable PTSD (PSTD). Participants were then interviewed by a clinician regarding their past month's drug use and the age at which they began using substances. Specifically, participants were asked about their use of tobacco, ethanol, cannabis, amphetamine, and MDMA.

The most common traumatic event experienced by both the TE and the PTSD groups was nondomestic violence (54% and 66%, respectively), followed by sexual abuse (25% and 40%, respectively). When comparing substance use across the three groups, Basedow et al. found that adolescents with probable PTSD were more likely to use MDMA (32%) compared to those that have not experienced a traumatic event (9%) and to those that have experienced a traumatic event (8%). No significant differences were observed across the three groups for the other substances.

Questions to consider:

(1) Why was the term probable PTSD used instead of stating that one group had PTSD?
(2) Why do you think adolescents with PTSD-like symptoms use MDMA more frequently compared to the other groups?

Answers to questions:

(1) Because the participants completed a self-report questionnaire to answer questions about PTSD symptoms, they cannot officially be diagnosed as having PTSD. To be diagnosed with a disorder, one needs to complete a clinical assessment with a doctor.
(2) Basedow et al. argue that MDMA use can act as a form of self-medication. Recall that MDMA belongs to a drug class called entactogens (or empathogens). These drugs have psychostimulant and hallucinogenic effects. More importantly, they increase feelings of empathy and sympathy. Because PTSD is characterized by reliving the traumatic experience, these individuals may use MDMA to improve their mood during these experiences. This leads to another question: if individuals use MDMA as a form of self-medication, why did the PTSD group not differ from the other groups when examining drugs like cannabis or ethanol? The most likely answer is that baseline use of tobacco, ethanol, and cannabis was already high (remember, this study used adolescents). Over 90% of the participants reported tobacco use during the past month, while over half of the participants reported using ethanol and cannabis. This could have led to a ceiling effect, where the researchers could not observe further increases in the use of these substances. On the flip side of this, amphetamine use was extremely low, with just 7% of the total sample reporting using amphetamine. Thirteen percent of the PTSD participants reported using amphetamine, compared to just 3% and 6% of the NoTE and TE groups. Due to the low numbers reported in each group, there was most likely not enough statistical power to detect a difference across groups.

As discussed in the Experiment Spotlight, one potential explanation for increased substance use in those with an anxiety disorder is self-medication.[37,45] In one study, 18.3% of those with GAD reported using ethanol to mitigate their symptoms.[50] Another study found that lifetime cannabis use is correlated with lifetime diagnosis and current diagnosis of panic disorder.[51] Perhaps individuals with panic disorder use cannabis as a way of coping with this specific anxiety disorder. While PTSD is not always directly related to cigarette smoking, there is an important moderating variable that influences this relationship: negative affect reduction expectancy. In other words, if individuals with PTSD expect smoking to improve their mood, they are more likely to smoke and to develop nicotine dependence.[52]

Animal models of stress

Clinical studies have provided ample evidence that stress and addiction are highly correlated. To better understand the neurobiological underpinnings linking stress to addiction, animal models of stress are invaluable. Researchers studying stress need to walk a fine line as ethical guidelines stipulate that stress needs to be minimized as much as possible. However, to study stress, one needs to subject an animal to a stressor. The key is to subject an animal to a stressor that will not cause permanent or severe physical pain. To this end, there are numerous stressors that can be used that are considered ethical. The most commonly used stressors will be detailed below. Because the previous chapter discussed how social factors influence drug-taking behaviors, I will begin this section by describing how researchers manipulate social interactions to increase stress.

Social isolation

You previously learned about environmental enrichment as an animal model of socioeconomic status (SES). Recall that animals that have access to novel objects and social peers are less likely to self-administer a drug. While environmental enrichment is protective against drug-taking behaviors, social isolation is a stressor in animals that can increase drug sensitivity.[53] Numerous studies have examined the effects of social isolation on volitional consumption of drugs, particularly ethanol. Evidence indicates that socially isolated animals consume more ethanol,[54] as well as a morphine solution,[55] compared to animals that live in pairs or in groups. Some research has shown that the timing of social isolation is a critical determinant of drug consumption, as social isolation during adolescence, but not during adulthood, increases ethanol consumption.[56] This finding makes sense as adolescents are highly sensitive to peers, as we covered in the previous chapter, and stress is known to increase drug-seeking behaviors in adolescents.

In addition to volitional drug consumption, socially isolated animals more readily acquire drug self-administration[57,58] (but see[59]), self-administer more drug[60–62] (but see[63]), show greater motivation for drug reinforcement as assessed with the progressive ratio schedule,[57,61,62] and show increased drug-seeking and relapse-like behaviors[60,64] (but see[57]). Research has shown mixed results concerning the dose-response curve for cocaine in isolated animals. In one study, the dose-response curve for cocaine was shifted to the right in isolated rats, meaning they self-administered greater amounts of higher doses of cocaine relative to lower doses.[59] In a later study conducted by the same research group, they found that isolated rats self-administered more cocaine at a lower dose but were less likely to acquire self-administration at a higher dose.[65] Despite this discrepancy, isolation appears to increase the direct reinforcing effects of drugs.

Isolation stress also potentiates the locomotor effects of drugs, particularly stimulants.[59,63,66–68] One interesting finding is that isolated animals chronically treated with ethanol have higher mortality rates compared to animals raised in groups. Even when the amount of ethanol is controlled such that isolated and group-housed animals drink the same amount of ethanol, isolated animals still show a higher mortality rate.[69] These results provide further support that isolation enhances the physiological effects of drugs. The increased behavioral sensitization observed in isolated animals provides one account as to why they self-administer more drugs.

In the previous chapter, you learned that the effects of environmental enrichment on CPP are mixed. While socially isolated rats develop greater CPP to MDMA compared to group-housed rats,[70] this finding appears to be the exception, not the rule. Most studies assessing the contribution of social isolation stress have found either decreased CPP in socially isolated animals[71] or no differences in CPP as a function of housing condition.[72] Why is this the case? Perhaps environmentally enriched animals are better at learning and forming associations between the interoceptive cues of the drug and an environmental context. Environmentally enriched animals have very different brains compared to isolated animals. They have larger brains[73] and have more dendritic spines in the hippocampus[74] compared to isolated animals. In other words, social isolation may reduce CPP by impairing Pavlovian conditioning without altering operant conditioning. Another possible explanation is that isolation increases the hedonic set point (see Chapter 5). That is, animals raised in isolation need more drug to achieve the "desired" effect. In drug self-administration studies, this leads to greater self-administration. In CPP, each animal receives the same dose of the drug. If the dose is too low, isolated animals may not reach their hedonic set point, thus deriving little "pleasure" from the drug. This could account for the blunted CPP that has been observed in some studies.

Maternal separation

One way to model early-life stress in rodents is to separate a young pup from the mother; this is known as *maternal separation*.[53] Maternal separation has been shown to alter physiological responses to drugs and withdrawal symptoms. Rats that have been maternally separated show greater withdrawal symptoms following chronic morphine treatment.[75] However, discrepant findings have been observed concerning the effects of maternal separation on drug-induced alterations in locomotor activity. Some studies have shown that maternal separation increases the locomotor stimulant effects of drugs,[76] but other studies have observed no differences in locomotor activity or even decreased activity in maternally separated animals.[68,77,78] Some factors have been identified that modulate the effects of maternal separation on locomotor activity and behavioral sensitization. An early study found that rat pups that experience maternal separation in a room-temperature environment (which is colder than what rodents normally experience) show increased activity following *d*-amphetamine treatments, whereas pups that experience maternal separation in a warmer environment are less sensitive to the locomotor-stimulant effects of *d*-amphetamine.[79] Another consideration is the dose of the drug. For example, when a lower dose of ethanol is used, maternally separated rats show enhanced locomotor activity; however, maternal separation leads to a blunted locomotor response following administration of a higher dose of ethanol.[80]

Although there are some conflicting findings regarding drug-induced alterations in locomotor activity, there is consistent evidence that maternally separated animals are more likely to acquire drug self-administration[81] and to self-administer more drug.[81–90] Maternal separation also impairs the extinction of drug-taking behavior.[81] In other words, maternally separated animals continue to respond on a manipulandum even when the drug is no longer available. Although maternal separation increases drug-seeking behavior during extinction, it does not always increase relapse-like behavior following extinction.[90]

Research has shown that the length of maternal separation can profoundly alter drug-taking behaviors. Experiencing brief periods of maternal separation (15 min) *decreases* consumption of higher concentrations of ethanol (8%–20%) but increases preference for a low concentration of ethanol (5%).[91,92] Rats that experience 15 min of maternal separation from days 2–15 of life self-administer significantly less cocaine compared to control groups, whereas pups that experience 180 min of maternal separation show greater self-administration at the lowest tested dose of cocaine.[93] Similarly, rats exposed to 180 min of maternal separation from days 2–14 acquire methamphetamine self-administration at a faster rate and self-administer more methamphetamine compared to rats that receive daily 15-min separations[94]; however, it is important to note that Lewis et al. did not test a control group that did not receive maternal separation.[94] Likewise, rats that receive longer periods of maternal separation develop greater nicotine CPP and consume more ethanol compared to rats that receive 15 min of separation, although no differences are observed between maternally separated animals and control animals.[95,96]

Social defeat

When discussing the role of social interaction on addiction-like behaviors in the previous chapter, I did not include how antagonistic social interactions impact drug self-administration in animals. We all have most likely had to deal with an antagonistic social interaction, whether getting into an argument with a friend, encountering a rude customer, or being harassed by a stranger. Antagonistic social interaction can be modeled in rodents using the resident-intruder paradigm, in which an "intruder" is placed in the home cage of a "resident" for a specified amount of time (typically 10 min). The researcher can then place a divider between the two animals to prevent physical contact. However, the resident will continue to engage in aggressive behaviors toward the intruder (e.g., biting the divider). Often, the researcher uses an intruder that is smaller than the resident. This will cause the intruder to experience **social defeat** as it has more difficulty defending itself from the larger resident[53] (see Fig. 11.3 for a schematic of the resident-intruder paradigm). Social defeat is a major risk factor for the development of addiction-like behaviors, including:

(1) enhanced locomotor responses to drugs,[97–103]
(2) increased ethanol consumption,[104–106]
(3) facilitated acquisition of drug self-administration[58,107,108] (but see[109,110]),
(4) increased drug self-administration,[58,89,97–101,105,107,111–115]

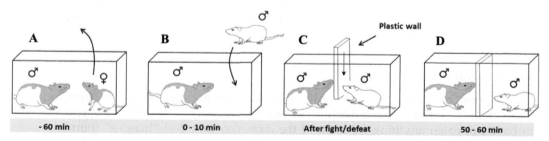

FIGURE 11.3 An example of the resident-intruder paradigm to examine social defeat. In this example, a smaller rat (all white) is placed in the home cage of a larger rat (white and gray). The two rats directly interact with one another for 10 min before being separated by a plastic wall. For approximately 1 h, the two rats are separated by the wall before the smaller rat is removed from the cage. *Image comes from Fig. 1 of Rajalingam D, Nymoen I, Jacobsen DP, et al. Repeated social defeat promotes persistent inflammatory changes in splenic myeloid cells; decreased expression of β-arrestin-2 (ARRB2) and increased expression of interleukin-6 (IL-6). BMC Neurosci. 2020;21(1):25. https://doi.org/10.1186/s12868-020-00574-4. This is an open-access article distributed under the terms of the Creative Commons Attribution 4.0 International License (https://creativecommons.org/licenses/by/4.0/).*

(5) greater motivation to self-administer drug[105,107,114,115] (but see[109,110]),
(6) elevated CPP,[102,103,105,116–119] although social defeat paradoxically decreases MDMA CPP,[120]
(7) delayed extinction of drug self-administration and CPP,[107,109] and
(8) increased reinstatement and incubation of drug-seeking behavior.[108,121]

The resident-intruder paradigm described above involves intermittent periods of social defeat. Another paradigm that can be used is *subordination stress*. An intruder lives in a protective cage in the home cage of the resident for an extended period of time (e.g., several weeks). In contrast to episodic social defeat, subordination stress *decreases* the reinforcing effects of drugs.[122] This finding may seem counterintuitive. However, being continuously subjected to social defeat may lead to anhedonia, in which animals lose motivation to work for access to drug. There is some evidence to support this hypothesis as animals subjected to long-term subordination stress consume less sucrose compared to both controls and dominant partners,[123] and rats subjected to daily bouts of social defeat across 5 weeks show reduced preference for sucrose.[124]

Environmental stressors

Beyond social contact, there are stimuli in our environment that can act as stressors. Imagine moving from a rural area to a heavily congested urban center. Adjusting to the environmental change can be difficult because of the increased light and noise pollution. Lights and sounds can be used as stressors in animals as well. In one study, mice were exposed to a cue that signaled the delivery of loud, aversive noise. When the cue signaling the noise was presented, mice engaged in increased ethanol-seeking behavior.[125]

Researchers can model childhood stress by placing rodent pups in cages with limited bedding and nesting. Rats raised in this type of environment acquire cocaine self-administration faster but show lower demand intensity in a threshold procedure.[126] Bolton et al. argue that the decreased demand intensity observed in these rats reflects increased drug use vulnerability.[126] How is this the case? Decreased demand intensity could indicate that the hedonic set point has been shifted such that animals do not need as much drug to achieve a reward threshold.

Food deprivation

While food restriction is commonly used to motivate animals to acquire food self-administration, chronic food restriction can act as a stressor. Just think how you feel when you have not eaten anything for most of the day. You may feel "hangry" (portmanteau of hungry and angry). Food deprivation increases the locomotor effects and the direct reinforcing effects of drugs.[127] Drug-seeking and reinstatement of drug-seeking behavior increase during periods of food deprivation.[128] Food restriction can even increase the conditioned rewarding effects of drugs as assessed in CPP.[129]

Restraint

Imagine being confined such that you are unable to move. You most likely would feel stressed (and most likely agitated as well). Restraint is a commonly used stressor in rodents. Rodents can be placed in a contraption that looks like a tube or a small box (see Fig. 11.4 for an example). The animal has no room to move, and there is no way for the animal to escape from the restraint. Chronic restraint stress increases behavioral sensitization,[130,131] and a single exposure to restraint stress increases cocaine-induced hyperactivity 3 weeks after restraint.[132] However, restraint stress can *inhibit* psychostimulant-induced locomotor sensitization if administered 5–30 min following drug treatments.[133]

Restraint stress increases the acquisition rate of drug self-administration[134] and the amount of self-administered drugs.[135] Rodents subjected to restraint stress consume more ethanol,[136] although stress-induced increases in ethanol consumption appear to be dependent on which mouse strain is used. 129SVEV mice consume more ethanol following chronic stress,[137] but C57BL/6J mice often show decreased ethanol consumption following stress[138] while WSC-1 mice are unaffected by restraint stress.[139] In addition to strain differences, stress differentially alters ethanol consumption in rats that have different basal levels of ethanol consumption. Restraint stress increases ethanol consumption in rats that have a moderate preference for ethanol, but stress decreases consumption in rats that have high baseline ethanol consumption.[140] These studies highlight the care that needs to be taken when selecting mouse strains, and they show the impact that individual differences can have on stress-induced changes in drug consumption.

Chronic restraint during extinction of drug self-administration increases drug-induced reinstatement of cocaine-seeking behavior,[141] and stress potentiates reacquisition of drug self-administration following extinction learning.[142] Similarly, chronic restraint stress during

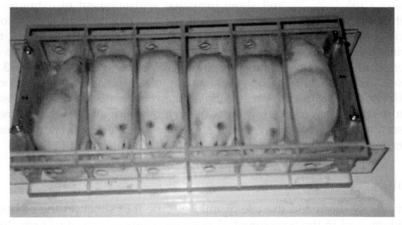

FIGURE 11.4 Restraint stress entails placing rodents in a contraption that prevents them from moving. *The image comes from Fig. 2 of Smith C. Using rodent models to simulate stress of physiologically relevant severity: when, why and how. In: Qian X, ed.* Glucocorticoids: New Recognition of Our Familiar Friend: *IntechOpen. https://doi.org/10.5772/52045. This is an open-access article distributed under the terms of the Creative Commons Attribution 3.0 International License (https://creativecommons.org/licenses/by/3.0/).*

drug withdrawal magnifies the incubation of cocaine craving.[143] An odor previously paired with restraint stress can reinstate heroin-seeking behavior on its own.[134,144]

The effects of restraint on CPP are largely dependent on how long restraint is administered. Whereas acute restraint stress increases CPP,[144–147] chronic restraint stress inhibits or fails to alter the acquisition of CPP.[130,143,145] Reconciling the differential effects of acute and chronic restraint stress on CPP is somewhat difficult. Neither alterations in hedonic set point nor increased anhedonia following chronic restraint stress can be used as viable explanations as both acute and chronic restraint stress potentiate drug self-administration. Chronic restraint stress is known to impair cognitive and memory functions, particularly spatial memories, in animals[148]; as such, animals subjected to chronic restraint stress may not be able to effectively associate the environmental cues of the CPP compartment with the drug.

Forced swimming

Stress can be measured with the *forced swim test* (also known as the *Porsolt test*, named after Dr. Roger Porsolt). As the name suggests, animals are placed in a container of water from which they cannot escape. Researchers measure how much time animals spend swimming and trying to climb the sides of the container to escape. Immobility time is an important variable that is often interpreted as a form of "behavioral despair". As such, the forced swim test has traditionally been used as a model of major depression and serves as a valuable screening tool for potential antidepressant-like drugs.[149] Because the forced swim test involves a stressful event (it should not be difficult to imagine how stressed you would be if you found yourself in the middle of the ocean with nothing to hold on to), some work has examined if forced swimming exacerbates addiction-like behaviors.

Animals subjected to forced swim stress are more sensitive to the sedative effects of ethanol[150] and consume more ethanol[151] (but see[150]), treat cocaine as an inelastic good,[152] develop greater CPP,[151,153–157] and show increased reinstatement of extinguished CPP.[158] Climbing behavior during the forced swim test has specifically been shown to predict demand elasticity of heroin and heroin-seeking behavior.[159] Like restraint stress, the duration of forced swim stress modulates the effects of stress on drug sensitivity. While a single exposure to forced swim stress potentiates morphine CPP, experiencing stress for 3 consecutive days decreases CPP.[160] Additionally, strain differences in ethanol consumption have been observed following forced swim stress, similar to what has been observed with restraint stress. In mouse strains that have a low baseline preference for ethanol, swim stress decreases ethanol consumption.[161] Conversely, swim stress has no effect on ethanol consumption in alcohol-preferring rats while increasing consumption in Wistar rats.[162]

Physical pain

I have previously discussed the use of mild electric foot shock in the compulsive drug-seeking paradigm and in the assessment of risky decision-making in rodents. In these paradigms, the animal can avoid shock by not responding on the manipulandum associated with shock. In inescapable shock paradigms, the researcher can directly shock a rodent's tail or place them in an operant chamber in which a shock is randomly administered to a rodent.

If a researcher does not have access to a shock generator, they can introduce a mildly painful stimulus to a rodent by pinching its tail. Physically painful stimuli increase psychostimulant-induced locomotor activity,[163] increase ethanol consumption,[104,162] increase acquisition of self-administration,[163] and increase CPP.[164,165] In a progressive ratio schedule, intermittent foot shock stress *decreases* nicotine self-administration in adolescents, but not in adults,[110] although this effect is marginal. It is also important to note that adolescent rats earn fewer infusions in the progressive ratio schedule compared to adults, approximately half of the infusions when compared to adults. The significant decrease reported by Zou et al.[110] may reflect an artifact of the low baseline levels of responding observed in adolescents.

Pharmacological stressors

Some drugs act as **anxiogenics**; that is, they increase anxiety levels. Recall that noradrenaline is an important component of the HPA axis and is released in response to stress. Noradrenergic α_2 receptor antagonists (e.g., yohimbine, a chemical derived from the bark of the yohimbe evergreen tree) are commonly used as pharmacological stressors. As α_2 receptors are coupled to G_i proteins, stimulating these receptors decreases norepinephrine levels. Conversely, by inhibiting α_2 receptors, drugs like yohimbine increase norepinephrine levels, thus mimicking the effects of ACTH on the adrenal glands. Noradrenergic α_2 receptor antagonists increase ethanol consumption,[166] increase self-administration,[167–169] increase motivation to self-administer drug,[110] increase CPP,[170] and decrease the rate of extinction.[171–181] One would think that higher doses of α_2 receptor antagonists would further potentiate drug-related behaviors, but that is not always the case. While a lower dose of yohimbine (2.5 mg/kg) and a higher dose of yohimbine (5 mg/kg) increases the reacquisition of oxycodone (0.25 mg/kg) CPP, the higher dose *decreases* the reacquisition of higher doses of oxycodone (2 and 5 mg/kg).[170]

Combining stressors

Realistically, we face multiple stressors at once. Perhaps you must study for a difficult exam while in the process of interviewing for a job. Others may be diagnosed with a serious illness shortly after losing a loved one. Some preclinical studies have determined how the combination of several stressors impacts drug sensitivity. In one study, rats were raised either in isolation, in pairs, or in groups with access to novel objects (i.e., environmental enrichment). Half of the rats were subjected to restraint stress for 2 h followed by placement in cold water for up to 15 min. Half of the rats served as a control, never being subjected to stress. Rats were then trained to self-administer cocaine. Rats exposed to restraint/cold stress self-administered more cocaine during acquisition; however, isolated rats increased their self-administration at a faster rate during acquisition compared to nonstressed isolated rats.[174] In another study, rats were first subjected to forced swim stress. Later, some rats were given restraint stress while others were not restrained. Rats were then tested for amphetamine CPP. Restraint stress increased amphetamine CPP, but only in rats that "gave up" sooner in the forced swim test (known as "passive" rats). Furthermore, passive rats showed greater CPP compared to "active" rats, regardless if they received restraint stress.[175] Other stressor combinations that increase drug sensitivity include early-life maternal separation and adolescent social isolation,[82] maternal separation and cold exposure,[176] social isolation and social defeat,[58] and pharmacological stress and restraint stress.[177]

Early-life stress and adult stress can interact in such a way to paradoxically blunt the direct reinforcing effects of drugs. For example, while maternal separation increases methamphetamine self-administration in rats, maternally separated rats that experience variable stressors (e.g., tail pinch, loud noise, forced swimming, restraint, etc.) during adulthood self-administer similar amounts of methamphetamine compared to control animals.[86] Similarly, restraint stress decreases ethanol consumption in animals that have experienced maternal separation.[92] While maternal separation and social defeat increase ethanol consumption when administered independently, animals exposed to both stressors do not increase ethanol consumption later in life.[89] Furthermore, social isolation has been shown to prevent social defeat stress-induced increases in cocaine self-administration in a progressive ratio schedule.[58] One argument that has been proposed to explain this seemingly paradoxical finding comes from Lewis et al.[86] While early-life stress activates the HPA axis, which can potentiate addiction-like behaviors, chronic stress administered during adulthood decreases the activation of the HPA axis in animals subjected to early-life stress. Previous studies have found that chronic variable stress in adulthood reverses HPA axis reactivity.[178] How does this translate to real life? Some individuals experience hardships early in life, which may better prepare them to handle adversity later in life. Thus, these individuals may habituate to stressful situations if they have previously experienced them.

Predictable and unpredictable stress

In addition to combining stressors, one method to increase ecological validity is to expose animals to unpredictable stress. So far, the studies I have cited in this section use predictable stress; that is, each animal receives the same type of stressor(s) throughout the experiment. In real life, stressors can occur when we do not expect them (e.g., the sudden loss of a loved one). Researchers can model unpredictable stress by varying the type of stressor the animal receives on a given day, and they can expose the animal to the stressor at random times during the day. Some research has shown that rats exposed to unpredictable stress are more sensitive to the locomotor stimulant and the conditioned rewarding effects of cocaine.[179] However, other studies have shown no effects of unpredictable stress on drug CPP.[180] Some studies have even shown that chronic exposure to mild unpredictable stress can reduce the rewarding effects of drugs.[181]

Inescapable versus escapable stress

In the paradigms that have been covered so far, stress is inescapable. The rodent cannot escape from restraint or from isolation housing. What if the animal could avoid a stressor? Animals that experience escapable stress show decreased CPP,[165] and they learn to extinguish drug self-administration faster than animals exposed to inescapable stress and control animals.[182] Why does it matter if stress is escapable or not? If they are both stressors, shouldn't they increase addiction-like behaviors? Imagine you have an upcoming exam. You may feel stressed about the situation. However, you know that you can prepare for the exam by studying, and you know you will be able to relax after the exam. Contrast this with the ongoing stress an individual faces when experiencing something like homelessness. These individuals have a much more difficult time escaping their situation. Getting a job is extremely difficult because having an address is typically a requirement for securing employment. Because there does not appear to be any recourse from the stress, these individuals may be at greater risk for substance use. Overall, stress, *per se*, is not a risk factor for SUDs; instead, one's perception of how much control they have over the stressor is an important determinant of subsequent drug use and relapse.

Stress-induced reinstatement

Previously, you learned about drug-induced and cue-induced reinstatement, and you have learned that animals subjected to stressors are often more likely to show increased relapse-like behaviors in these reinstatement tests. There is a third major type of reinstatement paradigm that can be used: *stress-induced reinstatement*. A researcher typically uses restraint, shock, or a pharmacological stressor (e.g., yohimbine) prior to the reinstatement session as opposed to using a priming injection or a drug-paired cue to reinstate drug-seeking behavior. Overall, the following stressors have been shown to reinstate drug-seeking behaviors: social defeat,[183–186] food deprivation,[187] restraint,[186,188–194] forced swimming,[147,187,188,195–205] physical pain,[126,173,186,203,206–212] and noradrenergic receptor antagonists.[166–169,173,205,208,209,213–222] Additionally, research has shown that stimuli that become paired with social defeat can reinstate drug-seeking behavior.[223] These results should not be too surprising as I have already mentioned that stress is a critical determinant of relapse.

Quiz 11.1

1. What is the mechanism of action of most pharmacological stressors?
    a. GABA receptor agonist
    b. Noradrenergic α_1 antagonist
    c. Noradrenergic α_2 agonist
    d. Noradrenergic α_2 antagonist
2. How do hormones differ from neurotransmitters?
3. What is the major stress hormone in humans?
    a. Corticosterone
    b. Corticotropin-releasing hormone
    c. Cortisol
    d. Mineralocorticoid
4. Which of the following is NOT commonly used as a stressor in animals?
    a. Antagonist social interaction
    b. Physical pain
    c. Sexual deprivation
    d. Social isolation

5. True/False: perceived lack of control over stress may be a better predictor of substance use compared to stress itself.

Answers to quiz 11.1

1. d — Noradrenergic α_2 antagonist
2. Hormones and neurotransmitters are highly similar. Hormones travel through the blood stream as opposed to traveling across a synapse. Hormones also bind to receptors located on glands/organs instead of receptors on neurons.
3. c — Cortisol
4. c — Sexual deprivation
5. True

Neuromechanisms mediating relationship between stress and addiction

Earlier, I described the HPA axis and how it regulates the release of glucocorticoids and noradrenaline following stress exposure. Naturally, research has examined how glucocorticoids mediate drug sensitivity. Yet, stress has wide-reaching effects on the nervous system. After discussing how stress and drugs can similarly activate the HPA axis, I will detail how stress can alter other neurotransmitter systems, thus further potentiating the effects of stress on addiction.

HPA axis

Stressors are not the sole activator of the HPA axis. In humans, acute drug administration typically increases the activity of the HPA axis and the autonomic nervous system; however, long-term drug use is associated with decreased basal levels of cortisol.[224] This finding is similar to what has been observed in animals exposed to chronic stress in adulthood after experiencing early-life stress. Dependent cocaine users have higher glucocorticoid concentrations compared to nonusers.[6] Heavy cannabis users show an increased startle response to unpredictable shock delivery, further suggesting that increased substance use leads to altered stress reactivity.[225] Withdrawal from nicotine also increases startle responses to shock delivery, but this startle response is blunted when participants are told they will be able to smoke a cigarette following the procedure.[226]

In animals, acute drug administration increases activation of the HPA axis and increases corticosterone levels,[227,228] but blunts the effects of stress on HPA activity.[229] Conversely, extended access to cocaine decreases plasma levels of corticosterone but potentiates stress-induced increases in corticosterone.[230] Acute withdrawal from cocaine increases plasma levels of corticosterone and CRF mRNA levels in the paraventricular nucleus of the hypothalamus during restraint.[231] Heroin withdrawal increases a startle response to an acoustic stimulus, which is blocked by a CRF receptor antagonist.[232]

Glucocorticoid levels are associated with treatment retention, as increased cortisol levels are linked to premature dropout from treatment.[233] Corticosterone and CRH levels rise during extinction and during reinstatement tests.[183,212,229] In particular, increased CRH is observed in the ventral tegmental area (VTA), which leads to increased glutamate levels that regulate the release of dopamine in the mesocorticolimbic pathway.[212] We will further discuss the effects of stress on the dopaminergic and the glutamatergic systems later in this chapter.

To further elucidate the role of the HPA axis on addiction-like behaviors, one can employ behavioral pharmacology experiments in which CRF or glucocorticoid receptor ligands are administered to an animal. Direct administration of CRH reinstates drug-seeking behavior[213] while blocking CRF receptors reduces ethanol consumption[234] and attenuates drug CPP.[177] Blocking these receptors also decreases stress-induced increases in drug self-administration,[97,98,111,113,169] CPP,[102,153] and reinstatement.[158,169,194,201]

The role of the noradrenergic system has already been covered somewhat in this chapter as pharmacological stressors commonly target this neurotransmitter system (e.g., yohimbine). Chapter 2 also discussed the role of the noradrenergic system in addiction-like behaviors. There is additional research that supports the role of the noradrenergic system in stress-induced increases in drug sensitivity. For example, animals that have had an adrenalectomy show blunted stress-induced behavioral sensitization.[129] Not only does stress cause the release of noradrenaline from the adrenal glands, but it has been shown to alter the noradrenergic system in the brain. A single exposure to restraint stress increases norepinephrine efflux in the medial prefrontal cortex (mPFC)[235] while long-term maternal separation decreases norepinephrine levels in the amygdala.[236] Isolation stress decreases presynaptic noradrenergic function in the hippocampus,[67] which may account for some of the memory deficits associated with long-term stress.

If noradrenergic α_2 receptor antagonists can increase stress-like responses, can agonists at these receptors protect against stress-induced increases in drug-seeking behaviors? There is support that α_2 receptor agonists are efficacious in blocking stress-induced drug craving and reinstatement,[167,209,217] and they can block withdrawal-induced startle responses.[232] While noradrenergic α_2 receptors are inhibitory, α_1 and β_2 receptors are excitatory. Antagonists at these receptors block stress-induced reinstatement.[146,201,205,209] I want to mention that although yohimbine is an antagonist at α_2 receptor receptors, there is some evidence that yohimbine's ability to modify drug-seeking behavior is mediated by a nonnoradrenergic mechanism. Mice lacking the α_2 receptor extinguish cocaine CPP at a similar rate as control animals.[172] Similarly, lesions to noradrenergic cell bodies fail to attenuate yohimbine-induced potentiation in ethanol self-administration or ethanol-seeking behavior.[168] These findings show the need to examine other neurotransmitter systems on stress-induced drug seeking and relapse-like behavior. Before discussing these other neurotransmitters, I want to discuss the neuroanatomical changes that can occur following stress exposure.

Shared neuroanatomical underpinnings of stress and addiction

Early life stress increases Fos expression in several regions implicated in SUDs, including the nucleus accumbens (NAc), the amygdala, and the lateral habenula.[84,125] Acute stressors also activate the NAc, the amygdala, the mPFC, the hypothalamus, and the hippocampus.[222,237,238] Stress potentiates cocaine-induced c-Fos expression in the NAc and the amygdala.[64] Repeated restraint stress during cocaine withdrawal increases the neuronal activity of the basolateral amygdala (BLA), an effect that is additive when compared to restraint stress or cocaine exposure alone.[239] However, isolation stress blunts c-Fos expression in the mPFC and the VTA during heroin withdrawal.[62] While stress can activate several areas associated with addiction, lesions to or temporary inactivation of these regions can reverse the effects of stress on addiction vulnerability. Specifically, lesions to the NAc core decrease reactivation of CPP following stress exposure,[240] and inhibiting the frontal cortex impairs the ability of stress to reinstate drug-seeking behavior.[206] Lesions to or inactivation of the lateral habenula block stress-induced ethanol seeking and cocaine seeking.[218,241]

In addition to increasing activity in the NAc, early-life stress increases dendritic length, dendritic head diameter, and dendritic spine density in this region[78,242] (Fig. 11.5), and social isolation increases long-term potentiation (LTP) in the VTA.[243] Although LTP is enhanced in the VTA, maternal separation decreases the number of neurons in the VTA.[83] These neural adaptations may explain why drugs are more reinforcing in animals subjected to early-life stress. In contrast to the NAc, early-life stress decreases dendritic length in the frontal cortex[242] (but see[78]). The altered functioning in the frontal cortex may account for the cognitive deficits associated with stress. Maternal separation increases the number of neurons in the amygdala,[83] which may explain increased anxiety-like behaviors observed in animals subjected to stress.

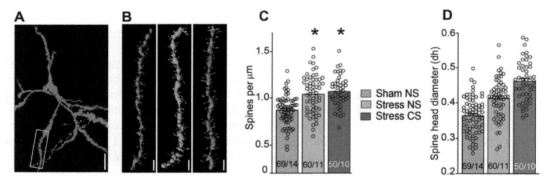

FIGURE 11.5 The effects of stress on dendritic spine morphology in medium spiny neurons of the NAc. (A) shows a representative neuron used in the analysis of dendritic spine morphology. (B) shows the three main conditions of the experiment. The left panel shows control animals that never encountered stress. To understand the next two panels, a description of the experimental design is needed. Two groups received restraint stress, which was paired with the scent of lemon. The middle panel shows a representative dendrite of a rat that was exposed to the scent of sandalwood following restraint stress. Sandalwood is a neutral stimulus that was never explicitly paired with restraint stress. The right panel shows a representative dendrite of a rat that was exposed to the lemon scent following restraint stress. Panels (C) and (D) show the number of dendritic spines and dendritic spine head diameter for each group, respectively. Experiencing stress increases the number of dendritic spines, regardless if rats were exposed to the neutral stimulus (sandalwood scent) or to the conditioned stimulus (lemon scent). Stress also increases dendritic spine head diameter, but this effect is further increased in rats exposed to the conditioned stimulus. *Image modified from Fig. 3 of Garcia-Keller C, Carter JS, Kruyer A, et al. Behavioral and accumbens synaptic plasticity induced by cues associated with restraint stress. Neuropsychopharmacology. 2021;46(10):1848–1856. https://doi.org/10.1038/s41386-021-01074-7. Copyright 2021.*

The hippocampus appears to be highly sensitive to the effects of stress. Individuals with PTSD often experience memory deficits resulting from a damaged hippocampus.[244] At a cellular level, stress reverses morphine-induced LTP in the hippocampus, leading to long-term depression (LTD).[245] Early-life stress in the form of maternal separation increases dendritic arborization in the hippocampus, suggesting that exposure to stress during childhood can cause part of the brain to mature too quickly.[246] Related to this finding is adolescent social isolation increases activity of the ventral hippocampus.[247]

The bed nucleus of the stria terminalis (BNST) is an important structure that serves as an intersection between the stress and the reward systems. The BNST receives noradrenergic inputs from other regions.[248] Not surprisingly, this region mediates drug reinforcement. Various drugs increase noradrenergic activity in the BNST,[249] and chronic drug self-administration causes an upregulation of the norepinephrine transporter (NET) in the BNST.[250] Ethanol and stress affect intra-BNST CRF neurons in a similar fashion.[251] Inactivation of the BNST reduces stress-induced reinstatement[214] as does blocking noradrenergic β receptors[252] and CRF receptors[253] in this region.

Dopamine and serotonin

Beyond the HPA axis and glucocorticoids, stressors produce widespread changes in other neurotransmitter systems. Because dopamine is often linked to SUDs, I will start by discussing the effects of stress on the dopaminergic system. Acute stress increases dopamine release and reuptake in the NAc, an effect that is dependent on activation of the HPA axis,[254] while decreasing tyrosine hydroxylase (precursor to dopamine) levels.[190] The decreased tyrosine may result from the fact that increased dopamine is released during stress; that is, tyrosine is being converted into dopamine at such a quick rate that there is not enough time to replenish tyrosine hydroxylase levels. Increases in NAc dopamine may also arise from the increased neuronal firing observed in the VTA following stress exposure.[66] A single exposure to restraint stress also increases dopamine metabolism in the NAc shell,[255] further providing support for increased dopaminergic activity in this region. However, some discrepancies have been observed following chronic stress. While repeated stress decreases basal dopamine levels in the NAc shell,[256] the PFC,[257] and the amygdala,[236] social isolation stress increases dopamine levels in the NAc and the mPFC.[258,259] In contrast to acute stress, chronic stressors increase tyrosine hydroxylase levels in the NAc, as well as in the dorsal striatum, the hippocampus, and the hypothalamus.[85,260,261] This may occur because prolonged stress leads to compensatory changes in the dopaminergic system; as already covered in this book, prolonged drug exposure modifies neurotransmitter systems by downregulating or upregulating certain receptors.

Directly relevant to addiction, stress potentiates drug-induced increases in NAc and PFC dopamine[65,112,131,257,262,263] and increases dopamine turnover in the NAc following drug exposure.[77] However, stress has been shown to decrease cocaine-induced dopamine release in the NAc shell and the VTA.[256] These results show that stress can differentially alter drug-induced increases in dopamine across various brain regions.

Concerning alterations to dopamine receptors, both maternal separation and social isolation increase NAc and dorsal striatum dopamine D_2-like receptor expression while decreasing D_1-like receptor expression.[82,242,261] Drug administration can further potentiate stress-induced upregulation of D_2-like receptor expression in the NAc.[77] In contrast to the NAc, early-life stress downregulates dopamine D_2-like receptors in the frontal cortex,[242] perhaps due to increased presynaptic activity of dopaminergic neurons in the NAc.[67] Enhanced presynaptic dopamine functioning is also observed in the amygdala of stressed animals.[67]

Behavioral pharmacology/neuroscience studies provide additional evidence that dopamine receptors mediate the influence of stress on addiction-like behaviors. Blocking D_1-like and/or D_2-like receptors can blunt stress-induced increases in CPP,[118,143] as well as stress-induced reinstatement and reinstatement-induced increases in corticosterone levels.[140,183–185,198,207,211,213] Interestingly, while antagonizing D_1-like receptors blocks stress-induced reinstatement, chronic administration of a D_1-like receptor antagonist during extinction potentiates reinstatement.[139] Mice lacking D_2-like receptors show less cocaine-induced behavioral sensitization, cocaine seeking, and relapse-like behavior following stress.[264] However, mice with lower D_2-like receptor expression consume more ethanol following chronic stress.[265] Dopaminergic activity in the hippocampus mediates stress-induced reinstatement of morphine CPP, as blocking D_1-like or D_2-like receptors in either the dentate gyrus or the CA1 region attenuates stress-induced reinstatement.[266,267]

While maternal separation decreases basal levels of the dopamine transporter (DAT),[76] social isolation during adolescence increases striatal DAT expression.[261] Stress also increases DAT uptake in striatal regions,[268] mirroring what has been observed in the NAc. Stress upregulates DAT in animals treated with cocaine.[77] Ethanol consumption decreases monoamine oxidase (MAO) levels in the NAc of animals subjected to chronic maternal separation.[269] This means that MAO is not able to metabolize dopamine, allowing for excess dopamine levels. In sum, stress has major effects on the dopaminergic system, which may account for why stress increases the reinforcing effects of drugs and increases relapse-like behaviors.

In addition to dopamine, there is support for the role of serotonin in mediating the effects of stress on drug seeking. Maternal stress upregulates serotonin transporter (SERT) and serotonin 5-HT$_{1A}$ receptors while decreasing serotonin levels in the amygdala.[236,270] Isolation stress also decreases basal serotonin levels in the NAc,[271] which is consistent with the increased presynaptic serotonergic activity observed in this region following stress.[67] Remember, presynaptic receptors are autoreceptors that serve as negative feedback loops to limit the synthesis and release of additional neurotransmitter. Long-term maternal separation increases serotonin levels in the raphe nuclei, which is where most serotonergic cell bodies originate.[236] In addition to upregulating SERT and the inhibitory 5-HT$_{1A}$ receptor, stress increases SERT function[156] and increases 5-HT$_{1A}$ functioning in the raphe nuclei.[54] These results suggest that inhibited serotonergic activity resulting from stress increases drug sensitivity. Behavioral pharmacology studies have provided support for this hypothesis. Blocking 5-HT$_{1A}$ receptors prevents the ability of yohimbine to increase ethanol self-administration and ethanol reinstatement.[168] In contrast to 5-HT$_{1A}$ receptors, stimulating the excitatory 5-HT$_{2C}$ receptor decreases stress-induced reinstatement.[216] Finally, activating GABA receptors in the raphe nuclei, thus inhibiting serotonergic cell bodies, potentiates stress-induced reinstatement, an effect that is reversed by GABA receptor antagonism.[200]

Monoaminergic activity in the mPFC is important for stress-induced increases in drug-taking behaviors. Restraint stress increases dopamine release in the mPFC[235] and increases dopamine metabolism in the mPFC, and this effect is potentiated in animals that have received lesions to serotonergic cell bodies, similar to what has been observed in the NAc shell.[255] Long-term stress decreases dopamine and serotonin levels in the mPFC.[256] Blockade of intra-mPFC serotonin 5-HT$_{1A}$ receptors attenuates the effect of stress on cocaine CPP[272] while blocking dopamine D$_1$-like receptors in this region prevents stress-induced reinstatement.[206]

Glutamate

Like drugs, stress increases glutamatergic activity. A single exposure to restraint stress increases basal levels of NAc core and mPFC glutamate,[131,193] and repeated social defeat stress increases glutamatergic synaptic plasticity in the VTA.[119] Animals with a history of drug exposure show greater stress-induced glutamate release in the mPFC.[273] Restraint stress administered before being tested for stress-induced reinstatement of CPP increases NAc glutamate levels.[274] Isolation stress during adolescence alters glutamate release from the hippocampus to the NAc.[275]

Along with altered glutamate release, maternal separation alters the expression of AMPA receptor subunits in the NAc and the VTA.[81] Restraint stress administered before being tested for stress-induced reinstatement of CPP upregulates the GluN2B subunit of the NMDA receptor, as well as glutamate transporters, in the mPFC.[192] Recall that the GluN2B has been implicated as a major mediator of addiction-like behaviors as blocking GluN2B-containing NMDA receptors decreases relapse-like behavior. Outside of the mesocorticolimbic pathway, social stress downregulates NMDA receptor expression in the hippocampus,[276] specifically the GluN1 subunit.[120] This may partially account for the memory deficits observed in those that have experienced chronic or traumatic stressors.

Attenuating extracellular glutamate levels is one way to mitigate the effects of stress on drug use. NMDA receptor channel blockers reverse the effects of stress-induced increases in locomotor activity[263] and cocaine CPP.[116] Additionally, blocking NMDA receptors in the NAc core or in the VTA decreases stress-induced reinstatement of cocaine CPP.[144,277] In Chapter 7, I mentioned that N-acetylcysteine attenuates the reinforcing effects of drugs by restoring glutamate homeostasis. As N-acetylcysteine decreases striatal glutamate levels,[278] it is not surprising then that N-acetylcysteine prevents amphetamine-induced hyperactivity in mice exposed to isolation stress[279] and stress-induced reinstatement of ethanol or cocaine seeking.[280]

Acetylcholine

Earlier you learned about the fight-or-flight response, which is controlled by our sympathetic nervous system. Norepinephrine is not the only neurotransmitter that controls the sympathetic nervous system. Neurons exiting the central nervous system release acetylcholine, which then binds to nicotinic receptors located on noradrenergic neurons. Activation of nicotinic receptors causes the noradrenergic neuron to generate an action potential, sending a message to the target glands/muscles. In other words, when we encounter a stressor, acetylcholine levels increase. In the central nervous system, stress increases acetylcholine levels in the frontal cortex, the striatum, and the hippocampus.[272,281–283] CRH has recently been shown to activate cholinergic interneurons in the striatum and in the NAc.[284] The laterodorsal tegmental nucleus (LDT) contains a large collection of cholinergic neurons that project axons to several brain regions, including the VTA. Intra-LDT infusions of noradrenergic α_2 or β receptor antagonists prevent restraint stress from increasing cocaine CPP.[144]

Some research has determined if ligands that bind to acetylcholine nicotinic or muscarinic receptors are efficacious in preventing stress-induced increases in drug reinforcement. A partial agonist at nicotinic receptors reduces yohimbine-induced nicotine seeking.[285] Blocking nicotinic receptors also attenuates stress-induced reinstatement of ethanol CPP,[188] but not ethanol-seeking behavior.[215] In the VTA, blocking nicotinic and muscarinic acetylcholine receptors prevents restraint stress from increasing cocaine CPP.[144] Microinjections of a nicotinic receptor antagonist into the hippocampus or into the mPFC attenuates stress-induced potentiation of nicotine CPP, although intra-amygdalar nicotinic receptor blockade exacerbates the effects of stress on CPP.[286]

Endocannabinoids

The role of endocannabinoids on stress is an emerging area of research. Endocannabinoid levels are lower in the striatum, the PFC, and the amygdala of animals that have experienced stress.[87,287] Recall that endocannabinoids are metabolized by fatty acid amide hydrolase (FAAH). Inhibiting FAAH, leading to increased endocannabinoid levels, attenuates stress-induced reinstatement.[288] These results suggest that stress promotes a hypoactive endocannabinoid system, which promotes drug seeking. However, early-life stress decreases NAc and dorsal striatum FAAH levels, leading to greater concentrations of endocannabinoids in these regions.[242,289] This discrepancy may be due to when the stress is administered (early life vs. later in life). Prolonged stress may lead to a downregulation of FAAH to compensate for the reduced endocannabinoid levels resulting from stress.

Stressors also modify cannabinoid CB_1 receptor expression in various brain regions, with increases observed in the NAc and the PFC[82,88] (note: Romano-López et al.[88] found decreased CB_1 receptor expression in the PFC) and decreases observed in the dorsal striatum, the amygdala, and the hypothalamus.[289,290] The ability of corticosterone to increase cocaine seeking is largely controlled by CB_1 receptors,[291] specifically in the prelimbic region of the mPFC.[292] Directly stimulating intra-mPFC CB_1 receptors potentiates stress-induced cocaine-seeking whereas blocking these receptors, whether systemically or directly into the mPFC, blunts this effect.[292,293] Systemic and intra-NAc administration of a CB_1 receptor antagonist prevents stress-induced reinstatement of cocaine CPP and corresponding increases in extracellular glutamate levels.[204,274]

Recent research has examined the mediating role of *fatty acid binding proteins* (*FABPs*) on stress-induced CPP. FABPs are part of the endocannabinoid system and are proposed to act as intracellular transporters for endocannabinoids. Specifically, they transport endocannabinoids to FAAH for metabolism. Mice lacking the gene encoding for FABPs fail to develop CPP following stress.[294] These results provide further support that low endocannabinoid levels help promote stress-induced drug seeking.

Peptides

There are numerous peptides, with the most well-known one being the opioid peptide class. Drugs like morphine and heroin are opioids. The ability of stress to increase the rewarding valence of drugs may be partially mediated by opioid receptors. Early-life stress increases gene expression of mu, kappa, and delta opioid receptors in the striatum[295] and increases kappa opioid receptor expression in the amygdala.[296] Blocking mu and delta opioid receptors decreases the effects of stress on ethanol CPP,[164] and delta receptor antagonists decrease stress-induced reinstatement of ethanol seeking.[297] Although not a traditional opioid receptor, the nociceptin receptor blocks stress-induced reinstatement of ethanol-, but not cocaine-, seeking behavior.[210,298]

Kappa receptors have received considerable attention and have been implicated as a critical mediator of stress-induced increases in addiction vulnerability. Mice lacking the prodynorphin gene fail to increase ethanol consumption or show potentiated cocaine CPP following stress.[151,154] Administration of a kappa receptor agonist potentiates stress-induced cocaine CPP while a kappa receptor antagonist blocks this effect[117,154,155,157] (but see[164]) Kappa receptor antagonists also block stress-induced increases in drug self-administration[152,202] and stress-induced reinstatement.[199,203,222]

One potential mechanism underlying the role of kappa receptors on stress and drug sensitivity centers around serotonergic activity in the NAc. Kappa receptors regulate the activation of serotonin 5-HT$_{1B}$ receptors in the NAc. By stimulating kappa receptors, increased dynorphin release subsequently decreases the activity of 5-HT$_{1B}$ receptors in the NAc, leading to decreased serotonergic tone.[299] As already covered, decreased serotonergic activity is implicated in the effects of stress on drug use. Somewhat similarly, animals with a history of drug exposure show greater GABAergic activity in the raphe nuclei following pharmacological stress.[219] Additionally, blocking kappa receptors prevents stress from increasing SERT activity.[156]

While opioids are a major peptide class implicated in SUDs, there are other important peptides that we have not yet covered in this book. The first one is called *cocaine- and amphetamine-regulated transcript (CART) peptide*. CART gets its name because drugs like cocaine and amphetamine increase its activity.[300] What exactly is CART? This peptide has been shown to regulate dopamine neurotransmission in the brain.[301] CART levels can increase or decrease depending on what type of stressor is applied and which brain region is examined. Restraint stress decreases CART levels in the NAc, an effect that is observed in rats that stop swimming sooner in the forced swim task.[175] Maternal separation decreases CART levels in the PVN of the hypothalamus[302]; likewise, social isolation stress decreases CART expression in the hypothalamus, the NAc, and the VTA.[303,304] In the central nucleus of the amygdala (CeA), restraint stress increases CART peptide and dopamine levels,[175,305] and chronic stress increases CART peptide in the hippocampus.[305] As stress decreases CART levels in the NAc and the PVN, what do you think happens if CART is directly injected into either region? Intra-PVN injections of CART attenuate stress-induced ethanol consumption.[306]

One mechanism that may account for stress-induced relapse is *orexin* (also known as hypocretin), a neuropeptide best known for its role in arousal and the sleep-wake cycle (narcolepsy has been proposed to be an orexinergic disorder). Stress activates orexinergic neurons in the lateral hypothalamic nucleus, which increases orexin A levels in the VTA. Orexin A then binds to orexinergic receptors on dopaminergic neurons. Activation of these receptors causes 2-arachidonoylglycerol (2-AG) (an endocannabinoid) to inhibit GABAergic neurons that typically inhibit dopaminergic neurons.[307] More recent research has shown that *neuropeptide S* (also involved in arousal and the sleep-wake cycle) is released during stress, which then activates orexinergic neurons in the lateral hypothalamic nucleus.[308] Long-term stress also appears to dysregulate the orexinergic system, as rats exposed to early-life stress show a blunted orexinergic response following cocaine administration.[125] Further support for the role of orexin in stress-induced reinstatement comes from behavioral pharmacology studies assessing the effects of orexin receptor antagonists on this form of drug seeking. Systemic blockade of orexin receptors prevents stress-induced reinstatement.[309] Blocking orexin receptors in the NAc, the VTA, or the CA1 region (but not the dentate gyrus) of the hippocampus inhibits the ability of stressors to reinstate morphine CPP.[187,195–197] Inhibiting orexin receptors in the CeA also prevents yohimbine-induced reinstatement of cocaine-seeking behavior.[220] Blocking neuropeptide S receptors prevents stress-induced reinstatement.[221]

The orexinergic system also appears to mediate the effects of yohimbine on delaying the extinction of cocaine CPP. In particular, yohimbine decreases excitatory neurotransmission in the BNST,[171,172] an effect that is blocked by orexin receptor antagonists and absent in orexin knockout mice.[171] Conrad et al. also found that an orexin 1 receptor antagonist prevents yohimbine's ability to delay extinction.[171]

Finally, I want to mention a neuropeptide called *oxytocin*. Often referred to as the "love hormone", oxytocin plays a critical role in pair bonding in mammals. It is also involved in producing uterine contractions and is responsible for the letdown reflex (ejection of milk from the nipples). Food deprivation stress leads to a concomitant decrease in oxytocin levels and overexpression of oxytocin receptors.[128] Oxytocin administration reverses stress-induced increases in ethanol consumption,[106] drug CPP,[103] and reinstatement.[189,191–193] The beneficial effects of oxytocin may be partially due to its ability to inhibit stress-induced increases in mPFC glutamate.[193]

Hedonic allostasis model of addiction revisited

In Chapter 5, I introduced Dr. George Koob's hedonic allostasis model of addiction. Now that you understand the neurobiological underpinnings of stress, it is time to discuss this model in its entirety. According to Koob,[310] addiction consists of three stages: (1) binge/intoxication, (2) withdrawal/negative affect, and (3) preoccupation/anticipation. The first stage is largely driven by positive reinforcement, incentive salience, and habit formation. As such, the basal ganglia and the ventral striatum are highly involved during this initial stage. Recall that continued drug use leads to compensatory changes in the brain, leading to withdrawal symptoms upon discontinuation of drug use. Individuals often feel dysphoric and experience physical symptoms (e.g., flu-like symptoms during opioid withdrawal). Withdrawal symptoms can motivate an individual to continue drug use, even if they no longer enjoy using the drug. This is negative reinforcement, and according to Koob, this stage is largely controlled by the amygdala and its surrounding structures. The final stage reflects the influence conditioned stimuli have on drug seeking. These cues can precipitate intense cravings, causing the individual to exert much of their efforts in procuring more drugs. The preoccupation/anticipation stage is mediated by the PFC. You can think of these three stages as a circle. As individuals continuously use a substance, they are more likely to experience withdrawal symptoms upon cessation. At the same time, stimuli and environmental cues become paired with the drug, which can elicit withdrawal symptoms and cravings when presented alone. As the individual goes through withdrawal, they become focused on using more drugs. This ultimately leads back to the binge/intoxication stage, and the process repeats itself. Koob has often conceptualized his model as a downward spiral.

So, what does stress have to do with the hedonic allostasis model of addiction? Stress is proposed to play a critical role in each stage of the addiction cycle.[311] As already covered in this chapter, stress hormones increase the self-administration of drugs. CRH also increases incentive salience. For example, injecting CRH directly into the NAc increases operant responding when presented with a conditioned stimulus (e.g., tone).[312] During drug withdrawal, the HPA axis becomes dysregulated. Elevated stress hormones are observed during withdrawal, which can contribute to the negative affect individuals experience. Because the HPA axis becomes dysregulated, excess glucocorticoid levels impair cognitive processing in the PFC, leading to inhibitory control deficits. This then promotes the preoccupation that is frequently observed in individuals with SUDs. The upregulated glucocorticoid receptor levels observed during prolonged abstinence may explain why stress can serve as a major trigger for relapse. Fig. 11.6 shows the contributions of the HPA axis to each stage of the addiction cycle.

Methods for reducing stress

By now, I hope you appreciate the role stress can have on the addiction process. Because stress is a major risk factor for relapse, efforts are needed to reduce stress in individuals with a SUD. I will not be able to discuss every intervention that has been reported in the literature. Instead, I will focus on some of the major interventions that have been studied. Earlier in this chapter, you read that coping strategies can modulate the relationship between stress and drug use. In HIV-positive adults that experience childhood sexual abuse, receiving a coping intervention decreases ethanol and cocaine use to a greater extent compared to those that do not receive the intervention.[313] This makes sense, but the question is: "what are effective coping strategies for dealing with stress?". Some treatment interventions that can decrease stress and drug use have previously been covered in this book. One such treatment is cognitive-behavioral therapy (CBT). Recall that in CBT individuals identify problematic thought processes that can lead to maladaptive behaviors. In this case, individuals work on identifying the warning signs of too much stress, and they work with the therapist to develop coping strategies to reduce stress. In one study, Milosevic et al.[314] used *integrated group CBT* to treat individuals with a comorbid mood/anxiety disorder and SUD. Individuals received 12 sessions in which they learned about symptoms of mood disorders, anxiety disorders, and SUDs; costs and benefits of using substances to treat mood/anxiety symptoms; associating triggers with moods, physical sensations, and behaviors; cognitive

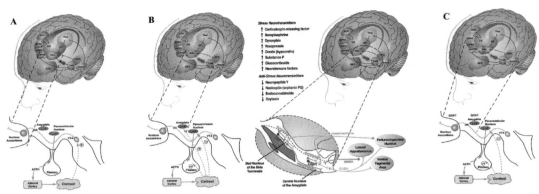

FIGURE 11.6 Influence of the hypothalamic-pituitary-adrenal (HPA) axis during each stage of the addiction cycle: binge/intoxication (A), withdrawal/negative affect (B), and preoccupation/anticipation (C). In (A), cortisol that is released from the adrenal cortex stimulates neurons in the ventral tegmental area (VTA). These neurons activate the nucleus accumbens (NAc). The HPA axis potentiates the activity of the basal ganglia (depicted in blue), leading to habitual drug use. In (B), the HPA axis is highly activated. Cortisol travels to the pituitary and the hypothalamus to control the release of additional cortisol. However, cortisol stimulates corticotropin-releasing factor (CRF) receptors in the amygdala. This increases the stress response in the extended amygdala (in red), which includes the bed nucleus of the stria terminalis (BNST) and the NAc. The right image of (B) shows a magnified view of the extended amygdala and its connections with other brain regions. The elevated stress hormone levels during withdrawal influence the negative affect individuals feel upon cessation of drug use, fueling the "need" to continue using. In (C), long-term abstinence leads to glucocorticoid receptor (GCR) upregulation, which has a major effect on frontal cortical structures, hippocampus, and insula (in green). The impairment observed in these regions may explain the preoccupation with drug use/drug cravings, inhibitory control deficits, and memory impairments observed in those with a history of chronic drug use. This finding may also account for why stress can serve as a trigger for relapse. *Image comes from Figs 2, 7, and 11 of Koob GF, Schulkin J. Addiction and stress: an allostatic view. Neurosci Biobehav Rev. 2019;106:245—262. https://doi.org/10.1016/j.neubiorev.2018.09.008, Copyright 2019, with permission from Elsevier.*

distortions; developing balanced thoughts; strategies for coping with urges and cravings; confronting fearful situations; and problem-solving. Individuals also completed homework assignments during treatment. Results showed that treatment decreased stress and excessive ethanol consumption, although drug use did not change following treatment. CBT is also effective in treating comorbid PTSD and alcohol use disorder.[315] When combined with contingency management, CBT reduces tobacco use in individuals exposed to trauma.[316]

Another treatment that has already been discussed is mindfulness training. Mindfulness training has been shown to reduce stress in smokers.[317] Additionally, in individuals with low SES, mindfulness has an indirect effect on smoking lapse during a quit attempt; that is, mindfulness mitigates the effects of stress on smoking.[318] African American smokers that have a higher mindfulness disposition are less likely to consume ethanol in response to stress.[319] Similarly, individuals with high levels of trait mindfulness show less attentional bias to ethanol cues and experience less stress than those low in mindfulness.[320] In college students, mindfulness is negatively correlated with ethanol problems and stress.[321] Breath counting, a form of mindfulness training, reduces stress-induced ethanol seeking in students.[322] Compared to CBT, individuals with cocaine and/or alcohol use disorder have lower anxiety following stress provocation when given mindfulness training, although drug craving is similar across treatment modalities.[323] Mindfulness can be effective in those that experience extreme stress. For example, firefighters face life-threatening situations and witness the aftermath of tragedies. Mindfulness reduces PTSD symptoms and ethanol problems in urban firefighters.[324]

One activity that has received attention is physical exercise. Anecdotally speaking, I like to exercise when I encounter stress from work. I am not alone, as exercise is a popular method for managing stress.[325] Empirical research shows that exercise improves anxiety and provides positive mental health for individuals with multiple disorders, including PTSD and SUDs.[326] Specific to those with a SUD, exercise alleviates stress, anxiety, drug craving, and withdrawal symptoms.[327,328] In animals, running on a wheel reverses stress-induced increases in ethanol self-administration[114] and morphine CPP.[329] Exercise also attenuates incubation of drug craving,[330] cue-induced reinstatement,[331] drug-induced reinstatement,[332] and stress-induced reinstatement.[333] In Chapter 1 you learned that drug exposure *in utero* can negatively impact a child (e.g., fetal alcohol syndrome). Some research has shown that exercise can reverse the effects of prenatal drug exposure in animals. Specifically, pregnant rats that receive swimming exercise have pups that show decreased anxiety- and depressive-like behaviors and consume less ethanol later in life.[334]

The beneficial effects of exercise on addiction-like behaviors may result from alterations to multiple neurotransmitter systems. Exercise prevents stress-induced activation of orexinergic neurons, although this effect is observed in male rats only.[335] Work has also shown that exercise enhances the sensitivity of CB_1 receptors, which allows for greater inhibition of neurons in the striatum.[336] Even when animals are forced to engage in chronic exercise, a stressor in itself, they show blunted stress-induced reinstatement.[333] This may occur because the chronic forced exercise causes adaptations that occur to the HPA axis, mitigating the effects of subsequent stress on drug seeking. While exercise has beneficial effects on stress and drug craving, this does not necessarily mean that exercise is an effective treatment for SUDs. One controlled trial found that individuals assigned to an exercise condition did not achieve greater smoking abstinence rates compared to a control group even though exercise improved stress and anxiety.[337]

A common treatment for individuals with anxiety disorders is exposure therapy. You read about exposure therapies in Chapter 6. To recap, during exposure therapy, individuals encounter the stimulus that provokes anxiety in a clinical setting. Individuals may be asked to directly face the anxiety-provoking stimulus, or they can imagine the stimulus. Technological advances now allow the therapist to expose individuals to the stimulus via virtual reality. Exposure therapy reduces drug cravings and/or use and PTSD symptoms across sessions.[338,339] Virtual exposure therapy (VRET) has been shown to be as effective as CBT in reducing nicotine use in those with nicotine dependence.[340] Exposure therapy can easily be used with other treatment interventions to decrease PTSD symptoms and ethanol/drug use, such as integrated coping skills therapy and pharmacological treatments.[341–343] An interesting dissociation has been observed concerning the benefits of exposure therapy on drug use and PTSD symptoms in veterans that have a single SUD or cooccurring SUDs. Individuals with cooccurring SUDs show more reductions in substance use during treatment; however, those with a single SUD show greater improvements in PTSD symptoms.[344]

Quiz 11.2

1. Which opioid receptor primarily mediates stress-induced alterations in drug-seeking behavior?
    a. Delta
    b. Kappa
    c. Mu
    d. Nociceptin
2. True/false: the stress that a pregnant individual experiences has no impact on drug sensitivity in the offspring.
3. Which of the following statements about exercise is true?
    a. Exercise can attenuate drug self-administration in animals but does not consistently decrease drug use in humans.
    b. Exercise decreases drug use without alleviating stress.
    c. Exercise is effective in reducing stress, but it has no beneficial effects on drug craving.
    d. Exercise reduces drug use, but only in those with PTSD.
4. How do CRF receptor antagonists affect drug-seeking behaviors?
    a. They decrease drug seeking.
    b. They have no effect on drug seeking.
    c. They increase drug seeking.
5. Which brain structure serves as a "bridge" between the stress and the reward systems?
    a. Amygdala
    b. BNST
    c. LDT
    d. Raphe nuclei

Answers to quiz 11.2

1. b — Kappa
2. False
3. a — Exercise can attenuate drug self-administration in animals but does not consistently decrease drug use in humans.
4. a — They decrease drug seeking
5. b — BNST

Chapter summary

Stress is something we all experience, with some experiencing stressors daily. Chronic stress has detrimental effects on one's health and is a major risk factor for substance use and SUDs. Individuals with anxiety disorders like PTSD are especially at risk for developing a comorbid SUD. Importantly, the perception of control over stress appears to be an important determinant of addiction. Given the large effects of stress on health, numerous animal models have been developed to study the effects of stress on substance use and relapse-like behaviors. Overall, these studies largely confirm what has been reported in epidemiological studies: stress exacerbates drug seeking and can act as a trigger for relapse. Animal studies have been valuable in elucidating the neuromechanisms of stress-induced potentiation of addiction-like behaviors. As you learned, stress affects virtually every neurotransmitter system implicated in SUDs. The HPA axis, which regulates stress responses, can also increase drug sensitivity, further showing how stress regulates substance use. Due to the wide-reaching effects of stress on neurobiological substrates, developing an effective treatment for the simultaneous reduction in stress and substance use can be difficult. However, some treatments have shown some promise in alleviating stress and blunting drug cravings, such as CBT, mindfulness, exercise, and exposure therapy. Even if one does not have a SUD, managing stress is an important skill we all need to adopt as it can greatly improve our physical and mental wellbeing.

Glossary

Anxiogenic — drug that increases anxiety.
Corticosteroid — class of steroid hormones produced by the adrenal glands involved in a range of physiological responses, including stress responses and salt retention.
Glucocorticoid — a specific type of corticosteroid involved in the stress response; cortisol is the primary glucocorticoid in humans, whereas corticosterone is the major glucocorticoid in other species like rodents.
Hormone — a type of molecule that travels through the bloodstream to reach a target receptor on a gland/organ.
Hypothalamic-pituitary-adrenal (HPA) axis — a system that regulates the production and release of stress hormones (like cortisol and adrenaline) from the adrenal glands.
Neuroendocrine system — system by which the nervous system and endocrine system work together to influence important physiological processes such as feeding/drinking, metabolism, blood pressure, mating, and stress responses.
Social defeat — an antagonistic form of social interaction in which one animal is repeatedly "defeated" (i.e., attacked) by another animal.
Stress — physical and mental responses to an external cause.

Supplementary data related to this chapter can be found online at https://educate.elsevier.com/9780323905787.

References

1. Hammer RR, Dingel MJ, Ostergren JE, Nowakowski KE, Koenig BA. The experience of addiction as told by the addicted: incorporating biological understandings into self-story. *Cult Med Psychiatr*. 2012;36(4):712−734. https://doi.org/10.1007/s11013-012-9283-x.
2. Argento E, Strathdee SA, Goldenberg S, Braschel M, Montaner J, Shannon K. Violence, trauma and living with HIV: longitudinal predictors of initiating crystal methamphetamine injection among sex workers. *Drug Alcohol Depend*. 2017;175:198−204. https://doi.org/10.1016/j.drugalcdep.2017.02.014.
3. DiMaggio C, Galea S, Li G. Substance use and misuse in the aftermath of terrorism. A Bayesian meta-analysis. *Addiction*. 2009;104(6):894−904. https://doi.org/10.1111/j.1360-0443.2009.02526.x.
4. Sinha R. Chronic stress, drug use, and vulnerability to addiction. *Ann N Y Acad Sci*. 2008;1141:105−130. https://doi.org/10.1196/annals.1441.030.
5. Zilberman N, Yadid G, Efrati Y, Rassovsky Y. Negative and positive life events and their relation to substance and behavioral addictions. *Drug Alcohol Depend*. 2019;204:107562. https://doi.org/10.1016/j.drugalcdep.2019.107562.

6. Fox HC, Jackson ED, Sinha R. Elevated cortisol and learning and memory deficits in cocaine dependent individuals: relationship to relapse outcomes. *Psychoneuroendocrinology*. 2009;34(8):1198−1207. https://doi.org/10.1016/j.psyneuen.2009.03.007.
7. Rougemont-Bücking A, Grazioli VS, Daeppen JB, Gmel G, Studer J. Family-related stress versus external stressors: differential impacts on alcohol and illicit drug use in young men. *Eur Addiction Res*. 2017;23(6):284−297. https://doi.org/10.1159/000485031.
8. Calcaterra SL, Beaty B, Mueller SR, Min SJ, Binswanger IA. The association between social stressors and drug use/hazardous drinking among former prison inmates. *J Subst Abuse Treat*. 2014;47(1):41−49. https://doi.org/10.1016/j.jsat.2014.02.002.
9. Johnson V, Pandina RJ. Alcohol problems among a community sample: longitudinal influences of stress, coping, and gender. *Subst Use Misuse*. 2000;35(5):669−686. https://doi.org/10.3109/10826080009148416.
10. Mahoney 3rd JJ, Newton TF, Omar Y, Ross EL, De La Garza 2nd R. The relationship between lifetime stress and addiction severity in cocaine-dependent participants. *Eur Neuropsychopharmacol*. 2013;23(5):351−357. https://doi.org/10.1016/j.euroneuro.2012.05.016.
11. Karlsgodt KH, Lukas SE, Elman I. Psychosocial stress and the duration of cocaine use in non-treatment seeking individuals with cocaine dependence. *Am J Drug Alcohol Abuse*. 2003;29(3):539−551. https://doi.org/10.1081/ada-120023457.
12. Douglas KR, Chan G, Gelernter J, et al. Adverse childhood events as risk factors for substance dependence: partial mediation by mood and anxiety disorders. *Addict Behav*. 2010;35(1):7−13. https://doi.org/10.1016/j.addbeh.2009.07.004.
13. Garami J, Valikhani A, Parkes D, et al. Examining perceived stress, childhood trauma and interpersonal trauma in individuals with drug addiction. *Psychol Rep*. 2019;122(2):433−450. https://doi.org/10.1177/0033294118764918.
14. Wills TA, Vaccaro D, McNamara G. The role of life events, family support, and competence in adolescent substance use: a test of vulnerability and protective factors. *Am J Community Psychol*. 1992;20(3):349−374. https://doi.org/10.1007/BF00937914.
15. Johnson V, Pandina RJ. A longitudinal examination of the relationships among stress, coping strategies, and problems associated with alcohol use. *Alcohol Clin Exp Res*. 1993;17(3):696−702. https://doi.org/10.1111/j.1530-0277.1993.tb00822.x.
16. Butters JE. Family stressors and adolescent cannabis use: a pathway to problem use. *J Adolesc*. 2002;25(6):645−654. https://doi.org/10.1006/jado.2002.0514.
17. Davis JP, Christie NC, Dworkin ER, et al. Influences of victimization and comorbid conditions on latency to illicit drug use among adolescents and young adults. *Drug Alcohol Depend*. 2020;206:107721. https://doi.org/10.1016/j.drugalcdep.2019.107721.
18. Bell TM, Raymond J, Vetor A, et al. Long-term prescription opioid utilization, substance use disorders, and opioid overdoses after adolescent trauma. *J Trauma & Acute Care Surg*. 2019;87(4):836−840. https://doi.org/10.1097/TA.0000000000002261.
19. Otten R, Mun CJ, Shaw DS, Wilson MN, Dishion TJ. A developmental cascade model for early adolescent-onset substance use: the role of early childhood stress. *Addiction*. 2019;114(2):326−334. https://doi.org/10.1111/add.14452.
20. McCabe SE, Cranford JA, Boyd CJ. Stressful events and other predictors of remission from drug dependence in the United States: longitudinal results from a national survey. *J Subst Abuse Treat*. 2016;71:41−47. https://doi.org/10.1016/j.jsat.2016.08.008.
21. Syan SK, Minhas M, Oshri A, et al. Predictors of premature treatment termination in a large residential addiction medicine program. *J Subst Abuse Treat*. 2020;117:108077. https://doi.org/10.1016/j.jsat.2020.108077.
22. Preston KL, Epstein DH. Stress in the daily lives of cocaine and heroin users: relationship to mood, craving, relapse triggers, and cocaine use. *Psychopharmacology*. 2011;218(1):29−37. https://doi.org/10.1007/s00213-011-2183-x.
23. Miranda Jr R, Wemm SE, Treloar Padovano H, et al. Weaker memory performance exacerbates stress-induced cannabis craving in youths' daily lives. *Clin Psychol Sci*. 2019;7(5):1094−1108. https://doi.org/10.1177/2167702619841976.

24. Preston KL, Kowalczyk WJ, Phillips KA, et al. Exacerbated craving in the presence of stress and drug cues in drug-dependent patients. *Neuropsychopharmacology*. 2018;43(4):859–867. https://doi.org/10.1038/npp.2017.275.
25. Jaremko KM, Sterling RC, Van Bockstaele EJ. Psychological and physiological stress negatively impacts early engagement and retention of opioid-dependent individuals on methadone maintenance. *J Subst Abuse Treat*. 2015;48(1):117–127. https://doi.org/10.1016/j.jsat.2014.08.006.
26. Panlilio LV, Stull SW, Kowalczyk WJ, et al. Stress, craving and mood as predictors of early dropout from opioid agonist therapy. *Drug Alcohol Depend*. 2019;202:200–208. https://doi.org/10.1016/j.drugalcdep.2019.05.026.
27. Tate SR, McQuaid JR, Brown SA. Characteristics of life stressors predictive of substance treatment outcomes. *J Subst Abuse Treat*. 2005;29(2):107–115. https://doi.org/10.1016/j.jsat.2005.05.003.
28. Kang SY, Deren S, Goldstein MF. Relationships between childhood abuse and neglect experience and HIV risk behaviors among methadone treatment drop-outs. *Child Abuse Neglect*. 2002;26(12):1275–1289. https://doi.org/10.1016/s0145-2134(02)00412-x.
29. Pasareanu AR, Vederhus JK, Opsal A, Kristensen Ø, Clausen T. Mental distress following inpatient substance use treatment, modified by substance use; comparing voluntary and compulsory admissions. *BMC Health Serv Res*. 2017;17(1):5. https://doi.org/10.1186/s12913-016-1936-y.
30. Hagen E, Erga AH, Hagen KP, et al. One-year sobriety improves satisfaction with life, executive functions and psychological distress among patients with polysubstance use disorder. *J Subst Abuse Treat*. 2017;76:81–87. https://doi.org/10.1016/j.jsat.2017.01.016.
31. Anderson KG, Ramo DE, Brown SA. Life stress, coping and comorbid youth: an examination of the stress-vulnerability model for substance relapse. *J Psychoact Drugs*. 2006;38(3):255–262. https://doi.org/10.1080/02791072.2006.10399851.
32. Alegría AA, Hasin DS, Nunes EV, et al. Comorbidity of generalized anxiety disorder and substance use disorders: results from the national epidemiologic survey on alcohol and related conditions. *J Clin Psychiatr*. 2010;71(9):1187–1253. https://doi.org/10.4088/JCP.09m05328gry.
33. Brook JS, Zhang C, Rubenstone E, Primack BA, Brook DW. Comorbid trajectories of substance use as predictors of antisocial personality disorder, major depressive episode, and generalized anxiety disorder. *Addict Behav*. 2016;62:114–121. https://doi.org/10.1016/j.addbeh.2016.06.003.
34. Goodwin RD, Lieb R, Hoefler M, et al. Panic attack as a risk factor for severe psychopathology. *Am J Psychiatr*. 2004;161(12):2207–2214. https://doi.org/10.1176/appi.ajp.161.12.2207.
35. Prior K, Mills K, Ross J, Teesson M. Substance use disorders comorbid with mood and anxiety disorders in the Australian general population. *Drug Alcohol Rev*. 2017;36(3):317–324. https://doi.org/10.1111/dar.12419.
36. Lemyre A, Gauthier-Légaré A, Bélanger RE. Shyness, social anxiety, social anxiety disorder, and substance use among normative adolescent populations: a systematic review. *Am J Drug Alcohol Abuse*. 2019;45(3):230–247. https://doi.org/10.1080/00952990.2018.1536882.
37. Wolitzky-Taylor K, Bobova L, Zinbarg RE, Mineka S, Craske MG. Longitudinal investigation of the impact of anxiety and mood disorders in adolescence on subsequent substance use disorder onset and vice versa. *Addict Behav*. 2012;37(8):982–985. https://doi.org/10.1016/j.addbeh.2012.03.026.
38. Bakken K, Landheim AS, Vaglum P. Substance-dependent patients with and without social anxiety disorder: occurrence and clinical differences. A study of a consecutive sample of alcohol-dependent and poly-substance-dependent patients treated in two counties in Norway. *Drug Alcohol Depend*. 2005;80(3):321–328. https://doi.org/10.1016/j.drugalcdep.2005.04.011.
39. Mancebo MC, Grant JE, Pinto A, Eisen JL, Rasmussen SA. Substance use disorders in an obsessive compulsive disorder clinical sample. *J Anxiety Disord*. 2009;23(4):429–435. https://doi.org/10.1016/j.janxdis.2008.08.008.
40. Gentes EL, Schry AR, Hicks TA, et al. Prevalence and correlates of cannabis use in an outpatient VA posttraumatic stress disorder clinic. *Psychol Addict Behav*. 2016;30(3):415–421. https://doi.org/10.1037/adb0000154.
41. Lipschitz DS, Rasmusson AM, Anyan W, et al. Posttraumatic stress disorder and substance use in inner-city adolescent girls. *J Nerv Ment Dis*. 2003;191(11):714–721. https://doi.org/10.1097/01.nmd.0000095123.68088.da.
42. Mergler M, Driessen M, Havemann-Reinecke U, et al. Differential relationships of PTSD and childhood trauma with the course of substance use disorders. *J Subst Abuse Treat*. 2018;93:57–63. https://doi.org/10.1016/j.jsat.2018.07.010.

43. Young-Wolff KC, Fromont SC, Delucchi K, Hall SE, Hall SM, Prochaska JJ. PTSD symptomatology and readiness to quit smoking among women with serious mental illness. *Addict Behav.* 2014;39(8):1231–1234. https://doi.org/10.1016/j.addbeh.2014.03.024.
44. Bowe A, Rosenheck R. PTSD and substance use disorder among veterans: characteristics, service utilization and pharmacotherapy. *J Dual Diagn.* 2015;11(1):22–32. https://doi.org/10.1080/15504263.2014.989653.
45. Basedow LA, Kuitunen-Paul S, Wiedmann MF, Roessner V, Golub Y. Self-reported PTSD is associated with increased use of MDMA in adolescents with substance use disorders. *Eur J Psychotraumatol.* 2021;12(1):1968140. https://doi.org/10.1080/20008198.2021.1968140.
46. Kok T, de Haan H, van der Meer M, Najavits L, de Jong C. Assessing traumatic experiences in screening for PTSD in substance use disorder patients: what is the gain in addition to PTSD symptoms? *Psychiatr Res.* 2015;226(1):328–332. https://doi.org/10.1016/j.psychres.2015.01.014.
47. Tripp JC, Worley MJ, Straus E, Angkaw AC, Trim RS, Norman SB. Bidirectional relationship of posttraumatic stress disorder (PTSD) symptom severity and alcohol use over the course of integrated treatment. *Psychol Addict Behav.* 2020;34(4):506–511. https://doi.org/10.1037/adb0000564.
48. Vogel L, Koller G, Ehring T. The relationship between posttraumatic stress symptoms and craving in patients with substance use disorder attending detoxification. *Drug Alcohol Depend.* 2021;223:108709. https://doi.org/10.1016/j.drugalcdep.2021.108709.
49. Najavits LM, Harned MS, Gallop RJ, et al. Six-month treatment outcomes of cocaine-dependent patients with and without PTSD in a multisite national trial. *J Stud Alcohol Drugs.* 2007;68(3):353–361. https://doi.org/10.15288/jsad.2007.68.353.
50. Robinson J, Sareen J, Cox BJ, Bolton J. Self-medication of anxiety disorders with alcohol and drugs: results from a nationally representative sample. *J Anxiety Disord.* 2009;23(1):38–45. https://doi.org/10.1016/j.janxdis.2008.03.013.
51. Zvolensky MJ, Cougle JR, Johnson KA, Bonn-Miller MO, Bernstein A. Marijuana use and panic psychopathology among a representative sample of adults. *Exp Clin Psychopharmacol.* 2010;18(2):129–134. https://doi.org/10.1037/a0019022.
52. Hruska B, Bernier J, Kenner F, et al. Examining the relationships between posttraumatic stress disorder symptoms, positive smoking outcome expectancies, and cigarette smoking in people with substance use disorders: a multiple mediator model. *Addict Behav.* 2014;39(1):273–281. https://doi.org/10.1016/j.addbeh.2013.10.002.
53. Miczek KA, Yap JJ, Covington 3rd HE. Social stress, therapeutics and drug abuse: preclinical models of escalated and depressed intake. *Pharmacol Ther.* 2008;120(2):102–128. https://doi.org/10.1016/j.pharmthera.2008.07.006.
54. Advani T, Hensler JG, Koek W. Effect of early rearing conditions on alcohol drinking and 5-HT1A receptor function in C57BL/6J mice. *Int J Neuropsychopharmacol.* 2007;10(5):595–607. https://doi.org/10.1017/S1461145706007401.
55. Raz S, Berger BD. Social isolation increases morphine intake: behavioral and psychopharmacological aspects. *Behav Pharmacol.* 2010;21(1):39–46. https://doi.org/10.1097/FBP.0b013e32833470bd.
56. Schenk S, Gorman K, Amit Z. Age-dependent effects of isolation housing on the self-administration of ethanol in laboratory rats. *Alcohol.* 1990;7(4):321–326. https://doi.org/10.1016/0741-8329(90)90090-y.
57. Baarendse PJ, Limpens JH, Vanderschuren LJ. Disrupted social development enhances the motivation for cocaine in rats. *Psychopharmacology.* 2014;231(8):1695–1704. https://doi.org/10.1007/s00213-013-3362-8.
58. Burke AR, Miczek KA. Escalation of cocaine self-administration in adulthood after social defeat of adolescent rats: role of social experience and adaptive coping behavior. *Psychopharmacology.* 2015;232(16):3067–3079. https://doi.org/10.1007/s00213-015-3947-5.
59. Phillips GD, Howes SR, Whitelaw RB, Wilkinson LS, Robbins TW, Everitt BJ. Isolation rearing enhances the locomotor response to cocaine and a novel environment, but impairs the intravenous self-administration of cocaine. *Psychopharmacology.* 1994;115(3):407–418. https://doi.org/10.1007/BF02245084.
60. Cortés-Patiño DM, Serrano C, Garcia-Mijares M. Early social isolation increases persistence of alcohol-seeking behavior in alcohol-related contexts. *Behav Pharmacol.* 2016;27(2–3 Spec Issue):185–191. https://doi.org/10.1097/FBP.0000000000000213.
61. Kosten TA, Zhang XY, Kehoe P. Heightened cocaine and food self-administration in female rats with neonatal isolation experience. *Neuropsychopharmacology.* 2006;31(1):70–76. https://doi.org/10.1038/sj.npp.1300779.

62. Singh A, Xie Y, Davis A, Wang ZJ. Early social isolation stress increases addiction vulnerability to heroin and alters c-Fos expression in the mesocorticolimbic system. *Psychopharmacology*. 2022;239(4):1081−1095. https://doi.org/10.1007/s00213-021-06024-1.
63. Schenk S, Robinson B, Amit Z. Housing conditions fail to affect the intravenous self-administration of amphetamine. *Pharmacol, Biochem Behav*. 1988;31(1):59−62. https://doi.org/10.1016/0091-3057(88)90311-5.
64. Fosnocht AQ, Lucerne KE, Ellis AS, Olimpo NA, Briand LA. Adolescent social isolation increases cocaine seeking in male and female mice. *Behav Brain Res*. 2019;359:589−596. https://doi.org/10.1016/j.bbr.2018.10.007.
65. Howes SR, Dalley JW, Morrison CH, Robbins TW, Everitt BJ. Leftward shift in the acquisition of cocaine self-administration in isolation-reared rats: relationship to extracellular levels of dopamine, serotonin and glutamate in the nucleus accumbens and amygdala-striatal FOS expression. *Psychopharmacology*. 2000;151(1):55−63. https://doi.org/10.1007/s002130000451.
66. Fabricius K, Helboe L, Fink-Jensen A, Wörtwein G, Steiniger-Brach B, Sotty F. Increased dopaminergic activity in socially isolated rats: an electrophysiological study. *Neurosci Lett*. 2010;482(2):117−122. https://doi.org/10.1016/j.neulet.2010.07.014.
67. Lapiz MD, Fulford A, Muchimapura S, Mason R, Parker T, Marsden CA. Influence of postweaning social isolation in the rat on brain development, conditioned behavior, and neurotransmission. *Neurosci Behav Physiol*. 2003;33(1):13−29. https://doi.org/10.1023/a:1021171129766.
68. Weiss IC, Domeney AM, Heidbreder CA, Moreau JL, Feldon J. Early social isolation, but not maternal separation, affects behavioral sensitization to amphetamine in male and female adult rats. *Pharmacol, Biochem Behav*. 2001;70(2−3):397−409. https://doi.org/10.1016/s0091-3057(01)00626-8.
69. Yanai J, Ginsburg BE. Increased sensitivity to chronic ethanol in isolated mice. *Psychopharmacologia*. 1976;46(2):185−189. https://doi.org/10.1007/BF00421390.
70. Meyer A, Mayerhofer A, Kovar KA, Schmidt WJ. Rewarding effects of the optical isomers of 3,4-methylenedioxy-methylamphetamine ('Ecstasy') and 3,4-methylenedioxy-ethylamphetamine ('Eve') measured by conditioned place preference in rats. *Neurosci Lett*. 2002;330(3):280−284. https://doi.org/10.1016/s0304-3940(02)00821-2.
71. Dow-Edwards D, Iijima M, Stephenson S, Jackson A, Weedon J. The effects of prenatal cocaine, post-weaning housing and sex on conditioned place preference in adolescent rats. *Psychopharmacology*. 2014;231(8):1543−1555. https://doi.org/10.1007/s00213-013-3418-9.
72. Grotewold SK, Wall VL, Goodell DJ, Hayter C, Bland ST. Effects of cocaine combined with a social cue on conditioned place preference and nucleus accumbens monoamines after isolation rearing in rats. *Psychopharmacology*. 2014;231(15):3041−3053. https://doi.org/10.1007/s00213-014-3470-0.
73. Bennett EL, Rosenzweig MR, Diamond MC. Rat brain: effects of environmental enrichment on wet and dry weights. *Science*. 1969;163(3869):825−826. https://doi.org/10.1126/science.163.3869.825.
74. Rojas JJ, Deniz BF, Miguel PM, et al. Effects of daily environmental enrichment on behavior and dendritic spine density in hippocampus following neonatal hypoxia-ischemia in the rat. *Exp Neurol*. 2013;241:25−33. https://doi.org/10.1016/j.expneurol.2012.11.026.
75. Kalinichev M, Easterling KW, Holtzman SG. Early neonatal experience of Long-Evans rats results in long-lasting changes in morphine tolerance and dependence. *Psychopharmacology*. 2001;157(3):305−312. https://doi.org/10.1007/s002130100806.
76. Brake WG, Zhang TY, Diorio J, Meaney MJ, Gratton A. Influence of early postnatal rearing conditions on mesocorticolimbic dopamine and behavioural responses to psychostimulants and stressors in adult rats. *Eur J Neurosci*. 2004;19(7):1863−1874. https://doi.org/10.1111/j.1460-9568.2004.03286.x.
77. Gracia-Rubio I, Martinez-Laorden E, Moscoso-Castro M, Milanés MV, Laorden ML, Valverde O. Maternal separation impairs cocaine-induced behavioural sensitization in adolescent mice. *PLoS One*. 2016;11(12):e0167483. https://doi.org/10.1371/journal.pone.0167483.
78. Muhammad A, Kolb B. Maternal separation altered behavior and neuronal spine density without influencing amphetamine sensitization. *Behav Brain Res*. 2011;223(1):7−16. https://doi.org/10.1016/j.bbr.2011.04.015.
79. Zimmerberg B, Shartrand AM. Temperature-dependent effects of maternal separation on growth, activity, and amphetamine sensitivity in the rat. *Dev Psychobiol*. 1992;25(3):213−226. https://doi.org/10.1002/dev.420250306.

80. Fernández M, Fabio MC, Nizhnikov ME, Spear NE, Abate P, Pautassi RM. Maternal isolation during the first two postnatal weeks affects novelty-induced responses and sensitivity to ethanol-induced locomotor activity during infancy. *Dev Psychobiol.* 2014;56(5):1070−1082. https://doi.org/10.1002/dev.21192.
81. Castro-Zavala A, Martín-Sánchez A, Luján MÁ, Valverde O. Maternal separation increases cocaine intake through a mechanism involving plasticity in glutamate signalling. *Addiction Biol.* 2021;26(2):e12911. https://doi.org/10.1111/adb.12911.
82. Amancio-Belmont O, Becerril Meléndez AL, Ruiz-Contreras AE, Méndez-Díaz M, Próspero-García O. Maternal separation plus social isolation during adolescence reprogram brain dopamine and endocannabinoid systems and facilitate alcohol intake in rats. *Brain Res Bull.* 2020;164:21−28. https://doi.org/10.1016/j.brainresbull.2020.08.002.
83. Bassey RB, Gondré-Lewis MC. Combined early life stressors: prenatal nicotine and maternal deprivation interact to influence affective and drug seeking behavioral phenotypes in rats. *Behav Brain Res.* 2019;359:814−822. https://doi.org/10.1016/j.bbr.2018.07.022.
84. Bertagna NB, Favoretto CA, Rodolpho BT, et al. Maternal separation stress affects voluntary ethanol intake in a sex dependent manner. *Front Physiol.* 2021;12:775404. https://doi.org/10.3389/fphys.2021.775404.
85. García-Gutiérrez MS, Navarrete F, Aracil A, et al. Increased vulnerability to ethanol consumption in adolescent maternal separated mice. *Addiction Biol.* 2016;21(4):847−858. https://doi.org/10.1111/adb.12266.
86. Lewis CR, Staudinger K, Tomek SE, Hernandez R, Manning T, Olive MF. Early life stress and chronic variable stress in adulthood interact to influence methamphetamine self-administration in male rats. *Behav Pharmacol.* 2016;27(2−3 Spec Issue):182−184. https://doi.org/10.1097/FBP.0000000000000166.
87. Portero-Tresserra M, Gracia-Rubio I, Cantacorps L, et al. Maternal separation increases alcohol-drinking behaviour and reduces endocannabinoid levels in the mouse striatum and prefrontal cortex. *Eur Neuropsychopharmacol.* 2018;28(4):499−512. https://doi.org/10.1016/j.euroneuro.2018.02.003.
88. Romano-López A, Méndez-Díaz M, Ruiz-Contreras AE, Carrisoza R, Próspero-García O. Maternal separation and proclivity for ethanol intake: a potential role of the endocannabinoid system in rats. *Neuroscience.* 2012;223:296−304. https://doi.org/10.1016/j.neuroscience.2012.07.071.
89. Thompson SM, Simmons AN, McMurray MS. The effects of multiple early life stressors on adolescent alcohol consumption. *Behav Brain Res.* 2020;380:112449. https://doi.org/10.1016/j.bbr.2019.112449.
90. Zhang XY, Sanchez H, Kehoe P, Kosten TA. Neonatal isolation enhances maintenance but not reinstatement of cocaine self-administration in adult male rats. *Psychopharmacology.* 2005;177(4):391−399. https://doi.org/10.1007/s00213-004-1963-y.
91. Leichtweis KS, Carvalho M, Morais-Silva G, Marin MT, Amaral V. Short and prolonged maternal separation impacts on ethanol-related behaviors in rats: sex and age differences. *Stress.* 2020;23(2):162−173. https://doi.org/10.1080/10253890.2019.1653847.
92. Roman E, Hyytiä P, Nylander I. Maternal separation alters acquisition of ethanol intake in male ethanol-preferring AA rats. *Alcohol Clin Exp Res.* 2003;27(1):31−37. https://doi.org/10.1097/01.ALC.0000047352.88145.80.
93. Moffett MC, Harley J, Francis D, Sanghani SP, Davis WI, Kuhar MJ. Maternal separation and handling affects cocaine self-administration in both the treated pups as adults and the dams. *J Pharmacol Exp Therapeut.* 2006;317(3):1210−1218. https://doi.org/10.1124/jpet.106.101139.
94. Lewis CR, Staudinger K, Scheck L, Olive MF. The effects of maternal separation on adult methamphetamine self-administration, extinction, reinstatement, and MeCP2 immunoreactivity in the nucleus accumbens. *Front Psychiatr.* 2013;4:55. https://doi.org/10.3389/fpsyt.2013.00055.
95. Dalaveri F, Nakhaee N, Esmaeilpour K, Mahani SE, Sheibani V. Effects of maternal separation on nicotine-induced conditioned place preference and subsequent learning and memory in adolescent female rats. *Neurosci Lett.* 2017;639:151−156. https://doi.org/10.1016/j.neulet.2016.11.059.
96. Jaworski JN, Francis DD, Brommer CL, Morgan ET, Kuhar MJ. Effects of early maternal separation on ethanol intake, GABA receptors and metabolizing enzymes in adult rats. *Psychopharmacology.* 2005;181(1):8−15. https://doi.org/10.1007/s00213-005-2232-4.
97. Boyson CO, Holly EN, Shimamoto A, et al. Social stress and CRF-dopamine interactions in the VTA: role in long-term escalation of cocaine self-administration. *J Neurosci.* 2014;34(19):6659−6667. https://doi.org/10.1523/JNEUROSCI.3942-13.2014.

98. Boyson CO, Miguel TT, Quadros IM, Debold JF, Miczek KA. Prevention of social stress-escalated cocaine self-administration by CRF-R1 antagonist in the rat VTA. *Psychopharmacology*. 2011;218(1):257–269. https://doi.org/10.1007/s00213-011-2266-8.
99. Burke AR, Forster GL, Novick AM, Roberts CL, Watt MJ. Effects of adolescent social defeat on adult amphetamine-induced locomotion and corticoaccumbal dopamine release in male rats. *Neuropharmacology*. 2013;67:359–369. https://doi.org/10.1016/j.neuropharm.2012.11.013.
100. Covington 3rd HE, Miczek KA. Repeated social-defeat stress, cocaine or morphine. Effects on behavioral sensitization and intravenous cocaine self-administration "binges". *Psychopharmacology*. 2001;158(4):388–398. https://doi.org/10.1007/s002130100858.
101. Covington 3rd HE, Miczek KA. Intense cocaine self-administration after episodic social defeat stress, but not after aggressive behavior: dissociation from corticosterone activation. *Psychopharmacology*. 2005;183(3):331–340. https://doi.org/10.1007/s00213-005-0190-5.
102. Ferrer-Pérez C, Reguilón MD, Manzanedo C, Aguilar MA, Miñarro J, Rodríguez-Arias M. Antagonism of corticotropin-releasing factor CRF1 receptors blocks the enhanced response to cocaine after social stress. *Eur J Pharmacol*. 2018;823:87–95. https://doi.org/10.1016/j.ejphar.2018.01.052.
103. Ferrer-Pérez C, Castro-Zavala A, Luján MÁ, et al. Oxytocin prevents the increase of cocaine-related responses produced by social defeat. *Neuropharmacology*. 2019;146:50–64. https://doi.org/10.1016/j.neuropharm.2018.11.011.
104. Funk D, Vohra S, Lê AD. Influence of stressors on the rewarding effects of alcohol in Wistar rats: studies with alcohol deprivation and place conditioning. *Psychopharmacology*. 2004;176(1):82–87. https://doi.org/10.1007/s00213-004-1859-x.
105. Montagud-Romero S, Reguilón MD, Pascual M, et al. Critical role of TLR4 in uncovering the increased rewarding effects of cocaine and ethanol induced by social defeat in male mice. *Neuropharmacology*. 2021;182:108368. https://doi.org/10.1016/j.neuropharm.2020.108368.
106. Reguilón MD, Ferrer-Pérez C, Miñarro J, Rodríguez-Arias M. Oxytocin reverses ethanol consumption and neuroinflammation induced by social defeat in male mice. *Horm Behav*. 2021;127:104875. https://doi.org/10.1016/j.yhbeh.2020.104875.
107. Riga D, Schmitz L, van Mourik Y, et al. Stress vulnerability promotes an alcohol-prone phenotype in a preclinical model of sustained depression. *Addiction Biol*. 2020;25(1):e12701. https://doi.org/10.1111/adb.12701.
108. Tidey JW, Miczek KA. Acquisition of cocaine self-administration after social stress: role of accumbens dopamine. *Psychopharmacology*. 1997;130(3):203–212. https://doi.org/10.1007/s002130050230.
109. Rodríguez-Arias M, Montagud-Romero S, Rubio-Araiz A, et al. Effects of repeated social defeat on adolescent mice on cocaine-induced CPP and self-administration in adulthood: integrity of the blood-brain barrier. *Addiction Biol*. 2017;22(1):129–141. https://doi.org/10.1111/adb.12301.
110. Zou S, Funk D, Shram MJ, Lê AD. Effects of stressors on the reinforcing efficacy of nicotine in adolescent and adult rats. *Psychopharmacology*. 2014;231(8):1601–1614. https://doi.org/10.1007/s00213-013-3314-3.
111. Burke AR, DeBold JF, Miczek KA. CRF type 1 receptor antagonism in ventral tegmental area of adolescent rats during social defeat: prevention of escalated cocaine self-administration in adulthood and behavioral adaptations during adolescence. *Psychopharmacology*. 2016;233(14):2727–2736. https://doi.org/10.1007/s00213-016-4336-4.
112. Han X, Albrechet-Souza L, Doyle MR, Shimamoto A, DeBold JF, Miczek KA. Social stress and escalated drug self-administration in mice II. Cocaine and dopamine in the nucleus accumbens. *Psychopharmacology*. 2015;232(6):1003–1010. https://doi.org/10.1007/s00213-014-3734-8.
113. Han X, DeBold JF, Miczek KA. Prevention and reversal of social stress-escalated cocaine self-administration in mice by intra-VTA CRFR1 antagonism. *Psychopharmacology*. 2017;234(18):2813–2821. https://doi.org/10.1007/s00213-017-4676-8.
114. Reguilón MD, Ferrer-Pérez C, Ballestín R, Miñarro J, Rodríguez-Arias M. Voluntary wheel running protects against the increase in ethanol consumption induced by social stress in mice. *Drug Alcohol Depend*. 2020;212:108004. https://doi.org/10.1016/j.drugalcdep.2020.108004.
115. Wang J, Bastle RM, Bass CE, Hammer Jr RP, Neisewander JL, Nikulina EM. Overexpression of BDNF in the ventral tegmental area enhances binge cocaine self-administration in rats exposed to repeated social defeat. *Neuropharmacology*. 2016;109:121–130. https://doi.org/10.1016/j.neuropharm.2016.04.045.

116. García-Pardo MP, Calpe-López C, Miñarro J, Aguilar MA. Role of N-methyl-D-aspartate receptors in the long-term effects of repeated social defeat stress on the rewarding and psychomotor properties of cocaine in mice. *Behav Brain Res*. 2019;361:95–103. https://doi.org/10.1016/j.bbr.2018.12.025.
117. McLaughlin JP, Li S, Valdez J, Chavkin TA, Chavkin C. Social defeat stress-induced behavioral responses are mediated by the endogenous kappa opioid system. *Neuropsychopharmacology*. 2006;31(6):1241–1248. https://doi.org/10.1038/sj.npp.1300872.
118. Montagud-Romero S, Reguilon MD, Roger-Sanchez C, et al. Role of dopamine neurotransmission in the long-term effects of repeated social defeat on the conditioned rewarding effects of cocaine. *Prog Neuro Psychopharmacol Biol Psychiatr*. 2016;71:144–154. https://doi.org/10.1016/j.pnpbp.2016.07.008.
119. Stelly CE, Pomrenze MB, Cook JB, Morikawa H. Repeated social defeat stress enhances glutamatergic synaptic plasticity in the VTA and cocaine place conditioning. *Elife*. 2016;5:e15448. https://doi.org/10.7554/eLife.15448.
120. García-Pardo MP, Miñarro J, Llansola M, Felipo V, Aguilar MA. Role of NMDA and AMPA glutamatergic receptors in the effects of social defeat on the rewarding properties of MDMA in mice. *Eur J Neurosci*. 2019;50(3):2623–2634. https://doi.org/10.1111/ejn.14190.
121. Holly EN, Boyson CO, Montagud-Romero S, et al. Episodic social stress-escalated cocaine self-administration: role of phasic and tonic corticotropin releasing factor in the anterior and posterior ventral tegmental area. *J Neurosci*. 2016;36(14):4093–4105. https://doi.org/10.1523/JNEUROSCI.2232-15.2016.
122. Tamashiro KL, Hegeman MA, Nguyen MM, et al. Dynamic body weight and body composition changes in response to subordination stress. *Physiol Behav*. 2007;91(4):440–448. https://doi.org/10.1016/j.physbeh.2007.04.004.
123. Rygula R, Abumaria N, Flügge G, Fuchs E, Rüther E, Havemann-Reinecke U. Anhedonia and motivational deficits in rats: impact of chronic social stress. *Behav Brain Res*. 2005;162(1):127–134. https://doi.org/10.1016/j.bbr.2005.03.009.
124. Iniguez SD, Riggs LM, Nieto SJ, et al. Social defeat stress induces a depression-like phenotype in adolescent male c57BL/6 mice. *Stress*. 2014;17(3):247–255. https://doi.org/10.3109/10253890.2014.910650.
125. Mollenauer S, Bryson R, Robison M, Sardo J, Coleman C. EtOH self-administration in anticipation of noise stress in C57BL/6J mice. *Pharmacol, Biochem Behav*. 1993;46(1):35–38. https://doi.org/10.1016/0091-3057(93)90313-i.
126. Bolton JL, Ruiz CM, Rismanchi N, et al. Early-life adversity facilitates acquisition of cocaine self-administration and induces persistent anhedonia. *Neurobiology of Stress*. 2018;8:57–67. https://doi.org/10.1016/j.ynstr.2018.01.002.
127. Schroff KC, Cowen MS, Koch S, Spanagel R. Strain-specific responses of inbred mice to ethanol following food shortage. *Addiction Biol*. 2004;9(3–4):265–271. https://doi.org/10.1080/13556210412331292596.
128. Shalev U, Morales M, Hope B, Yap J, Shaham Y. Time-dependent changes in extinction behavior and stress-induced reinstatement of drug seeking following withdrawal from heroin in rats. *Psychopharmacology*. 2001;156(1):98–107. https://doi.org/10.1007/s002130100748.
129. Mousavi A, Askari N, Vaez-Mahdavi MR. Augmentation of morphine-conditioned place preference by food restriction is associated with alterations in the oxytocin/oxytocin receptor in rat models. *Am J Drug Alcohol Abuse*. 2020;46(3):304–315. https://doi.org/10.1080/00952990.2019.1648483.
130. Deroche V, Piazza PV, Casolini P, Maccari S, Le Moal M, Simon H. Stress-induced sensitization to amphetamine and morphine psychomotor effects depend on stress-induced corticosterone secretion. *Brain Res*. 1992;598(1–2):343–348. https://doi.org/10.1016/0006-8993(92)90205-n.
131. Garcia-Keller C, Martinez SA, Esparza MA, Bollati F, Kalivas PW, Cancela LM. Cross-sensitization between cocaine and acute restraint stress is associated with sensitized dopamine but not glutamate release in the nucleus accumbens. *Eur J Neurosci*. 2013;37(6):982–995. https://doi.org/10.1111/ejn.12121.
132. Kuribara H. Inhibitory effect of restraint on induction of behavioral sensitization to methamphetamine and cocaine in mice. *Pharmacol Biochem Behav*. 1996;54(2):327–331. https://doi.org/10.1016/0091-3057(95)02026-8.
133. Carter JS, Kearns AM, Vollmer KM, et al. Long-term impact of acute restraint stress on heroin self-administration, reinstatement, and stress reactivity. *Psychopharmacology*. 2020;237(6):1709–1721. https://doi.org/10.1007/s00213-020-05486-z.
134. Paula Avalos M, Susana Guzmán A, Rigoni D, et al. Minocycline prevents chronic restraint stress-induced vulnerability to developing cocaine self-administration and associated glutamatergic mechanisms: a potential role of microglia. *Brain Behav Immun*. 2022;101:359–376. https://doi.org/10.1016/j.bbi.2022.01.014.

135. Lynch WJ, Kushner MG, Rawleigh JM, Fiszdon J, Carroll ME. The effects of restraint stress on voluntary ethanol consumption in rats. *Exp Clin Psychopharmacol*. 1999;7(4):318−323. https://doi.org//1064-1297.7.4.318.
136. Yang X, Wang S, Rice KC, Munro CA, Wand GS. Restraint stress and ethanol consumption in two mouse strains. *Alcohol Clin Exp Res*. 2008;32(5):840−852. https://doi.org/10.1111/j.1530-0277.2008.00632.x.
137. Lopez MF, Anderson RI, Becker HC. Effect of different stressors on voluntary ethanol intake in ethanol-dependent and nondependent C57BL/6J mice. *Alcohol*. 2016;51:17−23. https://doi.org/10.1016/j.alcohol.2015.11.010.
138. Tambour S, Brown LL, Crabbe JC. Gender and age at drinking onset affect voluntary alcohol consumption but neither the alcohol deprivation effect nor the response to stress in mice. *Alcohol Clin Exp Res*. 2008;32(12):2100−2106. https://doi.org/10.1111/j.1530-0277.2008.00798.x.
139. Rockman GE, Hall A, Glavin GB. Effects of restraint stress on voluntary ethanol intake and ulcer proliferation in rats. *Pharmacol, Biochem Behav*. 1986;25(5):1083−1087. https://doi.org/10.1016/0091-3057(86)90089-4.
140. Ball KT, Stone E, Best O, et al. Chronic restraint stress during withdrawal increases vulnerability to drug priming-induced cocaine seeking via a dopamine D1-like receptor-mediated mechanism. *Drug Alcohol Depend*. 2018;187:327−334. https://doi.org/10.1016/j.drugalcdep.2018.03.024.
141. Yu G, Sharp BM. Basolateral amygdala and ventral hippocampus in stress-induced amplification of nicotine self-administration during reacquisition in rat. *Psychopharmacology*. 2015;232(15):2741−2749. https://doi.org/10.1007/s00213-015-3911-4.
142. Glynn RM, Rosenkranz JA, Wolf ME, et al. Repeated restraint stress exposure during early withdrawal accelerates incubation of cue-induced cocaine craving. *Addiction Biol*. 2018;23(1):80−89. https://doi.org/10.1111/adb.12475.
143. Capriles N, Cancela LM. Effect of acute and chronic stress restraint on amphetamine-associated place preference: involvement of dopamine D(1) and D(2) receptors. *Eur J Pharmacol*. 1999;386(2−3):127−134. https://doi.org/10.1016/s0014-2999(99)00746-3.
144. Shinohara F, Asaoka Y, Kamii H, Minami M, Kaneda K. Stress augments the rewarding memory of cocaine via the activation of brainstem-reward circuitry. *Addiction Biol*. 2019;24(3):509−521. https://doi.org/10.1111/adb.12617.
145. del Rosario Capriles N, Cancela LM. Motivational effects mu- and kappa-opioid agonists following acute and chronic restraint stress: involvement of dopamine D(1) and D(2) receptors. *Behav Brain Res*. 2002;132(2):159−169. https://doi.org/10.1016/s0166-4328(01)00414-4.
146. Wada S, Yanagida J, Sasase H, et al. Acute restraint stress augments the rewarding memory of cocaine through activation of α1 adrenoceptors in the medial prefrontal cortex of mice. *Neuropharmacology*. 2020;166:107968. https://doi.org/10.1016/j.neuropharm.2020.107968.
147. Meng S, Quan W, Qi X, Su Z, Yang S. Effect of baclofen on morphine-induced conditioned place preference, extinction, and stress-induced reinstatement in chronically stressed mice. *Psychopharmacology*. 2014;231(1):27−36. https://doi.org/10.1007/s00213-013-3204-8.
148. Woo H, Hong CJ, Jung S, Choe S, Yu SW. Chronic restraint stress induces hippocampal memory deficits by impairing insulin signaling. *Mol Brain*. 2018;11(1):37. https://doi.org/10.1186/s13041-018-0381-8.
149. Slattery DA, Cryan JF. Using the rat forced swim test to assess antidepressant-like activity in rodents. *Nat Protoc*. 2012;7(6):1009−1014. https://doi.org/10.1038/nprot.2012.044.
150. Boyce-Rustay JM, Cameron HA, Holmes A. Chronic swim stress alters sensitivity to acute behavioral effects of ethanol in mice. *Physiol Behav*. 2007;91(1):77−86. https://doi.org/10.1016/j.physbeh.2007.01.024.
151. Sperling RE, Gomes SM, Sypek EI, Carey AN, McLaughlin JP. Endogenous kappa-opioid mediation of stress-induced potentiation of ethanol-conditioned place preference and self-administration. *Psychopharmacology*. 2010;210(2):199−209. https://doi.org/10.1007/s00213-010-1844-5.
152. Groblewski PA, Zietz C, Willuhn I, Phillips PE, Chavkin C. Repeated stress exposure causes strain-dependent shifts in the behavioral economics of cocaine in rats. *Addiction Biol*. 2015;20(2):297−301. https://doi.org/10.1111/adb.12123.
153. Kreibich AS, Briand L, Cleck JN, Ecke L, Rice KC, Blendy JA. Stress-induced potentiation of cocaine reward: a role for CRF R1 and CREB. *Neuropsychopharmacology*. 2009;34(12):2609−2617. https://doi.org/10.1038/npp.2009.91.

154. McLaughlin JP, Marton-Popovici M, Chavkin C. Kappa opioid receptor antagonism and prodynorphin gene disruption block stress-induced behavioral responses. *J Neurosci*. 2003;23(13):5674–5683. https://doi.org/10.1523/JNEUROSCI.23-13-05674.2003.
155. Schindler AG, Li S, Chavkin C. Behavioral stress may increase the rewarding valence of cocaine-associated cues through a dynorphin/kappa-opioid receptor-mediated mechanism without affecting associative learning or memory retrieval mechanisms. *Neuropsychopharmacology*. 2010;35(9):1932–1942. https://doi.org/10.1038/npp.2010.67.
156. Schindler AG, Messinger DI, Smith JS, et al. Stress produces aversion and potentiates cocaine reward by releasing endogenous dynorphins in the ventral striatum to locally stimulate serotonin reuptake. *J Neurosci*. 2012;32(49):17582–17596. https://doi.org/10.1523/JNEUROSCI.3220-12.2012.
157. Smith JS, Schindler AG, Martinelli E, Gustin RM, Bruchas MR, Chavkin C. Stress-induced activation of the dynorphin/κ-opioid receptor system in the amygdala potentiates nicotine conditioned place preference. *J Neurosci*. 2012;32(4):1488–1495. https://doi.org/10.1523/JNEUROSCI.2980-11.2012.
158. Karimi S, Attarzadeh-Yazdi G, Yazdi-Ravandi S, et al. Forced swim stress but not exogenous corticosterone could induce the reinstatement of extinguished morphine conditioned place preference in rats: involvement of glucocorticoid receptors in the basolateral amygdala. *Behav Brain Res*. 2014;264:43–50. https://doi.org/10.1016/j.bbr.2014.01.045.
159. Stafford NP, Kazan TN, Donovan CM, Hart EE, Drugan RC, Charntikov S. Individual vulnerability to stress is associated with increased demand for intravenous heroin self-administration in rats. *Front Behav Neurosci*. 2019;13:134. https://doi.org/10.3389/fnbeh.2019.00134.
160. Fatahi Z, Zeighamy Alamdary S, Khodagholi F, Shahamati SZ, Razavi Y, Haghparast A. Effect of physical stress on the alteration of mesolimbic system apoptotic factors in conditioned place preference paradigm. *Pharmacol Biochem Behav*. 2014;124:231–237. https://doi.org/10.1016/j.pbb.2014.06.017.
161. Boyce-Rustay JM, Janos AL, Holmes A. Effects of chronic swim stress on EtOH-related behaviors in C57BL/6J, DBA/2J and BALB/cByJ mice. *Behav Brain Res*. 2008;186(1):133–137. https://doi.org/10.1016/j.bbr.2007.07.031.
162. Vengeliene V, Siegmund S, Singer MV, Sinclair JD, Li TK, Spanagel R. A comparative study on alcohol-preferring rat lines: effects of deprivation and stress phases on voluntary alcohol intake. *Alcohol Clin Exp Res*. 2003;27(7):1048–1054. https://doi.org/10.1097/01.ALC.0000075829.81211.0C.
163. Piazza PV, Deminiere JM, le Moal M, Simon H. Stress- and pharmacologically-induced behavioral sensitization increases vulnerability to acquisition of amphetamine self-administration. *Brain Res*. 1990;514(1):22–26. https://doi.org/10.1016/0006-8993(90)90431-a.
164. Matsuzawa S, Suzuki T, Misawa M, Nagase H. Involvement of mu- and delta-opioid receptors in the ethanol-associated place preference in rats exposed to foot shock stress. *Brain Res*. 1998;803(1–2):169–177. https://doi.org/10.1016/s0006-8993(98)00679-9.
165. Will MJ, Watkins LR, Maier SF. Uncontrollable stress potentiates morphine's rewarding properties. *Pharmacol Biochem Behav*. 1998;60(3):655–664. https://doi.org/10.1016/s0091-3057(98)00027-6.
166. Bertholomey ML, Verplaetse TL, Czachowski CL. Alterations in ethanol seeking and self-administration following yohimbine in selectively bred alcohol-preferring (P) and high alcohol drinking (HAD-2) rats. *Behav Brain Res*. 2013;238:252–258. https://doi.org/10.1016/j.bbr.2012.10.030.
167. Lê AD, Harding S, Juzytsch W, Funk D, Shaham Y. Role of alpha-2 adrenoceptors in stress-induced reinstatement of alcohol seeking and alcohol self-administration in rats. *Psychopharmacology*. 2005;179(2):366–373. https://doi.org/10.1007/s00213-004-2036-y.
168. Lê AD, Funk D, Harding S, Juzytsch W, Fletcher PJ. The role of noradrenaline and 5-hydroxytryptamine in yohimbine-induced increases in alcohol-seeking in rats. *Psychopharmacology*. 2009;204(3):477–488. https://doi.org/10.1007/s00213-009-1481-z.
169. Marinelli PW, Funk D, Juzytsch W, et al. The CRF1 receptor antagonist antalarmin attenuates yohimbine-induced increases in operant alcohol self-administration and reinstatement of alcohol seeking in rats. *Psychopharmacology*. 2007;195(3):345–355. https://doi.org/10.1007/s00213-007-0905-x.
170. Campbell AT, Kwiatkowski D, Boughner E, Leri F. Effect of yohimbine stress on reacquisition of oxycodone seeking in rats. *Psychopharmacology*. 2012;222(2):247–255. https://doi.org/10.1007/s00213-012-2640-1.

171. Conrad KL, Davis AR, Silberman Y, et al. Yohimbine depresses excitatory transmission in BNST and impairs extinction of cocaine place preference through orexin-dependent, norepinephrine-independent processes. *Neuropsychopharmacology*. 2012;37(10):2253–2266. https://doi.org/10.1038/npp.2012.76.
172. Davis AR, Shields AD, Brigman JL, et al. Yohimbine impairs extinction of cocaine-conditioned place preference in an alpha2-adrenergic receptor independent process. *Learn Mem*. 2008;15(9):667–676. https://doi.org/10.1101/lm.1079308.
173. Kupferschmidt DA, Tribe E, Erb S. Effects of repeated yohimbine on the extinction and reinstatement of cocaine seeking. *Pharmacol, Biochem Behav*. 2009;91(3):473–480. https://doi.org/10.1016/j.pbb.2008.08.026.
174. Hofford RS, Prendergast MA, Bardo MT. Modified single prolonged stress reduces cocaine self-administration during acquisition regardless of rearing environment. *Behav Brain Res*. 2018;338:143–152. https://doi.org/10.1016/j.bbr.2017.10.023.
175. Wisłowska-Stanek A, Płaźnik A, Kołosowska K, et al. Differences in the dopaminergic reward system in rats that passively and actively behave in the Porsolt test. *Behav Brain Res*. 2019;359:181–189. https://doi.org/10.1016/j.bbr.2018.10.027.
176. Odeon MM, Yamauchi L, Grosman M, Acosta GB. Long-term effects of repeated maternal separation and ethanol intake on HPA axis responsiveness in adult rats. *Brain Res*. 2017;1657:193–201. https://doi.org/10.1016/j.brainres.2016.11.034.
177. Grakalic I, Schindler CW, Baumann MH, Rice KC, Riley AL. Effects of stress modulation on morphine-induced conditioned place preferences and plasma corticosterone levels in Fischer, Lewis, and Sprague-Dawley rat strains. *Psychopharmacology*. 2006;189(3):277–286. https://doi.org/10.1007/s00213-006-0562-5.
178. Ladd CO, Thrivikraman KV, Huot RL, Plotsky PM. Differential neuroendocrine responses to chronic variable stress in adult Long Evans rats exposed to handling-maternal separation as neonates. *Psychoneuroendocrinology*. 2005;30(6):520–533. https://doi.org/10.1016/j.psyneuen.2004.12.004.
179. Haile CN, GrandPre T, Kosten TA. Chronic unpredictable stress, but not chronic predictable stress, enhances the sensitivity to the behavioral effects of cocaine in rats. *Psychopharmacology*. 2001;154(2):213–220. https://doi.org/10.1007/s002130000650.
180. Miller LL, Ward SJ, Dykstra LA. Chronic unpredictable stress enhances cocaine-conditioned place preference in type 1 cannabinoid receptor knockout mice. *Behav Pharmacol*. 2008;19(5–6):575–581. https://doi.org/10.1097/FBP.0b013e32830ded11.
181. Papp M, Lappas S, Muscat R, Willner P. Attenuation of place preference conditioning but not place aversion conditioning by chronic mild stress. *J Psychopharmacol*. 1992;6(3):352–356. https://doi.org/10.1177/026988119200600302.
182. Baratta MV, Pomrenze MB, Nakamura S, Dolzani SD, Cooper DC. Control over stress accelerates extinction of drug seeking via prefrontal cortical activation. *Neurobiol Stress*. 2015;2:20–27. https://doi.org/10.1016/j.ynstr.2015.03.002.
183. Guerrero-Bautista R, Ribeiro Do Couto B, Hidalgo JM, et al. Modulation of stress- and cocaine prime-induced reinstatement of conditioned place preference after memory extinction through dopamine D3 receptor. *Prog Neuro Psychopharmacol Biol Psychiatr*. 2019;92:308–320. https://doi.org/10.1016/j.pnpbp.2019.01.017.
184. Guerrero-Bautista R, Franco-García A, Hidalgo JM, et al. Distinct regulation of dopamine D3 receptor in the basolateral amygdala and dentate gyrus during the reinstatement of cocaine cpp induced by drug priming and social stress. *Int J Mol Sci*. 2021;22(6):3100. https://doi.org/10.3390/ijms22063100.
185. Reguilón MD, Montagud-Romero S, Ferrer-Pérez C, et al. Dopamine D2 receptors mediate the increase in reinstatement of the conditioned rewarding effects of cocaine induced by acute social defeat. *Eur J Pharmacol*. 2017;799:48–57. https://doi.org/10.1016/j.ejphar.2017.01.039.
186. Ribeiro Do Couto B, Aguilar MA, Manzanedo C, Rodríguez-Arias M, Armario A, Miñarro J. Social stress is as effective as physical stress in reinstating morphine-induced place preference in mice. *Psychopharmacology*. 2006;185(4):459–470. https://doi.org/10.1007/s00213-006-0345-z.
187. Edalat P, Kavianpour M, Zarrabian S, Haghparast A. Role of orexin-1 and orexin-2 receptors in the CA1 region of hippocampus in the forced swim stress- and food deprivation-induced reinstatement of morphine seeking behaviors in rats. *Brain Res Bull*. 2018;142:25–32. https://doi.org/10.1016/j.brainresbull.2018.06.016.
188. Bhutada P, Mundhada Y, Ghodki Y, Dixit P, Umathe S, Jain K. Acquisition, expression, and reinstatement of ethanol-induced conditioned place preference in mice: effects of exposure to stress and modulation by mecamylamine. *J Psychopharmacol*. 2012;26(2):315–323. https://doi.org/10.1177/0269881111431749.

189. Cai J, Che X, Xu T, et al. Repeated oxytocin treatment during abstinence inhibited context- or restraint stress-induced reinstatement of methamphetamine-conditioned place preference and promoted adult hippocampal neurogenesis in mice. *Exp Neurol*. 2022;347:113907. https://doi.org/10.1016/j.expneurol.2021.113907.
190. Cruz FC, Leão RM, Marin MT, Planeta CS. Stress-induced reinstatement of amphetamine-conditioned place preference and changes in tyrosine hydroxylase in the nucleus accumbens in adolescent rats. *Pharmacol Biochem Behav*. 2010;96(2):160–165. https://doi.org/10.1016/j.pbb.2010.05.001.
191. Fan XY, Shi G, Zhao P. Reversal of oxycodone conditioned place preference by oxytocin: promoting global DNA methylation in the hippocampus. *Neuropharmacology*. 2019;160:107778. https://doi.org/10.1016/j.neuropharm.2019.107778.
192. Han WY, Du P, Fu SY, et al. Oxytocin via its receptor affects restraint stress-induced methamphetamine CPP reinstatement in mice: involvement of the medial prefrontal cortex and dorsal hippocampus glutamatergic system. *Pharmacol Biochem Behav*. 2014;119:80–87. https://doi.org/10.1016/j.pbb.2013.11.014.
193. Qi J, Yang JY, Wang F, Zhao YN, Song M, Wu CF. Effects of oxytocin on methamphetamine-induced conditioned place preference and the possible role of glutamatergic neurotransmission in the medial prefrontal cortex of mice in reinstatement. *Neuropharmacology*. 2009;56(5):856–865. https://doi.org/10.1016/j.neuropharm.2009.01.010.
194. Taslimi Z, Sarihi A, Haghparast A. Glucocorticoid receptors in the basolateral amygdala mediated the restraint stress-induced reinstatement of methamphetamine-seeking behaviors in rats. *Behav Brain Res*. 2018;348:150–159. https://doi.org/10.1016/j.bbr.2018.04.022.
195. Azizbeigi R, Haghparast A. Involvement of orexin-2 receptor in the ventral tegmental area in stress- and drug priming-induced reinstatement of conditioned place preference in rats. *Neurosci Lett*. 2019;696:121–126. https://doi.org/10.1016/j.neulet.2018.12.029.
196. Azizbeigi R, Farzinpour Z, Haghparast A. Role of orexin-1 receptor within the ventral tegmental area in mediating stress- and morphine priming-induced reinstatement of conditioned place preference in rats. *Basic Clin Neurosci*. 2019;10(4):373–382. https://doi.org/10.32598/bcn.9.10.130.
197. Ebrahimian F, Naghavi FS, Yazdi F, Sadeghzadeh F, Taslimi Z, Haghparast A. Differential roles of orexin receptors within the dentate gyrus in stress- and drug priming-induced reinstatement of conditioned place preference in rats. *Behav Neurosci*. 2016;130(1):91–102. https://doi.org/10.1037/bne0000112.
198. Farzinpour Z, Mousavi Z, Karimi-Haghighi S, Haghparast A. Antagonism of the D1- and D2-like dopamine receptors in the nucleus accumbens attenuates forced swim stress- and morphine priming-induced reinstatement of extinguished rats. *Behav Brain Res*. 2018;341:16–25. https://doi.org/10.1016/j.bbr.2017.12.010.
199. Graziane NM, Polter AM, Briand LA, Pierce RC, Kauer JA. Kappa opioid receptors regulate stress-induced cocaine seeking and synaptic plasticity. *Neuron*. 2013;77(5):942–954. https://doi.org/10.1016/j.neuron.2012.12.034.
200. Li C, Staub DR, Kirby LG. Role of GABAA receptors in dorsal raphe nucleus in stress-induced reinstatement of morphine-conditioned place preference in rats. *Psychopharmacology*. 2013;230(4):537–545. https://doi.org/10.1007/s00213-013-3182-x.
201. McReynolds JR, Vranjkovic O, Thao M, et al. Beta-2 adrenergic receptors mediate stress-evoked reinstatement of cocaine-induced conditioned place preference and increases in CRF mRNA in the bed nucleus of the stria terminalis in mice. *Psychopharmacology*. 2014;231(20):3953–3963. https://doi.org/10.1007/s00213-014-3535-0.
202. Polter AM, Bishop RA, Briand LA, Graziane NM, Pierce RC, Kauer JA. Poststress block of kappa opioid receptors rescues long-term potentiation of inhibitory synapses and prevents reinstatement of cocaine seeking. *Biol Psychiatr*. 2014;76(10):785–793. https://doi.org/10.1016/j.biopsych.2014.04.019.
203. Redila VA, Chavkin C. Stress-induced reinstatement of cocaine seeking is mediated by the kappa opioid system. *Psychopharmacology*. 2008;200(1):59–70. https://doi.org/10.1007/s00213-008-1122-y.
204. Vaughn LK, Mantsch JR, Vranjkovic O, et al. Cannabinoid receptor involvement in stress-induced cocaine reinstatement: potential interaction with noradrenergic pathways. *Neuroscience*. 2012;204:117–124. https://doi.org/10.1016/j.neuroscience.2011.08.021.
205. Vranjkovic O, Hang S, Baker DA, Mantsch JR. β-adrenergic receptor mediation of stress-induced reinstatement of extinguished cocaine-induced conditioned place preference in mice: roles for β1 and β2 adrenergic receptors. *J Pharmacol Exp Therapeut*. 2012;342(2):541–551. https://doi.org/10.1124/jpet.112.193615.

206. Capriles N, Rodaros D, Sorge RE, Stewart J. A role for the prefrontal cortex in stress- and cocaine-induced reinstatement of cocaine seeking in rats. *Psychopharmacology*. 2003;168(1−2):66−74. https://doi.org/10.1007/s00213-002-1283-z.
207. Figueroa-Guzman Y, Mueller C, Vranjkovic O, et al. Oral administration of levo-tetrahydropalmatine attenuates reinstatement of extinguished cocaine seeking by cocaine, stress or drug-associated cues in rats. *Drug Alcohol Depend*. 2011;116(1−3):72−79. https://doi.org/10.1016/j.drugalcdep.2010.11.023.
208. Hammerslag LR, Denehy ED, Carper B, Nolen TL, Prendergast MA, Bardo MT. Effects of the glucocorticoid receptor antagonist PT150 on stress-induced fentanyl seeking in male and female rats. *Psychopharmacology*. 2021;238(9):2439−2447. https://doi.org/10.1007/s00213-021-05865-0.
209. Lê AD, Funk D, Juzytsch W, et al. Effect of prazosin and guanfacine on stress-induced reinstatement of alcohol and food seeking in rats. *Psychopharmacology*. 2011;218(1):89−99. https://doi.org/10.1007/s00213-011-2178-7.
210. Martin-Fardon R, Ciccocioppo R, Massi M, Weiss F. Nociceptin prevents stress-induced ethanol- but not cocaine-seeking behavior in rats. *Neuroreport*. 2000;11(9):1939−1943. https://doi.org/10.1097/00001756-200006260-00026.
211. Shaham Y, Stewart J. Effects of opioid and dopamine receptor antagonists on relapse induced by stress and re-exposure to heroin in rats. *Psychopharmacology*. 1996;125(4):385−391. https://doi.org/10.1007/BF02246022.
212. Wang B, Shaham Y, Zitzman D, Azari S, Wise RA, You ZB. Cocaine experience establishes control of midbrain glutamate and dopamine by corticotropin-releasing factor: a role in stress-induced relapse to drug seeking. *J Neurosci*. 2005;25(22):5389−5396. https://doi.org/10.1523/JNEUROSCI.0955-05.2005.
213. Brown ZJ, Kupferschmidt DA, Erb S. Reinstatement of cocaine seeking in rats by the pharmacological stressors, corticotropin-releasing factor and yohimbine: role for D1/5 dopamine receptors. *Psychopharmacology*. 2012;224(3):431−440. https://doi.org/10.1007/s00213-012-2772-3.
214. Buffalari DM, See RE. Inactivation of the bed nucleus of the stria terminalis in an animal model of relapse: effects on conditioned cue-induced reinstatement and its enhancement by yohimbine. *Psychopharmacology*. 2011;213(1):19−27. https://doi.org/10.1007/s00213-010-2008-3.
215. Cippitelli A, Brunori G, Schoch J, et al. Differential regulation of alcohol taking and seeking by antagonism at α4β2 and α3β4 nAChRs. *Psychopharmacology*. 2018;235(6):1745−1757. https://doi.org/10.1007/s00213-018-4883-y.
216. Fletcher PJ, Rizos Z, Sinyard J, Tampakeras M, Higgins GA. The 5-HT2C receptor agonist Ro60-0175 reduces cocaine self-administration and reinstatement induced by the stressor yohimbine, and contextual cues. *Neuropsychopharmacology*. 2008;33(6):1402−1412. https://doi.org/10.1038/sj.npp.1301509.
217. Gass JT, Olive MF. Reinstatement of ethanol-seeking behavior following intravenous self-administration in Wistar rats. *Alcohol Clin Exp Res*. 2007;31(9):1441−1445. https://doi.org/10.1111/j.1530-0277.2007.00480.x.
218. King CE, Becker HC. Oxytocin attenuates stress-induced reinstatement of alcohol seeking behavior in male and female mice. *Psychopharmacology*. 2019;236(9):2613−2622. https://doi.org/10.1007/s00213-019-05233-z.
219. Liu J, Johnson B, Wu R, et al. TAAR1 agonists attenuate extended-access cocaine self-administration and yohimbine-induced reinstatement of cocaine-seeking. *Br J Pharmacol*. 2020;177(15):3403−3414. https://doi.org/10.1111/bph.15061.
220. Schmoutz CD, Zhang Y, Runyon SP, Goeders NE. Antagonism of the neuropeptide S receptor with RTI-118 decreases cocaine self-administration and cocaine-seeking behavior in rats. *Pharmacol Biochem Behav*. 2012;103(2):332−337. https://doi.org/10.1016/j.pbb.2012.09.003.
221. Simms JA, Bito-Onon JJ, Chatterjee S, Bartlett SE. Long-Evans rats acquire operant self-administration of 20% ethanol without sucrose fading. *Neuropsychopharmacology*. 2010;35(7):1453−1463. https://doi.org/10.1038/npp.2010.15.
222. Zhou Y, Leri F, Grella SL, Aldrich JV, Kreek MJ. Involvement of dynorphin and kappa opioid receptor in yohimbine-induced reinstatement of heroin seeking in rats. *Synapse*. 2013;67(6):358−361. https://doi.org/10.1002/syn.21638.
223. Manvich DF, Stowe TA, Godfrey JR, Weinshenker D. A method for psychosocial stress-induced reinstatement of cocaine seeking in rats. *Biol Psychiatr*. 2016;79(11):940−946. https://doi.org/10.1016/j.biopsych.2015.07.002.
224. Wemm SE, Sinha R. Drug-induced stress responses and addiction risk and relapse. *Neurobiol Stress*. 2019;10:100148. https://doi.org/10.1016/j.ynstr.2019.100148.
225. Hefner KR, Starr MJ, Curtin JJ. Heavy marijuana use but not deprivation is associated with increased stressor reactivity. *J Abnorm Psychol*. 2018;127(4):348−358. https://doi.org/10.1037/abn0000344.

226. Bradford DE, Curtin JJ, Piper ME. Anticipation of smoking sufficiently dampens stress reactivity in nicotine-deprived smokers. *J Abnorm Psychol*. 2015;124(1):128–136. https://doi.org/10.1037/abn0000007.
227. Glodosky NC, Cuttler C, Freels TG, et al. Cannabis vapor self-administration elicits sex- and dose-specific alterations in stress reactivity in rats. *Neurobiol Stress*. 2020;13:100260. https://doi.org/10.1016/j.ynstr.2020.100260.
228. Morse DE. Neuroendocrine responses to nicotine and stress: enhancement of peripheral stress responses by the administration of nicotine. *Psychopharmacology*. 1989;98(4):539–543. https://doi.org/10.1007/BF00441956.
229. Goeders NE, Clampitt DM. Potential role for the hypothalamo-pituitary-adrenal axis in the conditioned reinforcer-induced reinstatement of extinguished cocaine seeking in rats. *Psychopharmacology*. 2002;161(3):222–232. https://doi.org/10.1007/s00213-002-1007-4.
230. Mantsch JR, Cullinan WE, Tang LC, et al. Daily cocaine self-administration under long-access conditions augments restraint-induced increases in plasma corticosterone and impairs glucocorticoid receptor-mediated negative feedback in rats. *Brain Res*. 2007;1167:101–111. https://doi.org/10.1016/j.brainres.2007.05.080.
231. Mantsch JR, Taves S, Khan T, et al. Restraint-induced corticosterone secretion and hypothalamic CRH mRNA expression are augmented during acute withdrawal from chronic cocaine administration. *Neurosci Lett*. 2007;415(3):269–273. https://doi.org/10.1016/j.neulet.2007.01.036.
232. Park PE, Vendruscolo LF, Schlosburg JE, Edwards S, Schulteis G, Koob GF. Corticotropin-releasing factor (CRF) and α2 adrenergic receptors mediate heroin withdrawal-potentiated startle in rats. *Int J Neuropsychopharmacol*. 2013;16(8):1867–1875. https://doi.org/10.1017/S1461145713000308.
233. Ligabue KP, Schuch JB, Scherer JN, et al. Increased cortisol levels are associated with low treatment retention in crack cocaine users. *Addict Behav*. 2020;103:106260. https://doi.org/10.1016/j.addbeh.2019.106260.
234. Lodge DJ, Lawrence AJ. The CRF1 receptor antagonist antalarmin reduces volitional ethanol consumption in isolation-reared fawn-hooded rats. *Neuroscience*. 2003;117(2):243–247. https://doi.org/10.1016/s0306-4522(02)00793-5.
235. Shinohara F, Arakaki S, Amano T, Minami M, Kaneda K. Noradrenaline enhances the excitatory effects of dopamine on medial prefrontal cortex pyramidal neurons in rats. *Neuropsychopharmacol Rep*. 2020;40(4):348–354. https://doi.org/10.1002/npr2.12135.
236. Arborelius L, Eklund MB. Both long and brief maternal separation produces persistent changes in tissue levels of brain monoamines in middle-aged female rats. *Neuroscience*. 2007;145(2):738–750. https://doi.org/10.1016/j.neuroscience.2006.12.007.
237. Briand LA, Vassoler FM, Pierce RC, Valentino RJ, Blendy JA. Ventral tegmental afferents in stress-induced reinstatement: the role of cAMP response element-binding protein. *J Neurosci*. 2010;30(48):16149–16159. https://doi.org/10.1523/JNEUROSCI.2827-10.2010.
238. Razavi Y, Karimi S, Karimi-Haghighi S, Hesam S, Haghparast A. Changes in c-fos and p-CREB signaling following exposure to forced swim stress or exogenous corticosterone during morphine-induced place preference are dependent on glucocorticoid receptor in the basolateral amygdala. *Can J Physiol Pharmacol*. 2020;98(11):741–752. https://doi.org/10.1139/cjpp-2019-0712.
239. Munshi S, Rosenkranz JA, Caccamise A, Wolf ME, Corbett CM, Loweth JA. Cocaine and chronic stress exposure produce an additive increase in neuronal activity in the basolateral amygdala. *Addiction Biol*. 2021;26(1):e12848. https://doi.org/10.1111/adb.12848.
240. Wang J, Zhao Z, Liang Q, et al. The nucleus accumbens core has a more important role in resisting reactivation of extinguished conditioned place preference in morphine-addicted rats. *J Int Med Res*. 2008;36(4):673–681. https://doi.org/10.1177/147323000803600408.
241. Gill MJ, Ghee SM, Harper SM, See RE. Inactivation of the lateral habenula reduces anxiogenic behavior and cocaine seeking under conditions of heightened stress. *Pharmacol Biochem Behav*. 2013;111:24–29. https://doi.org/10.1016/j.pbb.2013.08.002.
242. Romano-López A, Méndez-Díaz M, García FG, Regalado-Santiago C, Ruiz-Contreras AE, Próspero-García O. Maternal separation and early stress cause long-lasting effects on dopaminergic and endocannabinergic systems and alters dendritic morphology in the nucleus accumbens and frontal cortex in rats. *Develop Neurobiol*. 2016;76(8):819–831. https://doi.org/10.1002/dneu.22361.
243. Whitaker LR, Degoulet M, Morikawa H. Social deprivation enhances VTA synaptic plasticity and drug-induced contextual learning. *Neuron*. 2013;77(2):335–345. https://doi.org/10.1016/j.neuron.2012.11.022.

244. Samuelson KW. Post-traumatic stress disorder and declarative memory functioning: a review. *Dialogues Clin Neurosci.* 2011;13(3):346–351. https://doi.org/10.31887/DCNS.2011.13.2/ksamuelson.
245. Yang Y, Zheng X, Wang Y, et al. Stress enables synaptic depression in CA1 synapses by acute and chronic morphine: possible mechanisms for corticosterone on opiate addiction. *J Neurosci.* 2004;24(10):2412–2420. https://doi.org/10.1523/JNEUROSCI.5544-03.2004.
246. Bianco CD, Hübner IC, Bennemann B, de Carvalho CR, Brocardo PS. Effects of postnatal ethanol exposure and maternal separation on mood, cognition and hippocampal arborization in adolescent rats. *Behav Brain Res.* 2021;411:113372. https://doi.org/10.1016/j.bbr.2021.113372.
247. Almonte AG, Ewin SE, Mauterer MI, Morgan JW, Carter ES, Weiner JL. Enhanced ventral hippocampal synaptic transmission and impaired synaptic plasticity in a rodent model of alcohol addiction vulnerability. *Sci Rep.* 2017;7(1):12300. https://doi.org/10.1038/s41598-017-12531-z.
248. Flavin SA, Winder DG. Noradrenergic control of the bed nucleus of the stria terminalis in stress and reward. *Neuropharmacology.* 2013;70:324–330. https://doi.org/10.1016/j.neuropharm.2013.02.013.
249. Jadzic D, Bassareo V, Carta AR, Carboni E. Nicotine, cocaine, amphetamine, morphine, and ethanol increase norepinephrine output in the bed nucleus of stria terminalis of freely moving rats. *Addiction Biol.* 2021;26(1):e12864. https://doi.org/10.1111/adb.12864.
250. Macey DJ, Smith HR, Nader MA, Porrino LJ. Chronic cocaine self-administration upregulates the norepinephrine transporter and alters functional activity in the bed nucleus of the stria terminalis of the rhesus monkey. *J Neurosci.* 2003;23(1):12–16. https://doi.org/10.1523/JNEUROSCI.23-01-00012.2003.
251. Snyder AE, Salimando GJ, Winder DG, Silberman Y. Chronic intermittent ethanol and acute stress similarly modulate BNST CRF neuron activity via noradrenergic signaling. *Alcohol Clin Exp Res.* 2019;43(8):1695–1701. https://doi.org/10.1111/acer.14118.
252. Vranjkovic O, Gasser PJ, Gerndt CH, Baker DA, Mantsch JR. Stress-induced cocaine seeking requires a beta-2 adrenergic receptor-regulated pathway from the ventral bed nucleus of the stria terminalis that regulates CRF actions in the ventral tegmental area. *J Neurosci.* 2014;34(37):12504–12514. https://doi.org/10.1523/JNEUROSCI.0680-14.2014.
253. Erb S, Stewart J. A role for the bed nucleus of the stria terminalis, but not the amygdala, in the effects of corticotropin-releasing factor on stress-induced reinstatement of cocaine seeking. *J Neurosci.* 1999;19(20):RC35. https://doi.org/10.1523/JNEUROSCI.19-20-j0006.1999.
254. Rougé-Pont F, Deroche V, Le Moal M, Piazza PV. Individual differences in stress-induced dopamine release in the nucleus accumbens are influenced by corticosterone. *Eur J Neurosci.* 1998;10(12):3903–3907. https://doi.org/10.1046/j.1460-9568.1998.00438.x.
255. Morrow BA, Roth RH. Serotonergic lesions alter cocaine-induced locomotor behavior and stress-activation of the mesocorticolimbic dopamine system. *Synapse.* 1996;23(3):174–181. https://doi.org/10.1002/(SICI)1098-2396(199607)23:3<174::AID-SYN6>3.0.CO;2-5.
256. Mangiavacchi S, Masi F, Scheggi S, Leggio B, De Montis MG, Gambarana C. Long-term behavioral and neurochemical effects of chronic stress exposure in rats. *J Neurochem.* 2001;79(6):1113–1121. https://doi.org/10.1046/j.1471-4159.2001.00665.x.
257. Fabricius K, Steiniger-Brach B, Helboe L, Fink-Jensen A, Wörtwein G. Socially isolated rats exhibit changes in dopamine homeostasis pertinent to schizophrenia. *Int J Dev Neurosci.* 2011;29(3):347–350. https://doi.org/10.1016/j.ijdevneu.2010.09.003.
258. Hall FS, Wilkinson LS, Humby T, et al. Isolation rearing in rats: pre- and postsynaptic changes in striatal dopaminergic systems. *Pharmacol Biochem Behav.* 1998;59(4):859–872. https://doi.org/10.1016/s0091-3057(97)00510-8.
259. Jones GH, Hernandez TD, Kendall DA, Marsden CA, Robbins TW. Dopaminergic and serotonergic function following isolation rearing in rats: study of behavioural responses and postmortem and in vivo neurochemistry. *Pharmacol Biochem Behav.* 1992;43(1):17–35. https://doi.org/10.1016/0091-3057(92)90635-s.
260. Karkhanis AN, Leach AC, Yorgason JT, et al. Chronic social isolation stress during peri-adolescence alters presynaptic dopamine terminal dynamics via augmentation in accumbal dopamine availability. *ACS Chem Neurosci.* 2019;10(4):2033–2044. https://doi.org/10.1021/acschemneuro.8b00360.
261. Lampert C, Arcego DM, de Sá Couto-Pereira N, et al. Short post-weaning social isolation induces long-term changes in the dopaminergic system and increases susceptibility to psychostimulants in female rats. *Int J Dev Neurosci.* 2017;61:21–30. https://doi.org/10.1016/j.ijdevneu.2017.05.003.

262. Miczek KA, Nikulina EM, Shimamoto A, Covington 3rd HE. Escalated or suppressed cocaine reward, tegmental BDNF, and accumbal dopamine caused by episodic versus continuous social stress in rats. *J Neurosci*. 2011;31(27):9848−9857. https://doi.org/10.1523/JNEUROSCI.0637-11.2011.
263. Pacchioni AM, Gioino G, Assis A, Cancela LM. A single exposure to restraint stress induces behavioral and neurochemical sensitization to stimulating effects of amphetamine: involvement of NMDA receptors. *Ann N Y Acad Sci*. 2002;965:233−246. https://doi.org/10.1111/j.1749-6632.2002.tb04165.x.
264. Sim HR, Choi TY, Lee HJ, et al. Role of dopamine D2 receptors in plasticity of stress-induced addictive behaviours. *Nat Commun*. 2013;4:1579. https://doi.org/10.1038/ncomms2598.
265. Delis F, Thanos PK, Rombola C, et al. Chronic mild stress increases alcohol intake in mice with low dopamine D2 receptor levels. *Behav Neurosci*. 2013;127(1):95−105. https://doi.org/10.1037/a0030750.
266. Nazari-Serenjeh F, Rezaee L, Zarrabian S, Haghparast A. Comparison of the role of D1- and D2-like receptors in the CA1 region of the hippocampus in the reinstatement induced by a subthreshold dose of morphine and forced swim stress in extinguished morphine-cpp in rats. *Neurochem Res*. 2018;43(11):2092−2101. https://doi.org/10.1007/s11064-018-2631-7.
267. Norozpour Y, Zarrabian S, Rezaee L, Haghparast A. D1- and D2-like receptors in the dentate gyrus region of the hippocampus are involved in the reinstatement induced by a subthreshold dose of morphine and forced swim stress in extinguished morphine-CPP in rats. *Behav Neurosci*. 2019;133(6):545−555. https://doi.org/10.1037/bne0000335.
268. Yorgason JT, Calipari ES, Ferris MJ, et al. Social isolation rearing increases dopamine uptake and psychostimulant potency in the striatum. *Neuropharmacology*. 2016;101:471−479. https://doi.org/10.1016/j.neuropharm.2015.10.025.
269. Bendre M, Comasco E, Nylander I, Nilsson KW. Effect of voluntary alcohol consumption on Maoa expression in the mesocorticolimbic brain of adult male rats previously exposed to prolonged maternal separation. *Transl Psychiatr*. 2015;5(12):e690. https://doi.org/10.1038/tp.2015.186.
270. Vicentic A, Francis D, Moffett M, et al. Maternal separation alters serotonergic transporter densities and serotonergic 1A receptors in rat brain. *Neuroscience*. 2006;140(1):355−365. https://doi.org/10.1016/j.neuroscience.2006.02.008.
271. Kosten TA, Zhang XY, Kehoe P. Infant rats with chronic neonatal isolation experience show decreased extracellular serotonin levels in ventral striatum at baseline and in response to cocaine. *Brain Res. Develop Brain Res*. 2004;152(1):19−24. https://doi.org/10.1016/j.devbrainres.2004.05.005.
272. Smith DG, Davis RJ, Gehlert DR, Nomikos GG. Exposure to predator odor stress increases efflux of frontal cortex acetylcholine and monoamines in mice: comparisons with immobilization stress and reversal by chlordiazepoxide. *Brain Res*. 2006;1114(1):24−30. https://doi.org/10.1016/j.brainres.2006.07.058.
273. Hamamura T, Fibiger HC. Enhanced stress-induced dopamine release in the prefrontal cortex of amphetamine-sensitized rats. *Eur J Pharmacol*. 1993;237(1):65−71. https://doi.org/10.1016/0014-2999(93)90094-x.
274. Guzman AS, Avalos MP, De Giovanni LN, et al. CB1R activation in nucleus accumbens core promotes stress-induced reinstatement of cocaine seeking by elevating extracellular glutamate in a drug-paired context. *Sci Rep*. 2021;11(1):12964. https://doi.org/10.1038/s41598-021-92389-4.
275. Deutschmann AU, Kirkland JM, Briand LA. Adolescent social isolation induced alterations in nucleus accumbens glutamate signalling. *Addiction Biol*. 2022;27(1):e13077. https://doi.org/10.1111/adb.13077.
276. Bellinger FP, Davidson MS, Bedi KS, Wilce PA. Ethanol prevents NMDA receptor reduction by maternal separation in neonatal rat hippocampus. *Brain Res*. 2006;1067(1):154−157. https://doi.org/10.1016/j.brainres.2005.09.067.
277. De Giovanni LN, Guzman AS, Virgolini MB, Cancela LM. NMDA antagonist MK 801 in nucleus accumbens core but not shell disrupts the restraint stress-induced reinstatement of extinguished cocaine-conditioned place preference in rats. *Behav Brain Res*. 2016;315:150−159. https://doi.org/10.1016/j.bbr.2016.08.011.
278. Durieux AM, Fernandes C, Murphy D, et al. Targeting glia with N-acetylcysteine modulates brain glutamate and behaviors relevant to neurodevelopmental disorders in C57BL/6J mice. *Front Behav Neurosci*. 2015;9:343. https://doi.org/10.3389/fnbeh.2015.00343.
279. Herrmann AP, Benvenutti R, Pilz LK, Elisabetsky E. N-acetylcysteine prevents increased amphetamine sensitivity in social isolation-reared mice. *Schizophr Res*. 2014;155(1−3):109−111. https://doi.org/10.1016/j.schres.2014.03.012.

280. Garcia-Keller C, Smiley C, Monforton C, Melton S, Kalivas PW, Gass J. N-Acetylcysteine treatment during acute stress prevents stress-induced augmentation of addictive drug use and relapse. *Addiction Biol.* 2020;25(5):e12798. https://doi.org/10.1111/adb.12798.
281. Imperato A, Puglisi-Allegra S, Casolini P, Angelucci L. Changes in brain dopamine and acetylcholine release during and following stress are independent of the pituitary-adrenocortical axis. *Brain Res.* 1991;538(1):111−117. https://doi.org/10.1016/0006-8993(91)90384-8.
282. Mark GP, Rada PV, Shors TJ. Inescapable stress enhances extracellular acetylcholine in the rat hippocampus and prefrontal cortex but not the nucleus accumbens or amygdala. *Neuroscience.* 1996;74(3):767−774. https://doi.org/10.1016/0306-4522(96)00211-4.
283. Stengård K. Tail pinch increases acetylcholine release in rat striatum even after toluene exposure. *Pharmacol Biochem Behav.* 1995;52(2):261−264. https://doi.org/10.1016/0091-3057(95)00090-j.
284. Lemos JC, Shin JH, Alvarez VA. Striatal Cholinergic interneurons are a novel target of corticotropin releasing factor. *J Neurosci.* 2019;39(29):5647−5661. https://doi.org/10.1523/JNEUROSCI.0479-19.2019.
285. Yuan M, Malagon AM, Yasuda D, Belluzzi JD, Leslie FM, Zaveri NT. The α3β4 nAChR partial agonist AT-1001 attenuates stress-induced reinstatement of nicotine seeking in a rat model of relapse and induces minimal withdrawal in dependent rats. *Behav Brain Res.* 2017;333:251−257. https://doi.org/10.1016/j.bbr.2017.07.004.
286. Javadi P, Rezayof A, Sardari M, Ghasemzadeh Z. Brain nicotinic acetylcholine receptors are involved in stress-induced potentiation of nicotine reward in rats. *J Psychopharmacol.* 2017;31(7):945−955. https://doi.org/10.1177/0269881117707745.
287. Serrano A, Pavon FJ, Buczynski MW, et al. Deficient endocannabinoid signaling in the central amygdala contributes to alcohol dependence-related anxiety-like behavior and excessive alcohol intake. *Neuropsychopharmacology.* 2018;43(9):1840−1850. https://doi.org/10.1038/s41386-018-0055-3.
288. Chauvet C, Nicolas C, Thiriet N, Lardeux MV, Duranti A, Solinas M. Chronic stimulation of the tone of endogenous anandamide reduces cue- and stress-induced relapse in rats. *Int J Neuropsychopharmacol.* 2014;18(1):pyu025. https://doi.org/10.1093/ijnp/pyu025.
289. Malone DT, Kearn CS, Chongue L, Mackie K, Taylor DA. Effect of social isolation on CB1 and D2 receptor and fatty acid amide hydrolase expression in rats. *Neuroscience.* 2008;152(1):265−272. https://doi.org/10.1016/j.neuroscience.2007.10.043.
290. Sciolino NR, Bortolato M, Eisenstein SA, et al. Social isolation and chronic handling alter endocannabinoid signaling and behavioral reactivity to context in adult rats. *Neuroscience.* 2010;168(2):371−386. https://doi.org/10.1016/j.neuroscience.2010.04.007.
291. Kupferschmidt DA, Klas PG, Erb S. Cannabinoid CB1 receptors mediate the effects of corticotropin-releasing factor on the reinstatement of cocaine seeking and expression of cocaine-induced behavioural sensitization. *Br J Pharmacol.* 2012;167(1):196−206. https://doi.org/10.1111/j.1476-5381.2012.01983.x.
292. McReynolds JR, Doncheck EM, Li Y, et al. Stress promotes drug seeking through glucocorticoid-dependent endocannabinoid mobilization in the prelimbic cortex. *Biol Psychiatr.* 2018;84(2):85−94. https://doi.org/10.1016/j.biopsych.2017.09.024.
293. McReynolds JR, Doncheck EM, Vranjkovic O, et al. CB1 receptor antagonism blocks stress-potentiated reinstatement of cocaine seeking in rats. *Psychopharmacology.* 2016;233(1):99−109. https://doi.org/10.1007/s00213-015-4092-x.
294. Hamilton J, Marion M, Figueiredo A, et al. Fatty acid binding protein deletion prevents stress-induced preference for cocaine and dampens stress-induced corticosterone levels. *Synapse.* 2018;72(6):e22031. https://doi.org/10.1002/syn.22031.
295. Granholm L, Todkar A, Bergman S, Nilsson K, Comasco E, Nylander I. The expression of opioid genes in non-classical reward areas depends on early life conditions and ethanol intake. *Brain Res.* 2017;1668:36−45. https://doi.org/10.1016/j.brainres.2017.05.006.
296. Nakamoto K, Taniguchi A, Tokuyama S. Changes in opioid receptors, opioid peptides and morphine antinociception in mice subjected to early life stress. *Eur J Pharmacol.* 2020;881:173173. https://doi.org/10.1016/j.ejphar.2020.173173.
297. Nielsen CK, Simms JA, Bito-Onon JJ, Li R, Ananthan S, Bartlett SE. The delta opioid receptor antagonist, SoRI-9409, decreases yohimbine stress-induced reinstatement of ethanol-seeking. *Addiction Biol.* 2012;17(2):224−234. https://doi.org/10.1111/j.1369-1600.2010.00295.x.

298. Sartor GC, Powell SK, Wiedner HJ, Wahlestedt C, Brothers SP. Nociceptin receptor activation does not alter acquisition, expression, extinction and reinstatement of conditioned cocaine preference in mice. *Brain Res.* 2016;1632:34—41. https://doi.org/10.1016/j.brainres.2015.11.044.
299. Fontaine HM, Silva PR, Neiswanger C, et al. Stress decreases serotonin tone in the nucleus accumbens in male mice to promote aversion and potentiate cocaine preference via decreased stimulation of 5-HT1B receptors. *Neuropsychopharmacology.* 2022;47(4):891—901. https://doi.org/10.1038/s41386-021-01178-0.
300. Douglass J, McKinzie AA, Couceyro P. PCR differential display identifies a rat brain mRNA that is transcriptionally regulated by cocaine and amphetamine. *J Neurosci.* 1995;15(3 Pt 2):2471—2481. https://doi.org/10.1523/JNEUROSCI.15-03-02471.1995.
301. Hubert GW, Jones DC, Moffett MC, Rogge G, Kuhar MJ. CART peptides as modulators of dopamine and psychostimulants and interactions with the mesolimbic dopaminergic system. *Biochem Pharmacol.* 2008;75(1):57—62. https://doi.org/10.1016/j.bcp.2007.07.028.
302. Younes-Rapozo V, de Moura EG, da Silva Lima N, et al. Early weaning is associated with higher neuropeptide Y (NPY) and lower cocaine- and amphetamine-regulated transcript (CART) expressions in the paraventricular nucleus (PVN) in adulthood. *Br J Nutr.* 2012;108(12):2286—2295. https://doi.org/10.1017/S0007114512000487.
303. Dandekar MP, Singru PS, Kokare DM, Subhedar NK. Cocaine- and amphetamine-regulated transcript peptide plays a role in the manifestation of depression: social isolation and olfactory bulbectomy models reveal unifying principles. *Neuropsychopharmacology.* 2009;34(5):1288—1300. https://doi.org/10.1038/npp.2008.201.
304. Somalwar AR, Choudhary AG, Balasubramanian N, Sakharkar AJ, Subhedar NK, Kokare DM. Cocaine- and amphetamine-regulated transcript peptide promotes reward seeking behavior in socially isolated rats. *Brain Res.* 2020;1728:146595. https://doi.org/10.1016/j.brainres.2019.146595.
305. Hunter RG, Bellani R, Bloss E, Costa A, Romeo RD, McEwen BS. Regulation of CART mRNA by stress and corticosteroids in the hippocampus and amygdala. *Brain Res.* 2007;1152:234—240. https://doi.org/10.1016/j.brainres.2007.03.042.
306. Liu J, Hu X-M, Li X-J, Zhao H. Traumatic stress affects alcohol-drinking behavior through cocaine- and amphetamine-regulated transcript 55-102 in the paraventricular nucleus in rats. *Mol Med Rep.* 2018;17(1):1157—1165. https://doi.org/10.3892/mmr.2017.7989.
307. Tung LW, Lu GL, Lee YH, et al. Orexins contribute to restraint stress-induced cocaine relapse by endocannabinoid-mediated disinhibition of dopaminergic neurons. *Nat Commun.* 2016;7:12199. https://doi.org/10.1038/ncomms12199.
308. Chou YH, Hor CC, Lee MT, et al. Stress induces reinstatement of extinguished cocaine conditioned place preference by a sequential signaling via neuropeptide S, orexin, and endocannabinoid. *Addiction Biol.* 2021;26(3):e12971. https://doi.org/10.1111/adb.12971.
309. Richards JK, Simms JA, Steensland P, et al. Inhibition of orexin-1/hypocretin-1 receptors inhibits yohimbine-induced reinstatement of ethanol and sucrose seeking in Long-Evans rats. *Psychopharmacology.* 2008;199(1):109—117. https://doi.org/10.1007/s00213-008-1136-5.
310. Koob GF. Neurobiology of opioid addiction: opponent process, hyperkatifeia, and negative reinforcement. *Biol Psychiatr.* 2020;87(1):44—53. https://doi.org/10.1016/j.biopsych.2019.05.023.
311. Koob GF, Schulkin J. Addiction and stress: an allostatic view. *Neurosci Biobehav Rev.* 2019;106:245—262. https://doi.org/10.1016/j.neubiorev.2018.09.008.
312. Peciña S, Schulkin J, Berridge KC. Nucleus accumbens corticotropin-releasing factor increases cue-triggered motivation for sucrose reward: paradoxical positive incentive effects in stress? *BMC Biol.* 2006;4:8. https://doi.org/10.1186/1741-7007-4-8.
313. Meade CS, Drabkin AS, Hansen NB, Wilson PA, Kochman A, Sikkema KJ. Reductions in alcohol and cocaine use following a group coping intervention for HIV-positive adults with childhood sexual abuse histories. *Addiction.* 2010;105(11):1942—1951. https://doi.org/10.1111/j.1360-0443.2010.03075.x.
314. Milosevic I, Chudzik SM, Boyd S, McCabe RE. Evaluation of an integrated group cognitive-behavioral treatment for comorbid mood, anxiety, and substance use disorders: a pilot study. *J Anxiety Disord.* 2017;46:85—100. https://doi.org/10.1016/j.janxdis.2016.08.002.
315. Sannibale C, Teesson M, Creamer M, et al. Randomized controlled trial of cognitive behaviour therapy for co-morbid post-traumatic stress disorder and alcohol use disorders. *Addiction.* 2013;108(8):1397—1410. https://doi.org/10.1111/add.12167.

316. Japuntich SJ, Lee LO, Pineles SL, et al. Contingency management and cognitive behavioral therapy for trauma-exposed smokers with and without posttraumatic stress disorder. *Addict Behav*. 2019;90:136–142. https://doi.org/10.1016/j.addbeh.2018.10.042.
317. Tang YY, Tang R, Posner MI. Mindfulness meditation improves emotion regulation and reduces drug abuse. *Drug Alcohol Depend*. 2016;163(suppl 1):S13–S18. https://doi.org/10.1016/j.drugalcdep.2015.11.041.
318. Cambron C, Hopkins P, Burningham C, Lam C, Cinciripini P, Wetter DW. Socioeconomic status, mindfulness, and momentary associations between stress and smoking lapse during a quit attempt. *Drug Alcohol Depend*. 2020;209:107840. https://doi.org/10.1016/j.drugalcdep.2020.107840.
319. Adams CE, Cano MA, Heppner WL, et al. Testing a moderated mediation model of mindfulness, psychosocial stress, and alcohol use among African American smokers. *Mindfulness*. 2015;6(2):315–325. https://doi.org/10.1007/s12671-013-0263-1.
320. Garland EL, Boettiger CA, Gaylord S, Chanon VW, Howard MO. Mindfulness is Inversely associated with alcohol attentional bias among recovering alcohol-dependent adults. *Cogn Ther & Res*. 2012;36(5):441–450. https://doi.org/10.1007/s10608-011-9378-7.
321. Bodenlos JS, Noonan M, Wells SY. Mindfulness and alcohol problems in college students: the mediating effects of stress. *J Am Coll Health*. 2013;61(6):371–378. https://doi.org/10.1080/07448481.2013.805714.
322. Shuai R, Bakou AE, Hardy L, Hogarth L. Ultra-brief breath counting (mindfulness) training promotes recovery from stress-induced alcohol-seeking in student drinkers. *Addict Behav*. 2020;102:106141. https://doi.org/10.1016/j.addbeh.2019.106141.
323. Brewer JA, Sinha R, Chen JA, et al. Mindfulness training and stress reactivity in substance abuse: results from a randomized, controlled stage I pilot study. *Subst Abuse*. 2009;30(4):306–317. https://doi.org/10.1080/08897070903250241.
324. Smith BW, Ortiz JA, Steffen LE, et al. Mindfulness is associated with fewer PTSD symptoms, depressive symptoms, physical symptoms, and alcohol problems in urban firefighters. *J Consult Clin Psychol*. 2011;79(5):613–617. https://doi.org/10.1037/a0025189.
325. Cairney J, Kwan MY, Veldhuizen S, Faulkner GE. Who uses exercise as a coping strategy for stress? Results from a national survey of Canadians. *J Phys Activ Health*. 2014;11(5):908–916. https://doi.org/10.1123/jpah.2012-0107.
326. Ashdown-Franks G, Firth J, Carney R, et al. Exercise as medicine for mental and substance use disorders: a meta-review of the benefits for neuropsychiatric and cognitive outcomes. *Sports Med*. 2020;50(1):151–170. https://doi.org/10.1007/s40279-019-01187-6.
327. Daniel JZ, Cropley M, Fife-Schaw C. The effect of exercise in reducing desire to smoke and cigarette withdrawal symptoms is not caused by distraction. *Addiction*. 2006;101(8):1187–1192. https://doi.org/10.1111/j.1360-0443.2006.01457.x.
328. Read JP, Brown RA, Marcus BH, et al. Exercise attitudes and behaviors among persons in treatment for alcohol use disorders. *J Subst Abuse Treat*. 2001;21(4):199–206. https://doi.org/10.1016/s0740-5472(01)00203-3.
329. Rozeske RR, Greenwood BN, Fleshner M, Watkins LR, Maier SF. Voluntary wheel running produces resistance to inescapable stress-induced potentiation of morphine conditioned place preference. *Behav Brain Res*. 2011;219(2):378–381. https://doi.org/10.1016/j.bbr.2011.01.030.
330. Zlebnik NE, Carroll ME. Prevention of the incubation of cocaine seeking by aerobic exercise in female rats. *Psychopharmacology*. 2015;232(19):3507–3513. https://doi.org/10.1007/s00213-015-3999-6.
331. Lynch WJ, Piehl KB, Acosta G, Peterson AB, Hemby SE. Aerobic exercise attenuates reinstatement of cocaine-seeking behavior and associated neuroadaptations in the prefrontal cortex. *Biol Psychiatr*. 2010;68(8):774–777. https://doi.org/10.1016/j.biopsych.2010.06.022.
332. Zlebnik NE, Anker JJ, Gliddon LA, Carroll ME. Reduction of extinction and reinstatement of cocaine seeking by wheel running in female rats. *Psychopharmacology*. 2010;209(1):113–125. https://doi.org/10.1007/s00213-010-1776-0.
333. Robison LS, Alessi L, Thanos PK. Chronic forced exercise inhibits stress-induced reinstatement of cocaine conditioned place preference. *Behav Brain Res*. 2018;353:176–184. https://doi.org/10.1016/j.bbr.2018.07.009.
334. Torabi M, Pooriamehr A, Bigdeli I, Miladi-Gorji H. Maternal swimming exercise during pregnancy attenuates anxiety/depressive-like behaviors and voluntary morphine consumption in the pubertal male and female rat offspring born from morphine dependent mothers. *Neurosci Lett*. 2017;659:110–114. https://doi.org/10.1016/j.neulet.2017.08.074.

335. James MH, Campbell EJ, Walker FR, et al. Exercise reverses the effects of early life stress on orexin cell reactivity in male but not female rats. *Front Behav Neurosci*. 2014;8:244. https://doi.org/10.3389/fnbeh.2014.00244.
336. De Chiara V, Errico F, Musella A, et al. Voluntary exercise and sucrose consumption enhance cannabinoid CB1 receptor sensitivity in the striatum. *Neuropsychopharmacology*. 2010;35(2):374–387. https://doi.org/10.1038/npp.2009.141.
337. Ussher M, West R, McEwen A, Taylor A, Steptoe A. Efficacy of exercise counselling as an aid for smoking cessation: a randomized controlled trial. *Addiction*. 2003;98(4):523–532. https://doi.org/10.1046/j.1360-0443.2003.00346.x.
338. Brady KT, Dansky BS, Back SE, Foa EB, Carroll KM. Exposure therapy in the treatment of PTSD among cocaine-dependent individuals: preliminary findings. *J Subst Abuse Treat*. 2001;21(1):47–54. https://doi.org/10.1016/s0740-5472(01)00182-9.
339. Peirce JM, Schacht RL, Brooner RK. The effects of prolonged exposure on substance use in patients with posttraumatic stress disorder and substance use disorders. *J Trauma Stress*. 2020;33(4):465–476. https://doi.org/10.1002/jts.22546.
340. Park CB, Choi JS, Park SM, et al. Comparison of the effectiveness of virtual cue exposure therapy and cognitive behavioral therapy for nicotine dependence. *Cyberpsychol Behav Soc Netw*. 2014;17(4):262–267. https://doi.org/10.1089/cyber.2013.0253.
341. Foa EB, Yusko DA, McLean CP, et al. Concurrent naltrexone and prolonged exposure therapy for patients with comorbid alcohol dependence and PTSD: a randomized clinical trial. *JAMA*. 2013;310(5):488–495. https://doi.org/10.1001/jama.2013.8268.
342. Foa EB, Asnaani A, Rosenfield D, Zandberg LJ, Gariti P, Imms P. Concurrent varenicline and prolonged exposure for patients with nicotine dependence and PTSD: a randomized controlled trial. *J Consult Clin Psychol*. 2017;85(9):862–872. https://doi.org/10.1037/ccp0000213.
343. Norman SB, Trim R, Haller M, et al. Efficacy of integrated exposure therapy vs integrated coping skills therapy for comorbid posttraumatic stress disorder and alcohol use disorder: a randomized clinical trial. *JAMA Psychiatr*. 2019;76(8):791–799. https://doi.org/10.1001/jamapsychiatry.2019.0638.
344. Jeffirs SM, Jarnecke AM, Flanagan JC, Killeen TK, Laffey TF, Back SE. Veterans with PTSD and comorbid substance use disorders: does single versus poly-substance use disorder affect treatment outcomes? *Drug Alcohol Depend*. 2019;199:70–75. https://doi.org/10.1016/j.drugalcdep.2019.04.001.

CHAPTER 12

Gender and sex differences in addiction

Introduction

One popular saying is "men are from Mars, women are from Venus". This statement is used to indicate fundamental differences between men and women. This term comes from John Gray's 1992 book *Men Are from Mars, Women Are from Venus*.[1] The purpose of Gray's book was to provide a way to improve communication between partners, not to study gender differences in addiction. Gray has received criticism as others have pointed out that men and women are more similar than what is depicted in his book. So, does that mean men and women are similar when it comes to drug use patterns and addiction vulnerability? Up until now, I have largely avoided mentioning any differences found between men and women or male and female animals. This was a deliberate choice as I wanted to avoid adding another layer of complexity to the previous chapters. The role of gender/sex on addiction is an important topic that needs to be covered in its own chapter. However, this is a topic that has historically not received as much attention as it should. For example, using PubMed, if I search terms related to gender/sex differences in cocaine self-administration, I find just three articles published in 2000. As you are about to learn, men and women differ in self-reported drug use, and animal studies report large sex differences in a variety of addiction-like behaviors. These findings are important because they suggest that treatments designed to reduce drug use may be more effective for men compared to women and vice versa.

Learning objectives

By the end of this chapter, you should be able to …

(1) Differentiate sex from gender.
(2) Explain how drug use patterns differ between men and women/male and female animals.
(3) Discuss how sex hormones account for differences in drug sensitivity observed between males and females.

(4) Identify some of the differential neurobiological responses to drugs that have been observed in males and females.
(5) Discuss how sociocultural factors influence gender differences in drug use.
(6) Describe how drug use differs between sexual/gender minority individuals and sexual/gender normative individuals.

Differentiating sex from gender

Often, individuals use the terms **sex** and **gender** interchangeably. However, these are not isomorphic terms. Sex refers to the biological and the physiological characteristics of an individual that are determined by our chromosomes. Biological *males* inherit one X sex-based chromosome and one Y sex-based chromosome, whereas biological *females* inherit two X chromosomes. The presence of the Y chromosome causes reproductive organs to develop into the penis, testes, and vas deferens. The absence of the Y chromosome leads to the development of the vagina, ovaries, and fallopian tubes. Gender is not based on physiology. Instead, gender is a psychosocial construct that includes norms and roles associated with *men* and *women*. While sex is often dichotomized, gender is measured on a continuum. Individuals may not identify entirely as a man or as a woman. We will come back to this concept later in the chapter. When reading preclinical studies, you will encounter terms related to sex (males vs. females) only. When reading clinical studies, you may encounter the terms men/women or male/female.

Gender differences in drug use patterns

In Chapter 1, I provided some statistics on drug use patterns as detailed in the *National Survey on Drug Use and Health (NSDUH)*.[2] In 2020, men aged 12 and older reported more past month tobacco use (25.1% vs. 16.5%), ethanol use (52.6% vs. 47.6%), and cannabis use (13.3% vs. 10.4%) compared to women. A higher percentage of men also reported binge drinking (24.9% vs. 19.7%) and heavy ethanol use (8.3% vs. 4.6%) during the past month. While men report more past month illicit drug use (14.9% vs. 12.1%), this discrepancy is smaller compared to what was reported the year prior. In 2019, 15.5% of men reported past-month illicit drug use compared to 10.7% of women. As noted by the *NSDUH*, caution is needed when interpreting the 2020 statistics as the methodology for collecting data changed from 2019 to 2020.

The *NSDUH* did report some instances in which drug use was higher in women compared to men.[2] Slightly more women (0.7%) reported past month use of methamphetamine compared to men (0.6%), although lifetime use of methamphetamine was considerably higher in men (6.7% vs. 4.5%). Women are also more likely to misuse prescription pain medications and sedative-hypnotic medications (1.0% vs. 0.8% and 0.9% vs. 0.7%, respectively) compared to men. One potential explanation to account for higher misuse rates observed for prescription pain medications and sedative-hypnotic medications in women is that they are more likely to be prescribed these medications relative to men.[3] We will return to this concept when we discuss the sociocultural factors underlying gender differences in substance use.

Epidemiological studies largely support the findings reported in the *NSDUH*. Men are more likely to smoke cigarettes,[4] use cannabis,[5] use cocaine,[6] and use methamphetamine,[7] with men being more likely to begin using cocaine at an earlier age compared to women.[8] Men are also more frequently diagnosed with substance use disorders (SUDs), including those for ethanol, cannabis, cocaine, opioids, sedative-hypnotics, and hallucinogens.[9-11] Men with cannabis use disorder meet more criteria for cannabis dependence and have longer episodes of cannabis use disorder compared to women.[12] Men entering treatment for a SUD often report using more ethanol compared to women.[13] Laboratory-based studies have also shown that men self-administer more ethanol intravenously, which is considered a form of binge-like "drinking".[14] Gender differences are also observed in adolescents. Adolescent boys are more likely to engage in moderate ethanol consumption and are more likely to smoke cigarettes, including E-cigarettes, compared to girls.[15,16] Additionally, male undergraduate students are more likely to report drug use and misuse,[17] particularly ethanol use.[18]

One factor that could account for increased drug use in men is that they have more opportunities to use drugs, particularly for cannabis, cocaine, heroin, and hallucinogens.[19] Importantly, Van Etten et al. found that women were just as likely as men to eventually begin using these drugs when given the opportunity to do so.[19] In fact, some work has shown that women may be at *greater* risk of developing problems related to drug use. For example, women are more likely to use heroin.[5] While men are more likely to smoke, women are more sensitive to the subjective effects of nicotine[20] and report higher levels of dependence.[21] Adolescent girls transition from first cigarette use to daily smoking at a faster rate compared to boys.[22] Similarly, women report using methamphetamine at an earlier age compared to men.[23] Although men and women self-administer similar amounts of cocaine in a laboratory, women report feeling less high and less stimulated when given higher doses of cocaine,[24] and women have lower systolic blood pressure following administration of cocaine.[25] Because women experience fewer subjective and physiological responses to cocaine, they may need to use more cocaine to feel the same effects as men. Indeed, women are more likely to use crack cocaine and amphetamine compared to men and report greater cocaine dependence compared to men,[8,26] including reporting greater cocaine use upon entry to treatment for a SUD.[13] An interesting interaction has been observed between gender and dose when examining D-amphetamine self-administration in men and women. A higher dose is more reinforcing in men, but a lower dose is more reinforcing in women.[27] This is an important concept we will come back to when discussing animal studies assessing sex differences in addiction-like behaviors. Finally, women progress from initial drug use to treatment entry for a SUD at a *faster* rate compared to men,[28] and women entering treatment for cocaine use disorder have more severe drug problems.[29]

One consistent finding in the literature is that women experience more adverse drug reactions compared to men.[30] Women report larger unpleasant effects following oxycodone administration, such as feeling nauseous, compared to men.[31] Women often report experiencing worse withdrawal symptoms upon cessation of substance use. This effect has also been observed following cessation of cigarette smoking,[4,21,32] cannabis use,[33] and ketamine use.[34] While ethanol withdrawal has been reported to be worse in men,[35] women with depression and ethanol dependence experience withdrawal for a longer period of time.[36]

Given that women experience more severe withdrawal symptoms, one would expect that women have a higher chance of relapse. There is some evidence that women are more likely to relapse due to nicotine withdrawal symptoms.[32] Following 2 weeks of abstinence, women work harder to receive ethanol, as assessed in a progressive ratio schedule.[37] This finding suggests that women may be more motivated to use ethanol following a period of abstinence. One interesting phenomenon has been reported concerning cocaine relapse. During a 5-month period, men are more likely to transition from abstinence to cocaine use. At first glance, these results would suggest that men are more likely to relapse. However, men are also more likely to transition from drug use to abstinence during a 5-month period.[38] In other words, cocaine use is a more stable behavior in women compared to men.

In addition to withdrawal, one factor that can influence relapse is conditioned responses when exposed to drug-paired stimuli (see Chapter 6). Gender differences have been observed in how individuals respond to these drug-paired stimuli. For example, when asked to imagine scenes depicting heroin use, women report greater craving for heroin, but they also report greater sadness; these changes are accompanied by greater increases in systolic blood pressure.[39] These conditioned responses may serve as a risk factor for subsequent relapse in women.

Animal studies assessing sex differences in addiction-like behaviors

Considering that differences are observed in substance use patterns between men and women, animal studies need to take sex into consideration when conducting research on the neurobehavioral mechanisms underlying addiction. Historically, studies often used males only. To improve the quality of preclinical, as well as clinical, studies, the National Institutes of Health (NIH) implemented a policy stating that sex needs to be examined as a biological variable.[40] Introduced in 2016, this policy mandates that any researcher applying for federal grant money from the NIH must examine gender/sex differences. This entails more than just including a sample that contains both men/males and women/females. Researchers need to include sex as a biological variable, meaning they need to include sex in their statistical analyses and need to discuss these differences (if observed) in their research. Granted, there are some instances in which including sex as a biological variable is unnecessary. Imagine that a researcher wants to study the long-term effects of cannabinoid treatment in men receiving chemotherapy for testicular cancer or wants to study SUDs in National Football League (NFL) players. Using women in these studies is not practical as women do not get testicular cancer nor play in the NFL (at least not at the time of this writing). As I mentioned in the Introduction, using search terms to find research articles examining gender/sex differences in cocaine self-administration yields just three papers published in the year 2000. In 2021, at least 20 papers were published examining sex differences in cocaine self-administration.

Before discussing the literature examining sex differences in addiction-like behaviors, it is important to note that the presence or the absence of such differences is largely dependent on the drug dose(s) used in the experiment. Recall that in humans, women are more sensitive to a low dose of amphetamine, whereas men are more sensitive to a higher dose. In animal studies, if too low of a dose is used, neither sex will work to receive that dose or will not show a preference for an environment paired with that dose. At certain doses, sex differences may not be observed because of a ceiling effect (i.e., there is an upper limit to the number of responses an animal can emit for drug reinforcement, just as there is an upper limit to the duration of time an animal can spend in a drug-paired environment).

Physiological and behavioral drug effects

Drugs can differentially alter physiological responses in males and females. Following an intoxicating dose of ethanol, males take longer to recover their "righting reflex" (i.e., ability to get back on all four legs after being placed backside down on a surface), and a higher percentage of males develop acute ethanol tolerance.[41] Conversely, females show less conditioned taste avoidance when given ethanol.[42,43] However, pre-exposing animals to ethanol during adolescence attenuates conditioned taste aversion to a greater extent in males relative to females.[43] Like ethanol, females show less avoidance of cocaine-paired saccharin.[44] Additionally, females show less anxiety-like behavior following administration of cocaine.[45] Whereas males experience more analgesia following opioid administration,[46] females experience greater hypothermia following morphine, as well as ethanol, administration that lasts for a longer period of time.[41,47]

Somewhat consistent with human studies, withdrawal symptoms are more pronounced in females following cessation of ethanol[48] and nicotine[49] administration. Following acute ethanol withdrawal, females display greater anxiety-like behavior.[50] While heroin withdrawal symptoms are similar in males and females,[51] males experience worse morphine withdrawal symptoms during early abstinence, but females experience more persistent withdrawal symptoms.[52] Cannabis withdrawal is similar across sex, but females show greater retropulsion (i.e., walking backwards).[53]

One common behavioral measure following drug administration is locomotor activity, particularly locomotor sensitization (i.e., increased locomotor responses following repeated drug administration). Locomotor sensitization is a predictor of subsequent drug-taking behavior.[54] Thus, sex differences in locomotor sensitization can shed insights into the potential risk of developing addiction-like behaviors in male and female animals. Research has consistently shown that females are more active following acute psychostimulant administration relative to males and show greater psychostimulant-induced locomotor sensitization.[55–63] This effect is not isolated to psychostimulants as females are also more sensitive to the acute locomotor stimulant effects of opioids and show greater sensitization to fentanyl.[46,64–66]

Collectively, findings such as decreased conditioned taste aversion, increased withdrawal symptoms, and increased locomotor sensitization suggest that female animals may be at heightened risk of developing addiction-like behaviors. Is this the case though? Keep reading to find out.

Voluntary oral consumption

Because ethanol is primarily consumed as a beverage, studies often measure voluntary consumption of ethanol in rodents. Female rats consume more ethanol compared to males.[42,50,67–70] While most studies measuring oral consumption use ethanol, studies have shown that females consume more nicotine[71] and cocaine.[72] When studying sex differences in drug consumption, researchers need to consider size differences between males and females. For example, I use Sprague Dawley rats (a type of albino strain) in my research. By 12 weeks of age, males are ~350 g while females are ~220 g. Although male and female mice consume a similar amount of a nicotine solution, females consume more nicotine when adjusted for body weight.[73] Similarly, some studies have observed higher ethanol consumption in males relative to females[74] or no sex differences in consumption[75] (Fig. 12.1), but when body weight is taken into consideration, females self-administer more ethanol.

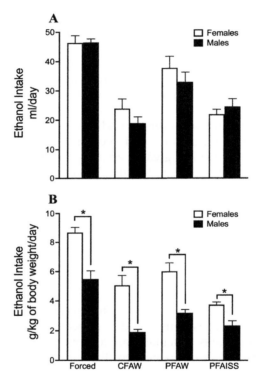

FIGURE 12.1 Ethanol consumption across four different conditions for female (*white bars*) and male (*black bars*) rats. During the forced phase of the experiment, rats had continuous access to a water bottle containing 6% ethanol only. During the continuous free choice to alcohol and water (CFAW) phase, rats had continuous access to two water bottles, one containing ethanol and the other containing water. The periodic free choice to alcohol and water (PFAW) was similar to the CFAW phase, except rats were deprived of ethanol for 3-day periods. The final phase was the period free-choice to alcohol and isocaloric-sweetened solution (PFAISS). This condition was identical to the PFAW phase, except that when rats had access to ethanol, they also had access to a sweetened solution. When absolute consumption was measured (A), males and females consumed similar amounts of ethanol. However, when body weight was taken into account (B), females consumed significantly more ethanol (g/kg of body weight) compared to males. *Image is recreated and modified from Fig. 1 of Juárez J, Barrios de Tomasi E. Sex differences in alcohol drinking patterns during forced and voluntary consumption in rats. Alcohol. 1999;19(1):15–22. https://doi.org/10.1016/s0741-8329(99)00010-5.*

While females often consume more drug, particularly ethanol, compared to males, exposing male rats to chronic intermittent ethanol for 6 weeks increases their ethanol consumption to what is observed in females given either vehicle or ethanol.[69] Similar research has shown that males pre-exposed to ethanol consume more ethanol compared to females.[42] This mechanism may account for increased ethanol consumption often reported in men compared to women. As many individuals are first exposed to ethanol during adolescence, men may develop a greater preference for ethanol over time relative to women.

Acquisition and maintenance of drug self-administration

Acquisition of drug self-administration is typically quantified by measuring the percentage of animals that meet specific criteria (e.g., X number of active responses by X number of sessions), the rate at which animals meet the specific criteria, and the total number of infusions earned during the acquisition period. Overall, the percentage of animals acquiring drug self-administration tends to be higher in females compared to males[76-78] (but see[79,80]). Females also tend to acquire psychostimulant and opioid self-administration at a faster rate[45,77,78,81-85] (but see[79,86,87]). However, discrepant results are observed concerning the rate of nicotine self-administration, with studies observing faster acquisition in females,[88] faster acquisition in males,[79] and no differences in rate of acquisition.[89] Additionally, Chellian et al.[90] found no sex differences in initial self-administration of nicotine but observed greater nicotine intake in females by the end of training.

Females may be more likely to acquire drug self-administration and to do so at a faster rate, but many studies have observed no sex differences in the amount of responding for drug reinforcement during acquisition and during maintenance of drug self-administration.[79,89,91-101] However, when sex differences are observed, results typically show greater drug self-administration in females, especially for psychostimulants[78,79,86,102-104] and opioids.[51,87] Similar to research examining ethanol consumption, males previously exposed to morphine self-administer more oxycodone during acquisition compared to females,[105] once again suggesting that early-life exposure to drugs can differentially increase drug sensitivity later in life.

Like the acquisition rate of nicotine self-administration, discrepancies in responding for nicotine have been observed. Some studies have reported greater nicotine self-administration in female rats[106] while others have shown greater self-administration in males.[79] Exposure to tobacco smoke or to nicotine injections during adolescence increases nicotine self-administration in females, but not in males,[107] a finding that is opposite to what has been observed with ethanol and oxycodone. Conversely, concurrent nicotine increases ethanol self-administration in adolescent males only.[108] This last finding suggests that men are more likely to drink when smoking a cigarette compared to women.

As with consumption studies, one important consideration is the size of males and females. When active lever responses are analyzed, males and females respond similarly for oxycodone; however, when intake is expressed as mg/kg, females consume more oxycodone.[109] Another important consideration is the drug dose that researchers use in self-administration studies, something I mentioned when introducing the section on sex differences in animal research. When measuring sex differences in drug self-administration, careful consideration needs to be taken when selecting the dose. For example, when mice self-administer 30 μg/kg of heroin, females show elevated responding, but this effect disappears when a higher dose (60 μg/kg) is used.[51] Differences in drug dose may account for the large discrepancies observed in nicotine self-administration. Donny et al. found greater self-administration in females at a low dose of nicotine (0.02 mg/kg/infusion), an effect that disappears at higher doses.[88] Swalve et al.[79] observed greater self-administration in males at a slightly higher dose than what Donny et al.[88] used (0.03 mg/kg). Finally, the schedule of reinforcement may influence sex differences in drug self-administration. When a fixed ratio schedule is used, females respond more for ethanol compared to males,[99,110] but females respond less for ethanol when a fixed interval schedule is used.[111] Remember that in a fixed interval schedule, animals must wait for a certain period before a response leads to reinforcement. Thus, the decreased responding in females observed in this schedule does not necessarily mean they are less motivated to earn ethanol. Instead, they may just be better at learning the contingencies of reinforcement associated with the fixed interval schedule. To measure differences in motivation for drug reinforcement, the progressive ratio schedule can be used.

Motivation to self-administer drug

The progressive ratio schedule is often used to measure the motivation to self-administer drugs (i.e., a drug's reinforcing efficacy) and constitutes one of the dimensions of the 3-CRIT model of addiction. Most experiments have used cocaine to compare progressive ratio responding between males and females. These experiments have largely shown increased responding in females[45,55,82,84,102,112,113] (but see[92,114,115]). While Lynch and Taylor did not observe differences in responding for cocaine during initial training, they found that giving rats access to cocaine for 24 h increases motivation for cocaine to a greater extent in females compared to males.[112] During cocaine withdrawal, females increase their responding at a faster rate relative to males.[113]

Females are also more motivated to earn infusions of *d*-amphetamine[114] and nicotine,[76,88] but results have been more mixed when other drugs have been used. While an early report found enhanced motivation for methamphetamine in females,[78] more recent studies have failed to replicate this finding.[80,91] Similarly, the reinforcing efficacy of ethanol can either be equivalent in males and females[74,93] or be higher in females relative to males.[115] One study found that males respond more for ethanol, but when weight is taken into account, females are more motivated to earn ethanol.[116] While no sex differences have been observed for morphine-maintained responding in the progressive ratio schedule,[96] early exposure to morphine *decreases* the reinforcing efficacy of oxycodone in females.[105]

Escalation of drug self-administration

Escalation of drug self-administration is often modeled by providing animals extended access to drug self-administration, usually 6 h, and is another defining feature of the 3-CRIT model of addiction. Not as many studies have compared escalation of drug intake between males and females compared to other drug self-administration paradigms. However, females show greater escalation of cocaine[55] and methamphetamine self-administration.[117,118] There is some evidence that sex differences in escalation of self-administration, at least for methamphetamine, are somewhat dependent on age. Westbrook et al. recently observed greater escalation of methamphetamine self-administration in adolescent, but not adult, female rats compared to males,[118] demonstrating that sex differences in dependence-related behavior are sensitive to developmental periods. To better model how humans take methamphetamine (characterized by binge-like consumption), Cornett and Goeders[119] allowed rats to self-administer methamphetamine for 96-hour periods across 5 weeks (note: rats had access to water while in the operant chambers and were fed each day). Both males and females escalated their methamphetamine self-administration, but no sex differences were observed in this model. While escalation has historically been measured with psychostimulant drugs, recent studies have observed greater escalation of heroin self-administration in females relative to males.[51,87]

Behavioral economic measures of self-administration

Similar to escalation experiments, few studies have applied behavioral economic analyses to examine sex differences in drug self-administration. In Chapter 5, you learned that demand intensity refers to consumption of a reinforcer when the price is minimally constrained while demand elasticity indicates how much an animal is willing to defend the price of a reinforcer as the price increases. While males are more willing to respond for ethanol as the response requirement increases,[120] inconsistencies have been reported when other drugs have been tested. A couple of studies have failed to observe sex differences in demand elasticity of nicotine;[121,122] however, one study found that males treat nicotine as more inelastic when higher doses are used, but females treat it as more inelastic when lower doses are used.[123] Females have also been shown to have higher demand intensity for nicotine compared to males.[122] Interestingly, nicotine makes ethanol more inelastic in females, but not in males.[120] Cocaine demand intensity and/or elasticity typically do not differ as a function of sex,[95,124,125] but there is evidence that females given intermittent access to cocaine are more likely to treat cocaine as an inelastic good compared to males.[126]

One potential explanation for the vast discrepancies reported in the literature relates to how price of the reinforcer is manipulated. The response requirement can increase across sessions (somewhat akin to a progressive ratio schedule), or the drug dose can decrease across sessions or within a single session. For example, Grebenstein et al. did not observe sex differences in demand elasticity or intensity for nicotine when the dose of nicotine decreased across sessions.[121] Likewise, Powell et al.[122] decreased the dose of nicotine across sessions and within sessions and did not observe sex differences in demand intensity/elasticity. The one study to observe sex differences in behavioral economic measures of nicotine consumption increased the response requirement across sessions while keeping dose constant.[123]

Conditioned place preference (CPP)

There is one interpretational issue when comparing males and females in drug self-administration paradigms. As females are more sensitive to the locomotor-stimulant effects of drugs like cocaine, the increased self-administration observed in females could merely reflect increased hyperactivity. This could also account for why studies using nicotine have been more inconsistent compared to other drugs like cocaine. Because conditioned place preference (CPP) relies on Pavlovian conditioning, using this paradigm to measure sex differences is important because it can provide insights on how environmental cues drive addiction-like behaviors in males and females. While females consistently consume more ethanol, sex differences are often not observed in ethanol CPP.[127,128] One study did find that male mice develop ethanol CPP when lower doses are used (1.0 and 1.5 g/kg), but females develop CPP to higher doses (2.5 and 3.0 g/kg).[129] It is important to note that the results of this study appear to be an exception as opposed to the rule as other studies have observed similar ethanol CPP in males and females when 1.0 and 1.5 g/kg are used.[127,128] There is evidence for sex differences in ethanol CPP in late adolescent mice, as late adolescent females develop CPP while late adolescent males do not.[130] This suggests that late adolescence may be a critical period in which females are more sensitive to the rewarding effects of ethanol, which is similar to what has been observed for escalation of methamphetamine self-administration.

While most CPP experiments give animals multiple exposures to a drug and a specific environmental context, Nentwig et al.[131] used a variation of CPP to study the conditioned rewarding effects of ethanol. Rats received a single injection of ethanol on either Day 1 or Day 3 of the experiment and an injection of saline on either Day 1 or Day 3. On Days 2 and 4, rats were not tested. On Day 5, all rats were tested for CPP. Females developed CPP to the compartment paired with ethanol, regardless of when ethanol was administered. Interestingly, males developed CPP when ethanol was administered on Day 1 only. Why could this be the case? One potential explanation is that ethanol is not as rewarding *per se* in males when ethanol is given on Day 1. Instead, males are sensitive to the novelty of the apparatus. Remember, when males are given ethanol on Day 1, this is the first time they have experienced the CPP chamber. If males receive ethanol on Day 3, they are familiar with the apparatus; thus, there is not as much novelty, even if they are placed in a separate compartment compared to when they received saline.

Concerning nicotine, males develop greater CPP to lower doses of nicotine compared to females,[132] although the relationship between sex and nicotine CPP appears to be dependent on how many times nicotine is paired with an environmental context. For example, when nicotine is paired once with a CPP compartment, *females* develop greater CPP to a low dose of nicotine.[133] After pairing nicotine with the environmental context a second time, males and females show similar CPP at each dose of nicotine tested. This finding suggests that females are more sensitive to the initial rewarding effects of nicotine, but males ultimately develop a greater preference for nicotine over time. This is similar to the finding that women are more sensitive to the subjective effects of nicotine despite the fact that more men smoke cigarettes.[20]

Psychostimulant CPP is generally similar between males and females,[58,60,62,134] but when sex differences are observed, females are more sensitive to the conditioned reward effects of cocaine,[63] *d*-amphetamine,[56] and methamphetamine.[135] In contrast to psychostimulants, sex differences in opioid CPP have been more variable. Some studies have failed to observe sex differences[46,66] while others have observed greater morphine and oxycodone CPP in females.[65] However, fentanyl CPP is higher in males compared to females.[64] This last result is interesting as Gaulden et al. observed that females are more sensitive to the locomotor-stimulant effects of fentanyl.[64] More work is needed to determine if the increased fentanyl CPP observed in males is a unique characteristic of fentanyl compared to other opioids.

One limitation to using CPP to examine sex differences is that this paradigm requires the use of numerous groups. In drug self-administration experiments, the drug dose can be manipulated within the same group of animals. In CPP, one drug dose must be tested in a single group of animals. To effectively measure differential sensitivity to the conditioned rewarding effects of a drug in males and females, multiple doses need to be used. This can make examining sex differences in CPP difficult.

Extinction of drug preference

Extinction can be measured in both drug self-administration and CPP experiments. Understanding potential sex differences in the extinction of self-administration or CPP is valuable because this knowledge can inform us of difficulties therapists can encounter when treating men or women with a SUD. Many studies have failed to detect sex differences during extinction of self-administration.[74,79,85,98,100,136–139] However, when sex differences are observed, females show greater responding compared to males, with these differences being observed for ethanol,[140] nicotine,[141] cocaine,[103,104,142,143] and the cannabinoid WIN55,212-2 (somewhat similar to THC, the psychoactive compound in cannabis).[144] Taken at face value, these results indicate that women may be at higher risk of continued drug use during a treatment intervention, meaning extra efforts should be taken to help women abstain from multiple substances.

There is one limitation to comparing extinction responses between males and females. In studies in which females self-administer more drugs at the end of acquisition, their responses will start off higher during extinction compared to males. For example, Bertholomey et al. found increased responding for ethanol in females during the first extinction session, but this effect was most likely due to the increased ethanol responding observed during training.[110] One way to control for this caveat is to calculate extinction responses as the percentage of responses emitted at the end of training. When responding is represented as a percentage of responding at the end of baseline training, there are no sex differences in extinction of nicotine self-administration.[121] This finding questions if sex is an important factor when measuring extinction of drug-seeking behavior.

Relapse-like behavior

If females are more resistant to extinction, one could assume that they are more likely to show relapse-like behavior during abstinence. Although females are consistently more likely to develop relapse-like behavior following methamphetamine self-administration[91,121,141,149] (but see[101,150]), results with other drugs have been more mixed. Multiple studies have not observed sex differences when examining cocaine reinstatement,[126,138,142,145,146] although some have found that females show greater reinstatement[134,136,147] and greater incubation of cocaine craving.[148] So far, there is not much evidence for sex differences in nicotine reinstatement,[141] and studies with ethanol have been widely mixed, with some reporting greater reinstatement in males[74,93] and at least one reporting greater reinstatement in females.[110] Like ethanol CPP, ethanol reinstatement is influenced by age. Late adolescent females show reinstatement of ethanol CPP while late adolescent males do not.[130] Heroin relapse-like behavior has been shown to be greater in females compared to males,[101,149] but other studies have failed to replicate this finding.[87,98] Very few studies have examined reinstatement to cannabinoids, but one study found that females show greater reinstatement of WIN55,212-2-seeking behavior.[150]

Recall that reinstatement can be induced by exposure to a priming injection of the drug, exposure to cues previously paired with drug administration, or exposure to a stressor. There is some evidence that the type of reinstatement being measured can modulate sex differences in relapse-like behavior. Males and females show similar drug-induced reinstatement of fentanyl-seeking behavior, but females show less reinstatement compared to males when presented with cues.[85] Malone et al.[85] also demonstrated that self-administration history can influence sex-dependent effects on cue-induced reinstatement. Sex differences were observed when animals had extended access to fentanyl during self-administration, but not when given 1-hour sessions.

Taken together, when sex differences are observed in preclinical studies, females consume greater amounts of drug, are more sensitive to the reinforcing and the conditioned rewarding effects of drugs, take longer to extinguish drug preferences, and are more likely to "relapse". After completing the quiz below, you will learn how gender/sex can act as a moderating variable. That is, relationships between certain behaviors and drug-related behaviors can depend on which gender/sex is being examined.

Quiz 12.1

(1) In animal studies, females consistently self-administer more drug, except for _____.
   a. cocaine
   b. ethanol
   c. morphine
   d. nicotine
(2) What is the major limitation when interpreting self-administration data between male and female animals, especially for psychostimulant drugs?
(3) Which of the following statements regarding the NIH Sex as a Biological Variable mandate is true?
   a. Any researcher applying for NIH funding must use both men/males and women/females; no exceptions will be given.
   b. Individuals that do not apply for NIH funding are still required to include both men/males and women/females in their research studies.
   c. While researchers must include men/males and women/females in their studies, they do not need to directly compare each other.
   d. None of the above.
(4) Although men often report using more drugs, women have marginally higher misuse rates of _____.

a. cocaine
   b. ethanol
   c. nicotine
   d. sedative-hypnotics
(5) True/False: the number of studies examining sex differences has increased since the early 2000s.

Answers to quiz 12.1

(1) d — nicotine
(2) Females are more hyperactive following psychostimulant drugs compared to males. During self-administration, females could become more hyperactive, leading to greater responses on the lever/nose poke aperture, leading to greater hyperactivity.
(3) d — None of the above
(4) d — sedative-hypnotics
(5) True

Gender/sex as a moderating variable

In this book, you have learned about multiple factors that are associated or predictive of drug use. When discussing factors such as maladaptive decision making, peer influences, and stress, there was hardly any mention of gender or sex. Importantly, there are situations in which one such relationship is observed in men only and vice versa. As detailed in Chapter 8, impulsive decision making is consistently predictive of drug self-administration in animals. Interestingly, high impulsive females show greater reinstatement of cocaine-seeking behavior than low impulsive females and high/low impulsive males.[143] This implies that high impulsive women may be at greater risk for relapse compared to high impulsive men and to low impulsive women. Methylphenidate (Ritalin) is often used to treat attention-deficit/hyperactivity disorder (ADHD). Preclinical work has shown that male rats treated with methylphenidate during early life self-administer more cocaine during adulthood, particularly in a progressive ratio schedule, a finding not observed in females.[92]

Speaking of impulsivity, clinical studies have shown that drugs can differentially alter impulsivity in men and women. Women that engage in heavy ethanol consumption show elevated motor impulsivity compared to heavy drinking men.[151] Decision making is also more impaired in cocaine/methamphetamine-dependent women compared to dependent men,[152] although ethanol-dependent men show greater impulsive decision making relative to women.[153]

Gender/sex differences are also observed in risky decision making. Biological males are typically more risk seeking, an effect that is observed in humans[154] and in animals.[155] One study found that prenatal exposure to cocaine interacts with sex. While adolescent boys and girls in the control condition showed similar risk taking in the balloon analog risk task (BART), adolescent males prenatally exposed to cocaine engaged in more risk taking while risky decision making was nearly abolished in females exposed to cocaine.[156] However, adult men acutely treated with *d*-amphetamine become more risk taking, whereas women become more risk averse.[157] While risky decision making has been implicated in addiction, one study found that risk-taking behavior in female rats is negatively associated with cocaine self-administration.[158] However, my laboratory has shown that increased risky decision making in females is positively correlated with cocaine CPP.[63]

In Chapter 10, you learned about social influences on drug use. I described one experiment I conducted examining the preference for social interaction and *d*-amphetamine in male rats.[159] Recall that males prefer social interaction over amphetamine, but adult males show greater preference for amphetamine. In a follow-up study assessing concurrent choice of social interaction and amphetamine in adolescent females, we found that females do not prefer social interaction over amphetamine.[160] Sex differences have also been observed when measuring social facilitation of nicotine self-administration as males show greater self-administration of a low dose of nicotine in the presence of a partner while females respond less for nicotine.[161] These results are interesting because they show how social interaction can act as a protective factor against drug use and as a risk factor for drug use in males.

Gender/sex can modulate the relationship between stress and addiction. Stress is associated with nicotine withdrawal symptoms, and this association is more pronounced in women compared to men.[162] Yohimbine administered to cocaine-dependent individuals increases feelings of anxiety and craving to a greater extent in women compared to men.[163] Trauma history is a significant predictor of relapse in women, but not in men.[164] In animals, maternal separation increases cocaine CPP in males, but not in females,[165] but females that have experienced separation are more sensitive to the locomotor-stimulant effects of ethanol and methamphetamine.[166,167] The duration of maternal separation can interact with sex. Whereas brief bouts of maternal separation decrease ethanol consumption in both males and females, longer periods of separation differentially affect ethanol intake, with increased consumption observed in males and decreased consumption observed in females.[168]

Importantly, treatment interventions can be differentially effective across sex. Earlier, I mentioned that women entering treatment for cocaine use disorder show more severe drug problems. Yet, they respond better to treatment after 6 months.[29] Wheel running, a form of alternative reinforcement in rodents, attenuates ethanol and cocaine self-administration to a greater extent in females compared to males.[169,170] However, while wheel running is more effective at suppressing reinstatement of cocaine seeking in male rats,[171] it is more effective in attenuating reinstatement of heroin seeking in females.[98] How do these results apply to real life? Exercise has been shown to be a protective factor against drug use (as we covered in the previous chapter), but this effect is typically more common in women compared to men.[172]

Similar to exercise, potential pharmacotherapies have been shown to differentially influence drug-taking behaviors in males and females. I will highlight just some of these findings here. The $GABA_B$ receptor agonist baclofen decreases the number of rats acquiring cocaine self-administration to a greater extent in females compared to males.[81] A drug called nicotinamide selectively attenuates relapse-like behavior in males,[139] but the hormone oxytocin reduces methamphetamine-seeking behavior in females only.[91] The buprenorphine analog BU08028 is efficacious at reducing incubation of heroin seeking in both males and females, but it also *increases* reacquisition of heroin self-administration in females only.[173] When combined with environmental enrichment, a glycine transporter-1 inhibitor increases the rate of extinction of cocaine self-administration and attenuates cocaine-seeking behavior in males, but this combined therapeutic approach increases the rate of extinction learning without altering relapse-like behavior in females.[174] These findings are critically important because they can inform clinicians on how to best treat men and women with a SUD. While one treatment can be effective in reducing drug use in men, it can exacerbate cravings in females.

Potential causes of gender/sex differences in drug sensitivity

By now, you know that gender/sex is an important variable to consider when studying addiction-like behaviors. However, this chapter has not covered the *mechanisms* underlying gender/sex differences that have been observed in previous studies. This is the focus of the current section. We will first discuss some of the biological differences of males and females that may account for the differential findings observed in preclinical studies.

Biological factors

Hormonal Influences

In the last chapter you learned about the neuroendocrine system and the hypothalamic—pituitary—adrenal (HPA) axis as related to stress. In addition to the HPA axis, there is the **hypothalamic—pituitary—gonadal (HPG) axis**. The hypothalamus releases *gonadotropin-releasing hormone* (*GnRH*). This causes the pituitary gland to release important hormones that regulate our reproductive organs (see next paragraph). Our reproductive organs also secrete hormones. Testes produce **androgens**, with testosterone being the prototypic androgen. Ovaries produce **estrogens**, which include estradiol (also spelled as oestradiol).

In women, hormone levels fluctuate during the **menstrual cycle**. The menstrual cycle commences with *menstruation* (hence menstrual cycle). During menstruation, the endometrium (inner lining of uterus) begins to shed. Estradiol levels are lower compared to *follicle-stimulating hormone* (*FSH*) during menstruation. FSH is secreted by the pituitary gland and regulates the functions of both ovaries and testes. Menstruation occurs at the beginning of the *follicular phase*, which gets its name from FSH. The follicular phase lasts approximately 2 weeks. During this phase, FSH helps stimulate estradiol production, and it is critically involved in the development of a follicle, a sac-like structure containing an unfertilized egg. While estradiol levels increase during the second half of the follicular phase, FSH levels stay constant until the very end of this phase. The second major phase of the menstrual cycle is called the *luteal phase*, which begins with *ovulation*. During ovulation, FSH levels increase sharply, along with *luteinizing hormone* (which gives the luteal phase its name). The unfertilized egg cell (called an ovum) is released from the follicle. The ovum then travels through the fallopian tubes, where it can be fertilized if sperm are present. After the ovum has been released from the follicle, a structure called the *corpus luteum* is formed. This structure acts as a temporary gland that secretes **progesterone**, a type of a progestogen hormone. Progesterone helps the uterus accept a fertilized egg. If the ovum is unfertilized, progesterone helps prepare the uterine wall to shed during menstruation. As progesterone levels increase, estradiol, FSH, and luteinizing hormone levels decrease. Fig. 12.2 shows the stages of the menstrual cycle.

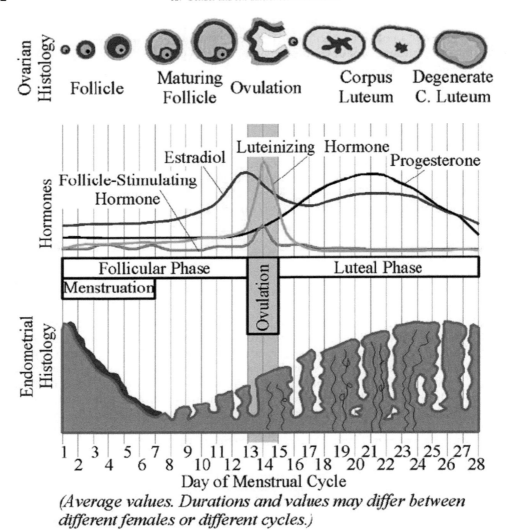

FIGURE 12.2 A schematic of a woman's menstrual cycle. The menstrual cycle is composed of two main phases: the follicular phase and the luteal phase. The follicular phase begins with menstruation and ends as ovulation begins. The top portion of the figure shows what happens to the ovaries across the menstrual cycle. The middle portion shows how hormone levels fluctuate during the cycle. The bottom portion shows how the endometrium changes during the cycle. During menstruation, the endometrium sheds; the endometrium thickens following menstruation. *Figure was modified from Wikimedia Commons user Chris 73 (https://commons.wikimedia.org/wiki/File:MenstrualCycle.png).*

Overall, many studies have failed to observe differences in drug-taking behaviors across the menstrual cycle in women.[175,176] However, some studies have indicated a role for estrogen in ethanol consumption.[177] Specifically, chronic ethanol consumption is more common in the luteal phase of monkeys.[178] Likewise, women have elevated positive subjective ratings of ethanol during the luteal phase.[179] Women receiving treatment for an alcohol use disorder drink more during the luteal phase and during menstruation compared to the other phases.[180] In addition to experiencing more positive ethanol effects during the luteal phase, women metabolize ethanol at a faster rate in the luteal phase,[181] which may account for the increased ethanol consumption observed during this phase.

An interesting dissociation has been observed concerning menstrual cycle influences on ethanol intake/subject effects compared to other drugs. While ethanol consumption/subjective ethanol experiences are associated with the luteal phase, responding for cocaine is higher during the follicular phase in monkeys,[182] and women report greater feelings of being high and stimulated and experiencing more "good drug effects" following intravenous nicotine, oral d-amphetamine, and smoked cocaine during this phase.[175,183,184] Giving estradiol, which is elevated during the follicular phase, to women increases subjective ratings of pleasant stimulation following d-amphetamine administration.[185] As estradiol levels begin to decrease during the late follicular phase, women experience more unpleasant stimulation following d-amphetamine administration.[186] As progesterone levels increase, women have a lower urge to smoke cigarettes[187] and are less likely to smoke.[188] Progesterone administration during the follicular phase tends to decrease smoking.[189] Related to this finding, women with higher progesterone levels show lower stress-induced and cue-induced cocaine craving.[190]

Rodents have their own cycle, called the **estrous cycle**. The estrous cycle is composed of four different phases: diestrus, proestrus, estrus, and metestrus. This cycle progresses more rapidly compared to the menstrual cycle. In rats, no phase lasts more than 3 days, and the metestrus phase lasts for just 6–8 h. The entire estrous cycle occurs over the span of approximately 5 days, compared to ~28 days for a human's menstrual cycle. Diestrus/metestrus is somewhat similar to the follicular phase, with elevated estrogens being observed at the end of this phase. However, unlike the follicular phase, progesterone levels are also elevated during diestrus/metestrus. Estrogen levels are highest during proestrus, while progesterone levels spike at the end of the proestrus phase and at the beginning of the estrus phase.

Like humans, the rodent estrous cycle affects drug sensitivity in females, with estrogens purportedly underlying the increased addiction-like behaviors observed in females compared to males.[191] Females self-administer more cocaine[148,192] and heroin[193] and are willing to work harder to receive nicotine and cocaine as assessed in a progressive ratio schedule[76,84,193–195] during the estrus phase. Incubation of cocaine craving and reinstatement of cocaine-seeking behavior are highest during estrus.[104,148,192] The role of the estrous cycle on behavioral economic indices of self-administration needs more clarification as some inconsistencies have been reported in the literature. One study found that cocaine and remifentanil demand intensity, but not demand elasticity, are increased during the estrus phase compared to the diestrus phase.[124] However, a separate study found that cocaine demand elasticity, but not demand intensity, is lower during the proestrus/estrus phase compared to the diestrus phase.[125] Although somewhat discrepant, these results provide further support that increased estrogen levels are associated with addiction-like behaviors.

Drugs that selectively target estrogen receptors can be used to further elucidate the role of hormones on maintaining addiction-like behaviors. Estrogen receptor activation in the prelimbic cortex is an important mediator of relapse-like behavior.[196] Blocking estrogen receptors abolishes locomotor sensitization, drug CPP, and motivation to self-administer drugs.[197,198] To better isolate the contribution of estrogens and progesterone on drug sensitivity in animals, researchers can surgically remove the ovaries in a procedure called an *ovariectomy*. Ovariectomized rats have decreased levels of both estradiol and progesterone.[199] Ovariectomized females show attenuated drug-induced increases in locomotor activity and behavior sensitization,[61,200] drug CPP,[201] drug preference over natural reinforcement,[202] and reinstatement of drug-seeking behavior.[150,203,204]

After a rat has received an ovariectomy, researchers can administer estradiol or progesterone to determine the contributions of each individual hormone to addiction-like behaviors. Ovariectomized females that receive estradiol show:

(1) increased stimulant-induced hyperactivity and locomotor sensitization,[57,200,205]
(2) faster acquisition of cocaine self-administration,[83,206-209]
(3) greater drug self-administration,[106,206,208,209]
(4) enhanced motivation for drug reinforcement,[210]
(5) greater escalation of drug self-administration,[211]
(6) elevated drug CPP,[135]
(7) greater drug-induced anxiety-like behavior,[212]
(8) increased relapse-like behavior,[203,204] and
(9) greater drug choice over natural reinforcement.[202]

Estradiol even increases preference for cocaine over food in castrated male rats.[213] Interestingly, the effects of estradiol on cocaine self-administration follow an inverted U-shape curve. That is, high doses of estradiol, as well as chronic administration of estradiol, fail to increase self-administration.[206] Another interesting finding is that estradiol *increases* extinction of CPP.[214]

In contrast to estradiol, progesterone blunts the locomotor-stimulant effects[215] and the rewarding/reinforcing effects of drugs in females.[200,201,207,211] Progesterone and allopregnanolone (metabolite of progesterone) prevent the escalation of drug self-administration[216] and decrease relapse-like behavior in females,[192,217,218] but not in males.[218] If progesterone can attenuate drug-taking behaviors, does this mean that individuals can receive hormone treatment to reduce substance use? Progesterone decreases cocaine-induced increases in diastolic blood pressure and attenuates the positive subjective effects of cocaine (e.g., feeling high).[219,220] In cocaine-dependent individuals, progesterone decreases drug cravings.[221] Recently, a randomized clinical trial showed that progesterone can be used to treat cannabis withdrawal symptoms.[222] Unfortunately, progesterone fails to modify cocaine drug self-administration in women.[176,220]

To complicate matters, most of the research I have cited in this section has focused on the relationship between hormones and psychostimulant drugs. While rodent studies largely show that increased estradiol levels increase the reinforcing effects of psychostimulants, there are some exceptions. The estrous cycle does not appear to influence nicotine self-administration or ethanol consumption.[70,88] Additionally, recent research has shown that females are *less* sensitive to the conditioned rewarding effects of fentanyl during estrus[64] and that administration of estradiol can *decrease* opioid self-administration.[223] Collectively, these results suggest that estradiol can act as a protective factor against opioid use but can potentiate psychostimulant use.

While estrogens directly affect how females respond to drugs, cues that are present during drug self-administration training can modulate the influence of the estrous cycle on cocaine reinforcement. Recall that female rats self-administer more drug during the estrus phase compared to the diestrus/metestrus phases. However, this effect occurs only when discrete cues are paired with drug delivery.[224] Interestingly, cues presented during reinforcement during the estrus phase increase activation of the immediate early gene c-Fos in striatal regions, which may account for the estrus-dependent increases in cocaine self-administration. This result highlights the complex interaction between the environment and biology.

Not only can environmental cues modulate the influence of the estrous cycle on drug-related behaviors, but the estrous cycle can act as a modulator in itself. To provide one example, the antibiotic ceftriaxone attenuates reinstatement of cocaine seeking in both males and females, except for when females are in the estrus phase.[146] This finding is important because it may partially explain why women are more susceptible to relapse. During treatment, the intervention may become less effective as the individual approaches the late follicular phase of the menstrual cycle. As estradiol levels begin to increase, the individual may become more sensitive to stimuli previously paired with the drug, which can then precipitate cravings and relapse.

Neurobiological Differences

Given that some inconsistencies have been observed regarding hormonal effects on the development and maintenance of drug-taking behaviors, research has determined if there are other neurobiological differences between males and females that can account for sex differences reported in the literature. The ventral tegmental area (VTA) contains a large collection of dopaminergic cell bodies that project axons to the nucleus accumbens (NAc) and to the frontal cortex. One major sex difference is that females have a greater *proportion* of dopaminergic cell bodies in the VTA compared to males.[225] Basal firing of VTA dopamine neurons is influenced by the estrous cycle, as firing is highest during estrus and lowest during proestrus.[226] In the NAc, females have elevated protein kinase A (PKA) levels in the NAc,[59] and estradiol has a facilitatory role on cocaine- and amphetamine-regulated transcript (CART) peptide in the NAc and in the dorsal striatum.[227] As NAc PKA activation is important for drug reinforcement,[228] the increased PKA levels may account for the increased drug-taking behaviors observed in females. Like PKA, CART peptide is implicated in the reinforcing effects of drugs.[229] As females enter a phase marked by high estradiol levels, they have elevated CART peptide levels, which can increase the desire to use drugs.

Drugs also seem to differentially increase dopamine levels in the NAc in males and females. For example, low doses of ethanol (e.g., 0.25 and 0.50 g/kg) greatly increase dopamine levels of females, whereas only modest increases in dopamine are observed in males; furthermore, a higher dose of ethanol (e.g., 2 g/kg) *decreases* intra-NAc dopamine levels in males, but not in females.[67,68] Additionally, females exposed to heroin show greater dopamine release compared to drug-naïve rats, an effect that is absent in males.[87] In addition to differences in dopamine release, prolonged drug exposure differentially alters protein expression throughout the brain. Chronic ethanol administration downregulates dopamine D_1-like receptor expression and serotonin 5-HT_{2A} receptor expression in both the frontal cortex and in the hippocampus, an effect observed in females only.[230] Drug withdrawal can also cause sex-specific alterations in neurons. Female, but not male, rodents subjected to 20 days of forced abstinence from cocaine show decreased levels of the early growth response 3 (ERG3) transcription factor in D_2-like-containing medium spiny neurons (MSNs) in the NAc.[231] Remember that transcription factors are involved in transcribing DNA into RNA.

What's even more interesting is that Engeln et al. found that overexpressing ERG3 increases drug-induced reinstatement in males but blocks reinstatement in females.[231]

Sex differences are also observed in the dorsal striatum. Females in the proestrus phase show a greater percentage of Fos activation in the striatum following stimulant administration compared to males and to ovariectomized females.[232] The ability of estradiol to increase drug self-administration may be linked to its ability to potentiate dopamine release in the dorsal striatum[233] and dopamine/serotonin levels in the VTA.[201] Drug-induced increases in striatal dopamine levels are highest during the estrus phase.[234] Estradiol administration potentiates psychostimulant-induced increases in intra-dorsal striatum and intra-NAc dopamine[205,233,235,236] and increases striatal dopamine transporter (DAT) expression.[236] Fig. 12.3 shows how the estrous cycle influences dopamine release. Females have higher levels of *DA and cAMP-regulated phosphoprotein of 32 kDa* (*DARPP-32*) in the dorsal striatum, as well as in the NAc, compared to males following cocaine self-administration.[238] Like PKA and CART peptide, DARPP-32 is an important mediator of addiction-like behaviors.[239]

Dopamine is not the only neurotransmitter system that differs between males and females. Differences in glutamatergic activity have been observed in males and females following drug administration. Specifically, methamphetamine self-administration selectively increases evoked excitatory potentials (i.e., increases NMDA receptor activation) in the prefrontal cortex (PFC) of females, but not males.[97] Chronic ethanol administration selectively downregulates GluN2B-containing NMDA receptor protein expression in the hippocampus of females.[230] Somewhat similarly, during reinstatement of methamphetamine-seeking behavior, females show greater excitation in the dentate gyrus of the hippocampus and show an increased GluN2A/2B ratio in this region.[240] As an increased GluN2A/2B ratio is important for preserving memories,[241] exposing animals to cues previously associated with drugs seemingly activates drug-related memories via this mechanism. In Chapter 7, you learned about the glutamate homeostasis hypothesis addiction. Recall that administration of drugs can alter the ratio of AMPA and NMDA receptors by upregulating AMPA receptors. *Protein interacting with C kinase 1* (*PICK1*) is important for removing AMPA receptors from the synapse. Deleting PICK1 from the medial PFC (mPFC) attenuates cocaine-seeking behavior in male mice, but *increases* relapse-like behavior in females.[242] Beyond ionotropic glutamate receptors, the mGluR5 appears to be required for estradiol to increase drug self-administration as blocking these receptors inhibits estradiol-induced increases in cocaine self-administration.[243]

Excitatory neurotransmission is influenced by more than just glutamate. There is a neuropeptide called *substance P* that regulates excitatory neurotransmission. At least in the NAc, substance P can depress excitation.[244] An interesting finding is that there are increased substance P levels in the NAc core of males, but decreased levels in the NAc core of females.[89] Because there is less substance P to decrease excitatory neurotransmission, this can account for why females show greater drug-induced activation in dopaminergic and glutamatergic systems.

During drug withdrawal, sex differences are observed in brain activation beyond what is observed in D_2-like receptors in the NAc. In particular, during ethanol withdrawal, both males and females show increased Fos expression in the anterior cingulate cortex (ACC), the amygdala, and the lateral habenula; however, females also show increased Fos expression in the PFC and in the NAc.[50] Likewise, females show greater brain activation during

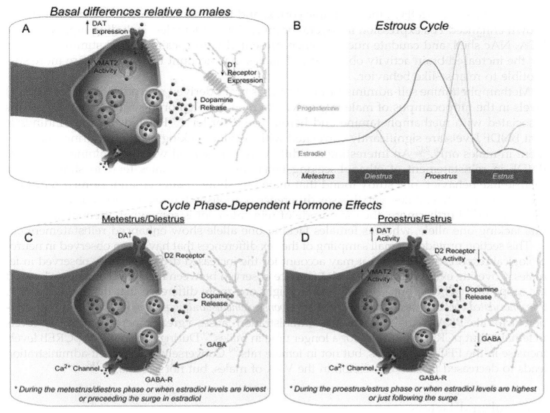

FIGURE 12.3 A schematic depicting basal differences in dopamine release in the striatum of females relative to males (A). Recall that medium spiny neurons (MSNs) are highly abundant in the striatum and regulate the release of dopamine from dopaminergic neurons. Females have decreased dopamine D_1-like receptors on MSNs, which causes greater disinhibition of the dopamine neuron, leading to greater dopamine release. Additionally, females have increased vesicular monoamine transporter 2 ($VMAT_2$) activity and dopamine transporter (DAT) activity. Panel (B) shows how estradiol and progesterone levels fluctuate across each phase of the estrous cycle. Notice that estradiol and progesterone levels are elevated during the proestrus phase. Panels (C) and (D) compare dopaminergic activity of female animals during metestrus/diestrus and proestrus/estrus. During metestrus/diestrus, when estradiol/progesterone levels are low, dopaminergic activity is similar to what is observed in male rats and female rats that have received an ovariectomy. During proestrus, ovarian hormone levels increase, which leads to alterations in the striatal system, including decreased GABAergic activity, decreased dopamine D_2 autoreceptor activity, and increased dopamine release. This altered activity is observed in the estrus phase even though ovarian hormone levels begin to decrease to levels observed in the metestrus/diestrus phases. *Image comes from Fig. 2 of Zachry JE, Nolan SO, Brady LJ, Kelly SJ, Siciliano CA, Calipari ES. Sex differences in dopamine release regulation in the striatum.* Neuropsychopharmacology. 2021;46(3):491–499. https://doi.org/10.1038/s41386-020-00915-1, Copyright 2021.

relapse-like behaviors. Earlier I mentioned that females show increased excitation in the dentate gyrus during methamphetamine relapse-like behavior. Females also show elevated activity in multiple brain regions during reinstatement of methamphetamine-seeking behavior, such as the dorsal striatum, the NAc (both core and shell regions), the amygdala, the lateral orbitofrontal cortex (lOFC), the prelimbic cortex, and the cingulate cortex.[137] Other work has shown enhanced Fos expression in several brain regions of females, including hippocampus, VTA, NAc shell, and caudate nucleus when presented with cocaine-paired stimuli.[245] Overall, the increased brain activity observed in females may account for why they are more susceptible to relapse-like behavior.

Methamphetamine self-administration increases brain-derived neurotrophic factor (BDNF) levels in the hippocampus of male rats only.[94] In men, a polymorphism of the *BDNF* gene is associated with methamphetamine and heroin use disorder.[246] Related to these findings is that BNDF levels are significantly correlated with cocaine-seeking behavior during reinstatement in males only.[247] An interesting sex difference is observed when examining the role of BNDF in reinstatement of ethanol-seeking behavior. While females tend to show greater relapse-like behavior, one study found that male rats show greater ethanol-primed reinstatement compared to females; yet, this sex difference is abolished in rats lacking one allele of the *BDNF* gene.[93] Furthermore, the magnitude of reinstatement is somewhat attenuated in male rats lacking one allele, whereas females lacking one allele show enhanced reinstatement.

This section provides a small sampling of the sex differences that have been observed in neurobiological responses to drugs that may account for the increased drug sensitivity observed in females. Covering every neurobiological difference observed between males and females is beyond the scope of this book, but I do want to highlight one more difference that has been reported. *Phosphorylated cyclic adenosine monophosphate response element-binding protein (pCREB)* is a transcription factor like PKA. In the NAc, cocaine administration causes a greater increase in pCREB levels in females, but pCREB is induced for a longer time in males.[59] During cocaine CPP, pCREB levels increase in the PFC of male rats, but not in female rats.[60] Conversely, cocaine self-administration leads to decreased pCREB expression in the VTA of males, but not in females.[248]

Sociocultural factors

If females are more sensitive to the reinforcing effects of drugs, why do epidemiological studies often show increased substance use in men? To answer this question, we need to discuss nonbiological factors accounting for gender differences in drug use, particularly sociocultural factors. As summarized by Becker et al., culture is a major factor influencing differential drug use in men and women.[249] For example, drug use is more normalized in men. There is some evidence to this when we examine state incarceration rates for drug offenses between 1986 and 1996 (near the peak of the "War on Drugs"). In 1986, ~8% of incarcerated men were in state prison for a drug offense compared to ~12% of incarcerated women. By 1996, drug-related offenses accounted for ~22% of men in state prison compared to ~37.5% of women in prison. Thus, during a 10-year period, drug-related incarcerations increased by 522% in men and 888% in women.[250] Even during the mid-2010s, approximately 35% of female prisoners were incarcerated due to a drug offense compared to only 19% of male prisoners.[251] When looking at these statistics, it is important to note that there are more men in prison for drug offenses compared to women; these statistics just highlight the disproportionate increases in prison sentences for drug offenses between men and women. These data seemingly show that women are punished more harshly for committing a drug offense compared to men, possibly because drug use is more socially acceptable in men.

At the beginning of this chapter, I mentioned that men have more opportunities to use drugs. From a historical perspective, this has not always been quite the case. During the 1800s, "medicines" containing ethanol and opiates were often sold to women to treat multiple health conditions, including the pain and discomfort associated with menstruation.[249] Because these drugs were targeted at women more so than men, opiate dependence was higher in women during the late 1800s. Access to opiates decreased dramatically during the 1900s as the United States created the Food and Drug Administration (FDA) and started to place prohibitive taxes on certain narcotics like opiates. Consequently, the number of women using drugs containing ethanol/opiates declined dramatically. Decades later, sedative-hypnotic drugs, particularly diazepam, became highly associated with women during the 1960 and 1970s. Why? Diazepam is a benzodiazepine, which is often used to treat anxiety disorders, and women are more likely to be diagnosed with an anxiety disorder.[252] In the 1980s and 1990s, physicians began prescribing psychostimulant drugs to women at a higher rate relative to men. Psychostimulant drugs are often linked to ADHD; however, this drug class has been used for weight loss. This makes sense as one effect of drugs like amphetamine and methylphenidate is appetite suppression. In each of these cases, increased drug use was observed in women during periods in which doctors increased their prescriptions for these medications. When these drugs are not widely prescribed (i.e., antidepressant drugs are now often prescribed for anxiety disorders instead of benzodiazepines), women do not report the same level of drug use as men. Becker et al. propose that women may be more sensitive to the availability of drugs compared to men; that is, legal/regulatory constraints on drug availability seem to impact women more so than men.[249]

Drug use in sexual/gender minority individuals

The primary focus of this chapter has been to discuss how biological sex impacts drug use and drug sensitivity. However, I would be remiss if I did not discuss how one's **sexual identity** or **gender identity** influence drug-taking behaviors. Sexual identity refers to how one thinks of oneself in terms of to whom they are romantically or sexually attracted.[253] Terms like *gay* (attraction to the same sex), *bisexual* (attraction to either sex), or *asexual* (no attraction to either sex) are often used. Individuals can also identify as *pansexual*. To understand what the term pansexual means, we need to discuss gender identity. Even if an individual is born biologically male (XY chromosomes), that person may not identify as a man. Instead, they may identify as a woman. These individuals are *transgendered*, which contrasts with *cisgender* individuals whose gender aligns with their biological sex.[253] Other individuals are *nonbinary*, which can be further subdivided. These individuals have a gender identity that fluctuates over time (*gender fluid*), or they do not identify as a man or as a woman (*agender*).[254] Going back to pansexual individuals: these individuals are not limited in sexual choice with regard to biological sex, gender, or gender identity.

Research has shown that sexual/gender minority individuals are more likely to use tobacco products, ethanol (including more binge/heavy drinking episodes), and illicit drugs,[255-259] and they are more likely to begin using substances at an earlier age.[260] Consistent with the finding that adolescent boys are more likely to smoke cigarettes compared to girls, transgendered boys smoke more compared to transgendered girls, and transgendered boys begin smoking at an earlier age compared to cisgender boys,[261] with one study finding that transgendered students are more than 3 times as likely to use cigarettes.[256] Approximately one quarter of 16–18-year-old sexual minority individuals binge drink, and one-fifth use drugs.[262] High ethanol dependence (~15%) and alcohol use disorder (ranging from 10% to 18% across early adulthood) are observed in sexual minority individuals.[257,263] Sexual minority women are at increased risk for developing drug-related problems. They smoke more frequently, smoke more cigarettes, consume more ethanol, and are more likely to use tobacco and ethanol concurrently compared to heterosexual women;[264,265] furthermore, sexual minority women show greater escalation of tobacco and ethanol use as they transition from adolescence into adulthood.[266] SUD rates have been reported to be highest in sexual minority women, with one study reporting that the percentage of these women meeting criteria for alcohol use disorder, tobacco use disorder, and other SUDs being 27%, 29%, and 11%, respectively.[267]

Goldbach et al. identified six factors that increase the risk of substance use in sexual minority individuals: psychological stress, housing status, victimization, lack of supportive environments, negative disclosure reactions, and internalizing/externalizing problem behavior (e.g., anxiety, depression, suicidal ideation, and conduct problems like truancy).[268] As covered in the previous chapter, stress is a major risk factor for substance use; this is true of sexual minority individuals as well. Compounding the role of stress on drug use in sexual minorities is inadequate housing. Being homeless or living in a shelter impacts drug use in both sexual normative and sexual minority individuals, but sexual/gender minorities without adequate housing show greater increases in drug use.[269] Sexual/gender minorities experience high levels of discrimination, domestic violence, and urban violence,[257,270,271] which promotes increased ethanol and drug use/misuse.[272,273] See the Experiment Spotlight below for an example of how **sexual minority stress** influences ethanol use in sexual minority women.

Experiment Spotlight: Minority stress and alcohol use in sexual minority women's daily lives (Lewis et al.[272])

In addition to dealing with the same stressors covered in the previous chapter, sexual and gender minority individuals face additional types of stress, termed sexual minority stress. As mentioned above, these individuals can encounter discrimination, hostility, and/or violence due to their sexual/gender identity (*distal stressors*). They also experience *proximal stressors*, such as fears of "coming out" and internalized homophobia/transphobia. By now, you know that sexual/gender minority individuals are at increased risk of using substances, especially ethanol. Lewis et al.[272] used an interesting approach to further determine how sexual minority stress influences ethanol use (drinking likelihood, drinking quantity, and binge drinking) and ethanol-related consequences on the same day as the stressor's occurrence and the day following the stressful event.

The study included 321 sexual minority women between the ages of 18 and 35. Participants had to meet multiple inclusion criteria: (a) be a cisgender woman, (b) be in a same-sex relationship for at least 3 months in which they see their partner at least once per week, and (c) be able to complete a daily diary survey each morning. Either the participant or her partner had to meet additional criteria: (a) be mostly attracted to other women; (b) consume ethanol on at least 3 days during the past 2 weeks; and (c) binge drink at least once in the past 2 weeks.

Four types of sexual minority stressors were measured in the study. (1) Global negative sexual minority stress entailed asking participants the question "Did any negative events occur yesterday that were related to your being a sexual minority or to sexual orientation issues?". (2) Sexual minority events include specific experiences individuals encountered the previous day (e.g., being verbally harassed, being told that they are being oversensitive about sexual minority issues, etc.). (3) Concealment of sexual identity refers to avoiding discussions that would reveal the individual's sexual identity to other individuals, including immediate family, friends, acquaintances, classmates/coworkers, and strangers. (4) Sexual minority discrimination was assessed with a single question related to being treated unfairly or being rejected because of one's sexual identity.

During the experiment, participants, as well as their partners, completed a daily "diary" survey (online) answering questions related to the sexual minority stressors described above. Additionally, individuals identified if they consumed ethanol the previous day and how much ethanol they consumed. Binge drinking was defined as consuming four or more standard drinks (e.g., four 12-oz servings of beer). Participants indicated if they experienced any ethanol consequences like doing embarrassing things, taking risks, getting sick, or feeling hung over. Data were analyzed one of two ways. First, Lewis et al. performed within-person correlations, which compare an individual's level of sexual minority stress *relative* to the sexual minority stress they normally face. Second, a between-person correlation was performed, which determines if greater levels of sexual minority stress *across* individuals are associated with ethanol-related behaviors.

Across the 14-day experiment, approximately 36% of individuals in the study experienced global sexual minority stress, but 71% experienced at least one sexual minority stress event. Over half of the sample (51%) reported concealment, and one-third (~34%) faced discrimination. Lewis et al. found that individuals experiencing global or specific sexual minority stress, as well as discrimination, were more likely to consume ethanol that same day. Furthermore, experiencing a specific stressor increased next-day binge drinking/drinking quantity. These significant associations were observed when the within-person analysis was used only. When examining between-person correlations, Lewis et al. found that experiencing a specific sexual minority stress event was positively associated with drinking quantity and experiencing a negative ethanol consequence. Additionally, global negative sexual minority stress and discrimination were positively associated with negative ethanol consequences.

Questions to consider:

(1) When discussing the results of their study, Lewis et al. state that one limitation is the correlative nature of the data. In other words, Lewis et al. found that sexual minority stress is associated with drinking behaviors. Why is this a potential limitation?
(2) Given that discrepant correlations were obtained when using a within-person and a between-person approach, how should these data be interpreted? Is one analytic approach better than the other? Why or why not?

Answers to questions:

(1) One challenge with correlations is determining causality. Do sexual minority stressors lead to increased drinking, or does drinking lead an individual to experience sexual minority stress? Imagine the following scenario: an individual goes to a bar and begins drinking. While at the bar, the individual encounters someone that uses a derogatory term. If the individual had not been at the bar, they would not have experienced this stressor. Therefore, the increased drinking that occurred on that date was not due to exposure to sexual minority stress.

(2) One approach is not necessarily better than the other. They provide different pieces of information. As noted by Lewis et al., the within-person analysis indicates that experiencing more sexual minority stressors than what is typically experienced is correlated with ethanol use. That is, sexual minority individuals often face discrimination and hostility. However, there may be some days in which they experience even greater levels of harassment. Sexual minority women may be at greater risk of consuming ethanol on days in which they experience heightened discrimination/hostility. The between-person approach simply asks "are greater levels of sexual minority stress, regardless of one's baseline levels of sexual minority stress, associated with ethanol use/ethanol-related behaviors?"

Just as social factors can be used in a therapeutic setting to help those with a SUD, social support can act as a protective factor against drug use in sexual/gender minority individuals. This makes sense as social rejection can serve as a stressor in itself. Sexual/gender minority adolescents living in an area that is more accepting are less likely to use illicit drugs, and adolescent girls have lower odds of using cannabis and cigarettes.[274] Likewise, sexual minority students attending a school with a gay-straight alliance are less likely to use cannabis, cocaine, or hallucinogens.[275]

Similar to what we covered in Chapter 10 when comparing drug use between racial minorities to White individuals, when other factors are statistically controlled, the increased drug used observed in sexual minority individuals decrease. For example, before controlling for additional variables, Lowry et al. found that gay and bisexual high school students use more drugs like cocaine, methamphetamine, and heroin, but when factors such as social stressors, age, sex, and race/ethnicity were controlled, many of these differences disappeared.[259] Furthermore, some research has shown that the relationship between transgender identity and cigarette smoking disappears when controlling for other variables like socioeconomic status, age, presence of depressive symptoms, and binge drinking.[258] However, other work has shown greater drug use in sexual minorities even after controlling for other demographic variables.[268] Despite these discrepancies, more work is needed to understand substance use in sexual and gender minority individuals. Because these individuals are more likely to be victims of violent or sexual acts, additional services are needed to help those at increased risk of developing a SUD.

Quiz 12.2

(1) During the menstrual cycle, when are estradiol levels highest?
  a. Early follicular phase
  b. Late follicular phase
  c. Luteal phase
  d. Menstruation

(2) How does progesterone administration affect drug sensitivity in females?
  a. Decreases drug sensitivity
  b. Has no effect on drug sensitivity
  c. Increases drug sensitivity
  d. Increases drug sensitivity at low doses but decreases drug sensitivity at high doses

(3) Which of the following is an example of how sex acts as a moderating variable?
  a. Being homeless increases the risk of drug use in transgendered women, but not necessarily in transgendered men.
  b. Female rats show more relapse-like behaviors compared to male rats.
  c. Gay and bisexual individuals have more drinking-related problems compared to heterosexuals.
  d. Transgender individuals smoke more cigarettes compared to cisgender individuals.

(4) Which of the following statements regarding neurobiological differences between males and females is true?
  a. CART peptide levels increase as progesterone increases in females, which accounts for the increased drug sensitivity observed in females.
  b. Drug-induced increases in dopamine levels are higher in females compared to males.
  c. Substance P levels are higher in the NAc of females compared to males.
  d. All of the above.

(5) According to Dr. Jill Becker, what factor can explain why women report less drug use if they are more sensitive to the reinforcing effects of drugs?
  a. They are more impacted by laws/norms that influence the availability of drugs.
  b. They are more sensitive to alternative reinforcers.
  c. They have greater self-control.
  d. They receive more social support.

Answers to quiz 12.2

(1) b — Late follicular phase
(2) a — Decreases drug sensitivity
(3) a — Being homeless increases the risk of drug use in transgendered women, but not necessarily in transgendered men.
(4) b — Drug-induced increases in dopamine levels are higher in females compared to males.
(5) a — They are more impacted by laws/norms that influence the availability of drugs.

Chapter summary

Clinical and preclinical research has demonstrated a complex relationship between gender/sex and substance use. While females tend to be more sensitive to the reinforcing/conditioned rewarding effects of drugs, men often report greater drug use. The greater drug sensitivity observed in females can be largely attributed to hormonal differences. The estrogen estradiol appears to potentiate the effects of drugs, whereas the progestogen progesterone attenuates these effects. The ability of estradiol to increase drug effects in females can be partially explained by its actions on the dopaminergic system in striatal regions. Females tend to have greater dopamine release when estradiol levels are highest. Other neurobiological differences have been observed between males and females. Collectively, research examining gender/sex differences in drug dependence vulnerability has revealed that environmental factors modulate the influence of sex on drug use, as women may be more sensitive to the availability of drugs as laws and regulations change. With cannabis legalization now in place in several U.S. states, I will be interested to see if/how cannabis use rates will differ between men and women.

Glossary

Androgen — type of steroid hormone that regulates the development and maintenance of male characteristics; includes testosterone.
Estrogen — type of steroid hormone that regulates the development and maintenance of female characteristics; includes estradiol.
Estrous cycle — the recurring reproductive cycle of specific female mammals, including rats and mice; typically composed of proestrus, estrus, diestrus, and metestrus phases.
Gender identity — one's self-perceived gender.
Gender — a sociocultural construct that encompasses the norms and behaviors associated with one's sex.
Hypothalamic—pituitary—gonadal (HPG) axis — system that regulates the production and release of steroidal sex hormones from the testes and the ovaries.
Menstrual cycle — a series of changes in hormone production that affect the uterus and ovaries of a biological female that allow for reproduction to occur; primarily composed of the follicular and luteal phases, with menstruation occurring at the beginning of the follicular phase and ovulation occurring at the end of the follicular phase and at the beginning of the luteal phase.
Progesterone — a type of progestogen steroid hormone that is produced during the luteal phase of the menstrual cycle; helps prepare a fertilized egg cell to embed itself in the wall of the uterus.
Sex — a trait that denotes one's reproductive organs (testes/penis for males and ovaries/vagina for females).
Sexual identity — how one thinks of oneself in terms of to whom one is romantically or sexually attracted (not to be confused with gender identity).
Sexual minority stress — stressors that are specific to sexual minority individuals (e.g., discrimination based on sexual identity).

References

1. Gray J. *Men Are from Mars, Women Are from Venus.* HarperCollins Publishers; 1992.
2. Substance Abuse and Mental Health Services Administration. *Key Substance Use and Mental Health Indicators in the United States: Results from the 2019 National Survey on Drug Use and Health (HHS Publication No. PEP20-07-01-001, NSDUH Series H-55).* Rockville, MD: Center for Behavioral Health Statistics and Quality, Substance Abuse and Mental Health Services Administration; 2020. Retrieved from: https://www.samhsa.gov/data/.
3. Svarstad BL, Cleary PD, Mechanic D, Robers PA. Gender differences in the acquisition of prescribed drugs: an epidemiological study. *Med Care.* 1987;25(11):1089—1098. https://doi.org/10.1097/00005650-198711000-00007.
4. Pirie PL, Murray DM, Luepker RV. Gender differences in cigarette smoking and quitting in a cohort of young adults. *Am J Publ Health.* 1991;81(3):324—327. https://doi.org/10.2105/ajph.81.3.324.

5. Dean AJ, McBride M, Macdonald EM, Connolly Y, McDermott BM. Gender differences in adolescents attending a drug and alcohol withdrawal service. *Drug Alcohol Rev.* 2010;29(3):278−285. https://doi.org/10.1111/j.1465-3362.2009.00152.x.
6. Palamar JJ, Davies S, Ompad DC, Cleland CM, Weitzman M. Powder cocaine and crack use in the United States: an examination of risk for arrest and socioeconomic disparities in use. *Drug Alcohol Depend.* 2015;149:108−116. https://doi.org/10.1016/j.drugalcdep.2015.01.029.
7. Uhlmann S, Debeck K, Simo A, Kerr T, Montaner JS, Wood E. Crystal methamphetamine initiation among street-involved youth. *Am J Drug Alcohol Abuse.* 2014;40(1):31−36. https://doi.org/10.3109/00952990.2013.836531.
8. Pope SK, Falck RS, Carlson RG, Leukefeld C, Booth BM. Characteristics of rural crack and powder cocaine use: gender and other correlates. *Am J Drug Alcohol Abuse.* 2011;37(6):491−496. https://doi.org/10.3109/00952990.2011.600380.
9. Lev-Ran S, Le Strat Y, Imtiaz S, Rehm J, Le Foll B. Gender differences in prevalence of substance use disorders among individuals with lifetime exposure to substances: results from a large representative sample. *Am J Addict.* 2013;22(1):7−13. https://doi.org/10.1111/j.1521-0391.2013.00321.x.
10. Palmer RH, Young SE, Hopfer CJ, et al. Developmental epidemiology of drug use and abuse in adolescence and young adulthood: evidence of generalized risk. *Drug Alcohol Depend.* 2009;102(1−3):78−87. https://doi.org/10.1016/j.drugalcdep.2009.01.012.
11. Sher KJ, Grekin ER, Williams NA. The development of alcohol use disorders. *Annu Rev Clin Psychol.* 2005;1:493−523. https://doi.org/10.1146/annurev.clinpsy.1.102803.144107.
12. Khan SS, Secades-Villa R, Okuda M, et al. Gender differences in cannabis use disorders: results from the national epidemiologic survey of alcohol and related conditions. *Drug Alcohol Depend.* 2013;130(1−3):101−108. https://doi.org/10.1016/j.drugalcdep.2012.10.015.
13. Wechsberg WM, Craddock SG, Hubbard RL. How are women who enter substance abuse treatment different than men?: a gender comparison from the Drug Abuse Treatment Outcome Study (DATOS). *Drugs Soc.* 1998;13(1−2):97−115. https://doi.org/10.1300/J023v13n01_06.
14. Gowin JL, Sloan ME, Stangl BL, Vatsalya V, Ramchandani VA. Vulnerability for alcohol use disorder and rate of alcohol consumption. *Am J Psychiatr.* 2017;174(11):1094−1101. https://doi.org/10.1176/appi.ajp.2017.16101180.
15. Kong G, Kuguru KE, Krishnan-Sarin S. Gender differences in U.S. adolescent E-cigarette use. *Curr Addic Rep.* 2017;4(4):422−430. https://doi.org/10.1007/s40429-017-0176-5.
16. Thamotharan S, Hahn H, Fields S. Drug use status in youth: the role of gender and delay discounting. *Subst Use Misuse.* 2017;52(10):1338−1347. https://doi.org/10.1080/10826084.2017.1280831.
17. McCabe SE, Morales M, Cranford JA, Delva J, McPherson MD, Boyd CJ. Race/ethnicity and gender differences in drug use and abuse among college students. *J Ethn Subst Abuse.* 2007;6(2):75−95. https://doi.org/10.1300/J233v06n02_06.
18. O'Malley PM, Johnston LD. Epidemiology of alcohol and other drug use among American college students. *J Stud Alcohol.* 2002;(14):23−39. https://doi.org/10.15288/jsas.2002.s14.23.
19. Van Etten ML, Neumark YD, Anthony JC. Male-female differences in the earliest stages of drug involvement. *Addiction.* 1999;94(9):1413−1419. https://doi.org/10.1046/j.1360-0443.1999.949141312.x.
20. Myers CS, Taylor RC, Moolchan ET, Heishman SJ. Dose-related enhancement of mood and cognition in smokers administered nicotine nasal spray. *Neuropsychopharmacology.* 2008;33(3):588−598. https://doi.org/10.1038/sj.npp.1301425.
21. Panday S, Reddy SP, Ruiter RA, Bergström E, de Vries H. Nicotine dependence and withdrawal symptoms among occasional smokers. *J Adolesc Health.* 2007;40(2):144−150. https://doi.org/10.1016/j.jadohealth.2006.09.001.
22. Thorner ED, Jaszyna-Gasior M, Epstein DH, Moolchan ET. Progression to daily smoking: is there a gender difference among cessation treatment seekers? *Subst Use Misuse.* 2007;42(5):829−835. https://doi.org/10.1080/10826080701202486.
23. He J, Xie Y, Tao J, et al. Gender differences in socio-demographic and clinical characteristics of methamphetamine inpatients in a Chinese population. *Drug Alcohol Depend.* 2013;130(1−3):94−100. https://doi.org/10.1016/j.drugalcdep.2012.10.014.
24. Lynch WJ, Kalayasiri R, Sughondhabirom A, et al. Subjective responses and cardiovascular effects of self-administered cocaine in cocaine-abusing men and women. *Addiction Biol.* 2008;13(3−4):403−410. https://doi.org/10.1111/j.1369-1600.2008.00115.x.

25. Singha AK, McCance-Katz EF, Petrakis I, Kosten TR, Oliveto A. Sex differences in self-reported and physiological response to oral cocaine and placebo in humans. *Am J Drug Alcohol Abuse*. 2000;26(4):643−657. https://doi.org/10.1081/ada-100101900.
26. Staton-Tindall M, Oser CB, Duvall JL, et al. Male and female stimulant use among rural Kentuckians: the contribution of spirituality and religiosity. *J Drug Issues*. 2008;38(3):863−882. https://doi.org/10.1177/002204260803800310.
27. Vansickel AR, Stoops WW, Rush CR. Human sex differences in d-amphetamine self-administration. *Addiction*. 2010;105(4):727−731. https://doi.org/10.1111/j.1360-0443.2009.02858.x.
28. Hernandez-Avila CA, Rounsaville BJ, Kranzler HR. Opioid-, cannabis- and alcohol-dependent women show more rapid progression to substance abuse treatment. *Drug Alcohol Depend*. 2004;74(3):265−272. https://doi.org/10.1016/j.drugalcdep.2004.02.001.
29. Kosten TA, Gawin FH, Kosten TR, Rounsaville BJ. Gender differences in cocaine use and treatment response. *J Subst Abuse Treat*. 1993;10(1):63−66. https://doi.org/10.1016/0740-5472(93)90100-g.
30. Franconi F, Brunelleschi S, Steardo L, Cuomo V. Gender differences in drug responses. *Pharmacol Res*. 2007;55(2):81−95. https://doi.org/10.1016/j.phrs.2006.11.001.
31. Zacny JP, Drum M. Psychopharmacological effects of oxycodone in healthy volunteers: roles of alcohol-drinking status and sex. *Drug Alcohol Depend*. 2010;107(2−3):209−214. https://doi.org/10.1016/j.drugalcdep.2009.10.012.
32. Weinberger AH, Platt JM, Shuter J, Goodwin RD. Gender differences in self-reported withdrawal symptoms and reducing or quitting smoking three years later: a prospective, longitudinal examination of U.S. adults. *Drug Alcohol Depend*. 2016;165:253−259. https://doi.org/10.1016/j.drugalcdep.2016.06.013.
33. Herrmann ES, Weerts EM, Vandrey R. Sex differences in cannabis withdrawal symptoms among treatment-seeking cannabis users. *Exp Clin Psychopharmacol*. 2015;23(6):415−421. https://doi.org/10.1037/pha0000053.
34. Chen W-Y, Huang M-C, Lin S-K. Gender differences in subjective discontinuation symptoms associated with ketamine use. *Subst Abuse Treat Prev Pol*. 2014;9:39. https://doi.org/10.1186/1747-597X-9-39.
35. Deshmukh A, Rosenbloom MJ, Sassoon S, O'Reilly A, Pfefferbaum A, Sullivan EV. Alcoholic men endorse more DSM-IV withdrawal symptoms than alcoholic women matched in drinking history. *J Stud Alcohol*. 2003;64(3):375−379. https://doi.org/10.15288/jsa.2003.64.375.
36. Weinberger AH, Maciejewski PK, McKee SA, Reutenauer EL, Mazure CM. Gender differences in associations between lifetime alcohol, depression, panic disorder, and posttraumatic stress disorder and tobacco withdrawal. *Am J Addict*. 2009;18(2):140−147. https://doi.org/10.1080/10550490802544888.
37. Plawecki MH, White K, Kosobud A, et al. Sex differences in motivation to self-administer alcohol after 2 weeks of abstinence in young-adult heavy drinkers. *Alcohol Clin Exp Res*. 2018;42(10):1897−1908. https://doi.org/10.1111/acer.13860.
38. Gallop RJ, Crits-Christoph P, Ten Have TR, et al. Differential transitions between cocaine use and abstinence for men and women. *J Consult Clin Psychol*. 2007;75(1):95−103. https://doi.org/10.1037/0022-006X.75.1.95.
39. Yu J, Zhang S, Epstein DH, et al. Gender and stimulus difference in cue-induced responses in abstinent heroin users. *Pharmacol Biochem Behav*. 2007;86(3):485−492. https://doi.org/10.1016/j.pbb.2007.01.008.
40. National Institutes of Health. *Consideration of Sex as a Biological Variable in NIH-Funded Research*. Notice number NOT-OD-15-102; 2016. https://grants.nih.gov/grants/guide/notice-files/not-od-15-102.html.
41. Webb B, Burnett PW, Walker DW. Sex differences in ethanol-induced hypnosis and hypothermia in young Long-Evans rats. *Alcohol Clin Exp Res*. 2002;26(5):695−704.
42. de la Torre ML, Escarabajal MD, Agüero Á. Sex differences in adult Wistar rats in the voluntary consumption of ethanol after pre-exposure to ethanol-induced flavor avoidance learning. *Pharmacol Biochem Behav*. 2015;137:7−15. https://doi.org/10.1016/j.pbb.2015.07.011.
43. Sherrill LK, Berthold C, Koss WA, Juraska JM, Gulley JM. Sex differences in the effects of ethanol pre-exposure during adolescence on ethanol-induced conditioned taste aversion in adult rats. *Behav Brain Res*. 2011;225(1):104−109. https://doi.org/10.1016/j.bbr.2011.07.003.
44. Jenney CB, Dasalla J, Grigson PS. Female rats exhibit less avoidance than male rats of a cocaine-, but not a morphine-paired, saccharin cue. *Brain Res Bull*. 2018;138:80−87. https://doi.org/10.1016/j.brainresbull.2017.09.001.
45. Martini M, Pinto AX, Valverde O. Estrous cycle and sex affect cocaine-induced behavioural changes in CD1 mice. *Psychopharmacology*. 2014;231(13):2647−2659. https://doi.org/10.1007/s00213-014-3433-5.

46. Collins D, Reed B, Zhang Y, Kreek MJ. Sex differences in responsiveness to the prescription opioid oxycodone in mice. *Pharmacol, Biochem Behav*. 2016;148:99–105. https://doi.org/10.1016/j.pbb.2016.06.006.
47. Kest B, Adler M, Hopkins E. Sex differences in thermoregulation after acute and chronic morphine administration in mice. *Neurosci Lett*. 2000;291(2):126–128. https://doi.org/10.1016/s0304-3940(00)01393-8.
48. Matzeu A, Terenius L, Martin-Fardon R. Exploring sex differences in the attenuation of ethanol drinking by naltrexone in dependent rats during early and protracted abstinence. *Alcohol Clin Exp Res*. 2018;42(12):2466–2478. https://doi.org/10.1111/acer.13898.
49. Nesil T, Kanit L, Ugur M, Pogun S. Nicotine withdrawal in selectively bred high and low nicotine preferring rat lines. *Pharmacol, Biochem Behav*. 2015;131:91–97. https://doi.org/10.1016/j.pbb.2015.02.009.
50. Li J, Chen P, Han X, et al. Differences between male and female rats in alcohol drinking, negative affects and neuronal activity after acute and prolonged abstinence. *Int J Phys Pathophys & Pharmac*. 2019;11(4):163–176.
51. Towers EB, Tunstall BJ, McCracken ML, Vendruscolo LF, Koob GF. Male and female mice develop escalation of heroin intake and dependence following extended access. *Neuropharmacology*. 2019;151:189–194. https://doi.org/10.1016/j.neuropharm.2019.03.019.
52. Bobzean S, Kokane SS, Butler BD, Perrotti LI. Sex differences in the expression of morphine withdrawal symptoms and associated activity in the tail of the ventral tegmental area. *Neurosci Lett*. 2019;705:124–130. https://doi.org/10.1016/j.neulet.2019.04.057.
53. Marusich JA, Lefever TW, Antonazzo KR, Craft RM, Wiley JL. Evaluation of sex differences in cannabinoid dependence. *Drug Alcohol Depend*. 2014;137:20–28. https://doi.org/10.1016/j.drugalcdep.2014.01.019.
54. Piazza PV, Deminiere JM, le Moal M, Simon H. Stress- and pharmacologically-induced behavioral sensitization increases vulnerability to acquisition of amphetamine self-administration. *Brain Res*. 1990;514(1):22–26. https://doi.org/10.1016/0006-8993(90)90431-a.
55. Algallal H, Allain F, Ndiaye NA, Samaha AN. Sex differences in cocaine self-administration behaviour under long access versus intermittent access conditions. *Addiction Biol*. 2020;25(5):e12809. https://doi.org/10.1111/adb.12809.
56. Brown RW, Perna MK, Noel DM, Whittemore JD, Lehmann J, Smith ML. Amphetamine locomotor sensitization and conditioned place preference in adolescent male and female rats neonatally treated with quinpirole. *Behav Pharmacol*. 2011;22(4):374–378. https://doi.org/10.1097/FBP.0b013e328348737b.
57. Hu M, Becker JB. Effects of sex and estrogen on behavioral sensitization to cocaine in rats. *J Neurosci*. 2003;23(2):693–699. https://doi.org/10.1523/JNEUROSCI.23-02-00693.2003.
58. Mathews IZ, McCormick CM. Female and male rats in late adolescence differ from adults in amphetamine-induced locomotor activity, but not in conditioned place preference for amphetamine. *Behav Pharmacol*. 2007;18(7):641–650. https://doi.org/10.1097/FBP.0b013e3282effbf5.
59. Nazarian A, Sun WL, Zhou L, Kemen LM, Jenab S, Quinones-Jenab V. Sex differences in basal and cocaine-induced alterations in PKA and CREB proteins in the nucleus accumbens. *Psychopharmacology*. 2009;203(3):641–650. https://doi.org/10.1007/s00213-008-1411-5.
60. Nygard SK, Klambatsen A, Hazim R, et al. Sexually dimorphic intracellular responses after cocaine-induced conditioned place preference expression. *Brain Res*. 2013;1520:121–133. https://doi.org/10.1016/j.brainres.2013.04.060.
61. Savageau MM, Beatty WW. Gonadectomy and sex differences in the behavioral responses to amphetamine and apomorphine of rats. *Pharmacol Biochem Behav*. 1981;14(1):17–21. https://doi.org/10.1016/0091-3057(81)90097-6.
62. Schindler CW, Bross JG, Thorndike EB. Gender differences in the behavioral effects of methamphetamine. *Eur J Pharmacol*. 2002;442(3):231–235. https://doi.org/10.1016/s0014-2999(02)01550-9.
63. Yates JR, Horchar MJ, Kappesser JL, Broderick MR, Ellis AL, Wright MR. The association between risky decision making and cocaine conditioned place preference is moderated by sex. *Drug Alcohol Depend*. 2021;228:109079. https://doi.org/10.1016/j.drugalcdep.2021.109079.
64. Gaulden AD, Burson N, Sadik N, et al. Effects of fentanyl on acute locomotor activity, behavioral sensitization, and contextual reward in female and male rats. *Drug Alcohol Depend*. 2021;229(Pt A):109101. https://doi.org/10.1016/j.drugalcdep.2021.109101.
65. Karami M, Zarrindast MR. Morphine sex-dependently induced place conditioning in adult Wistar rats. *Eur J Pharmacol*. 2008;582(1–3):78–87. https://doi.org/10.1016/j.ejphar.2007.12.010.

66. Randall CK, Kraemer PJ, Bardo MT. Morphine-induced conditioned place preference in preweanling and adult rats. *Pharmacol Biochem Behav*. 1998;60(1):217−222. https://doi.org/10.1016/s0091-3057(97)00585-6.
67. Blanchard BA, Glick SD. Sex differences in mesolimbic dopamine responses to ethanol and relationship to ethanol intake in rats. *Recent Dev Alcohol*. 1995;12:231−241. https://doi.org/10.1007/0-306-47138-8_15.
68. Blanchard BA, Steindorf S, Wang S, Glick SD. Sex differences in ethanol-induced dopamine release in nucleus accumbens and in ethanol consumption in rats. *Alcohol Clin Exp Res*. 1993;17(5):968−973. https://doi.org/10.1111/j.1530-0277.1993.tb05650.x.
69. Henricks AM, Berger AL, Lugo JM, et al. Sex differences in alcohol consumption and alterations in nucleus accumbens endocannabinoid mRNA in alcohol-dependent rats. *Neuroscience*. 2016;335:195−206. https://doi.org/10.1016/j.neuroscience.2016.08.032.
70. Priddy BM, Carmack SA, Thomas LC, Vendruscolo JC, Koob GF, Vendruscolo LF. Sex, strain, and estrous cycle influences on alcohol drinking in rats. *Pharmacol Biochem Behav*. 2017;152:61−67. https://doi.org/10.1016/j.pbb.2016.08.001.
71. Bagdas D, Diester CM, Riley J, et al. Assessing nicotine dependence using an oral nicotine free-choice paradigm in mice. *Neuropharmacology*. 2019;157:107669. https://doi.org/10.1016/j.neuropharm.2019.107669.
72. Grathwohl C, Dadmarz M, Vogel WH. Oral self-administration of ethanol and cocaine in rats. *Pharmacology*. 2001;63(3):160−165. https://doi.org/10.1159/000056128.
73. Klein LC, Stine MM, Vandenbergh DJ, Whetzel CA, Kamens HM. Sex differences in voluntary oral nicotine consumption by adolescent mice: a dose-response experiment. *Pharmacol Biochem Behav*. 2004;78(1):13−25. https://doi.org/10.1016/j.pbb.2004.01.005.
74. Randall PA, Stewart RT, Besheer J. Sex differences in alcohol self-administration and relapse-like behavior in Long-Evans rats. *Pharmacol Biochem Behav*. 2017;156:1−9. https://doi.org/10.1016/j.pbb.2017.03.005.
75. Juárez J, Barrios de Tomasi E. Sex differences in alcohol drinking patterns during forced and voluntary consumption in rats. *Alcohol*. 1999;19(1):15−22. https://doi.org/10.1016/s0741-8329(99)00010-5.
76. Lynch WJ. Sex and ovarian hormones influence vulnerability and motivation for nicotine during adolescence in rats. *Pharmacol Biochem Behav*. 2009;94(1):43−50. https://doi.org/10.1016/j.pbb.2009.07.004.
77. Lynch WJ, Carroll ME. Sex differences in the acquisition of intravenously self-administered cocaine and heroin in rats. *Psychopharmacology*. 1999;144(1):77−82. https://doi.org/10.1007/s002130050979.
78. Roth ME, Carroll ME. Sex differences in the acquisition of IV methamphetamine self-administration and subsequent maintenance under a progressive ratio schedule in rats. *Psychopharmacology*. 2004;172(4):443−449. https://doi.org/10.1007/s00213-003-1670-0.
79. Swalve N, Smethells JR, Carroll ME. Sex differences in the acquisition and maintenance of cocaine and nicotine self-administration in rats. *Psychopharmacology*. 2016;233(6):1005−1013. https://doi.org/10.1007/s00213-015-4183-8.
80. Hankosky ER, Westbrook SR, Haake RM, Marinelli M, Gulley JM. Reduced sensitivity to reinforcement in adolescent compared to adult Sprague-Dawley rats of both sexes. *Psychopharmacology*. 2018;235(3):861−871. https://doi.org/10.1007/s00213-017-4804-5.
81. Campbell UC, Morgan AD, Carroll ME. Sex differences in the effects of baclofen on the acquisition of intravenous cocaine self-administration in rats. *Drug Alcohol Depend*. 2002;66(1):61−69. https://doi.org/10.1016/s0376-8716(01)00185-5.
82. Carroll ME, Morgan AD, Lynch WJ, Campbell UC, Dess NK. Intravenous cocaine and heroin self-administration in rats selectively bred for differential saccharin intake: phenotype and sex differences. *Psychopharmacology*. 2002;161(3):304−313. https://doi.org/10.1007/s00213-002-1030-5.
83. Hu M, Crombag HS, Robinson TE, Becker JB. Biological basis of sex differences in the propensity to self-administer cocaine. *Neuropsychopharmacology*. 2004;29(1):81−85. https://doi.org/10.1038/sj.npp.1300301.
84. Lynch WJ. Acquisition and maintenance of cocaine self-administration in adolescent rats: effects of sex and gonadal hormones. *Psychopharmacology*. 2008;197(2):237−246. https://doi.org/10.1007/s00213-007-1028-0.
85. Malone SG, Keller PS, Hammerslag LR, Bardo MT. Escalation and reinstatement of fentanyl self-administration in male and female rats. *Psychopharmacology*. 2021;238(8):2261−2273. https://doi.org/10.1007/s00213-021-05850-7.
86. Anker JJ, Zlebnik NE, Navin SF, Carroll ME. Responding during signaled availability and nonavailability of iv cocaine and food in rats: age and sex differences. *Psychopharmacology*. 2011;215(4):785−799. https://doi.org/10.1007/s00213-011-2181-z.
87. George BE, Barth SH, Kuiper LB, et al. Enhanced heroin self-administration and distinct dopamine adaptations in female rats. *Neuropsychopharmacology*. 2021;46(10):1724−1733. https://doi.org/10.1038/s41386-021-01035-0.

88. Donny EC, Caggiula AR, Rowell PP, et al. Nicotine self-administration in rats: estrous cycle effects, sex differences and nicotinic receptor binding. *Psychopharmacology*. 2000;151(4):392–405. https://doi.org/10.1007/s002130000497.
89. Pittenger ST, Swalve N, Chou S, et al. Sex differences in neurotensin and substance P following nicotine self-administration in rats. *Synapse*. 2016;70(8):336–346. https://doi.org/10.1002/syn.21907.
90. Chellian R, Behnood-Rod A, Wilson R, Bruijnzeel AW. Rewarding effects of nicotine self-administration increase over time in male and female rats. *Nicotine Tob Res*. 2021;23(12):2117–2126. https://doi.org/10.1093/ntr/ntab097.
91. Cox BM, Young AB, See RE, Reichel CM. Sex differences in methamphetamine seeking in rats: impact of oxytocin. *Psychoneuroendocrinology*. 2013;38(10):2343–2353. https://doi.org/10.1016/j.psyneuen.2013.05.005.
92. Crawford CA, Baella SA, Farley CM, et al. Early methylphenidate exposure enhances cocaine self-administration but not cocaine-induced conditioned place preference in young adult rats. *Psychopharmacology*. 2011;213(1):43–52. https://doi.org/10.1007/s00213-010-2011-8.
93. Hogarth SJ, Jaehne EJ, van den Buuse M, Djouma E. Brain-derived neurotrophic factor (BDNF) determines a sex difference in cue-conditioned alcohol seeking in rats. *Behav Brain Res*. 2018;339:73–78. https://doi.org/10.1016/j.bbr.2017.11.019.
94. Johansen A, McFadden LM. The neurochemical consequences of methamphetamine self-administration in male and female rats. *Drug Alcohol Depend*. 2017;178:70–74. https://doi.org/10.1016/j.drugalcdep.2017.04.011.
95. López AJ, Johnson AR, Euston TJ, et al. Cocaine self-administration induces sex-dependent protein expression in the nucleus accumbens. *Commun Biol*. 2021;4(1):883. https://doi.org/10.1038/s42003-021-02358-w.
96. Neelakantan H, Ward SJ, Walker EA. Effects of paclitaxel on mechanical sensitivity and morphine reward in male and female C57Bl6 mice. *Exp Clin Psychopharmacol*. 2016;24(6):485–495. https://doi.org/10.1037/pha0000097.
97. Pena-Bravo JI, Penrod R, Reichel CM, Lavin A. Methamphetamine self-administration elicits sex-related changes in postsynaptic glutamate transmission in the prefrontal cortex. *eNeuro*. 2019;6(1). https://doi.org/10.1523/ENEURO.0401-18.2018.
98. Smethells JR, Greer A, Dougen B, Carroll ME. Effects of voluntary exercise and sex on multiply-triggered heroin reinstatement in male and female rats. *Psychopharmacology*. 2020;237(2):453–463. https://doi.org/10.1007/s00213-019-05381-2.
99. Sneddon EA, Ramsey OR, Thomas A, Radke AK. Increased responding for alcohol and resistance to aversion in female mice. *Alcohol Clin Exp Res*. 2020;44(7):1400–1409. https://doi.org/10.1111/acer.14384.
100. Swalve N, Smethells JR, Carroll ME. Sex differences in attenuation of nicotine reinstatement after individual and combined treatments of progesterone and varenicline. *Behav Brain Res*. 2016;308:46–52. https://doi.org/10.1016/j.bbr.2016.04.023.
101. Venniro M, Zhang M, Shaham Y, Caprioli D. Incubation of methamphetamine but not heroin craving after voluntary abstinence in male and female rats. *Neuropsychopharmacology*. 2017;42(5):1126–1135. https://doi.org/10.1038/npp.2016.287.
102. Westenbroek C, Perry AN, Becker JB. Pair housing differentially affects motivation to self-administer cocaine in male and female rats. *Behav Brain Res*. 2013;252:68–71. https://doi.org/10.1016/j.bbr.2013.05.040.
103. Fuchs RA, Evans KA, Mehta RH, Case JM, See RE. Influence of sex and estrous cyclicity on conditioned cue-induced reinstatement of cocaine-seeking behavior in rats. *Psychopharmacology*. 2005;179(3):662–672. https://doi.org/10.1007/s00213-004-2080-7.
104. Kippin TE, Fuchs RA, Mehta RH, et al. Potentiation of cocaine-primed reinstatement of drug seeking in female rats during estrus. *Psychopharmacology*. 2005;182(2):245–252. https://doi.org/10.1007/s00213-005-0071-y.
105. Mavrikaki M, Lintz T, Constantino N, Page S, Chartoff E. Chronic opioid exposure differentially modulates oxycodone self-administration in male and female rats. *Addiction Biol*. 2021;26(3):e12973. https://doi.org/10.1111/adb.12973.
106. Flores RJ, Pipkin JA, Uribe KP, Perez A, O'Dell LE. Estradiol promotes the rewarding effects of nicotine in female rats. *Behav Brain Res*. 2016;307:258–263. https://doi.org/10.1016/j.bbr.2016.04.004.
107. Chellian R, Behnood-Rod A, Wilson R, et al. Adolescent nicotine and tobacco smoke exposure enhances nicotine self-administration in female rats. *Neuropharmacology*. 2020;176:108243. https://doi.org/10.1016/j.neuropharm.2020.108243.
108. Lárraga A, Belluzzi JD, Leslie FM. Nicotine increases alcohol intake in adolescent male rats. *Front Behav Neurosci*. 2017;11:25. https://doi.org/10.3389/fnbeh.2017.00025.

109. Fulenwider HD, Nennig SE, Hafeez H, et al. Sex differences in oral oxycodone self-administration and stress-primed reinstatement in rats. *Addiction Biol.* 2020;25(6):e12822. https://doi.org/10.1111/adb.12822.
110. Bertholomey ML, Nagarajan V, Torregrossa MM. Sex differences in reinstatement of alcohol seeking in response to cues and yohimbine in rats with and without a history of adolescent corticosterone exposure. *Psychopharmacology.* 2016;233(12):2277−2287. https://doi.org/10.1007/s00213-016-4278-x.
111. van Haaren F, Anderson K. Sex differences in schedule-induced alcohol consumption. *Alcohol.* 1994;11(1):35−40. https://doi.org/10.1016/0741-8329(94)90009-4.
112. Lynch WJ, Taylor JR. Sex differences in the behavioral effects of 24-h/day access to cocaine under a discrete trial procedure. *Neuropsychopharmacology.* 2004;29(5):943−951. https://doi.org/10.1038/sj.npp.1300389.
113. Towers EB, Bakhti-Suroosh A, Lynch WJ. Females develop features of an addiction-like phenotype sooner during withdrawal than males. *Psychopharmacology.* 2021;238(8):2213−2224. https://doi.org/10.1007/s00213-021-05846-3.
114. Shahbazi M, Moffett AM, Williams BF, Frantz KJ. Age- and sex-dependent amphetamine self-administration in rats. *Psychopharmacology.* 2008;196(1):71−81. https://doi.org/10.1007/s00213-007-0933-6.
115. Nieto SJ, Kosten TA. Female Sprague-Dawley rats display greater appetitive and consummatory responses to alcohol. *Behav Brain Res.* 2017;327:155−161. https://doi.org/10.1016/j.bbr.2017.03.037.
116. Hogarth SJ, Djouma E, van den Buuse M. 7,8-Dihydroxyflavone enhances cue-conditioned alcohol reinstatement in rats. *Brain Sci.* 2020;10(5):270. https://doi.org/10.3390/brainsci10050270.
117. Reichel CM, Chan CH, Ghee SM, See RE. Sex differences in escalation of methamphetamine self-administration: cognitive and motivational consequences in rats. *Psychopharmacology.* 2012;223(4):371−380. https://doi.org/10.1007/s00213-012-2727-8.
118. Westbrook SR, Dwyer MR, Cortes LR, Gulley JM. Extended access self-administration of methamphetamine is associated with age- and sex-dependent differences in drug taking behavior and recognition memory in rats. *Behav Brain Res.* 2020;390:112659. https://doi.org/10.1016/j.bbr.2020.112659.
119. Cornett EM, Goeders NE. 96-hour methamphetamine self-administration in male and female rats: a novel model of human methamphetamine addiction. *Pharmacol Biochem Behav.* 2013;111:51−57. https://doi.org/10.1016/j.pbb.2013.08.005.
120. Barrett ST, Thompson BM, Emory JR, et al. Sex differences in the reward-enhancing effects of nicotine on ethanol reinforcement: a reinforcer demand analysis. *Nicotine Tob Res.* 2020;22(2):238−247. https://doi.org/10.1093/ntr/ntz056.
121. Grebenstein P, Burroughs D, Zhang Y, LeSage MG. Sex differences in nicotine self-administration in rats during progressive unit dose reduction: implications for nicotine regulation policy. *Pharmacol Biochem Behav.* 2013;114−115:70−81. https://doi.org/10.1016/j.pbb.2013.10.020.
122. Powell GL, Cabrera-Brown G, Namba MD, et al. Economic demand analysis of within-session dose-reduction during nicotine self-administration. *Drug Alcohol Depend.* 2019;201:188−196. https://doi.org/10.1016/j.drugalcdep.2019.03.033.
123. Chellian R, Wilson R, Polmann M, Knight P, Behnood-Rod A, Bruijnzeel AW. Evaluation of sex differences in the elasticity of demand for nicotine and food in rats. *Nicotine Tob Res.* 2020;22(6):925−934. https://doi.org/10.1093/ntr/ntz171.
124. Lacy RT, Austin BP, Strickland JC. The influence of sex and estrous cyclicity on cocaine and remifentanil demand in rats. *Addiction Biol.* 2020;25(1):e12716. https://doi.org/10.1111/adb.12716.
125. Sun W, Yuill MB, Fan M. Behavioral economics of cocaine self-administration in male and female rats. *Behav Pharmacol.* 2021;32(1):21−31. https://doi.org/10.1097/FBP.0000000000000598.
126. Kawa AB, Robinson TE. Sex differences in incentive-sensitization produced by intermittent access cocaine self-administration. *Psychopharmacology.* 2019;236(2):625−639. https://doi.org/10.1007/s00213-018-5091-5.
127. Cunningham CL, Shields CN. Effects of sex on ethanol conditioned place preference, activity and variability in C57BL/6J and DBA/2J mice. *Pharmacol Biochem Behav.* 2018;173:84−89. https://doi.org/10.1016/j.pbb.2018.07.008.
128. Grisel JE, Beasley JB, Bertram EC, et al. Initial subjective reward: single-exposure conditioned place preference to alcohol in mice. *Front Neurosci.* 2014;8:345. https://doi.org/10.3389/fnins.2014.00345.
129. Barros-Santos T, Libarino-Santos M, Anjos-Santos A, et al. Sex differences in the development of conditioned place preference induced by intragastric alcohol administration in mice. *Drug Alcohol Depend.* 2021;229(Pt A):109105. https://doi.org/10.1016/j.drugalcdep.2021.109105.

130. Roger-Sánchez C, Aguilar MA, Rodríguez-Arias M, Aragon CM, Miñarro J. Age- and sex-related differences in the acquisition and reinstatement of ethanol CPP in mice. *Neurotoxicol Teratol*. 2012;34(1):108–115. https://doi.org/10.1016/j.ntt.2011.07.011.
131. Nentwig TB, Myers KP, Grisel JE. Initial subjective reward to alcohol in Sprague-Dawley rats. *Alcohol*. 2017;58:19–22. https://doi.org/10.1016/j.alcohol.2016.11.005.
132. Yararbas G, Keser A, Kanit L, Pogun S. Nicotine-induced conditioned place preference in rats: sex differences and the role of mGluR5 receptors. *Neuropharmacology*. 2010;58(2):374–382. https://doi.org/10.1016/j.neuropharm.2009.10.001.
133. Edwards AW, Konz N, Hirsch Z, Weedon J, Dow-Edwards DL. Single trial nicotine conditioned place preference in pre-adolescent male and female rats. *Pharmacol Biochem Behav*. 2014;125:1–7. https://doi.org/10.1016/j.pbb.2014.07.016.
134. Morris Bobzean SA, Dennis TS, Addison BD, Perrotti LI. Influence of sex on reinstatement of cocaine-conditioned place preference. *Brain Res Bull*. 2010;83(6):331–336. https://doi.org/10.1016/j.brainresbull.2010.09.003.
135. Chen HH, Yang YK, Yeh TL, et al. Methamphetamine-induced conditioned place preference is facilitated by estradiol pretreatment in female mice. *Chin J Physiol*. 2003;46(4):169–174.
136. Lynch WJ, Carroll ME. Reinstatement of cocaine self-administration in rats: sex differences. *Psychopharmacology*. 2000;148(2):196–200. https://doi.org/10.1007/s002130050042.
137. Pittenger ST, Chou S, Murawski NJ, et al. Female rats display higher methamphetamine-primed reinstatement and c-Fos immunoreactivity than male rats. *Pharmacol Biochem Behav*. 2021;201:173089. https://doi.org/10.1016/j.pbb.2020.173089.
138. Weber RA, Logan CN, Leong KC, Peris J, Knackstedt L, Reichel CM. Regionally specific effects of oxytocin on reinstatement of cocaine seeking in male and female rats. *Int J Neuropsychopharmacol*. 2018;21(7):677–686. https://doi.org/10.1093/ijnp/pyy025.
139. Witt EA, Reissner KJ. The effects of nicotinamide on reinstatement to cocaine seeking in male and female Sprague Dawley rats. *Psychopharmacology*. 2020;237(3):669–680. https://doi.org/10.1007/s00213-019-05404-y.
140. Logrip ML, Gainey SC. Sex differences in the long-term effects of past stress on alcohol self-administration, glucocorticoid sensitivity and phosphodiesterase 10A expression. *Neuropharmacology*. 2020;164:107857. https://doi.org/10.1016/j.neuropharm.2019.107857.
141. Feltenstein MW, Ghee SM, See RE. Nicotine self-administration and reinstatement of nicotine-seeking in male and female rats. *Drug Alcohol Depend*. 2012;121(3):240–246. https://doi.org/10.1016/j.drugalcdep.2011.09.001.
142. Kohtz AS, Lin B, Smith ME, Aston-Jones G. Attenuated cocaine-seeking after oxytocin administration in male and female rats. *Psychopharmacology*. 2018;235(7):2051–2063. https://doi.org/10.1007/s00213-018-4902-z.
143. Perry JL, Nelson SE, Carroll ME. Impulsive choice as a predictor of acquisition of IV cocaine self-administration and reinstatement of cocaine-seeking behavior in male and female rats. *Exp Clin Psychopharmacol*. 2008;16(2):165–177. https://doi.org/10.1037/1064-1297.16.2.165.
144. Fattore L, Spano MS, Altea S, Angius F, Fadda P, Fratta W. Cannabinoid self-administration in rats: sex differences and the influence of ovarian function. *Br J Pharmacol*. 2007;152(5):795–804. https://doi.org/10.1038/sj.bjp.0707465.
145. Fosnocht AQ, Lucerne KE, Ellis AS, Olimpo NA, Briand LA. Adolescent social isolation increases cocaine seeking in male and female mice. *Behav Brain Res*. 2019;359:589–596. https://doi.org/10.1016/j.bbr.2018.10.007.
146. Bechard AR, Hamor PU, Schwendt M, Knackstedt LA. The effects of ceftriaxone on cue-primed reinstatement of cocaine-seeking in male and female rats: estrous cycle effects on behavior and protein expression in the nucleus accumbens. *Psychopharmacology*. 2018;235(3):837–848. https://doi.org/10.1007/s00213-017-4802-7.
147. Smith MA, Pennock MM, Walker KL, Lang KC. Access to a running wheel decreases cocaine-primed and cue-induced reinstatement in male and female rats. *Drug Alcohol Depend*. 2012;121(1–2):54–61. https://doi.org/10.1016/j.drugalcdep.2011.08.006.
148. Kerstetter KA, Aguilar VR, Parrish AB, Kippin TE. Protracted time-dependent increases in cocaine-seeking behavior during cocaine withdrawal in female relative to male rats. *Psychopharmacology*. 2008;198(1):63–75. https://doi.org/10.1007/s00213-008-1089-8.
149. Venniro M, Russell TI, Zhang M, Shaham Y. Operant social reward decreases incubation of heroin craving in male and female rats. *Biol Psychiatr*. 2019;86(11):848–856. https://doi.org/10.1016/j.biopsych.2019.05.018.

150. Fattore L, Spano MS, Altea S, Fadda P, Fratta W. Drug- and cue-induced reinstatement of cannabinoid-seeking behaviour in male and female rats: influence of ovarian hormones. *Br J Pharmacol*. 2010;160(3):724−735. https://doi.org/10.1111/j.1476-5381.2010.00734.x.
151. Weafer J, De Arcangelis J, de Wit H. Sex differences in behavioral impulsivity in at-risk and non-risk drinkers. *Front Psychiatr*. 2015;6:72. https://doi.org/10.3389/fpsyt.2015.00072.
152. van der Plas EA, Crone EA, van den Wildenberg WP, Tranel D, Bechara A. Executive control deficits in substance-dependent individuals: a comparison of alcohol, cocaine, and methamphetamine and of men and women. *J Clin Exp Neuropsychol*. 2009;31(6):706−719. https://doi.org/10.1080/13803390802484797.
153. Myerson J, Green L, van den Berk-Clark C, Grucza RA. Male, but not female, alcohol-dependent African Americans discount delayed gains more steeply than propensity-score matched controls. *Psychopharmacology*. 2015;232(24):4493−4503. https://doi.org/10.1007/s00213-015-4076-x.
154. Korucuoglu O, Harms MP, Kennedy JT, et al. Adolescent decision-making under risk: neural correlates and sex differences. *Cerebr Cortex*. 2020;30(4):2690−2706. https://doi.org/10.1093/cercor/bhz269.
155. Orsini CA, Willis ML, Gilbert RJ, Bizon JL, Setlow B. Sex differences in a rat model of risky decision making. *Behav Neurosci*. 2016;130(1):50−61. https://doi.org/10.1037/bne0000111.
156. Allen JW, Bennett DS, Carmody DP, Wang Y, Lewis M. Adolescent risk-taking as a function of prenatal cocaine exposure and biological sex. *Neurotoxicol Teratol*. 2014;41:65−70. https://doi.org/10.1016/j.ntt.2013.12.003.
157. White TL, Lejuez CW, de Wit H. Personality and gender differences in effects of d-amphetamine on risk taking. *Exp Clin Psychopharmacol*. 2007;15(6):599−609. https://doi.org/10.1037/1064-1297.15.6.599.
158. Orsini CA, Blaes SL, Dragone RJ, et al. Distinct relationships between risky decision making and cocaine self-administration under short- and long-access conditions. *Prog Neuro Psychopharmacol Biol Psychiatr*. 2020;98:109791. https://doi.org/10.1016/j.pnpbp.2019.109791.
159. Yates JR, Beckmann JS, Meyer AC, Bardo MT. Concurrent choice for social interaction and amphetamine using conditioned place preference in rats: effects of age and housing condition. *Drug Alcohol Depend*. 2013;129(3):240−246. https://doi.org/10.1016/j.drugalcdep.2013.02.024.
160. Weiss VG, Hofford RS, Yates JR, Jennings FC, Bardo MT. Sex differences in monoamines following amphetamine and social reward in adolescent rats. *Exp Clin Psychopharmacol*. 2015;23(4):197−205. https://doi.org/10.1037/pha0000026.
161. Peartree NA, Hatch KN, Goenaga JG, et al. Social context has differential effects on acquisition of nicotine self-administration in male and female rats. *Psychopharmacology*. 2017;234(12):1815−1828. https://doi.org/10.1007/s00213-017-4590-0.
162. Lawless MH, Harrison KA, Grandits GA, Eberly LE, Allen SS. Perceived stress and smoking-related behaviors and symptomatology in male and female smokers. *Addict Behav*. 2015;51:80−83. https://doi.org/10.1016/j.addbeh.2015.07.011.
163. Moran-Santa Maria MM, McRae-Clark A, Baker NL, Ramakrishnan V, Brady KT. Yohimbine administration and cue-reactivity in cocaine-dependent individuals. *Psychopharmacology*. 2014;231(21):4157−4165. https://doi.org/10.1007/s00213-014-3555-9.
164. Hyman SM, Paliwal P, Chaplin TM, Mazure CM, Rounsaville BJ, Sinha R. Severity of childhood trauma is predictive of cocaine relapse outcomes in women but not men. *Drug Alcohol Depend*. 2008;92(1−3):208−216. https://doi.org/10.1016/j.drugalcdep.2007.08.006.
165. Ganguly P, Honeycutt JA, Rowe JR, Demaestri C, Brenhouse HC. Effects of early life stress on cocaine conditioning and AMPA receptor composition are sex-specific and driven by TNF. *Brain Behav Immun*. 2019;78:41−51. https://doi.org/10.1016/j.bbi.2019.01.006.
166. Kawakami SE, Quadros IM, Takahashi S, Suchecki D. Long maternal separation accelerates behavioural sensitization to ethanol in female, but not in male mice. *Behav Brain Res*. 2007;184(2):109−116. https://doi.org/10.1016/j.bbr.2007.06.023.
167. Sprowles J, Vorhees CV, Williams MT. Impact of preweaning stress on long-term neurobehavioral outcomes in Sprague-Dawley rats: differential effects of barren cage rearing, pup isolation, and the combination. *Neurotoxicol Teratol*. 2021;84:106956. https://doi.org/10.1016/j.ntt.2021.106956.
168. Roman E, Gustafsson L, Hyytiä P, Nylander I. Short and prolonged periods of maternal separation and voluntary ethanol intake in male and female ethanol-preferring AA and ethanol-avoiding ANA rats. *Alcohol Clin Exp Res*. 2005;29(4):591−601. https://doi.org/10.1097/01.alc.0000158933.70242.fc.

169. Cosgrove KP, Hunter RG, Carroll ME. Wheel-running attenuates intravenous cocaine self-administration in rats: sex differences. *Pharmacol Biochem Behav*. 2002;73(3):663–671. https://doi.org/10.1016/s0091-3057(02)00853-5.
170. Gallego X, Cox RJ, Funk E, Foster RA, Ehringer MA. Voluntary exercise decreases ethanol preference and consumption in C57BL/6 adolescent mice: sex differences and hippocampal BDNF expression. *Physiol Behav*. 2015;138:28–36. https://doi.org/10.1016/j.physbeh.2014.10.008.
171. Peterson AB, Hivick DP, Lynch WJ. Dose-dependent effectiveness of wheel running to attenuate cocaine-seeking: impact of sex and estrous cycle in rats. *Psychopharmacology*. 2014;231(13):2661–2670. https://doi.org/10.1007/s00213-014-3437-1.
172. Korhonen T, Kujala UM, Rose RJ, Kaprio J. Physical activity in adolescence as a predictor of alcohol and illicit drug use in early adulthood: a longitudinal population-based twin study. *Twin Res Hum Genet*. 2009;12(3):261–268. https://doi.org/10.1375/twin.12.3.261.
173. Bossert JM, Townsend EA, Altidor LK, et al. Sex differences in the effect of chronic delivery of the buprenorphine analogue BU08028 on heroin relapse and choice in a rat model of opioid maintenance. *Br J Pharmacol*. 2021. https://doi.org/10.1111/bph.15679. Advance online publication.
174. Kantak KM, Gauthier JM, Mathieson E, Knyazhanskaya E, Rodriguez-Echemendia P, Man HY. Sex differences in the effects of a combined behavioral and pharmacological treatment strategy for cocaine relapse prevention in an animal model of cue exposure therapy. *Behav Brain Res*. 2020;395:112839. https://doi.org/10.1016/j.bbr.2020.112839.
175. Evans SM, Haney M, Foltin RW. The effects of smoked cocaine during the follicular and luteal phases of the menstrual cycle in women. *Psychopharmacology*. 2002;159(4):397–406. https://doi.org/10.1007/s00213-001-0944-7.
176. Reed SC, Evans SM, Bedi G, Rubin E, Foltin RW. The effects of oral micronized progesterone on smoked cocaine self-administration in women. *Horm Behav*. 2011;59(2):227–235. https://doi.org/10.1016/j.yhbeh.2010.12.009.
177. Frydenberg H, Flote VG, Larsson IM, et al. Alcohol consumption, endogenous estrogen and mammographic density among premenopausal women. *Breast Cancer Res*. 2015;17(1):103. https://doi.org/10.1186/s13058-015-0620-1.
178. Dozier BL, Stull CA, Baker EJ, et al. Chronic ethanol drinking increases during the luteal menstrual cycle phase in rhesus monkeys: implication of progesterone and related neurosteroids. *Psychopharmacology*. 2019;236(6):1817–1828. https://doi.org/10.1007/s00213-019-5168-9.
179. Evans SM, Levin FR. Response to alcohol in women: role of the menstrual cycle and a family history of alcoholism. *Drug Alcohol Depend*. 2011;114(1):18–30. https://doi.org/10.1016/j.drugalcdep.2010.09.001.
180. Hayaki J, Holzhauer CG, Epstein EE, et al. Menstrual cycle phase, alcohol consumption, alcohol cravings, and mood among women in outpatient treatment for alcohol use disorder. *Psychol Addict Behav*. 2020;34(6):680–689. https://doi.org/10.1037/adb0000576.
181. Sutker PB, Goist Jr KC, King AR. Acute alcohol intoxication in women: relationship to dose and menstrual cycle phase. *Alcohol Clin Exp Res*. 1987;11(1):74–79. https://doi.org/10.1111/j.1530-0277.1987.tb01266.x.
182. Mello NK, Knudson IM, Mendelson JH. Sex and menstrual cycle effects on progressive ratio measures of cocaine self-administration in cynomolgus monkeys. *Neuropsychopharmacology*. 2007;32(9):1956–1966. https://doi.org/10.1038/sj.npp.1301314.
183. DeVito EE, Herman AI, Waters AJ, Valentine GW, Sofuoglu M. Subjective, physiological, and cognitive responses to intravenous nicotine: effects of sex and menstrual cycle phase. *Neuropsychopharmacology*. 2014;39(6):1431–1440. https://doi.org/10.1038/npp.2013.339.
184. Justice AJ, de Wit H. Acute effects of d-amphetamine during the follicular and luteal phases of the menstrual cycle in women. *Psychopharmacology*. 1999;145(1):67–75. https://doi.org/10.1007/s002130051033.
185. Lile JA, Kendall SL, Babalonis S, Martin CA, Kelly TH. Evaluation of estradiol administration on the discriminative-stimulus and subject-rated effects of d-amphetamine in healthy pre-menopausal women. *Pharmacol, Biochem Behav*. 2007;87(2):258–266. https://doi.org/10.1016/j.pbb.2007.04.022.
186. Justice AJ, de Wit H. Acute effects of d-amphetamine during the early and late follicular phases of the menstrual cycle in women. *Pharmacol, Biochem Behav*. 2000;66(3):509–515. https://doi.org/10.1016/s0091-3057(00)00218-5.
187. Allen AM, Lunos S, Heishman SJ, al'Absi M, Hatsukami D, Allen SS. Subjective response to nicotine by menstrual phase. *Addict Behav*. 2015;43:50–53. https://doi.org/10.1016/j.addbeh.2014.12.008.

188. Ethier AR, McKinney TL, Tottenham LS, Gordon JL. The effect of reproductive hormones on women's daily smoking across the menstrual cycle. *Biol Sex Differ*. 2021;12(1):41. https://doi.org/10.1186/s13293-021-00384-1.
189. Sofuoglu M, Babb DA, Hatsukami DK. Progesterone treatment during the early follicular phase of the menstrual cycle: effects on smoking behavior in women. *Pharmacol Biochem Behav*. 2001;69(1−2):299−304. https://doi.org/10.1016/s0091-3057(01)00527-5.
190. Sinha R, Fox H, Hong KI, Sofuoglu M, Morgan PT, Bergquist KT. Sex steroid hormones, stress response, and drug craving in cocaine-dependent women: implications for relapse susceptibility. *Exp Clin Psychopharmacol*. 2007;15(5):445−452. https://doi.org/10.1037/1064-1297.15.5.445.
191. Lynch WJ, Roth ME, Mickelberg JL, Carroll ME. Role of estrogen in the acquisition of intravenously self-administered cocaine in female rats. *Pharmacol Biochem Behav*. 2001;68(4):641−646. https://doi.org/10.1016/s0091-3057(01)00455-5.
192. Feltenstein MW, Byrd EA, Henderson AR, See RE. Attenuation of cocaine-seeking by progesterone treatment in female rats. *Psychoneuroendocrinology*. 2009;34(3):343−352. https://doi.org/10.1016/j.psyneuen.2008.09.014.
193. Lacy RT, Strickland JC, Feinstein MA, Robinson AM, Smith MA. The effects of sex, estrous cycle, and social contact on cocaine and heroin self-administration in rats. *Psychopharmacology*. 2016;233(17):3201−3210. https://doi.org/10.1007/s00213-016-4368-9.
194. Quigley JA, Logsdon MK, Graham BC, Beaudoin KG, Becker JB. Activation of G protein-coupled estradiol receptor 1 in the dorsolateral striatum enhances motivation for cocaine and drug-induced reinstatement in female but not male rats. *Biol Sex Differ*. 2021;12(1):46. https://doi.org/10.1186/s13293-021-00389-w.
195. Datta U, Martini M, Sun WL. Sex differences in the motivational contrast between sucrose and cocaine in rats. *J Drug Design & Res*. 2017;4(3):1042.
196. Doncheck EM, Anderson EM, Konrath CD, et al. Estradiol regulation of the prelimbic cortex and the reinstatement of cocaine seeking in female rats. *J Neurosci*. 2021;41(24):5303−5314. https://doi.org/10.1523/JNEUROSCI.3086-20.2021.
197. Bakhti-Suroosh A, Nesil T, Lynch WJ. Tamoxifen blocks the development of motivational features of an addiction-like phenotype in female rats. *Front Behav Neurosci*. 2019;13:253. https://doi.org/10.3389/fnbeh.2019.00253.
198. Segarra AC, Torres-Díaz YM, Silva RD, et al. Estrogen receptors mediate estradiol's effect on sensitization and CPP to cocaine in female rats: role of contextual cues. *Horm Behav*. 2014;65(2):77−87. https://doi.org/10.1016/j.yhbeh.2013.12.007.
199. Alagwu EA, Nneli RO. Effect of ovariectomy on the level of plasma sex hormones in albino rats. *Niger J Physiol Sci*. 2005;20(1−2):90−94.
200. Souza MF, Couto-Pereira NS, Freese L, et al. Behavioral effects of endogenous or exogenous estradiol and progesterone on cocaine sensitization in female rats. *Braz J Med Biol Res*. 2014;47(6):505−514. https://doi.org/10.1590/1414-431x20143627.
201. Russo SJ, Festa ED, Fabian SJ, et al. Gonadal hormones differentially modulate cocaine-induced conditioned place preference in male and female rats. *Neuroscience*. 2003;120(2):523−533. https://doi.org/10.1016/s0306-4522(03)00317-8.
202. Kerstetter KA, Ballis MA, Duffin-Lutgen S, Carr AE, Behrens AM, Kippin TE. Sex differences in selecting between food and cocaine reinforcement are mediated by estrogen. *Neuropsychopharmacology*. 2012;37(12):2605−2614. https://doi.org/10.1038/npp.2012.99.
203. Anker JJ, Larson EB, Gliddon LA, Carroll ME. Effects of progesterone on the reinstatement of cocaine-seeking behavior in female rats. *Exp Clin Psychopharmacol*. 2007;15(5):472−480. https://doi.org/10.1037/1064-1297.15.5.472.
204. Larson EB, Roth ME, Anker JJ, Carroll ME. Effect of short- vs. long-term estrogen on reinstatement of cocaine-seeking behavior in female rats. *Pharmacol Biochem Behav*. 2005;82(1):98−108. https://doi.org/10.1016/j.pbb.2005.07.015.
205. Becker JB, Beer ME. The influence of estrogen on nigrostriatal dopamine activity: behavioral and neurochemical evidence for both pre- and postsynaptic components. *Behav Brain Res*. 1986;19(1):27−33. https://doi.org/10.1016/0166-4328(86)90044-6.
206. Hu M, Becker JB. Acquisition of cocaine self-administration in ovariectomized female rats: effect of estradiol dose or chronic estradiol administration. *Drug Alcohol Depend*. 2008;94(1−3):56−62. https://doi.org/10.1016/j.drugalcdep.2007.10.005.

207. Jackson LR, Robinson TE, Becker JB. Sex differences and hormonal influences on acquisition of cocaine self-administration in rats. *Neuropsychopharmacology*. 2006;31(1):129–138. https://doi.org/10.1038/sj.npp.1300778.
208. Perry AN, Westenbroek C, Becker JB. Impact of pubertal and adult estradiol treatments on cocaine self-administration. *Horm Behav*. 2013;64(4):573–578. https://doi.org/10.1016/j.yhbeh.2013.08.007.
209. Roth ME, Casimir AG, Carroll ME. Influence of estrogen in the acquisition of intravenously self-administered heroin in female rats. *Pharmacol Biochem Behav*. 2002;72(1–2):313–318. https://doi.org/10.1016/s0091-3057(01)00777-8.
210. Ramôa CP, Doyle SE, Naim DW, Lynch WJ. Estradiol as a mechanism for sex differences in the development of an addicted phenotype following extended access cocaine self-administration. *Neuropsychopharmacology*. 2013;38(9):1698–1705. https://doi.org/10.1038/npp.2013.68.
211. Larson EB, Anker JJ, Gliddon LA, Fons KS, Carroll ME. Effects of estrogen and progesterone on the escalation of cocaine self-administration in female rats during extended access. *Exp Clin Psychopharmacol*. 2007;15(5):461–471. https://doi.org/10.1037/1064-1297.15.5.461.
212. Rauhut AS, Curran-Rauhut MA. 17 β-Estradiol exacerbates methamphetamine-induced anxiety-like behavior in female mice. *Neurosci Lett*. 2018;681:44–49. https://doi.org/10.1016/j.neulet.2018.05.025.
213. Bagley JR, Adams J, Bozadjian RV, Bubalo L, Ploense KL, Kippin TE. Estradiol increases choice of cocaine over food in male rats. *Physiol Behav*. 2019;203:18–24. https://doi.org/10.1016/j.physbeh.2017.10.018.
214. Twining RC, Tuscher JJ, Doncheck EM, Frick KM, Mueller D. 17β-estradiol is necessary for extinction of cocaine seeking in female rats. *Learn Mem*. 2013;20(6):300–306. https://doi.org/10.1101/lm.030304.113.
215. Quiñones-Jenab V, Perrotti LI, Mc Monagle J, Ho A, Kreek MJ. Ovarian hormone replacement affects cocaine-induced behaviors in ovariectomized female rats. *Pharmacol Biochem Behav*. 2000;67(3):417–422. https://doi.org/10.1016/s0091-3057(00)00381-6.
216. Anker JJ, Zlebnik NE, Carroll ME. Differential effects of allopregnanolone on the escalation of cocaine self-administration and sucrose intake in female rats. *Psychopharmacology*. 2010;212(3):419–429. https://doi.org/10.1007/s00213-010-1968-7.
217. Holtz NA, Lozama A, Prisinzano TE, Carroll ME. Reinstatement of methamphetamine seeking in male and female rats treated with modafinil and allopregnanolone. *Drug Alcohol Depend*. 2012;120(1–3):233–237. https://doi.org/10.1016/j.drugalcdep.2011.07.010.
218. Anker JJ, Holtz NA, Zlebnik N, Carroll ME. Effects of allopregnanolone on the reinstatement of cocaine-seeking behavior in male and female rats. *Psychopharmacology*. 2009;203(1):63–72. https://doi.org/10.1007/s00213-008-1371-9.
219. Evans SM, Foltin RW. Exogenous progesterone attenuates the subjective effects of smoked cocaine in women, but not in men. *Neuropsychopharmacology*. 2006;31(3):659–674. https://doi.org/10.1038/sj.npp.1300887.
220. Sofuoglu M, Mitchell E, Kosten TR. Effects of progesterone treatment on cocaine responses in male and female cocaine users. *Pharmacol Biochem Behav*. 2004;78(4):699–705. https://doi.org/10.1016/j.pbb.2004.05.004.
221. Milivojevic V, Fox HC, Sofuoglu M, Covault J, Sinha R. Effects of progesterone stimulated allopregnanolone on craving and stress response in cocaine dependent men and women. *Psychoneuroendocrinology*. 2016;65:44–53. https://doi.org/10.1016/j.psyneuen.2015.12.008.
222. Sherman BJ, Caruso MA, McRae-Clark AL. Exogenous progesterone for cannabis withdrawal in women: feasibility trial of a novel multimodal methodology. *Pharmacol Biochem Behav*. 2019;179:22–26. https://doi.org/10.1016/j.pbb.2019.01.008.
223. Sharp JL, Ethridge SB, Ballard SL, Potter KM, Schmidt KT, Smith MA. The effects of chronic estradiol treatment on opioid self-administration in intact female rats. *Drug Alcohol Depend*. 2021;225:108816. https://doi.org/10.1016/j.drugalcdep.2021.108816.
224. Johnson AR, Thibeault KC, Lopez AJ, et al. Cues play a critical role in estrous cycle-dependent enhancement of cocaine reinforcement. *Neuropsychopharmacology*. 2019;44(7):1189–1197. https://doi.org/10.1038/s41386-019-0320-0.
225. Kritzer MF, Creutz LM. Region and sex differences in constituent dopamine neurons and immunoreactivity for intracellular estrogen and androgen receptors in mesocortical projections in rats. *J Neurosci*. 2008;28(38):9525–9535. https://doi.org/10.1523/JNEUROSCI.2637-08.2008.
226. Zhang D, Yang S, Yang C, Jin G, Zhen X. Estrogen regulates responses of dopamine neurons in the ventral tegmental area to cocaine. *Psychopharmacology*. 2008;199(4):625–635. https://doi.org/10.1007/s00213-008-1188-6.

227. Shieh KR, Yang SC. Effects of estradiol on the stimulation of dopamine turnover in mesolimbic and nigrostriatal systems by cocaine- and amphetamine-regulated transcript peptide in female rats. *Neuroscience*. 2008;154(4):1589−1597. https://doi.org/10.1016/j.neuroscience.2008.01.086.
228. Self DW, Genova LM, Hope BT, Barnhart WJ, Spencer JJ, Nestler EJ. Involvement of cAMP-dependent protein kinase in the nucleus accumbens in cocaine self-administration and relapse of cocaine-seeking behavior. *J Neurosci*. 1998;18(5):1848−1859. https://doi.org/10.1523/JNEUROSCI.18-05-01848.1998.
229. Couceyro PR, Evans C, McKinzie A, et al. Cocaine- and amphetamine-regulated transcript (CART) peptides modulate the locomotor and motivational properties of psychostimulants. *J Pharmacol Exp Therapeut*. 2005;315(3):1091−1100. https://doi.org/10.1124/jpet.105.091678.
230. Marco EM, Peñasco S, Hernández MD, et al. Long-term effects of intermittent adolescent alcohol exposure in male and female rats. *Front Behav Neurosci*. 2017;11:233. https://doi.org/10.3389/fnbeh.2017.00233.
231. Engeln M, Mitra S, Chandra R, et al. Sex-specific role for Egr3 in nucleus accumbens D2-medium spiny neurons following long-term abstinence from cocaine self-administration. *Biol Psychiatr*. 2020;87(11):992−1000. https://doi.org/10.1016/j.biopsych.2019.10.019.
232. Castner SA, Becker JB. Sex differences in the effect of amphetamine on immediate early gene expression in the rat dorsal striatum. *Brain Res*. 1996;712(2):245−257. https://doi.org/10.1016/0006-8993(95)01429-2.
233. Becker JB. Direct effect of 17 beta-estradiol on striatum: sex differences in dopamine release. *Synapse*. 1990;5(2):157−164. https://doi.org/10.1002/syn.890050211.
234. Becker JB, Ramirez VD. Sex differences in the amphetamine stimulated release of catecholamines from rat striatal tissue in vitro. *Brain Res*. 1981;204(2):361−372. https://doi.org/10.1016/0006-8993(81)90595-3.
235. Tobiansky DJ, Will RG, Lominac KD, et al. Estradiol in the preoptic area regulates the dopaminergic response to cocaine in the nucleus accumbens. *Neuropsychopharmacology*. 2016;41(7):1897−1906. https://doi.org/10.1038/npp.2015.360.
236. Yu PL, Wu CI, Lee TS, Pan WH, Wang PS, Wang SW. Attenuation of estradiol on the reduction of striatal dopamine by amphetamine in ovariectomized rats. *J Cell Biochem*. 2009;108(6):1318−1324. https://doi.org/10.1002/jcb.22361.
237. Zachry JE, Nolan SO, Brady LJ, Kelly SJ, Siciliano CA, Calipari ES. Sex differences in dopamine release regulation in the striatum. *Neuropsychopharmacology*. 2021;46(3):491−499. https://doi.org/10.1038/s41386-020-00915-1.
238. Lynch WJ, Kiraly DD, Caldarone BJ, Picciotto MR, Taylor JR. Effect of cocaine self-administration on striatal PKA-regulated signaling in male and female rats. *Psychopharmacology*. 2007;191(2):263−271. https://doi.org/10.1007/s00213-006-0656-0.
239. Mahajan SD, Aalinkeel R, Reynolds JL, et al. Therapeutic targeting of "DARPP-32": a key signaling molecule in the dopiminergic pathway for the treatment of opiate addiction. *Int Rev Neurobiol*. 2009;88:199−222. https://doi.org/10.1016/S0074-7742(09)88008-2.
240. Takashima Y, Tseng J, Fannon MJ, et al. Sex differences in context-driven reinstatement of methamphetamine seeking is associated with distinct neuroadaptations in the dentate gyrus. *Brain Sci*. 2018;8(12):208. https://doi.org/10.3390/brainsci8120208.
241. Holehonnur R, Phensy AJ, Kim LJ, et al. Increasing the GluN2A/GluN2B ratio in neurons of the mouse basal and lateral amygdala inhibits the modification of an existing fear memory trace. *J Neurosci*. 2016;36(36):9490−9504. https://doi.org/10.1523/JNEUROSCI.1743-16.2016.
242. Wickens MM, Kirkland JM, Knouse MC, McGrath AG, Briand LA. Sex-specific role for prefrontal cortical protein interacting with C kinase 1 in cue-induced cocaine seeking. *Addiction Biol*. 2021;26(5):e13051. https://doi.org/10.1111/adb.13051.
243. Martinez LA, Gross KS, Himmler BT, et al. Estradiol facilitation of cocaine self-administration in female rats requires activation of mGluR5. *eNeuro*. 2016;3(5). https://doi.org/10.1523/ENEURO.0140-16.2016.
244. Kombian SB, Ananthalakshmi KV, Parvathy SS, Matowe WC. Substance P depresses excitatory synaptic transmission in the nucleus accumbens through dopaminergic and purinergic mechanisms. *J Neurophysiol*. 2003;89(2):728−737. https://doi.org/10.1152/jn.00854.2002.
245. Zhou L, Pruitt C, Shin CB, Garcia AD, Zavala AR, See RE. Fos expression induced by cocaine-conditioned cues in male and female rats. *Brain Struct Funct*. 2014;219(5):1831−1840. https://doi.org/10.1007/s00429-013-0605-8.
246. Cheng C-Y, Hong C-J, Yu Y-W, Chen T-J, Wu H-C, Tsai S-J. Brain-derived neurotrophic factor (Val66Met) genetic polymorphism is associated with substance abuse in males. *Brain Res Mol Brain Res*. 2005;140(1−2):86−90. https://doi.org/10.1016/j.molbrainres.2005.07.008.

247. Jordan CJ, Andersen SL. Working memory and salivary brain-derived neurotrophic factor as developmental predictors of cocaine seeking in male and female rats. *Addiction Biol.* 2018;23(3):868−879. https://doi.org/10.1111/adb.12535.
248. Castro-Zavala A, Martín-Sánchez A, Valverde O. Sex differences in the vulnerability to cocaine's addictive effects after early-life stress in mice. *Eur Neuropsychopharmacol.* 2020;32:12−24. https://doi.org/10.1016/j.euroneuro.2019.12.112.
249. Becker JB, McClellan M, Reed BG. Sociocultural context for sex differences in addiction. *Addiction Biol.* 2016;21(5):1052−1059. https://doi.org/10.1111/adb.12383.
250. Mauer M, Potler C, Wolf R. *Gender and Justice: Women, Drugs, and Sentencing Policy*. The Sentencing Project; 1999.
251. United Nations. *World Drug Report*. United Nations publication; 2018. Sales No. E.18.XI.9.
252. McLean CP, Asnaani A, Litz BT, Hofmann SG. Gender differences in anxiety disorders: prevalence, course of illness, comorbidity and burden of illness. *J Psychiatr Res.* 2011;45(8):1027−1035. https://doi.org/10.1016/j.jpsychires.2011.03.006.
253. Moleiro C, Pinto N. Sexual orientation and gender identity: review of concepts, controversies and their relation to psychopathology classification systems. *Front Psychol.* 2015;6:1511. https://doi.org/10.3389/fpsyg.2015.01511.
254. Suen LW, Lunn MR, Katuzny K, et al. What sexual and gender minority people want researchers to know about sexual orientation and gender identity questions: a qualitative study. *Arch Sex Behav.* 2020;49(7):2301−2318. https://doi.org/10.1007/s10508-020-01810-y.
255. Coulter RW, Blosnich JR, Bukowski LA, Herrick AL, Siconolfi DE, Stall RD. Differences in alcohol use and alcohol-related problems between transgender- and nontransgender-identified young adults. *Drug Alcohol Depend.* 2015;154:251−259. https://doi.org/10.1016/j.drugalcdep.2015.07.006.
256. De Pedro KT, Gilreath TD, Jackson C, Esqueda MC. Substance use among transgender students in California public middle and high schools. *J Sch Health.* 2017;87(5):303−309. https://doi.org/10.1111/josh.12499.
257. Diehl A, Pillon SC, Caetano R, Madruga CS, Wagstaff C, Laranjeira R. Violence and substance use in sexual minorities: data from the Second Brazilian National Alcohol and Drugs Survey (II BNADS). *Arch Psychiatr Nurs.* 2020;34(1):41−48. https://doi.org/10.1016/j.apnu.2019.11.003.
258. Hoffman L, Delahanty J, Johnson SE, Zhao X. Sexual and gender minority cigarette smoking disparities: an analysis of 2016 Behavioral Risk Factor Surveillance System data. *Prev Med.* 2018;113:109−115. https://doi.org/10.1016/j.ypmed.2018.05.014.
259. Lowry R, Johns MM, Robin LE, Kann LK. Social stress and substance use disparities by sexual orientation among high school students. *Am J Prev Med.* 2017;53(4):547−558. https://doi.org/10.1016/j.amepre.2017.06.011.
260. Talley AE, Turner B, Foster AM, Phillips 2nd G. Sexual minority youth at risk of early and persistent alcohol, tobacco, and marijuana use. *Arch Sex Behav.* 2019;48(4):1073−1086. https://doi.org/10.1007/s10508-018-1275-7.
261. Wheldon CW, Watson RJ, Fish JN, Gamarel K. Cigarette smoking among youth at the intersection of sexual orientation and gender identity. *LGBT Health.* 2019;6(5):235−241. https://doi.org/10.1089/lgbt.2019.0005.
262. Robbins T, Wejnert C, Balaji AB, et al. Binge drinking, non-injection drug use, and sexual risk behaviors among adolescent sexual minority males, 3 US cities, 2015. *J Urban Health.* 2020;97(5):739−748. https://doi.org/10.1007/s11524-020-00479-x.
263. Peralta RL, Victory E, Thompson CL. Alcohol use disorder in sexual minority adults: age- and sex- specific prevalence estimates from a national survey, 2015−2017. *Drug Alcohol Depend.* 2019;205:107673. https://doi.org/10.1016/j.drugalcdep.2019.107673.
264. Hequembourg AL, Blayney JA, Bostwick W, Van Ryzin M. Concurrent daily alcohol and tobacco use among sexual minority and heterosexual women. *Subst Use Misuse.* 2020;55(1):66−78. https://doi.org/10.1080/10826084.2019.1656252.
265. Krueger EA, Braymiller JL, Barrington-Trimis JL, Cho J, McConnell RS, Leventhal AM. Sexual minority tobacco use disparities across adolescence and the transition to young adulthood. *Drug Alcohol Depend.* 2020;217:108298. https://doi.org/10.1016/j.drugalcdep.2020.108298.
266. Marshal MP, King KM, Stepp SD, et al. Trajectories of alcohol and cigarette use among sexual minority and heterosexual girls. *J Adolesc Health.* 2012;50(1):97−99. https://doi.org/10.1016/j.jadohealth.2011.05.008.
267. Kahle EM, Veliz P, McCabe SE, Boyd CJ. Functional and structural social support, substance use and sexual orientation from a nationally representative sample of US adults. *Addiction.* 2020;115(3):546−558. https://doi.org/10.1111/add.14819.

268. Goldbach JT, Tanner-Smith EE, Bagwell M, Dunlap S. Minority stress and substance use in sexual minority adolescents: a meta-analysis. *Prev Sci.* 2014;15(3):350–363. https://doi.org/10.1007/s11121-013-0393-7.
269. Fletcher JB, Kisler KA, Reback CJ. Housing status and HIV risk behaviors among transgender women in Los Angeles. *Arch Sex Behav.* 2014;43(8):1651–1661. https://doi.org/10.1007/s10508-014-0368-1.
270. Connolly D, Aldridge A, Davies E, et al. Comparing transgender and cisgender experiences of being taken advantage of sexually while under the influence of alcohol and/or other drugs. *J Sex Res.* 2021;58(9):1112–1117. https://doi.org/10.1080/00224499.2021.1912692.
271. Walters MA, Paterson J, Brown R, McDonnell L. Hate crimes against trans people: assessing emotions, behaviors, and attitudes toward criminal justice agencies. *J Interpers Violence.* 2020;35(21–22):4583–4613. https://doi.org/10.1177/0886260517715026.
272. Lewis RJ, Romano KA, Ehlke SJ, et al. Minority stress and alcohol use in sexual minority women's daily lives. *Exp Clin Psychopharmacol.* 2021;29(5):501–510. https://doi.org/10.1037/pha0000484.
273. Wolford-Clevenger C, Flores LY, Bierma S, Cropsey KL, Stuart GL. Minority stress and drug use among transgender and gender diverse adults: a daily diary study. *Drug Alcohol Depend.* 2021;220:108508. https://doi.org/10.1016/j.drugalcdep.2021.108508.
274. Watson RJ, Park M, Taylor AB, et al. Associations between community-level LGBTQ-supportive factors and substance use among sexual minority adolescents. *LGBT Health.* 2020;7(2):82–89. https://doi.org/10.1089/lgbt.2019.0205.
275. Heck NC, Livingston NA, Flentje A, Oost K, Stewart BT, Cochran BN. Reducing risk for illicit drug use and prescription drug misuse: high school gay-straight alliances and lesbian, gay, bisexual, and transgender youth. *Addict Behav.* 2014;39(4):824–828. https://doi.org/10.1016/j.addbeh.2014.01.007.

SECTION V

Epilogue

CHAPTER 13

Beyond substance use disorders: Behavioral addictions

Introduction

The focus of this book has been on substance use disorders (SUDs). During the past 12 chapters, you have learned about neurobiological, behavioral, cognitive, and sociocultural factors associated with SUDs. Although the term addiction is often used synonymously with SUDs, there are other forms of addiction that do not involve drugs. The purpose of this chapter is to discuss **behavioral addictions**. Behavioral addiction is an activity (other than drug use) an individual continuously engages in despite negative consequences. Many behavioral addictions revolve around behaviors that are perfectly normal when done in moderation. The issue is that individuals with a behavioral addiction have a difficult time refraining from the activity even if it causes them personal distress, negatively impacts their school/work performance, and/or damages their relationships with family, friends, and/or romantic partners. Historically, the term addiction only applied to individuals that excessively and compulsively used drugs; in fact, up until 2013, the *Diagnostic and Statistical Manual of Mental Disorders* (*DSM*) considered SUDs as the only addiction. With the release of the fifth edition of the *DSM*, the term addiction is no longer exclusively used to denote SUDs.[1] As you are about to learn, behavioral addictions and SUDs are quite similar. In fact, many of the concepts that you have learned in this textbook can be applied to behavioral addictions.

Learning objectives

By the end of this chapter, you should be able to …

(1) Identify the officially recognized behavioral addictions and recognize other addictive-like behaviors.
(2) Identify the psychiatric conditions that often cooccur with behavioral addictions.
(3) Discuss the shared neurobehavioral mechanisms of behavioral addictions and SUDs.
(4) Compare and contrast the various treatments for behavioral addictions.

What's considered a behavioral addiction?

Currently, the *DSM-5* officially recognizes one behavioral addiction: gambling disorder (formally called pathological gambling). Gambling disorder has been recognized since before the *DSM-5*. With the release of the *DSM-5*, gambling disorder was moved from the impulse-control disorder section to the newly created "Substance-Related and Addictive Disorders" category.[1] To be diagnosed with gambling disorder, an individual must meet four of the following nine symptoms:

(1) preoccupation with gambling, such that they are constantly planning on how to get more money to gamble,
(2) gambling with greater amounts of money to receive the same thrill,
(3) repeatedly failing to control or to stop gambling,
(4) restlessness or irritability when trying to stop gambling,
(5) difficulty functioning in everyday life activities,
(6) engaging in gambling to escape from a dysphoric (e.g., unhappy) state,
(7) gambling to regain lost money due to gambling (known as **chasing losses** or loss-chasing),
(8) lying about gambling to loved ones, and
(9) relying on others to fund gambling.

Although gambling disorder is the only formally recognized nonsubstance-related behavioral addiction in the *DSM-5*, the *International Statistical Classification of Diseases and Related Health Problems* (*ICD*) recognizes gambling disorders and gaming disorders as behavioral addictions. The *DSM-5* also includes internet gaming disorder in Section III of the manual, meaning that additional research is needed before it can be classified as a behavioral addiction. According to the *ICD-11*,[2] individuals with gaming disorder show the following characteristics:

(1) impaired control over gaming (i.e., they have difficulty moderating how much time they spend playing games or have difficulty stopping the game),
(2) giving games increased priority over other life activities, and
(3) continued or escalated gaming behavior despite negative consequences.

To provide a real-life example of how a gaming disorder can negatively impact one's life: when I was in college, one of my friends excessively played the game *World of Warcraft*. He spent so much time playing the game that he never attended class. He assumed that his professors would drop him from their classes (note, this is not how college classes work). By the end of his first semester in college, my friend had a 0.00 GPA. If you are reading this as a college student, do not be like my friend.

The *DSM-5* and the *ICD-11* have a section for eating disorders that are separate from behavioral addictions.[1,2] There are multiple eating disorders, with *anorexia nervosa* being the most well-known eating disorder. Individuals with anorexia fear weight gain and will use extreme methods to control their weight. Beyond anorexia, there are other eating disorders that share similar features as SUDs, and as such, will be discussed in more detail here: **binge eating disorder** and **bulimia nervosa**. In both conditions, individuals experience recurring episodes in which they eat large amounts of food in a short period of time (∼2 h), more than what would normally be expected during this time frame. These individuals also will eat beyond fullness and will report a lack of control over their eating. How do these conditions differ? To answer this question, let's first look at the *DSM-5* criteria for binge eating disorder (note, the *ICD-11* criteria are similar to those listed in the *DSM-5*). To be diagnosed with binge eating disorder, one must meet the following criteria:

(1) engage in at least three of the following behaviors:
   (a) unusual rapid consumption (typically measured within a 2-h period),
   (b) eating beyond fullness,
   (c) eating large quantities of food, even when not hungry,
   (d) eating alone to avoid embarrassment, and
   (e) feeling disgusted, depressed, or guilty after a binge eating episode;
(2) experience distress due to binge eating; and
(3) engage in binge eating at least once a week over the span of 3 months.

What distinguishes bulimia nervosa from binge eating disorder is that individuals with bulimia will participate in maladaptive behaviors to control their weight following a binge. For example, these individuals may force themselves to vomit or will take a laxative following a meal. Other individuals may exercise excessively or will fast for prolonged periods of time. Individuals with bulimia nervosa are overly concerned with their appearance, similar to what is observed in those with anorexia nervosa.

The *ICD-11* recognizes compulsive sexual behavior disorder as an impulse-control disorder,[2] but the symptoms of compulsive sexual behavior mirror those of other behavioral addictions. Like gambling and gaming disorders, individuals with compulsive sexual behavior have difficulty controlling intense sexual urges/impulses, make sex the central focus of one's life to the point that other responsibilities are ignored, have difficulty abstaining from sex, and continue to engage in sexual activity despite negative consequences associated with the activity or deriving no pleasure/satisfaction from it.

While not officially recognized by the *DSM* or the *ICD*, additional behavioral addictions related to internet use, shopping, social media use, exercise, plastic surgery, pornography, and work may be included in the future. For simplicity, I will use the word addiction to describe these behaviors even if they have not been formally recognized as behavioral addictions. What is preventing something like "exercise addiction" from being formally recognized as a behavioral addiction? First, not enough research has been conducted on these behaviors to verify that they share most of the same features as SUDs. Second, there has been some debate as to whether these behaviors should be classified as their own mental disorder or if they are the result of an already existing disorder such as obsessive-compulsive disorder (OCD) and/or other anxiety disorders and impulse-control disorders like attention-deficit/hyperactivity disorder (ADHD).[3] With that said, this is a good time to discuss psychiatric disorders that are often observed in those that engage in addictive-like behaviors.

Comorbid disorders observed in behavioral addictions

You have already learned that individuals with a SUD commonly have a comorbid disorder, such as schizophrenia, major depressive disorder, bipolar disorder, ADHD, and/or one of several anxiety disorders. SUDs and behavioral addictions have high comorbidity rates as well.[4] Specifically, estimates of comorbidity between behavioral addictions and SUDs range between 21% and 64%.[5] One study found that almost 75% of individuals with gambling disorder have comorbid alcohol use disorder, and over 60% have a cooccurring nicotine use disorder[6]. Behavioral addictions can also cooccur with one another, with one study finding that 45% of those with gambling disorder meet diagnostic criteria for a second behavioral addiction.[7] Individuals with a binge eating disorder show greater gambling behavior.[8,9] One study found that 1.49% of the entire sample had a lifetime diagnosis of gambling disorder; however, 5.7% of individuals with binge eating disorder displayed pathological gambling.[10] The reverse is true as 8.1% of those with a gambling disorder binge eat.[7] Finally, compulsive sexual behavior is associated with pathological gambling and compulsive shopping.[11]

Like SUDs, individuals with behavioral addictions are more likely to be diagnosed with other psychiatric conditions. A large proportion of individuals with a gambling disorder also have a mood disorder (~50%), an anxiety disorder (~41%), or a personality disorder (~61%).[9] Even when compared to recreational gamblers, individuals that display pathological gambling score higher in depression.[12] Likewise, binge eating, addictive gaming, excessive social media use, compulsive sex, and compulsive exercise are associated with elevated anxiety and depression.[8,11,13–19] In Chapter 11, you learned that exercise can be used to alleviate stress; yet, when individuals compulsively exercise, they experience greater depression and anxiety.[20] This finding exemplifies the distress that individuals feel when engaging in compulsive behaviors such as excessive exercising. Furthermore, individuals who exercise professionally score higher in depression compared to those who exercise recreationally.[20]

Approximately 4.5% of individuals with gambling disorder are diagnosed with schizophrenia.[21] This may not seem significant, but you need to take into consideration that schizophrenia is present in just 0.5%–1.0% of the general population, meaning that individuals with gambling disorder are over 4 times as likely to be diagnosed with schizophrenia. Schizophrenia is also associated with binge eating disorders[22] and problematic internet use.[23] Schizophrenia is primarily treated with atypical antipsychotic medications, which are associated with weight gain. One research question has centered around binge eating as a potential mechanism for the weight gain observed in individuals prescribed such medications. There is some evidence that binge eating is a cooccurring condition in those with schizophrenia taking atypical antipsychotic medications.[24]

Lifetime prevalence of ADHD in those with gambling disorder has been reported to be approximately 29%,[25] and ADHD is associated with severity of gambling disorder.[26] ADHD symptomology is also significantly associated with eating disorders, including bulimia nervosa and binge eating disorder[27] (but see[28]). A recent study found that self-reported ADHD symptoms are correlated with binge eating and food addiction in patients with an alcohol use disorder,[29] demonstrating the complex associations between SUDs, behavioral addictions, and impulse-control disorders. Other behavioral addictions associated with ADHD include internet addiction,[30] gaming (including internet gaming) disorder,[17,18] social media addiction,[17] sex addiction,[11] and work addiction.[31] As you will learn later, impulsivity is a defining feature of behavioral addictions.

Learning and cognitive mechanisms underlying behavioral addictions

In Chapters 5–7, you learned about the behavioral and the cognitive bases of SUDs. Just like SUDs, phenomena such as operant and Pavlovian conditioning, cognitive biases, decision making, and memory contribute to the development and the maintenance of behavioral addictions and/or have been shown to be distorted in individuals with such addictions. This section focuses on the learning and the cognitive mechanisms associated with behavioral addictions.

Operant and Pavlovian conditioning

Going to a casino, playing a video game, eating something like cake, having sex, and browsing the internet/social media share an important feature with ethanol and drugs: they are reinforcing! As you learned in Chapter 5, behaviors that are reinforcing are more likely to be repeated in the future. Recall that reinforcement can be divided into positive reinforcement and negative reinforcement. With positive reinforcement, something pleasurable is added after performing a behavior; however, with negative reinforcement, something aversive is removed following a behavior. Think of social media: receiving a bunch of "likes" or upvotes after posting something is positively reinforcing. Hundreds of millions of individuals use social media and enjoy video games, but the percentage of individuals that develops a social media addiction or gaming disorder ranges from less than 1% to approximately 37%.[32,33] These results indicate that positive reinforcement on its own does not fully account for why someone develops a behavioral addiction, which is correct. However, individuals with gambling disorder, binge eating disorders, internet addiction, and gaming disorders are more sensitive to rewards[34-40] (but see[8]). Furthermore, individuals with gambling disorder are more sensitive to monetary rewards compared to other rewards.[41] Going back to social media, someone at risk of developing social media addiction is more sensitive to the attention they receive for their posts. Receiving this attention will motivate these individuals to continue posting messages, photos, and/or videos, and these individuals may compulsively check their social media to see if they have received more attention or to post additional content.

Not only are behavioral addictions associated with increased reward sensitivity, but they are linked to altered sensitivity to punishment, particularly in gambling disorder and internet gaming disorder.[38] Recall that punishers decrease the frequency of a behavior. Intuitively, if an individual is more sensitive to rewards, they should be less sensitive to punishment. However, individuals with behavioral addictions, particularly binge eating disorders, are generally more sensitive to punishment compared to normal controls[37,39] (but see[8,35]). This finding contrasts with what is typically observed in SUDs as substance dependence is often associated with decreased punishment sensitivity. In SUDs, individuals are less sensitive to the aversive effects of a drug (e.g., hangover), which may account for why these individuals continue to use the substance. Conversely, someone with a gambling disorder may be more sensitive to losing on a slot machine. To alleviate these negative feelings, the individual continues to gamble (chasing losses). Individuals with a binge eating disorder are especially sensitive to social punishment[39] and may respond to this social stressor by engaging in a binge. Increased sensitivity to social punishment may also exacerbate social media addiction, as individuals may be more motivated to make another post if a previous post fails to garner the expected attention. Overall, individuals with a behavioral addiction may continue engaging in maladaptive behavior as a way of dealing with stressors, which is a form of negative reinforcement.

Even though behavioral addictions are associated with decreased punishment sensitivity, individuals that pathologically gamble consistently underestimate the probability of a loss compared to those that do not gamble.[42] In other words, gamblers may be more sensitive to punishment, but they are not as effective at predicting the likelihood of losing when playing a slot machine or playing a round of Blackjack. Related to this point, *near misses* are subjectively rated as more unpleasant than a complete miss (i.e., receiving no jackpot symbols or just one jackpot symbol), but individuals report a greater desire to continue playing following a near miss.[43] This finding directly supports what I have already mentioned about punishment sensitivity in behavioral addictions. That is, continued gambling following a near miss may alleviate the frustration of missing the jackpot previously (i.e., a form of negative reinforcement).

In addition to operant processes, Pavlovian conditioning is involved in behavioral addictions. Numerous stimuli can become paired with a particular behavior. For example, casinos are flooded with lights and sounds. These stimuli can elicit strong cravings, just as conditioned stimuli can elicit drug cravings in those with a SUD. Individuals with gambling disorder report greater craving when presented with cues related to gambling, but they do not show increases in craving when presented with neutral cues or shown pictures of palatable food.[44] Additionally, individuals that engage in problem gambling show increased attentional bias to gambling-related cues.[38,45–51] Even animals display increased sign-tracking (the animal equivalent of attentional bias) to a stimulus that predicts probabilistic food delivery compared to a stimulus that signals consistent food delivery.[52] Enhanced attentional bias is observed in other behavioral addictions like binge eating disorder and internet gaming disorder.[38,53,54] See the Experiment Spotlight below to see how attentional bias toward high-calorie foods can be measured.

Experiment Spotlight: Attentional bias to high-calorie food in binge eaters with high shape/weight concern (Seo and Lee[53])

In Chapter 7 you learned about several methods for measuring attentional bias. One of these methods is to track how much time an individual spends looking at specific stimuli on a computer screen. Seo and Lee[53] used eye-tracking equipment to determine if binge eaters spend more time looking at high-calorie foods. Additionally, Seo and Lee wanted to determine if individuals with high shape and/or weight concern (SWC) show greater attentional bias to high-calorie foods. This is somewhat paradoxical, but individuals with SWC are more likely to binge eat.

To complete their study, Seo and Lee recruited 716 Korean women. Before being eligible to complete the experiment, participants had to complete the Eating Disorder Diagnostic Scale (EDDS) and the Eating Disorder Examination Questionnaire (EDE-Q) to ensure they met the criteria for binge eating disorder. A subscale of the EDE-Q was used to categorize participants as having high SWC or low SWC. Participants were ineligible for the experiment if they had a diagnosis of another eating disorder like anorexia nervosa, used maladaptive methods for controlling their weight (e.g., vomiting after eating), or were currently taking pharmacotherapies. Out of the original 716 participants that were recruited, 120 completed the experiment. Participants were categorized into one of four groups: (1) binge eater with high SWC, (2) binge eater with low SWC, (3) healthy control with high SWC, and (4) healthy control with low SWC.

Participants were instructed to fast for 6 h before arriving to the laboratory. After arriving to the lab, participants completed a hunger visual analog scale in which they indicated how hungry they felt and how much they wanted to eat. They were also asked to rate how full/satiated they felt. Participants then completed questionnaires about depression and anxiety before completing a questionnaire to assess food cravings. After completing these questionnaires, participants completed an eye-tracking task. Participants were given 54 trials in which they viewed a pair of images presented on the computer screen. Seo and Lee used a total of 27 images throughout the experiment: nine high-calorie foods, nine low-calorie foods, and nine neutral household items. Each pair was matched as much as possible according to their color, shape, and size. For example, one food pairing consisted of a pizza and a watermelon cut in half (see figure below). Other food pairings included sausages and carrots, chocolate truffles and walnuts, cookies and grapes, and potato wedges and bananas. Participants were presented with each pairing for 4 s. Each cue was presented twice during the experiment, which allowed for the image to be presented on both sides of the screen.

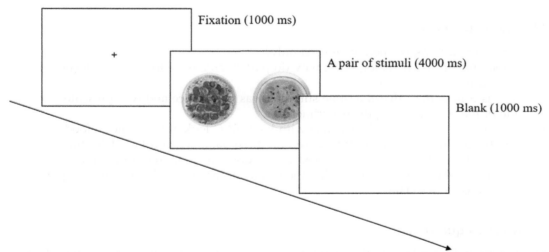

Figure comes from Fig. 1 of Seo CL, Lee JH. Attentional bias to high-calorie food in binge eaters with high shape/weight concern. Front Psychiatr. 2021;12:606296. https://doi.org/10.3389/fpsyt.2021.606296. This is an open-access article distributed under the terms of the Creative Commons Attribution 4.0 International License (https://creativecommons.org/licenses/by/4.0/).

The figure on the next page shows the results of the eye-tracking portion of the experiment. High SWC individuals who binge eat took less time to begin looking at high-calorie foods compared to low SWC individuals who binge eat. Additionally, high SWC/binge eaters spent more time looking at high-calorie foods compared to low SWC/binge eaters. Seo and Lee argue that these results support the view that high SWC exacerbates binge eating as these individuals have difficulty directing their attention away from highly palatable foods.

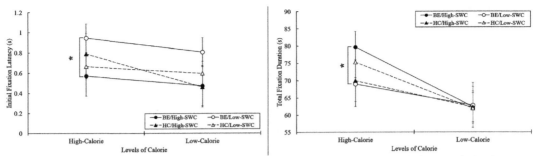

Figure comes from Fig. 2 (left panel) and 3 (right panel) of Seo CL, Lee JH. Attentional bias to high-calorie food in binge eaters with high shape/weight concern. Front Psychiatr. 2021;12:606296. https://doi.org/10.3389/fpsyt.2021.606296. This is an open-access article distributed under the terms of the Creative Commons Attribution 4.0 International License (https://creativecommons.org/licenses/by/4.0/).

Questions to consider:

(1) Why do you think women were used only?
(2) In the hunger visual analog scale, why do participants have to rate their level of hunger and their level of satiation?
(3) Why is it important that the same stimulus was used twice and was presented on each side of the screen during the experiment?
(4) Seo and Lee found that binge eaters with high SWC look at high-calorie foods faster than binge eaters with low SWC. Based on these data, Seo and Lee claim that binge eaters with high SWC show enhanced orientation bias to high-calorie foods. Looking at the left panel of the figure depicting the results of the study, do you see any potential limitations to this claim?

Answers to questions:

(1) As you will learn later in this chapter, eating disorders, including binge eating disorders, are more common in women.
(2) Having participants answer questions that measure opposing constructs is an important *manipulation check* to ensure that participants are paying attention to the questions and not just randomly selecting answers. An individual cannot simultaneously be hungry and full. If a participant states that they are both of these things, the researcher knows that the participant's responses are invalid and that the participant should be excluded from the study. Similar types of manipulation checks are included in most questionnaires, including ones that measure constructs such as depression and anxiety (as in the current experiment).
(3) This ensures that attentional bias toward one stimulus reflects a true attentional bias and is not an artifact of side preference. Theoretically, an individual may always start by looking to their left and then scan to the right side of the screen. If binge-eating individuals with high SWC show increased attentional bias to high-calorie foods, they

should look at these foods faster compared to binge-eating low SWC individuals, regardless of which side the high-calorie food is presented.
(4) If you look at the latency to initial fixation for both high-calorie and low-calorie foods, you will notice that high SWC binge-eating individuals start viewing these images faster compared to the low SWC binge-eating individuals. Even though the difference between these two groups was not statistically significant when examining low-calorie foods, the results are largely similar across both food types. These results suggest that high SWC binge eaters are more sensitive to food cues in general. This is not a major limitation to the study as Seo and Lee found that the total amount of time spent looking at high-calorie foods is significantly higher in high SWC binge eaters compared to low SWC binge eaters, an effect that was absent for low-calorie foods.

One paradigm you have previously learned about is conditioned place preference (CPP). CPP measures the conditioned rewarding effects of a stimulus. CPP has been used to examine college students' preference for a virtual environment paired with candy.[55] In one virtual room, participants received chocolate dispensed in a cup beside them, but they did not receive candy when exploring the other virtual room. When given a preference test, participants spent more time in the room paired with chocolate. Although these experiments did not test individuals with a binge eating disorder, Astur et al. found significant correlations between dieting and CPP scores for the chocolate-paired environment in women.[55] In other words, increased dieting is associated with greater preference for an environment previously paired with an indulgent treat. These findings can be used to explain how environmental stimuli can elicit approach behaviors toward palatable foods in individuals with bulimia nervosa or binge eating disorder. Theoretically, individuals with a binge eating disorder should show enhanced CPP compared to healthy controls. Future work is needed to test this hypothesis.

Cognitive biases

Attentional bias is one cognitive bias that greatly influences addictive behaviors. There are multiple cognitive biases that have been proposed to exacerbate gambling.[56] This section will cover some of these gambling-specific biases. **Overconfidence** is one such bias. As related to gambling, individuals with gambling disorder often display overconfidence in their ability to win in situations that are purely probability based. Overconfidence can be measured in the Georgia Gambling Task (GGT) (see Fig. 13.1 for a description of the task).[57] Pathological gamblers are more confident in their responses, place more bets, and lose more points compared to nonpathological gamblers.[58] This is known as *paradoxical betting*. The near miss effect covered in the previous section is a direct example of **gambler's fallacy**. Gambler's fallacy is a cognitive bias in which individuals believe independent events are influenced by previous events.[59] For instance, someone rolling a die may erroneously predict that they will roll a six soon because they have not rolled a six in a while. However, this is not how probability works. On a six-sided die, the probability of rolling any number is 0.1667 (1/6). One more cognitive bias I want to mention is the *illusory correlation*. An illusory correlation is a "fake" correlation; for example, an individual with pathological gambling may believe that their luck will increase the likelihood of winning. These biases may account for why individuals with gambling disorder continue to gamble even though they are more sensitive to punishment.

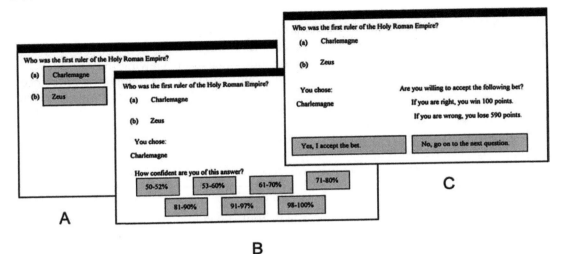

FIGURE 13.1 A schematic of the Georgia gambling task (GGT). Participants first answer a general knowledge question and then rate how confident they are in their answer. Individuals then indicate if they want to accept a bet or not. If individuals are overly confident in their responses, the number of points they can earn is substantially less than the number of points they can lose. *Image comes from Fig. 1 of Goodie AS. The effects of control on betting: paradoxical betting on items of high confidence with low value.* J Exp Psychol Learn Mem Cogn. 2003;29(4):598−610. https://doi.org/10.1037/0278-7393.29.4.598. Copyright © 2003 by the American Psychological Association.

Maladaptive decision making and memory impairments

Related to the cognitive biases described above, deficits in decision making, particularly in the Iowa Gambling Task (IGT), and cognitive control are defining features of behavioral addictions.[60,61] Decision-making inflexibility is often observed in those with gambling disorder.[62,63] There are a couple of ways to measure cognitive inflexibility. First, a reversal-learning paradigm measures how well individuals can alter their responses after contingencies have reversed.[64] For example, individuals can be given two choices, one being more advantageous than the other. Once individuals have learned which choice is more advantageous, the researcher switches the options such that the advantageous option is now disadvantageous. The researcher then measures the time required for individuals to modify their responses accordingly. Individuals with pathological gambling tendencies take longer to switch their response strategy in this task.[62] A second way to measure cognitive inflexibility is with the Wisconsin card sorting task (WCST).[65] Participants are shown several cards that contain different geometric shapes on them. One card may have two yellow crosses on it while another card may have four blue circles on it. The participant is then handed a separate card and is instructed to match it to one of the cards in front of him/her. However, the participant is not told *how* to match the card. They are merely told whether a particular match is correct or incorrect. As the experimental session progresses, the "rules" for matching the cards change. The researcher determines how well individuals can shift their matching strategy. In short, individuals with pathological gambling tendencies perform worse on the WCST.[63] Binge eating is also associated with cognitive inflexibility, as they perform worse in tasks like the WCST.[66]

In addition to impaired cognitive flexibility, steeper delay discounting (i.e., devaluation of larger, delayed rewards in favor of small, immediate rewards) is characteristic of several behavioral disorders, including gambling disorder,[67,68] binge eating disorder,[69] and internet gaming disorder.[70] Furthermore, gamblers that show steeper delay discounting are more likely to chase their losses.[71] Animal research has shown some correlations between delay discounting and binge eating, albeit inconsistencies have been reported. If rats are assigned to a condition in which they are allowed to binge eat, they show greater delay discounting.[72] Another study found increased binge eating in rats that show greater delay discounting,[73] but this was not replicated in a later study.[74]

As gambling involves risky decision making, one could reasonably expect to observe increased risky decision making in gamblers. Indeed, gambling is associated with shallower probability discounting (i.e., increased risky choice)[75] and increased risk taking in the balloon analog risk task (BART).[76] Risky decision making is not necessarily a defining feature of all behavioral addictions, as women with a binge eating disorder either do not differ from controls in the BART[77] or show increased risk aversion in this task.[78] These results highlight dissociations across behavioral addictions and highlight the potential need to selectively target risky decision making in pathological/problematic gamblers during therapy to better increase treatment outcomes.

Working memory is also impaired in individuals with a behavioral addiction,[66,79] and individuals with a binge eating disorder experience deficits in immediate recall and delayed recall in a visuospatial memory task.[80] Long-term memory distortions have been proposed to influence pathological gambling, as individuals with gambling disorder are more likely to remember wins more than losses.[56] This hypothesis may be extended to account for other behavioral addictions and may explain why behavioral addictions persist despite enhanced punishment sensitivity often observed in these individuals. For example, an individual with a binge eating disorder often experiences distress due to their compulsive eating, but because their working memory and long-term memory are impaired, they may have more difficulty linking their distress back to the behavior that caused it originally.

Quiz 13.1

(1) Which of the following is officially recognized as a behavioral addiction in the *DSM-5*?
   a. Binge eating disorder
   b. Gambling disorder
   c. Internet gaming disorder
   d. All of the above
(2) Which gambling task captures how overconfidence leads to impaired decision making?
   a. Cambridge gambling task
   b. Georgia gambling task
   c. Iowa gambling task
   d. Slot machine task
(3) Which of the following statements about cognitive/decision-making impairments and behavioral addictions is true?

a. Behavioral addictions are commonly associated with working memory deficits.
   b. Binge eating disorder is consistently linked to maladaptive risky decision making.
   c. Individuals with gambling disorder show greater discounting of both delayed and probabilistic reinforcers.
   d. All of the above
(4) True/False: individuals with a behavioral addiction are insensitive to punishment

Answers to quiz 13.1

(1) b — Gambling disorder
(2) b — Georgia gambling task
(3) a — Behavioral addictions are commonly associated with working memory deficits.
(4) False

Risk factors associated with behavioral addictions

Chapters 9—12 discussed specific factors linked to SUDs, such as personality traits, familial/peer influences, sociocultural factors, gender/sex, and stress. Similar to SUDs, research has shown (1) personality characteristics that are more common in individuals with a behavioral addiction, (2) specific ways in which family members and/or social peers influence addictive-like behaviors, (3) differential rates of behavioral addictions across income and race/ethnicity, (4) gender differences in prevalence of behavioral addictions, and (5) deleterious effects of stress on the maintenance of behavioral addictions and increased likelihood of relapse following stress exposure. These findings will be discussed below.

Personality traits

Many of the personality traits associated with SUDs have been linked to behavioral addictions. Individuals with behavioral addictions often score higher in neuroticism and harm avoidance, but lower in agreeableness, conscientiousness, self-directedness, cooperativeness, and mindfulness.[81–86] Another personality trait identified as a risk factor for behavioral addictions is impulsivity.[12,38,79,84,87–91] Specific facets of impulsivity that are correlated with addictive-like behaviors include sensation seeking[83,89,92] (but see[93]), urgency,[68,92–97] lack of premeditation[93,94,97] (but see[96]), and lack of perseverance[93,95,98,97] (but see[94,96]). Eating disorders also are associated with attentional deficits,[91] even when no differences are observed in ADHD diagnoses between binge-eating and nonbinge-eating individuals.[28] There is some evidence that impulsivity severity differs across behavioral addictions. For example, individuals with internet addiction score higher on impulsivity measures compared to those with gambling disorder.[79]

While sensation seeking has been identified as a risk factor for behavioral addictions, some inconsistent findings have been observed concerning the relationship between sensation seeking and gambling disorder (see[87,96,99] for positive associations; see[84,94,97] for negative associations). These inconsistencies could be due, at least in part, by the type of gambling being studied. Sensation seeking is predictive of gambling, but only in those that engage in *strategic* gambling.[100] Another important consideration is that sensation seeking is composed of four major domains: thrill and adventure seeking, experience seeking, disinhibition, and boredom susceptibility. There is evidence that disinhibition and boredom susceptibility, but not thrill adventure seeking or experience seeking, are elevated in gamblers.[101] These results show the need to examine individual subscales of impulsivity when determining potential personality constructs that are predictive of addictions.

Peer and familial influences

Social influences have a direct impact on behavioral addictions. Simply put, living with a gambler increases the probability of gambling.[102] In one experiment, children between the ages of 9 and 14 answered questions about gambling behavior. Amazingly, 86% of the participants that identified regular gambling did so with a family member, and 75% reported gambling with friends; conversely, only 18% of the participants identified gambling alone.[103] While adolescents are more likely to gamble if their parents gamble, adolescent gambling problems are associated with paternal (i.e., father) gambling severity, but not maternal gambling severity.[104] Adolescents and young adults that identify excessive gambling in peers or concerns over familial gambling are more likely to engage in pathological gambling.[105,106] Additionally, adolescents that are classified as probable pathological gamblers are more sensitive to peer pressure.[107] Problematic gambling during adolescence is also associated with family problems and lack of familial and peer support.[108] Related to this point, a recent study found that gambling behavior is increased in those that experience more pressure from family and peers to stop gambling.[99] The increased gambling observed in these individuals could be a form of negative reinforcement as they try to escape the criticisms/concerns of family/peers.

Social influences are a major determinant of binge eating disorders. In particular, the development of binge eating disorders is influenced by negative familial and peer comments about one's weight. Adolescent girls with a binge eating disorder like bulimia nervosa are more likely to have a mother who has criticized their weight and physical appearance.[109] Family members and peers also influence bulimia in women by reinforcing the **thin ideal**, the concept of an ideally slim female body, and by modeling abnormal eating behaviors.[110,111] These findings are somewhat similar to what was covered in the Experiment Spotlight. Binge eaters that are highly concerned about their body shape/weight show enhanced attentional bias to high-calorie foods. If other individuals critique one's weight and/or image, this could increase the shape/weight concerns of those prone to a binge eating disorder, which then can exacerbate cravings for "forbidden" foods.

One way in which parents can influence addictive behaviors is through their parenting style. Recall that parents can be authoritative (high warmth/high supervision), authoritarian (low warmth/high supervision), permissive (high warmth/low supervision), or uninvolved (low warmth/low supervision). Increased parental monitoring/supervision that is characteristic of authoritative parenting decreases the likelihood of gambling in adolescents[104] and is protective against compulsive internet use[112] and smartphone addiction.[113] Conversely, more authoritarian-like parenting styles are associated with gambling disorder,[114] binge eating disorders,[115] internet addiction,[112] and smartphone addiction.[113]

Sociocultural factors

Beyond immediate family members and peers, sociocultural factors can greatly contribute to behavioral addictions. Low socioeconomic status (SES) is predictive of problematic gambling[116,117] and pathological gambling.[118] This finding makes sense as low SES individuals may be more tempted to gamble to improve their financial situation. Low SES individuals or individuals from disadvantaged neighborhoods play the lottery more frequently.[116,119]

SES is not necessarily a risk factor for binge eating disorder, but the risk factors for binge eating vary across SES groups. While being overweight and dieting are associated with binge eating in both high SES and low SES adolescents, body dissatisfaction and family teasing are risk factors for high SES individuals, whereas food insecurity is predictive of binge eating in low SES individuals.[120] Gender can also act as a moderating variable between SES and binge eating. SES is not associated with binge eating in men, but binge eating is observed more frequently in low SES women.[121]

The media is often blamed for disordered eating, especially in women. Television, movies, music videos, magazines, and other forms of media often feature women that meet the thin ideal. The reinforcement of the thin ideal by media is correlated with bulimia nervosa in women,[110] particularly in women with high levels of body dysphoria.[122] With the rise of social media, research has shown a link between social media usage and elevations in body dissatisfaction in women that binge eat.[123]

Racial differences have been observed in behavioral addictions. Gambling disorder is more prevalent in African Americans and Asian Americans compared to Whites,[124] and African Americans with gambling disorder report more symptoms compared to Whites.[125] Even in a sample of recreational gamblers, African Americans, as well as Hispanics and Asians, display more behaviors indicative of a gambling disorder compared to Whites.[117,118] It is important to note that when SES is taken into consideration, the increased gambling observed in African Americans disappears.[119] This is similar to what has been reported for SUDs. Being African American, in itself, does not necessarily increase one's chances of gambling. Instead, African Americans are more likely to live in disadvantaged areas, which have higher gambling rates. In a college sample, Asians meet more criteria for pathological gambling and are more likely to: (1) report lying about losses, (2) report having a family history of pathological gambling, and (3) report feeling a loss of control over gambling.[126] There is some evidence for cultural differences in gambling rates. Compared to Americans, Brazilians progress from recreational gambling to gambling disorder at a faster rate and are more likely to endorse loss chasing.[127] This cultural difference could be explained by SES differences in the United States and Brazil, as Brazilians are more likely to be classified as low SES compared to Americans.

In contrast to gambling disorder, binge eating disorder and internet addiction are more prominent in Whites compared to African Americans,[128,129] although Asian women report greater binge eating compared to White women.[130] Whites develop binge eating disorder at an earlier age compared to African Americans.[131] One potential explanation for the decreased binge eating observed in African Americans is differences in *weight bias*, or negative attitudes held toward one's body weight. In a sample of obese individuals, African Americans showed reduced weight bias compared to Whites.[132] Barnes et al. also found that weight bias was significantly correlated with binge eating episodes.[132] Another explanation that has been proposed is that African American men have different beauty standards compared to White men. Research has shown that African American men prefer larger female body silhouettes compared to White men.[133] There is some contradictory evidence to this hypothesis as some have shown no differences in preferences between African Americans and Whites.[134]

Gender and sexual/gender identity

Some behavioral addictions are more commonly observed in men, including gambling disorder[135] (but see[136]), internet addiction,[30,129,137] and gaming disorder.[17,31] Increased betting is even observed in boys entering adolescence.[138] Men are also more likely to have a comorbid eating disorder and gambling disorder,[10] and men with gambling disorder are more likely to engage in compulsive sexual behavior.[81] Conversely, binge eating disorders and addictive-like use of social media are more associated with women.[17,128,139] Female children also score higher in mobile phone addiction.[137] The influence of food craving on binge eating-like behavior is stronger in women compared to men.[141] Even though men are more likely to experience gambling problems,[118] women with gambling disorder have worse symptoms compared to men[142] and begin showing symptoms of pathological gambling at a faster rate.[143] This mirrors what has been observed for SUDs. Men often report increased drug use, but women tend to be more sensitive to the reinforcing effects of drugs. One factor that could account for gender differences in internet gaming disorder is differential sensitivity to reward and punishment. Men are more sensitive to rewards but less sensitive to punishment compared to women.[144]

Just as what has been reported for SUDs, sexual and gender minority individuals are at increased risk of developing an eating disorder.[145] According to Nagata et al., eating disorders in sexual minority individuals may stem from sexual minority stress,[145] which you read about in the previous chapter. They also propose that body dissatisfaction is a major stressor that gender minority individuals experience, which can promote disordered eating. In addition to eating disorders, sexual minority individuals are more likely to show symptoms of gambling disorder,[146] and they may be more at risk for developing problematic gaming and internet use.[147] The increased incidence rates of gambling and gaming/internet use may also be related to sexual minority stress, as sexual/gender minority individuals engage in these activities to mitigate the stress associated with harassment, discrimination, and/or physical violence.

Stress

Many individuals with a behavioral addiction report experiencing increased cravings following stress.[11,99,136,148] Stress increases the desire to gamble in those with gambling disorder.[149] Early life stress and adversity are more common in individuals with gambling disorder and binge eating disorders.[140,148,150] Similarly, individuals with posttraumatic stress disorder (PTSD) or PTSD-like symptoms are more likely to engage in activities like pathological gambling and binge eating.[151,152] One hypothesis is that PTSD influences pathological gambling by enhancing cognitive distortions related to gambling behavior.[153] This is highly plausible as memory deficits are frequently observed in individuals with PTSD.

Even if you do not have a SUD or a behavioral addiction, you may have noticed that stress can increase your desire for "comfort foods", or those that are often calorically dense. *Stress eating* can be particularly problematic in those with a binge eating disorder. Indeed, individuals with binge eating disorder or bulimia nervosa show greater stress reactivity[85] and are more likely to eat in response to stress.[154–158] Women that experience stress at work, sexual/aggravated assault, or intimate partner violence are at risk of developing a binge eating disorder,[159,160] and racial minority or sexual minority individuals subjected to discrimination are more likely to binge eat.[161,162]

Neurobiological underpinnings of behavioral addictions

There is abundant evidence that behaviors like pathological gambling and binge eating share many of the same features as SUDs. If pathological gambling and binge eating are addictions like SUDs, one would expect to see similar neurobiological changes during the development of a behavioral addiction. As you are about to learn, behavioral addictions are controlled by the same neural mechanisms as those that govern SUDs. The purpose of this section is to discuss the major neurobiological underpinnings of behavioral addictions.

Brain regions implicated in behavioral addictions

Because behaviors like gambling, eating, and sex are reinforcing, they activate the mesocorticolimbic pathway just as drugs do. The ventral striatum, which includes the nucleus accumbens (NAc), has received considerable attention when studying the neurobiology of pathological gambling and binge eating. Activation of the ventral striatum is correlated with gambling severity,[163] and near misses activate brain structures implicated in reward.[43] Individuals with binge eating disorder show greater activation of the ventral striatum in response to food cues[62] while individuals with gaming disorder or internet pornography addiction show greater neural activation following exposure to gaming-related cues or pornographic images, respectively.[164,165] Just as drugs hijack the mesocorticolimbic pathway of the brain, other reinforcers can work in a similar fashion. Cues paired with monetary rewards activate the ventral striatum in individuals with gambling disorder, but this region shows decreased activity following presentation of cues that are predictive of other types of rewards.[41] Eventually, ventral striatal activation diminishes when individuals are exposed to the reinforcer related to the behavioral addiction[166] just as prolonged drug use elicits less activation of areas like the NAc.

Remember, as individuals continuously use ethanol and/or drugs, the basal ganglia, particularly the dorsal striatum (caudate and putamen), become an important locus of addiction. Therefore, it should not be surprising to learn that altered striatal function has been observed in individuals with a behavioral addiction.[167] Gambling disorder and internet gaming disorder are associated with dysfunction in the dorsal striatum,[168,169] and morphological abnormalities have been observed in the striatum of individuals with a behavioral addiction.[88,170] Similarly, individuals with internet gaming disorder show reduced functional connectivity between the ventral striatum and the frontal cortex, but they show increased connectivity between the dorsal striatum and the frontal cortex.[171] These findings reflect the compulsive nature of addictions, whether substance related or not. Because the dorsal striatum is critical for habit formation, continued gambling or binge eating occurs despite the distress one may experience while engaging in these activities.

As activity of the dorsal striatum increases during the development of an addiction, individuals with a behavioral addiction show decreased activity in the prefrontal cortex (PFC),[166,168] particularly when making decisions.[172] Inactivating the medial PFC (mPFC) impairs decision making in the rat gambling task (rGT)[173] and increases consumption of a palatable food in an animal model of binge eating.[174] These findings are consistent with data showing a hypoactive frontal cortex in substance-dependent individuals. However, stimuli associated with gambling, food, or gaming cause greater activation of the frontal cortex and the anterior cingulate cortex (ACC).[40,44,175] Similarly, greater frontal cortical activation is observed in individuals with binge eating disorder as they wait for food delivery.[176] The heightened frontal cortical activation in response to cues/environmental stimuli maps onto the attentional bias that is observed in those with a behavioral addiction.

Brain regions involved in memory are altered in individuals with a behavioral addiction. Individuals with gambling disorder show decreased hippocampal and amygdalar volume.[177,178] This finding could help explain some of the memory deficits observed in those with gambling disorder. However, individuals with internet gaming disorder have a larger hippocampus and amygdala.[179] Yoon et al. argue that a larger hippocampus/amygdala observed in these individuals may be due to increased sensitivity to gaming-related memories these individuals experience.[179] More work is needed to understand why opposite findings have been observed concerning hippocampal volume across behavioral addictions.

Neurotransmitters implicated in behavioral addictions

Dopamine has received considerable attention when examining behavioral addictions as it is the major neurotransmitter of the mesocorticolimbic pathway. Individuals with gambling disorder show greater striatal dopamine release in response to uncertainty compared to nongamblers,[180] and dopamine release correlates significantly with gambling severity.[181] One important finding is that individuals with gambling disorder experience greater dopamine release after losing money, but not when winning money.[182] Similarly, gambling severity is correlated with dopamine release following a near miss.[183] These findings may explain why individuals with gambling disorder chase their losses. As you learned above, individuals with a behavioral addiction show altered striatal activity. One alteration that has been observed in individuals with gambling disorder is reduced dopamine transporter (DAT) expression in the striatum.[184] Because DAT is responsible for removing excess dopamine from the synapse, downregulated DAT expression alters the amount of dopamine in the synapse, which can alter the activity of this region.

The role of dopamine in gambling disorder is further strengthened by research with individuals with Parkinson's disease. Parkinson's disease results from the death of dopaminergic neurons in the substantia nigra, which sends dopaminergic neurons to the basal ganglia. This is what leads to the motor control issues observed in these individuals. What does this have to do with gambling? A major treatment for Parkinson's disease is dopamine receptor agonists, which increase dopamine levels in the brain, particularly the basal ganglia. One observation that has been reported is that individuals on dopamine receptor agonist treatment are more likely to exhibit pathological gambling.[185] Individuals with gambling disorder and Parkinson's disease have increased dopamine release in the NAc.[186] Additionally, dopamine receptor agonists decrease activity of the frontal cortex following a loss, leading to decreased loss sensitivity.[187]

In animals, acute systemic administration of dopamine D_2-like receptor ligands does not affect decision making in the rGT.[188] However, acute administration of a D_2 or a D_4 receptor agonist causes rats to treat non-wins as wins in a rodent analog of a slot machine task,[189,190] and the D_3 receptor agonist pramipexole increases preference for a variable ratio schedule (i.e., gambling-like schedule) over a fixed ratio schedule.[191] Chronic treatment of the dopamine $D_{2/3}$ receptor agonist ropinirole increases gambling-like behavior.[192] These findings are similar to the increased gambling behavior observed in individuals receiving dopamine receptor agonist treatment for Parkinson's disease.

Like gambling disorder, the dopaminergic system is altered in those with binge eating disorder. Dopamine D_2-like receptors are downregulated in the striatum.[193] In rats, binge eating decreases dopaminergic signaling in the ventral tegmental area (VTA) and in the PFC,[194] decreases dopamine D_2-like receptors in the NAc,[195] and decreases dopamine D_1-like receptors in the striatum.[196] The decreased dopaminergic activity is similar to what has been observed in those with a SUD and can explain some of the long-term memory deficits reported in those with a binge eating disorder.

Beyond dopamine, there is evidence that the serotonergic system is dysregulated in those with a behavioral addiction.[197,198] Individuals with gambling disorder have decreased serotonin transporter (SERT) levels.[199] The decreased SERT expression observed in these individuals could result from a compensatory response to decreased serotonin levels. In rodents, a $5-HT_{1A}$ receptor agonist impairs performance in the rGT, as rats respond less for the most optimal option.[200] As the $5-HT_{1A}$ receptor is inhibitory, stimulating these receptors leads to decreased activity of serotonergic neurons.

SERT is implicated in eating disorders, as decreased SERT binding is associated with increased binge eating.[201] Individuals recovering from bulimia nervosa have decreased SERT binding in the midbrain, but they have *increased* SERT binding in the ACC.[202] Individuals with bulimia nervosa have enhanced $5-HT_{1A}$ receptor activity and binding,[203] and they show enhanced $5-HT_{1A}$ receptor binding following recovery from bulimia.[204] In contrast to $5-HT_{1A}$ receptors, women that have recovered from bulimia nervosa have decreased binding of the serotonin $5-HT_{2A}$ receptor in the frontal cortex.[205] Upregulation of serotonin $5-HT_{2C}$ receptors has been observed in the PFC of rodents that binge eat.[195] Interestingly, activating $5-HT_{2C}$ receptors decreases binge eating behavior in rodents.[206] Recall from Chapter 3 that $5-HT_{2C}$ receptor agonists reduce the reinforcing effects of drugs.

Although dopamine and serotonin have received considerable attention, multiple neurotransmitter systems have been implicated in behavioral addictions. Both major amino acid neurotransmitters, glutamate and GABA, contribute to behavioral addictions. Gambling severity is associated with decreased glutamate-glutamine levels,[207] and individuals with gambling disorder have increased $GABA_A$ receptor availability in the hippocampus.[208]

Pathological gambling is associated with blunted dopamine-induced GABA release in the frontal cortex, which may partially explain the inhibitory failures observed in individuals that excessively gamble.[209] Somewhat similarly, disrupting GABAergic signaling in the mPFC impairs decision making in the rGT.[210] The drug N-acetylcysteine, which is important for maintaining glutamate homeostasis, and the Alzheimer's treatment memantine, which blocks NMDA receptors, have shown some promise in reducing binge eating.[211] The $GABA_B$ receptor agonist baclofen also decreases binge eating[212] and transiently decreases gambling behaviors.[213]

Opioids are an important mediator of appetite and the hedonic effects of food.[214] As such, studying the opioid system in relation to binge eating disorders makes sense. While pathological gambling is associated with decreased mu opioid receptor availability in the ACC, binge eating disorder is linked to widespread decreases in mu opioid receptor availability.[215] Although opioid receptor availability is decreased in those with a behavioral addiction, direct stimulation of mu opioid receptors in the NAc increases binge eating in rodents.[216] Conversely, deleting mu opioid receptors decreases consumption of a palatable food in mice.[217] Likewise, blocking mu opioid receptors can reduce attentional bias in individuals with binge eating disorder[218] and can reduce binge eating in both humans and rodents.[219,220] Mu opioid receptor antagonists also block gambling behavior in humans and in rats.[221,222]

In addition to opioids, the endocannabinoid system regulates appetite and has received attention as a potential mediator of binge eating disorders. Using the intermittent access model of binge eating, Satta et al. observed altered endocannabinoid levels in the striatum, the amygdala, and the hippocampus of binge eating rats, as well as a downregulation of cannabinoid CB_1 receptors in the PFC.[223] In Chapter 2, I discussed the drug rimonabant, which was used in Europe as a smoking cessation aid before being discontinued due to concerns about its side effect profile. Even though rimonabant cannot be used therapeutically, this drug is useful in experimental studies aiming to elucidate the role of the endocannabinoid system on binge eating behaviors. Rimonabant reduces binge eating in rodents.[224] Future work is needed to determine if cannabinoid receptors are involved solely in eating-related disorders or if they mediate other behavioral addictions.

Incentive sensitization theory revisited

Originally developed by Robinson and Berridge[225] to explain SUDs, the incentive sensitization theory has been applied to behavioral addictions.[226] Recall that the incentive sensitization theory differentiates "liking" from "wanting". Liking refers to the pleasure one experiences when engaging in an activity such as gambling or eating a palatable food, particularly during the early stages of engaging in the activity (e.g., gambling for the first time). Wanting is more akin to negative reinforcement. Individuals perform the activity to alleviate anxiety. Here, the individual continues engaging in an activity even if they no longer enjoy it. At a behavioral level, wanting is an important mechanism that drives addictions.

The incentive sensitization theory also argues that the reward pathway in the brain becomes overly responsive to a particular activity and the stimuli associated with that activity. Individuals with gambling disorder show greater drug-induced increases in striatal dopamine release.[227] Animal models have also shown that chronic exposure to probabilistic reinforcement, similar to what is observed in gambling, leads to a sensitized dopaminergic system.[228] Specifically, rats trained in a variable ratio schedule of reinforcement show greater amphetamine-induced locomotor activity and have more dopamine D_2-like receptors in the dorsal striatum[228] (Fig. 13.2).

A recent hypothesis has been proposed to explain why gambling disorder and SUDs are often observed together.[229] This hypothesis posits that chronic exposure to unpredictable reinforcement leads to a sensitized reward system which then increases responses to drugs and to stress (Fig. 13.3). Recall that anxiety disorders are often observed in individuals with a SUD and with a behavioral addiction. The evidence for cross sensitization between gambling and substance use has already been covered in this chapter as both humans and animals exposed to gambling-like conditions show enhanced dopaminergic responses to drugs. You have learned previously that acute drug use activates the stress pathway whereas

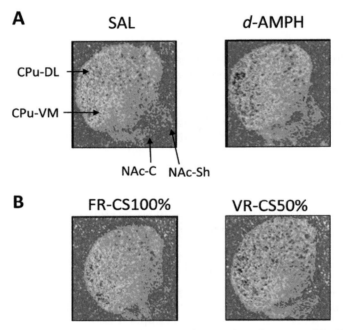

FIGURE 13.2 Dopamine D_2-like receptor distribution in the dorsolateral striatum (CPu-DL), the ventromedial striatum (CPu-VM), the nucleus accumbens core (NAc-C), and the nucleus accumbens shell (NAc-Sh) in rats exposed to one of four conditions. In (A), rats were chronically treated with either saline (SAL) or d-amphetamine (d-AMPH). In (B), rats were trained to respond for reinforcement that was paired with a conditioned stimulus (CS). Some rats responded according to a fixed ratio schedule in which presentation of the CS always predicted delivery of reinforcement (FR-CS100%). Another group of rats responded according to a variable ratio schedule of reinforcement in which the CS predicted reinforcement delivery 50% of the time (VR-CS50%). Rats trained on the VR-CS50% schedule had elevated D_2-like receptor density in the dorsal striatum, which mimicked the increased receptor density observed in rats chronically treated with d-AMPH. *Image comes from Fig. 5 of Fugariu V, Zack MH, Nobrega JN, Fletcher PJ, Zeeb FD. Effects of exposure to chronic uncertainty and a sensitizing regimen of amphetamine injections on locomotion, decision-making, and dopamine receptors in rats.* Neuropsychopharmacology. 2020;45(5):811−822. https://doi.org/10.1038/s41386-020-0599-x. Copyright 2020.

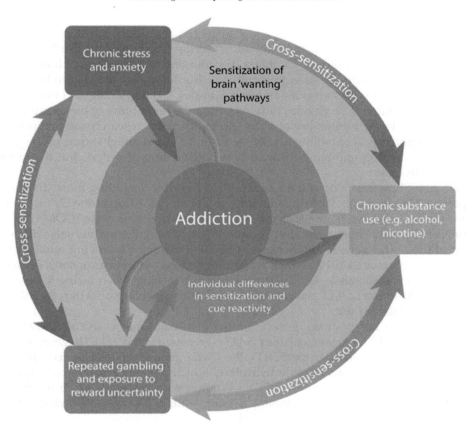

FIGURE 13.3 Proposed framework for cross sensitization that occurs between gambling-like conditions, substance use, and stress. *Image comes from Fig. 1 of Hellberg SN, Russell TI, Robinson M. Cued for risk: evidence for an incentive sensitization framework to explain the interplay between stress and anxiety, substance abuse, and reward uncertainty in disordered gambling behavior. Cognit Affect Behav Neurosci. 2019;19(3):737–758. https://doi.org/10.3758/s13415-018-00662-3. Copyright 2019.*

chronic drug use blunts the activity of the stress pathway in response to drugs. Likewise, regular gamblers experience increased heart rate and elevated salivary cortisol levels while gambling, but pathological gamblers experience no increases in cortisol during gambling.[230] Explaining this effect is somewhat difficult as discrepant findings have been observed concerning baseline cortisol levels in pathological gamblers. There is some evidence that individuals with gambling disorder have higher baseline cortisol levels,[12] but other studies have observed lower baseline cortisol levels in these individuals.[231] Additionally, in individuals with gambling disorder, ventral striatum activation in response to monetary rewards is positively correlated with cortisol levels.[232] While Hellberg et al.[229] focused on gambling, binge eating may lead to cross sensitization of the stress and the reward systems as an overactive stress system is observed in individuals with binge eating disorders.[150]

Genetic associations

There is some evidence that behavioral addictions are heritable just like SUDs, although studies have primarily focused on gambling disorder and binge eating/bulimia. Compared to control individuals, individuals with gambling disorder are more likely to have an immediate family member that also has a gambling disorder.[84] Binge eating and bulimia nervosa are also highly heritable,[233] with one study reporting the heritability of binge eating disorder to be 57%.[234] Due to the heritability of behavioral addictions, research has examined the genetic links of pathological gambling and binge eating. I will highlight some of the genes that have been implicated in gambling disorder and binge eating disorders, but by no means is this an exhaustive list.

Specific genes have been implicated in gambling disorder, including those that encode for the dopamine D_1 receptor,[235] the dopamine D_2 receptor,[236] the dopamine D_4 receptor,[237] the serotonin 5-HT_{2A} receptor,[238] the enzyme catechol-O-methyltransferase (COMT),[239] and the α_4 subunit of the acetylcholine nicotinic receptor.[240] Additionally, individuals with a specific variation of the *DRD4* gene show increased gambling behavior following administration of the dopamine precursor L-DOPA, which is commonly used to treat Parkinson's disease.[241] Individuals with a specific allele for SERT are more risk seeking when facing losses.[242] While genetic variations in the *COMT* gene are associated with gambling disorder, they do not predict problematic internet use.[78]

Binge eating also has a strong genetic basis.[243] Polymorphisms in the *DRD2* gene, the *COMT* gene, the *HTR1A* gene (encodes 5-HT_{1A} receptor), the *HTR1B* gene, the *HTR2A* gene, the *SLC6A4* gene (encodes for SERT), the *CNR1* gene (encodes CB_1 receptor), and the *FAAH* gene (encodes fatty acid amide hydrolase, which metabolizes cannabinoids) are also associated with binge eating disorder.[244–250] Women with a specific allele for the *SLC6A4* gene are at increased risk of developing a binge eating disorder following adverse life events.[251] Although Akkermann et al. failed to detect a relationship between the *SLC6A4* gene and binge eating, they found that a polymorphism in this gene is associated with more severe symptoms in those with an eating disorder.[252]

Treatment interventions for behavioral addictions

As behavioral addictions share many of the same features as SUDs, the treatment interventions you have previously learned about are used to combat behavioral addictions. Cognitive-behavioral therapy (CBT) is widely used in the treatment of behavioral addictions, including gambling disorder,[253] binge eating disorders,[254,257] compulsive buying,[256] internet/social media addiction,[129,257] and gaming disorder.[258] One advantage of CBT is that it can be easily included with other treatment approaches like pharmacotherapy,[259,260] motivational enhancement therapy (MET),[261] virtual reality,[262,265] music therapy,[264] and nutritional counseling.[265] Another advantage of CBT is that it can be successfully implemented with individuals with a comorbid condition like schizophrenia.[266] One benefit of CBT, especially when combined with MET, is that it can reduce physiological arousal to cues paired with gambling.[267] Receiving CBT or pharmacotherapy for binge eating disorder decreases impulsivity, which remains lower at follow-up; the reductions in impulsivity are associated with improvements in binge eating and depression.[268]

In addition to CBT, mindfulness training has received some support for the treatment of behavioral addictions.[269] Mindfulness treatment decreases problematic gambling behavior,[253] binge eating,[270] internet/smartphone addiction,[139,271] and gaming disorder.[272] Directly related to binge eating, individuals that engage in more mindful eating practices are less likely to binge eat.[13] As noted by Omiwole et al., mindfulness-based practices for the treatment of problematic eating behaviors in adolescents are still in its infancy[273]; thus, more work is needed to determine the effectiveness of mindfulness training in this population. One disadvantage to mindfulness training is that long-term abstinence is not always observed.[274] However, mindfulness-based practices may be useful in reducing problematic behaviors early during an intervention such as CBT.

Because conditioned stimuli promote problematic behaviors like pathological gambling or binge eating, treatment interventions need to address the heightened attention individuals place on these stimuli. Individuals in treatment for gambling disorder show less attentional bias for gambling-paired stimuli.[90] One study examined the role of bias modification training in a group of individuals with binge eating disorder.[275] Participants were assigned to one of two conditions. In one condition, participants were instructed to attend to food stimuli. In another condition, participants were instructed to avoid food stimuli. Individuals that were instructed to avoid food stimuli showed reduced attentional bias to food stimuli, but this effect was transient; the intervention also failed to decrease food craving. A separate clinical-based study used an attentional bias modification program that lasted for 8 weeks.[276] The results from this trial were more promising, as participants decreased the frequency of binge episodes, even after 3 months of discontinuing the program and despite no changes in food cravings. Interpreting the results of this study is difficult as Boutelle et al. did not include a control group, and they experienced high attrition rates throughout the course of the experiment.[276]

Another intervention that can be used to ameliorate the impact of conditioned stimuli on craving is cue exposure and response prevention treatment (i.e., extinction training; see Chapter 6).[255] Although individual studies have found beneficial effects of cue exposure therapies for binge eating disorders,[255] one literature review indicated that this therapy provides little advantages over CBT,[277] and another review noted that most studies examining exposure therapy for binge eating have been underpowered and have been methodologically weak.[278] Another problem is that combining cue exposure and prevention treatment with CBT *worsens* treatment outcomes in those with gambling disorder.[279] Cue exposure therapies can be combined with virtual reality therapy to produce greater long-term abstinence, with one clinical trial reporting a 70% abstinence rate for binge eating in individuals given virtual reality/exposure therapy compared to 26% given CBT.[280]

One intervention that has been used primarily for binge eating disorders is exercise. In Chapter 11 you learned that exercise can be used to treat SUDs. There is some evidence that aerobic exercise and yoga reduce the number of binge episodes and reduce depressive symptoms.[281,284] One study found that exercise decreases participants' self-reported wanting of high-fat foods without altering their liking of such foods, which also results in a reduction in binge eating.[283] Exercise combined with dietary therapy is well received by patients[284] and quickly improves individuals' self-reported eating when compared to CBT, but this beneficial effect disappears at treatment follow-up,[285] suggesting that exercise therapy should be used as an adjunct treatment.

Another treatment that has been used nearly exclusively for binge eating disorders is **brief strategic therapy (BST)**, which has been proposed as an alternative to CBT. Previously used to treat phobic disorders, BST does not focus on cognitive distortions that drive psychological disorders; instead, BST operates under the assumption that disorders are maintained because of faulty attempted solutions for controlling the disorder. For example, an individual with a binge eating disorder may try to abstain from certain food by implementing a strict diet. Anyone that has ever attempted to diet knows how challenging it can be. After a while, cravings can intensify, leading to relapse. This is the same challenge individuals attempting to abstain from drug use face. The first stage of BST is understanding the relationship between perceptions of potential solutions and reactions to the attempted solutions (i.e., dieting leads to increased hunger which can lead to compensatory bingeing). The second stage of treatment is to modify how one thinks about the solutions that they have attempted in the past. During this stage, an individual that has relied on dieting will learn that their dieting is what increases future vulnerability to a binge episode. During the third and fourth stages, individuals evaluate effective strategies and adopt a healthier relationship with food.[286] BST appears to provide better long-term benefits to binge eating disorders compared to CBT.[286] Overall, BST appears to be an effective strategy for reducing addictive behaviors, but work is needed to determine its effectiveness in other addictions like gambling disorder.

Less commonly used interventions for behavioral addictions include MET administered alone,[261,287] twelve (12)-step facilitated treatments,[288] and virtual reality administered alone.[258] While MET appears to be effective in individuals with gambling disorder,[261] this intervention does not reduce binge eating; instead, MET is effective in encouraging those with binge eating disorder to change their behaviors.[287] The effectiveness of 12-step programs like Gambler's Anonymous has been inconsistent.[289]

Multiple variables weaken the effectiveness of treatment interventions for behavioral addictions, such as low education levels,[290,291] race,[291] reward sensitivity,[292] impaired cognitive flexibility,[293] poor response inhibition,[294] harm avoidance,[295] lack of perseverance,[98] and impulsivity,[86] specifically sensation seeking and negative urgency.[86,293] Stress is a major risk factor for relapse during treatment and following treatment for a behavioral addiction. Indeed, individuals seeking treatment for a binge eating disorder report that stress is a primary trigger for a binge and for causing relapse,[296] and individuals with PTSD have higher treatment dropout rates.[297] In women, being divorced is a predictor of gambling relapse.[290] Related to stress, low SES is a predictor of poor treatment outcomes following therapy.[82]

Quiz 13.2

(1) Which behavioral addiction is primarily observed in women?
  a. Gambling disorder
  b. Gaming disorder
  c. Internet addiction
  d. Social media addiction
(2) Which therapy is often considered the "gold standard" for behavioral addictions?
  a. Brief strategic therapy

b. Cognitive-behavioral therapy
   c. Mindfulness-based therapy
   d. Motivational enhancement therapy
(3) Which of the following medical disorders has strengthened the argument that dopamine is an important mediator of gambling disorder?
   a. Alzheimer's disease
   b. Bipolar disorder
   c. Parkinson's disease
   d. Schizophrenia
(4) Which of the following personality traits is NOT linked to behavioral addictions?
   a. Harm avoidance
   b. Impulsivity
   c. Introversion
   d. Neuroticism
(5) Which parenting style is most closely linked to behavioral addictions?
   a. Authoritarian
   b. Authoritative
   c. Permissive
   d. Uninvolved

Answers to quiz 13.2

(1) d — Social media addiction
(2) b — Cognitive-behavioral therapy
(3) c — Parkinson's disease
(4) c — Introversion
(5) a — Authoritarian

Chapter summary

There is strong evidence that the term addiction can be extended to behaviors beyond substance use. In particular, gambling disorder and binge eating disorders share many of the same features as SUDs. Positive reinforcement is important for the initiation of activities like gambling or gaming, and stimuli can become paired with these activities, which can then elicit strong cravings when the individual is not currently engaged in the activity. Individuals with a behavioral addiction show many of the same deficits in decision making and memory as what has been observed in those with a SUD. SUDs and behavioral addictions share many of the same risk factors, including harm avoidance, neuroticism, and impulsivity; having authoritarian parents and having family members and friends that engage in the addictive behavior; coming from a disadvantaged background; and stress. Behavioral addictions and SUDs also share many of the same underlying neurobiological mechanisms, as a dysfunctional mesocorticolimbic pathway is observed in conditions like gambling disorder and binge eating disorders. While dopamine and serotonin have received the most attention, other neurotransmitter systems mediate behavioral addictions. More work is needed to further elucidate the neurobiological, behavioral, cognitive, and sociocultural factors involved in behavioral addictions. As we learn more about behaviors such as compulsive social media use or excessive online shopping, we will be better equipped to treat individuals that find themselves engaging in these behaviors.

Glossary

Behavioral addiction — a psychological disorder in which an individual has a compulsion to engage in a nonsubstance-related behavior despite aversive outcomes to one's physical and/or mental well-being.

Binge eating disorder — a psychological disorder characterized by periods of excessive eating that cause distress and negatively impact one's physical and/or mental well-being.

Brief strategic therapy (BST) — a form of treatment most associated with binge eating disorders that does not focus on cognitive distortions that drive psychological disorders, but instead operates under the assumption that disorders are maintained because of faulty attempted solutions for controlling the disorder.

Bulimia nervosa — a psychological disorder characterized by periods of excessive eating followed by purges that cause distress and negatively impact one's physical and/or mental well-being.

Chasing losses — increased gambling frequency that occurs following a string of losses in an effort to recoup the lost money (also called **loss chasing**).

Gambler's fallacy — a cognitive bias in which an individual believes that a probabilistic outcome that has not occurred frequently in the past is now more likely to occur even though the probability of the outcome never changes (e.g., "I have not won the lottery the past 10 times I have played. I'm due to win soon").

Overconfidence — a cognitive bias characterized by excessive confidence in one's cognitive and/or physical abilities.

Thin ideal — concept of an ideally slim female body.

Supplementary data related to this chapter can be found online at https://educate.elsevier.com/9780323905787.

References

1. American Psychiatric Association. *Diagnostic and Statistical Manual of Mental Disorders*. 5th ed. 2013. https://doi.org/10.1176/appi.books.9780890425596.
2. World Health Organization. *International Statistical Classification of Diseases and Related Health Problems*. 11th ed.; 2019. https://icd.who.int/.
3. Pinna F, Dell'Osso B, Nicola MD, et al. Behavioural addictions and the transition from DSM-IV-TR to DSM-5. *J Psychopathol*. 2015;21:380—389.
4. Jiménez-Murcia S, Del Pino-Gutiérrez A, Fernández-Aranda F, et al. Treatment outcome in male gambling disorder patients associated with alcohol use. *Front Psychol*. 2016;7:465. https://doi.org/10.3389/fpsyg.2016.00465.
5. Grant JE. *Impulse Control Disorders: a Clinician's Guide to Understanding and Treating Behavioral Addictions*. Norton Press; 2008.
6. Petry NM, Stinson FS, Grant BF. Comorbidity of DSM-IV pathological gambling and other psychiatric disorders: results from the national epidemiologic survey on alcohol and related conditions. *J Clin Psychiatr*. 2005;66(5):564—574. https://doi.org/10.4088/jcp.v66n0504.
7. Tang K, Kim HS, Hodgins DC, McGrath DS, Tavares H. Gambling disorder and comorbid behavioral addictions: demographic, clinical, and personality correlates. *Psychiatr Res*. 2020;284:112763. https://doi.org/10.1016/j.psychres.2020.112763.
8. Gat-Lazer S, Geva R, Gur E, Stein D. Reward dependence and harm avoidance among patients with binge-purge type eating disorders. *Eur Eat Disord Rev*. 2017;25(3):205—213. https://doi.org/10.1002/erv.2505.
9. Yip SW, White MA, Grilo CM, Potenza MN. An exploratory study of clinical measures associated with subsyndromal athological gambling in patients with binge eating disorder. *J Gambl Stud*. 2011;27(2):257—270. https://doi.org/10.1007/s10899-010-9207-z.

10. Jiménez-Murcia S, Steiger H, Israël M, et al. Pathological gambling in eating disorders: prevalence and clinical implications. *Compr Psychiatr*. 2013;54(7):1053–1060. https://doi.org/10.1016/j.comppsych.2013.04.014.
11. Odlaug BL, Lust K, Schreiber LR, et al. Compulsive sexual behavior in young adults. *Ann Clin Psychiatr*. 2013;25(3):193–200.
12. Wohl MJA, Matheson K, Young MM, Anisman H. Cortisol rise following awakening among problem gamblers: dissociation from comorbid symptoms of depression and impulsivity. *J Gambl Stud*. 2008;24(1):79–90. https://doi.org/10.1007/s10899-007-9080-6.
13. Giannopoulou I, Kotopoulea-Nikolaidi M, Daskou S, Martyn K, Patel A. Mindfulness in eating is inversely related to binge eating and mood disturbances in university students in health-related disciplines. *Nutrients*. 2020;12(2):396. https://doi.org/10.3390/nu12020396.
14. Hussain Z, Griffiths MD. Problematic social networking site use and comorbid psychiatric disorders: a systematic review of recent large-scale studies. *Front Psychiatr*. 2018;9:686. https://doi.org/10.3389/fpsyt.2018.00686.
15. Jo YS, Bhang SY, Choi JS, Lee HK, Lee SY, Kweon YS. Clinical characteristics of diagnosis for internet gaming disorder: comparison of DSM-5 IGD and ICD-11 GD diagnosis. *J Clin Med*. 2019;8(7):945. https://doi.org/10.3390/jcm8070945.
16. Meyer M, Sattler I, Schilling H, et al. Mental disorders in individuals with exercise addiction-a cross-sectional study. *Front Psychiatr*. 2021;12:751550. https://doi.org/10.3389/fpsyt.2021.751550.
17. Schou Andreassen C, Billieux J, Griffiths MD, et al. The relationship between addictive use of social media and video games and symptoms of psychiatric disorders: a large-scale cross-sectional study. *Psychol Addict Behav*. 2016;30(2):252–262. https://doi.org/10.1037/adb0000160.
18. Vadlin S, Åslund C, Hellström C, Nilsson KW. Associations between problematic gaming and psychiatric symptoms among adolescents in two samples. *Addict Behav*. 2016;61:8–15. https://doi.org/10.1016/j.addbeh.2016.05.001.
19. Zlot Y, Goldstein M, Cohen K, Weinstein A. Online dating is associated with sex addiction and social anxiety. *J Behav Addic*. 2018;7(3):821–826. https://doi.org/10.1556/2006.7.2018.66.
20. Weinstein A, Maayan G, Weinstein Y. A study on the relationship between compulsive exercise, depression and anxiety. *J Behav Addic*. 2015;4(4):315–318. https://doi.org/10.1556/2006.4.2015.034.
21. Granero R, Fernández-Aranda F, Pino-Gutierrez AD, et al. The prevalence and features of schizophrenia among individuals with gambling disorder. *J Psychiatr Res*. 2021;136:374–383. https://doi.org/10.1016/j.jpsychires.2021.02.025.
22. Kouidrat Y, Amad A, Lalau JD, Loas G. Eating disorders in schizophrenia: implications for research and management. *Schizophr Res & Treat*. 2014;2014:791573. https://doi.org/10.1155/2014/791573.
23. Lee JY, Chung YC, Song JH, et al. Contribution of stress and coping strategies to problematic Internet use in patients with schizophrenia spectrum disorders. *Compr Psychiatr*. 2018;87:89–94. https://doi.org/10.1016/j.comppsych.2018.09.007.
24. de Beaurepaire R. Binge eating disorders in antipsychotic-treated patients with schizophrenia: prevalence, antipsychotic specificities, and changes over time. *J Clin Psychopharmacol*. 2021;41(2):114–120. https://doi.org/10.1097/JCP.0000000000001357.
25. Retz W, Ringling J, Retz-Junginger P, Vogelgesang M, Rösler M. Association of attention-deficit/hyperactivity disorder with gambling disorder. *J Neural Transm*. 2016;123(8):1013–1019. https://doi.org/10.1007/s00702-016-1566-x.
26. Aymamí N, Jiménez-Murcia S, Granero R, et al. Clinical, psychopathological, and personality characteristics associated with ADHD among individuals seeking treatment for gambling disorder. *BioMed Res Int*. 2015;2015:965303. https://doi.org/10.1155/2015/965303.
27. Nickel K, Maier S, Endres D, et al. Systematic review: overlap between eating, autism spectrum, and attention-deficit/hyperactivity disorder. *Front Psychiatr*. 2019;10:708. https://doi.org/10.3389/fpsyt.2019.00708.
28. Halevy-Yosef R, Bachar E, Shalev L, et al. The complexity of the interaction between binge-eating and attention. *PLoS One*. 2019;14(4):e0215506. https://doi.org/10.1371/journal.pone.0215506.
29. El Ayoubi H, Barrault S, Gateau A, et al. Adult attention-deficit/hyperactivity disorder among alcohol use disorder inpatients is associated with food addiction and binge eating, but not BMI. *Appetite*. 2022;168:105665. https://doi.org/10.1016/j.appet.2021.105665.
30. Wang BQ, Yao NQ, Zhou X, Liu J, Lv ZT. The association between attention deficit/hyperactivity disorder and internet addiction: a systematic review and meta-analysis. *BMC Psychiatr*. 2017;17(1):260. https://doi.org/10.1186/s12888-017-1408-x.

31. Schou Andreassen C, Griffiths MD, Sinha R, Hetland J, Pallesen S. The relationships between workaholism and symptoms of psychiatric disorders: a large-scale cross-sectional study. *PLoS One.* 2016;11(5):e0152978. https://doi.org/10.1371/journal.pone.0152978.
32. Ramesh Masthi NR, Pruthvi S, Phaneendra MS. A comparative study on social media usage and health status among students studying in pre-university colleges of urban Bengaluru. *Indian J Community Med.* 2018;43(3):180–184. https://doi.org/10.4103/ijcm.IJCM_285_17.
33. Sussman CJ, Harper JM, Stahl JL, Weigle P. Internet and video game addictions: diagnosis, epidemiology, and neurobiology. *Child & Adolesc Psychiatr Clin North Am.* 2018;27(2):307–326. https://doi.org/10.1016/j.chc.2017.11.015.
34. Brevers D, Koritzky G, Bechara A, Noël X. Cognitive processes underlying impaired decision-making under uncertainty in gambling disorder. *Addict Behav.* 2014;39(10):1533–1536. https://doi.org/10.1016/j.addbeh.2014.06.004.
35. Dong G, Huang J, Du X. Enhanced reward sensitivity and decreased loss sensitivity in Internet addicts: an fMRI study during a guessing task. *J Psychiatr Res.* 2011;45(11):1525–1529. https://doi.org/10.1016/j.jpsychires.2011.06.017.
36. Eichen DM, Chen EY, Schmitz MF, Arlt J, McCloskey MS. Addiction vulnerability and binge eating in women: exploring reward sensitivity, affect regulation, impulsivity & weight/shape concerns. *Pers Indiv Differ.* 2016;100:16–22. https://doi.org/10.1016/j.paid.2016.03.084.
37. Eneva KT, Murray S, O'Garro-Moore J, et al. Reward and punishment sensitivity and disordered eating behaviors in men and women. *J. Eating Disord.* 2017;5:6. https://doi.org/10.1186/s40337-017-0138-2.
38. Fauth-Bühler M, Mann K. Neurobiological correlates of internet gaming disorder: similarities to pathological gambling. *Addict Behav.* 2017;64:349–356. https://doi.org/10.1016/j.addbeh.2015.11.004.
39. Fussner LM, Luebbe AM, Smith AR. Social reward and social punishment sensitivity in relation to dietary restraint and binge/purge symptoms. *Appetite.* 2018;127:386–392. https://doi.org/10.1016/j.appet.2018.05.133.
40. Schienle A, Schäfer A, Hermann A, Vaitl D. Binge-eating disorder: reward sensitivity and brain activation to images of food. *Biol Psychiatr.* 2009;65(8):654–661. https://doi.org/10.1016/j.biopsych.2008.09.028.
41. Sescousse G, Barbalat G, Domenech P, Dreher JC. Imbalance in the sensitivity to different types of rewards in pathological gambling. *Brain.* 2013;136(Pt 8):2527–2538. https://doi.org/10.1093/brain/awt126.
42. Ojala KE, Janssen LK, Hashemi MM, et al. Dopaminergic drug effects on probability weighting during risky decision making. *eNeuro.* 2018;5(2). https://doi.org/10.1523/ENEURO.0330-18.2018.
43. Clark L, Lawrence AJ, Astley-Jones F, Gray N. Gambling near-misses enhance motivation to gamble and recruit win-related brain circuitry. *Neuron.* 2009;61(3):481–490. https://doi.org/10.1016/j.neuron.2008.12.031.
44. Limbrick-Oldfield EH, Mick I, Cocks RE, et al. Neural substrates of cue reactivity and craving in gambling disorder. *Transl Psychiatr.* 2017;7(1):e992. https://doi.org/10.1038/tp.2016.256.
45. Brevers D, Cleeremans A, Bechara A, et al. Time course of attentional bias for gambling information in problem gambling. *Psychol Addict Behav.* 2011;25(4):675–682. https://doi.org/10.1037/a0024201.
46. Brevers D, Cleeremans A, Tibboel H, et al. Reduced attentional blink for gambling-related stimuli in problem gamblers. *J Behav Ther Exp Psychiatr.* 2011;42(3):265–269. https://doi.org/10.1016/j.jbtep.2011.01.005.
47. Ciccarelli M, Nigro G, Griffiths MD, Cosenza M, D'Olimpio F. Attentional bias in non-problem gamblers, problem gamblers, and abstinent pathological gamblers: an experimental study. *J Affect Disord.* 2016;206:9–16. https://doi.org/10.1016/j.jad.2016.07.017.
48. Hønsi A, Mentzoni RA, Molde H, Pallesen S. Attentional bias in problem gambling: a systematic review. *J Gambl Stud.* 2013;29(3):359–375. https://doi.org/10.1007/s10899-012-9315-z.
49. McGrath DS, Meitner A, Sears CR. The specificity of attentional biases by type of gambling: an eye-tracking study. *PLoS One.* 2018;13(1):e0190614. https://doi.org/10.1371/journal.pone.0190614.
50. van Holst RJ, Lemmens JS, Valkenburg PM, Peter J, Veltman DJ, Goudriaan AE. Attentional bias and disinhibition toward gaming cues are related to problem gaming in male adolescents. *J Adolesc Health.* 2012;50(6):541–546. https://doi.org/10.1016/j.jadohealth.2011.07.006.
51. Vizcaino EJV, Fernandez-Navarro P, Blanco C, et al. Maintenance of attention and pathological gambling. *Psychol Addict Behav.* 2013;27(3):861–867. https://doi.org/10.1037/a0032656.
52. Anselme P, Robinson M. From sign-tracking to attentional bias: implications for gambling and substance use disorders. *Prog Neuro Psychopharmacol Biol Psychiatr.* 2020;99:109861. https://doi.org/10.1016/j.pnpbp.2020.109861.

53. Seo CL, Lee JH. Attentional bias to high-calorie food in binge eaters with high shape/weight concern. *Front Psychiatr.* 2021;12:606296. https://doi.org/10.3389/fpsyt.2021.606296.
54. Stojek M, Shank LM, Vannucci A, et al. A systematic review of attentional biases in disorders involving binge eating. *Appetite.* 2018;123:367–389. https://doi.org/10.1016/j.appet.2018.01.019.
55. Astur RS, Palmisano AN, Hudd EC, et al. Pavlovian conditioning to food reward as a function of eating disorder risk. *Behav Brain Res.* 2015;291:277–282. https://doi.org/10.1016/j.bbr.2015.05.016.
56. Fortune EE, Goodie AS. Cognitive distortions as a component and treatment focus of pathological gambling: a review. *Psychol Addict Behav.* 2012;26(2):298–310. https://doi.org/10.1037/a0026422.
57. Goodie AS. The effects of control on betting: paradoxical betting on items of high confidence with low value. *J Exp Psychol Learn Mem Cogn.* 2003;29(4):598–610. https://doi.org/10.1037/0278-7393.29.4.598.
58. Goodie AS. The role of perceived control and overconfidence in pathological gambling. *J Gambl Stud.* 2005;21(4):481–502. https://doi.org/10.1007/s10899-005-5559-1.
59. Tversky A, Kahneman D. Belief in the law of small numbers. *Psychol Bull.* 1971;76(2):105–110. https://doi.org/10.1037/h0031322.
60. Kim BM, Lee J, Choi AR, et al. Event-related brain response to visual cues in individuals with Internet gaming disorder: relevance to attentional bias and decision-making. *Transl Psychiatr.* 2021;11(1):258. https://doi.org/10.1038/s41398-021-01375-x.
61. Ledgerwood DM, Orr ES, Kaploun KA, et al. Executive function in pathological gamblers and healthy controls. *J Gambl Stud.* 2012;28(1):89–103. https://doi.org/10.1007/s10899-010-9237-6.
62. Lee JE, Namkoong K, Jung YC. Impaired prefrontal cognitive control over interference by food images in binge-eating disorder and bulimia nervosa. *Neurosci Lett.* 2017;651:95–101. https://doi.org/10.1016/j.neulet.2017.04.054.
63. Izquierdo A, Jentsch JD. Reversal learning as a measure of impulsive and compulsive behavior in addictions. *Psychopharmacology.* 2012;219(2):607–620. https://doi.org/10.1007/s00213-011-2579-7.
64. Grant DA, Berg E. A behavioral analysis of degree of reinforcement and ease of shifting to new responses in a Weigl-type card-sorting problem. *J Experiment Psychol.* 1948;38(4):404–411. https://doi.org/10.1037/h0059831.
65. Eneva KT, Arlt JM, Yiu A, Murray SM, Chen EY. Assessment of executive functioning in binge-eating disorder independent of weight status. *Int J Eat Disord.* 2017;50(8):942–951. https://doi.org/10.1002/eat.22738.
66. Albein-Urios N, Martinez-González JM, Lozano Ó, Verdejo-Garcia A. Monetary delay discounting in gambling and cocaine dependence with personality comorbidities. *Addict Behav.* 2014;39(11):1658–1662. https://doi.org/10.1016/j.addbeh.2014.06.001.
67. Michalczuk R, Bowden-Jones H, Verdejo-Garcia A, Clark L. Impulsivity and cognitive distortions in pathological gamblers attending the UK National Problem Gambling Clinic: a preliminary report. *Psychol Med.* 2011;41(12):2625–2635. https://doi.org/10.1017/S003329171100095X.
68. Carr MM, Wiedemann AA, Macdonald-Gagnon G, Potenza MN. Impulsivity and compulsivity in binge eating disorder: a systematic review of behavioral studies. *Prog Neuro Psychopharmacol Biol Psychiatr.* 2021;110:110318. https://doi.org/10.1016/j.pnpbp.2021.110318.
69. Cheng Y-S, Ko H-C, Sun C-K, Yeh P-Y. The relationship between delay discounting and Internet addiction: a systematic review and meta-analysis. *Addict Behav.* 2021;114:106751. https://doi.org/10.1016/j.addbeh.2020.106751.
70. Ciccarelli M, Cosenza M, D'Olimpio F, Griffiths MD, Nigro G. An experimental investigation of the role of delay discounting and craving in gambling chasing behavior. *Addict Behav.* 2019;93:250–256. https://doi.org/10.1016/j.addbeh.2019.02.002.
71. Vickers SP, Goddard S, Brammer RJ, Hutson PH, Heal DJ. Investigation of impulsivity in binge-eating rats in a delay-discounting task and its prevention by the d-amphetamine prodrug, lisdexamfetamine. *J Psychopharmacol.* 2017;31(6):784–797. https://doi.org/10.1177/0269881117691672.
72. Cano AM, Murphy ES, Lupfer G. Delay discounting predicts binge-eating in Wistar rats. *Behav Process.* 2016;132:1–4. https://doi.org/10.1016/j.beproc.2016.08.011.
73. Moore CF, Blasio A, Sabino V, Cottone P. Impulsive choice does not predict binge-like eating in rats. *Behav Pharmacol.* 2018;29(8):726–731. https://doi.org/10.1097/FBP.0000000000000446.
74. Kyonka E, Schutte NS. Probability discounting and gambling: a meta-analysis. *Addiction.* 2018;113(12):2173–2181. https://doi.org/10.1111/add.14397.
75. Ciccarelli M, Malinconico R, Griffiths MD, Nigro G, Cosenza M. Reward preferences of pathological gamblers under conditions of uncertainty: an experimental study. *J Gambl Stud.* 2016;32(4):1175–1189. https://doi.org/10.1007/s10899-016-9593-y.

76. Leslie M, Leppanen J, Paloyelis Y, Nazar BP, Treasure J. The influence of oxytocin on risk-taking in the balloon analogue risk task among women with bulimia nervosa and binge eating disorder. *J Neuroendocrinol*. 2019;31(8):e12771. https://doi.org/10.1111/jne.12771.
77. Neveu R, Fouragnan E, Barsumian F, et al. Preference for safe over risky options in binge eating. *Front Behav Neurosci*. 2016;10:65. https://doi.org/10.3389/fnbeh.2016.00065.
78. Ioannidis K, Redden SA, Valle S, Chamberlain SR, Grant JE. Problematic internet use: an exploration of associations between cognition and COMT rs4818, rs4680 haplotypes. *CNS Spectr*. 2020;25(3):409–418. https://doi.org/10.1017/S1092852919001019.
79. Zhou Z, Zhou H, Zhu H. Working memory, executive function and impulsivity in Internet-addictive disorders: a comparison with pathological gambling. *Acta Neuropsychiatr*. 2016;28(2):92–100. https://doi.org/10.1017/neu.2015.54.
80. Eneva KT, Murray SM, Chen EY. Binge-eating disorder may be distinguished by visuospatial memory deficits. *Eat Behav*. 2017;26:159–162. https://doi.org/10.1016/j.eatbeh.2017.04.001.
81. Cowie ME, Kim HS, Hodgins DC, McGrath DS, Scanavino M, Tavares H. Demographic and psychiatric correlates of compulsive sexual behaviors in gambling disorder. *J Behav Addic*. 2019;8(3):451–462. https://doi.org/10.1556/2006.8.2019.35.
82. Granero R, Valero-Solis S, Fernández-Aranda F, et al. Response trajectories of gambling severity after cognitive behavioral therapy in young-adult pathological gamblers. *J Behav Addic*. 2020;9(1):140–152. https://doi.org/10.1556/2006.2020.00008.
83. Liao Z, Huang Q, Huang S, et al. Prevalence of internet gaming disorder and its association with personality traits and gaming characteristics among Chinese adolescent gamers. *Front Psychiatr*. 2020;11:598585. https://doi.org/10.3389/fpsyt.2020.598585.
84. Mann K, Lemenager T, Zois E, et al. Comorbidity, family history and personality traits in pathological gamblers compared with healthy controls. *Eur Psychiatr*. 2017;42:120–128. https://doi.org/10.1016/j.eurpsy.2016.12.002.
85. Peterson CB, Thuras P, Ackard DM, et al. Personality dimensions in bulimia nervosa, binge eating disorder, and obesity. *Compr Psychiatr*. 2010;51(1):31–36. https://doi.org/10.1016/j.comppsych.2009.03.003.
86. Ramos-Grille I, Gomà-i-Freixanet M, Aragay N, Valero S, Vallès V. Predicting treatment failure in pathological gambling: the role of personality traits. *Addict Behav*. 2015;43:54–59. https://doi.org/10.1016/j.addbeh.2014.12.010.
87. Barrault S, Varescon I. Online and live regular poker players: do they differ in impulsive sensation seeking and gambling practice? *J Behav Addic*. 2016;5(1):41–50. https://doi.org/10.1556/2006.5.2016.015.
88. Grant JE, Isobe M, Chamberlain SR. Abnormalities of striatal morphology in gambling disorder and at-risk gambling. *CNS Spectr*. 2019;24(6):609–615. https://doi.org/10.1017/S1092852918001645.
89. Greenberg NR, Zhai ZW, Hoff RA, Krishnan-Sarin S, Potenza MN. Problematic shopping and self-injurious behaviors in adolescents. *J Behav Addic*. 2020;9(4):1068–1078. https://doi.org/10.1556/2006.2020.00093.
90. Schmidt C, Gleesborg C, Schmidt H, et al. A bias towards natural rewards away from gambling cues in gamblers undergoing active treatment. *Brain Res*. 2021;1764:147479. https://doi.org/10.1016/j.brainres.2021.147479.
91. Steadman KM, Knouse LE. Is the relationship between ADHD symptoms and binge eating mediated by impulsivity? *J Atten Disord*. 2016;20(11):907–912. https://doi.org/10.1177/1087054714530779.
92. Kotbagi G, Morvan Y, Romo L, Kern L. Which dimensions of impulsivity are related to problematic practice of physical exercise? *J Behav Addic*. 2017;6(2):221–228. https://doi.org/10.1556/2006.6.2017.024.
93. Claes L, Islam MA, Fagundo AB, et al. The relationship between non-suicidal self-injury and the UPPS-P impulsivity facets in eating disorders and healthy controls. *PLoS One*. 2015;10(5):e0126083. https://doi.org/10.1371/journal.pone.0126083.
94. Maclaren VV, Fugelsang JA, Harrigan KA, Dixon MJ. The personality of pathological gamblers: a meta-analysis. *Clin Psychol Rev*. 2011;31(6):1057–1067. https://doi.org/10.1016/j.cpr.2011.02.002.
95. Rømer Thomsen K, Callesen MB, Hesse M, et al. Impulsivity traits and addiction-related behaviors in youth. *J Behav Addic*. 2018;7(2):317–330. https://doi.org/10.1556/2006.7.2018.22.
96. Savvidou LG, Fagundo AB, Fernández-Aranda F, et al. Is gambling disorder associated with impulsivity traits measured by the UPPS-P and is this association moderated by sex and age? *Compr Psychiatr*. 2017;72:106–113. https://doi.org/10.1016/j.comppsych.2016.10.005.
97. Shakeel MK, Hodgins DC, Goghari VM. A Comparison of self-reported impulsivity in gambling disorder and bipolar disorder. *J Gambl Stud*. 2019;35(1):339–350. https://doi.org/10.1007/s10899-018-9808-5.

98. Mallorquí-Bagué N, Vintró-Alcaraz C, Verdejo-García A, et al. Impulsivity and cognitive distortions in different clinical phenotypes of gambling disorder: profiles and longitudinal prediction of treatment outcomes. *Eur Psychiatr*. 2019;61:9–16. https://doi.org/10.1016/j.eurpsy.2019.06.006.
99. Chow CF, Cheung C, So L. Factors influencing gambling behavior among employees in Macau gambling industry. *J Gambl Stud*. 2022;38(1):87–121. https://doi.org/10.1007/s10899-021-10034-1.
100. Bonnaire C, Bungener C, Varescon I. Sensation seeking in a community sample of French gamblers: comparison between strategic and non-strategic gamblers. *Psychiatr Res*. 2017;250:1–9. https://doi.org/10.1016/j.psychres.2017.01.057.
101. Fortune EE, Goodie AS. The relationship between pathological gambling and sensation seeking: the role of subscale scores. *J Gambl Stud*. 2010;26(3):331–346. https://doi.org/10.1007/s10899-009-9162-8.
102. Lutter M, Tisch D, Beckert J. Social explanations of lottery play: new evidence based on national survey data. *J Gambl Stud*. 2018;34(4):1185–1203. https://doi.org/10.1007/s10899-018-9748-0.
103. Gupta R, Derevensky J. Familial and social influences on juvenile gambling behavior. *J Gambl Stud*. 1997;13(3):179–192. https://doi.org/10.1023/A:1024915231379.
104. Vachon J, Vitaro F, Wanner B, Tremblay RE. Adolescent gambling: relationships with parent gambling and parenting practices. *Psychol Addict Behav*. 2004;18(4):398–401. https://doi.org/10.1037/0893-164X.18.4.398.
105. Gay J, Gill PR, Corboy D. Parental and peer influences on emerging adult problem gambling: does exposure to problem gambling reduce stigmatizing perceptions and increase vulnerability? *J Gambl Issues*. 2016;33:30–51. https://doi.org/10.4309/jgi.2016.33.3.
106. Zhai ZW, Yip SW, Steinberg MA, et al. Relationships between perceived family gambling and peer gambling and adolescent problem gambling and binge-drinking. *J Gambl Stud*. 2017;33(4):1169–1185. https://doi.org/10.1007/s10899-017-9670-x.
107. Langhinrichsen-Rohling J, Rohde P, Seeley JR, Rohling ML. Individual, family, and peer correlates of adolescent gambling. *J Gambl Stud*. 2004;20(1):23–46. https://doi.org/10.1023/B:JOGS.0000016702.69068.53.
108. Hardoon KK, Gupta R, Derevensky JL. Psychosocial variables associated with adolescent gambling. *Psychol Addict Behav*. 2004;18(2):170–179. https://doi.org/10.1037/0893-164X.18.2.170.
109. Pike KM, Rodin J. Mothers, daughters, and disordered eating. *J Abnorm Psychol*. 1991;100(2):198–204. https://doi.org/10.1037//0021-843x.100.2.198.
110. Stice E. Modeling of eating pathology and social reinforcement of the thin-ideal predict onset of bulimic symptoms. *Behav Res Ther*. 1998;36(10):931–944. https://doi.org/10.1016/s0005-7967(98)00074-6.
111. Young EA, Clopton JR, Bleckley MK. Perfectionism, low self-esteem, and family factors as predictors of bulimic behavior. *Eat Behav*. 2004;5(4):273–283. https://doi.org/10.1016/j.eatbeh.2003.12.001.
112. Li S, Lei H, Tian L. A meta-analysis of the relationship between parenting style and Internet addiction among mainland Chinese teenagers. *SBP (Soc Behav Pers): Int J*. 2018;46(9):1475–1488. https://doi.org/10.2224/sbp.7631.
113. Lian L, You X, Huang J, Yang R. Who overuses Smartphones? Roles of virtues and parenting style in Smartphone addiction among Chinese college students. *Comput Hum Behav*. 2016;65:92–99. https://doi.org/10.1016/j.chb.2016.08.027.
114. Villalta L, Arévalo R, Valdepérez A, Pascual JC, de los Cobos JP. Parental bonding in subjects with pathological gambling disorder compared with healthy controls. *Psychiatr Q*. 2015;86(1):61–67. https://doi.org/10.1007/s11126-014-9336-0.
115. Amianto F, Martini M, Olandese F, et al. Affectionless control: a parenting style associated with obesity and binge eating disorder in adulthood. *Eur Eat Disord Rev*. 2021;29(2):178–192. https://doi.org/10.1002/erv.2809.
116. Welte JW, Barnes GM, Wieczorek WF, Tidwell MC, Parker J. Gambling participation in the U.S.–results from a national survey. *J Gambl Stud*. 2002;18(4):313–337. https://doi.org/10.1023/a:1021019915591.
117. Welte JW, Barnes GM, Wieczorek WF, Tidwell MC, Parker JC. Risk factors for pathological gambling. *Addict Behav*. 2004;29(2):323–335. https://doi.org/10.1016/j.addbeh.2003.08.007.
118. Pasternak AV, Fleming MF. Prevalence of gambling disorders in a primary care setting. *Arch Fam Med*. 1999;8(6):515–520. https://doi.org/10.1001/archfami.8.6.515, 4th.
119. Barnes GM, Welte JW, Tidwell MC, Hoffman JH. Gambling on the lottery: sociodemographic correlates across the lifespan. *J Gambl Stud*. 2011;27(4):575–586. https://doi.org/10.1007/s10899-010-9228-7.
120. West CE, Goldschmidt AB, Mason SM, Neumark-Sztainer D. Differences in risk factors for binge eating by socioeconomic status in a community-based sample of adolescents: findings from Project EAT. *Int J Eat Disord*. 2019;52(6):659–668. https://doi.org/10.1002/eat.23079.

121. Reagan P, Hersch J. Influence of race, gender, and socioeconomic status on binge eating frequency in a population-based sample. *Int J Eat Disord*. 2005;38(3):252−256. https://doi.org/10.1002/eat.20177.
122. Young EA, McFatter R, Clopton JR. Family functioning, peer influence, and media influence as predictors of bulimic behavior. *Eat Behav*. 2001;2(4):323−337. https://doi.org/10.1016/s1471-0153(01)00038-1.
123. Srivastava P, Felonis CR, Clancy OM, Wons OB, Abber SR, Juarascio AS. Real-time predictors of body dissatisfaction in females with binge eating: an ecological momentary assessment study. *Eat Weight Disord*. 2022;27(4):1547−1553. https://doi.org/10.1007/s40519-021-01296-0.
124. Alegria AA, Petry NM, Hasin DS, Liu SM, Grant BF, Blanco C. Disordered gambling among racial and ethnic groups in the US: results from the national epidemiologic survey on alcohol and related conditions. *CNS Spectr*. 2009;14(3):132−142. https://doi.org/10.1017/s1092852900020113.
125. Chamberlain SR, Leppink E, Redden SA, Odlaug BL, Grant JE. Racial-ethnic related clinical and neurocognitive differences in adults with gambling disorder. *Psychiatr Res*. 2016;242:82−87. https://doi.org/10.1016/j.psychres.2016.05.038.
126. Rinker DV, Rodriguez LM, Krieger H, Tackett JL, Neighbors C. Racial and ethnic differences in problem gambling among college students. *J Gambl Stud*. 2016;32(2):581−590. https://doi.org/10.1007/s10899-015-9563-9.
127. Medeiros GC, Leppink EW, Yaemi A, Mariani M, Tavares H, Grant JE. Electronic gaming machines and gambling disorder: a cross-cultural comparison between treatment-seeking subjects from Brazil and the United States. *Psychiatr Res*. 2015;230(2):430−435. https://doi.org/10.1016/j.psychres.2015.09.032.
128. Udo T, Grilo CM. Prevalence and correlates of DSM-5-defined eating disorders in a nationally representative sample of U.S. adults. *Biol Psychiatr*. 2018;84(5):345−354. https://doi.org/10.1016/j.biopsych.2018.03.014.
129. Young KS. Cognitive behavior therapy with Internet addicts: treatment outcomes and implications. *Cyberpsychol Behav*. 2007;10(5):671−679. https://doi.org/10.1089/cpb.2007.9971.
130. Monterubio GE, Fitzsimmons-Craft EE, Balantekin KN, et al. Eating disorder symptomatology, clinical impairment, and comorbid psychopathology in racially and ethnically diverse college women with eating disorders. *Int J Eat Disord*. 2020;53(11):1868−1874. https://doi.org/10.1002/eat.23380.
131. Udo T, White MA, Lydecker JL, et al. Biopsychosocial correlates of binge eating disorder in Caucasian and African American women with obesity in primary care settings. *Eur Eat Disord Rev*. 2016;24(3):181−186. https://doi.org/10.1002/erv.2417.
132. Barnes RD, Ivezaj V, Grilo CM. An examination of weight bias among treatment-seeking obese patients with and without binge eating disorder. *Gen Hosp Psychiatr*. 2014;36(2):177−180. https://doi.org/10.1016/j.genhosppsych.2013.10.011.
133. Rosen EF, Brown A, Braden J, et al. African-American males prefer a larger female body silhouette than do Whites. *Bull Psychonomic Soc*. 1993;31(6):599−601. https://doi.org/10.3758/BF03337366.
134. Freedman RE, Carter MM, Sbrocco T, Gray JJ. Do men hold African-American and Caucasian women to different standards of beauty? *Eat Behav*. 2007;8(3):319−333. https://doi.org/10.1016/j.eatbeh.2006.11.008.
135. Cheung NW. Social strain, couple dynamics and gender differences in gambling problems: evidence from Chinese married couples. *Addict Behav*. 2015;41:175−184. https://doi.org/10.1016/j.addbeh.2014.10.013.
136. Edens EL, Rosenheck RA. Rates and correlates of pathological gambling among VA mental health service users. *J Gambl Stud*. 2012;28(1):1−11. https://doi.org/10.1007/s10899-011-9239-z.
137. Menendez-García A, Jiménez-Arroyo A, Rodrigo-Yanguas M, et al. Internet, video game and mobile phone addiction in children and adolescents diagnosed with ADHD: a case-control study. *Adicciones*. 2022;34(3):208−217. https://doi.org/10.20882/adicciones.1469.
138. Hardoon KK, Derevensky JL. Social influences involved in children's gambling behavior. *J Gambl Stud*. 2001;17(3):191−215. https://doi.org/10.1023/a:1012216305671.
139. Lan Y, Ding JE, Li W, et al. A pilot study of a group mindfulness-based cognitive-behavioral intervention for smartphone addiction among university students. *J Behav Addic*. 2018;7(4):1171−1176. https://doi.org/10.1556/2006.7.2018.103.
140. Afifi TO, Sareen J, Fortier J, et al. Child maltreatment and eating disorders among men and women in adulthood: results from a nationally representative United States sample. *Int J Eat Disord*. 2017;50(11):1281−1296. https://doi.org/10.1002/eat.22783.
141. Chao AM, Grilo CM, Sinha R. Food cravings, binge eating, and eating disorder psychopathology: exploring the moderating roles of gender and race. *Eat Behav*. 2016;21:41−47. https://doi.org/10.1016/j.eatbeh.2015.12.007.

142. Grant JE, Chamberlain SR, Schreiber LR, Odlaug BL. Gender-related clinical and neurocognitive differences in individuals seeking treatment for pathological gambling. *J Psychiatr Res*. 2012;46(9):1206–1211. https://doi.org/10.1016/j.jpsychires.2012.05.013.
143. González-Ortega I, Echeburúa E, Corral P, Polo-López R, Alberich S. Predictors of pathological gambling severity taking gender differences into account. *Eur Addiction Res*. 2013;19(3):146–154. https://doi.org/10.1159/000342311.
144. Zhang J, Hu Y, Wang Z, Wang M, Dong GH. Males are more sensitive to reward and less sensitive to loss than females among people with internet gaming disorder: fMRI evidence from a card-guessing task. *BMC Psychiatr*. 2020;20(1):357. https://doi.org/10.1186/s12888-020-02771-1.
145. Nagata JM, Ganson KT, Austin SB. Emerging trends in eating disorders among sexual and gender minorities. *Curr Opin Psychiatr*. 2020;33(6):562–567. https://doi.org/10.1097/YCO.0000000000000645.
146. Richard J, Martin-Storey A, Wilkie E, Derevensky JL, Paskus T, Temcheff CE. Variations in gambling disorder symptomatology across sexual identity among college student-athletes. *J Gambl Stud*. 2019;35(4):1303–1316. https://doi.org/10.1007/s10899-019-09838-z.
147. Broman N, Hakansson A. Problematic gaming and internet use but not gambling may be overrepresented in sexual minorities - a pilot population web survey study. *Front Psychol*. 2018;9:2184. https://doi.org/10.3389/fpsyg.2018.02184.
148. Roberts A, Sharman S, Coid J, et al. Gambling and negative life events in a nationally representative sample of UK men. *Addict Behav*. 2017;75:95–102. https://doi.org/10.1016/j.addbeh.2017.07.002.
149. Tschibelu E, Elman I. Gender differences in psychosocial stress and in its relationship to gambling urges in individuals with pathological gambling. *J Addict Dis*. 2011;30(1):81–87. https://doi.org/10.1080/10550887.2010.531671.
150. Chami R, Monteleone AM, Treasure J, Monteleone P. Stress hormones and eating disorders. *Mol Cell Endocrinol*. 2019;497:110349. https://doi.org/10.1016/j.mce.2018.12.009.
151. Braun J, El-Gabalawy R, Sommer JL, Pietrzak RH, Mitchell K, Mota N. Trauma exposure, DSM-5 posttraumatic stress, and binge eating symptoms: results from a nationally representative sample. *J Clin Psychiatr*. 2019;80(6):19m12813. https://doi.org/10.4088/JCP.19m12813.
152. Moore 3rd LH, Grubbs JB. Gambling Disorder and comorbid PTSD: a systematic review of empirical research. *Addict Behav*. 2021;114:106713. https://doi.org/10.1016/j.addbeh.2020.106713.
153. Grubbs JB, Chapman H, Shepherd KA. Post-traumatic stress and gambling related cognitions: analyses in inpatient and online samples. *Addict Behav*. 2019;89:128–135. https://doi.org/10.1016/j.addbeh.2018.09.035.
154. Laessle RG, Schulz S. Stress-induced laboratory eating behavior in obese women with binge eating disorder. *Int J Eat Disord*. 2009;42(6):505–510. https://doi.org/10.1002/eat.20648.
155. Leung SL, Barber JA, Burger A, Barnes RD. Factors associated with healthy and unhealthy workplace eating behaviours in individuals with overweight/obesity with and without binge eating disorder. *Obesity Sci & Practi*. 2018;4(2):109–118. https://doi.org/10.1002/osp4.151.
156. Lyu Z, Jackson T. Acute stressors reduce neural inhibition to food cues and increase eating among binge eating disorder symptomatic women. *Front Behav Neurosci*. 2016;10:188. https://doi.org/10.3389/fnbeh.2016.00188.
157. Schulz S, Laessle RG. Stress-induced laboratory eating behavior in obese women with binge eating disorder. *Appetite*. 2012;58(2):457–461. https://doi.org/10.1016/j.appet.2011.12.007.
158. Wonderlich JA, Breithaupt L, Thompson JC, Crosby RD, Engel SG, Fischer S. The impact of neural responses to food cues following stress on trajectories of negative and positive affect and binge eating in daily life. *J Psychiatr Res*. 2018;102:14–22. https://doi.org/10.1016/j.jpsychires.2018.03.005.
159. Breland JY, Donalson R, Li Y, Hebenstreit CL, Goldstein LA, Maguen S. Military sexual trauma is associated with eating disorders, while combat exposure is not. *Psychological Trauma*. 2018;10(3):276–281. https://doi.org/10.1037/tra0000276.
160. Huston JC, Grillo AR, Iverson KM, Mitchell KS, VA Boston Healthcare System. Associations between disordered eating and intimate partner violence mediated by depression and posttraumatic stress disorder symptoms in a female veteran sample. *Gen Hosp Psychiatr*. 2019;58:77–82. https://doi.org/10.1016/j.genhosppsych.2019.03.007.
161. Beccia AL, Jesdale WM, Lapane KL. Associations between perceived everyday discrimination, discrimination attributions, and binge eating among Latinas: results from the National Latino and Asian American Study. *Ann Epidemiol*. 2020;45:32–39. https://doi.org/10.1016/j.annepidem.2020.03.012.

162. Grunewald W, Convertino AD, Safren SA, et al. Appearance discrimination and binge eating among sexual minority men. *Appetite*. 2021;156:104819. https://doi.org/10.1016/j.appet.2020.104819.
163. Brevers D, Noël X, He Q, Melrose JA, Bechara A. Increased ventral-striatal activity during monetary decision making is a marker of problem poker gambling severity. *Addiction Biol*. 2016;21(3):688–699. https://doi.org/10.1111/adb.12239.
164. Brand M, Snagowski J, Laier C, Maderwald S. Ventral striatum activity when watching preferred pornographic pictures is correlated with symptoms of Internet pornography addiction. *Neuroimage*. 2016;129:224–232. https://doi.org/10.1016/j.neuroimage.2016.01.033.
165. Liu L, Yip SW, Zhang JT, et al. Activation of the ventral and dorsal striatum during cue reactivity in Internet gaming disorder. *Addiction Biol*. 2017;22(3):791–801. https://doi.org/10.1111/adb.12338.
166. Reuter J, Raedler T, Rose M, Hand I, Gläscher J, Büchel C. Pathological gambling is linked to reduced activation of the mesolimbic reward system. *Nat Neurosci*. 2005;8(2):147–148. https://doi.org/10.1038/nn1378.
167. Tomasi D, Volkow ND. Striatocortical pathway dysfunction in addiction and obesity: differences and similarities. *Crit Rev Biochem Mol Biol*. 2013;48(1):1–19. https://doi.org/10.3109/10409238.2012.735642.
168. Lee D, Namkoong K, Lee J, Jung YC. Dorsal striatal functional connectivity changes in Internet gaming disorder: a longitudinal magnetic resonance imaging study. *Addiction Biol*. 2021;26(1):e12868. https://doi.org/10.1111/adb.12868.
169. Meng YJ, Deng W, Wang H-y, et al. Reward pathway dysfunction in gambling disorder: a meta-analysis of functional magnetic resonance imaging studies. *Behav Brain Res*. 2014;275:243–251. https://doi.org/10.1016/j.bbr.2014.08.057.
170. Cai C, Yuan K, Yin J, et al. Striatum morphometry is associated with cognitive control deficits and symptom severity in internet gaming disorder. *Brain Imaging & Behav*. 2016;10(1):12–20. https://doi.org/10.1007/s11682-015-9358-8.
171. Dong G-H, Dong H, Wang M, et al. Dorsal and ventral striatal functional connectivity shifts play a potential role in internet gaming disorder. *Commun Biol*. 2021;4(1):866. https://doi.org/10.1038/s42003-021-02395-5.
172. Tanabe J, Thompson L, Claus E, Dalwani M, Hutchison K, Banich MT. Prefrontal cortex activity is reduced in gambling and nongambling substance users during decision-making. *Hum Brain Mapp*. 2007;28(12):1276–1286. https://doi.org/10.1002/hbm.20344.
173. Zeeb FD, Baarendse PJ, Vanderschuren LJ, Winstanley CA. Inactivation of the prelimbic or infralimbic cortex impairs decision-making in the rat gambling task. *Psychopharmacology*. 2015;232(24):4481–4491. https://doi.org/10.1007/s00213-015-4075-y.
174. Sinclair EB, Klump KL, Sisk CL. Reduced medial prefrontal control of palatable food consumption is associated with binge eating proneness in female rats. *Front Behav Neurosci*. 2019;13:252. https://doi.org/10.3389/fnbeh.2019.00252.
175. Zhang Y, Lin X, Zhou H, Xu J, Du X, Dong G. Brain activity toward gaming-related cues in internet gaming disorder during an addiction Stroop task. *Front Psychol*. 2016;7:714. https://doi.org/10.3389/fpsyg.2016.00714.
176. Simon JJ, Skunde M, Walther S, Bendszus M, Herzog W, Friederich HC. Neural signature of food reward processing in bulimic-type eating disorders. *Soc Cognit Affect Neurosci*. 2016;11(9):1393–1401. https://doi.org/10.1093/scan/nsw049.
177. Fuentes D, Rzezak P, Pereira FR, et al. Mapping brain volumetric abnormalities in never-treated pathological gamblers. *Psychiatr Res*. 2015;232(3):208–213. https://doi.org/10.1016/j.pscychresns.2015.04.001.
178. Rahman AS, Xu J, Potenza MN. Hippocampal and amygdalar volumetric differences in pathological gambling: a preliminary study of the associations with the behavioral inhibition system. *Neuropsychopharmacology*. 2014;39(3):738–745. https://doi.org/10.1038/npp.2013.260.
179. Yoon EJ, Choi JS, Kim H, et al. Altered hippocampal volume and functional connectivity in males with Internet gaming disorder comparing to those with alcohol use disorder. *Sci Rep*. 2017;7(1):5744. https://doi.org/10.1038/s41598-017-06057-7.
180. Linnet J, Mouridsen K, Peterson E, Møller A, Doudet DJ, Gjedde A. Striatal dopamine release codes uncertainty in pathological gambling. *Psychiatr Res*. 2012;204(1):55–60. https://doi.org/10.1016/j.pscychresns.2012.04.012.
181. Joutsa J, Johansson J, Niemelä S, et al. Mesolimbic dopamine release is linked to symptom severity in pathological gambling. *Neuroimage*. 2012;60(4):1992–1999. https://doi.org/10.1016/j.neuroimage.2012.02.006.
182. Linnet J, Peterson E, Doudet DJ, Gjedde A, Møller A. Dopamine release in ventral striatum of pathological gamblers losing money. *Acta Psychiatr Scand*. 2010;122(4):326–333. https://doi.org/10.1111/j.1600-0447.2010.01591.x.

183. Chase HW, Clark L. Gambling severity predicts midbrain response to near-miss outcomes. *J Neurosci*. 2010;30(18):6180−6187. https://doi.org/10.1523/JNEUROSCI.5758-09.2010.
184. Pettorruso M, Martinotti G, Cocciolillo F, et al. Striatal presynaptic dopaminergic dysfunction in gambling disorder: a I-FP-CIT SPECT study. *Addiction Biol*. 2019;24(5):1077−1086. https://doi.org/10.1111/adb.12677.
185. Clark CA, Dagher A. The role of dopamine in risk taking: a specific look at Parkinson's disease and gambling. *Front Behav Neurosci*. 2014;8:196. https://doi.org/10.3389/fnbeh.2014.00196.
186. Steeves TD, Miyasaki J, Zurowski M, et al. Increased striatal dopamine release in Parkinsonian patients with pathological gambling: a [11C] raclopride PET study. *Brain*. 2009;132(Pt 5):1376−1385. https://doi.org/10.1093/brain/awp054.
187. van Eimeren T, Ballanger B, Pellecchia G, Miyasaki JM, Lang AE, Strafella AP. Dopamine agonists diminish value sensitivity of the orbitofrontal cortex: a trigger for pathological gambling in Parkinson's disease? *Neuropsychopharmacology*. 2009;34(13):2758−2766. https://doi.org/10.1038/npp.2009.124.
188. Di Ciano P, Pushparaj A, Kim A, et al. The impact of selective dopamine D_2, D_3 and D_4 ligands on the rat gambling task. *PLoS One*. 2015;10(9):e0136267. https://doi.org/10.1371/journal.pone.0136267.
189. Cocker PJ, Le Foll B, Rogers RD, Winstanley CA. A selective role for dopamine D_4 receptors in modulating reward expectancy in a rodent slot machine task. *Biol Psychiatr*. 2014;75(10):817−824. https://doi.org/10.1016/j.biopsych.2013.08.026.
190. Cocker PJ, Hosking JG, Murch WS, Clark L, Winstanley CA. Activation of dopamine D4 receptors within the anterior cingulate cortex enhances the erroneous expectation of reward on a rat slot machine task. *Neuropharmacology*. 2016;105:186−195. https://doi.org/10.1016/j.neuropharm.2016.01.019.
191. Johnson PS, Madden GJ, Brewer AT, Pinkston JW, Fowler SC. Effects of acute pramipexole on preference for gambling-like schedules of reinforcement in rats. *Psychopharmacology*. 2011;213(1):11−18. https://doi.org/10.1007/s00213-010-2006-5.
192. Tremblay M, Silveira MM, Kaur S, et al. Chronic $D_{2/3}$ agonist ropinirole treatment increases preference for uncertainty in rats regardless of baseline choice patterns. *Eur J Neurosci*. 2017;45(1):159−166. https://doi.org/10.1111/ejn.13332.
193. Michaelides M, Thanos PK, Volkow ND, Wang GJ. Dopamine-related frontostriatal abnormalities in obesity and binge-eating disorder: emerging evidence for developmental psychopathology. *Int Rev Psychiatr*. 2012;24(3):211−218. https://doi.org/10.3109/09540261.2012.679918.
194. Corwin RL, Wojnicki FH, Zimmer DJ, et al. Binge-type eating disrupts dopaminergic and GABAergic signaling in the prefrontal cortex and ventral tegmental area. *Obesity*. 2016;24(10):2118−2125. https://doi.org/10.1002/oby.21626.
195. Chawla A, Cordner ZA, Boersma G, Moran TH. Cognitive impairment and gene expression alterations in a rodent model of binge eating disorder. *Physiol Behav*. 2017;180:78−90. https://doi.org/10.1016/j.physbeh.2017.08.004.
196. Heal DJ, Hallam M, Prow M, et al. Dopamine and μ-opioid receptor dysregulation in the brains of binge-eating female rats - possible relevance in the psychopathology and treatment of binge-eating disorder. *J Psychopharmacol*. 2017;31(6):770−783. https://doi.org/10.1177/0269881117699607.
197. Kaye W. Neurobiology of anorexia and bulimia nervosa. *Physiol Behav*. 2008;94(1):121−135. https://doi.org/10.1016/j.physbeh.2007.11.037.
198. Pallanti S, Bernardi S, Quercioli L, DeCaria C, Hollander E. Serotonin dysfunction in pathological gamblers: increased prolactin response to oral m-CPP versus placebo. *CNS Spectr*. 2006;11(12):956−964. https://doi.org/10.1017/s1092852900015145.
199. Marazziti D, Golia F, Picchetti M, et al. Decreased density of the platelet serotonin transporter in pathological gamblers. *Neuropsychobiology*. 2008;57(1−2):38−43. https://doi.org/10.1159/000129665.
200. Kuikka JT, Tammela L, Karhunen L, et al. Reduced serotonin transporter binding in binge eating women. *Psychopharmacology*. 2001;155(3):310−314. https://doi.org/10.1007/s002130100716.
201. Pichika R, Buchsbaum MS, Bailer U, et al. Serotonin transporter binding after recovery from bulimia nervosa. *Int J Eat Disord*. 2012;45(3):345−352. https://doi.org/10.1002/eat.20944.
202. Tiihonen J, Keski-Rahkonen A, Löppönen M, et al. Brain serotonin 1A receptor binding in bulimia nervosa. *Biol Psychiatr*. 2004;55(8):871−873. https://doi.org/10.1016/j.biopsych.2003.12.016.
203. Bailer UF, Bloss CS, Frank GK, et al. 5-HT_1A receptor binding is increased after recovery from bulimia nervosa compared to control women and is associated with behavioral inhibition in both groups. *Int J Eat Disord*. 2011;44(6):477−487. https://doi.org/10.1002/eat.20843.

204. Kaye WH, Frank GK, Meltzer CC, et al. Altered serotonin 2A receptor activity in women who have recovered from bulimia nervosa. *Am J Psychiatr*. 2001;158(7):1152–1155. https://doi.org/10.1176/appi.ajp.158.7.1152.
205. Price AE, Anastasio NC, Stutz SJ, Hommel JD, Cunningham KA. Serotonin 5-HT$_{2C}$ receptor activation suppresses binge intake and the reinforcing and motivational properties of high-fat food. *Front Pharmacol*. 2018;9:821. https://doi.org/10.3389/fphar.2018.00821.
206. Weidacker K, Johnston SJ, Mullins PG, Boy F, Dymond S. Impulsive decision-making and gambling severity: the influence of γ-amino-butyric acid (GABA) and glutamate-glutamine (Glx). *Eur Neuropsychopharmacol*. 2020;32:36–46. https://doi.org/10.1016/j.euroneuro.2019.12.110.
207. Mick I, Ramos AC, Myers J, et al. Evidence for GABA-A receptor dysregulation in gambling disorder: correlation with impulsivity. *Addiction Biol*. 2017;22(6):1601–1609. https://doi.org/10.1111/adb.12457.
208. Møller A, Rømer Thomsen K, Brooks DJ, et al. Attenuation of dopamine-induced GABA release in problem gamblers. *Brain and Behavior*. 2019;9(3):e01239. https://doi.org/10.1002/brb3.1239.
209. Paine TA, O'Hara A, Plaut B, Lowes DC. Effects of disrupting medial prefrontal cortex GABA transmission on decision-making in a rodent gambling task. *Psychopharmacology*. 2015;232(10):1755–1765. https://doi.org/10.1007/s00213-014-3816-7.
210. Hurley MM, Resch JM, Maunze B, Frenkel MM, Baker DA, Choi S. N-acetylcysteine decreases binge eating in a rodent model. *Int J Obes*. 2016;40(7):1183–1186. https://doi.org/10.1038/ijo.2016.31.
211. Popik P, Kos T, Zhang Y, Bisaga A. Memantine reduces consumption of highly palatable food in a rat model of binge eating. *Amino Acids*. 2011;40(2):477–485. https://doi.org/10.1007/s00726-010-0659-3.
212. Corwin RL, Boan J, Peters KF, Ulbrecht JS. Baclofen reduces binge eating in a double-blind, placebo-controlled, crossover study. *Behav Pharmacol*. 2012;23(5–6):616–625. https://doi.org/10.1097/FBP.0b013e328357bd62.
213. Dannon PN, Rosenberg O, Schoenfeld N, Kotler M. Acamprosate and baclofen were not effective in the treatment of pathological gambling: preliminary blind rater comparison study. *Front Psychiatr*. 2011;2:33. https://doi.org/10.3389/fpsyt.2011.00033.
214. Nathan PJ, Bullmore ET. From taste hedonics to motivational drive: central μ-opioid receptors and binge-eating behaviour. *Int J Neuropsychopharmacol*. 2009;12(7):995–1008. https://doi.org/10.1017/S146114570900039X.
215. Majuri J, Joutsa J, Johansson J, et al. Dopamine and opioid neurotransmission in behavioral addictions: a comparative PET study in pathological gambling and binge eating. *Neuropsychopharmacology*. 2017;42(5):1169–1177. https://doi.org/10.1038/npp.2016.265.
216. Blumenthal SA, Pratt WE. d-Fenfluramine and lorcaserin inhibit the binge-like feeding induced by μ-opioid receptor stimulation of the nucleus accumbens in the rat. *Neurosci Lett*. 2018;687:43–48. https://doi.org/10.1016/j.neulet.2018.09.028.
217. Awad G, Roeckel LA, Massotte D, Olmstead MC, Befort K. Deletion of mu opioid receptors reduces palatable solution intake in a mouse model of binge eating. *Behav Pharmacol*. 2020;31(2&3):249–255. https://doi.org/10.1097/FBP.0000000000000496.
218. Chamberlain SR, Mogg K, Bradley BP, et al. Effects of mu opioid receptor antagonism on cognition in obese binge-eating individuals. *Psychopharmacology*. 2012;224(4):501–509. https://doi.org/10.1007/s00213-012-2778-x.
219. Giuliano C, Robbins TW, Nathan PJ, Bullmore ET, Everitt BJ. Inhibition of opioid transmission at the μ-opioid receptor prevents both food seeking and binge-like eating. *Neuropsychopharmacology*. 2012;37(12):2643–2652. https://doi.org/10.1038/npp.2012.128.
220. Stancil SL, Adelman W, Dietz A, Abdel-Rahman S. Naltrexone reduces binge eating and purging in adolescents in an eating disorder program. *J Child Adolesc Psychopharmacol*. 2019;29(9):721–724. https://doi.org/10.1089/cap.2019.0056.
221. Di Ciano P, Le Foll B. Evaluating the impact of naltrexone on the rat gambling task to test its predictive validity for gambling disorder. *PLoS One*. 2016;11(5):e0155604. https://doi.org/10.1371/journal.pone.0155604.
222. Kraus SW, Etuk R, Potenza MN. Current pharmacotherapy for gambling disorder: a systematic review. *Expet Opin Pharmacother*. 2020;21(3):287–296. https://doi.org/10.1080/14656566.2019.1702969.
223. Satta V, Scherma M, Piscitelli F, et al. Limited access to a high fat diet alters endocannabinoid tone in female rats. *Front Neurosci*. 2018;12:40. https://doi.org/10.3389/fnins.2018.00040.
224. Scherma M, Fattore L, Satta V, et al. Pharmacological modulation of the endocannabinoid signalling alters binge-type eating behaviour in female rats. *Br J Pharmacol*. 2013;169(4):820–833. https://doi.org/10.1111/bph.12014.

225. Robinson TE, Berridge KC. The neural basis of drug craving: an incentive-sensitization theory of addiction. *Brain Res Brain Res Rev*. 1993;18(3):247−291. https://doi.org/10.1016/0165-0173(93)90013-p.
226. Berridge KC, Robinson TE. Liking, wanting, and the incentive-sensitization theory of addiction. *Am Psychol*. 2016;71(8):670−679. https://doi.org/10.1037/amp0000059.
227. Boileau I, Payer D, Chugani B, et al. In vivo evidence for greater amphetamine-induced dopamine release in pathological gambling: a positron emission tomography study with [(11)C]-(+)-PHNO. *Mol Psychiatr*. 2014;19(12):1305−1313. https://doi.org/10.1038/mp.2013.163.
228. Fugariu V, Zack MH, Nobrega JN, Fletcher PJ, Zeeb FD. Effects of exposure to chronic uncertainty and a sensitizing regimen of amphetamine injections on locomotion, decision-making, and dopamine receptors in rats. *Neuropsychopharmacology*. 2020;45(5):811−822. https://doi.org/10.1038/s41386-020-0599-x.
229. Hellberg SN, Russell TI, Robinson M. Cued for risk: evidence for an incentive sensitization framework to explain the interplay between stress and anxiety, substance abuse, and reward uncertainty in disordered gambling behavior. *Cognit Affect Behav Neurosci*. 2019;19(3):737−758. https://doi.org/10.3758/s13415-018-00662-3.
230. Paris JJ, Franco C, Sodano R, Frye CA, Wulfert E. Gambling pathology is associated with dampened cortisol response among men and women. *Physiol Behav*. 2010;99(2):230−233. https://doi.org/10.1016/j.physbeh.2009.04.002.
231. Maniaci G, Goudriaan AE, Cannizzaro C, van Holst RJ. Impulsivity and stress response in pathological gamblers during the trier social stress test. *J Gambl Stud*. 2018;34(1):147−160. https://doi.org/10.1007/s10899-017-9685-3.
232. Li Y, Sescousse G, Dreher JC. Endogenous cortisol levels are associated with an imbalanced striatal sensitivity to monetary versus non-monetary cues in pathological gamblers. *Front Behav Neurosci*. 2014;8:83. https://doi.org/10.3389/fnbeh.2014.00083.
233. Bulik CM, Sullivan PF, Kendler KS. Heritability of binge-eating and broadly defined bulimia nervosa. *Biol Psychiatr*. 1998;44(12):1210−1218. https://doi.org/10.1016/s0006-3223(98)00280-7.
234. Javaras KN, Laird NM, Reichborn-Kjennerud T, Bulik CM, Pope Jr HG, Hudson JI. Familiality and heritability of binge eating disorder: results of a case-control family study and a twin study. *Int J Eat Disord*. 2008;41(2):174−179. https://doi.org/10.1002/eat.20484.
235. da Silva Lobo DS, Vallada HP, Knight J, et al. Dopamine genes and pathological gambling in discordant sib-pairs. *J Gambl Stud*. 2007;23(4):421−433. https://doi.org/10.1007/s10899-007-9060-x.
236. Comings DE, Rosenthal RJ, Lesieur HR, et al. A study of the dopamine D2 receptor gene in pathological gambling. *Pharmacogenetics*. 1996;6(3):223−234. https://doi.org/10.1097/00008571-199606000-00004.
237. Comings DE, Gonzalez N, Wu S, et al. Studies of the 48 bp repeat polymorphism of the DRD4 gene in impulsive, compulsive, addictive behaviors: tourette syndrome, ADHD, pathological gambling, and substance abuse. *Am J Med Genet*. 1999;88(4):358−368. https://doi.org/10.1002/(sici)1096-8628(19990820)88:4<358::aid-ajmg13>3.0.co;2-g.
238. Wilson D, da Silva Lobo DS, Tavares H, Gentil V, Vallada H. Family-based association analysis of serotonin genes in pathological gambling disorder: evidence of vulnerability risk in the 5HT-2A receptor gene. *J Mol Neurosci*. 2013;49(3):550−553. https://doi.org/10.1007/s12031-012-9846-x.
239. Grant JE, Leppink EW, Redden SA, Odlaug BL, Chamberlain SR. COMT genotype, gambling activity, and cognition. *J Psychiatr Res*. 2015;68:371−376. https://doi.org/10.1016/j.jpsychires.2015.04.029.
240. Jeong J-E, Rhee J-K, Kim T-M, et al. The association between the nicotinic acetylcholine receptor α4 subunit gene (CHRNA4) rs1044396 and Internet gaming disorder in Korean male adults. *PLoS One*. 2017;12(12):e0188358. https://doi.org/10.1371/journal.pone.0188358.
241. Eisenegger C, Knoch D, Ebstein RP, Gianotti LR, Sándor PS, Fehr E. Dopamine receptor D4 polymorphism predicts the effect of L-DOPA on gambling behavior. *Biol Psychiatr*. 2010;67(8):702−706. https://doi.org/10.1016/j.biopsych.2009.09.021.
242. Neukam PT, Kroemer NB, Deza Araujo YI, et al. Risk-seeking for losses is associated with 5-HTTLPR, but not with transient changes in 5-HT levels. *Psychopharmacology*. 2018;235(7):2151−2165. https://doi.org/10.1007/s00213-018-4913-9.
243. Munn-Chernoff MA, Grant JD, Agrawal A, et al. Are there common familial influences for major depressive disorder and an overeating-binge eating dimension in both European American and African American female twins? *Int J Eat Disord*. 2015;48(4):375−382. https://doi.org/10.1002/eat.22280.

244. Hernández S, Camarena B, González L, Caballero A, Flores G, Aguilar A. A family-based association study of the HTR1B gene in eating disorders. *Brazilian J Psychiatr*. 2016;38(3):239−242. https://doi.org/10.1590/1516-4446-2016-1936.
245. Koren R, Duncan AE, Munn-Chernoff MA, et al. Preliminary evidence for the role of HTR2A variants in binge eating in young women. *Psychiatr Genet*. 2014;24(1):28−33. https://doi.org/10.1097/YPG.0000000000000014.
246. Davis C, Levitan RD, Yilmaz Z, Kaplan AS, Carter JC, Kennedy JL. Binge eating disorder and the dopamine D2 receptor: genotypes and sub-phenotypes. *Prog Neuro Psychopharmacol Biol Psychiatr*. 2012;38(2):328−335. https://doi.org/10.1016/j.pnpbp.2012.05.002.
247. Frieling H, Römer KD, Wilhelm J, et al. Association of catecholamine-O-methyltransferase and 5-HTTLPR genotype with eating disorder-related behavior and attitudes in females with eating disorders. *Psychiatr Genet*. 2006;16(5):205−208. https://doi.org/10.1097/01.ypg.0000218620.50386.f1.
248. Lim SW, Ha J, Shin DW, Woo HY, Kim KH. Associations between the serotonin-1A receptor C(-1019)G polymorphism and disordered eating symptoms in female adolescents. *J Neural Transm*. 2010;117(6):773−779. https://doi.org/10.1007/s00702-010-0412-9.
249. Monteleone P, Tortorella A, Castaldo E, Maj M. Association of a functional serotonin transporter gene polymorphism with binge eating disorder. *Am J Med Genet Part B Neuropsychiatr Genet*. 2006;141B(1):7−9. https://doi.org/10.1002/ajmg.b.30232.
250. Monteleone P, Bifulco M, Di Filippo C, et al. Association of CNR1 and FAAH endocannabinoid gene polymorphisms with anorexia nervosa and bulimia nervosa: evidence for synergistic effects. *Gene Brain Behav*. 2009;8(7):728−732. https://doi.org/10.1111/j.1601-183X.2009.00518.x.
251. Richardson J, Steiger H, Schmitz N, et al. Relevance of the 5-HTTLPR polymorphism and childhood abuse to increased psychiatric comorbidity in women with bulimia-spectrum disorders. *J Clin Psychiatr*. 2008;69(6):981−990. https://doi.org/10.4088/jcp.v69n0615.
252. Akkermann K, Nordquist N, Oreland L, Harro J. Serotonin transporter gene promoter polymorphism affects the severity of binge eating in general population. *Prog Neuro Psychopharmacol Biol Psychiatr*. 2010;34(1):111−114. https://doi.org/10.1016/j.pnpbp.2009.10.008.
253. McIntosh CC, Crino RD, O'Neill K. Treating problem gambling samples with cognitive behavioural therapy and mindfulness-based interventions: a clinical trial. *J Gambl Stud*. 2016;32(4):1305−1325. https://doi.org/10.1007/s10899-016-9602-1.
254. Palavras MA, Hay P, Filho CA, Claudino A. The efficacy of psychological therapies in reducing weight and binge eating in people with bulimia nervosa and binge eating disorder who are overweight or obese-A critical synthesis and meta-analyses. *Nutrients*. 2017;9(3):299. https://doi.org/10.3390/nu9030299.
255. Wilson GT, Eldredge KL, Smith D, Niles B. Cognitive-behavioral treatment with and without response prevention for bulimia. *Behav Res Ther*. 1991;29(6):575−583. https://doi.org/10.1016/0005-7967(91)90007-p.
256. Kellett S, Oxborough P, Gaskell C. Treatment of compulsive buying disorder: comparing the effectiveness of cognitive behavioural therapy with person-centred experiential counselling. *Behav Cognit Psychother*. 2021;49(3):370−384. https://doi.org/10.1017/S1352465820000521.
257. Zhou X, Rau PP, Yang CL, Zhou X. Cognitive behavioral therapy-based short-term abstinence intervention for problematic social media use: improved well-being and underlying mechanisms. *Psychiatr Q*. 2021;92(2):761−779. https://doi.org/10.1007/s11126-020-09852-0.
258. Park SY, Kim SM, Roh S, et al. The effects of a virtual reality treatment program for online gaming addiction. *Comput Methods Progr Biomed*. 2016;129:99−108. https://doi.org/10.1016/j.cmpb.2016.01.015.
259. Choi SW, Shin YC, Youn H, Lim SW, Ha J. Pharmacotherapy and group cognitive behavioral therapy enhance follow-up treatment duration in gambling disorder patients. *Ann Gen Psychiatr*. 2016;15:20. https://doi.org/10.1186/s12991-016-0107-1.
260. Quilty LC, Allen TA, Davis C, Knyahnytska Y, Kaplan AS. A randomized comparison of long acting methylphenidate and cognitive behavioral therapy in the treatment of binge eating disorder. *Psychiatr Res*. 2019;273:467−474. https://doi.org/10.1016/j.psychres.2019.01.066.
261. Petry NM, Weinstock J, Morasco BJ, Ledgerwood DM. Brief motivational interventions for college student problem gamblers. *Addiction*. 2009;104(9):1569−1578. https://doi.org/10.1111/j.1360-0443.2009.02652.x.
262. Bouchard S, Robillard G, Giroux I, et al. Using virtual reality in the treatment of gambling disorder: the development of a new tool for cognitive behavior therapy. *Front Psychiatr*. 2017;8:27. https://doi.org/10.3389/fpsyt.2017.00027.

263. Cesa GL, Manzoni GM, Bacchetta M, et al. Virtual reality for enhancing the cognitive behavioral treatment of obesity with binge eating disorder: randomized controlled study with one-year follow-up. *J Med Internet Res.* 2013;15(6):e113. https://doi.org/10.2196/jmir.2441.
264. Bong SH, Won GH, Choi TY. Effects of cognitive-behavioral therapy based music therapy in Korean adolescents with smartphone and internet addiction. *Psychiatr Investig.* 2021;18(2):110–117. https://doi.org/10.30773/pi.2020.0155.
265. Friederich HC, Schild S, Wild B, et al. Treatment outcome in people with subthreshold compared with full-syndrome binge eating disorder. *Obesity.* 2007;15(2):283–287. https://doi.org/10.1038/oby.2007.545.
266. Echeburúa E, Gómez M, Freixa M. Cognitive-behavioural treatment of pathological gambling in individuals with chronic schizophrenia: a pilot study. *Behav Res Ther.* 2011;49(11):808–814. https://doi.org/10.1016/j.brat.2011.08.009.
267. Freidenberg BM, Blanchard EB, Wulfert E, Malta LS. Changes in physiological arousal to gambling cues among participants in motivationally enhanced cognitive-behavior therapy for pathological gambling: a preliminary study. *Appl Psychophysiol Biofeedback.* 2002;27(4):251–260. https://doi.org/10.1023/a:1021057217447.
268. Boswell RG, Gueorguieva R, Grilo CM. Change in impulsivity is prospectively associated with treatment outcomes for binge-eating disorder. In: *Psychological Medicine.* Advance online publication; 2021:1–9. https://doi.org/10.1017/S003329172100475X.
269. Sancho M, De Gracia M, Rodríguez RC, et al. Mindfulness-based interventions for the treatment of substance and behavioral addictions: a systematic review. *Front Psychiatr.* 2018;9:95. https://doi.org/10.3389/fpsyt.2018.00095.
270. Mercado D, Robinson L, Gordon G, Werthmann J, Campbell IC, Schmidt U. The outcomes of mindfulness-based interventions for Obesity and Binge Eating Disorder: a meta-analysis of randomised controlled trials. *Appetite.* 2021;166:105464. https://doi.org/10.1016/j.appet.2021.105464.
271. Liu X, Jiang J, Zhang Y. Effects of logotherapy-based mindfulness intervention on internet addiction among adolescents during the COVID-19 pandemic. *Iran J Public Health.* 2021;50(4):789–797. https://doi.org/10.18502/ijph.v50i4.6005.
272. Li W, Garland EL, McGovern P, O'Brien JE, Tronnier C, Howard MO. Mindfulness-oriented recovery enhancement for internet gaming disorder in U.S. adults: A stage I randomized controlled trial. *Psychol Addict Behav.* 2017;31(4):393–402. https://doi.org/10.1037/adb0000269.
273. Omiwole M, Richardson C, Huniewicz P, Dettmer E, Paslakis G. Review of mindfulness-related interventions to modify eating behaviors in adolescents. *Nutrients.* 2019;11(12):2917. https://doi.org/10.3390/nu11122917.
274. Grohmann D, Laws KR. Two decades of mindfulness-based interventions for binge eating: a systematic review and meta-analysis. *J Psychosom Res.* 2021;149:110592. https://doi.org/10.1016/j.jpsychores.2021.110592.
275. Schmitz F, Svaldi J. Effects of bias modification training in binge eating disorder. *Behav Ther.* 2017;48(5):707–717. https://doi.org/10.1016/j.beth.2017.04.003.
276. Boutelle KN, Monreal T, Strong DR, Amir N. An open trial evaluating an attention bias modification program for overweight adults who binge eat. *J Behav Ther Exp Psychiatr.* 2016;52:138–146. https://doi.org/10.1016/j.jbtep.2016.04.005.
277. Butler RM, Heimberg RG. Exposure therapy for eating disorders: a systematic review. *Clin Psychol Rev.* 2020;78:101851. https://doi.org/10.1016/j.cpr.2020.101851.
278. Magson NR, Handford CM, Norberg MM. The empirical status of cue exposure and response prevention treatment for binge eating: a systematic review. *Behav Ther.* 2021;52(2):442–454. https://doi.org/10.1016/j.beth.2020.06.005.
279. Jimenez-Murcia S, Aymamí N, Gómez-Peña M, et al. Does exposure and response prevention improve the results of group cognitive-behavioural therapy for male slot machine pathological gamblers? *Br J Clin Psychol.* 2012;51(1):54–71. https://doi.org/10.1111/j.2044-8260.2011.02012.x.
280. Ferrer-Garcia M, Pla-Sanjuanelo J, Dakanalis A, et al. A randomized trial of virtual reality-based cue exposure second-level therapy and cognitive behavior second-level therapy for bulimia nervosa and binge-eating disorder: outcome at six-month followup. *Cyberpsychol Behav Soc Netw.* 2019;22(1):60–68. https://doi.org/10.1089/cyber.2017.0675.
281. Sundgot-Borgen J, Rosenvinge JH, Bahr R, Schneider LS. The effect of exercise, cognitive therapy, and nutritional counseling in treating bulimia nervosa. *Med Sci Sports Exerc.* 2002;34(2):190–195. https://doi.org/10.1097/00005768-200202000-00002.

282. Vancampfort D, Vanderlinden J, De Hert M, et al. A systematic review on physical therapy interventions for patients with binge eating disorder. *Disabil Rehabil.* 2013;35(26):2191−2196. https://doi.org/10.3109/09638288.2013.771707.
283. Beaulieu K, Hopkins M, Gibbons C, et al. Exercise training reduces reward for high-fat food in adults with overweight/obesity. *Med Sci Sports Exerc.* 2020;52(4):900−908. https://doi.org/10.1249/MSS.0000000000002205.
284. Bakland M, Rosenvinge JH, Wynn R, et al. Patients' views on a new treatment for Bulimia nervosa and binge eating disorder combining physical exercise and dietary therapy (the PED-t). A qualitative study. *Eat Disord.* 2019;27(6):503−520. https://doi.org/10.1080/10640266.2018.1560847.
285. Mathisen TF, Rosenvinge JH, Friborg O, et al. Is physical exercise and dietary therapy a feasible alternative to cognitive behavior therapy in treatment of eating disorders? A randomized controlled trial of two group therapies. *Int J Eat Disord.* 2020;53(4):574−585. https://doi.org/10.1002/eat.23228.
286. Jackson JB, Pietrabissa G, Rossi A, Manzoni GM, Castelnuovo G. Brief strategic therapy and cognitive behavioral therapy for women with binge eating disorder and comorbid obesity: a randomized clinical trial one-year follow-up. *J Consult Clin Psychol.* 2018;86(8):688−701. https://doi.org/10.1037/ccp0000313.
287. Dunn EC, Neighbors C, Larimer ME. Motivational enhancement therapy and self-help treatment for binge eaters. *Psychol Addict Behav.* 2006;20(1):44−52. https://doi.org/10.1037/0893-164X.20.1.44.
288. Bray B, Rodríguez-Martín BC, Wiss DA, Bray CE, Zwickey H. Overeaters Anonymous: an overlooked intervention for binge eating disorder. *Int J Environ Res Publ Health.* 2021;18(14):7303. https://doi.org/10.3390/ijerph18147303.
289. Schuler A, Ferentzy P, Turner NE, et al. Gamblers Anonymous as a recovery pathway: a scoping review. *J Gambl Stud.* 2016;32(4):1261−1278. https://doi.org/10.1007/s10899-016-9596-8.
290. Baño M, Mestre-Bach G, Granero R, et al. Women and gambling disorder: assessing dropouts and relapses in cognitive behavioral group therapy. *Addict Behav.* 2021;123:107085. https://doi.org/10.1016/j.addbeh.2021.107085.
291. Thompson-Brenner H, Franko DL, Thompson DR, et al. Race/ethnicity, education, and treatment parameters as moderators and predictors of outcome in binge eating disorder. *J Consult Clin Psychol.* 2013;81(4):710−721. https://doi.org/10.1037/a0032946.
292. Mestre-Bach G, Granero R, Steward T, et al. Reward and punishment sensitivity in women with gambling disorder or compulsive buying: implications in treatment outcome. *J Behav Addic.* 2016;5(4):658−665. https://doi.org/10.1556/2006.5.2016.074.
293. Mallorquí-Bagué N, Mestre-Bach G, Lozano-Madrid M, et al. Trait impulsivity and cognitive domains involving impulsivity and compulsivity as predictors of gambling disorder treatment response. *Addict Behav.* 2018;87:169−176. https://doi.org/10.1016/j.addbeh.2018.07.006.
294. Goudriaan AE, Oosterlaan J, De Beurs E, Van Den Brink W. The role of self-reported impulsivity and reward sensitivity versus neurocognitive measures of disinhibition and decision-making in the prediction of relapse in pathological gamblers. *Psychol Med.* 2008;38(1):41−50. https://doi.org/10.1017/S0033291707000694.
295. Granero R, Fernández-Aranda F, Mestre-Bach G, et al. Cognitive behavioral therapy for compulsive buying behavior: predictors of treatment outcome. *Eur Psychiatr.* 2017;39:57−65. https://doi.org/10.1016/j.eurpsy.2016.06.004.
296. Moghimi E, Davis C, Bonder R, Knyahnytska Y, Quilty L. Exploring women's experiences of treatment for binge eating disorder: methylphenidate vs. cognitive behavioural therapy. *Prog Neuro Psychopharmacol Biol Psychiatr.* 2022;114:110492. https://doi.org/10.1016/j.pnpbp.2021.110492.
297. Maniaci G, La Cascia C, Picone F, Lipari A, Cannizzaro C, La Barbera D. Predictors of early dropout in treatment for gambling disorder: the role of personality disorders and clinical syndromes. *Psychiatr Res.* 2017;257:540−545. https://doi.org/10.1016/j.psychres.2017.08.003.

Index

Note: 'Page numbers followed by "f" indicate figures and "t" indicate tables.'

A

Abstinence, 5
Acamprosate, 76, 158–159
Acetylcholine, 117–118, 459–460
Acquisition, 103–105
Action potential, 52–53
Addiction, 140–154, 146f, 148f
   cognitive model of, 270–271
   compensatory response theory of, 239–241, 242f–243f
   dopamine depletion hypothesis of, 110
   glutamate homeostasis hypothesis of, 293–296
   hedonic allostasis model of, 462–463
   homeostasis hypothesis, 293–296, 295f
   incentive sensitization theory of, 241–246
   opponent process theory of, 195–197
   substance use disorders (SUDs). *See* Substance use disorders (SUDs)
   threshold procedure, 202–203, 203f
Addiction-like behaviors, 99–101, 490–498
   altered gene expression on, 142–144
   choice procedure, 204
   compulsive drug seeking, 205–206
   drug self-administration, 99, 493–495
   long-access model, 201f
   paradigm, overview of, 99
   progressive ratio schedule, 202f
Adolescence, 441
Adolescent community reinforcement approach (A-CRA), 215
Agonist, 48
Agonist treatment, 71–72, 79, 116, 155, 339, 547–548
Alcohol mixed with energy drink (AmED), 21, 23
Alternative five factor model, 360
AMPA receptor to NMDA receptor ratio, 292
Amygdala, 97–98
Androgens, 501
Anhedonia, 10–11
Antagonist, 48
Anterior cingulate cortex (ACC), 98, 335–336
Anxiety disorders, 442–443
Appetitive stimulus, 188
Astrocytes, 164
Attentional and memory processes
   attention and automaticity, drug effects on, 271–275
   cellular mechanisms of, 290–293, 291f
   cognitive-based treatments, 296–299
      cognitive-behavioral therapy (CBT), 296–299, 298f
      cognitive functions in individuals, 299
   cognitive impairments, 281
   conditioned place preference (CPP), 286–287
   drug effects on, 277–281
   drug warnings, attention to, 275–276
   modeling relapse-like behavior, 282–286, 283f
      extinction, 282–284
      incubation of drug craving, 284–285
      reinstatement of drug-seeking behavior, 284
      renewal, 285–286
      resurgence, 286
   neurobiology of, 288–290
Attentional bias, 234–238, 236f
   reducing, 256–257
Attention deficit/hyperactivity disorder
   addiction, relationship with, 374–375
   psychostimulant treatment for, 375
   rodent model, 375–377
Augmented reality, 253
Automaticity, 271
Autoreceptors, 58
Aversion therapy, 253–254
Axon, 51

B

Balloon analog risk task (BART), 541
Barbiturate, 13f, 15, 68–69
Barratt Impulsiveness Scale (BIS-11), 370
Bed nucleus of the stria terminalis (BNST), 457
Behavioral addictions, 532–533
   binge eaters, high-calorie food in, 536–538
   binge eating disorder, 532–533, 536–538
   bulimia nervosa, 532–533
   cognitive biases associated with, 539
   comorbid disorders observed in, 533–534
   Gambler's fallacy, 539

Behavioral addictions (*Continued*)
  gender and sexual/gender identity, 545
  learning and cognitive mechanisms of, 534–541
  maladaptive decision making, 540–541
  memory impairments observed in, 540–541
  neurotransmitters implicated in, 547–549
  operant conditioning, role of, 535–536
  overconfidence, 539
  Pavlovian conditioning, role of, 535–536
  peer and familial influences on, 543
  personality traits associated with, 542–543
  risk factors associated with, 542–546
  sociocultural factors associated with, 544–545
  stress, 545–546
  treatment interventions for, 552–554
Behavioral pharmacology, 103
Benzodiazepine, 13f, 15, 23, 25–26, 68–69, 204, 322–323, 509
Big-Five Factors Inventory (BFI), 364
Blood–brain barrier (BBB), 55–56
Brain regions involved in reward, 96–99
Brief strategic therapy (BST), 554
Buprenorphine, 77, 339, 368, 500–501

C

Cannabis, 18–19
Catechol-*O*-methyltransferase (COMT), 62, 136, 339–340, 552
Cathinone, 11–12
Cell body, 47–48
Cellular pathways associated with addiction, 162–164, 163f
Chemogenetics, 147–148, 148f
Cigarette, 8–10, 21, 28, 31, 72, 137–138, 194, 232, 272, 281, 319, 324, 339, 364–365, 393, 442, 509–510
Cocaine- and amphetamine-regulated transcript (CART), 505
Cognitive distortions, 270
Commonly used drugs
  drug mechanisms of action, 66–70
    cannabis, 70
    cathinone, 11–12
    cocaine, 10
    depressants, 67–69, 69f
    entactogens (empathogens), 17–18
    ethanol, 12–14
    gamma-hydroxybutyrate (GHB), 15
    hallucinogens, 69–70, 70f
    inhalants, 70
    methamphetamine, 10
    nicotine, 8–10
    opioids, 14
    psychostimulants, 66–67, 67f
    sedative-hypnotics, 15
    synthetic cannabinoids, 18–19
    synthetic cathinones, 11–12
  glial cells, 55–56
  nervous system, 46
  neurons, 47–55
    action potentials, 52–54
    synapses and chemical transmission, 54–55
  neurotransmitters, 46–47
    acetylcholine, 57–58, 59f–60f
    endocannabinoids, 64–65
    endogenous opioids, 64
    GABA, 63–64
    glutamate, 62–63
    monoamines, 58–62
Community reinforcement and family training (CRAFT), 191, 213–214
Community reinforcement approach (CRA), 212–213
Conditioned place preference (CPP), 100–101, 247–251, 286–287, 373–374, 495–497
  acquisition *vs.* expression, 250
  advantages and disadvantages of, 251
  biased *versus* unbiased design, 249f, 250
  humans in, 250
Conditioned response (CR), 228–230
Conditioned stimulus (CS), 228–230
Conditioned tolerance, 240
Conditioned withdrawal symptoms, 241
Consolidation, 286–287
Conspecific, 399
Contiguity, 188
Contingency management, 188, 209–210
Corticosteroid, 438
Counter conditioning, 253–254
COVID-19 pandemic, 30–31
Craving, 108–109
Cue exposure therapy, 251–252
Cysteine/glutamate transporter, 294

D

Delay discounting, 320, 321f, 322–324, 326, 330, 333–334, 339
Delirium tremens, 13–14
Dendrites, 48
Dendritic spines, 48
Depolarization, 54
Designer receptor exclusively activated by a designer drug (DREADD), 147–148
Discriminative stimulus, 189
Dissociative hallucinogens, 15, 17
Disulfiram, 75–76, 137, 253–254
Divided attention steering simulator (DASS), 274

DNA methylation, 152–154
Dominance hierarchies, 406
Dopamine, 457–459
   receptor levels, 156–157
   receptors, genes encoding for, 135–136
Dopaminergic signaling, 102–110
Dopamine transporter (DAT), 137
Dorsal raphe nuclei (DRN), 147
Dorsal striatum, 96
Dot-probe task, 235
Downregulation, 141
Drug abuse, 4
Drug discrimination, 207–208, 207f
Drug Enforcement Administration (DEA), 6–7
Drug misuse, 4, 335
Drug overdose, 5
Drug seeking, 108–109
Drug self-administration, 99, 198–207, 199f–201f, 493–495
Drug stroop task, 237
Drug-taking behavior, 105–106

E

Early growth response protein 1 (EGR1), 400
E-cigarettes, 28
Endocannabinoids, 117, 460
Endogenous opioids, 115–117
Entactogens (empathogens), 17–18, 18f
Environmental enrichment, 408–410, 409f, 445, 500–501
Epigenetics, 152–154
Estrogens, 501
Estrous cycle, 503
Ethanol (alcohol), 4–5, 12–14, 67–68, 112, 136, 141–142, 194, 233, 272, 292, 318, 366, 394, 415, 441, 488
Ethnicity, 410
Euphoria, 10–11, 105, 160–161
Excitatory amino acid transporters (EAATs), 63
Excitatory neurotransmission, 506
Extinction burst, 101, 282
Eye-tracking experiments, 235
Eysenck Personality Questionnaire (EPQ), 358, 359f

F

Fatty acid amide hydrolase (FAAH), 460
Fatty acid binding proteins (FABPs), 460
Fetal alcohol syndrome (FAS), 30, 31f, 274
Five-choice serial reaction time task (5CSRTT), 369–370
Five-factor model, 360, 361t
Fos-related antigen 1 (FRA1), 140–141

G

GABA, 112
Gambler's fallacy, 539
Gambling disorder, 335, 532, 535, 542, 544–547, 552, 554
Gamma-hydroxybutyrate (GHB), 15
Gender identity, 509
Gender/sex differences
   acquisition and maintenance of, 493–494
   addiction-like behaviors, 490–498
      conditioned place preference (CPP), 495–497
      drug preference, extinction of, 497
      drug self-administration, acquisition and maintenance of, 493–494
      drug self-administration, escalation, 494–495
      physiological and behavioral drug effects, 491
      relapse-like behavior, 497–498
      self-administer drug, motivation to, 494
      self-administration, behavioral economic measures of, 495
      voluntary oral consumption, 491–493, 492f
   differentiating sex from gender, 488
   drug sensitivity, 501–509
      biological factors, 501–508, 502f, 507f
      sociocultural factors, 508–509
   drug use, 488–490, 509–512
   escalation, 494–495
   moderating variable, as, 499–501
   sexual minority women's daily lives
      alcohol use, 510–511
      minority stress, 510–511
Generalized anxiety disorder (GAD), 442
Genetic associations, 552
Genome-wide association study, 135
Genotype, 152
Glial cells, 55–56
Glial fibrillary acidic protein (GFAP), 164
Glucocorticoids, 438
Glutamate, 62–63, 113, 459
   homeostasis hypothesis, 293–296, 295f
Glutamatergic neurons, 147
Goal-tracking, 232–233

H

Habit formation, 96
Hallucinogens, 15–17, 16f
Haplotype, 135
Harm reduction, 71–72
Hedonic allostasis model of addiction, 197, 462–463, 464f
Heritability, 134
Heteroreceptors, 58
Higher-order (second-order) conditioning, 230

Hippocampus, 98–99
Histone, 154
Histone acetylation, 154
Histone deacetylation, 154
Histone modification, 154
Homeostasis hypothesis, 293–296, 295f
Hormonal influences, 501
Hormone, 438
Human drug self-administration, 206–207
Human gene association studies, 139
Hyperpolarization, 54
Hypothalamic–pituitary–adrenal (HPA) axis, 438, 454–455
Hypothalamic–pituitary–gonadal (HPG) axis, 501

I
Immediate early gene, 140–141
Impulsive action, 369
   addiction relationship to, 371
Impulsive choice, 369
Impulsive decision making, 319–325, 321f
   acute alcohol effects, 321
   acute drug effects, 320
   drug use, long-term consequences of, 323–325
Impulsivity
   behavioral measures of, 369–370
   novelty seeking/sensation seeking, 373–374
   trait impulsivity, 370–373
Incentive salience, 95
Incentive sensitization theory, 241–246, 244f–245f, 549–551, 550f–551f
Incubation of craving effect, 101
Institutional Animal Care and Use Committee (IACUC), 33
Insula, 98
Intoxication, 4–5, 272, 322
Ionotropic receptor, 63
Iowa gambling task (IGT), 316

L
Lateral habenula, 99
Laterodorsal tegmental nucleus (LDT), 459–460
Limbic structures, 334–335
Long-term depression (LTD), 290
Long-term memory (LTM), 277
Long-term potentiation (LTP), 290

M
Maladaptive decision making
   addiction, relationship to, 336–337
   drug-induced impairments, 338–340
   drug use, impaired decision making resulting from, 318–319
   measuring maladaptive decision making, 316–318, 317f
   metacognition, 336–337
   mindfulness, 336–337
   overview of, 316–319
   shared neuromechanisms with addiction, 332–336
      anterior cingulate cortex (ACC), 335–336
      limbic structures, 334–335
      prefrontal cortex, 332–334
Maternal separation, 446–447
MDMA, 443
Medial prefrontal cortex (mPFC), 455
Medium spiny neurons (MSN), 95
Menstrual cycle, 501
Mesocorticolimbic pathway, primary brain structures of, 92–96
   amygdala, 97–98
   anterior cingulate cortex (ACC), 98
   brain regions, 96–99
   dorsal striatum, 96
   hippocampus, 98–99
   insula, 98
   lateral habenula, 99
   nucleus accumbens (NAc), 95
   prefrontal cortex (PFC), 96
   temporal prediction of reward, 93–94
   ventral tegmental area (VTA), 93
Messenger RNA (mRNA), 141
Metaplasticity, 292–293
Methamphetamine, 29–30, 30f
Microdialysis, 103
Mindfulness, 336–337
Minnesota Multiphasic Personality Inventory (MMPI), 362–364
Molecular and cellular mechanisms of addiction
   addiction in animals, genetic approaches to, 140–152
      addiction-like behaviors, altered gene expression on, 142–144
      chemogenetics, 147–148, 148f
      DNA methylation, 152–154
      epigenetics, 152–154
      histone modification, 154
      mRNA expression, drug-induced changes in, 141–142
      optogenetics, 145–147, 146f
      transcription factor expression, drug-induced changes in, 140–141
      ventral tegmental area dopamine neurons, chemogenetic manipulations of, 149
   cellular changes following, 156–164
      abrupt drug cessation, altered receptor levels during, 158–159
      astrocytes, 164

cellular pathways associated with, 162–164, 163f
dendritic spines, alterations to, 159
dopamine receptor levels, 156–157
other neurotransmitter systems, drugs affect receptor levels of, 157–158
receptor levels, changes to, 156–159
transporter expression and function, 159–162
drug addiction in humans, genes associated with, 135–139
acetylcholine nicotinic receptor, genes encoding for, 137–138
dopamine receptors, genes encoding for, 135–136
ethanol metabolism, genes encoding for, 138
monoamine clearance, genes encoding for, 136–137
drug administration, cellular changes following, 156–164
abrupt drug cessation, altered receptor levels during, 158–159
dendritic spines, alterations to, 159
dopamine receptor levels, 156–157
other neurotransmitter systems, drugs affect receptor levels of, 157–158
receptor levels, changes to, 156–159
transporter expression and function, 159–162
human gene association studies, 139
nicotinic receptor, genes encoding for, 137–138
pharmacogenetic approach to, 155
substance use disorders, heritability of, 134
Monoamine oxidase (MAO), 62
Motivating operation, 188–189

N

N-acetylcysteine, 75, 80, 296, 459, 548–549
Naloxone, 45
Naltrexone, 76, 116, 282–284, 371
Negative punishment, 191
Negative reinforcement, 188, 195
Nervous system, 46
Neuroanatomical and neurochemical substrates, addiction
addiction-like behaviors, 99–101
conditioned place preference (CPP), 100–101
drug self-administration paradigm, 99
behavioral effects of drugs, sensitization to, 106–108
craving and drug seeking, 108–109
dopamine depletion hypothesis of addiction, 110
dopaminergic signaling, 102–110
drug-taking behavior, 105–106
drug use initiation, 103–105
relapse, 109–110
tolerance and withdrawal, 108
Neuroendocrine system, 438

Neuron, 47
Neurotransmitter systems, 47, 112–118
acetylcholine, 117–118
endocannabinoids, 117
endogenous opioids, 115–117
GABA, 112
glutamate, 113
norepinephrine, 115
serotonin, 114
Nicotine, 8–9
NMDA receptors, 63
Norepinephrine, 115
Nucleus accumbens (NAc), 95, 455–456, 505

O

Obsessive-compulsive disorder (OCD), 442
Operant conditioning, 535–536
choice procedure, 204
combat substance use disorders, punishment to, 215–216
compulsive drug seeking, 205–206
definition, 186–192
drug discrimination, 207–208, 207f
drug self-administration, 198–207, 199f–201f
Skinner, Burrhus Frederick, 186–187, 187f
hedonic allostasis model of addiction, 197
human drug self-administration, 206–207
long-access model, 201f
negative reinforcement, 195
opponent-process theory, 195–197, 196f
positive reinforcement, 194–195
progressive ratio schedule, 202f
punishment, 198
reinforcement, 187–189
schedules of reinforcement, 189–191, 189f
punishment, 191–192, 193f
second-order schedule, 205
self-administration paradigms, neurobiological differences, 208–216
short-access model, 199, 200f
threshold procedure, 202–203, 203f
Opiates, 14
Opioids, 14
Opponent-process theory, 195–197, 196f
Optogenetics, 145–147, 146f
Overconfidence, 539

P

Parental monitoring., 394
Parenting styles, 394–395
Parkinson's disease, 60–61, 108, 547–548
Pavlovian conditioning, 232–234, 233f, 535–536
attentional bias, 234–238, 236f

Pavlovian conditioning (*Continued*)
　reducing, 256–257
　augmented reality, 253
　aversion therapy, 253–254
　conditioned place preference (CPP), 247–251
　　acquisition *vs.* expression, 250
　　advantages and disadvantages of, 251
　　biased *versus* unbiased design, 249f, 250
　　humans, 250
　counter conditioning, 253–254
　cue exposure therapy, 251–252
　methamphetamine cue-induced craving, 254–256
　overview of, 228
　Pavlovian conditioned approach (PCA), 232–234, 233f
　Pavlovian-to-instrumental transfer (PIT), 238–239
　　compensatory response theory, 239–241, 242f–243f
　　incentive sensitization theory of addiction, 241–246, 244f–245f
　principles of, 228–231, 229f, 231f
　virtual reality exposure therapy (VRET), 252–253
Peer and familial influences, 543
Peptides, 461–462
Personality traits, 542–543
　alternative five factor model, 360
　attention deficit/hyperactivity disorder
　　addiction, 374–375
　　psychostimulant treatment for, 375
　　rodents, 375–377
　character inventory, 361–362, 363t
　drug use and addiction, 364–368
　　Big Five traits, 364
　　psychopathology, 367–368
　　psychotic proneness, 364–365
　　temperaments, 367
　Eysenck personality questionnaire, 358, 359f
　five-factor model of, 360, 361t
　impulsivity
　　action and addiction, 371
　　behavioral measures of, 369–370
　　novelty seeking/sensation seeking, 373–374
　　trait impulsivity, 372–373
　　trait measures of, 370–371
　measuring, 358–364
　Minnesota Multiphasic Personality Inventory (MMPI), 362–364
　sixteen personality factor questionnaire, 359–360
　temperament, 361–362
Pharmacogenetics, 155
Pharmacological stressors, 451
Pharmacodynamics, 66
Pharmacokinetics, 66

Phasic firing, 93
Phenotype, 152
Phobic disorders, 442
Phosphorylated cyclic adenosine monophosphate response element-binding protein (pCREB), 508
Physically dependence, 5
Physical pain, 450–451
Polydrug use, 21
Positive punishment, 191
Positive reinforcement, 188, 194–195
Posttraumatic stress disorder (PTSD), 443
Predictable stress, 452
Prefrontal cortex (PFC), 96, 332–334, 506
Primary reinforcers, 188
Probability discounting, 326, 328–329, 335, 541
Progesterone, 501
Protein interacting with C kinase 1 (PICK1), 506
Protein kinase, 162
Psychedelic (classic) hallucinogens, 15–16
Psychological dependence, 5
Psychostimulants, 8–12, 29
Punishment, 191–192, 193f

R

Race, 410
Rat gambling task (rGT), 316, 317f, 547
Receptor, 48
Reciprocal determinism, 396–397
Reconsolidation, 286–287
Reinforcer efficacy, 185
Reinstatement model, 101
Relapse, 5, 109–110, 497–498
　cocaine, in rats, 400–402
Religion/spirituality, 411–412
Renewal, 285
Residential treatment center, 413–414
Resting membrane potential, 52–53
Restraint, 449–450, 449f
Resurgence, 286
Rimonabant, 74–75, 79, 117, 549
Risky decision task (RDT), 326
Routes of administration, 7–8

S

Scalloping, 190–191
Schedule II psychostimulants, 10
Schedules of reinforcement, 189–191, 189f
Schizophrenia, 26–27, 60–61, 273, 364, 533–534
Schizotypal Personality Questionnaire-Brief (SPQ-B), 364
Secondary reinforcers, 188
Second messenger, 49

Second-order schedule, 205
Sedation, 12, 232, 331
Sedative-hypnotics, 15
Self-efficacy, 270–271
Sensitization, 106
Sensory preconditioning, 230
Serotonergic receptors, 62
Serotonin, 114, 457–459
Serotonin transporter (SERT), 137, 458
Sexual identity, 509
Sexual minority stress, 510
Short-access model, 199, 200f
Short-term/working memory (STM), 277
Sign-tracking, 232–233
Single-nucleotide polymorphism (SNP), 135
Sixteen personality factor questionnaire, 359–360
Smoking-related illnesses, 28
Social and sociocultural factors
  drug use in animal models, 399–406
    dominance hierarchies, 406
    observational learning, 404
    social facilitation, 404–406, 405f
  drug use, peer and familial influences on, 394–396
  drug use, social influence on, 396–399
    social control theory, 398–399
    social learning theory, 396–398, 397f
  racial and ethnic differences, 410–411
  religion/spirituality, 411–412
  socially based treatments, 412–418
    family therapy, 413
    halfway houses, 414–415
    residential treatment centers, 413–414
    12-Step programs, 416–418, 417f
    therapeutic communities, 415–416
  socioeconomic status (SES), 407–410
    addiction in animals, 408–410, 409f
Social anxiety disorder (SAD), 442
Social control theory, 398–399
Social defeat, 447–448, 447f
Social facilitation, 404–406, 405f
Social isolation, 445–446
Social learning theory, 396–398, 397f
Sociocultural factors, 544–545
Socioeconomic status (SES), 407–410
  addiction in animals, 408–410, 409f
Stable-step function opsins (SSFOs), 147
Stereotypy, 29, 108, 116
Stimulus substitution theory of conditioning, 228–230
Stress, 437, 545–546
  animal models, 444–453
    combining stressors, 451–452
    environmental stressors, 448

    food deprivation, 448
    forced swimming, 450
    inescapable *vs.* escapable stress, 452–453
    maternal separation, 446–447
    pharmacological stressors, 451
    physical pain, 450–451
    predictable stress, 452
    restraint, 449–450, 449f
    social defeat, 447–448, 447f
    social isolation, 445–446
    stress-induced reinstatement, 453
    unpredictable stress, 452
  corticosteroid, 438
  drug use patterns in humans, 440–444
  environmental stressors, 448
  food deprivation, 448
  forced swimming, 450
  glucocorticoids, 438
  hedonic allostasis model of addiction, 462–463, 464f
  hypothalamic-pituitary-adrenal (HPA) axis, 438
  inescapable *vs.* escapable, 452–453
  neurobiology of, 438–440, 439f
  neuroendocrine system, 438
  neuromechanisms, 454–462
    acetylcholine, 459–460
    dopamine, 457–459
    endocannabinoids, 460
    glutamate, 459
    HPA axis, 454–455
    peptides, 461–462
    serotonin, 457–459
    shared neuroanatomical underpinnings, 455–457, 456f
  reducing stress methods, 463–465
Stress-induced reinstatement, 453
Substance Abuse and Mental Health Services Administration (SAMHSA), 6–7
Substance use disorders (SUD), 443
  Cannabis, 18–19
  cathinone, 11–12
  commonly used drugs, 6–20
  depressants, 12–15, 13f
  Diagnostic and Statistical Manual (DSM-5), 5–6
  drug combinations, 21–25
  economic impacts, 31–32
  entactogens (empathogens), 17–18, 18f
  hallucinogens, 15–17, 16f
  health and economic impacts, 27–32
  health impacts, 27–31
  heritability of, 134
  inhalants, 20–21
  International Classification of Diseases (ICD-11), 6
  key aspects, 4–6

Substance use disorders (SUD) (*Continued*)
  neurobiological underpinnings of, 546–552
    brain regions implicated in, 546–547
    genetic associations, 552
    incentive sensitization theory, 549–551, 550f–551f
  nicotine, 8–9
  pharmacological treatments for, 71–80
    alcohol use disorder, 75–77
    cannabis use disorder, 79–80
    nicotine use disorder, 72–73
    opioid use disorders, 77–78
    psychostimulant use disorders, 78–79
  prevalence of, 25–27
  psychiatric conditions, comorbidity with, 26–27
  psychostimulants, 8–12
  risk factor for, 330–331
  routes of administration, 7–8
  schedule II psychostimulants, 10
  synthetic cannabinoids, 18–19
  synthetic cathinones, 11–12
  terminology, 4–5
  translational approach, 32–33
  United States, in the, 25–26
  world, across the, 26
Sympathetic nervous system, 439–440
Synapse, 54–55
Synaptic plasticity, 290
Synaptic vesicles, 54–55
Synthetic cannabinoids, 18–19
Synthetic cathinones, 11–12

T
Temperament, 361–362
Temperament and Character Inventory (TCI), 361–362, 363t

Temporal prediction of reward, 93–94
Terminal buttons, 51–52
  transporters, 51–52
Tobacco, 8–9, 8f, 24, 26–28, 72, 235, 273, 275, 367–368, 394, 442, 488
Tolerance withdrawal, 5, 108
Tonic firing, 93
Transcription factors, 140–141
  expression, drug-induced changes in, 140–141
Translational approach, 32–33
Transporter expression and function, 159–162
Transporters, 51–52

U
Unconditioned response (UR), 228–230
Unconditioned stimuli (US), 228–230
Unpredictable stress, 452
Upregulation, 141

V
Ventral tegmental area (VTA), 93, 145, 245–246
Vesicular glutamate transporter (VGLUT), 63
Vesicular monoamine transporter 2 (VMAT-2), 62
Vicarious punishment, 396
Vicarious reinforcement, 396
Virtual reality exposure therapy (VRET), 252–253
Voltammetry, 103
Voluntary oral consumption, 491–493, 492f

W
Wason selection task, 316–319, 317f
Withdrawal, 5, 108
  symptoms of, 5